S0-EAJ-189

PHYSICS
Concepts and Consequences

RAYMOND L. MURRAY

Department of Nuclear Engineering
North Carolina State University at Raleigh

GROVER C. COBB

Department of Physics
North Carolina State University at Raleigh

PHYSICS
Concepts and Consequences

PRENTICE-HALL, INC., ENGLEWOOD CLIFFS, NEW JERSEY

PHYSICS: Concepts and Consequences
BY RAYMOND L. MURRAY AND GROVER C. COBB

13-672501-5

Library of Congress Catalog Card Number: 79-118661

Printed in the United States of America

Current Printing (*last digit*):
10 9 8 7 6 5 4 3 2 1

PRENTICE-HALL INTERNATIONAL, INC., LONDON
PRENTICE-HALL OF AUSTRALIA, PTY. LTD., SYDNEY
PRENTICE-HALL OF CANADA, LTD., TORONTO
PRENTICE-HALL OF INDIA PRIVATE LTD., NEW DELHI
PRENTICE-HALL OF JAPAN, INC., TOKYO

To Our Children

TO THE INSTRUCTOR

The explosion of scientific knowledge in the twentieth century has been accompanied by serious concern about our educational methods. Members of the "two cultures"—scientific and literary—have given serious thought to the problems of understanding and responsibility in the transition to a technological age. Experiments in the teaching of mathematics, languages, and science involve new viewpoints, content, and devices. The authors and the publisher were convinced of the need for a change in the approach to the subject of physics as taught in liberal arts schools. We recognized five requirements of a new textbook:

1. It should emphasize the unity of the subject, in view of trends toward applying all aspects of science to problems of the modern world.
2. It should reflect the progress of technical knowledge in this century, giving special attention to the roles of basic particles, nuclei, atoms, and molecules.
3. It should take advantage of the student's daily association with science and technology and his awareness of new developments.
4. It should be concise, conceptual, and challenging, in accord with demands on our future intellectual leaders.
5. It should provide the student with an appreciation of the role physics plays in understanding both the natural and manmade environment, and should suggest the beneficial implications of science and technology for our society.

To accomplish these objectives, a significant revision of both content and order of presentation from that of the traditional physics text was necessary. The flow of ideas is described below to provide the instructor with an overview of the text.

The concepts of particles and atomic structure are reviewed immediately. This permits a demonstration of the laws of motion by examples using a variety of forces, including gravitational, electrical, and mechanical. The ideas of classical and of special relativity are introduced, and emphasis is placed on the meanings of description and measurements. Fields, potential, and mass-energy are developed as unifying concepts applicable to all physical phenomena. The principles of conservation of linear and angular momentum are readily illustrated by particle interactions and by the motion, under a central force, of planets, spacecraft, and charges. The study of periodic motion leads naturally to a discussion of waves, including those of mechanical, acoustical, and electromagnetic origin. An approach to the properties of matter from the molecular viewpoint is followed by a statistical view of kinetic theory and thermodynamics. The idea of transport of particles, momentum, thermal energy, and electric charge facilitates the microscopic view of electrical and magnetic properties. Circuits, instruments, and devices for generating electromagnetic radiation lead to the phenomena of geometrical and physical optics. The concepts of the quantum, the uncertainty principle, and wave mechanics can then be established, followed by a survey of nuclear physics. Finally, the relation of physics to biology is discussed and some of the problems on the frontier of science and technology are outlined.

To avoid a ponderous volume, historical material has been kept to a minimum. Algebra, geometry, and trigonometry serve as principal mathematical tools, with a nominal use of vectors and an implicit allusion to calculus as needed. Many problems are provided at the ends of chapters, both as exercises and as indications of the breadth of the subject of physics.

The approach, although unconventional, helps to unify the historically well-insulated branches of physics. Young people who have grown up in an age of the atomic bomb, the computer, and travel to the Moon take keen interest in modern developments. The introduction of new and relevant material can possibly stimulate many able students who would otherwise react negatively to physics. The text seeks to give the liberal arts student an appreciation of the physical universe, some insight into the philosophical implications of the subject, and experience in the analysis of real problems. It is hoped that a balance between the intellectual and practical aspects of physics has been achieved.

The authors wish to express appreciation to their colleagues for thoughtful reviews of various drafts. Special thanks are due to Dr. Joseph W. Straley for encouragement and most helpful suggestions.

RAYMOND L. MURRAY / GROVER C. COBB
Raleigh, North Carolina

TO THE STUDENT

What is physics? The science of physics seeks to explain the fundamentals of all phenomena of nature. The objects it investigates range from electrons to inventions to galaxies. The success of physics depends on accurate measurements of distance, time, weight, electrical charge, and the many other properties needed to deduce and verify theories of nature. By virtue of its descriptions in terms of basic and submicroscopic processes, physics underlies other physical and biological sciences and practical applications.

The end of the nineteenth century marked the peak of what is called "classical physics," a view of nature then thought to be complete, consistent, and useful; all that remained to be done, it was thought, was to improve the accuracy of measurements. The spectacular growth of "modern physics" and its uses in the twentieth century suggest instead that our understanding of nature will never be complete.

We are living in a rapidly changing era and are all familiar with ideas never dreamed of by the scientists of the past. Through science courses, newspapers, magazines, radio, television, and daily living with modern devices and scientific discoveries, we are already prepared to appreciate the meaning and challenge of physics.

Students will encounter a few surprises, however, where ideas conflict with common observation. They shall come to appreciate better that the senses of hearing, sight, and touch are inadequate for the very small, such as atoms; the abstract, such as energy and fields; the very fast, such as nuclear particles; or the very large, such as stars. Words fail us in describing many processes; we shall have to rely on mathematics to express many models of physics. In this role, mathematics goes beyond its value as a "shorthand" to record and remember findings, or to help solve problems, or to make new predictions.

When chemistry and physics are viewed on an atomic basis, there is no difference between the two sciences. The valence electrons of chemical processes become, in physics, the electrons of an electric current; chemical combining weights are closely related to the physical structure of the nucleus; gas laws arise as a consequence of kinetic theory; the heat of a chemical reaction is merely one form of energy as described by physics; the bonds of compounds are interatomic forces of matter. The language and the purpose of the two sciences often differ, but their basis is the same.

What are physicists? The popular picture of physicist (in fact, of most scientists) is of a solitary and often eccentric individual. In reality, the complexity of modern physics requires that most research involve a close-knit team effort, with regular discussion and a division of labor. The physicist's work is believed to be coldly methodical and orderly in all respects. In fact, good luck or accident plays a considerable part in scientific research, and an outcome may be quite unrelated to what was sought. The physicist is said to be aloof, living in an "ivory tower," lacking concern for the needs of society. The truth is that there is a growing participation of scientists in community and national affairs. In addition, many physicists are engaged in research aimed at creating a better and safer world—in environmental problems, manufacturing, communications, power, transportation, nuclear processes, and military protection. The only significant feature that distinguishes a physicist from the average person is the combination of a facility with the abstract, a driving curiosity, a logical mind, and the ability to derive esthetic satisfaction from understanding.

What can physics do? Through the accumulation and study of *data* on diverse processes, an *order* is noted. Mathematical formulas are then derived to express this order in terms of *principles*—that is, unifying concepts. Through the use of principles, analyses and checks on other phenomena are possible. Predictions of new effects are made, tested, and either verified or refuted. Some of the conclusions of such research can add to the understanding of our surroundings while others help in practical engineering or industrial applications. As new concepts become commonplace, physics moves on to new boundaries, leading to an ever-growing body of knowledge. Basic to man's character is the urge to explore the unknown, to brave the hazards that accompany progress, and to render mental order of of apparent mystery and chaos. That the search is often frustrating and apparently unending does not deter him from the quest. The more the physicist finds out through his combined attack with measurements and theory, the more clearly does he realize how limited man really is in the scheme of the universe. Man sees more dimly and comprehends less about the stars and galaxies than a tiny insect understands a mountain. On the opposite end of the scale he is a blind and clumsy giant, striving to fathom the complexities of the atom and the nucleus. Even on the Earth, he has barely begun to understand the behavior of the sea, the land, and the atmosphere, and knows little about the interior of our world.

The task of physics is to relate all processes and entities logically, to provide wherever possible a means for predicting consistent behavior. We must keep in mind, however, that the descriptions and explanations of physics involve the "How" of our world, but not the "Why." For instance, physics can tell us, through mathematical formulas, *how* the amount of force between two electrical charges depends on the distance between them, but cannot answer the question "*What* is charge?" The even more fundamental question "Why is nature as it is?" is beyond the scope of physics; attempts to answer this question must be made within the realms of philosophy, metaphysics, and theology.

Why study physics? The reasons for seeking a knowledge of the concepts and methods of physics are different for the student of science, the student of engineering, and the student of the liberal arts. Clearly enough, the professional physicist requires a knowledge of the subject in order to be able to experiment and explain, and the engineer finds physics basic to his methods of analysis and design. The graduate in liberal arts, on the other hand, usually studies the subject to broaden his cultural horizons and to enhance his intellectual contributions.

Much has been said in recent years of the tremendous influence that the physical sciences and technology have had on our society. Indeed, these fields tend to be regarded as insidious invaders acting independently of man. The authors believe, however, that they are an intimate part of the structure of our civilization, subject to use by man for his benefit or detriment as he chooses, just as are literature, the knowledge of human relations, and the principles governing the behavior of political institutions. We are convinced that an appreciation of what science is and of what technology can do is essential to the decision-making processes that men employ to develop an advanced culture. An awareness of the sciences that help explain all processes is vital to the complete education of a person living in a technologically oriented world—one which emphasizes exploration of the atom, the body, and the universe, and which seeks to improve the convenience, comfort, and health of its people.

How difficult is physics? For years, students have complained bitterly that physics is a difficult subject. Such comments often reflect the student's natural desire for respect or sympathy from associates. Nevertheless, everyone agrees that the study of physics is demanding and not to be approached as one reads a novel. The mastery of physics can be satisfying, as are all achievements. It can be an exciting subject, as we see new and almost unbelievable ideas unfold, as we realize its power to analyze advanced technical problems in behalf of society's needs. There is some truth in the student's observation "Everything in physics depends on everything else." For instance, to understand motion, one needs to be familiar with basic particles, which are in turn described in terms of motion! It would be wonderful if one could be born with the necessary knowledge, or could have it all painlessly instilled by "sleep-teaching," but these are idle thoughts. There is no substitute for patient study and reflection, which will eventually yield familiarity with physics, and certainly some degree of understanding.

RLM / GCC

CONTENTS

7

ELECTROMAGNETISM 170

8

MATTER AND ENERGY 192

9

MOTION OF MANY-PARTICLE SYSTEMS 237

10

INTERACTIONS AND CONSERVATION LAWS 268

11

VIBRATION AND WAVE MOTION 301

12

WAVE PHENOMENA 329

13

MECHANICAL AND THERMAL PROPERTIES OF MATTER 351

FROM ATOMS TO STARS

1

The question "Of what is the world composed?" has always interested man deeply. It took a very long time for him to recognize the fact that all objects, materials, and processes could be explained by the existence of basic particles. When at first he applied his senses to the land, waters, and air, to the Sun, Moon, and stars, to plants and animals—all with a great variety of forms, colors, and behavior—it was natural for man to consider each item as having independent character. The first major step out of the era of superstition was the distinction between living and nonliving things. Centuries passed before the ancient Greek philosophers proposed that the four basic ingredients of nature were earth, water, air, and fire. Viewed in a modern context, these would correspond to the three basic states of matter—solid, liquid, and gas—plus radiant energy. A wealth of facts and detailed descriptions have since become available through careful scientific research methods, of which the Greeks never dreamed.

Observations on chemical reactions and the behavior of gases led to the conclusion in the nineteenth century that matter is composed of molecules and atoms. Around the beginning of the twentieth century, it was found that the atoms themselves were formed from even smaller particles—nuclei and electrons. Still later, through investigations of nuclear reactions, the nucleus was proved to contain protons and neutrons. Even though the explanation of the nature of matter was satisfying at each of these stages of deeper investigation, further scrutiny always led to the need for description by a still finer structure. Unusual subparticles continue to be found and are not yet completely explained. Man may well be fated to pursue this elusive question throughout his existence.

We shall try to illustrate the great scope of physics in the next few sections. Starting with the presently known basic particles, we shall show how nuclei, atoms, ions, and molecules are formed. Then we shall review some of the devices invented by man, which are objects more on the scale of his size. Proceeding to the still larger realm, we review natural phenomena in, on, and near our Earth. Finally, we look outward to space with its planets, stars, and galaxies.

In our presentation of the concepts and laws of physics, we plan to use the fundamental particle description that is capable of explaining *most* of the observed phenomena of nature. To start above the atomic level would hamper our efforts to do justice to modern nuclear problems; to leap to the frontier of physics would expose us to the dilemmas presently facing scientists. Our minds find it easier to visualize an object if we can compare it to something familiar. Thus we might picture atoms as very small marbles—spherical and uniform in composition. For many purposes the atom's actual shape and structure is not important, so this simple model is worth keeping. Furthermore, models of processes involving more than one particle can then be devised. For instance, the interaction of two atomic particles may be likened roughly to the collision of two billiard balls. As we proceed to describe these models of matter, the reader must remember that they serve as conveniences for thought, not as ultimately correct descriptions.

1-1 FOUR BASIC PARTICLES

It would be convenient for our study if all matter were constructed from particles of only one kind, and if differences in materials depended only on arrangement. Such is not the case, as far as we know; in fact, there are dozens of elementary particles. To describe most matter and many processes, however, *four* will suffice—the *electron*, the *proton*, the *neutron*, and the *photon*. These "building blocks" are distinguished by several characteristics, of which we shall select three—size, mass, and charge. *Size* or dimension is measured in meters (recall that 1 meter = 39.4 inches, 1 inch = 2.54 centimeters). *Mass*, a more technical term, is a measure of *inertia:* the tendency of an object to keep moving or to resist a change in motion. Our familiarity with pounds, grams, and kilograms is enough for the moment to appreciate this property (recall that 1 kilogram = 2.2 pounds, 1 pound = 454 grams). The third characteristic, electrical *charge*, is familiar in that the current used in the home is due to the flow of charges, but mysterious from a fundamental standpoint. All evidence points to the existence of indivisible units of electricity, one sort called positive, the other negative, of such nature that like charges repel and unlike charges attract. Also, in any reaction of particles, charge is neither created nor destroyed. This example of *conservation* is one of several we encounter in the study of physics.

Matter in different situations sometimes acts like chunks of material or *particles,* and sometimes as vibratory disturbances or *waves.* We shall see later that all matter in motion can be viewed either way, which implies that its nature is more complicated than either simplified version. We shall now identify our four basic particles in terms of mass, size, and sign of charge.

Electron. This particle is *negatively* charged, bearing the unit of smallest known amount of electricity. It has a very small mass, 9.11×10^{-31} kilograms (kg), such that around 10^{30} electrons would have to be accumulated to amount to 1 kilogram. Its size is correspondingly minute, having an apparent radius of about 2.8×10^{-15} meter. We can show that its density (mass per unit volume) is exceedingly high. If the electron is assumed to be spherical, with volume $(\frac{4}{3})\pi(2.8 \times 10^{-15})^3 = 9.2 \times 10^{-44}$ cubic meter, its density is $(9.1 \times 10^{-31})/(9.2 \times 10^{-44}) \cong 10^{13}$ kilograms per cubic meter. This is billions of times greater than the density of any ordinary material. We also call the electron a *beta* particle when discussing its emission from a radioactive substance.

Proton. This particle is *positively* charged with the same smallest magnitude of charge as has the electron. It is about 1836 times as massive as the electron, however, having a mass of 1.67×10^{-27} kg. In spite of its greater mass, it is somewhat smaller than the electron, of radius around 1.4×10^{-15} meter. One can easily compute that its density—mass per unit volume—is enormously larger than that of even the densest metals.

Neutron. This particle has only slightly greater mass than that of the proton but has *no electrical charge.* This neutrality allows it to wander through matter practically free from electric or magnetic effects and to reside in the nucleus of an atom without contributing to the internal electrical forces.

Photon. This entity is a very small amount of radiant energy, found in the structure of light, X rays, radio waves, and other radiations that we generally call *electromagnetic.* It is electrically neutral but is able to interact with electrical charges. It exists only when it is in motion, at a speed around 3×10^8 meters per second. It can be said to have mass because its inertia can be imparted to other objects. We classify it as a particle, but it behaves as a wave in certain situations, and thus is said to exhibit a *dual character.* In a photoelectric cell, photons acting as if they were particles have the ability to dislodge electrons from metal plates. On the other hand, the bending of light in water can best be described by a wave analogy.

Although there are many other distinct particles in nature,* the above are the building blocks in the structure of ordinary matter and of energy-transfer processes (for instance, light transmission).

1-2 CONSTRUCTION OF NUCLEI AND ATOMS

The *periodic table of chemical elements,* devised by the nineteenth-century Russian chemist Mendeleev, contained 92 different entries until the discovery of the elements beyond uranium (such as plutonium). At this writing, there is a total of 104 elements (see Appendix C). Each of these is distinguished by its chemical properties; for example, hydrogen reacts chemically with other elements quite differently from oxygen. A given element may have atoms of different mass called *isotopes;* for example, hydrogen has three isotopes. All the similarities and differences of element construction can be explained in terms of three of our building blocks—protons, electrons, and neutrons. The following three facts concerning these particles and the nucleus have been obtained from experimental measurements:

1. An atom under normal conditions is electrically neutral. This means that it has an equal number of protons and electrons.

2. The constituent atomic material is generally separated into two regions: the central positive *nucleus,* where most of the mass resides; and the surrounding negatively charged "cloud" of *electrons* with very small mass.

3. Electrons can be produced as the result of processes in the nucleus, but there is good evidence that they do not exist there.

The conclusion to be drawn from these facts is that the nucleus contains only protons and neutrons (together called *nucleons*) and the outer region only electrons.

Let us assign symbols to represent the numbers of each type of particle in an individual atom:

$$\begin{aligned}
\mathrm{Z} &= \text{number of protons or electrons (atomic number)} \\
\mathrm{N} &= \text{number of neutrons} \\
\mathrm{A} &= \mathrm{Z} + \mathrm{N} = \text{number of nucleons (mass number)}
\end{aligned}$$

We may represent any element or isotope, using a shorthand notation for these numbers, as follows:

$$_{\mathrm{Z}}(\text{symbol of element})^{\mathrm{A}}$$

* Other particles include the *positron* (essentially identical to the electron except that it bears positive charge), the *antiproton,* and *antineutron.* These are examples of what is called "antimatter"—of rare occurrence, on our world at least.

Hydrogen, the first element in the periodic table, has three isotopes. One of these, called *ordinary hydrogen*, is represented by

$$_1H^1$$

meaning that, for ordinary hydrogen, Z = 1, A = 1, N = A − Z = 0; or, in words, one proton in its nucleus, one electron outside, and no neutrons (see Figure 1-1). The chemical properties of hydrogen are associated with its one removable electron. The mass of the hydrogen atom is very close to the proton mass of 1.67×10^{-27} kg, since the electron mass is extremely small, 0.0009×10^{-27} kg. The electron (negative) and the proton (positive) in hydrogen form a relatively stable structure because of the electrical attraction mentioned in Section 1-1. It is convenient to imagine the electron as moving in a roughly circular path or *orbit* about the proton, much as a planet moves about the Sun. Such pictures should not be taken literally, however. A more modern view is that the orbit is a general region where the electron is *most likely* to be. A given electron cannot be in two places at the same time, but if one looks at many atoms the electrons are so distributed as to offer a cloud-like appearance.

Another isotope of hydrogen is *heavy hydrogen* (*deuterium*)

$$_1H^2$$

meaning Z = 1, A = 2, N = 1. The nucleus (called the *deuteron*) contains a neutron and a proton tightly bound together. The nucleus of deuterium has the same single charge as does the nucleus of ordinary hydrogen, and each has one external electron. The isotopes $_1H^1$ and $_1H^2$ thus react chemically with the same elements, and their electrical properties are the same, but their masses are different.

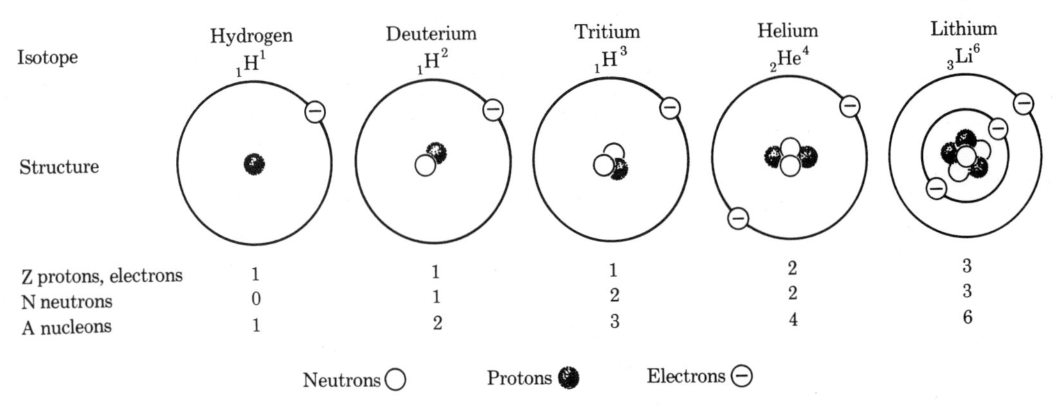

FIGURE 1-1

Building the light elements from neutrons, protons, and electrons

Still another isotope of hydrogen is *tritium*

$$_1H^3$$

with Z = 1, A = 3, N = 2. It has a nucleus (called the *triton*) that is radioactive—i.e., it disintegrates in the course of time.

The next element in the periodic table is *helium*, with two protons in its nucleus. The natural element is

$$_2He^4$$

with Z = 2, A = 4, N = 2. The nucleus contains two neutrons and two protons, and there are two (external) electrons. If one more proton is now

TABLE 1-1
COMPOSITION OF NATURAL LIGHT ELEMENTS

Element	Symbol	Z	A	N	Atomic Mass*
Hydrogen	H	1	1	0	1.00797
			2	1	
Helium	He	2	3	1	4.0026
			4	2	
Lithium	Li	3	6	3	6.939
			7	4	
Beryllium	Be	4	9	5	9.0122
Boron	B	5	10	5	10.811
			11	6	
Carbon	C	6	12	6	12.01115
			13	7	
Nitrogen	N	7	14	7	14.0067
Oxygen	O	8	16	8	15.9994
			17	9	
			18	10	
Fluorine	F	9	19	10	18.9984
Neon	Ne	10	20	10	20.183
Sodium	Na	11	23	12	22.9898
Magnesium	Mg	12	24	12	24.312
			25	13	
			26	14	
Aluminum	Al	13	27	14	26.9815

* Often called atomic weight. The figures given are the average for each *element*, on the scale in which C^{12} is exactly 12. For example, the natural abundances of C^{12} and C^{13} in the chemical carbon form a blend with average atomic mass 12.01115. A more complete table is given in Appendix C.

added to this nucleus, another electron must be put outside the nucleus to maintain electrical neutrality. One isotope of the new element thus formed—*lithium*—is

$$_3Li^6$$

where Z = 3, A = 6, N = 3. This isotope's nucleus has three neutrons and three protons. An important new feature is that the last electron added goes into an orbit, or *shell*, of its own, as sketched in Figure 1-1. According to quantum theory, no more than two electrons may reside in the orbit nearest the nucleus. Once that number is reached, the shell is said to be *closed* or complete. The chemical behavior of matter is strongly dependent on this situation. As we proceed up the periodic table with increasing numbers of particles, the atoms become increasingly complicated to sketch in the manner we have used for the lightest elements. Therefore, we shall merely use numbers to describe the structure, as in Table 1-1, where the most frequently found isotopes of a number of elements are listed. The elements from Z = 3 to Z = 10 are formed by successively adding protons to the nucleus and electrons to the second shell, up to a maximum of eight electrons. The second shell is complete at this point—at the element neon, Z = 10. This procedure continues up through the periodic table, with A and Z increasing, and more and more orbits and shells. The complexity becomes very great, as can be seen in the structure of the main isotope of the heaviest stable element, uranium—Figure 1-2 shows the schematic arrangement of $_{92}U^{238}$.

The elementary particles such as the neutron and electron can also be described by their A- and Z-values. Thus $_0n^1$ means that the neutron has no charge and has one heavy particle. The electron is represented by $_{-1}e^0$, which means that it has negative charge but that its mass is insignificant in comparison with those of atoms.

Sizes of atoms. From our calculations, we found the density of the electron to be exceedingly high: 10^{13} kilograms per cubic meter (kg/m^3). The proton is about the same size as the electron, but its mass and density are much greater than those of the electron. However, we know that ordinary matter has a much lower density; for example, water has a value of 1000 kg/m^3. The immediate conclusion to be made is that the atom is practically all empty space, with the matter highly concentrated in a few locations, in analogy to our solar system. Atoms vary in size according to their complexity, but most have dimensions in the range 10^{-10} to 10^{-9} meter, thousands of times larger than the sizes of nuclei.

Atomic masses. Constructed as they are of protons and neutrons, the masses (or mass numbers) of atoms of all elements would be expected to be multiples of the masses (or mass numbers) of nucleons. Inspection of the entries under "Atomic Mass" in Table 1-1 shows this to be *nearly* true, but not quite. There are several reasons for the differences. The slight excess in mass of the neutron (0.14% greater than that of the proton)

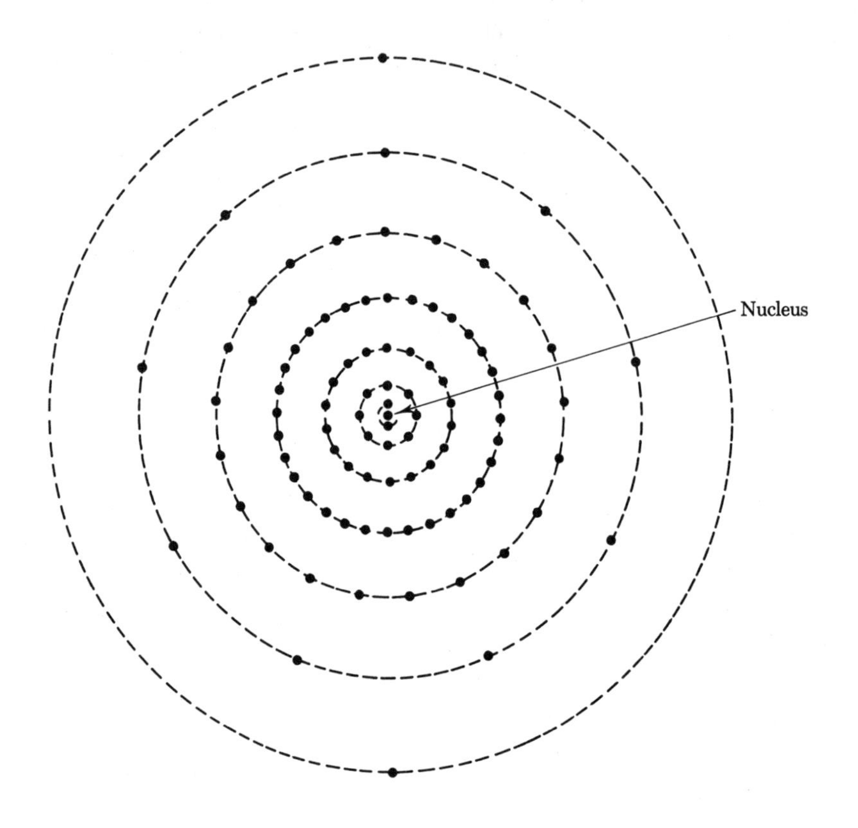

FIGURE 1-2

Electron orbits in uranium, U^{238} $Z = 92$, $N = 146$, $A = 238$

accounts for only part of the discrepancy. A natural element is usually a mixture of isotopes; for example, boron is about 20% isotope ${}_5B^{10}$ and 80% isotope ${}_5B^{11}$. By proportion, this would give an average mass of $(0.2)(10) + (0.8)(11) = 10.8$, intermediate between 10 and 11. A second effect has to do with a loss of mass on formation of nuclei from basic particles, a phenomenon explained by Einstein's formula $E = mc^2$, to be discussed later.

The scale of atomic masses (see Table 1-1) is based on the isotope C^{12}, which is conventionally taken to have an atomic mass of exactly 12. For many years the element oxygen was used for the reference; later, the isotope O^{16} was adopted; the present system dates from 1961.

1-3 ATOMIC AND NUCLEAR PROCESSES

As we have learned in chemistry, a *molecule* is a neutral particle containing two or more atoms, held together by forces that depend on the *sharing* of electrons. Because the effects of such sharing do not involve the internal nuclear structure to any important degree, we shall ignore this detail in sketching a few simple molecules (Figure 1-3).

The simplest *diatomic* molecule is that of hydrogen H_2, in which two hydrogen atoms are bound together by the combination of attractions and repulsions of the protons and electrons. The two electrons may be visualized as moving in a certain path about both nuclei, and thus as being shared by both. (Our diagrams are to be regarded as useful for thinking about molecules, but we continue to emphasize that they are rough models rather than final or exact descriptions.)

Possible combinations of atoms to form molecules by chemical reaction obey a simple rule: *atoms seek closed shells.* As discussed earlier, the maximum number of electrons in the first shell is 2. The second shell has a maximum occupancy of 8, the third shell 18, and so on. Whenever two atoms can share electrons in such a way that each atom appears to have a closed shell, these atoms have a natural affinity for each other. In the compound lithium hydride, LiH, the external electrons of lithium and hydrogen cooperate to give a closed shell of 2, as sketched. In hydrogen fluoride, HF, the 1 electron of hydrogen cooperates with the 7 external electrons of fluorine to give a closed shell of 8. Since closed shells are favored, vigorous chemical reactions between hydrogen and fluorine take place. From the point of view of each partner in the electron-sharing process, the shell is closed. The favorable combination of 2 hydrogen atoms and 1 oxygen atom to form water, H_2O, is easily pictured. The oxygen atom falls 2 electrons short of forming a complete second shell of 8 (which resembles neon). Two hydrogen atoms provide the electrons to fill this need. Long before the above atomic description was developed, the idea of *valence* was used. An element such as H or Na with 1 excess electron

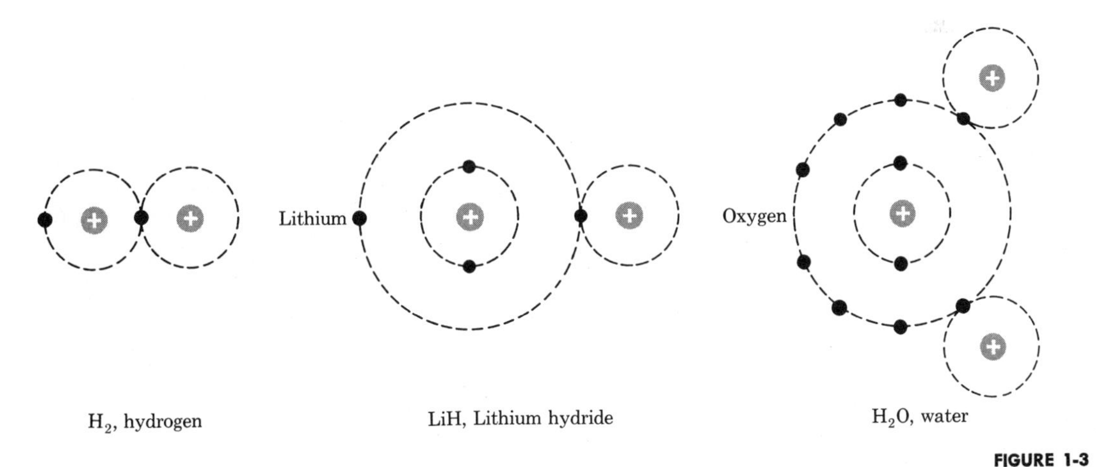

FIGURE 1-3

Electron sharing in molecules

in its outermost shell is said to have a valence of +1; an element such as F that is 1 electron short of a full shell has a valence of −1. Thus the chemical reaction equation for hydrogen and fluorine can be written

$$H^{(+1)} + F^{(-1)} \rightarrow HF$$

where the sum of valences on each side is zero. Elements that naturally have complete external shells such as He (2) and Ne (8) rarely combine permanently with other substances, since they have neither an excess nor a deficit of electrons; these are the "noble" gases, so named because they remain aloof from associations with other elements, with Z-values as follows: helium, 2; neon, 10; argon, 18; krypton, 36; xenon, 54; and radon, 86.*

Production of ions. An *ion* is a molecule or an atom that has an excess or deficiency of electrons compared with the neutral atom or molecule. The simplest example is the ion H^+, which is a hydrogen atom with its electron removed, leaving a net positive (proton) charge. Ions can be

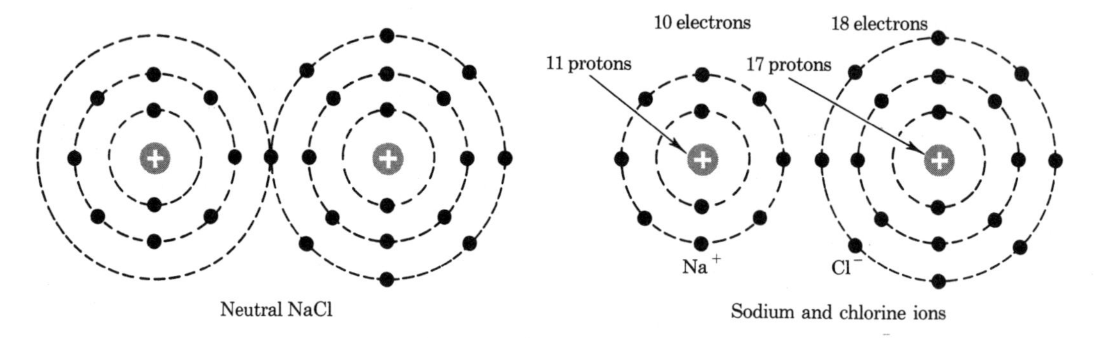

FIGURE 1-4

Ionization in salt-water solution

created in many ways—for example, by bombarding neutral atoms with photons or high-speed electrons, which remove electrons from the atom. They may also be produced by raising the temperature of a gas so high that mutual collisions of atoms remove electrons. On the average, most of the matter in the Earth and the universe is in the ionic state. Crystals of solids in the Earth consist of collections of positive ions with electrons moving through the array. The salts dissolved in sea water appear as ions. In Figure 1-4, the Na^+ and Cl^- ions of table salt in chemical solutions are compared with the molecule NaCl. High-temperature stars are almost

* The sum of the terms—up to a specific point—in a formula found by Rydberg will yield each of the foregoing Z-values:

$$Z = 2(1^2 + 2^2 + 2^2 + 3^2 + 3^2 + 4^2 + \ldots)$$

entirely composed of ions. The dissociation of electrons from atoms results from bombardment by other atoms, electrons, and photons.

Absorption and production of light by an atom. A simple explanation of processes involving light can be made using our atomic model. The photon, as a particle of radiant energy, interacts with an electron, say of hydrogen, and changes its state of motion. The light has been absorbed by the atom, and the electron has been shifted to a new orbit. Shortly thereafter the electron spontaneously returns to its normal path, and stored energy is released as light. This process takes place on a large scale in a very hot gas, giving off enough photons to be sensed by the human eye. Photons also are produced when the motion of high-speed charged particles is suddenly changed by electrical means, either in the atom or in X-ray generators, cyclotrons, and other devices.

Nuclear reactions and radioactivity. When we balance an equation for a *chemical reaction*, we make sure that the number of atoms and the sum of valences are the same on both sides, and thus use the idea of conservation of particles and electric charge. Thus

$$2[H^{(+1)}]_2 + [O^{(-2)}]_2 \rightarrow 2[H_2O]^{(0)}$$

expresses the fact that the four hydrogen atoms and the two oxygen atoms are preserved, and that the total valence (or charge) is zero before and after the reaction. Similar descriptions may be written for *nuclear reactions*. Suppose that a neutron reacts with a hydrogen atom to form heavy hydrogen, as in a nuclear reactor. Then we write

$${}_0n^1 + {}_1H^1 \rightarrow {}_1H^2$$

Before and after the reaction, the number of nucleons (A) is 2 and the number of positive charges (Z) is 1.

The phenomenon of *radioactivity*, in which particles are released by certain isotopes, is analogous to chemical dissociation. We are aware of radioactive cobalt-60(${}_{29}Co^{60}$); strontium-90(${}_{38}Sr^{90}$); and plutonium-239(${}_{94}Pu^{239}$). These are but three of the hundreds of radioactive isotopes found in nature, produced through bombardment of elements with accelerated charged particles, or created in a nuclear reactor, or resulting from an atomic-bomb explosion as fallout.

A particle as elementary as the neutron can disintegrate radioactively, according to

$${}_0n^1 \rightarrow {}_1H^1 + {}_{-1}e^0$$

We note that the total A of 1 and the total Z of zero is conserved in this process. This transformation of a neutron into a proton forms a basis of other spontaneous changes—for example, the "decay" of tritium into an isotope of helium plus an electron,

$$_1H^3 \rightarrow {}_2He^3 + {}_{-1}e^0$$

Here, the total A remains at 3 and the total Z at 1. Each isotope decays at its own definite rate; that is, the decay of a certain fraction of the nuclei takes place over each second of time. The time for the number of nuclei to drop to one-half the original value is called the element's *half-life.* It varies from fractions of a second to millions of years, depending on the element.

1-4 THE INVENTIONS OF MAN

We now turn from the submicroscopic realm of nuclei and atoms to those larger-scale objects built and used by man. Employing materials that are composed of very large numbers of atomic particles, and with a knowledge of physical principles, he has been able to improve upon nature, at least for his own comfort and convenience. The accomplishments in science and technology in the last two centuries exceed those in all of man's previous existence. Several factors seem to have contributed to this astounding growth. The first is the intellectual climate. Prehistoric man made slow progress because survival was his principal interest. He discovered the use of fire, metal weapons, building materials, the wheel and the lever, but little else. In this early period, the environment presumably did not encourage scientific spirit or interest. In classical Greek culture, between 600 B.C. and 200 B.C., philosophy and science flourished under such cultural leaders as Plato, Aristotle, and Archimedes; but there was little experimentation or practical use of physical ideas. Many inventions —such as the pump, the siphon, and an elementary steam engine—were developed by Hero of Alexandria, but most of them remained only interesting toys. During the thousand years preceding the Renaissance, men were preoccupied with political and religious matters, and until the fifteenth century little was done in science. Since then, the development of the scientific method of inquiry and the growth in curiosity about nature have resulted in an overwhelming accumulation of ideas and information.

A second factor that has caused science to grow is the appearance of men of genius; among those foremost in relatively recent times are Leonardo da Vinci, Isaac Newton, Clerk Maxwell, and Albert Einstein. A third factor is the impact of inventions and industry upon each other. The early machines of the Industrial Revolution led to an improvement in the economic status of many countries. The commercial value of invention was recognized, and new devices led to greater industry, increased resources, a higher standard of living, and more facilities for research.

A brief list of some of the inventions that have played an important role in this process is now presented; in this list we have taken the opportunity to preview some of the phenomena and principles of physics to be discussed later on. During the nineteenth century many of the key developments were mechanical in nature, while those of the twentieth century have been largely electronic and atomic. The most notable trend in recent times is a rapid increase in the number, size, and complexity of technical devices.

Spectacles (Bullet, 1282)

By the bending of light as it passes through a transparent substance, the images of objects are brought into sharper focus.

Printing (Gutenberg, 1454)

The molding of metal and the use of machinery forms the basis of printing.

Microscope (Hooke, 1600) and Telescope (Galileo, 1609)

These optical devices bring into better view the realms of the very small and the very distant, respectively.

Steam engine (Savery, 1698; Newcomen, 1705; Watt, 1783)

Processes involving the production, transfer, and practical use of heat are involved.

Electric battery (Galvani, 1780; Volta, 1782)

Chemical reactions take place to give electrical currents.

Cotton gin (Whitney, 1794)

This is an example of moving machinery designed to reduce man's labor.

Gasoline engine (Cecil, 1820; Brown, 1823)

Chemical energy from natural products is converted into motion and propulsion.

Generator (Faraday, 1831)

The motion of metal wires in magnetic fields creates useful electric currents.

Electromagnet (Henry, 1832)

The passage of an electric current through a wire gives rise to magnetic forces.

Harvester (McCormick, 1833)

This mechanical device has revolutionized the production of food.

Photograph (Daguerre, 1836)

The interaction of light with atoms in the film produces a permanent chemical change.

Telegraph (Morse, 1838)

Electrical signals are transmitted at very high speed through the motion of electrons in a wire.

Sewing machine (Howe, 1845) and Typewriter (Sholes, 1867)

These are other examples of assemblies of moving parts—levers, gears, and wheels—that transfer energy of motion.

Telephone (Bell, 1876)

The changes in electrical currents in magnets cause sound vibrations that we recognize as speech.

Phonograph (Edison, 1878)

The mechanical variations produced by sound are stored for future reproduction.

Electric light (Edison, 1879)

The flow of electricity in a wire creates heat and illumination.

Vacuum tube (Edison, 1883; Fleming, 1904; De Forest, 1906)

The control of charge flow in a vacuum or a gas is provided by electrical signals.

Motion pictures (Edison, 1888; Jenkins and Armat, 1890)

These consist of photographic processes and projection of light by lenses.

Automobile (Duryea, 1892)

The modern car is a complex device that uses chemical energy to provide heat, mechanical motion, electricity, light, and sound.

X-ray tube (Roentgen, 1895)

The acceleration of electrons by electric forces produces extremely penetrating light for photographic records.

Radio (Marconi, 1895)

Electrical charges are accelerated to produce electromagnetic waves that travel through space for communication purposes.

Airplane (Wright brothers, 1903)

The forces of molecules in the air on a high-speed object provide a suspension of this heavier-than-air machine.

Jet engine (Whittle, 1928)

The burning of fuel provides energy of gas motion that is transmitted to rotating machinery, which compresses and expels the heated gas.

Television (Zworykin, 1928)

Electrons controlled by incoming electromagnetic waves interact with atoms in the screen to produce visible light signals.

Nuclear reactor (Fermi and others, 1939)

The collision of neutrons with the element uranium produces other particles, heat, and light.

Transistor (Bardeen and Brattain, 1948)

Electrical charges in materials are controlled efficiently in a very small volume.

1-5 THE EARTH AND THE UNIVERSE

The tremendous scope of the subject of physics can be appreciated more fully as we proceed still higher up in the scale of size. The Earth, with its land, seas, and atmosphere, is vast compared with man and his apparatus; in turn, the universe is on a scale that is incomprehensibly larger. The branches of physics devoted to these subjects are *geophysics*, related closely to geology, oceanography, and meteorology; and *astrophysics*, related to astronomy.

The Earth. Little is known about the interior of the Earth, except that its 2000-mile-radius core is at a very high temperature, and is presumably in a plastic or fluid state. Surrounding the core is a mantle of rock 1800 miles thick and, outermost, a 30-mile-thick crust of rock with a variety of minerals, but containing mainly oxygen and silicon. The mountain peaks are as high as 5.5 miles—Mount Everest in Nepal is 29,028 feet; the lowest land depression is the Dead Sea in Israel–Jordan, 1292 feet below sea level. Occasional eruptions of volcanos disturb the Earth's surface, as do the more frequent tremors or earthquakes.

The salt-water seas cover more than 70 per cent of the area of the globe, and run to a maximum depth of 35,630 feet, off Hawaii. At the boundary between land and sea, the rise and fall of tides takes place.

Above the Earth is the *atmosphere*, a thin layer of air composed mainly of nitrogen and oxygen. By a distance of 100 miles, which is small compared with the Earth's radius of about 4000 miles, the atmosphere is down to one-billionth of its surface density. The air has many functions in addition to providing oxygen for breathing: it protects us from the glare of the sun; it contains and releases moisture in the form of rain and snow; it is the source of winds, tornadoes, hurricanes, thunder, and lightning; it provides a blanket to prevent heat loss from the globe; and it stores energy from the Sun. The presence of the atmosphere accounts for the blue of the sky and the red of the sunset; in it we see rainbows after a rain and mirages on the heated desert.

The universe. Man's knowledge of the space surrounding his Earth is now being expanded by measurements taken by space probes and by various spacecraft in orbit around the Earth. Through the centuries, the nature and movements of the Sun, the Moon, and the planets have been a subject of fascination, and through the use of the telescope and other devices, some degree of understanding of the orbiting bodies and the "fixed" stars has been gained.

From earliest times, the Earth was believed to be the center of the universe—a natural conclusion, since the stars and planets *seem* to revolve about the Earth. The elaborate theory of Ptolemy dating from A.D. 140 appeared to explain the motions satisfactorily, but only because time measurements were poor. The theory of Copernicus in the sixteenth century that the Sun is one of many stars and that planets orbit about the Sun was highly controversial because of religious prejudices. Many years elapsed until his view was accepted.

The nine planets of the solar system move in elliptical paths whose planes are nearly parallel. The central object, the Sun, provides the gravitational attraction making these orbits possible. Figure 1-5 shows the average distances of the planets from the Sun and indicates their sizes.

The Sun is a massive star containing more than 300,000 times as much material as the Earth, and of a volume 1 million times as great. The Sun accounts for all but $\frac{1}{7}$ of 1% of the total mass of the solar system. Its temperature, even at the cooler surface, is around 6000°C. The radiation received by our planet is a byproduct of the nuclear reactions going on within the interior of the Sun. It also provides the light that is reflected from the planets, permitting us to see them.

The planet nearest the Sun is *Mercury.* Its year is approximately the same as its day, causing its two sides to differ greatly in temperature. If it ever had an atmosphere, it has long since been boiled away by the heat from the sun.

The next planet, and the one whose orbit is nearest ours, is *Venus;* little of its surface details can be seen because of a thick cloud layer (presumably carbon dioxide). The recent explorations by the space probe Mariner, which sent back information when 10,000 miles from Venus, verified several earlier conclusions—its temperature is much higher than that on Earth: about 600°F; it does not have a significant daily rotation; it has no magnetic field as does the Earth.

Our Moon, about 240,000 miles away, is dry and airless, alternates from extremes of hot and cold, and is thus probably unfit for life. Its surface is found to be dusty and pocked with large craters. Its diameter is about $\frac{1}{4}$ and its mass around $\frac{1}{80}$ that of the Earth, while its density is roughly the same as the Earth's, suggesting that the two bodies may have had a common origin.

Rotating in an orbit larger than that of our planet is *Mars,* on the average 49 million miles distant from Earth. Its surface was once thought

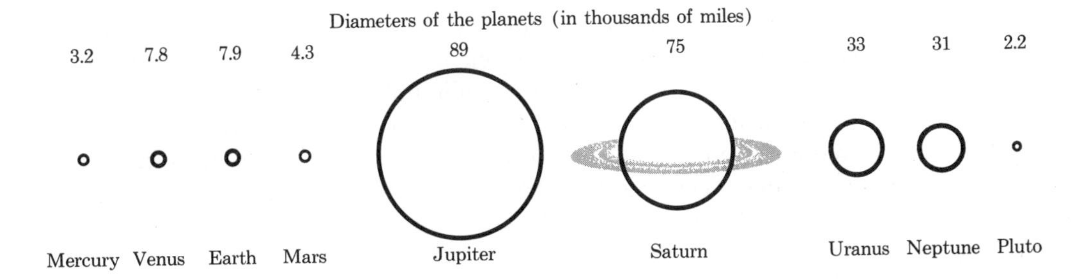

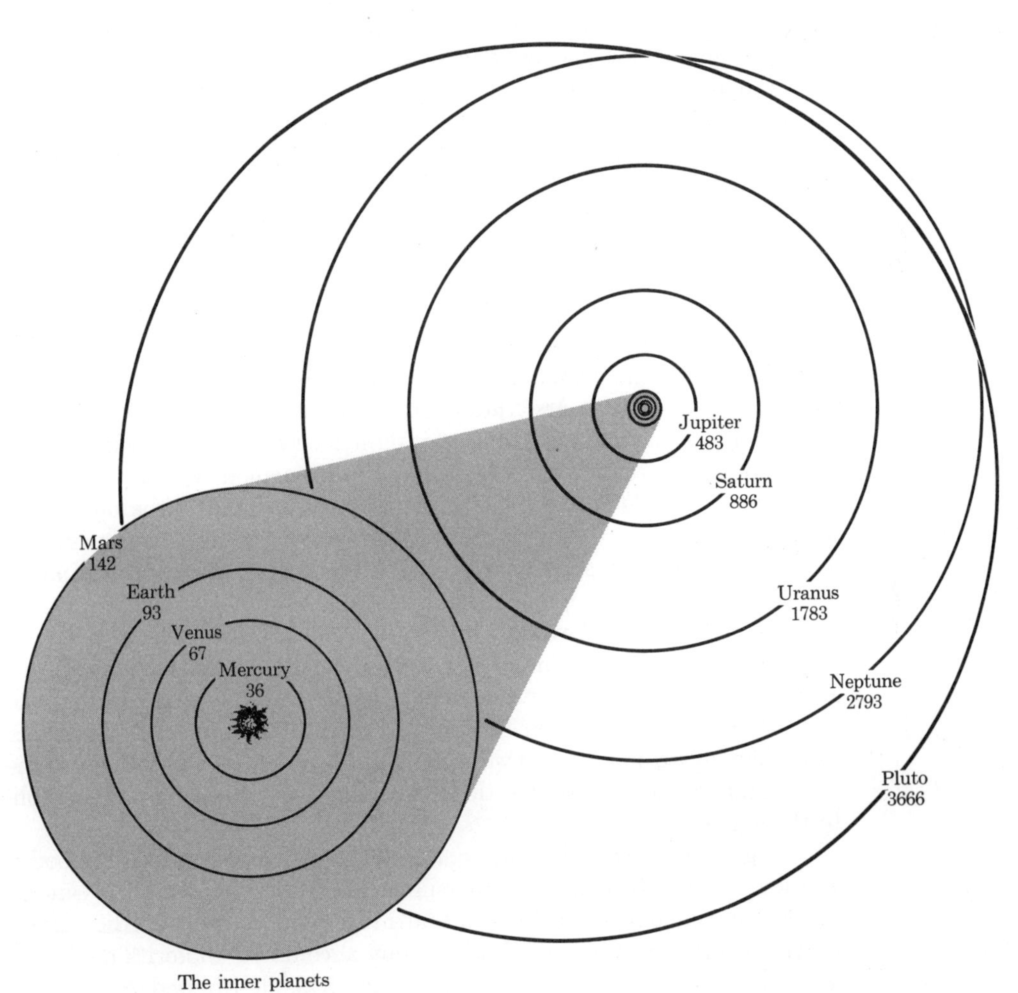

FIGURE 1-5

The solar system (average distances from the Sun in millions of miles are noted)

to be traversed by "canals," suggesting civilization. More recent information sent back by Mariner shows no canals and a moon-like surface. Mars has two small satellites only a few miles across. Its temperature is lower than ours, and its atmosphere is thin, but it may be capable of sustaining some low form of vegetable or animal life. Between Mars and Jupiter are the *asteroids*, small orbiting masses that may have been due to the breakup of a planet. The largest, Ceres, is only 480 miles in diameter; the smallest are mere grains of sand. There are probably billions of such particles, many of which strike the Earth or burn up in its atmosphere as meteorites. The asteroids are very strongly affected by Jupiter, and many accompany it, somewhat like moons.

Jupiter is the largest planet, more than 20 times the diameter of the Earth. It has twelve moons, one as large as Mercury. The planet spins very rapidly, with a short day of only 10 hours. It is extremely cold, −200°F, and thus is probably composed of condensed gases such as hydrogen, ammonia, and methane. The atmosphere is very deep because of Jupiter's strong gravity.

Saturn is the last planet that is visible to the naked eye. It rotates on its axis once every 30 years. The unusual rings about its equator may be due to sand, ice, or perhaps wind-driven gases. Its composition and temperature appear to be similar to those of Jupiter.

Uranus turns on an axis that is parallel to the plane of its orbit. Thus one-half of the planet is in continual darkness, the other half in light. Over a period of 42 years, the situation reverses. *Neptune* is very cold because of the small amount of radiation it receives from the Sun. It was discovered after its effect on the orbits of other planets was noticed.

Finally, *Pluto* was the last discovered (1930) and is the outermost planet of the solar system. From its vantage point, the Sun is merely a large star. Pluto has a peculiar orbit that sometimes goes within that of Neptune, and may have once been a moon of that planet. There appear to be no more planets beyond Pluto.

In addition to the nine planets with their total of 31 moons there are a few periodic *comets*. They have dense heads and long tails of luminous gas that stream away from the Sun because of the pressure of photons from that source of light. As we look at the sky on a clear night, the number of stars seems almost countless. Actually, only about 7000 are visible to the unaided eye.

A great void exists between our system and the nearest celestial body, which is a star about 25 million million miles away. Our Sun is but an average star among the 40 trillion that make up our *galaxy*, a disc-shaped collection of stars rotating slowly about a common center. The Milky Way (what we see of the rest of the galaxy) is about 1 billion billion miles in diameter. It is only one of an uncountable number of galaxies that reach into space for distances (visible to telescopes) up to 10^{22} miles, a figure that staggers the imagination. Among the questions that astrophysicists and cosmologists seek to answer are: How and when was the universe created? Is it finite or infinite in size? Serious work is also underway listening for signals that would show that there is intelligent life elsewhere.

PROBLEMS†

1-1 Express your height and weight in metric units.

1-2 A unit of time in microscopic physics is the *shake*, which is 10^{-8} second. How many shakes are there in a day? A year?

1-3 How many electrons would have to be lined up beside each other to span 1 inch?

1-4 Find the number of electrons in 1 kilogram of pure negative charge.

1-5 Calculate the density of the proton in kilograms per cubic meter, assuming it to be spherical. How many pounds per cubic inch is this?

1-6 Sketch the atom of beryllium, showing neutrons, protons, and electrons. How many total particles does it contain?

1-7 How many electrons, protons, and neutrons has a ${}_{92}U^{235}$ atom? How many has the ${}_{94}Pu^{239}$ atom?

1-8 Assuming that the human body is 95% water, H_2O, how much weight would a 150-pound person increase if all atoms of H were replaced by the isotope H^2?

1-9 What is the atomic mass of the electron if the proton has a value of 1.007277?

1-10 If the spacing between atoms in table salt, NaCl, is 2.8×10^{-8} centimeter, how many atoms are there in a cubic centimeter? (Let the atoms be located at the corners of cubes.)

1-11 Sketch and explain the molecule MgO in terms of electron-sharing and closed shells.

1-12 Explain, in terms of their structures, why fluorine (F) and sodium (Na) are very active chemicals. When they interact, what compound do they form?

1-13 Estimate the atomic mass of chlorine, 75% Cl^{35}, 25% Cl^{37}, using mass numbers of the isotopes.

1-14 A typical atom of the Earth occupies a cube 2.5×10^{-8} centimeter on a side. How many atoms are there in the Earth, whose radius is 6.4×10^{8} centimeters?

1-15 Complete the blanks in these nuclear reactions:

$$_{(\ )}N^{14} + {}_{2}(\ )^{4} \rightarrow {}_{8}O^{(\ )} + {}_{1}H^{1}$$

$$_{(\ )}B^{10} + {}_{0}n^{1} \rightarrow {}_{2}He^{(\ )} + {}_{3}(\ )^{7}$$

† Slide-rule accuracy—two or three significant figures—is usually adequate. Answers to selected problems are given in Appendix G. Those that are more difficult are marked with an asterisk (*).

1-16 How does the structure of radioactive carbon $_6C^{14}$ (used for finding the age of prehistoric wood) differ from the isotope into which it decays, $_7N^{14}$? What particle is emitted?

1-17 Assuming that the orbits of Earth, Mars, and Venus lie on the same plane and are nearly circles, calculate the closest distances our two neighboring planets come to us.

1-18 Compare the intensity of sunlight falling on Neptune with that falling on the Earth, assuming that it varies inversely as the square of the distance from the Sun.

1-19 Explain the different "phases" of the Moon. *Suggestion:* Sketch the position of the Moon in a circular path about the Earth, assuming the Sun to be in the same plane and a great distance away.

MEASUREMENT AND MOTION

2

To "understand" the physical universe, we use a combination of experimental observations and applied mathematical theory. The *measurements* or results of the experimental observations give us facts and clues on which to base our working *formulas* or *laws*. Since measurement plays such an important part in the systematic study of the physical world, it is necessary to understand both the idea and method that is involved.

2-1 FROM OBSERVATIONS TO LAWS

The beginning of each science has been marked by extensive, careful *observations*. Each of us, during his life experience, has made many observations which are catalogued for future use. We know, for example, that most objects fall toward the ground if released from our hand. Or, again, common experience tells us that it is usually harder to move the larger of two objects of the same composition. The weather bureau has provided us with the idea that before a storm, the pressure of the air drops. All of these are observational facts which are useful for everyday living, but do not quite fall into the category of science.

Casual observations can be refined into scientific observations. "It is a windy day" is the result of a casual observation. On the other hand, in a statement such as "wind velocity, 12–15 miles per hour southerly"—also the result of an observation—the actual speed and direction of the wind has been noted, which raises the observation out of the casual into the scientific. Many scientific observations are at first casual and become more quantitative as instruments are developed. When we make *quantitative* observations—those which include numbers—we are taking the first step toward a descriptive science. Further, to make an observation and relay this information to another requires a common language or scale. This raises the degree of sophistication from observation to *measurement.*

Measurement is the result of a scientific observation using a scale known to different observers. Scientific observations may be *passive,* which means that we merely accept the information coming to us and have no control over its source. Examination of the night sky with the eye or with a telescope is a good example of a passive observation. When the observations and measurements are controlled, we regard the process as an *experiment.* Thus experimental observations are the basis for any scientific discipline.

The results of measurement, called *data,* find two ultimate uses. First, the data become part of man's general body of knowledge; in this form, they may be found in tables and reference files. Second, the data are used as clues leading to scientific *theories,* which when tested become our physical *laws.* This use of data is very important to the theorist or theoretical scientist. He must have a certain supply of data to develop his mathematical equations, and then may require much more to test them.

We see, then, that casual observation leads to careful observation, careful observation to measurement, measurement to experiment, and experiment to theory, which in turn may become physical law. The importance of measurement in the development of science cannot be overemphasized; let us, then, examine some of its necessary features.

Every measurement—whether it be of a distance, a weight, an interval of time, or anything else—must include two things if it is to be understood: first, a *number*; and second, a *unit.* Table 2-1 gives a few familiar measurements including units that are used. For example, some results of measurements of distance might be expressed as 6 feet, 20 miles, 2 meters; or the result of measurements of weight might be 20 tons, 165 pounds, 50 grams. The result of making electrical measurements might yield 115 volts, 60 watts, or 90 horsepower.

Although there are many different units, every one may be expressed in terms of not more than three special units, sometimes called *fundamental* units: *length, time,* and *mass.* The first two are familiar to us; we might even argue persuasively that they are more basic than mass. As we shall see, these two units—length and time—lead to ideas of velocity, acceleration, and the description of motion.

The word "measurement" implies some precision—or better stated, imprecision—in the quantity being measured. In physics, as well as other sciences, precision is usually expressed by the *number of significant figures* used in recording the data. Thus 12.4 cm is a more precise figure than 12 cm, whereas 0.12 m and 12 cm express the same degree of precision.

In our technological age, many quantities are measured for which the degree of precision is stated. For a simple example, the volume of gasoline purchased at a service station is ordinarily expressed accurately within $\frac{1}{10}$ gal; whereas the gas gauge in the car allows an estimate only to about ± 1 gal at best. An expression such as "13.6 gallons" thus implies a precision to within $\frac{1}{10}$ gal. If the accuracy of measurement were made to within $\frac{1}{100}$ gal, one could express the value as "13.60 gallons." The additional significant figure (here, the zero) implies additional accuracy. A *significant figure*, although it is of some "significance," is not necessarily a *certainty*. The basic rule about usage of significant figures may be expressed: *The last figure written should be the first uncertain figure.*

TABLE 2-1
FAMILIAR MEASUREMENTS

Quantity Being Measured	Device	Units
Gasoline volume	Gas pump	gallons
Speed	Speedometer	miles per hour
Temperature	Thermometer	degrees (Fahrenheit or Celsius)
Air pressure	Barometer	inches of mercury
Rotational speed	Tachometer	revolutions per minute
Oil pressure	Pressure gauge	pounds per square inch
Electrical energy	Watt-hour meter	kilowatt-hours
Light intensity	Photoelectric light meter	lumens per square foot
Weight	Spring balance	pounds (U.S.)
Floor space	Tape measure	square foot
Distance between cities	Odometer	miles

A simple example will demonstrate how uncertain figures combine in arithmetical calculations. Suppose one is calculating gasoline mileage in an automobile; that the last trip taken was for 103 miles; and that 5.1 gal of gasoline were used. The last figure is based on the readings on the gasoline pump and is thus uncertain. It may be, say, 5.0 or 5.2, although 5.1 is an equally likely value.

Dividing 103 by 5.1 yields 20.20 miles per gallon (mi/gal); dividing 103 by 5.0 yields 20.60 mi/gal; and dividing 103 by 5.2 yields 19.81 mi/gal. Obviously the "best" answer is 20 mi/gal—any more figures than this would be misleading. That is, since the original data contained a factor with two significant figures, the most accurate answer can contain no more than two significant figures.

The rule we shall use when combining numbers is: When multiplying or dividing, the number of significant figures in the result shall be the same as that of the factor with the least number of significant figures.

2-2 BASIC DIMENSIONS—LENGTH AND TIME

Length

The measurement of length or distance is very basic and important in physics. The process is already familiar to us—to find our height, we place a mark on the wall opposite the top of our head and use a foot ruler to obtain its distance from the floor. We use a standard of length agreed upon among people in order to be able to discuss the result sensibly. We know that the ruler must be accurate—if a fraction of an inch has been cut off, or if it has shrunk, or if its marks have been poorly printed, our answer would be wrong. Also, it is clear that there is a limit on precision set by our ability to locate the mark correctly and by the size of the smallest division on the ruler. The science of measurement must account for all of these factors in careful detail.

At this point we should consider the problem of the units in which quantities are expressed. The *British system* of units, used in the commerce and industry of English-speaking countries, includes the familiar inch, foot, yard, and mile:

$$1 \text{ mi} = 5280 \text{ ft} = 1760 \text{ yd}$$
$$1 \text{ yd} = 3 \text{ ft} = 36 \text{ in.}$$

For scientific work, the preferable *metric system* is used. Based on the *meter*, (m) this sytem employs decimal fractions and multiples of the meter, as follows:

$$1 \text{ kilometer (km)} = 10^3 \text{ m}$$
$$1 \text{ centimeter (cm)} = 10^{-2} \text{ m}$$
$$1 \text{ millimeter (mm)} = 10^{-3} \text{ m}$$
$$1 \text{ micron } (\mu) = 10^{-6} \text{ m}$$
$$1 \text{ nanometer* (nm)} = 10^{-9} \text{ m}$$

The use of exponents of 10 in preference to whole numbers is recommended because of the tremendous range of sizes with which physics has to deal, as seen in Figure 2-1. It is curious, one may note, that four important objects—the atom, man, a star, and a galaxy—bear the same ratio of sizes to each other, about 1 to 10 billion.

When the metric system was first devised, a platinum–iridium bar of length thought to be one–ten-millionth of the distance from the North Pole to the Equator was selected as the length of 1 meter. In spite of improved geographic information, this bar, kept in France, remained for many years the reference for all other lengths. More recently (1960) the meter was redefined for convenience in the laboratory as 1,650,763.73 times the wavelength of orange light from an isotope of krypton, Kr^{86}. It is a nuisance to have to convert back and forth between the British and metric systems,

* Formerly *millimicron*.

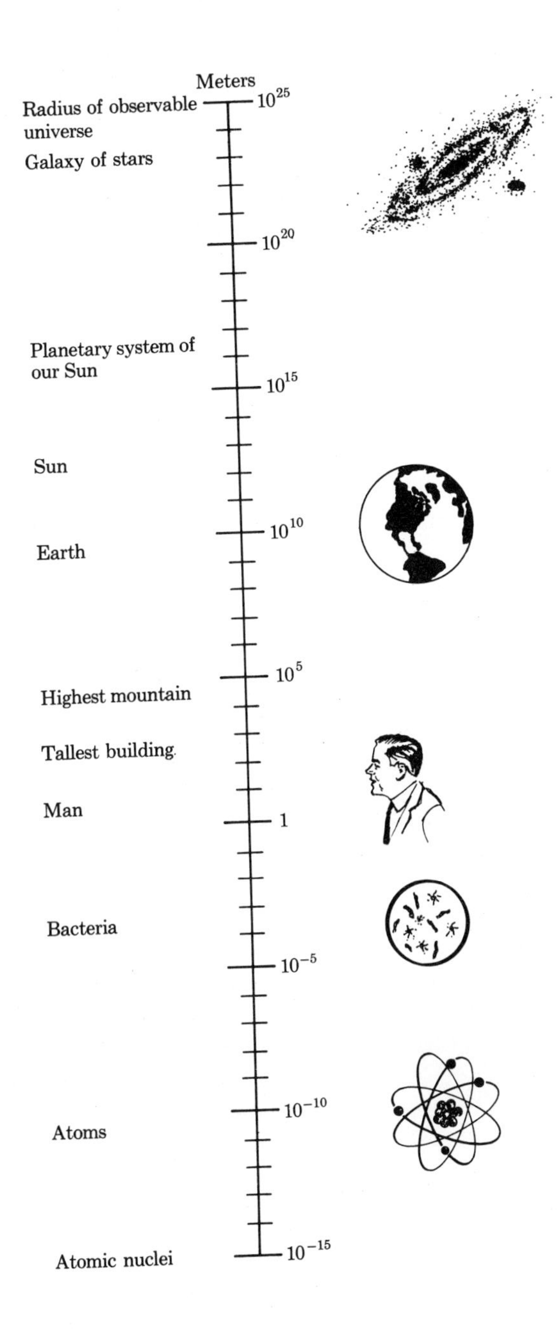

FIGURE 2-1

Range of sizes in the universe

but until the latter is universally adopted we might as well face the existence of a dual system and learn the conversions. Those most often used are the following:

1 in. = 2.54 cm
1 ft = 30.5 cm
1 mi = 1.61 km
1 m = 39.4 in.

Conversion Between Units

We must discuss very carefully the process by which we *convert* from one unit to another, to help avoid future confusion. To illustrate, let us again consider the measurement of our height. We must distinguish between the real physical distance, which is definite and fixed, and what we choose to *call* that distance, which depends on the units of measure we choose. For instance, one's height could be expressed in several ways:

5 ft, 10 in.
70 in.
178 cm
1.78 m
1.94 yd

All of these refer to the same actual distance. This list could go on and on, including such absurd figures as 0.0011 mi. The *number* we get in making a measurement thus depends on the *unit of measure* we adopt: *both* must be stated for the measurement to mean anything. Suppose we have obtained, by use of some device, a number for a certain length. To be specific, suppose we find someone's height—by use of a tape measure—to be 70 inches, and that we want to convert this into another number, as would be obtained if a different device were used—say, a stick marked off in centimeters. A *conversion factor* is need. Starting with the knowledge that

$$1 \text{ in.} = 2.54 \text{ cm}$$

then clearly 70 in. is (70)(2.54) = 178 cm. Another way to get the same answer is to express the length as a symbol with its units in parentheses; e.g.,

$$L \text{ (in.)} = 70 \text{ in.}$$

Now, since there are 2.54 centimeters in each inch—or 2.54 centimeters *per* inch—we can designate that number a *conversion factor:*

$$f = \frac{2.54 \text{ cm}}{1 \text{ in.}}$$

If we want to know the length of an object in centimeters, we *multiply* by that factor, doing the necessary arithmetic on the numbers *and* the units:

$$L\text{ (cm)} = (70\text{ in.})\left(\frac{2.54\text{ cm}}{1\text{ in.}}\right) = 178\text{ cm}$$

(Cancellation lines have been shown for this example.*)

In most of our illustrative calculations, three digits will be retained, as can be read from a slide rule. Numbers may be known to greater accuracy, and all available decimals would be used in a careful research problem. To illustrate a few conversions, let us find the following:

1. The height h in centimeters of the Empire State Building, which is 1250 feet high.

$$h = (1.25 \times 10^3\text{ ft})(30.5\text{ cm/ft}) = 3.81 \times 10^4\text{ cm}$$

2. The circumference s of the Earth at the Equator in meters, using a radius $R = 3960$ miles. Now,

$$s = 2\pi R = (6.28)(3.96 \times 10^3) = 2.49 \times 10^4\text{ miles}$$

Thus

$$s = (2.49 \times 10^4\text{ mi})(1.61 \times 10^3\text{ m/mi}) = 4.01 \times 10^7\text{ m}$$

3. The diameter d in inches of a 5-μ bacterium.

$$d = (5 \times 10^{-6}\text{ m})(39.4\text{ in./m}) = 1.97 \times 10^{-4}\text{ in.}$$

The most useful metric units of area are the *square centimeter* (cm^2) and the *square meter* (m^2). Note the conversion from the British square inch ($in.^2$):

$$1\text{ in.}^2 = 6.45\text{ cm}^2$$

For measurement of volume, the cubic centimeter (cm^3) and cubic meter (m^3) are needed, where

$$1\text{ in.}^3 = 16.4\text{ cm}^3$$

The liter, 1000 cm^3 or 10^{-3} m^3, is occasionally needed.

The unit used for measuring astronomical distances is the *light year*, the distance light travels in one year. The speed of light or of radio signals is 186,000 mi/sec or 3.00×10^8 m/sec, while a year is 3.16×10^7 sec; thus

$$1\text{ light year} = 5.89 \times 10^{12}\text{ mi} = 9.48 \times 10^{15}\text{ m}$$

* Note that the conversion factor is dimensionless, since 2.54 centimeters *is* 1 inch. For convenience, we write such expressions as 2.54 cm/in., with the same meaning intended.

The distance to the nearest star, Alpha Centauri, is 4.3 light years; to the galaxy Hydra, around 3 billion light years.

Angular Measure

We are quite familiar, from the study of geometry, with the use of angles. One common unit of angular measure is the *degree* (deg), $\frac{1}{360}$ of the circle, such that the right angle is 90 deg, the straight angle 180 deg, and so on. The degree in turn is divided into 60 *minutes*, each of which contains 60 *seconds*. The necessity for use of angular measure to supplement linear distances is brought out clearly by asking someone the question "How big does the Moon look to you?" A typical answer is "A few inches." Obviously, the apparent size depends on what object serves as comparison—one's hand in front of his face or a building in the distance.* A more reliable figure is that the Moon's diameter always occupies 0.52 deg out of the 180 deg between horizons. It is conventional to use a symbol such as θ to represent angles. Thus the hour hand at 2 P.M. makes an angle $\theta = 60$ deg with the vertical. An alternate unit for angular measure is the radian, which is an angle describing a circular arc equal in length to the circle's radius. The radian (rad) corresponds to about 57.3 deg, as can be verified by proportions: the angle θ is proportional to the arc s on the circle; i.e.,

$$\frac{\theta}{360 \text{ deg}} = \frac{s}{2\pi R}$$

Now, if $s = R$, $\theta = 360 \text{ deg}/2\pi = 57.3$ deg. Two useful relationships are $360 \text{ deg} = 2\pi$ rad and $1 \text{ deg} = 0.0175$ rad, from which any translation can be made. For any arc s on a circle of radius R, the angle in radians is

$$\theta = \frac{s}{R} \quad \text{or} \quad s = R\theta$$

Time

The astronomical and mechanical devices by which we note the passage of time are familiar to everyone. A succession of equal time intervals is formed by the beat of the seconds by a clock pendulum; by the movement of the second, minute, and hour hands on a watch or a clock run by 60-cycle-per-second electrical supply; by the daily cycle of the Earth's rotation and the yearly passage of the Earth about the Sun. All of these make use of periodic movements for which natural markers exist. It is fortunate that such devices are available, since humans are notoriously poor at judging

* It turns out, surprisingly, that a quarter coin just covers the Moon at 9 feet from the eye.

time. An hour at a party may seem like a minute, while an hour in class may seem like a day. Although there are fundamental questions about the exact meaning of time and the definition of simultaneous events, we shall temporarily follow Isaac Newton's poetic statement: "Absolute, true and mathematical time, of itself, and from its own nature flows equably without regard to anything external . . . ."

On the chart in Figure 2-2, we see the remarkable range of time intervals of interest.

Units of time The *second* (sec) is the principal working unit of time, and is defined as a certain fraction of a year. Other familiar units are related to each other in the following ways:

$$
\begin{aligned}
1 \text{ min} &= 60 \text{ sec} \\
1 \text{ hr} &= 60 \text{ min} = 3600 \text{ sec} \\
1 \text{ day} &= 24 \text{ hr} = 86{,}400 \text{ sec} \\
1 \text{ year}^* &\cong 365 + \text{days} \cong 3.16 \times 10^7 \text{ sec}
\end{aligned}
$$

Fractions of a second that are useful are the following:

$$
\begin{aligned}
1 \text{ millisecond (msec)} &= 10^{-3} \text{ sec} \\
1 \text{ microsecond } (\mu\text{sec}) &= 10^{-6} \text{ sec} \\
1 \text{ nanosecond (nsec)} &= 10^{-9} \text{ sec} \\
1 \text{ picosecond (psec)} &= 10^{-12} \text{ sec}
\end{aligned}
$$

Devices for obtaining the most accurate measurements of time employ *atomic vibrations*, which are remarkably constant. A crystal of quartz is placed in an electrical circuit to impose a natural frequency on it. Fine adjustments are made by use of molecules of ammonia gas NH_3, which make 2.3870×10^{10} vibrations each second. The error is less than 1 part in 1 billion. An even more accurate device is the *cesium clock*. The single electron in the valence shell of the Cs atom provides a magnetism that can be oscillated by radio waves 9,192,631,770 times each second. This device limits errors to 1 in 10 billion, which amounts to a gain or loss of only 1 second in every 300 years.

The measurement of times of the order of many centuries is possible by the use of radioactive isotopes such as those of thorium, uranium, and carbon. By calculating the amount of any of these isotopes present in a sample and employing the amount of product material and the well-known half-lives, ages of rocks or fossils may be deduced.

* The time required for the Earth to complete its revolution about the Sun (as of the first instant of the year 1900) is 365 days, 5 hours, 48 minutes, and 45.9747 seconds. This is reduced 0.53 second per century. The Gregorian calendar assumes instead 49 minutes and 12 seconds. If we round off to 365 days per year, an extra day (February 29) must be added to the calendar every fourth or "leap" year (except at each century year that is not divisible by 400—i.e., 1800, 1900, 2100, etc.).

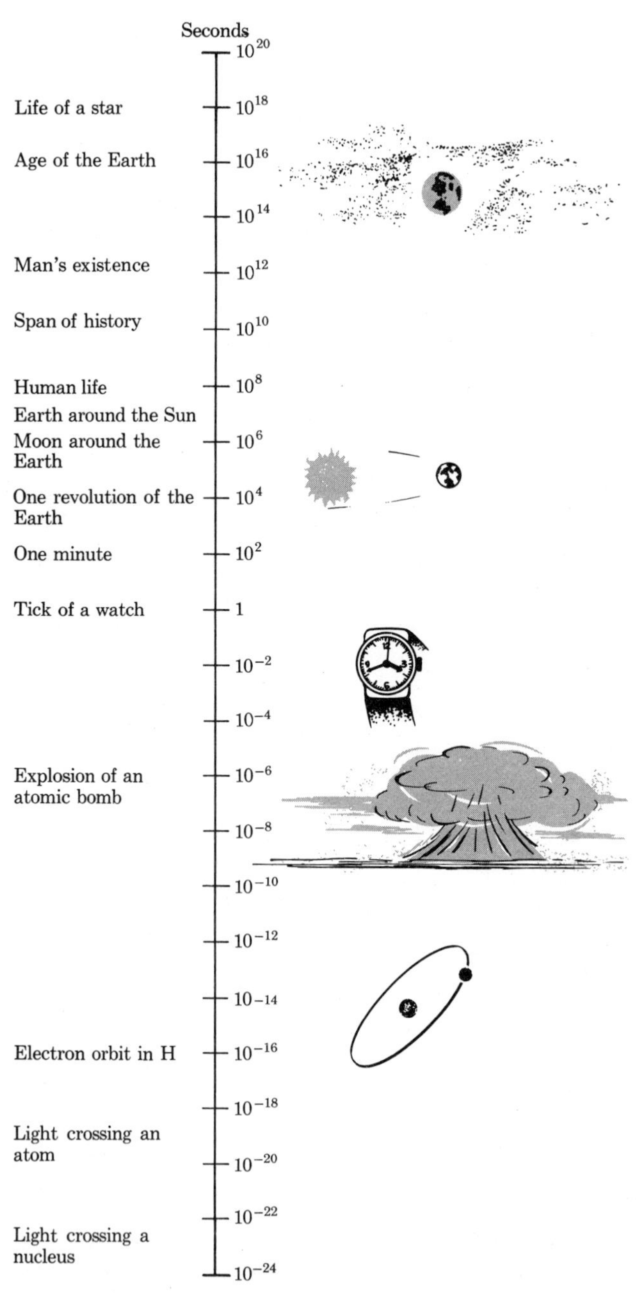

FIGURE 2-2

Range of times in the universe

2-3 MOTION, SPEED, AND ACCELERATION

The subjects of physics and athletic sports would seem to be very far apart. However, the problem of determining where the ball, the player, or the vehicle is located and how they move is much the same as that for an atom, part of a machine, or a planet. Some motions are obviously simpler than others. For example, the runner in the 100-yard dash, if he hopes to win, must move only in a *straight line* down the track. The driver in an automobile race moves in lines and curves over a nearly *plane surface.* The golfball travels horizontally, vertically, and, in a hook or slice, sidewise; thus it has *three-dimensional* motion. In each case we require a *frame of reference* and a *coordinate system.* The track or course is the frame, and the markers along the way are the coordinates. Physics serves to formalize the description of location and to reveal which rules govern the motion.

Motion in a Straight Line—Speed

Each point along a line is assigned a symbol x, to which we can attach identifying numbers, as required. Here we shall use subscripts 0, 1, 2, etc. to indicate specific points. The distance $x - x_0$ is called a *displacement.* (We often set x_0 equal to zero as a starting point or *origin.*) Times may be treated similarly, with the symbol t given to each instant, and $t - t_0$ representing the elapsed time. (Conveniently, t_0 is set equal to zero.) The *average speed* of an object moving along a line between two points is

$$v_{\text{av}} = \frac{\text{distance}}{\text{elapsed time}} = \frac{x - x_0}{t - t_0}$$

Two measurements are made—of distance, using a meter stick or equivalent; and of time, using a clock or stopwatch. Let us consider several simple "experiments."

A gun is fired at time zero to start the 100-meter Olympic run (Figure 2-3). Suppose the runners' starting point is labeled $x_0 = 0$. When the winner breaks the tape at $x_1 = 100$ meters, we stop the watch and read off time 10.1 seconds. We find the average speed of the winner to be

$$v_{\text{av}} = \frac{x_1 - x_0}{t_1 - t_0} = \frac{100 \text{ m}}{10.1 \text{ sec}} = 9.9 \text{ m/sec}$$

The three most common units of speed are miles per hour (mi/hr), feet per second (ft/sec), and meters per second (m/sec). Useful relationships are

$$60 \text{ mi/hr} = 88 \text{ ft/sec} = 26.8 \text{ m/sec}$$

Suppose an astronaut far from the Earth dons his spacesuit and leaves his ship with a relative speed of 6 m/sec. The distance he travels in a straight line is easily found if the speed and time are known. In a

FIGURE 2-3

The Olympic 100-meter run

period of 1 hour he will have drifted away a distance $x = vt = (6 \text{ m/sec})(3600 \text{ sec}) = 21{,}600$ m or about 13 mi.

Two wires separated by 3 meters are laid across the highway to detect speeders (Figure 2-4). As the front tires of a fast-moving automobile cross the wires, two signals 0.1 second apart are recorded electrically. Letting distance and time be $\Delta x = 3$, $\Delta t = 0.1$ (where Δ represents "difference"), we find*

$$v_{\text{av}} = \frac{\Delta x}{\Delta t} = \frac{3 \text{ m}}{0.1 \text{ sec}} = 30 \text{ m/sec}$$

Since speed is a ratio, it is common to state that it is the *rate of change* of position with time. A little later on, we shall introduce the related quantity—*velocity*—which is also a rate of change of position, but with the added feature of specified direction.

We saw in Section 2-1 that the ranges of sizes of objects in the universe and of times for events to happen were enormous. There appeared to be no limits large or small, however. Speed is peculiar in that it is *bounded*. An

* If we had sufficiently precise equipment, we could reduce Δx toward zero separation, while Δt would also approach zero. In the limit $\Delta x = 0$ and $\Delta t = 0$, the *instantaneous* speed is obtained:

$$v = \lim_{\Delta t \to 0} \frac{\Delta x}{\Delta t}$$

Using the abbreviating notation of calculus,

$$v = \frac{dx}{dt}$$

the derivative of position with respect to time.

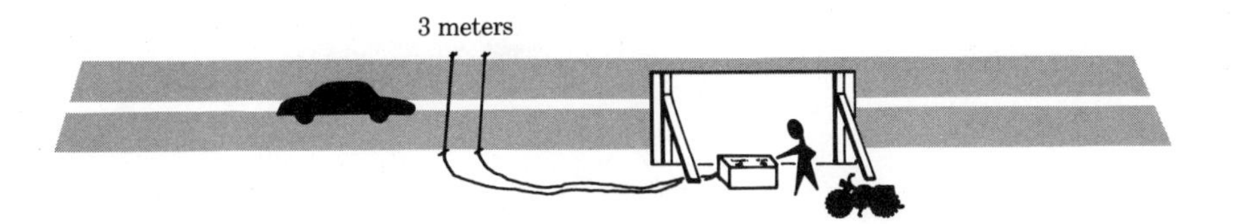

FIGURE 2-4

The speed control

object at rest obviously has zero speed. The highest speed possible is that of light, 3.0×10^8 m/sec, according to Einstein's well-verified theory. Figure 2-5 shows some typical values within these limits, all expressed in meters per second.

Acceleration

When we remark that a car has a good *acceleration*, we mean that it can gain speed rapidly. Suppose it can get from 30 mi/hr (13.4 m/sec) to 60 mi/hr (26.8 m/sec) in a time of 4.0 seconds. For any object, the average acceleration over a given interval of time is

$$a_{av} = \frac{\text{change in speed}}{\text{elapsed time}} = \frac{v - v_0}{t - t_0}$$

For the car,

$$a_{av} = \frac{26.8 - 13.4}{4.0 - 0} = 3.3 \text{ m/sec}^2$$

That is, it gains 3.3 m/sec every second. In words, the units of acceleration are "change in meters per second each second" or, abbreviated, "meters per second per second" (m/sec²). By comparing a difference in speeds Δv at two times separated by an interval Δt, we can form the average acceleration:*

$$a_{av} = \frac{\Delta v}{\Delta t}$$

The simplest type of acceleration found in nature is constant in time: near the Earth's surface, the effect of gravity gives just such motion to falling bodies. For example, an object is dropped from the roof of a tall building, as in Figure 2-6. Legend says that this experiment was performed by

* The instantaneous acceleration is

$$a = \lim_{\Delta t \to 0} \frac{\Delta v}{\Delta t}$$

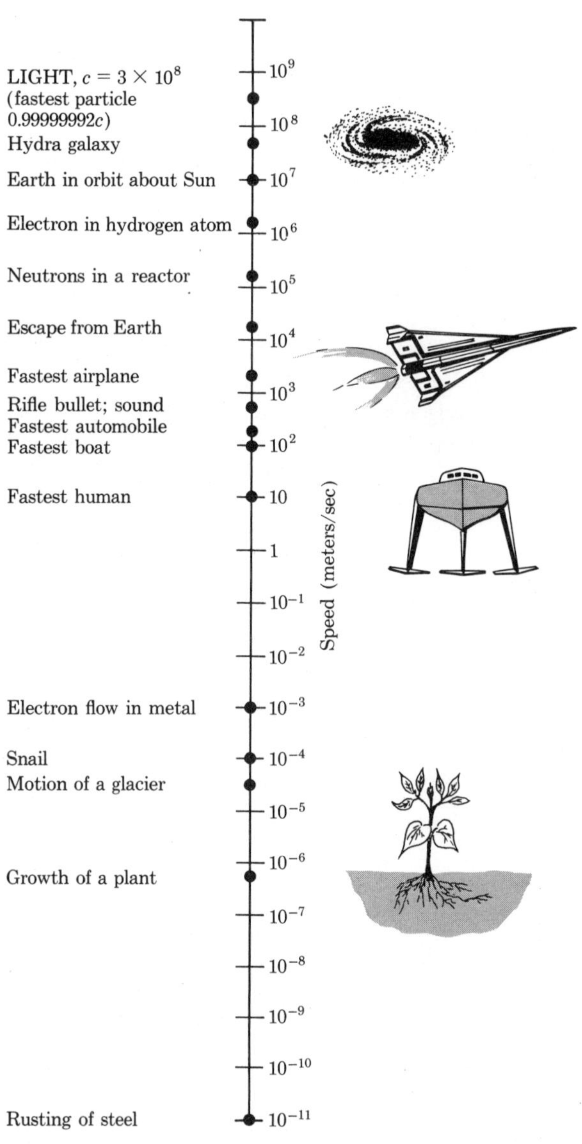

FIGURE 2-5

Range of speeds in the universe

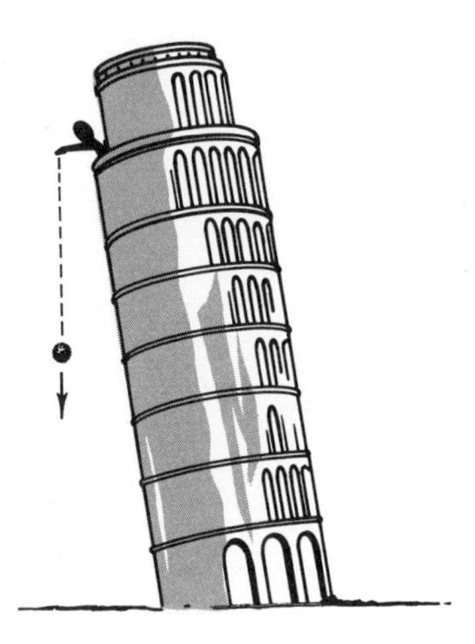

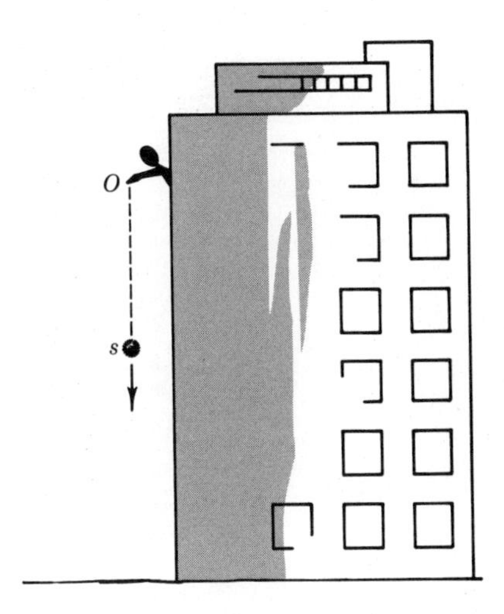

FIGURE 2-6

The falling object

Galileo* from the top of the Leaning Tower of Pisa in the sixteenth century. Its speed at any instant (neglecting air resistance) is found by observation to be

$$v = gt$$

and the distance it travels downward to be

$$s = \tfrac{1}{2}gt^2$$

where g is 9.8 m/sec^2 or 32 ft/sec^2, the acceleration of gravity. Figure 2-7 shows that, with acceleration held constant, speed increases linearly and distance increases quadratically. These relations refer to a special case of the more general type of motion under constant acceleration, for which we now give a mathematical derivation. Suppose that at time zero a body has a speed v_0, and picks up speed to reach a value v at a later time t. The average speed† is then $v_{av} = \dfrac{v + v_0}{2}$. During the time interval, the distance traveled is $s = v_{av}t$ or $s = \left(\dfrac{v + v_0}{2}\right) t$. Also, the acceleration is the increase

* Galileo Galilei, 1564–1642, an Italian known principally for his use of the telescope for astronomical discoveries, for his laws of falling bodies, and for his support of the theory of Copernicus that the planets go around the Sun. The heresy of his views involved him in a serious controversy with leaders of the Inquisition.

† This simple averaging is good only for uniformly accelerated motion—that is, when a = a constant.

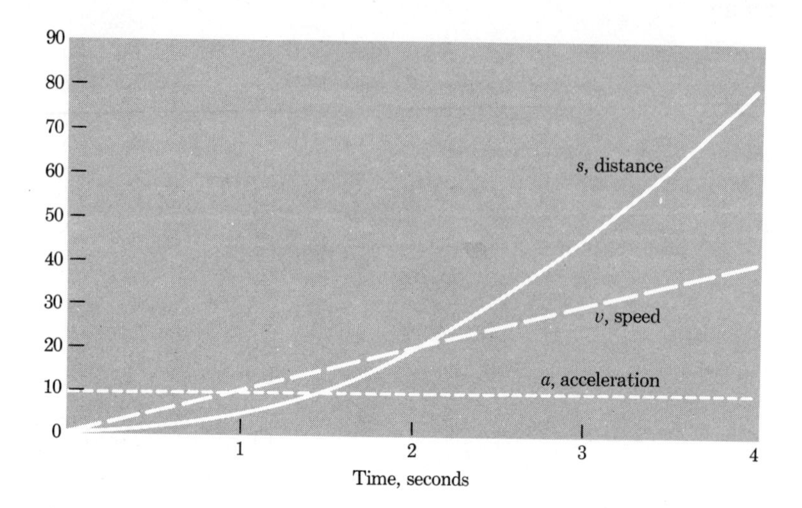

FIGURE 2-7

Graph of Galileo's data on falling object

in speed divided by the time: $a = \frac{v - v_0}{t}$. Now, if we multiply these last two equations together and rearrange, we get $v^2 = v_0^2 + 2as$. Then, if we insert $v = v_0 + at$ into the formula for s, we find $s = v_0 t + \frac{1}{2}at^2$. We then have four equations for uniformly accelerated motion:

$$a = \frac{v - v_0}{t}$$

$$s = \left(\frac{v + v_0}{2}\right) t$$

$$v^2 = v_0^2 + 2as$$

$$s = v_0 t + \tfrac{1}{2}at^2$$

There are four variables in these equations: a, s, v, and t (v_0 is a constant for any problem). One of the variables is missing from each of these equations: s from the first, a from the second, t from the third, and v from the fourth. The selection of the proper equation will thus save the student some time in the solution of a given problem.

Let us examine numerically the motion of a projectile. A bullet is fired downward from the top of a building 150 ft high with initial speed 400 ft/sec. We should like to know how long it takes the bullet to reach the ground and its speed on impact. Now, the acceleration is $a = 32$ ft/sec^2, and since $v^2 = v_0^2 + 2as$, its final speed is

$$v = \sqrt{(400)^2 + 2(32)(150)} = 412 \text{ ft/sec}$$

The time required is

$$t = \frac{2s}{v + v_0} = \frac{2(150)}{412 + 400} = 0.37 \text{ sec}$$

As a second example, suppose a racing car moving at 90 km/hr has 1 kilometer to go in the race and must make the distance in 30 seconds to break the speed record for the course. What constant acceleration is required, and what final speed does the car have? With an initial speed of

$$v_0 = \frac{9 \times 10^4 \text{ m/hr}}{3600 \text{ sec/hr}} = 25 \text{ m/sec}$$

the distance is $s = v_0 t + \frac{1}{2}at^2$, or

$$a = \frac{2(s - v_0 t)}{t^2} = \frac{2[1000 - (25)(30)]}{900} = 0.56 \text{ m/sec}^2$$

$$v = v_0 + at = 25 + (0.56)(30) = 42 \text{ m/sec}$$

Another example of uniform acceleration is the motion of an electron between two oppositely charged plates. The electrons that provide the image in a television picture tube are given such accelerations between the "electron gun" and the screen.

We must keep in mind that the acceleration is uniform only if there is no friction. A leaf and a rock do not fall with the same acceleration in air, but would if they were in a complete vacuum. To describe the firing of a space rocket from the ground, as in Figure 2-8, we must take account of air resistance, which is roughly proportional to the speed of the projectile.

Geometric Interpretation of Accelerated Motion

The algebraic derivation of the relationships between acceleration, speed, and position just given holds only for the case of constant acceler-

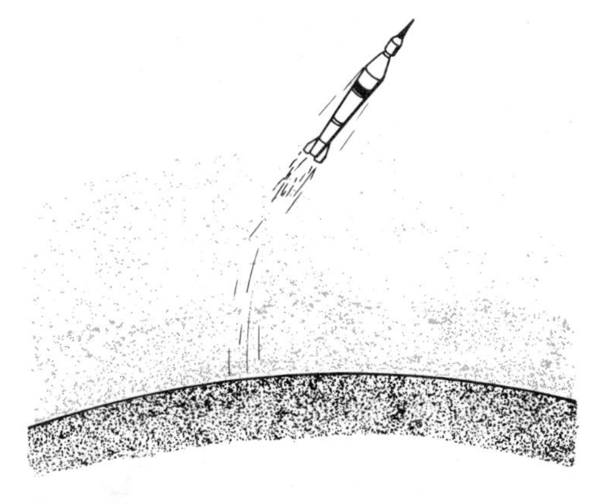

FIGURE 2-8

The rocket with air resistance

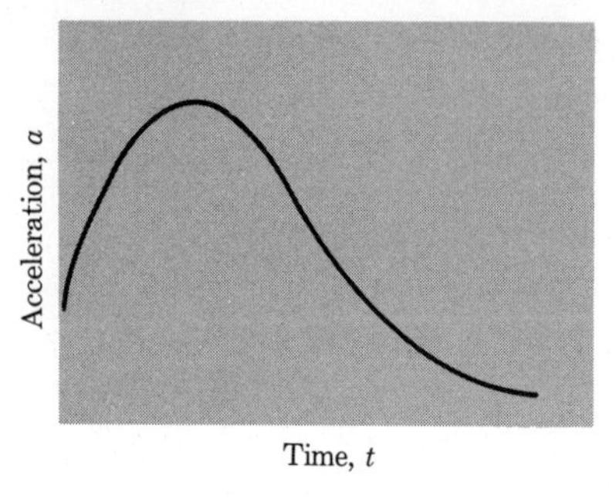

(a) Accurate graph of acceleration

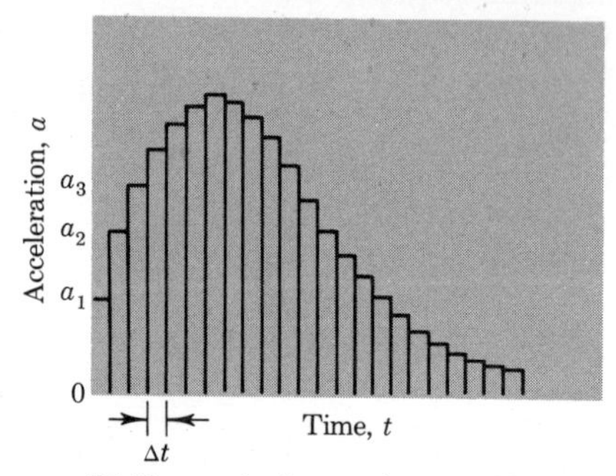

(b) Bar graph of approximate acceleration

FIGURE 2-9

Graphical description of accelerated motion

ation. It is thus desirable to have a more general method, applicable to cases with variable acceleration. Suppose that the acceleration a varies as plotted on the graph in Figure 2-9(a). This might represent the pickup of a car from rest until it reaches full speed. We first replace the smooth curve by a bar graph—Figure 2-9(b), where the acceleration is constant for equal short time intervals Δt. If we made these small enough, the plotted motion would be indistinguishable from the real motion.

Now let us find the gain in speed in the first interval. It is $v_1 = a_1\Delta t$; in the second interval it is $v_2 = a_2\Delta t$; and so on. The total gain in speed is clearly

$$v = a_1\Delta t + a_2\Delta t + a_3\Delta t + \ldots$$

which is also the sum of the areas of the bars in Figure 2-9(b). We see at once that the speed is given by the *area* under the bar graph—or more

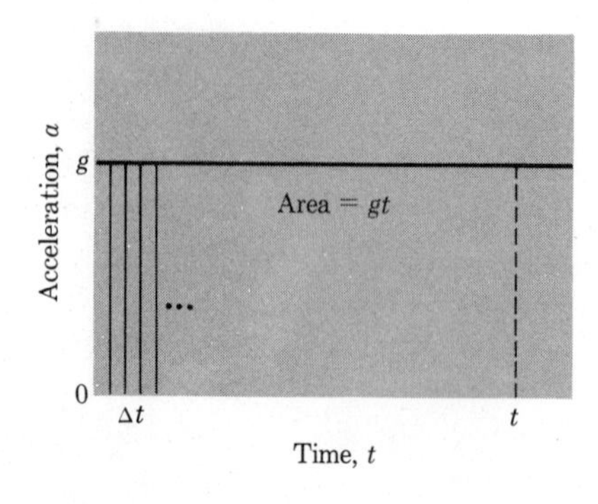

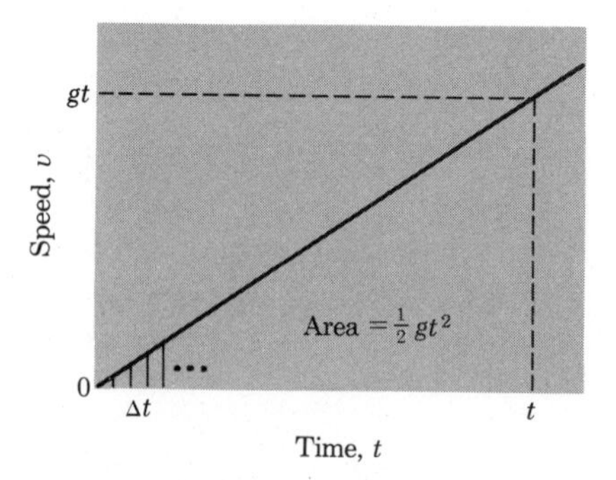

FIGURE 2-10

Motion with constant acceleration

accurately, if we wish, the area under the smooth curve up to time t. Let us test this method for the case of constant acceleration g, as in Figure 2-10. The area under the acceleration curve is $v = gt$, the correct answer. We may now proceed to find the distance traveled, since $s_1 = v_1\Delta t$, $s_2 = v_2\Delta t$, etc. The area under the triangular curve $v = gt$ is one-half the base times the height: $\frac{1}{2}t(gt)$ or $s = \frac{1}{2}gt^2$, as found by other methods.

The foregoing method may be viewed in either of two ways: (1) as the basic procedure of the mathematics of the integral calculus, or (2) as the technique for use of a high-speed electronic computer for solving motion problems.

Galileo and the Scientific Method

The description of the motion of uniformly accelerated objects appears rather straightforward to the modern physicist. However, its revelation by Galileo was of profound significance, not so much for the content of his discovery, but for the means he used and the intellectual consequences. Galileo was probably the first person effectively to apply the *scientific method*, based on measurement and mathematical interpretation. In order to appreciate this point fully, we must review the context of his work. The technique of measurements was known to the Babylonians and Egyptians, but the purpose of their work was principally utilitarian, and they made no attempt to develop physical laws. The early Greeks were imbued with the concept of *natural* motions. For example, Aristotle stated flatly that heavy bodies fall because the surface of the Earth is their "natural place." His idea that the larger the body the quicker the descent was revived and accepted at the end of the Dark Ages, an era whose intellectual effort was dominated by Aristotle's doctrines and those of the Church. Interpretations were made solely on the basis of reason. To make observations was considered unthinkable.

Galileo was trained in such an environment, but so great was his genius that he was able to conceive of making quantitative measurements on motion. He became convinced, perhaps by experiments at Pisa, that if there were no air resistance, bodies as dissimilar as a coin and a feather would fall at the same rate. Turning to the study of motion on a smooth incline as another example of uniformly accelerated motion, his first discoveries were that the time an object takes to fall a given height is independent of the slope of the incline, and that the horizontal speed remains uniform. Most importantly, he made measurements of distances traveled as a function of time, developed tables of data, prepared graphs, and by geometric methods showed that speed is proportional to time and that distance traveled is proportional to the square of the time. Such investigations appear innocuous enough now, but he was literally defying tradition by questioning the accepted views, and thus implying opposition to the whole structure of belief. His work in astronomy was of a similar character. By the use of the telescope, he discovered an eighth planet, Jupiter, which broke the astronomers' tradition that the number of planets was seven (a magic number).

He agreed with the view of Copernicus that the planets moved about the Sun, not vice versa, and thus contradicted the belief that the Earth and man were the focal points of the universe. The environment had always been thought to be designed for the benefit or punishment of man; Galileo and Copernicus relegated man to the role of an insignificant spectator. It is easy to see that the implicit degradation of man would be regarded as heretical. The natural response of the Church was resentment and a sense of danger to their beliefs.

Motion in a Plane

Many motions take place on horizontal plane surfaces, such as limited portions of the Earth's surface, or on vertical planes along the direction of gravity. Common examples we can visualize are the flight of airplanes, the voyage of a ship, the trajectory of a shell, and the path of electrons in a television picture tube. To plot these we need two coordinates. The most familiar coordinate system is the *rectangular* or *cartesian*, with its x- and y-axis. Each point on the plane is represented by a pair of numbers (x, y); straight-line distances between pairs of points (x_1, y_1) and (x_2, y_2) are found by use of the *Pythagorean theorem.*

Polar Coordinates

Almost everyone is more proficient in using cartesian (x, y) coordinates than a *polar* (r, θ) system. It is often difficult to reorient ourselves and translate between the cartesian and polar languages. However, in the mathematical description of physical processes, it is important to use the coordinate system that most naturally and simply represents the motion being studied. Experience has shown that the x, y system is best for motion that is more nearly in straight lines, while the r, θ system is best for nearly circular motion.

An example using the motion of a proton in the *cyclotron* (an accelerator for charged particles) will illustrate the value of the polar coordinate system. Suppose that the proton moves at constant speed in a circle of radius r (Figure 2-11). The location of any point may be specified by either r and θ or by x and y. As seen from the figure,

$$\cos\theta = \frac{x}{r}, \qquad \sin\theta = \frac{y}{r}$$

or

$$x = r\cos\theta, \qquad y = r\sin\theta$$

The correctness of these expressions is verified by forming

$$r = \sqrt{x^2 + y^2} = \sqrt{r^2(\cos^2\theta + \sin^2\theta)} = r$$

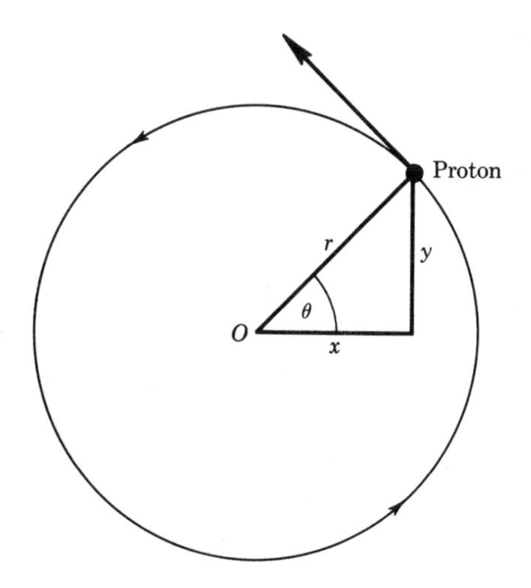

FIGURE 2-11

Circular motion and polar coordinates—proton in cyclotron

In passing, note that

$$\tan\theta = \frac{y}{x}$$

For instance, if x and y are both 1 meter, then

$$r = \sqrt{(1)^2 + (1)^2} = \sqrt{2} = 1.414\text{ m}$$

and $\tan\theta = 1/1$, $\theta = 45$ deg or $\pi/4$. When the proton is displaced along the circumference of a circle of radius R through a distance s, the angle θ is a fraction $s/(2\pi R)$ of the whole 2π, and thus

$$\theta = \frac{s}{2\pi R}(2\pi) = \frac{s}{R} \quad \text{or} \quad s = R\theta$$

Motion in a Circle

There is a great similarity between the mathematics of motion in a straight line and motion on a circular path, with r fixed at a value R, and θ changing. Examples include a proton in a magnetic field, an electron circulating around a nucleus (in a simple model), a point on a turning wheel, an astronaut in orbit, or a planet moving about the Sun.

In a time $t - t_0$ an angular displacement $\theta - \theta_0$ takes place, with an average angular speed, symbolized by ω_{av}, of

$$\omega_{av} = \frac{\text{displacement (angular)}}{\text{elapsed time}} = \frac{\theta - \theta_0}{t - t_0}$$

For example, if the miles-per-hour needle on a speedometer goes through an angle of 45 deg ($\pi/4$ rad) in 10 sec, its ω_{av} is $(\pi/4)/10 = 0.785$ rad/sec.

Look again at the proton moving with uniform speed v on the circle at radius R (Figure 2-11). The distance traveled along the arc increases according to $s = vt$. However, $s = R\theta$, so that $\theta = vt/R$. Now consider rates of change of angular speed. As a car starts up from rest, its wheels begin to turn on their axles, and the rate of rotation increases with time.

To describe how the angular speed changes with time, we define an angular acceleration,

$$\alpha_{av} = \frac{\text{change in angular speed}}{\text{elapsed time}} = \frac{\omega - \omega_0}{t - t_0} = \frac{\Delta\omega}{\Delta t}$$

The similarity that is noted between the variables of linear and angular motion suggests that a table be prepared. (It is assumed here that x_0 and θ_0 are zero at $t = 0$.)

	Linear	*Angular*
Displacement	x	θ
Speed	$v = \frac{\Delta x}{\Delta t}$	$\omega = \frac{\Delta\theta}{\Delta t}$
Acceleration	$a = \frac{\Delta v}{\Delta t}$	$\alpha = \frac{\Delta\omega}{\Delta t}$

Once a relation for the linear motion has been derived, we may borrow it for an angular problem. For instance, with constant angular acceleration α with an initial angle θ_0 and initial angular speed ω_0, we write at once

$$\omega = \omega_0 + \alpha t$$
$$\theta = \theta_0 + \omega_0 t + \tfrac{1}{2}\alpha t^2$$

Note that motion in a circle, even though it takes place in a plane, is really one-dimensional. Consider again the example of a rotating wheel. Suppose that the location at time zero of a spoke on the wheel is $\theta_0 = 0$, and that its angular speed then is $\omega_0 = 628$ rad/sec. If the angular acceleration is α, known to be constant at a value -31.4 rad/sec^2, when will the wheel stop? How many turns will it make before stopping? This reminds us of falling-body problems, and we may take advantage of the similarity: now,

$$\omega = 0 \quad \text{when} \quad t = -\frac{\omega_0}{\alpha} = \frac{(628 \text{ rad/sec})}{(31.4 \text{ rad/sec}^2)} = 20 \text{ sec}$$

and

$$\theta = 0 + 628(20) + \tfrac{1}{2}(-31.4)(20)^2 = 6280 \text{ rad}$$

which is 1000 turns.

A word of warning: a formula that does not properly describe a problem will obviously give meaningless answers. Full understanding of the principles and phenomena behind the formula and of the steps taken in deriving it will help give the needed assurance.

2-4 THE ALGEBRA OF VECTORS

When the measurement of a physical entity involves both direction and numerical value, the idea of *vectors* is necessary. For example, we should like to distinguish between the motion of two cars with identical speeds of 40 mi/hr but moving in different directions—say, one east and the other north. The velocity of one of the cars can be represented graphically, as on a map, by a line segment proportional to the magnitude—here the speed of 40 mi/hr—with an arrowhead pointing east to indicate direction.

Physical quantities fall into two classes: *scalar quantities* having only numerical values, which may be combined by the use of arithmetic; and *vector quantities* having both magnitude and direction, combined by the use of geometry. (Table 2-2 lists a few examples.) Let us now review the geometric method of treating vectors.

TABLE 2-2

SCALAR AND VECTOR QUANTITIES

SCALARS	VECTORS
Time	Displacement
Mass	Velocity
Volume	Force
Distance	Weight
Speed	Acceleration

Consider a displacement from point 1 to point 2 in Figure 2-12. (Point 1 could be labeled exactly by giving x_1 and y_1, and point 2 by giving x_2 and y_2). The displacement from one point to the other involves a change in the x-direction given by $x_2 - x_1$ and a change in the y-direction given by $y_2 - y_1$. We shall label those two *components* or parts of the displacement A_x and A_y; that is,

$$A_x = x_2 - x_1, \qquad A_y = y_2 - y_1$$

We denote the line connecting these two points as the displacement vector $\mathbf{A}$, and note that A is the distance between the two points. The absolute value of $\mathbf{A}$, often denoted by $|A|$, is given by $\sqrt{A_x^2 + A_y^2}$ according to the Pythagorean theorem for right triangles. It then follows that $\mathbf{A}$ itself can be specified completely by the knowledge of A_x and A_y, the components of the vector along the cartesian axes x and y. As a simple example, let us find the absolute value of the vector between the points $x_1 = 1$, $y_1 = 1$ and $x_2 = 3$, $y_2 = 2$. Now $A_x = x_2 - x_1 = 2$, and $A_y = y_2 - y_1 = 1$; hence

$$|A| = \sqrt{A_x^2 + A_y^2} = \sqrt{(2)^2 + (1)^2} = \sqrt{5}$$

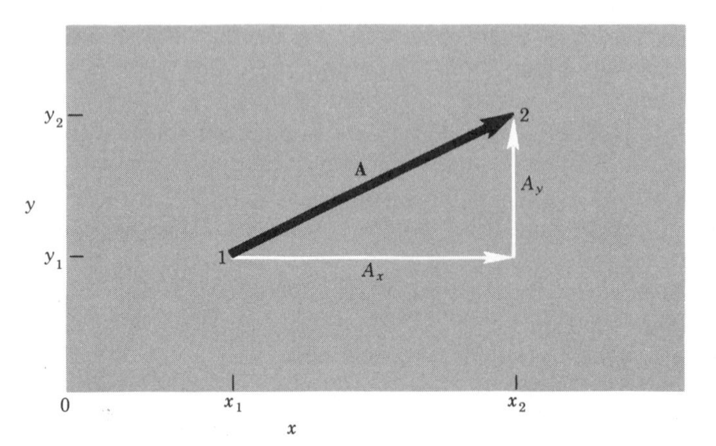

FIGURE 2-12

Components of a vector

Next consider a displacement **A** followed by a second displacement **B** (see Figure 2-13). The result of these two is the same as if the displacement **C** were carried out independently. A vector **C** that has same result as two other vectors **A** and **B** is called the *sum* of these vectors; thus in our example,

$$\mathbf{C} = \mathbf{A} + \mathbf{B}$$

Each of the three vectors is a sum of components along the x- and y-axes, e.g., $\mathbf{A} = \mathbf{A}_x + \mathbf{A}_y$. The vector **C** therefore has an x-component of magnitude $A_x + B_x$, and a y-component $A_y + B_y$. One procedure for obtaining the sum of two vectors can be stated as follows: Add the x-components of

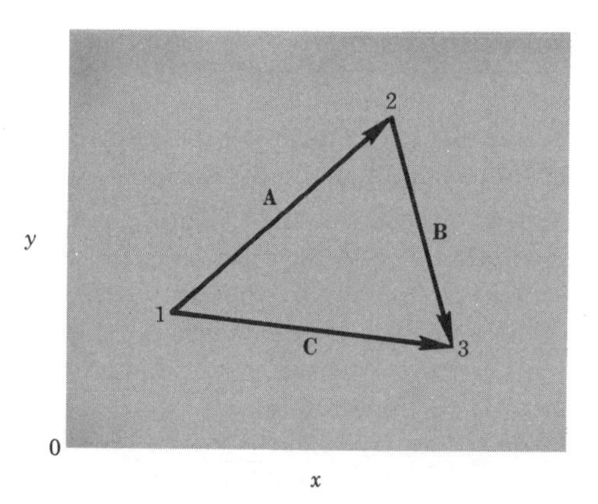

FIGURE 2-13

Addition of vectors

the two vectors, and the y-components of the two vectors, and combine the results using the Pythagorean theorem. To illustrate, suppose we travel on a road straight north for 6 miles and then northeast for 4 miles, as in Figure 2-14. We wish to know the distance, as the crow flies, from our starting point. The components of **A** are $A_x = 0$, $A_y = 6$, while those of B are $B_x = 2\sqrt{2}$ and $B_y = 2\sqrt{2}$. Then, $C_x = 2.83$, $C_y = 8.83$; hence

$$|C| = \sqrt{(2.83)^2 + (8.83)^2} = 9.27$$

This process may be extended to any number of vectors provided care is taken in noting the sign of each component. Just as algebra, geometry, and trigonometry are found to be extremely valuable tools for solving problems,

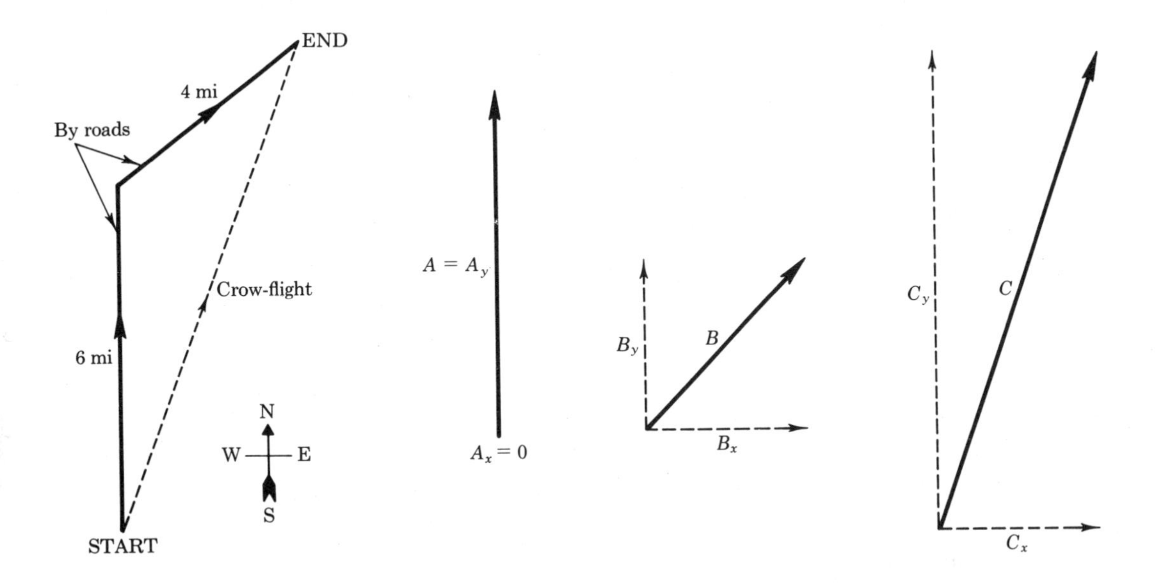

FIGURE 2-14

Vector addition by component method, $C_y = A_y + B_y$ and $C_x = A_x + B_x$

vector notation provides added convenience and power. Displacements in position in one, two, or three dimensions can be described geometrically by a vector, as a straight line of length proportional to the amount of the displacement, and with an arrowhead indicating the direction. Figure 2-15 shows examples in the different spaces.

Particle motion has both direction and magnitude, and thus lends itself to a vector treatment. Visualize, for instance, an electron traversing a cathode-ray tube—as in a television set—in a curved path. At any time t the location of the electric charge is given by the vector **r** (see Figure 2-16).

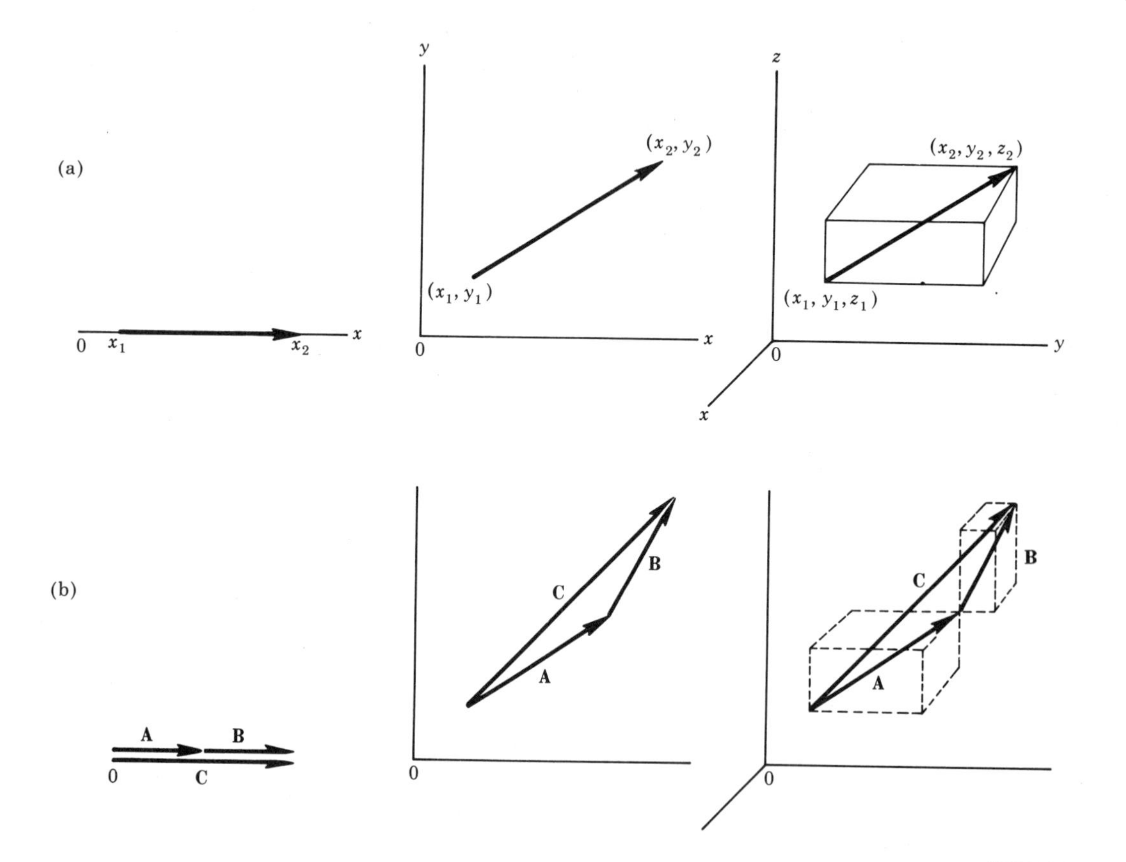

FIGURE 2-15

Vectors in one, two, and three dimensions

A very short time later, at $t + \Delta t$, it has moved to $\mathbf{r} + \Delta\mathbf{r}$. The displacement is seen to be the vector $\Delta\mathbf{r}$. The instantaneous *velocity* is defined as the ratio of this displacement and the time interval,

$$\mathbf{v} = \frac{\Delta\mathbf{r}}{\Delta t}$$

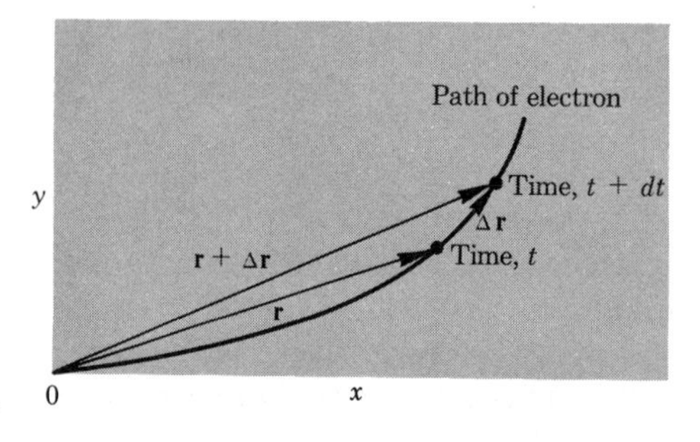

FIGURE 2-16

Vector displacement of electron in cathode-ray tube

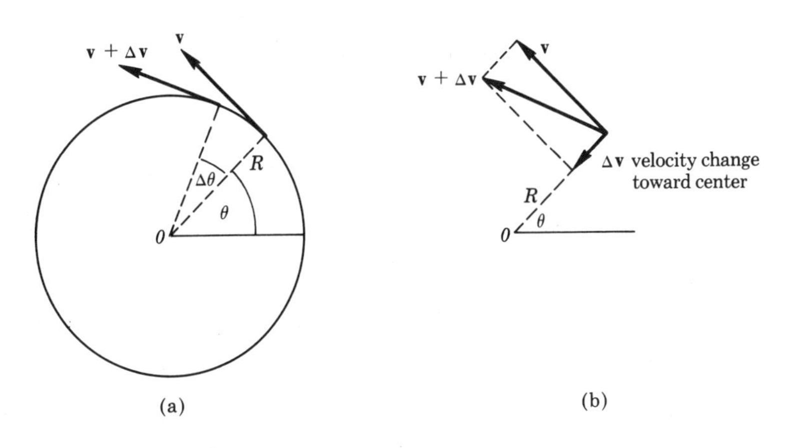

FIGURE 2-17

Motion in a circle at constant speed—centripetal acceleration

where it is imagined that $\Delta\mathbf{r}$ and Δt are both vanishingly small. The magnitude of the velocity is simply the speed, a scalar; the velocity vector is tangent to the curve at the point.

The instantaneous *vector acceleration* is similarly defined. At time t the velocity is $\mathbf{v}$, while at $t + \Delta t$ it is $\mathbf{v} + \Delta\mathbf{v}$, and the ratio of change in the velocity to elapsed time is

$$\mathbf{a} = \frac{\Delta\mathbf{v}}{\Delta t}$$

The magnitude of the vector acceleration is the ordinary acceleration. In summary, the new vectors $\mathbf{v}$ and $\mathbf{a}$ tell us not only the amount of speed and its rate of change, but also give their directions. For example, we can find the velocity and acceleration of the proton moving uniformly around a circular path. As seen in Figure 2-17, $\Delta\mathbf{r}$ and hence $\mathbf{v}$ are tangent to the circle, perpendicular to the radius. Since $\Delta r = R\Delta\theta$, the magnitude of the velocity is $v = \Delta r/\Delta t = R\Delta\theta/\Delta t$. Then, since $\omega = \Delta\theta/\Delta t$, we obtain

$$v = \omega R$$

Note that the velocity is changing, even though the speed is constant. Next, draw two velocity vectors for times t and $t + \Delta t$. The surprising result is that since $\Delta\mathbf{v}$ and thus $\mathbf{a}$ are perpendicular to $\mathbf{v}$, the acceleration is parallel to the radius—that is, toward the center of the circle. The magnitude of acceleration is $a = \Delta v/\Delta t = v\Delta\theta/\Delta t$; but since $\Delta\theta/\Delta t = v/R$,

$$a = \frac{v^2}{R} \qquad \text{(centripetal acceleration)}$$

This is constant, since R and v are constant. The term *centripetal* ("moving toward the center" in Latin) is applied to this particular acceleration. For

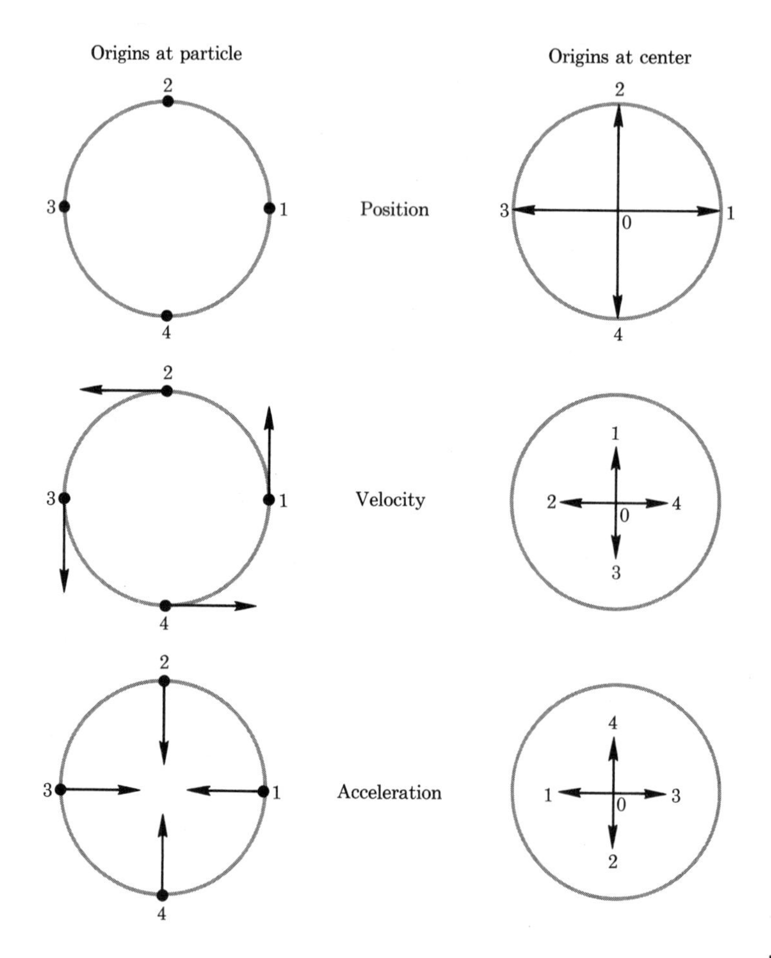

FIGURE 2-18

Changing vectors in circular motion

example, a car going at 20 m/sec rounds a curve of radius 100 m. Its centripetal acceleration must then be

$$a = \frac{v^2}{R} = \frac{(20)^2}{100} = 4.0 \text{ m/sec}^2$$

Recalling that the acceleration of gravity is $g = 9.8$ m/sec², we note that this result is about 0.4 g. An alternate form, using the formula $v = \omega R$, is

$$a = \omega^2 R$$

Note that the vector acceleration is changing even though the speed and the scalar acceleration are constant. An interesting pattern for the vectors

r, **v**, and **a** at different times is seen if their origins are put at the center of the circle (see Figure 2-18). The vectors stay 90 deg apart.

Frequency and period. Circular motion at constant speed is an example of *cyclic* or *periodic motion*. The time taken to make one revolution is called the *period* T, and the number of times a complete cycle is executed each second is called the *frequency* ν. As we shall see more fully in Chapter 11, there are many forms of periodic motion, not all necessarily circular, that can be characterized by T and ν. For instance, if we agitate the water in a pool, the level at some other point will rise and fall with a frequency ν. The alternating electric current in the home also has a frequency (usually 60 cycles per second). The different colors of light are regarded as corresponding to electromagnetic waves of different frequencies.

PROBLEMS

2-1 The world record for the mile run is 3:51 minutes. What would you expect to be the record time for the 1500-meter run?

2-2 A Frenchman by mistake measures the area of his plot of land as 2000 m² using a yardstick, thinking it to be a meter stick. What is the correct area?

2-3 How long does it take a light signal from the Earth to go to and come back from the Moon, 240,000 miles away? (See Figure P2-3.)

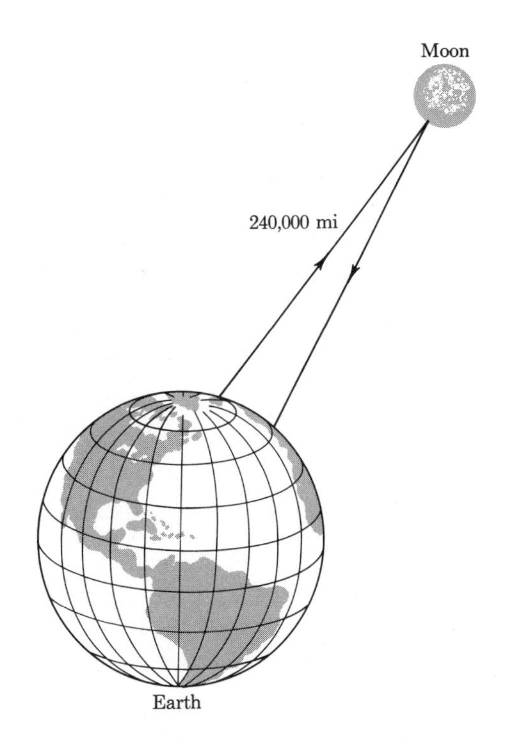

FIGURE P2-3

2-4 A person walks up a stalled escalator in 90 seconds. He rides up the same escalator when it is moving—and he is *not* walking—in 60 seconds. How long would it take him to *walk* up the moving escalator?

2-5 How long would it take light of speed 3×10^8 m/sec (or 186,000 mi/sec) to cross the following? (a) Our galaxy, of diameter 10^{18} mi. (Express also in years.) (b) An H atom, of diameter 10^{-10} m. (Express also in picoseconds.)

2-6 A single air molecule of typical speed 400 m/sec is put into a vacuum chamber of diameter 2 m. How many times per second will the particle cross the space between walls? How far will the molecule go per day?

2-7 Verify that 60 mi/hr = 88 ft/sec = 26.8 m/sec by use of unit-conversion methods.

2-8 Show that an acceleration of 9.8 m/sec^2 corresponds to 1.03 light years/(year)^2.

2-9 A motorist in a car moving at 60 mi/hr sees an obstacle in the road 300 feet ahead. With a reaction time of 2 seconds he slams on his brakes. What constant (negative) acceleration is required to avoid collision? What total time has elasped?

2-10* Find the "point of no return" for an airport runway of 1.5-mi length if a jet plane can accelerate at 10 ft/sec^2 and decelerate at 7 ft/sec^2. What length of time is available from the start of motion in which to decide on a course of action? Verify the latter answer by drawing a graph of v against t.

2-11 A baseball is thrown vertically with a speed of 50 ft/sec from a point that is 120 ft above the ground. Find the height to which it rises, the instant of time at which it passes by its original level, and its speed when it hits the ground.

2-12 A wall clock has a minute hand of length 0.16 m and an hour hand of length 0.12 m. Calculate the angular speed in rad/sec of each hand, and the linear speed in m/sec of the tips.

2-13 A wheel completes 600 revolutions per minute (rpm). What is the corresponding angular speed? If its rotation increases to 900 rpm in 5 sec, what is the average angular acceleration? How many turns did the wheel make in that length of time?

2-14 Find the average angular and linear speeds of (a) the center of the Earth about the Sun, a distance 92.9×10^6 mi away, and (b) a point on the Earth's surface at the equator as it turns about the axis between poles, assuming the Earth's radius to be 3960 mi. Express ω's in radians per second and v's in miles per hour.

2-15 A plane with the ability to travel in still air at a speed of 500 mi/hr encounters the jet stream with speed 300 miles per hour (see Figure P2-15). What speed relative to the ground does it have when riding the stream? When bucking the stream? What will its speed be if it moves at right angles to the stream?

FIGURE P2-15

2-16 The wind velocity is 15 m/sec, 30 deg south from east, as shown in Figure P2-16. Find its southerly and easterly components. Check the answer by use of the Pythagorean theorem.

2-17* A plane flies NE for 80 mi and then ENE for another 120 mi. Find its distance from the origin and angle with the east by the addition of vectors.

2-18 A plane with a speed of 100 mi/hr attempts to land on a field with a cross-wind of 60 mi/hr. At what angle to the landing strip should the pilot aim his plane, and with what speed does he hit the runway?

2-19* An airplane has a speed of 175 mi/hr in still air. It is flying due north along a north–south highway. The pilot hears a garbled message that there is a 50-mi/hr wind but doesn't catch the direction. However, he notes that he is traveling 150 mi/hr by watching the highway. What is the wind direction, and the direction that the plane is heading?

2-20 The Mariner spacecraft requires seven months to travel a 370-million-mile semicircular path to get to the planet Mars. (a) What is the

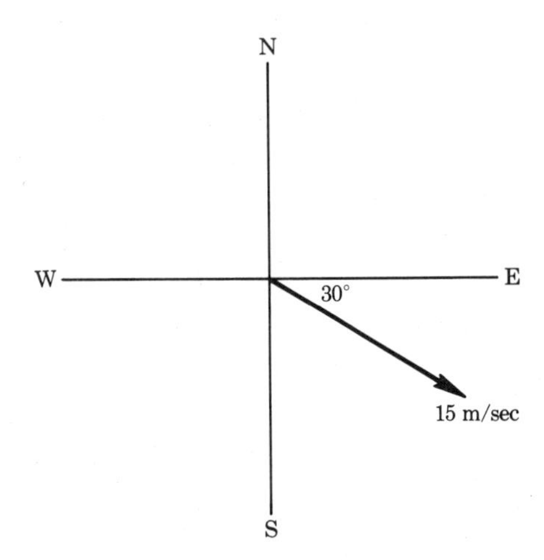

FIGURE P2-16

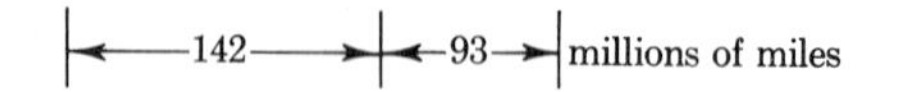

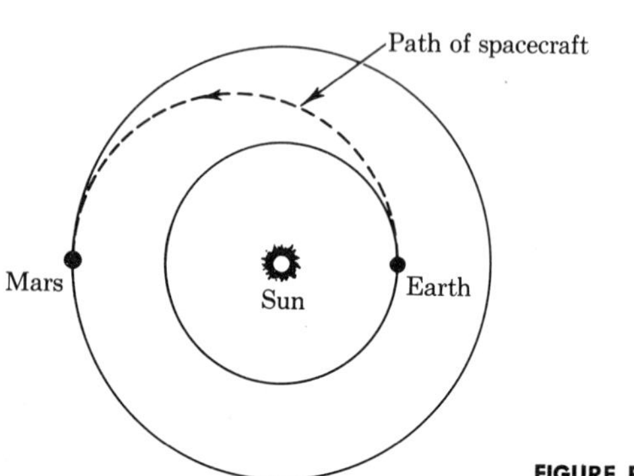

FIGURE P2-20

vehicle's average speed in miles per hour? (b) What is its centripetal acceleration in feet per second per second (ft/sec^2)?

2-21 Find the centripetal acceleration of a car traveling at a speed of 15 mi/hr rounding a corner formed by the intersection of two cross-streets if the radius of turn is 20 ft. How many times the acceleration of gravity is this?

2-22 Find the angular speed and centripetal acceleration of a point on the rim of a $33\frac{1}{3}$-rpm long-playing phonograph record of radius 0.15 m.

2-23 Find the magnitude of the acceleration of a particle on the tip of a fan blade 0.40-m in diameter and rotating at 600 rpm.

2-24* (a) Demonstrate that the way leap years are handled in the Gregorian calendar is consistent with the assumed length of the year. (b) By what year will it be necessary to make an additional correction to the calendar because of the error in the Gregorian year?

REALMS OF DESCRIPTION

3

We are now ready to elaborate on the idea that the explanations of physics involve a great deal of description and do not reveal ultimate reasons for natural phenomena. Observations are related by mathematical forms, and imagined simplified versions of the processes are proposed. Various approximate descriptions are found to be useful in treating certain ranges of size and speed of objects, while more complete views are needed in general. Among the important topics now to be reviewed are the role of mathematics, the meaning of models and approximations, and the ideas of frames of reference and relativity.

3-1 THE ROLE OF MATHEMATICS

Since the development of mathematics as a science by the Egyptians and the Greeks, the subject has enjoyed the respect of scientists in all disciplines. Galileo confided, "If I were again beginning my studies, I should follow the advice of Plato and start with mathematics, a science which proceeds very cautiously and admits nothing as established until it has been rigidly demonstrated." Galileo, in his utilization of abstract symbols for the description of nature, initiated the very fruitful line of reasoning known today as *mathematical physics*. This subject has made tremendous strides under the able hands of such intellectual leaders as Newton, Laplace, Maxwell, Einstein, and Schrödinger.

All of us are aware of the ability of a mathematical expression to state in a minimum of space the relationship between quantities. In physics the more common relationships are learned as working *formulas,* and the most general ones are called *physical laws.* It is important to note that some of the formulas given in the previous chapter are *definitions,* while others are only derived expressions. This role of mathematics—i.e., as the shorthand language of the scientist—is usually understood and accepted by the student without further explanation. However, an example will confirm this utility of mathematics.

Newton's law of gravitation may be stated in words as follows: "The force of attraction between two bodies is directly proportional to the product of their masses and inversely proportional to the square of the distance between their centers." This may be written mathematically as

$$F = K \frac{m_1 m_2}{r^2}$$

where K is a constant of proportionality, m_1 and m_2 stand for the masses, and r is the distance separating their centers. The formula contains all the information that the sentence has, is much easier to remember, and permits calculations of the force.

Once a physical fact has been put in mathematical form, the rules of arithmetic, algebra, and calculus fully apply to the resulting equation. This equation may be solved for any symbol, multiplied or divided by constants or variables or even other equations. In the realm of higher mathematics, it may be differentiated or integrated to obtain *new* equations.

Mathematics can thus be a creative tool in the development of a science. Following the "rules of the game," an investigator may derive previously unknown or unexplained relations. Suppose, for example, that early experimenters had found that finely divided soil dissolved in water more rapidly than chunks, and that a bed of small stones cooled off more quickly than one large piece. Knowing, as separate facts, that the sphere of radius r has a volume $\frac{4}{3}\pi r^3$ and a surface area $4\pi r^2$, they might seek to find a property of a body that increased as the size decreased. A little trial would show that the *ratio* of surface to volume, S/V, is simply $3/r$, which reflects the property sought. Thus we see that manipulation of mathematical expressions leads to new relations of description. Experimental tests may consequently be suggested, which on completion would prove or disprove the original assumptions. This type of application of mathematics forms much of the basis of the "modern physics" that has been developed in this century.

In the following chapters we shall use mathematical descriptions repeatedly as a shorthand method, as a tool, and as a form that reveals physical laws, selecting those descriptions that are most basic, general, and relevant.

Success in the study of physics depends a great deal on the manner with which one studies and remembers mathematical forms. To illustrate, note the several forms into which the expression relating speed v, time t, and distance s can be rearranged: $s = vt$, $v = s/t$, $t = s/v$, $s/vt = 1$, etc. Although some of these forms are handy for a particular problem, it is very unwise to try to remember all of them, or keep a list for reference. The important *physical principle* to keep in mind is that distance traveled depends directly on both speed and elapsed time. By the appreciation of the character of a relatively small number of physical principles it is possible to build up a very full description of the physical world. Rather than attempting to memorize a host of derived expressions, it is often preferable to be able to carry out the derivation from simpler basic relations, using mathematics as a tool rather than a crutch.

3-2 MODELS AND APPROXIMATIONS

In many of the realms of physical description, we observe directly the quantities involved. In throwing a stone, we can feel the stone's inertia, a property closely associated with its mass. We can see the stone move and observe or deduce its position, velocity, and acceleration. In other areas, however, the quantities are not observable by our senses. For example, the interaction of magnetic objects is clearly a force which cannot be described in the same way as the interaction of mechanical objects. The attempt to describe entities and events which transcend our physical senses usually requires the fabrication of a mental picture or image that is closely connected with some well-understood phenomenon. Such a structure is called a *model*. A model is a heuristic device—an aid to our imagination to help in understanding.

Many times in a particular area of physics, a model undergoes extensive changes as new knowledge is gained and applied. An evolving model frequently suggests experiments or tests that may lead to corrections or changes, or even abandonment of the model.

Physical Models

The behavior of even the simplest physical systems, such as a single molecule of water H_2O, is amazingly complicated, in part because of the many components—in this case, atoms, electrons, nuclei, and nucleons. The variety of possible interactions of components with each other and with the environment further complicates the situation. Even when we turn to the high-speed computer the problem of complete description may be hopeless—for instance, in trying to keep track of all the molecules in a glassful of water (some 10^{24} of them).

Thus, we devise *models* of the processes that contain the most important features. These might be called idealizations of the situation, with all of the less-significant details ignored in favor of the ones that dominate the process or which appear to lead to general truths or principles. Some examples will make the idea of models more specific:

1. A projectile such as a rocket, a bullet, or an electrically charged atom is often approximated as a *point mass*, an idealized object with material concentrated at a geometric point. For purposes of finding its basic motion, we ignore air resistance, internal motions, and changes of shape.

2. The parts of a mechanical device are assumed to be perfectly rigid, although it is well known that all materials can be stretched or compressed or twisted. This idealization permits a first-order approximation to be made of the machine's function. Naturally, when elastic properties become important, one re-examines the situation more closely.

3. We imagine an object to be isolated, or insulated from its environment, except for those disturbances we wish to apply. This technique is frequently used for processes where heat is involved. We realize that there is no perfect insulator for heat or electricity, but we know that the behavior of the system is dominated by internal effects rather than the outside environment.

4. Applied forces are taken to be uniform over the region of space in which an object moves. A good example is the force of gravity near the Earth. The force will actually change slightly with location, as will be discussed in Section 5-1. In the treatment of falling bodies, the inclusion of these deviations from uniformity would tend to obscure the meaningful or practical motion.

5. Atomic particles are visualized as being spherical in shape. For many measurements, this form is either sufficiently correct, or our ignorance of the more precise shape does not justify making any other assumption.

6. The internal structure of a nucleus is considered in terms of the water in a droplet, subject to possible distortion or division into smaller bits. Obviously this is not a "real" version of the nucleus, but is convenient for thinking, discussing, and even analyzing with some success.

One can continue at some length to list examples of physical idealizations that are made regularly and deliberately by physicists in their investigations. Among the virtues of this procedure are the ability to *visualize* processes, to perform mental predictions of performance, and to draw comparisons or analogies between different phenomena. One may quickly check measured results with those expected, and discover either errors in the experiment or weaknesses in the physical model that has been used. Communication with others is assisted by the use of common models. This is especially important in the learning process. It has been said that we do not learn the subject of physics, but merely become more familiar with it by long-continued association. Thus it is proper and necessary to study the simple, rough approximations to physical processes on first exposure to the subject, making refinements and improvements in the picture with further study.

There are hazards as well as benefits in using models. Having erected an image of a process, and having found it rather successful in giving agreement with observations, the human mind tends to believe that the model is real, and to resist criticisms or modifications of it. Part of the delay in accepting Einstein's theory of relativity or the modern quantum mechanics can be ascribed to such firm beliefs in models of nature. It is clear that one is more comfortable with a logical and reasonably accurate picture than being forced to revise his thinking toward a more complex or uncertain viewpoint. Progress toward the ultimate goal of a complete description of nature, however, is made only with open and flexible minds that are willing and able to accept new ideas.

Mathematical and Numerical Approximations

Simplifications in mathematical operations that are approximate fall into a different category from physical idealizations. Such approximations are made for convenience in handling problems. They may be based on physical knowledge, but are made after the general structure and principles of the subject have been set. Some familiar illustrations are given below.

The number of digits or significant figures that we use in a constant such as π depends on the accuracy desired. For quick guesses, the number 3 is enough; in many calculations, setting $\pi = 3.14$ will give results compatible with our ability to measure with simple instruments; for other problems, $\pi = 3.1416$ or even more figures of this irrational number* may be required.

Physical constants such as the speed of light fall in another class, because we have no method other than improved measurements for obtaining more decimal places. It is unnecessary, for most work, to use the present best value of 2.997925×10^8 m/sec. The value 3×10^8 is fortunately within 0.1 percent of the correct number. Sometimes there is a tendency to carry along more significant figures in a numerical calculation than the accuracy of known physical constants warrants. Generally, the answer is no better than that of the most important constant used in the solution, and it is misleading to quote any more figures than that limit sets. In a few situations, however, the nature of the algebraic operations requires that extra figures be carried along, particularly if comparisons of various cases are to be made.

As he proceeds into the study of physics, the student may be surprised to find that many laws or principles are of a *linear* form—that is, one variable increases linearly with another, as $y = bx$. Such relations may well be only the first *approximations* to the more correct description. It is reasonable to ask why we are willing to accept rough versions of the behavior of a system. The first reason is that linear expressions are simple and easy to manipulate, and that with them algebraic results, which are

* To twenty decimal places, $\pi = 3.14159\,26535\,89793\,23846$. A mathematician named Shanks once computed π by longhand in fifteen years to 707 decimal places. (Most of them turned out to be wrong.) An electronic computer can give a figure with 10,000 decimal places in a matter of hours.

preferable to mere numerical tables, may be obtained readily. The second is that linear expressions are frequently valid so long as one restricts the range of application. For example, the curve $y = 1 + x + x^2/2$ is very accurately approximated by $1 + x$ so long as x remains small—say, less than 0.01. From another viewpoint, we know that any smooth curve between two points can be nearly duplicated by a straight line, so long as the points are very near each other.

The use of the arithmetical average instead of detailed values of a mathematical function is very common. This will lead to sufficiently accurate results if the quantity being described varies slowly. The average in time of a changing electrical current can be used to represent heating effects, but obviously cannot be carried to the extreme case of a surge that blows out a fuse.

The science of physics makes use of numbers that are inconvenient to manipulate by ordinary arithmetic. An adequate number of significant figures is often attainable by use of only two terms in the binomial expansion of basic algebra:

$$(a + b)^n = a^n + na^{n-1}b + \frac{n(n-1)a^{n-2}}{2}b^2 + \ldots$$

Let us illustrate this point by asking for the value of $(1 - x)^{-1/2}$ when x is 0.0002. By ordinary means, it is $1/\sqrt{0.9998}$, an awkward number to compute by either slide rule or pencil and paper. Applying the binomial expansion, however, letting $a = 1$, $b = -x$, and $n = -\frac{1}{2}$, we have

$$(1 - x)^{-1/2} = 1 + \tfrac{1}{2}x + \tfrac{3}{8}x^2 + \ldots$$

which is also $1 + 0.0001 + 0.000000015 + \ldots$. The sum of two terms in the series gives 1.0001, which is within 1 part in 100 million of the exact answer.

3-3 DATA AND ERRORS

The process of measurement plays a key role in physics, since the development of detailed and correct conclusions depends vitally on reliable and accurate data. Let us proceed from some familiar crude examples of measurement toward a refined scientific approach. Suppose a new student asks us the question "How far is it from here to the library?" A glance allows us to respond "A block or so." Were we interested in verifying the answer, we might pace it off in what we hope are yard-long steps. Because of an accumulation of errors, the measurement may be high or low by a few feet. The use of a steel tape might bring us to within a fraction of an inch of the correct answer. Each of the above quotations has involved some uncertainty in length. In reporting our findings, we should quote what the distance is, as best we know it, but attach some further information on how sure (or unsure) we are of that answer. Thus we might state that the distance to the library is $x = 350.2 \pm 0.1$ ft, where the added ("plus-or-minus") number might be called our "uncertainty." In scientific measurements the first number is the *average* of the measurements and the second number is called the *error*. The science of statistics allows us to evaluate the error from our own data. The technique consists of analyzing the results of several measurements made under as nearly the same conditions as possible.

Suppose that we had collected the following set of eight measurements of the distance to the library, labeled x:

Trial	x
1	350.15
2	350.31
3	350.22
4	350.05
5	350.20
6	350.19
7	350.07
8	350.25

The *average* value, represented by $\bar{x}$, is easily obtained as the sum of the measured distances divided by the number of observations. (Note that here we can simply average the numbers after the decimal and add to 350.) Thus we find

$$\bar{x} = 350.18 \text{ ft}$$

We could stop at this point and say that this number is the distance of interest, but the question of reliability or precision has not yet been answered.

Experimental errors are of two general types—*systematic* errors and *statistical* fluctuations. An example of the first type might result from the use of a cloth tape measure that was washed and as a result is too short. Or systematic error in an electrical measurement can result from a drift in the voltage in the laboratory as the day goes on. In principle, such errors could be tracked down and eliminated, but it may be impossible in practice. The second type of error has to do with the fundamental variable behavior of physical events. For example, suppose we are measuring the radiation from a collection of radioactive atoms. In one interval of time, we might find that 4 particles disintegrate, in another 10, in still another 0, and so on. How the number observed varies depends on the mechanism of the particular measuring process.

Many events in physics obey what is called the *normal law of error*. A set of n ideal measurements of a general quantity yields values x_1, x_2, x_3, . . . with average value $\bar{x}$, and deviations from the average Δx_1, Δx_2, Δx_3, ... One-half the measurements will lie outside and one-half inside the range $\bar{x} - r$ to $\bar{x} + r$, where r is called the "probable error," given by

0.6745σ, where σ is the standard deviation, calculated from

$$\sigma = \sqrt{\frac{(\Delta x_1)^2 + (\Delta x_2)^2 + (\Delta x_3)^2 + \ldots}{n}}$$

Thus r serves as a measure of precision. Even if the measurement being performed is governed by another law besides the normal law, the latter is often used for convenience in error analysis. Let us return to our example of the distance to the library, seeking a value of the probable error. In the following table we first form the individual errors, as the difference between each observation and the average, neglecting signs. Then we square the errors and take their sum:

Trial	Δx	$(\Delta x)^2$
1	0.03	0.0009
2	0.13	0.0169
3	0.04	0.0016
4	0.13	0.0169
5	0.02	0.0004
6	0.01	0.0001
7	0.11	0.0121
8	0.07	0.0049
	Sum:	0.0538

Then, $\sigma = \sqrt{0.0538/8} = 0.0820$, and $r = (0.6745)(0.0820) = 0.0553$. We would thus state that the distance to the library is (in feet) 350.18 ± 0.06, where the error has been rounded off from 0.0553 to 0.06. Common sense suggests that the use of better instruments would give individual measurements that are closer to each other, corresponding to smaller probable error. This is correct *only up to a point,* as we shall see later. Should we attempt to reduce the uncertainties both in location and speed, a limit to the accuracy of our answers would be reached, regardless of the accuracy of the measuring devices employed.

3-4 FRAMES OF REFERENCE AND RELATIVITY

Everyone knows that several observers will give different accounts and interpretations of the same event. This fact is well verified by listening to conflicting stories of accidents in a court of law. Excluding deliberate lies, some of the variations are explained by physiological and psychological differences in people. The remainder are due to the vantage points of the observers.

Let us take a simple example: We are riding as passengers in an automobile that is going up a slight hill toward a stoplight, as in Figure 3-1. Suddenly it seems that the car behind crashes into our car. We turn around indignantly and blame the careless driver, only to have him berate us for

FIGURE 3-1

Illustration of relative motion

allowing our car to roll backward into his. Each of us is perfectly sincere in our interpretation of the collision. How can we know who is right? We could never tell unless we have been looking out the window at some fixed reference object, such as a nearby building, or there happened to be an impartial pedestrian on the street. The only certain fact was that the cars were moving *relative* to each other.

Next, imagine an object dropped from a moving airplane. To the pilot, the object seems to fall straight down, but to a person on the ground, it appears to follow a curved path. If asked to decide what the *real* motion was, we would be inclined to side with the observer on the ground and say it was a curve. However, remember that he is on the surface of a rotating Earth which itself is moving about the Sun. The object clearly must move in some different curved path as viewed from the Sun. Does the Sun represent the final fixed vantage point? Not necessarily, since astronomers tell us that the universe appears to be expanding. We are faced with the conclusion that there is probably *no* fixed object from which observation point we can form final judgments concerning motions—or, for that matter, concerning the laws of nature.

These examples point up the need to specify the point of view from which observations are made. The term *frame of reference* denotes a mathematical "structure" within which an observer sits. Examples might be sets of cartesian coordinates erected either by a person at the Earth's surface, or by a pilot of a moving plane (using, say, the tail of the plane as zero for his x-coordinate), or by hypothetical observers located at the surface of the Earth, at the center of the Earth, or at the Sun. Since the coordinate systems are constructions of our mind, they could be placed anywhere; but of course we are interested in what a real observer would see if located within any of them. New conceptual problems arise when two such frames of reference move with respect to each other. For example, a person on a moving train sets up a coordinate system that is fixed to the train, while a person at the station sets up one there. The questions to be asked are: What does each measure? How do they compare observations? Whose conclusions, if either one's, are preferable?

Putting the concept of relative motion into mathematical terms, we may be said to have a theory of *relativity*. Most people associate this word with Einstein, because his revolutionary view early in this century focused new attention on the subject. What is not generally realized is that Galileo and Newton had developed a classical theory of relativity centuries earlier. We shall review both, using examples of familiar moving objects to illustrate the ideas. It should be understood, however, that we expect the ideas to apply to all objects.

The Relativity of Newton

The sight of a jet airplane flying overhead is a familiar one. Examining its motion and observations by using some stationary markers on the Earth's surface, we can develop the meaning of relativity and extend the idea to all moving bodies. Visualize, as in Figure 3-2, the origin of the frame of an Earth observer, at the base of a tall radio-transmission tower. The second observer is in a plane, from which the tower can readily be seen. The plane is moving to the right with a known constant speed V. The two observers synchronize watches at time zero, when the origin at the tail of the plane coincides with the origin on the Earth's surface. By the time t seconds have elapsed, the plane is a distance Vt to the right. We now ask, What are the coordinates of a second radio tower in the distance, as measured by each observer? From the ground it is x; from the plane it is X (where we use small and large letters to distinguish each measurement). It is easy to see from Figure 3-1 that

$$x = X + Vt$$

or conversely,

$$X = x - Vt$$

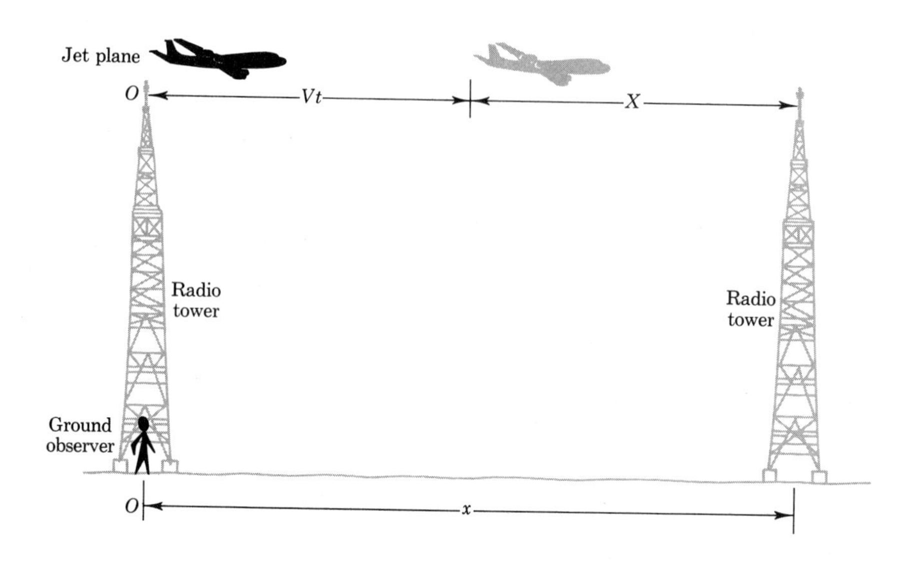

FIGURE 3-2

The Galilean transformation

FIGURE 3-3

Observation of a moving object from two frames of reference

These equations, which are based on everyday experience, are called the *Galilean transformation*, named after Galileo. It is very important to note that the equations would be the same whether the plane were moving with respect to the Earth or the Earth were moving backward under a "stationary" airplane (better, a hovering helicopter). Without a third object as standard, neither observer can know his absolute motion, in accord with our earlier discussion. The two forms of the equation can be used for mutual predictions of observations in the two frames. Suppose the observer on the plane measures a distance to the second tower X at time t. He deduces that the ground observer will find the tower at x. Conversely, if the person on the Earth sees x at t, he predicts that the person on the plane will find X.

The foregoing referred to the observation of a fixed object, in this case the second tower. Instead, let the two persons view an object in motion with some speed and acceleration—for example, a bird in flight, moving as sketched in Figure 3-3. At any instant t the bird's position is observed as x and X from the two frames, respectively; and after a time interval Δt its location is respectively $x + \Delta x$ and $X + \Delta X$. By use of one of the transformation formulas, we may find a relation between the two views of the bird's speed. Thus we write

$$X = x - Vt$$

$$X + \Delta X = x + \Delta x - V(t + \Delta t)$$

and subtract equations from each other to obtain

$$\Delta X = \Delta x - V\Delta t$$

We then divide by Δt and define speeds as $u = \Delta x/\Delta t$ as seen from the ground, and $U = \Delta X/\Delta t$ as seen from the plane. Then

$$U = u - V$$

which merely says that the plane is overtaking the bird and that the moving observer measures a speed that is lower than u by an amount V. A similar analysis will yield a relation between accelerations as seen in the two frames. After an elapsed time Δt, the respective speeds are $U + \Delta U$ and $u + \Delta u$, related by

$$U + \Delta U = u + \Delta u - V$$

We then subtract the previous equation, divide by Δt, and define observed accelerations $a = \Delta u/\Delta t$ as from the ground, and $A = \Delta U/\Delta t$ as from the plane. We find the simple result

$$A = a$$

which says that accelerations are identical, seen from two frames moving *uniformly* with respect to each other. In this classical description, it is thus immaterial where one observes the process of acceleration. In this sense, the two frames are said to be equivalent and are classed as *inertial frames*, referring to the measurement of inertia, force, and motion. We shall see later that, if the two frames of reference are *accelerating* with respect to each other, observations of acceleration appear different. In this event the frames are non-inertial, no longer equivalent, and require more careful examination.

Einstein's Theory of Relativity

For calculating the motion of familiar large-scale objects, Newton's concept of relativity is quite adequate, since the speeds of frames moving with relation to each other are low, at least in comparison with the speed of light, $c = 186{,}000$ mi/sec $= 3 \times 10^8$ m/sec. Even a modern jet plane with $V = 2000$ mi/hr $= 0.556$ mi/sec is slow, since $V/c = 0.556/186{,}000 = 0.000030$.

The Earth, in its yearly rotation about the Sun, has a ratio V/c somewhat larger. The circumference of its orbit is $2\pi r = 2\pi(9.3 \times 10^7)$ mi and the time taken to complete it is (365+ days)(86,400 sec/day) $= 3.16 \times 10^7$ sec. Thus $V = 18.5$ mi/sec, and since $c = 186{,}000$ mi/sec, $V/c = 0.0001$, still negligible compared with 1. The ratio becomes close to 1, however, for particles from nuclear reactions; or for charges accelerated by electrical forces, as in a cyclotron; or for stars receding from us at high speed. At this point Newton's relativity is inadequate and Einstein's* must be used. In order to understand the failure of the classical approach, let us return to the analogy of the jet plane and perform an experiment.

Suppose that a burst of radio waves or light waves is sent out from the first tower just as the plane passes over it. The observer in the plane and the observer on the ground have agreed in advance on the length of time they shall allow to elapse before noting how far the light has traveled. They plan to use this length and time to calculate the speed of light c. In the Newtonian picture, the Earth observer obtains an answer

$$c = \frac{x}{t}$$

and the plane observer

$$C = \frac{X}{t}$$

These speeds are those corresponding to u and U as we applied to the observations of the moving bird in a previous paragraph. They should be related by

$$C = c - V$$

since the plane is moving in the same direction as the light, and the plane's observer would expect to note the lower speed. The actual experimental fact is that *the two observers measure the same values for the speed of light.* This is not an approximate agreement due to the plane's low speed compared with that of light, but a real effect, verified in all situations.

Einstein's explanation for this seeming paradox was based on a new logic. First, he distinguished both distance *and time* as measured by the two observers in their own frames, letting (x, t) describe one and (X, T) the other. Next, he required that the connecting formula be a *linear* expression. This guarantees unique corresponding points in the two spaces. Then, as in classical physics, he assumed that the replacement of x by X and of t by T, along with the change of sign on V, would give the *same form* of transformation. This assures that the laws of physics are the same for each observer. All of these conditions are met by the expressions

$$x = k(X + VT)$$
$$X = k(x - Vt)$$

He supposed that the factor k should involve the speed of relative motion and the speed of light, and should reduce to 1 as V approaches zero. The

*Albert Einstein (1879–1955) was a genius who contributed new thoughts to science throughout his lifetime. He was an indifferent student, and was a clerk in the Swiss patent office when he made his first scientific discovery. He is known for his work on the molecular theory, the photoelectric effect, the theory of special relativity, the theory of general relativity, and unified field theory.

form of k can be found by reinterpreting the measurements of the speed of light. Consider again the results, which were $c = x/t$, $C = X/T$, but now insist that $c = C$ be found by measurements. It is a matter of straightforward algebra* to eliminate the times and distances to obtain

$$k = \frac{1}{\sqrt{1 - \left(\frac{V}{c}\right)^2}}$$

Next, let us find how the times t and T are related. Eliminating x or X between Einstein's equations, we easily obtain

$$t = k\left(T + \frac{VX}{c^2}\right)$$

$$T = k\left(t - \frac{Vx}{c^2}\right)$$

We see that the *times measured by the two observers are different.* The new set of four equations for x, X, t, and T is called the *Lorentz transformation,* necessary for frames moving with high relative but uniform speed. They form what is called Einstein's *theory of special relativity* ("special" refers to the restriction to cases of uniform relative speed). It is easy to see that if V/c is small, k approaches 1 and the Newtonian form results, with $T = t$.

There are some very startling consequences to Einstein's equations of relativity. The first has to do with measurements of *length*: Suppose that the pilot-observer carries a meter stick up with him, and measures its length while in flight at a speed V (see Figure 3-4). One end is placed on his origin $X_1 = 0$, and the other is found to be a distance X_2 away. The stick has a length $\Delta X = X_2 - X_1$ which is merely X_2 (1 meter). Now, what does the observer on the ground find the length to be? One end is at $x_1 = Vt$, the other at x_2, and $\Delta x = x_2 - x_1$. Using the Lorentz transformation that contains his time,

$$X_2 = k(x_2 - Vt)$$

or

$$\Delta X = k\Delta x$$

Thus the length of the meter stick on the plane as seen by the observer on the ground is, in meters,

$$\Delta x = \sqrt{1 - \left(\frac{V}{c}\right)^2}$$

This length is smaller than 1 meter for all V less than c, which means in general that *any object moving with respect to an observer is shorter than when at rest.* This called the *Lorentz contraction.* Let us turn the problem around,

* The reader is referred to Appendix D for the details.

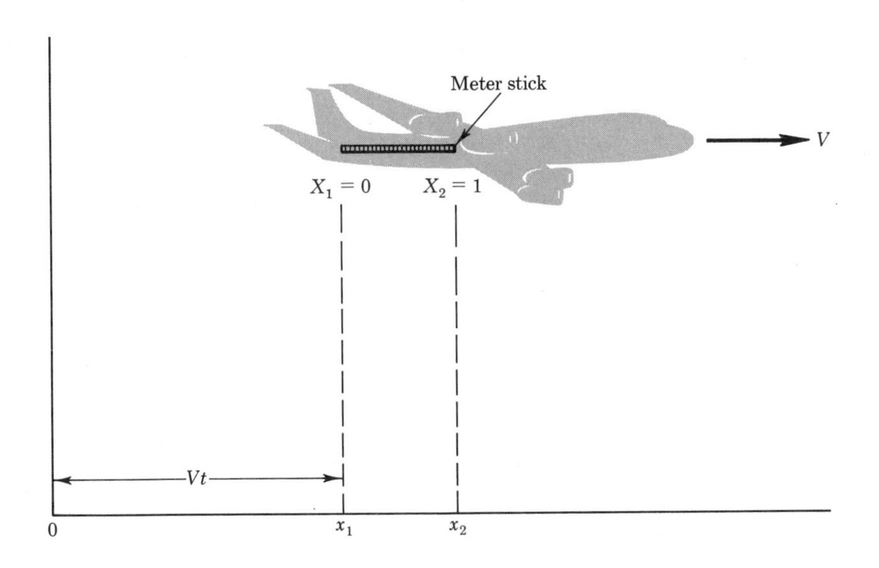

FIGURE 3-4

Measurements of length by stationary and moving observers

and ask what the length of a meter stick lying on the ground appears to be from the point of view of the pilot of the plane. By using the inverse transformation, we find that he too sees the stick to be contracted. This symmetry is to be expected, since the observers are moving relative to each other.

The second major consequence of Einstein's equations is that *from the viewpoint of an observer on the Earth, clocks on the moving frame run slow.* Imagine the plane to be equipped with a timing device such as a pendulum that marks seconds with each swing. A measurement is made when $x = X = 0$. The time for one swing on the plane is $\Delta T = 1$ second; but, from the transformation, the Earth sees $\Delta t = k\Delta T = \dfrac{1}{\sqrt{1 - \left(\frac{V}{c}\right)^2}}$ seconds, which is greater than 1 second. The plane's clock runs slow according to the Earth observer, a process called *time dilation*. Reversing the views, we find that the observer on the plane sees the Earth clock as also running slow. These concepts are at first mysterious and difficult to believe, since our personal experience is limited to motion at very low speeds compared with the speed of light. We can understand the reluctance of many scientists to accept the ideas at the time Einstein proposed them. Our appreciation of these relativistic effects can be improved by computing the length contraction of a meter stick and the time dilation of a 1-second tick of a clock for widely different speeds. The decrease in length is

$$1 - \sqrt{1 - \left(\frac{V}{c}\right)^2}$$

while the increase in time is

$$\frac{1}{\sqrt{1-\left(\frac{V}{c}\right)^2}} - 1$$

At one end of the scale are low or "nonrelativistic" speeds—for example, the Earth's motion around the Sun, calculated previously to be $V/c = 0.0001$; at the other end of the scale large objects such as a star receding rapidly from our solar system, or an electron in relativistic flight, say with speed $0.9999c$.

The calculations lead to the following results:

	V/c	*Length Contraction* (m)	*Time Dilation* (sec)
Low Speed	0.0001	5×10^{-9}	5×10^{-9}
High Speed	0.9999	0.986	69.7

The effects are negligible at low speed, but completely change the magnitude of length and time at high speed. Figure 3-5 shows the ratio of length of an object to its length at rest, as it depends on the ratio V/c. In general, the region of speed below $V/c = 0.1$ is considered "nonrelativistic," meaning that the effects of length contraction and time dilation are small and the use of classical formulas is adequate for many purposes.

Speeds of material objects add in a strange way according to special relativity. Suppose that a bullet is fired forward from the nose of the moving plane with a speed U relative to the plane. According to the *classical* concept of relativity, we would expect the speed measured from the ground to be

$$u = U + V$$

Einstein's relation, derived by comparing displacements Δx and ΔX in times Δt and ΔT, is quite different. As derived in Appendix D, the speed is

$$u = \frac{U + V}{1 + \frac{UV}{c^2}}$$

The sign on U is positive if the motion is in the same direction as V, negative if opposite.* If the moving frame and the projectile both approach

* If U and V are close to c, good approximations are:

$$u/c = 1 - \tfrac{1}{2}(1 - U/c)(1 - V/c)$$

for U positive, and

$$u/c = \frac{U + V}{2c + U - V}$$

for U negative.

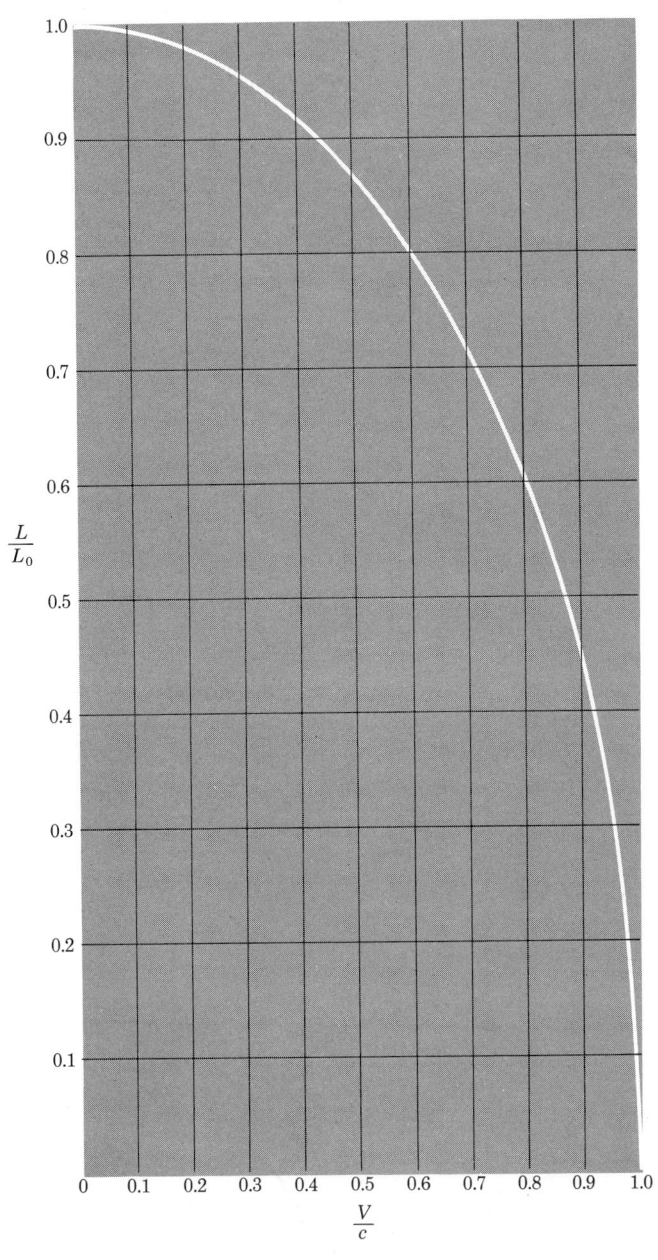

FIGURE 3-5

Dependence of length on speed

the speed of light—that is, if $V \to c$ and $U \to c$—the largest possible speed as seen from the ground is found from the formula to be $u = c$, *not* $2c$.

A numerical example is instructive. Suppose that the tritium nucleus (Section 1-3) of speed $V = 2.0 \times 10^8$ m/sec disintegrates and sends an electron forward with a speed (in the H^3 frame) $U = 2.9 \times 10^8$ m/sec. By classical analysis, the electron speed in the laboratory would be the simple sum 4.9×10^8 m/sec, which is in excess of the speed of light. By the theory of special relativity, however, it is

$$u = \frac{4.9 \times 10^8}{1 + \frac{(2.9)(2.0)}{(3)^2}} = 2.98 \times 10^8 \text{ m/sec}$$

In science-fiction stories we often read of the "fourth dimension" of relativity, a mysterious region into which a spaceship powered by some new principle can vanish. The basis for the phrase is that *time* has a new role in Einstein's relativity, being different for two observers, just as the space coordinates are. Instead of saying that a particle is at a point (x, y, z) at time t, it is preferable to say that it is at a point in a four-dimensional "space" (x, y, z, t).

3-5 EXPERIMENTAL TESTS OF SPECIAL RELATIVITY

The intimate relation of theory and experiment in physics is well illustrated by the investigations of Michelson and Morley in 1881. Their experiments were crucial in setting the stage for Einstein's theory of special relativity, by demonstrating the constancy of the speed of light. In order to see the meaning of the experiments, let us review the situation up to that time. The discovery of magnetism produced by electric currents had led to Maxwell's description of light as a vibration moving through space—that is, a wave with electric and magnetic components. The only waves previously known were those in air, water, or solids, and so it was presumed that there must also be some substance to carry light waves. This was difficult to imagine, since there was a vacuum between the Sun and the Earth, yet light was readily propagated through the space. The mechanistically oriented physicists of the time invented a medium called the *ether* that was supposed to pervade all of the universe. A close analogy was believed to hold between sound waves in air and electromagnetic waves in the ether. In the former case, the speed of sound measured will depend on the speed with which the observer passes through the air (the passengers on a plane traveling with supersonic speed outrun the sound generated by the rear jet engines). Similarly, observers on the Earth, which moves in the frame of reference of the stars, should experience a terrific "ether wind" that affects the speed of light. To resolve the controversy over the matter, Michelson and Morley set up a device to measure the speed of light both in the direction of motion of the Earth and at right angles to the motion. Their apparatus consisted of a set of mirrors, a light source, and a detector. As sketched in Figure 3-6(a), light could go by either path A or path B, since the plate was only partially covered with silver. For motion of light

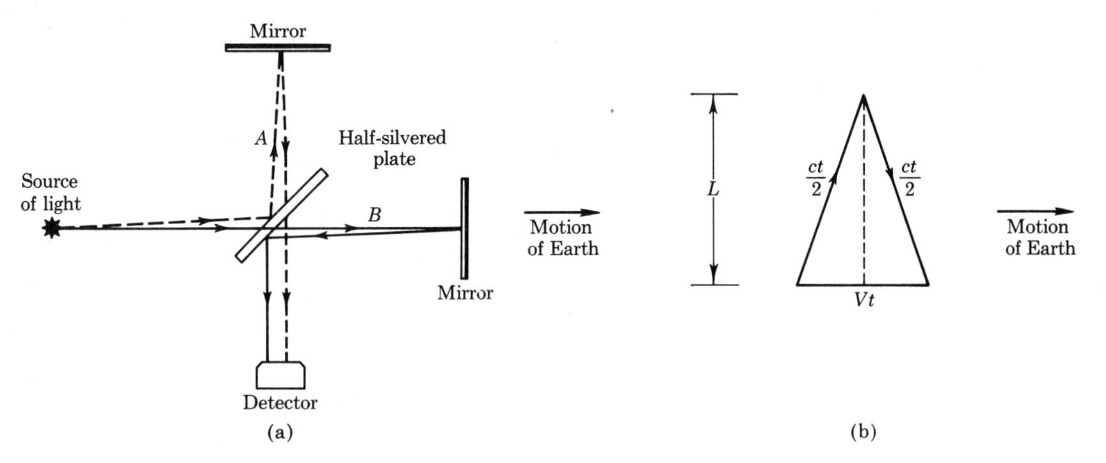

FIGURE 3-6

Michelson-Morley experiment

over a distance L in the same direction as the motion of the Earth of speed V, the time required should be $\frac{L}{c - V}$, while to return should take a time $\frac{L}{c + V}$. The total time was thus to be

$$t = \frac{L}{c + V} + \frac{L}{c - V} = \frac{2L}{c} \frac{1}{1 - \left(\frac{V}{c}\right)^2}$$

For motion at right angles, as seen in Figure 3-6(b), the observer has moved a distance Vt while the light goes a distance ct. From the right triangles, it is easily found that

$$t = \frac{2L}{c} \frac{1}{\sqrt{1 - \left(\frac{V}{c}\right)^2}}$$

One time formula has the square root, the other does not, so a time difference should be observed. Michelson and Morley's experiment revealed *no difference at all*, however, which should have told them that the ether wind does *not* exist and that the speed of light is *unaffected* by motion of the observer. Their interpretation, however, was that the Earth carries the ether along, and it was not until much additional work was done by Lodge, Lorentz, and Fitzgerald, and finally by Einstein, that the correct explanation was developed.

Other features of special relativity have also been tested. The slowing of moving clocks can be verified from the observation at the Earth's surface of the *meson*, a radioactive particle found in cosmic rays. The life of the meson in the laboratory is about 10^{-8} sec, but when it is moving with a speed near c its life is extended by a factor of about 7.

Although the theory of special relativity is complicated, it has the important property of guaranteeing that *the laws of nature are the same for all observers in uniform motion with respect to each other*. The symmetry of the observations on length contraction and time dilation are but two examples of this elegant property.

We noted that Newtonian relativity is not correct for frames accelerating with respect to each other. Similarly, Einstein's theory of special relativity cannot be used for such motions. His genius was such that he was able to extend the theory to describe observations in accelerated frames, in what is known as the theory of *general* relativity. As we shall discuss later (in Chapter 25), to develop this theory he combined concepts of geometry, motion, and gravity.

3-6 PHYSICS AS AN OPERATIONAL SCIENCE

The discovery of special relativity had a profound impact on scientific thinking, certainly because of its startling content. More importantly, it re-emphasized the necessity of continual examination of fundamental ideas. The concepts of length and time had remained unquestioned as obvious entities based on common experience. Einstein provided the impetus for the present *operational* viewpoint, which takes the approach that meaning cannot be attached to a concept unless legitimate physical measurements are possible. This does not say nature cannot be described by a formula or set of formulas, but it does say that we cannot assume it so without experimental demonstration. In order to see the importance of the operational method, let us examine the operations in measuring length—say, of an automobile at rest. This involves moving the measuring stick along in steps until the end of the car is reached. How do we know whether or not the way and speed with which we moved the stick affects its length? An increase in precision of measurement might reveal the need for specifying the exact technique. If the car were moving at a high rate of speed, there would be questions as to how we get from the ground to the automobile, and again whether the motion of the stick affects the measurement. If the object is a moving astronomical body or an elementary particle, the technique becomes even more complicated, and new definitions of operations must be found. For example, we are not able to *touch* a distant planet, and must rely on light signals to determine its features. Thus we begin to see how the concept of length may have different meanings corresponding to different operations. Strictly speaking, we cannot discuss a concept or pose sensible questions unless the operations are available, either in fact or by mental action. An example—Is the size of our solar system changing continuously in such a way that all things are equally affected? No operations appear to exist to answer the question; therefore, it must be classed as meaningless.

Use of the imagination to predict phenomena or principles is very valuable, however, in that it suggests experimental measurements to verify or deny a given supposition.

PROBLEMS

3-1 Find the average value of the speed of light, the average error, and the probable error, using the following measured data (each $\times 10^8$ and measured in m/sec): 2.99797, 2.99794, 2.99788, 2.99793.

3-2 A modern version of one of Galileo's experiments is performed by dropping a golfball from the observation deck of the Empire State Building. The measured positions at various times are as follows:

t, time (sec)	*s, distance fallen* (m)
0	0
2	19.7
4	78.5
6	176
8	315

Now, if the position s of a uniformly accelerated object is measured at various times t, the acceleration can be computed from adjacent points by

$$a = \frac{2\Delta s}{\Delta t^2}$$

where Δs and Δt^2 are the differences between positions and squares of times. (a) Compute four values for g using the above formula. (b) Find the average of the experimental results for g and the probable error. (c) Does the accepted value of g at New York City, 9.80267, fall within the range?

3-3 If the correct value of a number is C and an approximate value is A, then the *error* is $(A - C)$, the *fractional error* is $\frac{A - C}{C}$, and the *percentage error* is $100\left(\frac{A - C}{C}\right)$. Find each of these if $C = \frac{1}{1 - x}$ and $A = 1 + x$ (as the sum of the first two terms of the geometric series $1 + x + x^2 + x^3 + \ldots$), for $x = 0.01$, 0.1, and 0.9.

3-4 Find the percentage errors (see Problem 3-3) in the circumference of the Earth (a) if the radius is taken as 4000 mi rather than a more correct 3960 mi, (b) if π is taken as 3 rather than a more accurate value.

3-5 What percentage error (see Problem 3-3) is there in using the approximation $e^x \cong 1 + x + x^2/2$ when x is 0.2? When x is 1.0?

3-6 Calculate the ratio V/c for the motion of a point on the Earth's equator, radius 3960 mi $= 6.4 \times 10^6$ m, using the center of the Earth as a "fixed" reference. Would you class our motion as relativistic or nonrelativistic?

3-7 A rocket of height 20 ft on Earth is en route to a nearby galaxy and reaches a speed of 600,000,000 mi/hr. How much contraction in the length occurs?

3-8 An electron in a particular accelerator has a speed of 1.5×10^8 m/sec. By what fraction does its dimension in the direction of motion contract?

3-9 According to a person on the Earth, how many seconds a day does an astronaut's watch appear to lose when the ship is moving at a speed $\frac{1}{100}$ that of light? *Suggestion:* Use the approximation $(1 - x)^{-1/2} \cong 1 + x/2$.

3-10 We drive 800 miles from New York to Chicago in 20 hours. By how many nanoseconds does our watch appear to lose time during the trip, according to an observer at rest?

3-11* (a) Verify the approximate expression for adding speeds relativistically, $u/c = 1 - \frac{1}{2}(1 - U/c)(1 - V/c)$. *Hint:* Note that $(1 + x)^{-1} \cong 1 - x$ for small x. (b) A radioactive atom moving away from us with a speed of $0.99c$ emits an electron in the forward direction with speed $0.995c$. What is the electron's speed as measured in our laboratory?

3-12* A spaceship with speed $0.98c$ propels itself by shooting out protons at speed $0.99c$ relative to the rocket. How long would it take one of the particles released to reach the Earth, if the ship is 1 light year away?

FORCE, MASS, AND MOTION

4

The universe, the Earth, man's inventions, and the internal structure of matter appear to be hopelessly complex at first observation. The intricate relation of processes and objects has been reduced to reasonable order through careful study and analysis by scientists for several hundred years. One conclusion is that there is a limited number of basic kinds of *forces* in operation. We shall first define the concept of force and its relation to mass and motion, and in the following chapter identify and discuss carefully the major types of force—gravitational, electrical, magnetic, mechanical, frictional, atomic, and nuclear.

4-1 EXAMPLES OF FORCES

We may think of force as an entity that tends to cause or prevent motion of an object. Let us list a few familiar examples:

1. Through muscular action, we move our bodies in all our daily activities.

2. Any object that we attempt to push or pull tends to resist being moved.

3. The explosion of gases in the cylinders of an automobile engine provides forces on the pistons, and through gears and shafts the forces result in motion of the vehicle along the road.

4. The force of gravity causes all objects to be pulled toward or held on the Earth's surface.

5. The electrical charge on metal plates in a radio or television tube causes electrons to accelerate.

6. The internal forces in molecules and atoms prevent the component particles from separating.

7. The magnetic field of the Earth causes a compass needle to swing to a fixed direction.

Mathematically force is a vector quantity (see Section 2-4) having both magnitude and direction. Several forces acting on an object, even if they are from different sources, can be added by vector algebra to find the resultant.

4-2 NEWTON'S LAWS OF MOTION

The diverse processes listed in the previous section may all be brought into harmony in terms of three basic laws of *mechanics* first set down by Isaac Newton* in the seventeenth century. We shall state and explain the meaning and use of these powerful principles, taking some liberties with their phrasing in the hope of improving their clarity.

Before stating these laws, we must note that they apply only to motions relative to inertial frames of reference, that is, those moving uniformly—at constant speed—with respect to a fixed frame. They are also limited to the description of slowly-moving objects, the nonrelativistic region. From our discussion of relativity in Section 3-4, we would then conclude that Newton's laws will be only good *approximations* to reality, since there is no certainly known fixed frame. For most purposes, the view of the Earth's surface as a local fixed frame is both convenient and sufficiently accurate. There are, however, some problems in space mechanics for which this is not sufficient, and we must be careful to study Newton's laws with the above reservations in mind. The restriction to low speeds precludes their use for the description of some motions of atomic particles or of the motion of certain stars.

NEWTON'S FIRST LAW OF MOTION: IF NO FORCES ACT ON AN OBJECT, ITS MOTION WILL BE UNCHANGED.

There are several key words in this simple sentence. The word *object* refers to anything composed of matter. By *motion*, we mean the object's velocity, which has both magnitude and direction. *Unchanged* implies staying at constant velocity if moving, or staying at rest if originally in that

* Isaac Newton (1642–1727) was a man of genius whose contributions to many branches of physics were prodigious. Co-inventor of the calculus, he developed the theory of gravitation and explained the decomposition of white light into colors. His later years were devoted to English government service, as Master of the Mint.

condition. Finally, the statement also serves qualitatively to identify *force* as a *quantity that tends to change the motion of a material body*. The concept of force is basic to our present understanding of the laws of motion. We can think of a force as a "push" or "pull" on an object, and we need not consider the source of these forces at this time. If, however, we have such a force and let it act on an object, we can measure the change in its motion.

Let us perform an experiment with a compressed spring and two objects, which we may call A and B. Figure 4-1(a) indicates the action expected: when the bodies are released, A will move to the left and B to the right. The spring is used as the agent to cause the bodies to accelerate; but we could have used gravitational or electrical attractions, as shown in Figure 4-1(b).

If we measure the accelerations by observing changes in speed over a time interval, we shall find them to be of magnitude a_A and a_B respectively —directed oppositely, of course. Compressing the spring further before releasing it will yield greater accelerations. We find that the ratio of their magnitudes is

$$\frac{a_A}{a_B} = \text{a constant}$$

regardless of initial spring compression.

On finding constant ratios of acceleration when we cause body A to interact with a third body C (or B with C), we are prompted to suppose that there is some constant property of material objects involved. Thus we say that a body has *mass*—a measure of its inertia, which is resistance to being accelerated. If we assign the symbols m_A and m_B to the respective masses, the ratio should then be

$$\frac{a_A}{a_B} = \frac{m_B}{m_A} = \text{a constant}$$

which is compatible with our experience. The next step is to find numerical

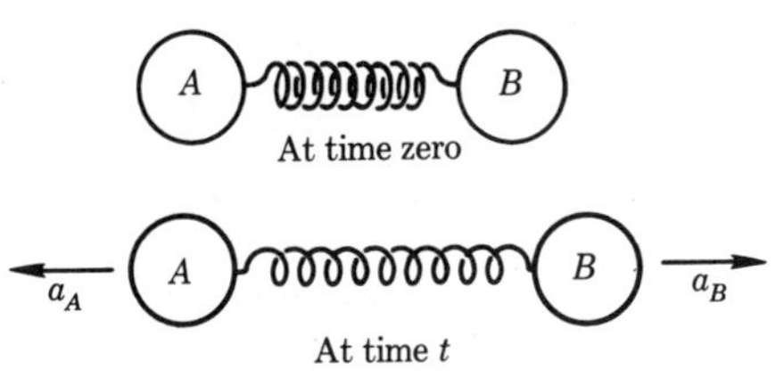

(a) Motion of objects connected by a spring

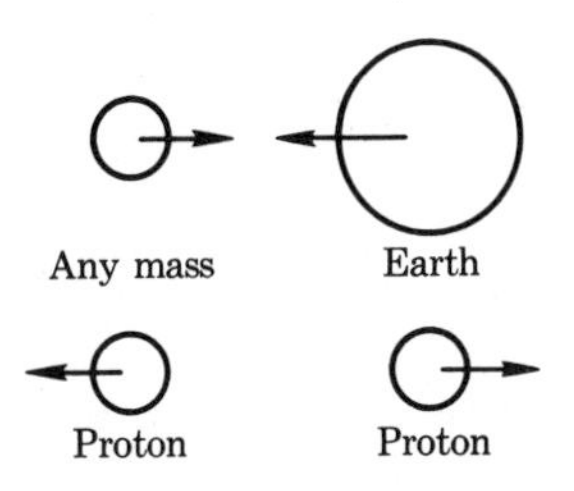

(b) Motion of objects with gravitational or electrical forces

FIGURE 4-1

Mutual accelerations of interacting objects

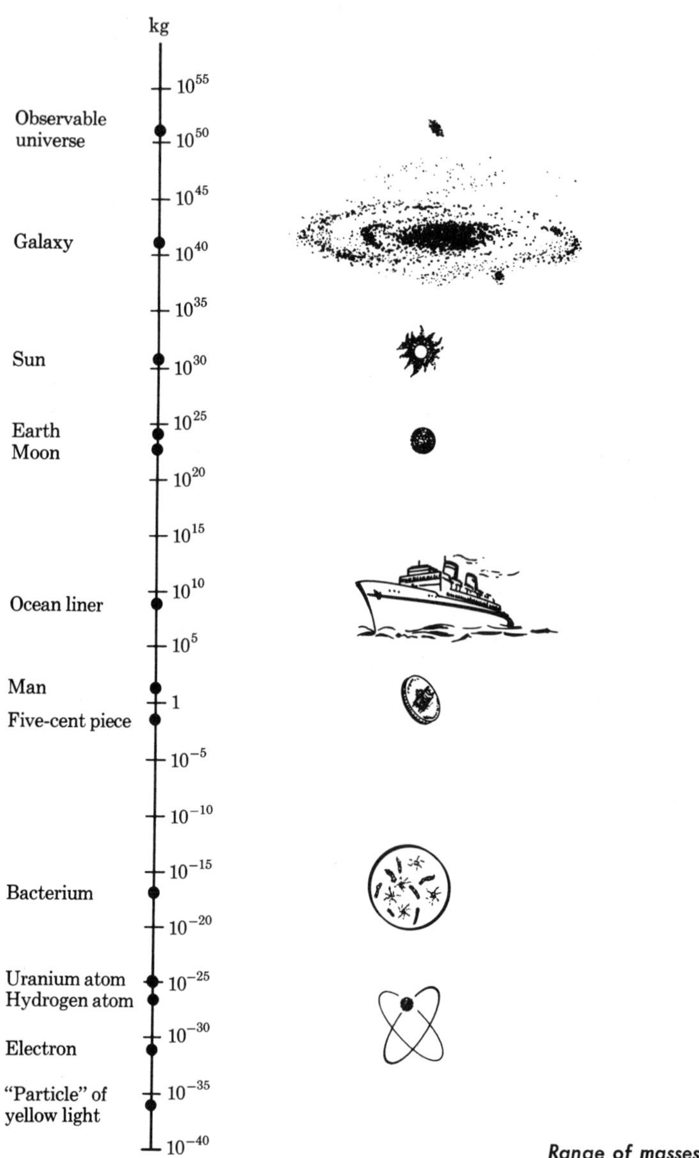

FIGURE 4-2

Range of masses in the universe

values. If we take body A as "standard" and let its mass be 1 kilogram,* then the measured ratio of accelerations tells us what m_B is. Mass values of all other objects could similarly be determined. Figure 4-2 shows the

* The actual standard of mass used in the metric system is a certain cylinder of platinum–iridium reposing in France.

great range of masses of objects in the universe. The results of our measurements and introduction of mass can be rearranged in the form

$$m_A \mathbf{a}_A = -m_B \mathbf{a}_B$$

where the directional aspect of acceleration is now included. The negative sign is required to express the fact that the accelerations are in opposite directions. We now *define* the force on body A as the product $m_A \mathbf{a}_B$, and label it $\mathbf{F}_A$. Thus

$$\mathbf{F}_A = m_A \mathbf{a}_A$$

Note that this quantitative statement is quite consistent with our previous notion that a stronger push gives a greater change in motion; and with our knowledge that the larger the mass, the smaller the acceleration. A *unit of force* is that force required to give an acceleration of 1 m/sec^2 to a mass of 1 kg. Thus force is measured in units of "kilogram-meters per second per second" (kg-m/sec^2). This combination of units is called the *newton*, a word that is easier to express and honors the famous scientist. Applying the above definition of force to any body under any acceleration yields Newton's second law, which we shall now state formally.

> NEWTON'S SECOND LAW OF MOTION: THE RATE OF CHANGE OF THE MOMENTUM OF A BODY IS EQUAL TO THE APPLIED FORCE.

Again, let us explain terms carefully. The *momentum* $\mathbf{p}$ is a vector quantity that is the product of the mass m of an object and its velocity $\mathbf{v}$. Thus

$$\mathbf{p} = m\mathbf{v}$$

For example, the momentum of an 800-kg automobile moving with speed 30 m/sec would be $p = mv = (800 \text{ kg})(30 \text{ m/sec}) = 24{,}000$ kg-m/sec. The *applied force* is provided by some agent external to the body under observation. The *rate of change* is found by comparing momenta at different times, and by forming

$$\frac{\mathbf{p}_2 - \mathbf{p}_1}{t_2 - t_1} \quad \text{or} \quad \frac{\Delta \mathbf{p}}{\Delta t}$$

Thus Newton's second law in mathematical form is

$$\mathbf{F} = \frac{\Delta \mathbf{p}}{\Delta t} \qquad \textit{(Newton's second law)}$$

It is easy to see that if $\mathbf{p} = m\mathbf{v}$ and m is constant, then $\Delta\mathbf{p} = m\Delta\mathbf{v}$ and, since $\Delta\mathbf{v} = \mathbf{a}\Delta t$, that $\Delta\mathbf{p}/\Delta t = m\mathbf{a}$, so that

$$\mathbf{F} = m\mathbf{a}$$

in accord with our earlier definition of force.

One more law can now be deduced:

NEWTON'S THIRD LAW OF MOTION: FOR EVERY FORCE EXERTED ON AN OBJECT THERE IS AN EQUAL AND OPPOSITE FORCE EXERTED BY THE OBJECT.

The proof of this third law follows from our experiment in which two objects interacted. Body B provides the force on body A, of amount $\mathbf{F}_{BA} = m_A\mathbf{a}_A$; while body A causes the force $\mathbf{F}_{AB} = m_B\mathbf{a}_B$. The vector relation yields

$$\mathbf{F}_{BA} = -\mathbf{F}_{AB}$$

that is, the forces of mutually interacting objects are equal in magnitude and opposite in direction. The truth of this statement may be made more plausible by giving a few familiar examples:

1. As we stand on the ground we exert a force downward on it due to our weight. The ground exerts a force upward on us of equal amount.
2. As we push on a wall with our hand, the wall pushes back in equal amount. No force can be applied to an object unless it pushes back, as we observe on punching a pillow.
3. A boy exerts a force on a toy wagon by pulling on the tongue. The wagon pulls back by exerting an equal force on the boy.

At first glance, Newton's laws do not seem to be of any great use in explaining motion. Contained within them, however, is a wealth of meaning and value for analyzing physical situations.

A few simple numerical examples will be sufficient at this stage to give us experience with units. (We abbreviate the word newton by the symbol N.)

1. Suppose that the electrical force on an electron of mass 9.11×10^{-31} kg between two charged plates is 1.82×10^{-20} N. Find the acceleration.

$$a = \frac{F(\text{N})}{m(\text{kg})} = \frac{1.82 \times 10^{-20}\ \text{kg-m/sec}^2}{9.11 \times 10^{-31}\ \text{kg}} = 2.0 \times 10^{10}\ \text{m/sec}^2$$

2. The force of gravity (weight) on a parachuter of mass 75 kg is known to be 735 N. Find his acceleration in free fall from an airplane.

$$a = \frac{F(\text{N})}{m(\text{kg})} = \frac{735\ \text{kg-m/sec}^2}{75\ \text{kg}} = 9.8\ \text{m/sec}^2$$

3. A section of current-carrying wire in an electric motor has a mass of 0.02 kg. Because of magnetic forces, it experiences an acceleration of 30 m/sec². Find the amount of force.

$$F = m(\text{kg})a(\text{m/sec}^2) = (0.02)(30) = 0.6\ \text{kg-m/sec}^2$$

$$F = 0.6\ \text{N}$$

The system of units based on the meter (for length), kilogram (for mass), and second (for time) is designated the mks system. We shall postpone discussion of the units of other systems.

Relation of Newton's first and second laws. Consider the mathematical form of the second law if no force is applied: if $\mathbf{F} = 0$, then

$$\frac{\Delta \mathbf{p}}{\Delta t} = 0 \quad \text{or} \quad \frac{m\Delta \mathbf{v}}{\Delta t} = 0$$

This can be true only if $m\mathbf{v}$ is a constant in time, which is consistent with the first law's statement that the velocity does not change under such conditions. A special case is $m\mathbf{v} = 0$, where zero is a perfectly good "constant," referring to a body at rest.

Relation of Newton's second and third laws. Some confusion can arise when we attempt to use the second and third laws at the same time. Suppose that we push against a door to open it. What other forces are involved? There may be some friction in the hinges, and the air behind the door tends to impede the motion. We apply a force, and the door

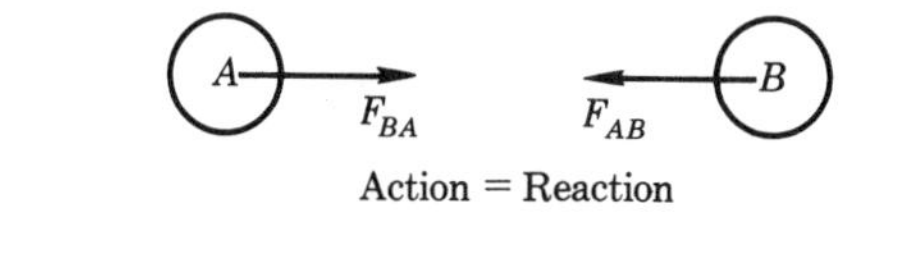

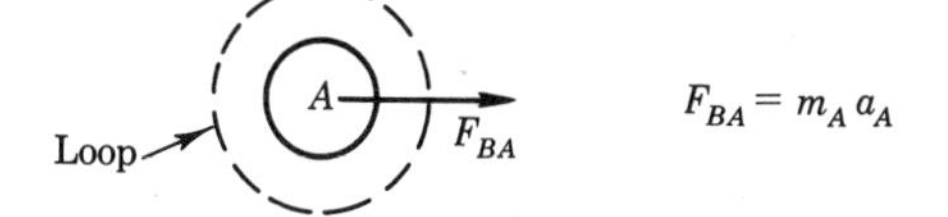

FIGURE 4-3

Relation of Newton's second and third laws

pushes back in accord with Newton's third law. To find the motion of the door we use Newton's second law, $F = ma$, where F consists of the force we apply, friction, and air resistance—but *not* the reaction force, since the latter acts on *us*, not on the door. The error can be avoided by a simple device. Suppose we are interested in the motion of body A, which is under the influence of body B, as in Figure 4-3. There are mutual forces (say, of attraction) of the objects on each other. By F_{BA} we mean the force of B on A, and by F_{AB} we mean the force of A on B. From Newton's third law,

$$\mathbf{F}_{AB} = -\mathbf{F}_{BA}$$

Now consider the forces acting on body A only. There is only one *externally applied* force—namely, F_{BA}, which will cause body A to be accelerated to the right, in the direction of the force, according to

$$F_{BA} = m_A a_A$$

We may now proceed to use this to study the behavior of object A. This technique of isolating bodies is very helpful in analyzing the motion of a system of objects with many parts: we may draw loops about parts of the system or about the whole system, keeping track of the external forces at each step, thus applying Newton's law appropriately.

4-3 APPLICATION OF NEWTON'S LAWS TO PARTICLES

Although Newton's laws appear rather straightforward when expressed as phrases or formulas, their use must be approached with care. We shall thus proceed from very simple problems toward the more complicated. Attention will be focused on the forces acting on the objects, not on the agent that "causes" the force. In all examples, we assume objects to be particles or "point masses," or represent larger bodies as points, in accord with our discussion of models in Section 3-2.

A Single Atom

Let us imagine one atom located in distant space, far from other atoms, planets, or stars, and thus experiencing no forces at all. If it happens to be at rest, it will not move, in accord with Newton's first law. On the other hand, if it were moving initially, it will continue to do so. In either case, the applied force F is zero; then, by Newton's second law, $ma = 0$, and the acceleration a is zero. Now let us change the situation. Light from a star falls on our atom, removes an electron, and the atom becomes an ion with net positive charge. Instruments carried by satellites and space probes show that there are electrical forces in space. The ion experiences a force F_E, where the subscript stands for "electrical." Then $F_E = ma$, and the particle speeds up with an acceleration $a = F_E/m$. Now, suppose instead that the atom progresses through empty space until it comes within the gravitational influence of a planet of mass M, which exerts a force F_G on the particle. By Newton's second law, the force of gravity on the atom of mass m is $F_G = ma$, and its acceleration is $a = F_G/m$. However, according to the third law the planet feels a force of magnitude F_G also, and is accelerated toward the atom by the amount F_G/M. From Figure 4-2 we see that M is larger than m by a factor of about 10^{50}, and so the acceleration of M is negligible. Near the Earth, the acceleration of gravity is the same for all objects, as discussed in Chapter 2. The value of g is approximately 9.81 m/sec^2 or 32.2 ft/sec^2, with some variation over the globe. A moving ion that comes near the Earth will also feel a magnetic force F_M that can produce an acceleration $a = F_M/m$, in addition to those due to electrical and gravitational forces. These separate examples show us that Newton's laws apply to quite a variety of situations, and that it is the mass of the object that tells what the acceleration will be.

Very rarely do we find any object under the influence of one type of force only. Let us examine some cases in which two or more forces act along the same line. The experiment that showed that there was an indivisible unit of electrical charge was performed by Millikan in the 1920's. His "oil-drop" experiment used a simple apparatus: essentially, two metal plates connected to a battery, as in Figure 4-4. In the space between the plates were two types of force—electrical, affecting charged particles; and gravitational, acting on the mass. Neglecting air resistance, when F_E is upward

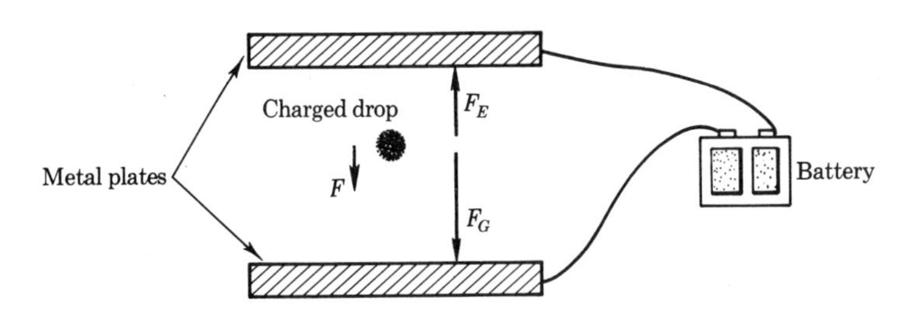

FIGURE 4-4

Forces in oil-drop experiment

and F_G is downward, the net force upward is $F = F_E - F_G$. Depending on which is larger, the particles of oil accelerate upward, downward, or remain at rest. Now, $a = F/m$, with F as the net force. When $F_E = F_G$, both F and a are zero.

Now let us study the forces acting on an object resting on the Earth's surface. The force of gravity acts "downward" on every mass, whether it be an atom, a stone, a person, or an automobile, as shown in Figure 4-5. The object is pressed by the force of gravity on the solid ground, which reacts with a force on the object, described by Newton's third law, as seen

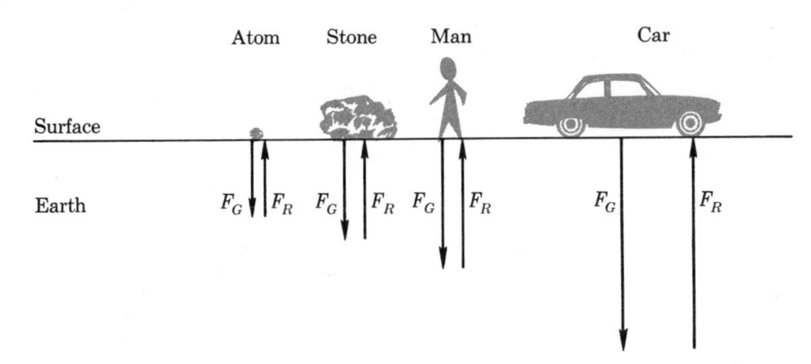

FIGURE 4-5

Force of gravity and reaction of the Earth's surface

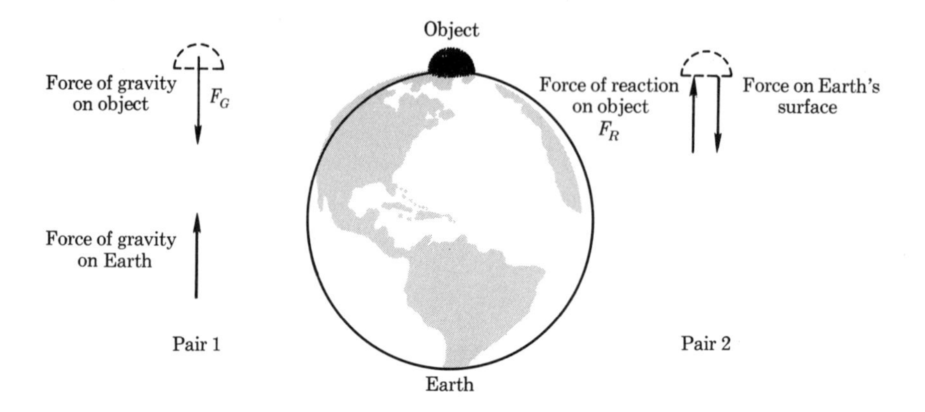

FIGURE 4-6

Action and reaction pairs, net force on object zero

in Figure 4-6. Again there is the gravity force on the Earth due to the object. Of the four forces, let us isolate those on the object—F_G is the force of gravity, or *weight*; and F_R is the reaction of the Earth. The net force is $F = F_G - F_R$, which is clearly zero, since there is no motion and a is zero. Where the surface of the Earth is not solid, an object tends to settle or sink. It feels the force of gravity and whatever resistance the medium offers, and will accelerate until the net force is zero—that is, until the reaction equals the weight.

An ice skater in motion. Motion on the Earth that does not involve gravity in one way or another is hard to imagine. A fair approximation is met by objects on smooth level surfaces, such as a sheet of ice on a lake in winter or on a skating rink. If, as in Figure 4-7, a skater of mass m is in horizontal motion on very smooth ice, with speed v, he experiences almost zero forces in the direction of motion; and, by Newton's law, if $F = 0$, then $a = 0$—and thus he maintains his forward speed. Suppose, now, that he hits a rough patch of ice. The metal of the skates scrapes on the surface and a frictional force F_F arises, which tends to push backward—that is, F_F is negative. Then, by Newton's law, $a = F_F/m$ is negative, and he decelerates, or slows down.

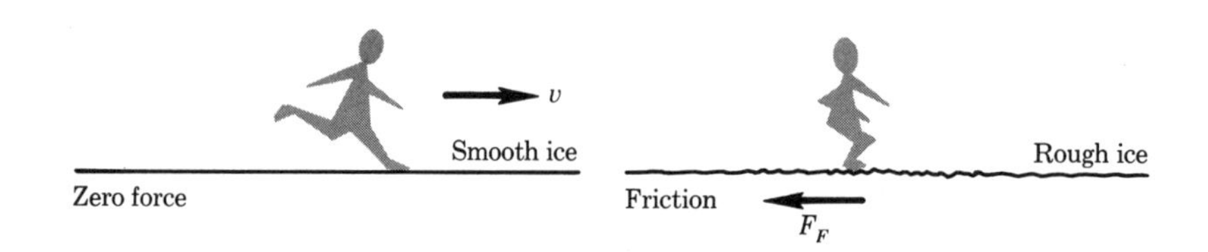

FIGURE 4-7

Ice skater in horizontal motion

A pulley lifting an object. Devices used in the home, in transportation, and in industry contain various structures that exert or experience forces. One of the simplest connections between parts of a device is a wire. If one pulls with a force F on a wire attached to an object, the reactions of atoms allow a tension force F_T to be transmitted to the mass. With this in mind, let us examine the forces acting when we use a simple pulley to lift an object from the ground, as in Figure 4-8. A mechanical force is applied of amount F_A, which creates the force of tension $F_T = F_A$ in the

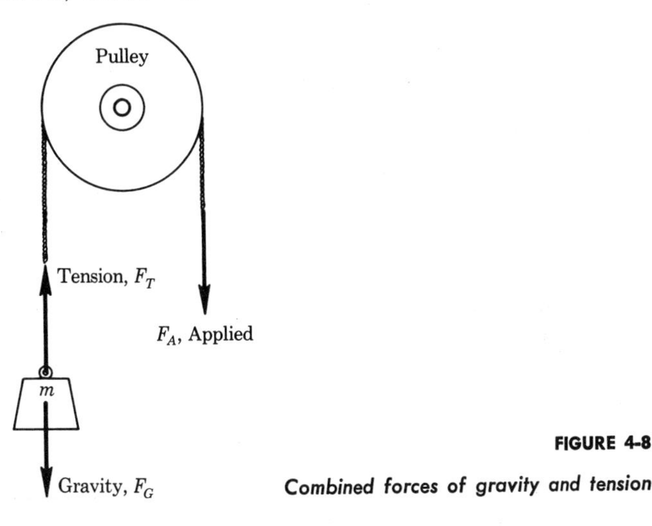

FIGURE 4-8

Combined forces of gravity and tension

rope. There are now two forces acting on the mass m: upward is F_T, downward is that of gravity, F_G. The net external force is thus $F = F_T - F_G$. The acceleration produced will be

$$a = \frac{F}{m} \quad \text{or} \quad a = \frac{F_A - F_G}{m}$$

A car on an incline. This example introduces forces that do not lie along the same line. Visualize, as in Figure 4-9, an automobile on a straight hill or incline. Neglecting friction, what forces are acting *on the car*? The gravity force is *downward* and of amount F_G. Let us abbreviate by labeling this force W, for weight. There are two directions of interest to us—those perpendicular and parallel to the plane on which the car is found. Resolve the weight vector into components $W_\perp$ (perpendicular) and $W_\parallel$ (parallel). From the diagram, $W_\perp = W \cos\theta$ and $W_\parallel = W \sin\theta$. As in the previous examples, there is a reaction force of the incline of amount F_R that balances the component $W_\perp$; that is, $W_\perp - F_R = 0$. There is no acceleration in that direction. Down the incline however, with no friction, there is an unbalanced force $W_\parallel$. By Newton's second law, the acceleration along the inclined plane is $a_\parallel = W_\parallel/m = (W/m)\sin\theta$. Notice that this is the same as $a \sin\theta$, where $a = W/m$ is the acceleration in free vertical fall.

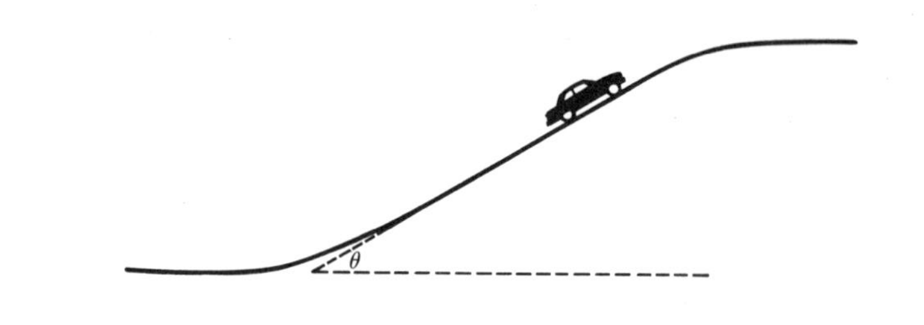

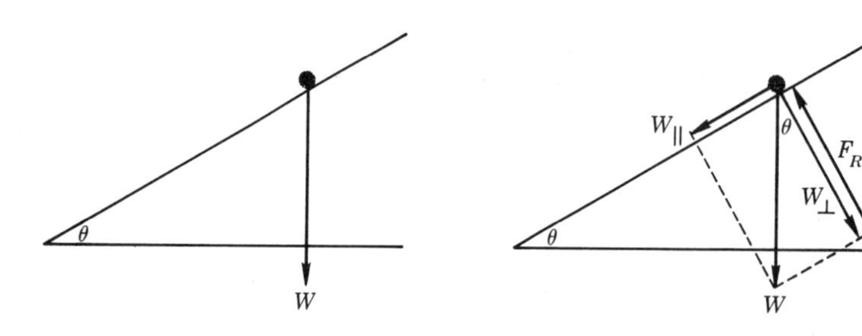

FIGURE 4-9

Forces and acceleration of car on an incline

For example, if a is 9.8 m/sec^2 and θ is 30 deg, $\sin\theta = 0.5$, then $a = 4.9$ m/sec^2.

A car held at rest on an incline by friction. Locking the emergency brake on the car of the previous example in order to prevent any motion, we then ask the question: How much frictional force F_F must be applied by the brakes? Again we resolve the weight W into components: the reaction force F_R balances the perpendicular component of W, while the frictional force up the incline balances the parallel component. The two components of acceleration are

$$a_\perp = \frac{W_\perp - F_R}{m} = 0$$

$$a_\parallel = \frac{W_\parallel - F_F}{m} = 0$$

The amount of frictional force must be

$$F_F = W_\parallel = W \sin\theta$$

If the car has a weight of 20,000 N and is at an angle of 45 deg, then $\sin\theta = \sqrt{2}/2 = 0.707$, and the frictional force must be 14,100 N.

Several Forces Acting on a Point

The vector method is useful when a number of forces are applied at different directions. Let us consider an example where the point is at rest (in equilibrium under the forces). Three fraternities, the Alphas, the Betas, and the Gammas, are engaged in a tug-of-war with ropes attached to a small ring, as in Figure 4-10. The object of the contest is to bring the ring outside the circle marked on the ground. Let us find of what force and at what angle the Alphas must pull to balance the pull of magnitude 1 ton by the Betas at 135 deg, and of 0.4 ton by the Gammas at 210 deg. If the ring is not to move, the vector force F on it must be zero. Let the forces be F_α, F_β, F_γ at angles α, β, γ from the zero axis, as in Figure 4-10(b). The x-components of force must add up to zero:

$$F_\alpha \sin \alpha + F_\beta \sin \beta + F_\gamma \sin \gamma = 0$$

Similarly the sum of the y-components of force is

$$F_\alpha \cos \alpha + F_\beta \cos \beta + F_\gamma \cos \gamma = 0$$

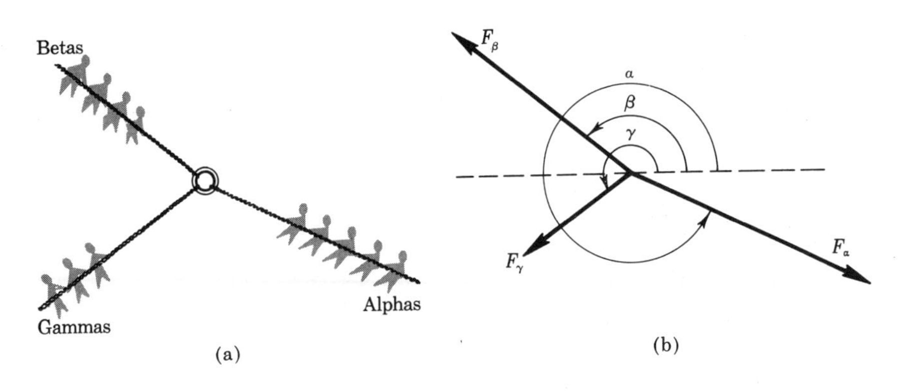

FIGURE 4-10

Forces in a tug-of-war match

We know everything but F_α and the angle α, but since there are two equations, it is now merely a matter of algebra to compute the solution.

$$F_\alpha \sin \alpha + 1.0 \sin 135° + 0.4 \sin 210° = 0$$
$$F_\alpha \cos \alpha + 1.0 \cos 135° + 0.4 \cos 210° = 0$$

Now, $\sin 135° = 0.707$, $\cos 135° = -0.707$, $\sin 210° = -0.500$, and $\cos 210° = -0.866$.

$$F_\alpha \sin \alpha = -0.507$$
$$F_\alpha \cos \alpha = 1.053$$

Dividing equations, $\tan \alpha = -0.481$ and $\alpha = -0.448$ or $334.3°$ $(-25.7°)$. Then, $\sin \alpha = -0.433$, and hence $F_\alpha = 1.17$ tons.

Falling mountain climbers. A mountain climber slips over the edge of a cliff and starts to drag his companion with him, as sketched in

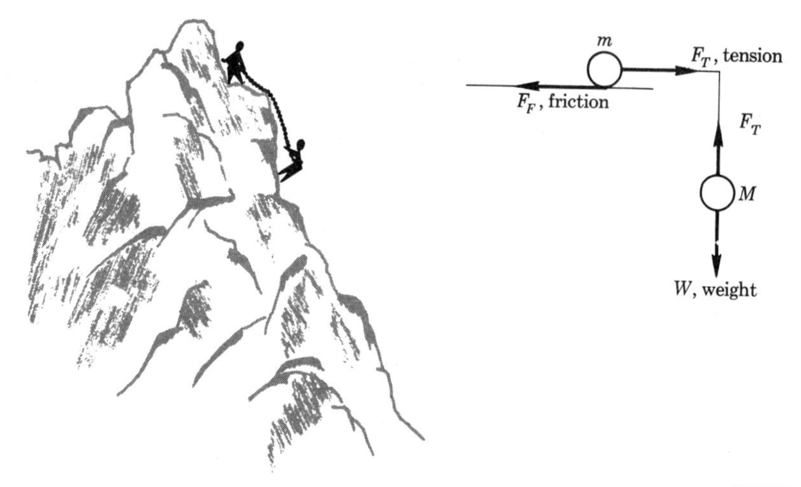

FIGURE 4-11

The mountain climbers

Figure 4-11. What acceleration will they experience, assuming that the rope does not stretch? Assume that the man on the level can develop a friction force F_F that is a fraction μ of his weight w. First we note that a tension force F_T is created in the rope that pulls m to the right, and tends to hold M up. We examine the motion of each man separately, applying Newton's second law. The net force to the right on the upper man is $F_T - F_F$. Thus

$$F_T - F_F = ma$$

The net force down on the lower man, of weight W, is

$$W - F_T = Ma$$

We can find a by adding equations and using the fact that $F_F = \mu w$:

$$F_T - \mu w + W - F_T = (m + M)a$$

$$a = \frac{W - \mu w}{M + m}$$

For example, suppose we are given $M = 75$ kg (165 lb), $m = 90$ kg (198 lb), $W = 735$ N, $w = 882$ N, and $\mu = 0.6$. Substituting,

$$a = \frac{735 - (0.6)(882)}{75 + 90} = 1.25 \text{ m/sec}^2$$

To prevent disaster, it is seen that μ must be as high as 0.83. If we want to find the tension in the rope, either of the two original equations can be used.

From the second,

$$F_T = W - Ma = 735 - (75)(1.25)$$
$$= 641 \text{ N}$$

Instead of using Newton's law separately, we could have looked at the whole system, with a net external force of $W - \mu w$, as if the two components were parallel. The total mass is $M + m$, and the acceleration is $a = \dfrac{W - \mu w}{M + m}$, as before.*

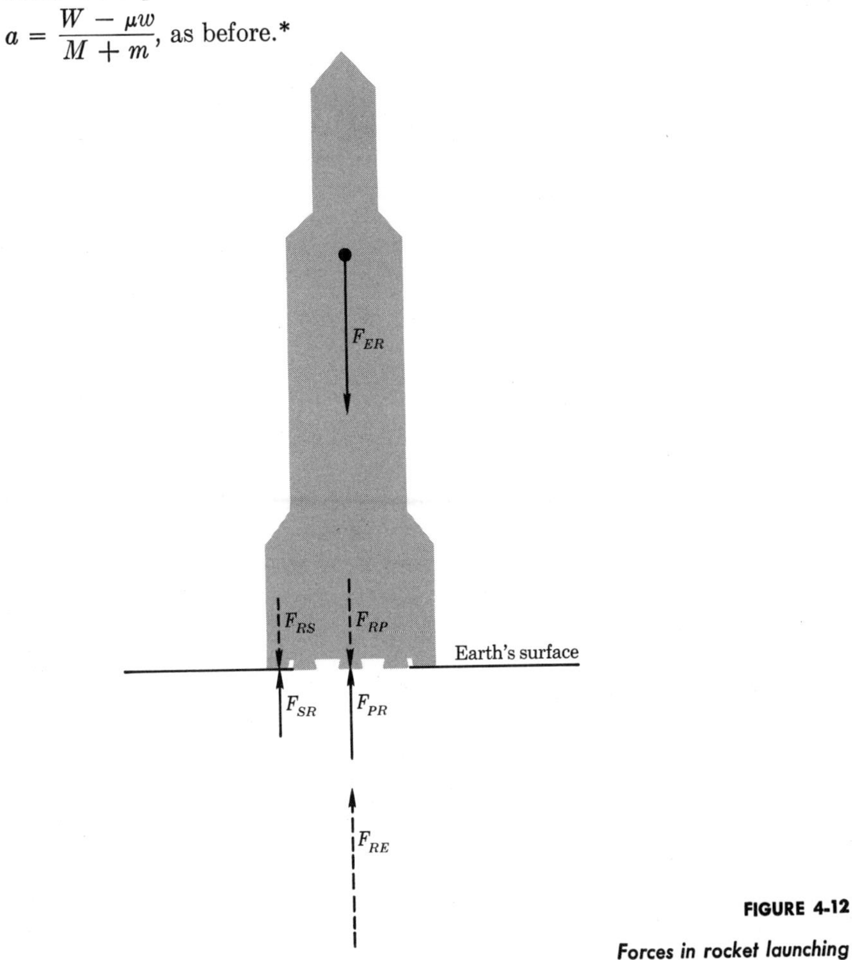

FIGURE 4-12

Forces in rocket launching

An Apollo rocket taking off vertically from a launching pad. The full description of this apparently simple situation requires consideration of six forces, as shown in Figure 4-12. These are

* Care must be exercised in such treatment to be sure *all* external forces are identified. Problem 4-11 illustrates the pitfalls of omission.

1. F_{RE} Rocket on Earth } (*gravity*)
2. F_{ER} Earth on Rocket }
3. F_{RS} Rocket on Surface } (*mechanical*)
4. F_{SR} Surface on Rocket }
5. F_{RP} Rocket on Propelling gases } (*momentum effects*)
6. F_{PR} Propelling gases on Rocket }

There are, we see, three action–reaction pairs, with three forces—numbers 2, 4 and 6—forming the net force on the projectile. This set of six forces acts as soon as the rocket motors are fired—even before the vehicle gets off the ground. Let us follow the whole launching and identify which of the forces act at various times.

Before the rocket engine is fired—Figure 4-13(a)—the only forces are the first four. Then all six come into play—Figure 4-13(b). After lift-off, the rocket's action–reaction with the Earth's surface is gone—Figure 4-13(c). After the fuel has all burned out, the only interactions are between rocket and Earth—Figure 4-13(d). Finally, when the rocket is very far from the Earth, the force of gravity between the objects is negligible—Figure 4-13(e).

There are several important features of this process. So long as the rocket is on (or near) the Earth, the gravity force F_{ER} is constant. As the thrust F_{PR} (due to propellant discharge) increases, the reaction of the Earth's surface to the rocket relaxes, but the vehicle does not take off, since $F_{PR} + F_{SR}$ does not yet exceed F_{ER}. Once the thrust is high enough, the rocket lifts off, but the ascent is slow (as we see in television or movies) because the net force $F_{PR} - F_{ER}$ is only a small fraction of the weight F_{ER}.

The Effect of Gravity

In our discussion of Newton's laws and the illustrations of their use, we emphasized that the acceleration of an object is determined by its mass, as an inertial property. The force may be one of various origins. From the standpoint of the principles of physics, no preferential attention should be given to any one of these types of force. Living as we do on a planet that provides a gravitational force, however, our daily experience places greater emphasis on *weight* than on other forces. We personally encounter electrical and magnetic forces on rather rare occasions, but sense the force of gravity with every motion of our body and observe its action every time something falls.

A second reason for the uniqueness of weight as a force is that the force of gravity on an object is directly proportional to the mass, the very property that determines acceleration under *any* force. From our present vantage point, this can only be viewed as remarkable coincidence. Later, we shall learn that the explanation for the equivalence of mass as an inertial property and mass as a gravitational property comes from the theory of general relativity.

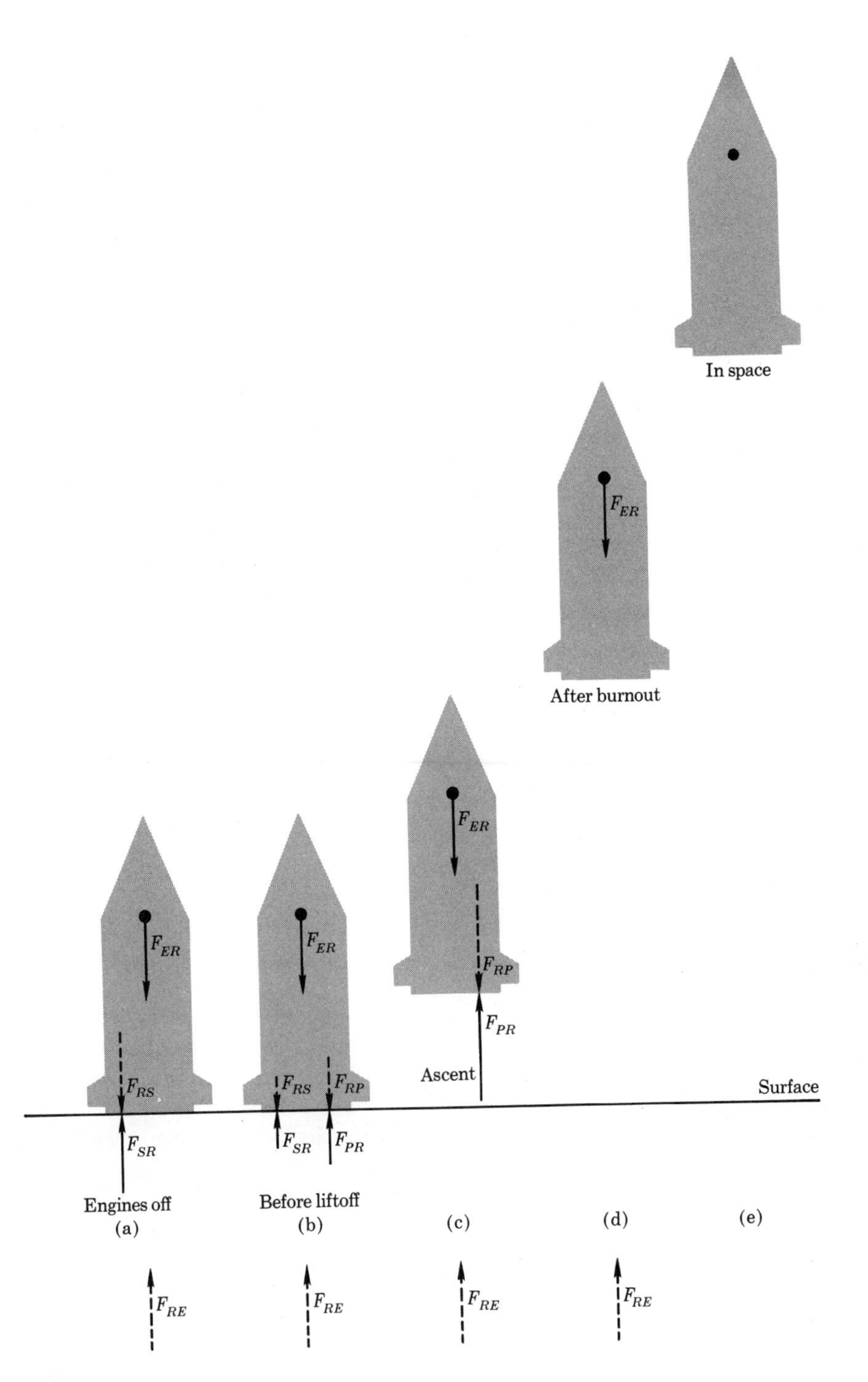

FIGURE 4-13

Force variations during rocket launching and flight

In order to bring out the full meaning of Newton's laws, it is convenient to study examples that reflect our experience. This requires that we anticipate the next chapter's general discussion of forces of all kinds by examining the relation of weight to mass in a little more detail. Experiments show that the force of gravity on an object is proportional to the mass, i.e.,

$$F_G \sim m$$

Further, the constant of proportionality—which we label g—is nearly the same everywhere on the Earth's surface:

$$F_G = mg$$

It is clear that g has the units of acceleration; hence the constant is called the acceleration of gravity, with numerical value on Earth around 32.2 ft/sec² or 9.81 m/sec², varying by a small percentage with location. In the analysis of forces in balance or of forces producing motion, we specify W (for weight) to be synonymous with F_G (for force of gravity). Thus, in a problem, we write

$$W = mg$$

A few calculations involving weight will be useful. To simplify, we use $g = 32$ ft/sec² and 9.8 m/sec².

1. The water molecule has a mass of 3.0×10^{-26} kg. What is its weight?

$$W = mg = (3.0 \times 10^{-26}\text{ kg})(9.8\text{ m/sec}^2) = 2.94 \times 10^{-25}\text{ N}$$

2. A man has a mass of 80 kg; thus his weight is

$$W = (80\text{ kg})(9.8\text{ m/sec}^2) = 784\text{ N}$$

Now let us turn to simple motion problems in which weight is involved.

1. A man rides in an ascending elevator that has an initial acceleration upward of $a = 10$ m/sec². Assuming his mass to be 60 kg, with what force is he pressed against the floor? There are two forces acting on the man—gravity pulling downward in amount W, and the floor pushing upward with force F. Thus the acceleration is produced by the net force

$$F - W = ma$$

but since $W = mg$, $F = m(g + a) = 60(9.8 + 10) = 1188$ N. This force is larger than the man's weight of $60(9.8) = 588$ N, and he appears to weigh more while accelerating upward, by an amount 600 N.

2. We allow our car of mass 1500 kg to coast down a long hill, a 5-deg incline. Neglecting friction, what acceleration will the car experience? The component of weight $W_{\parallel}$ along the incline (see Figure 4-9)

provides the acceleration. Thus, by Newton's second law,

$$W_{\parallel} = ma$$

but also $W_{\parallel} = W \sin \theta = mg \sin \theta$; hence $ma = mg \sin \theta$, $a = g \sin \theta = (9.8)(\sin 5 \text{ deg}) = 0.85 \text{ m/sec}^2$. It is interesting to note that the mass does not appear in the answer, since the m that determines the force of gravity is the same as the m appearing in Newton's law, and they cancel out.

4-4 DIMENSIONS AND SYSTEMS OF UNITS

We are now in a position to examine still more critically the meaning of the concept of measurement. Three basic *dimensions* used in the science of physics are mass (M), length (L), and time (T), which correspond to measurable quantities. Using these basic dimensions as algebraic quantities, one can construct more advanced quantities—such as speed, which can be formed as the ratio of length to time and abbreviated LT^{-1}; acceleration, LT^{-2}, and force, MLT^{-2}. One can invent as many such groups of dimensions as are needed or convenient to help describe nature. Mass, length, time, and the more advanced constructed dimensions should be considered as more fundamental than the *units* of measure which we select and agree on. In our present study we emphasize one consistent set of units—the meter, the kilogram, and the second—but realize that they are arbitrary standards. It has been jokingly said that one could express speed in furlongs per fortnight* just as well as in meters per second.

We have deliberately delayed the discussion of a painful subject, that of other units of mass, force, and acceleration besides the kilogram, the newton, and meters/(second)^2. Much of the difficulty that a student experiences with the subject of mechanics stems from three problems:

1. the confusion of weight and mass;
2. the conflict between everyday and scientific usages;
3. the variety of systems of units and conventions used.

These matters are unfortunately related.

In *A Description of the World*, written by Robert Morden in 1700, more than forty pages are devoted to the many different coins, weights, and measures used in various countries. At that early date, the units of weight used often depended on which commodity was being weighed. Even in our enlightened present we are faced with ounces, pounds, and tons in English-speaking countries and the metric weights in others. The three most important systems of units are the mks (meter–kilogram–second), as used throughout this book; the cgs (centimeter–gram–second), still used extensively in the literature of physics; and the British, employed mainly in the engineering and industry of English-speaking countries. Table 4-1

* 1 furlong = 160 ft, 1 fortnight = 2 weeks.

TABLE 4-1

SYSTEMS OF UNITS

Formula for Law	F	$= m$	a
Dimensions	MLT^{-2}	M	LT^{-2}
Units: mks	newton (N)	kilogram (kg)	m/sec^2
cgs	dyne	gram (g)	cm/sec^2
British	pound (lb)	slug	ft/sec^2

shows these systems in relation to Newton's second law and the dimensions of its physical quantities. Each system is consistent in that the same formula $F = ma$ can be used without any other constants of proportionality so long as the quantities F, m, and a are expressed in the correct units. Table 4-2 collects the necessary relation between the various units.

TABLE 4-2

UNITS OF LENGTH, MASS, AND FORCE

1 m	= 3.28 ft	1 ft	= 0.305 m
1 cm	= 10^{-2} m	1 m	= 10^2 cm
1 kg	= 10^3 g	1 g	= 10^{-3} kg
1 slug	= 14.6 kg	1 kg	= 0.0685 slug
1 N	= 10^5 dynes	1 dyne	= 10^{-5} N
1 lb	= 4.45 N	1 N	= 0.225 lb

Let us verify that each of these sets of units will give $F = ma$ consistently, starting with

$$F\ (\text{N}) = m\ (\text{kg})\ a\ (\text{m/sec}^2)$$

Applying factors to convert from mks to cgs,

$$[F\ (\text{N})]\ (10^5\ \text{dynes/N}) = [m\ (\text{kg})]\ (1000\ \text{g/kg})\ [a\ (\text{m/sec}^2)]\ (100\ \text{cm/m})$$

Cancellation of units and numbers gives

$$F\ (\text{dynes}) = [m\ (\text{g})][a\ (\text{cm/sec}^2)]$$

Instead, let us apply factors to convert from mks to British:

$$[F\ (\text{N})]\ (0.225\ \text{lb/N}) = [m\ (\text{kg})]\ (0.0685\ \text{slug/N})\ [a\ (\text{m/sec}^2)]\ (3.28\ \text{ft/m})$$

Cancellation leaves

$$F\ (\text{lb}) = [m\ (\text{slugs})][a\ (\text{ft/sec}^2)]$$

It is important to note that the pound is a unit of force, *not* a unit of mass. A "150-pound" person is one who produces a force of 150 pounds on scales where the acceleration of gravity is a standard* value $g_s = 32.2$ ft/sec². At another location where g is only 32.0 ft/sec², he would weigh less. The pound force in newtons is†

$$(0.454 \text{ kg})(9.81 \text{ m/sec}^2) = 4.45 \text{ N}$$

Thus

$$1 \text{ lb} = 4.45 \text{ N}$$
$$1 \text{ N} = 0.225 \text{ lb}$$

Suppose that we want to find the acceleration of a certain mass whose weight is W lb when a force of F lb is applied. There are two ways to get the answer: the first is correct, but involved; the second is a shortcut that gives the same answer. Newton's second law applies to the pound weight at a place where the acceleration is $g_s = 32.2$ ft/sec²:

$$W \text{ (lb)} = m \text{ (slugs)} \times (32.2 \text{ ft/sec}^2)$$

Thus

$$m \text{ (slugs)} = \frac{W \text{ (lb)}}{32.2 \text{ (ft/sec}^2)}$$

Now we apply Newton's law again to find the motion:

$$F \text{ (lb)} = [m \text{ (slugs)}][a \text{ (ft/sec}^2)]$$

or

$$F \text{ (lb)} = \frac{W \text{ (lb)}}{32.2 \text{ (ft/sec}^2)} a \text{ (ft/sec}^2)$$

Thus, from the formula, we see that a 1-lb force applied to an object of weight 1 lb gives an acceleration of 32.2 ft/sec². On the other hand, a 1-lb force on an object of weight 32.2 lb gives an acceleration of 1 ft/sec².

Now for the shortcut. If the object's weight is given to us in pounds, we merely divide by 32.2, which will give the proper *mass* in slugs to use in applying Newton's law. As an illustration, let us find the acceleration of a rocket of weight 10^5 lb (50 tons) when a thrust of 4×10^4 lb is applied. Its mass, using the shortcut, is $10^5/32.2 = 3100$ slugs. The acceleration is

$$a = \frac{4 \times 10^4 \text{ lb}}{3100 \text{ slugs}} = 12.8 \text{ ft/sec}^2$$

* By international agreement, the standard value is 32.17398, which we round off for calculations.

† More precisely, the numbers are 0.45359237, 9.80665, and 4.4482216.

Sometimes one finds forces expressed in units of *kilogram-force* (kgf). This is the force of standard gravity on a 1-kg mass, and hence

$$1 \text{ kgf} = (1 \text{ kg})(9.81 \text{ m/sec}^2) = 9.81 \text{ N}$$

We are frequently required to translate between weights in the British and metric systems. Suppose, for example, that a person weighs 180 lb and we want to find the corresponding weight in kgf. Now, 1 kgf = 2.2 lb, so the weight is

$$W = \frac{180 \text{ (lb)}}{2.2 \text{ (lb/kgf)}} = 81.8 \text{ kgf}$$

A mass of 81.8 kg would yield this weight in standard gravity.

A few numerical examples will help illustrate the use of units.

In mks: An automobile of mass 1000 kg is given an acceleration of 4 m/sec². How much force is required?

$$F \text{ (N)} = m \text{ (kg) } a \text{ (m/sec}^2)$$
$$F = (1000)(4) = 4000 \text{ N}$$

In cgs: A charged particle of mass 1.67×10^{-24} g experiences a force of 4.80×10^{-10} dynes. What acceleration occurs?

$$a \text{ (cm/sec}^2) = \frac{F \text{ (dynes)}}{m \text{ (g)}}$$

$$a = \frac{4.80 \times 10^{-10}}{1.67 \times 10^{-24}} = 2.87 \times 10^{14} \text{ cm/sec}^2$$

In British: An elevator lifts a man of weight 175 lb with an upward acceleration of 12 ft/sec². How much force is required?

$$m \text{ (slugs)} = \frac{W}{g_s} = \frac{175}{32.2} = 5.43 \text{ slugs}$$

Then

$$F \text{ (lb)} = m \text{ (slugs) } a \text{ (ft/sec}^2)$$
$$F = (5.43)(12) = 65.2 \text{ lb}$$

This is the *net* force over and above the weight. Note that the man seems to weigh 240.2 lb while going up in the elevator.

4-5 DENSITY AND ATOMIC STRUCTURE

An enormous number of compounds can be formed from the 104 known chemical elements. The mass of a certain sample of a compound is determined by the component particles, but the spacing of molecules in a substance depends on forces between them. The quantity of material can be expressed in terms of atoms or molecules. We learned in chemistry that a *gram-molecular weight* or *mole* of an element or compound is a weight that combined with or displaced 16 grams of oxygen. In the mks system one finds convenient the unit mass called the *kilomole* (kmole), which is 1000 moles. Thus, from Table 1-1, 1 kmole of hydrogen is 1.00797 kg, 1 kmole of oxygen is 15.9994 kg, etc. We recall from basic chemistry that the number of particles in a mole is the same for all substances. This fact is readily appreciated by an example. Each H atom has mass close to 1, each O atom has mass close to 16. Thus 1 gram of hydrogen has the same number of atoms as does 16 grams of oxygen. In the metric system, *Avogadro's number*, $N_a = 6.02 \times 10^{26}$, is the number of particles in a kilomole of any material. The mass of one particle—atom or molecule—of a material with molecular mass M (kg/kmole) is thus

$$m \text{ (kg)} = \frac{M \text{ (kg/kmole)}}{N_a \text{ (/kmole)}}$$

As a check, take hydrogen, $M \cong 1.008$. Then

$$m = \frac{1.008}{6.02 \times 10^{26}} = 1.67 \times 10^{-27} \text{ kg},$$

as was quoted in Section 1-1.

Density

Closely related to the idea of mass is the *density* (ρ) of a sample of material, defined as the mass (kg) in a unit volume (m³).* Density is a measure of an element's atomic composition and structure. Familiar formulas involving density are

$$\rho = \frac{M}{V}$$

and

$$M = \rho V$$

where M is in kilograms and V is in cubic meters; hence ρ is in kilograms per cubic meter (kg/m³). For example, if a piece of iron has a mass of 100 kg and its volume is 0.0127 m³, its density would be 7850 kg/m³. The spacing of molecules depends on the way the substance was prepared, and on the environment, especially in the case of gases. Densities specified at standard conditions (sea level, freezing temperature) are listed in Table 4-3

* The cubic meter of the mks system is an inconveniently large unit when we are discussing a system of small volume. So long as we are consistent in using other cgs units, volumes may be expressed in terms of the cubic centimeter, a more easily visualized size.

TABLE 4-3

SPECIFIC GRAVITY AND DENSITY*

(The density in kg/m³ is 10³ times the specific gravity)

	Substance	Specific Gravity	Density (lb/ft³)
Solids	Aluminum	2.70	168
	Brass	8.75	546
	Copper	8.93	558
	Iron	7.85	490
	Lead	11.3	708
	Osmium	22.5	1405
	Uranium	18.7	1167
	Gold	19.3	1205
	Ice	0.917	57.2
	Glass	≅2.6	≅162
	Wood: balsa	≅0.125	≅ 7.8
	oak	≅0.75	≅ 47
	ebony	≅1.22	≅ 76
	Cork	0.24	15
	Styrofoam	0.032	2.0
Liquids	Water	1.000	62.4
	Sea water	1.025	64.0
	Motor oil	≅0.90	≅56
	Mercury	13.6	849
	Milk	1.03	64.3
	Alcohol (ethyl)	0.791	49.4
Gases (at 0°C, atmospheric pressure)			
	Air	1.293×10^{-3}	0.0807
	Oxygen	1.429×10^{-3}	0.0892
	Nitrogen	1.251×10^{-3}	0.0781
	Hydrogen	0.090×10^{-3}	0.0056
	Helium	0.1785×10^{-3}	0.0111
	Tungsten fluoride	12.9×10^{-3}	0.805

* Main source: *Handbook of Chemistry and Physics*, Cleveland, Ohio: The Chemical Rubber Co., 1969.

for a few typical materials. The *weights* of a unit volume of material will be proportional to the mass density, so long as a standard gravity acceleration is used.

In the British system, densities are expressed in pounds per cubic foot.† For example, the densities of water and lead are 62.4 lb/ft³ and 708 lb/ft³, respectively. The *specific gravity* (really the specific mass) is the *ratio of density of a substance to the density of water*. This ratio is entirely independent of the units used. For lead in the mks system, the ratio is

† The density quoted in pounds per cubic foot (lb/ft³) must be considered as the *weight* per cubic foot in a *standard gravity*.

$\frac{11{,}300\ (\text{kg/m}^3)}{1000\ (\text{kg/m}^3)} = 11.3$; in the British system, $\frac{708\ (\text{lb/ft}^3)}{62.4\ (\text{lb/ft}^3)} = 11.3$.

The density of a substance depends directly on the masses of the individual atoms and molecules of which it is composed, and on the spacing of the particles. Let us picture a solid lattice, such as in Figure 4-14, in which the atoms are spaced regularly, as if at the corners of cubes with length of side d. Around each particle we may draw another box of side d, volume d^3, in which the particle resides. In a cubic meter there are $1/d^3$ such cubes, hence the number of atoms per unit volume is also

$$N = \frac{1}{d^3}$$

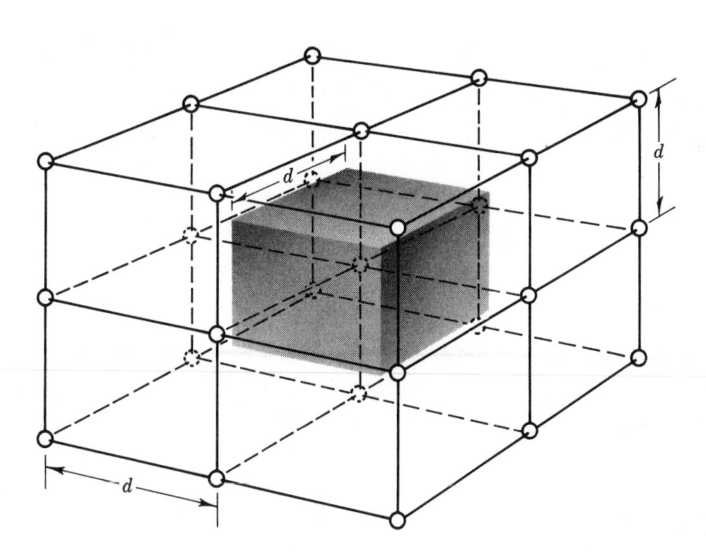

FIGURE 4-14

Lattice arrangement of atoms in a solid

Each of the particles has mass m; hence

$$\rho = Nm$$

The mass of one particle of molecular mass M is

$$m = \frac{M}{N_a}$$

where N_a is Avogadro's number. Then

$$\rho = \frac{NM}{N_a} \qquad \textit{(density)}$$

or conversely,

$$N = \frac{\rho N_a}{M} \qquad \textit{(particles per unit volume)}$$

The spacing of atoms is readily obtained from

$$d = \frac{1}{N^{1/3}}$$

Let us illustrate by finding the properties of a crystal of table salt, NaCl, which has equal numbers of the two particles of atomic masses 23 and 35 (average $M = 29$) arranged in a cubic lattice. With a density of 2.2×10^3 kg/m³, the number of atoms per cubic meter is

$$N = \frac{\rho N_a}{M} = \frac{(2.2 \times 10^3)(6.02 \times 10^{26})}{29} = 4.6 \times 10^{28} \text{ per cubic meter}$$

The atom spacing is

$$d = \frac{1}{N^{1/3}} = \frac{1}{3.6 \times 10^9} = 2.8 \times 10^{-10} \text{ m}$$

4-6 PRESSURE

The effect of a force on an object of some size may depend on the area over which it acts. We would find it most difficult, for example, to drive the *head* of a nail into a piece of wood, since the force is spread out over a large area compared with that of the sharp point. *Pressure is defined as the force* (say, in newtons) *on a unit area* (here, in square meters); or, in symbols,

$$P = \frac{F}{A}$$

Suppose a hydraulic press used to flatten metal exerts a force of 10^5 N on an area of 1 cm². The pressure is $P = \frac{F}{A} = \frac{10^5 \text{ N}}{10^{-4} \text{ m}^2} = 10^9 \text{ N/m}^2$. In the British system, the units of pressure are lb/in.² or lb/ft². Suppose that a piece of heavy machinery of weight 5000 lb and area of base 4 ft² is placed on a floor with uniform contact between surfaces. The pressure is (5000 lb)/(4 ft²) = 1250 lb/ft² or $\frac{1250 \text{ lb/ft}^2}{144 \text{ in.}^2/\text{ft}^2} = 8.68$ lb/in.². The blanket of air in the Earth's atmosphere exerts a pressure (at sea level) of 1.01×10^5 N/m², which is the same as 14.7 lb/in.². This pressure is also called 1 *atmosphere* (atm). Thus a pressure of 44.1 lb/in.² would be 3 atm. The fact that we live under an atmospheric pressure to which our bodies have become accustomed makes us forget how much force is involved. Riding in an airplane with cabin improperly "pressurized" is a painful way to realize the delicate balance of the body with respect to sudden changes in pressure. To take another instance, a deep-sea diver must ascend to the surface very slowly to avoid the "bends," essentially a reaction to a quick change in pressure. Obviously, the Apollo spacecraft and the astronauts' spacesuits were sealed with extraordinary care.

A simple experiment will reveal the magnitude of air pressure. A small amount of water is placed in a metal can such as that used for anti-freeze. The open can is then heated to form steam that drives out the air. The lid is replaced tightly, and the can immersed in cold water to condense the steam. The collapse of the walls of the can is startling—even if we know that the total force on the outside is about half a ton.

We must be careful in the interpretation of a measurement of pressure in an enclosed space. Consider the process of finding the pressure in the tire of a bicycle or automobile. A gauge that is connected to the valve registers the pressure *above atmospheric*. To verify this, note that if the tire is flat because of an opening to the atmosphere, the gauge reads zero. Then if the gauge reads 30 lb/in.^{2} when the tire is filled, the actual pressure inside is 30 + 14.7 or 44.7 lb/in.^{2}. This latter figure is the true or *absolute pressure*. The distinction is often found expressed by the units psig (pounds per square inch gauge) or psia (pounds per square inch absolute).

Relation of Density and Pressure

The pressure at the base of a column of material, such as a concrete pillar or a tall water-tower, is easily found if the density of the substance is known. Visualize cubes 1 meter on a side piled up on top of each other, as in Figure 4-15. Each has a density ρ: the total mass is ρh; ρhg is the weight applied to the base, which is of unit area. Thus for such a column the pressure is

$$P = \rho g h$$

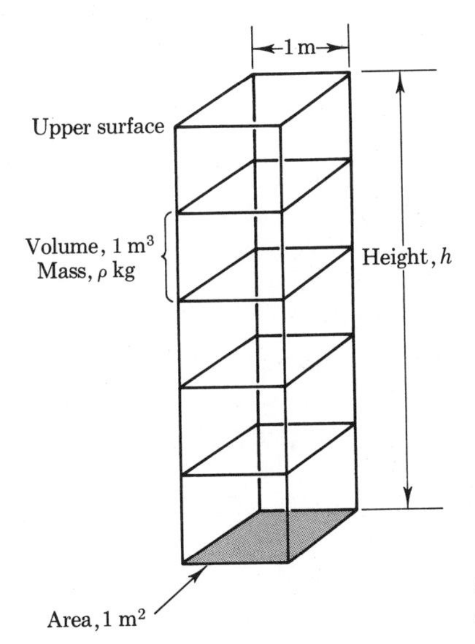

FIGURE 4-15

Pressure due to column of height h, $P = \rho g h$

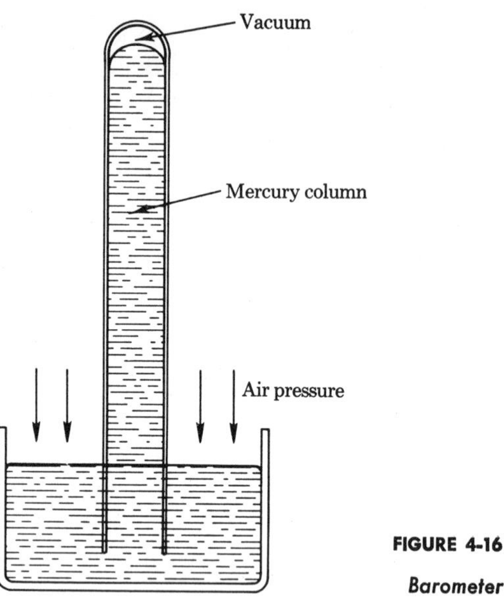

FIGURE 4-16
Barometer

The *barometer*, a device to measure air pressure, employs this principle. A glass tube with one end closed is filled with liquid mercury and inverted over a container, as shown in Figure 4-16. The level of mercury drops, leaving above the surface an almost perfect vacuum,* which exerts no pressure. The height of the column adjusts itself until the pressure of air on the surface in the container balances the pressure due to the column. Measurements of the height h thus yield P. At sea level, the column of Hg, density 13.6×10^3 kg/m³, has a height

$$h = \frac{P}{\rho g} = \frac{1.01 \times 10^5 \text{ N/m}^2}{(13.6 \times 10^3 \text{ kg/m}^3)(9.81 \text{ m/sec}^2)} = 0.760 \text{ m}$$

or 760 mm. This corresponds very closely to 30 in. The quotation of the daily barometric pressure, we may recall, is in "inches": for a "high" associated with fair weather, the pressure is above 30 inches of mercury; for a "low" (stormy weather), the pressure is below 30 inches.

4-7 RELATIVISTIC MASS

Although Newton's laws, including $\mathbf{F} = m\mathbf{a}$, are extremely useful in describing motions, we must remember that they are only approximately correct, applicable for speeds much less than that of light. We saw in Chapter 3 that the relativity of Einstein was more general and correct,

* The only molecules present in this space will be the mercury evaporated from the surface.

and that the phenomena of length contraction and time dilation were significant at high speeds. Einstein's discovery had a profound effect on viewpoints on the laws of dynamics—that is, on the relation between force, mass, and motion. He accepted Newton's law in the form $F = \Delta p/\Delta t$, with $p = mv$, but showed that *mass is dependent on speed,* and is not constant. In particular, if an object has speed v relative to a "fixed" frame, the mass observed from that frame is

$$m = \frac{m_0}{\sqrt{1 - \left(\frac{v}{c}\right)^2}}$$

where c is the speed of light 3×10^8 m/sec, and m_0 is the mass of the object at rest (the "rest mass"). The origin of the phenomenon can be traced to time dilation, as we can show* qualitatively through the following imaginary experiment. Visualize, as in Figure 4-17, an airplane passing over

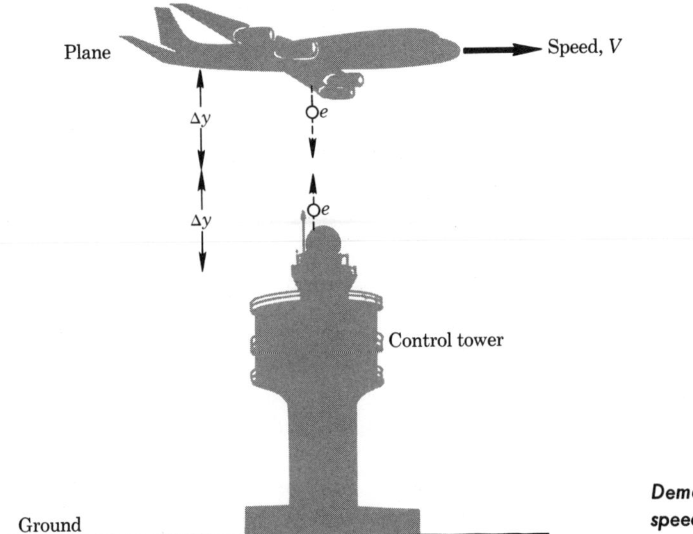

FIGURE 4-17

Demonstration of mass increase with speed

a control tower with speed V relative to the ground. The pilot and the observer on top of the tower have agreed prior to the flight that each will send out a particle, say an electron, with the same vertical component of velocity (as seen in his own frame of reference). The plan calls for the electrons to be released at the right instant and aimed in such a way that they collide at equal distances Δy from their original levels. These distances are the same for both observers, since the Lorentz transformation has to do only with x-motion. The idea is that the electrons will bounce off each other

* A more complete proof is found in C. W. Sherwin's *Basic Concepts of Physics*. New York: Holt, Rinehart, and Winston, 1961, p. 127.

and return to their sources, and by means of time and distance measurements speeds can be measured. From the point of view of each observer, the y-motion of his electron has merely been reversed by the collision after the particle went out a distance Δy for a certain length of time. The change of y-velocity is thus twice the initial value. We ignore the effect of gravity in order to get at the basic relativistic effects; we assume that the speeds in the y-direction are very small compared with c, so that Newton's classical laws may be applied to the collision—let us say, by the ground observer. He presumes that his own electron has a mass m_0, and seeks to determine the "unknown" mass m of the electron released from the plane by measuring accelerations—that is, by finding *changes* in the y-velocity over a time interval, as was described in Section 4-2. For his electron, he finds the change in velocity to be $\Delta v_0 = 2v_0$; for the plane's electron, he finds the change to be $\Delta v = 2v$. In order to obtain v it is necessary for him to measure the total time interval $2\Delta t$ required for that electron to travel a distance $2\Delta y$, and then form $v = \Delta y/\Delta t$. At this point, the symmetry of relativity comes into the picture. The plane's observer has measured the speed of this same electron as $v_0 = \Delta y/\Delta T$, using his own clock. From the last two expressions, we see that $v = v_0(\Delta T/\Delta t)$, and thus the ground observer's measurements yield a mass ratio

$$\frac{m}{m_0} = \frac{\Delta v_0}{\Delta v} = \frac{v_0}{v} = \frac{\Delta t}{\Delta T}$$

Recall from Section 3-4 that the ratio of time intervals as measured by "fixed" and moving frames is $\Delta t/\Delta T = k$, hence

$$\frac{m}{m_0} = k = \frac{1}{\sqrt{1 - (V/c)^2}}$$

The effect is associated with the high speed V of the moving frame, and is shared by all objects in that frame—the electron, the pilot, and the plane itself. Although the deduction of a larger mass by the observer on the "fixed" ground frame is in reality a consequence of time dilation, it is much simpler for us to think of the mass of any object as increasing with speed relative to the frame of observation.

There are several interesting consequences to the dependence of mass on speed (we return to the use of v instead of V for speed):

1. If v is zero, the denominator is 1, and $m = m_0$, which is of course correct.

2. As v increases, the denominator gets smaller, and m becomes larger than m_0.

3. As v approaches c, the denominator approaches zero, and m approaches infinity.

Let us contrast ordinary motions of relatively large-scale objects with the very high speeds of atomic particles obtainable in particle accelerators such as the cyclotron. Suppose the speed of a bullet from a rifle is 300 m/sec.

Then $v/c = 10^{-5}$, $(v/c)^2 = 10^{-10}$, which can surely be ignored compared with the 1 in the denominator of the mass formula. Then $m \cong m_0$ and $F = \Delta(m_0 v)/\Delta t \cong m_0 a$. On the other hand, let the speed of a proton in an accelerator be 1.5×10^8 m/sec. Then $v/c = \frac{1}{2}$, $(v/c)^2 = \frac{1}{4}$, and $m/m_0 = (1 - \frac{1}{4})^{-1/2} = 1.155$, which means that a 15% increase in mass has been produced.

We can now see that no object can move faster than the speed of light. As the speed approaches c, the force needed to continue to accelerate an increasingly massive object has to increase toward infinity. Alternatively, we can see that with a constant force, the object gains speed at a slower and slower rate as v approaches c.

The Significance of Relativity

Inspection of Einstein's formula relating mass and speed reveals the reason for the long delay in discovering the special theory of relativity. For the motion of objects of the scale of man, the effect is extremely small; and before the discovery of the electron and the development of modern electronics was impossible to detect. Until methods for making precise astronomical measurements were perfected, there was no means for knowing that some stars travel at speeds approaching the speed of light. For purposes of our daily living, relativity as a physical description has no significance. On the other hand, the *concept* of relativity has had considerable impact on philosophy, partly because of the popular attention the idea has evoked. The phrase "everything is relative" suggests the modern tendency to believe that truth is not absolute, but depends on one's point of view. It is valid to conjecture that the concept of relativity, along with the uncertainty principle of quantum mechanics, has had some indirect influence on moral philosophy and on the role of religion in man's life.

PROBLEMS

4-1 Find the momentum (in the metric system) of an 80-kg football player running at top speed if he can do the 100-yard dash in 9.5 seconds.

4-2 Calculate the momentum of a 1200-kg automobile moving with a speed of 20 m/sec; with a speed of 35 m/sec. What is the change in momentum? How much average force is required to cause this change in 4 seconds?

4-3 Calculate the force required to give an automobile of 1000-kg mass an acceleration of 5 m/sec^2. If that same force were applied to a 2500-kg car, what acceleration would result?

4-4 In Millikan's experiment, the gravitational force downward on a charged oil drop of mass 10^{-6} kg is 10^{-5} N and the electrical force upward is 0.998×10^{-5} N. (a) What is the magnitude and direction of the acceleration of the drop? (b) How long will it take to move, starting from rest, a distance of 3 cm?

4-5 An electrical force of 9×10^{-20} N acts on a proton. Compute the acceleration of the proton of mass 1.67×10^{-27} kg.

4-6 The force of gravity on a spaceship of mass 5×10^{4} kg approaching the surface of the planet Venus is 4.3×10^{5} N. What acceleration does the ship experience?

4-7 The Moon requires about 28 days to complete its orbit about the Earth at a radius of 3.84×10^{8} m. (a) Find its speed. (b) Find its radial acceleration toward the Earth. (c) Find the force required to give the Moon of mass 7.4×10^{22} kg this acceleration.

4-8 A small boy constructs from his Erector Set the arrangement of girders, wheels, and strings shown in Figure P4-8. Derive a formula for the acceleration, if the mass on the 45-deg incline is 50 grams, and the mass on the other end of the string is 30 grams. Which way will the system accelerate and how much, ignoring friction?

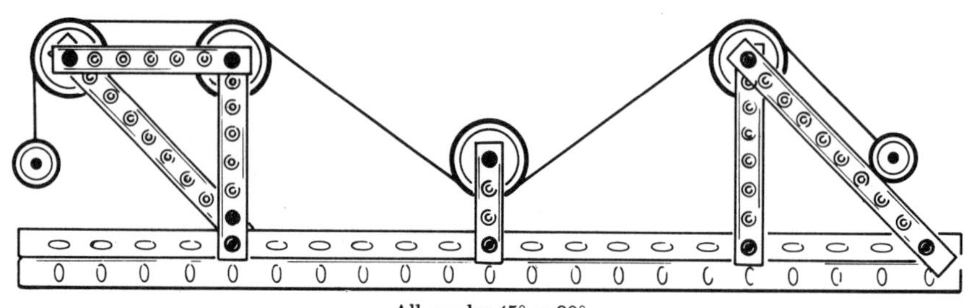

FIGURE P4-8

4-9 A bird of weight 0.8 lb perches on a telegraph wire between two poles 150 ft apart, and depresses the wire by 3 in. How much tension is produced in the wire? Assume that the wire is weightless and initially horizontal.

4-10* A tow truck is called to pull a 1200-kg car out of a ditch (see Figure P4-10). The bank has an angle of 30 deg. (a) What minimum force must be applied at the winding drum on the truck, neglecting friction in turning the car wheels? (b) The driver of the truck (of mass 1800 kg) releases his brakes by mistake just as the car reaches the top of the incline. What acceleration results? (c) Check the result in part (b) by finding the sum of the external forces and the total mass. Be sure all forces are accounted for.

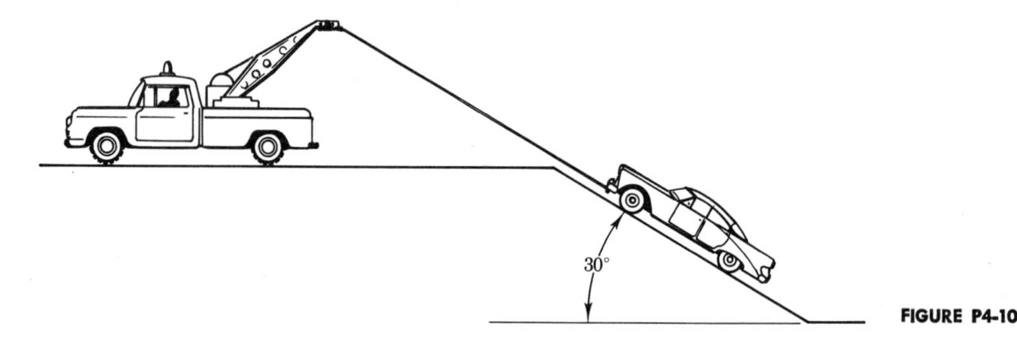

FIGURE P4-10

4-11 A boat is tied to the bank of a moving stream across which the wind blows (see Figure P4-11). (a) Write the equations stating that there is no net force along the stream or perpendicular to it. *Suggestion:* regard the boat as a point, with forces and angles as follows: wind, F_W, w; stream, F_S, s; rope, F_R, r. (b)* Solve the equations to find the angle the rope makes with the bank and the amount of force the rope experiences.

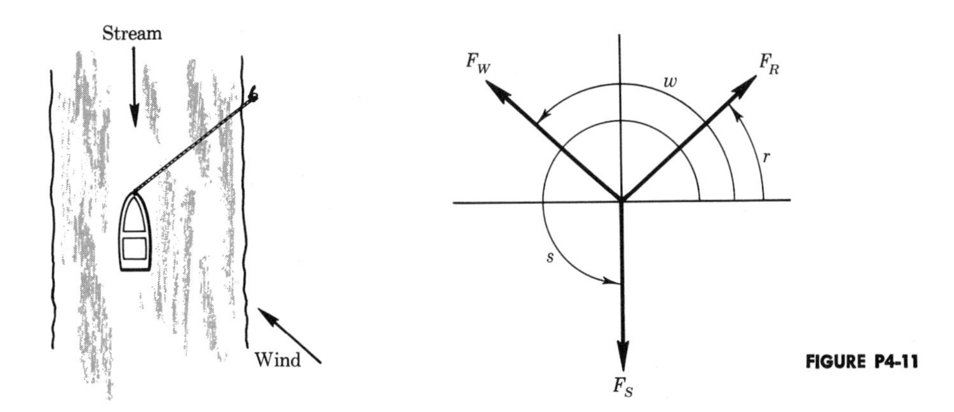

FIGURE P4-11

4-12 A water skier can swing far out to the side of the path of the boat towing him by changing the angle of his skis with the resisting water (see Figure P4-12). Analyze the forces that make this motion possible. Can the skier ever get ahead of the boat?

4-13 What horizontal force must be applied to the axle of a wheel, of radius r and weight w, to make the wheel "climb a step" of height h?

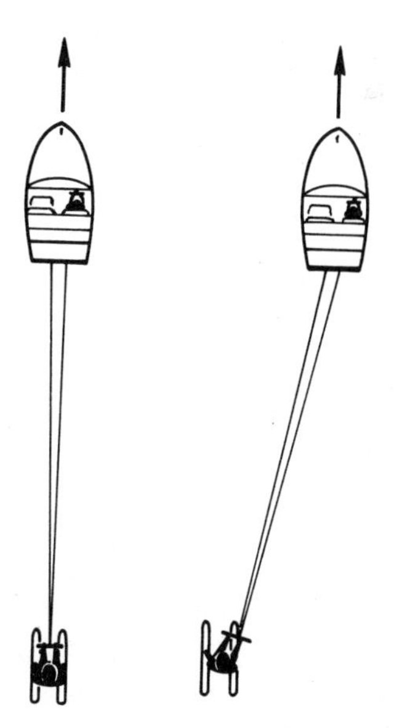

FIGURE P4-12

4-14 We hold firm on a rope connected to a pulley and to a mass of 25 kg, and then pull hard enough to give an upward acceleration of 9.81 m/sec^2. Compare the forces in the two cases.

4-15 What force must be provided, by an electromagnet that lifts scrap metal from cars, to lift a 2000-kg load with a vertical acceleration of 0.5 m/sec^2?

4-16 The inclined plane is often used to load ships. Neglecting friction, what force F must be used to accelerate a box of mass M up an incline of angle θ at 1 m/sec^2? Down the incline at 1 m/sec^2? At what angle is no force required to achieve that acceleration?

4-17 An elevator starts from the ground level with an acceleration upward of 10 ft/sec^2. With what force do the feet of a person weighing 165 lb press against the floor of the elevator?

4-18 Rope-climbing practice is a part of the training of the U. S. Marines. A rope is thrown over a pulley, and a Marine of mass 70 kg climbs with an acceleration of 2 m/sec^2. Neglecting friction, what happens to a mass of 70 kg used as counterbalance? How large is the force of tension in the rope?

4-19 Among the specifications accompanying a car shipped from Germany are the following: weight, 1100 kgf; maximum speed, 180 km/hr; fuel capacity, 22 liters; and tire pressure, 2.1×10^5 N/m^2. Translate these into lb, mi/hr, gal, and $lb/in.^2$.

4-20 (a) Verify the relation 1 kg = 0.0685 slug. (b) A force of 4×10^5 dynes acts on a mass of 2.5 g. Find the acceleration in cm/sec^2. Check the result by converting force and mass into the mks system, applying Newton's second law, and then converting from m/sec^2 to cm/sec^2.

4-21 A modern space rocket such as the hydrogen-burning J-2 can develop 200,000 lb of thrust. (a) What acceleration in ft/sec^2 will it give to a Moon spacecraft of weight 90,000 lb? (b) Convert force and mass into metric units, obtain the acceleration, and check to see if it agrees with the result from part a.

4-22 Using force, length, and time as fundamental dimensions, express the dimension of mass.

4-23* (a) Find the number of uranium atoms per cubic meter, if the mass of 1 kmole of U is 238.1 kg, and its density is 18.7×10^3 kg/m^3; (b) calculate the spacing of U atoms if it is assumed they are at the corners of cubes.

4-24 The diameter of a sphere of styrofoam is measured to be 4 in. and its weight is found to be 0.60 oz. Calculate its density in lb/ft^3 and its specific gravity.

4-25 NaK is an alloy of the metals sodium (percentage weight, 56) and potassium (percentage weight, 44) that serves as a cooling agent for high-temperature nuclear reactors. Estimate the specific gravity of NaK if those of its components are 0.93 and 0.84, respectively.

4-26 Verify that the standard air pressure of 14.7 $lb/in.^2$ is the same as 1.01×10^5 N/m^2.

4-27 Compute the magnitude of the force in tons due to air pressure on the sides of an evacuated container 20 cm × 15 cm × 5 cm.

4-28 Find the increase above atmospheric pressure, the total pressure, and total force on a skindiver of body area 1.5 m², 25 m below the fresh water level. Give answers in both metric and British systems.

4-29 In the seventeenth century, in Magdeburg, Germany, a public demonstration of the force of air was conducted. Teams of horses attempted unsuccessfully to pull apart two 1-ft-diameter hemispheres from which most of the air had been withdrawn (see Figure P4-29). The force needed to separate is found to be the same as that due to air pressure on a *disc* of the same radius. Calculate its value.

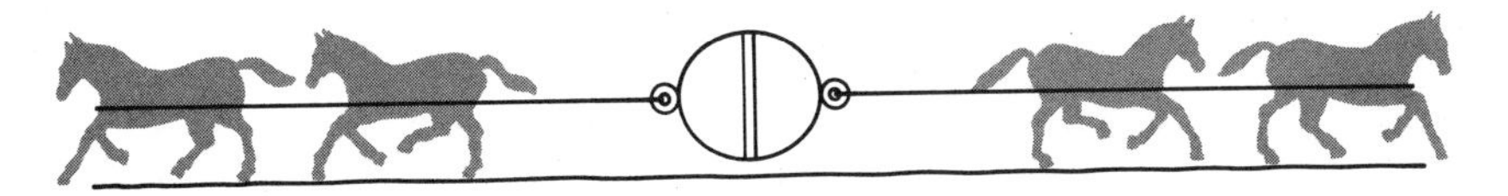

FIGURE P4-29

4-30 Find the unknown heights of liquids in the connecting tubes in Figure P4-30.

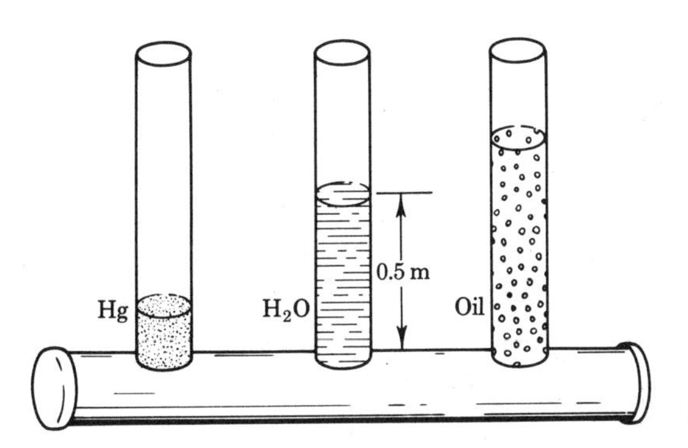

FIGURE P4-30

4-31 One end of a U-shaped tube filled with a liquid is attached to a tank of compressed air under pressure P, while the other end is open to atmospheric pressure P_a (see Figure P4-31). Show that the difference in pressure

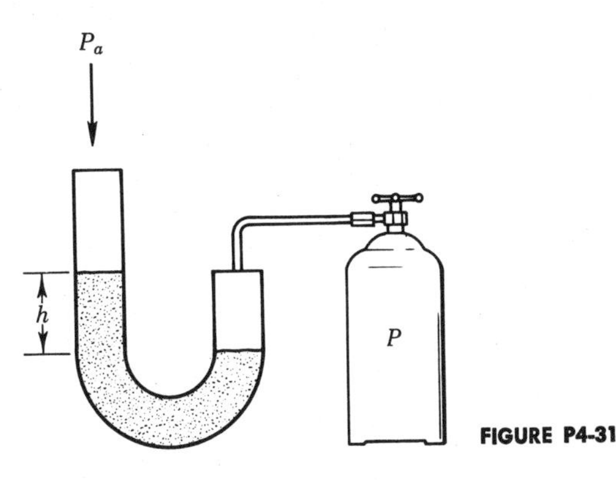

FIGURE P4-31

as measured by this *manometer* is $P - P_a = \rho g h$, and compute the height in centimeters if P is 1.25×10^5 N/m^2 and the fluid has specific gravity 2 at standard atmospheric pressure.

4-32 Explain why a water barometer would not be very practical.

4-33 Calculate the mass and momentum of an electron with rest mass $m_0 = 9.11 \times 10^{-31}$ kg, when it is moving at a speed 0.999 times the speed of light.

4-34 By what fraction larger than its rest mass is the mass of a bullet of speed 1200 m/sec? Should relativity be taken into account in aiming at a distant target?

4-35 Compute the percentage mass increase of a neutron of speed 5.1×10^7 m/sec, as produced by a nuclear reaction.

FORCES IN NATURE

5

We have introduced the properties of matter such as mass and charge, and defined motion in terms of distance and time.

Processes in nature have been shown to involve the ideas of force, matter, and motion. Our next step is to classify and describe the various *forces,* seeking to find their physical origin and mathematical statement. Experimental observations provide the basis for identifying and describing forces; and as noted before, the ultimate origin of forces remains unknown and perhaps unknowable.

5-1 CLASSIFICATION OF FORCES

Some time in the future, it may be possible to state that there is only one fundamental force, and that differences are only apparent. At the present stage of the development of physics, however, we are required to assume a variety of forces that appear to have distinct and different origins:

1. *Gravitational*—the interaction of objects by virtue of their *mass:* the existence of weight is evident on the Earth, and planetary motion is also governed by gravitation.

2. *Electrostatic* and *electromagnetic*—the interaction of objects bearing electrical *charge* at rest or in motion: the bonds in atoms and the operation of electrical devices depend on these forces.

3. *Nuclear*—the interaction of matter on the subatomic level: it provides the bonds that keep nuclei stable in spite of electrostatic repulsions.

4. *Frame-dependent* (*non-inertial*)—due to the relative motion of coordinate frames.

To this list we can add what may be called *macroscopic* (large-scale) forces: combinations of interacting forces involving many particles. Examples are elastic, cohesive, and frictional forces, and the force between bulk magnetic materials.

Arriving at the state of understanding of the forces that is depicted in the following pages was the consequence of the lifetime work of many scientists over a span of hundreds of years. The distilled essence of that understanding as we review it does not properly reveal the pattern of development of the subject. Omitted is much on the accumulation of experimental evidence, the misconceptions, the reasoning, and the flashes of inspiration that would give our account the depth of historical background that it deserves. As compensation, we shall aim to achieve a sense of clarity and unity in the complex science of physics in order to provide a basis for appreciation of recent and possible future advances.

5-2 GRAVITATIONAL ATTRACTION

All particles of matter are found to attract each other in a manner that depends on the masses of the objects and on the distance of their separation. Consider the following pairs of objects, also sketched in Figure 5-1:

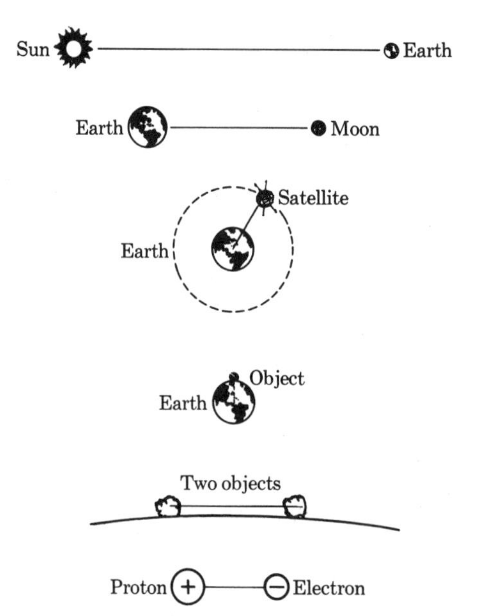

FIGURE 5-1
Examples of gravitational attraction

the Sun and the Earth
the Earth and the Moon
the Earth and a spacecraft or satellite
the Earth and any object on the Earth's surface
any two pieces of material on the Earth
a proton and an electron

Between any two particles in each of these objects acts a force that is represented by the formula discovered by Newton:*

$$F = k_G \frac{m_1 m_2}{R^2} \qquad \textit{(law of gravitation)}$$

where m_1 and m_2 are the masses in kilograms, R is the distance of separation of the particles in meters, F is the magnitude of the force (N), and k_G is a constant† of nature

$$k_G = 6.67 \times 10^{-11} \text{ N-m}^2/\text{kg}^2$$

These units may be verified by inspection of the formula. The term *inverse-square law* is applied to the relation because of the proportionality of force to $1/R^2$. A fuller meaning of this important law is now developed.

1. As stated, the formula applies only to particles, while the force between large-scale masses is the sum of forces between individual particles. We shall show later (in Section 6-2) that the force between two large spheres that are uniform in composition is given by the inverse-square formula if the total masses *are assumed to be located at their centers.*

2. The force of gravity on an object at a certain location (usually the Earth's surface) is called the *weight* of the body. Let us compute this force for a 1-kg mass, assuming that the Earth is a uniform sphere of radius 6.4×10^6 m (about 3960 mi) and mass 6.0×10^{24} kg.

$$F = \frac{(6.67 \times 10^{-11})(6.0 \times 10^{24})(1)}{(6.4 \times 10^6)^2} = 9.8 \text{ N}$$

The *weight* of a 1-kg mass is thus 9.8 N. Corresponding to this force is an acceleration of the mass m of amount

$$g = \frac{F}{m} = \frac{9.8 \text{ N}}{1 \text{ kg}} = 9.8 \text{ m/sec}^2$$

* Around 1660, Sir Isaac Newton had been studying the problem of what holds the Moon in its course around the Earth; and—as the story goes—on seeing an apple fall to the ground, suddenly thought of the law of universal gravitation that applies to all bodies. Verification of the law in the laboratory came much later (accomplished by Cavendish in 1798).

† This universal gravitational constant, often symbolized instead by the letter G, is measured experimentally. It has units chosen to conform with those of force. The units in this law agree with those in the law of inertia $F = ma$. Similar proportionality constants will be needed also for electricity and magnetism.

The symbol g instead of a is used here to denote the *acceleration of gravity.**

Two fine points should be noted. First, it has been assumed without proof that the mass that gives gravitational attraction is the same as the mass that provides inertia. The most precise measurements available reveal that these are identical. Second, the fact that the Earth is not really a uniform sphere, but has surface variations in height and density, and is spinning on its axis means that the value of g varies from place to place† on the Earth's surface by as much as 0.04 m/sec². The standard for use in fixing weights is 9.80665 m/sec². In the British system this is 32.174 ft/sec². For our numerical examples and problems, we use 9.8 m/sec² or 32.2 ft/sec² because they are easy numbers to work with.

3. The weight of an object decreases with distance above the Earth's surface. For small heights above the ground this is unimportant, as we can see by the following simple argument.

Form the ratio of forces at heights R and $R + \Delta R$. All of the constants cancel, leaving

$$\frac{F(R)}{F(R + \Delta R)} = \frac{(R + \Delta R)^2}{R^2} = \left(1 + \frac{\Delta R}{R}\right)^2$$

Now, if ΔR is 1000 m (0.62 mi), and R is 6.4×10^6 m (3960 mi), the ratio comes out 1.0003, implying a change of only $\frac{3}{100}\%$. The force of gravity on an artificial satellite at a height of 1000 mi is significantly smaller than the surface value, however. By the ratio of the squares of radii, the force ratio is

$$\left(\frac{3960}{4960}\right)^2 = 0.64$$

4. The weight of a body on the surface of the Moon with radius 0.272 times that of the Earth and mass 0.0123 times that of the Earth is smaller by the ratio

$$\frac{0.0123}{(0.272)^2} \cong \frac{1}{6}$$

This is the basis for the familiar statement that a man could jump 6 times higher on the Moon than on the Earth (Figure 5-2). Through muscular action, a person could give himself the same vertical speed v_0 at either location. Then the heights to which he rises are $h_M = v_0^2/2g_M$ and $h_E = v_0^2/2g_E$. Their ratio is $h_M/h_E = g_E/g_M \cong 6$. The difference between mass and weight is clearly revealed by the study of motion at the surface of the Moon. The weight of or the force of gravity on an object there is $\frac{1}{6}$ of its

* The phrase "the force was 7 g's" stems from this symbol. An acceleration of $7g$ would require such a force.

† For example, in Greenland it is 9.825, in New York City 9.803, and in Liberia 9.782.

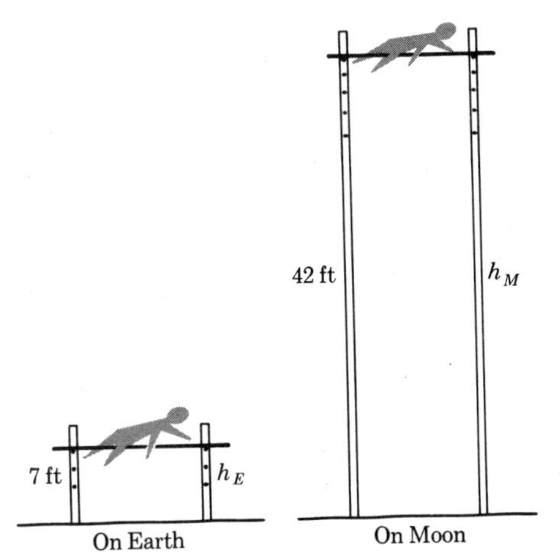

FIGURE 5-2
High jumping against gravity

Earth value, but the mass of the object is the *same* as on the Earth. It is easier, on the Moon, to hold a body from falling, but just as hard to accelerate it sideways.

Let us consider the launching of a landing craft from the surface of the Moon using rocket thrust. In a vertical lift-off, to overcome gravity and attain a certain acceleration a, the net force required would be

$$F_M = ma - mg_M$$

If the launching were performed instead on the Earth, the vertical force would have to be

$$F_E = ma - mg_E$$

However, on either planetary body the force needed for horizontal acceleration would be merely

$$F = ma$$

5-3 ELECTROSTATIC FORCES

Particles attract or repel each other if they are electrically charged. Objects that have a negative (−) charge are the following:

the electron
an atom or molecule with extra electrons
any body with an excess of electrons

Those having positive (+) charges are the following:

the proton
any nucleus
an atom or molecule with electrons removed
any body with a deficit of electrons

Simple experiments may be performed to generate "static electricity," or what physicists call electrostatic charges. By rubbing a glass rod with a cloth such as nylon, a positive charge may be left on the rod—that is, electrons will have been removed from it; by rubbing a hard rubber rod with fur, a negative charge due to electrons is deposited on the rod. Early experimenters assigned the plus and minus signs before the relation of electricity to basic particles was known as well as it is now. If we had it to do over again, we would probably assign + to the charge on the electron, since it is more mobile than the other charge. The first simple experimental rule regarding charges is *like charges repel, unlike charges attract* (Figure 5-3).

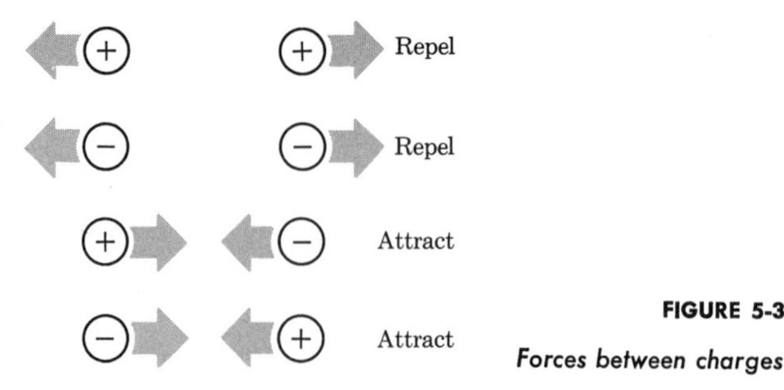

FIGURE 5-3
Forces between charges

For example, in the neutral hydrogen atom the attraction of proton and electron is responsible for maintaining a stable atom. The magnitude of the electrostatic force between any two particles can be written qualitatively as a proportionality:

$$F \backsim \frac{n_1 n_2}{R^2}$$

where n_1 and n_2 are the number of electrons carried by (or absent from) the objects and R is the distance of separation of centers. The similarity to the law of gravitation is obvious. We shall show shortly, however, that the strength of electrical forces is enormously larger than that of gravitational forces.

We have expressed the force formula in terms of numbers of electrons, which places it on a particle basis. Experimentally, the constant of proportionality is 2.3×10^{-28} N/m². For convenience in working with practical electrical circuits, it is desirable to choose a unit of electricity that is far larger than the minute electron charge. Named after the eighteenth-century French physicist C. A. Coulomb, the *coulomb*, abbreviated C, is chosen as the *unit of charge*, such that the charge on the electron is

$$e = 1.60 \times 10^{-19} \text{ C}$$

or

$$1 \text{ coulomb} = 6.25 \times 10^{18} \text{ electron charges}$$

Coulomb's law of electrostatic force is then written

$$F = k_E \frac{q_1 q_2}{R^2} \qquad (\textit{Coulomb's law})$$

where q_1 and q_2 are in coulombs. The *electrical* proportionally constant k_E is found experimentally to be approximately* 9×10^9. Since this number is simple and thus easy to remember, we shall use it in illustrative calculations. The unit of measurement of k_E is found to be N-m^2/C^2, by inspection of the force law.

Charge is a rather mysterious entity, since we cannot say what it is but only how it acts. The smallest known amount of charge is a definite number, 1.60×10^{-19} C, of the *same* magnitude on the electron (negative) and on the proton (positive), in spite of the fact that the proton's mass is 1836 times that of the electron.

Let us determine, using Coulomb's law, the electrostatic force between the electron and the proton in the hydrogen atom. The magnitude of charge is the same on each of the particles; $q_1 = q_2 = e = 1.60 \times 10^{-19}$ C. The known average separation is $R = 5.3 \times 10^{-11}$ m. Then,

$$F = \frac{(9 \times 10^9)(1.60 \times 10^{-19})^2}{(5.3 \times 10^{-11})^2} = 8.2 \times 10^{-8}\ \text{N}$$

This force is just the correct value to keep the electron on a circular orbit of radius 5.3×10^{-11} m about the proton.

Now let us compare the strength of electrical and gravitational forces between two electrons, separated by a distance R. Their masses are m_1 and m_2, both 9.11×10^{-31} kg; their charges are q_1 and q_2, both 1.60×10^{-19} C. Then,

$$\begin{aligned} \frac{F(\text{electric})}{F(\text{gravitational})} &= \frac{k_E q_1 q_2/R^2}{k_G m_1 m_2/R^2} \\ &= \frac{(9 \times 10^9)(1.60 \times 10^{-19})^2}{(6.67 \times 10^{-11})(9.11 \times 10^{-31})^2} \\ &= 4.2 \times 10^{42} \end{aligned}$$

In words, the electrical force between electrons is enormously greater than the gravitational force. Even when we compare the two types of force for particles as massive as two ionized uranium atoms, $m_1 = m_2 \cong 395 \times 10^{-27}$ kg, the ratio F_E/F_G is still more than 2×10^{31}. Although the gravitational and electrical forces are on an entirely different scale for the same separations, each plays a most important role in nature—the first at the astronomical level, the second at the atomic level. The masses of suns, planets, and satellites are large enough so that gravitational interactions are significant; such bodies have little or no *net* electrical charge, and thus

* More accurately, 8.987554×10^9. Our k_E is written $1/4\pi\epsilon_0$ in many texts.

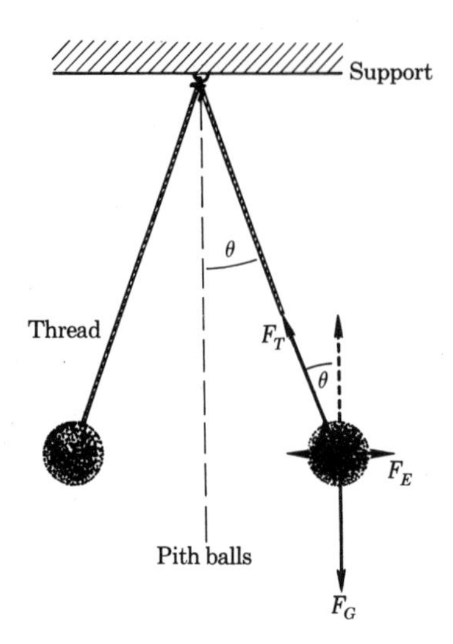

FIGURE 5-4

Repulsion of bodies with like charge

there are no appreciable electrical forces. In the atom, the component parts are held very close together by electrical forces, and gravitation is unimportant.

An early device used to show that like charges repel consisted of two lightweight spheres (made of pith from plant stalks) attached by threads to a common support (Figure 5-4). The spheres were charged by touching them with a rod that had previously been "electrified" by rubbing with cloth, just as we pick up electrical charge by walking on a rug in dry weather. The pith balls then stood apart with the threads describing an angle θ with the vertical, its value depending on the amount of charge deposited. The problem of finding the amount of charge on each ball is similar to Coulomb's experiment conducted in 1785, in which he established the law now bearing his name. Also, the example is interesting in that it combines three types of force—those due to gravity (F_G), to electricity

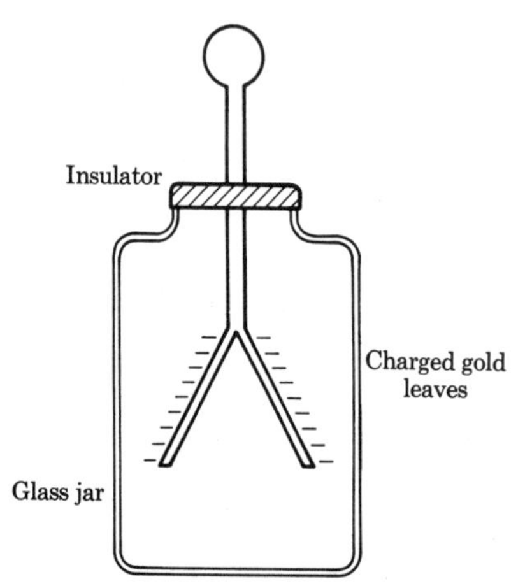

FIGURE 5-5

Electroscope

(F_E), and to mechanical tension (F_T). From the force diagram for one sphere we see that

$$F_T \sin \theta = F_E, \qquad F_T \cos \theta = F_G$$

Combining equations, $F_E = F_G \tan \theta$, where $F_E = k_E q^2/R^2$ and $F_G = mg$. The reader can verify that for $R = 0.02$ m, pith balls of mass 10^{-4} kg, $\theta =$ 30 deg, each ball carries a charge of 5×10^{-9} C, corresponding to about 30 billion electrons. A similar principle is applied in the electroscope, a device used to measure the amount of charge deposited by noting the degree of separation of the gold leaves (Figure 5-5).

The phenomenon of electrostatic *induction* can be easily demonstrated: if one runs a comb through his hair (on a dry day) and brings it near a small scrap of paper, the paper will leap to the comb and stick there briefly. The explanation is that the excess negative charge on the electrified comb tends to repel the electrons on the paper; then, the positive charges on the paper are, on the average, nearer the negatively charged comb—that is, there is an attractive force between the objects. The paper remains attached only for a moment because upon contact it receives electrons from the comb and acquires a net negative charge, at which point the like-charged bodies repel each other.

5-4 ELECTROMAGNETIC INTERACTIONS

Whenever we switch on an electrical circuit, whether it be for a flashlight, a TV set, an automobile ignition, or the lights in a room, there is a flow of charge which we call a *current*. A current consists of very large numbers of electrons or other charged particles in motion which exert forces on each other of *electromagnetic* orgin.

Electrical charges *at rest*, we saw in the previous section, exert electrical forces on each other. There are additional forces between steadily moving charges, as we can show by a simple experiment involving the interaction of only two positive charges. Visualize two accelerators for charged particles that can project positive ions by the use of suitable electrodes (Figure 5-6). One accelerator sends out hydrogen ions (protons) with charge q_1, the other helium ions (*alpha* particles) with charge q_2. The forces between charges causes them to move in curved paths, as shown in the figure. Since the charges are in motion, a force in addition to the electrostatic repulsion is operative. Let us examine the situation at the instant thc particles are moving parallel to each other, with speeds v_1 and v_2. If we could measure the force between the particles at that moment, we would find the electrostatic force, *plus* the electromagnetic force of magnitude

$$F = k_M \frac{(q_1 v_1)(q_2 v_2)}{R^2}$$

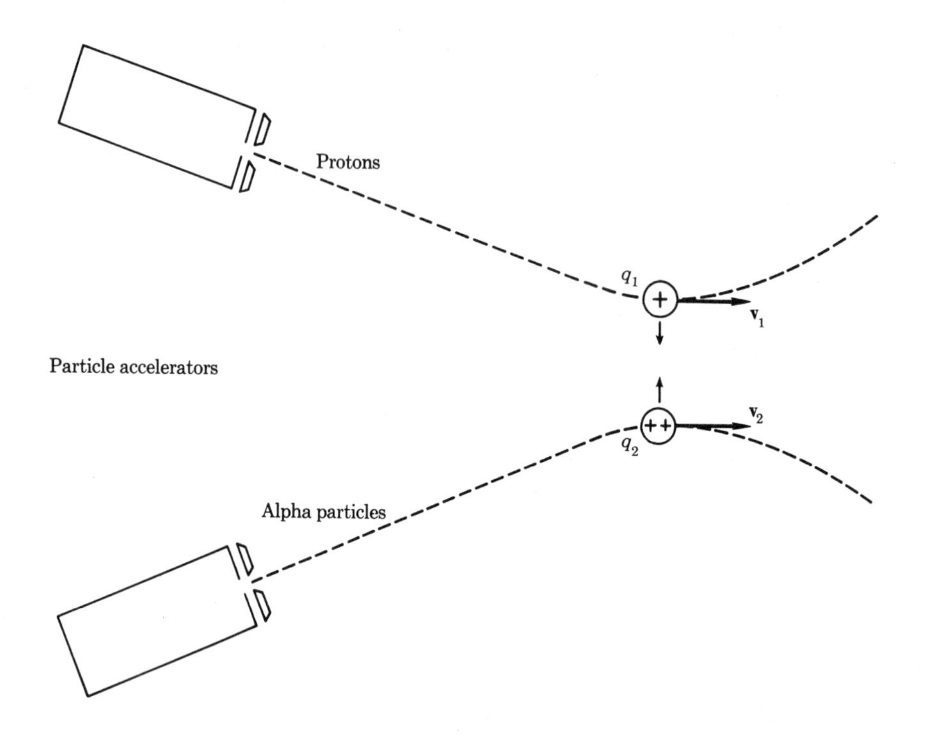

FIGURE 5-6

Magnetic force between moving electric charges

This is a special form of a description formulated by Ampere, a fellow-countryman and contemporary of Coulomb. The electromagnetic force is on a line joining the two positive charges, which *attract* each other—the reverse of the electrostatic case. In contrast with the electrical-force formula, where only q_1 and q_2 appear, this expression contains the products of charge and speed, q_1v_1 and q_2v_2. The inverse-square dependence on distance of separation is still present, however; and the *magnetic* constant of proportionality (for the mks system) is

$$k_M = 10^{-7} \text{ N-sec}^2/\text{C}^2$$

The units can be verified by substituting units for all other quantities. It should be emphasized that we have selected special directions of relative motions, for which the electromagnetic-force expression is simple and its structure readily understood. In a later section, we shall discuss the more general nature of forces between moving charges and currents.

Because of dependence on velocity, the magnitudes of electromagnetic forces cannot be easily compared with those of static electricity and gravity. At modest speeds of the charges involved, electromagnetic forces are generally smaller than electrostatic forces. To illustrate, let us calculate the magnetic force between two electrons of charge $e = 1.60 \times 10^{-19}$ C moving parallel to each other at speeds of 10^6 m/sec and separated by a typical atomic distance of 10^{-10} m.

$$F = k_M \left(\frac{ev}{R}\right)^2 = 10^{-7}\left[\frac{(1.60 \times 10^{-19})(10^6)}{10^{-10}}\right]^2$$
$$= 2.6 \times 10^{-13}\ \text{N}$$

Note that this is very small in comparison to typical electrostatic forces. For example, the electrostatic force of attraction between the electron and the proton in the hydrogen atom, as computed in Section 5-3, is 8.2×10^{-8} N, a factor of about 3×10^5 larger. Thus the electromagnetic force due to slowly moving charges in atoms tends to be masked by the much larger electrostatic forces. The gravitational force is also acting, but can be neglected.

As a second illustration, let us compute the force between a proton ($Z = 1$, $q_1 = Z\,e = 1.60 \times 10^{-19}$ C) and an alpha particle ($Z = 2$, $q_2 = 3.20 \times 10^{-19}$ C) when they pass at a distance of 10^{-3} m with speeds of 10^7 m/sec and 3×10^7 m/sec, respectively. Then,

$$F = 10^{-7}\,\frac{(1.60 \times 10^{-19})(10^7)(3.20 \times 10^{-19})(3 \times 10^7)}{(10^{-3})^2}$$
$$= 1.54 \times 10^{-24}\ \text{N}$$

This force is extremely small because the separation (1 mm) is very large on the scale of atomic dimensions.

5-5 NUCLEAR FORCES

The attractive forces that hold particles together in the nucleus are not yet as well understood as are the forces at the atomic level. Many qualitative statements have been made, but a complete mathematical description is not available. One of the known facts is that the forces between neutrons and protons are negligible unless the particles are within very short distances of each other. Interaction sets in at a separation of about 10^{-15} m. These "short-range forces" are capable of holding positively charged protons together in spite of the strong electrostatic repulsion that still appears to exist within nuclei. That nuclear forces between nucleons are very strong can be demonstrated numerically. We recall that the electrostatic force in the hydrogen atom was around 10^{-7} N, at separations of the proton and electron of about 10^{-10} m. In the nucleus of helium, the two protons are separated by about 10^{-15} m, so the electrical repulsion force is larger than the atomic attraction in hydrogen by a factor of $(10^{-10}/10^{-15})^2$ or 10^{10}. The nuclear-attraction force must certainly be even larger, since the helium nucleus is a stable structure. We thus deduce that the force is about 10^3 N. Next, we find that the forces are essentially independent of charge, since there appears to be similar interactions between protons and neutrons, protons and protons, and neutrons and neutrons. The forces have a property of being "saturable," which means that each particle interacts mainly with its neighbor and not with all the other particles in the nucleus. We can thus visualize strong bonds linking pairs of particles and weaker bonds connecting the pairs with each other. From the above combination of properties we can conclude that it is not possible for nuclei of mass number much greater than 250 to exist. Although the nuclear forces would be very strong—even at that point—the mutual electrical repulsion of the protons would prevent the structure from remaining together.

A mechanism explaining the force between two nucleons in terms of a sharing of another particle, the *meson*, has been advanced, but the description is not sufficient to explain the apparent existence of a host of other subnuclear particles that are observed when nuclei are bombarded by high-speed particles. This "particle problem" remains one of the challenges of modern physics, and we shall reserve further discussion until a later chapter.

5-6 FRAME-DEPENDENT FORCES

We now turn from the fundamental forces—related to gravity, electricity, and nuclear interaction—to another class that has its origin in the motion of frames of reference. We have already seen in Chapters 3 and 4 that observations of events from different viewpoints yield different answers, and that application of Einstein's theory of relativity is required when very high speeds are involved. In the discussion to follow, we shall limit our attention to low-speed motion such that special relativity is not required; however, we shall still find unusual effects associated with acceleration.

Consider again the concept and restrictions of Newton's laws of motion. The relation $\mathbf{F} = m\mathbf{a}$ holds only in frames of reference that are moving with *uniform speed* relative to each other. In our present "space age," uniform speeds are almost the exception rather than the rule, and accelerated motion is experienced everywhere—for example in driving an automobile, in taking off in an airplane, and in orbiting the Earth or the Moon in a spacecraft. It is not legitimate to use the strict form of Newton's laws within such *accelerated* frames of reference. To bring the situation into focus, let us compare two situations. First, suppose we are riding in a plane that is cruising along at a constant velocity. We experience no unusual effects due to motion: everything is normal in terms of experience on the ground. If we choose, we can perform simple experiments such as dropping an object, which appears to fall vertically in complete accord with Newton's laws.

Conditions are markedly different when we are instead riding in a jet plane that is accelerating down the runway. It would be hazardous indeed to drop an object from any height, since it would fly backward and might well hit us or a passenger behind us. This behavior is easily explained by our knowledge that in reality the object is dropping as a freely falling body with respect to the Earth and that our frame of reference is accelerating. Suppose, however, that we had no knowledge of our actual situation or that we were asked to set up the laws of motion for the accelerating frame. It is clear that we could no longer use Newton's laws in their standard form. One approach would be to say that there is a constant force acting along the length of the plane. Thus when we drop an object, we consider that a combination of gravity F_G and a "special force" F_S is acting. To describe the motion, we would write

$$\mathbf{F} = \mathbf{F}_G + \mathbf{F}_S$$

and then apply

$$\mathbf{F} = m\mathbf{a}$$

One's attitude toward this special force depends on his point of view or frame of reference. A person on the ground would say that the special force did not exist, but that the motion seen was merely a consequence of the observer's acceleration. On the other hand, being vigorously thrust back in his seat as the plane takes off can hardly be interpreted as other than a real force by the person on the plane.

A similar difficulty arises whenever one's frame of reference is moving in a curved or circular path with respect to one in which Newton's laws are believed to hold. Examples of such frames range from electrons in atomic orbits and electrical charges moving in circular paths in an accelerator to rotating wheels and shafts, automobiles rounding a curve, the Earth's rotation on its axis, and the Earth's yearly path around the Sun. In each of these cases, the object that contains the frame of reference is being accelerated.

Before examining the nature of forces as observed on the vehicle, we consider the view from what is assumed to be the "fixed" frame. This motion will be treated as "real." A particle moving with uniform speed v on the circumference of a circle of radius R was shown earlier (Section 2-4) to have a radial acceleration $a = v^2/R$ directed toward the center. To achieve such a centripetal acceleration, a force $F = ma = mv^2/R$ or $F = m\omega^2 R$ must be applied perpendicular to the path. Otherwise, the object would continue in a straight line. Such a force is designated the *centripetal force*. Let us examine several examples.

A small boy whirls his toy airplane of mass 0.15 kg on the end of a string of length $R = 0.8$ m, at a speed of 5 m/sec. The boy pulling the string provides the necessary inward force on the plane of amount

$$F = \frac{mv^2}{R} = \frac{(0.15\text{ kg}) \times (5\text{ m/sec})^2}{0.8\text{ m}} = 4.7\text{ N}$$

Or, let us compute the amount of force on the tires of a car of weight 3220 lb that must be provided by friction to cause the car to go around a curve with radius 400 feet at a speed of 60 mi/hr. As noted in Section 4-9, we should divide the weight by $g_s = 32.2$ if the force is wanted in pounds. Also, 60 mi/hr = 88 ft/sec. Then,

$$F = \frac{(3220/32.2 \text{ slugs})(88 \text{ ft/sec})^2}{400 \text{ ft}} = 1940 \text{ lb}$$

Next, let us find the force that will keep an electron of mass 9.1×10^{-31} kg moving in the hydrogen atom in an orbit of radius 5.3×10^{-11} m with a speed of 2.2×10^6 m/sec.

$$F_c = \frac{mv^2}{R} = \frac{(9.1 \times 10^{-31})(2.2 \times 10^6)^2}{5.3 \times 10^{-11}}$$
$$= 8.2 \times 10^{-8} \text{ N}$$

This force is just that provided by the electrostatic attraction between the electron and proton, as computed in Section 5-3,

$$F_E = \frac{k_E q^2}{R^2} = \frac{(9 \times 10^9)(1.60 \times 10^{-19})^2}{(5.3 \times 10^{-11})^2} = 8.2 \times 10^{-8} \text{ N}$$

These examples would appear to be straightforward applications of Newton's laws. Conceptual difficulties arise, however, when we attempt to describe the motion from the viewpoint of the circulating object. Suppose we are riding in the back seat of an automobile as it rounds a sharp curve. From our point of view, there is a sideways force that causes us to slide along the seat and that presses us against the inside of the car door. The explanation is very simple from the viewpoint of an observer *outside*. The car turns, but inertia keeps our body moving in a straight line. Then the door of the car provides a centripetal force causing us to move on the same curve as the vehicle, and we react to that force. This is correct, of course, but not very satisfying, because our long personal experience with forces and accelerations gives us the impression that there is an external force on our body tending to throw us outward. From our frame of reference, we feel what we call a *centrifugal** force. It is natural to want to continue using Newton's laws to describe our observations from the rotating frame, but their standard form does not apply to such a non-inertial (accelerated) reference system. Suppose, though, that we persist in using the form. We note first from Figure 5-7 that the passenger's frame of reference is rotating with the same angular speed ω as that of the car on the circle. We introduce a "special force" of amount $m\omega^2 R$, as a centrifugal correction. If a force F is applied to any object in the rotating frame, its motion is governed by

$$F - m\omega^2 R = ma$$

* From the Latin *centrum* ("center") and *fugare* ("to flee").

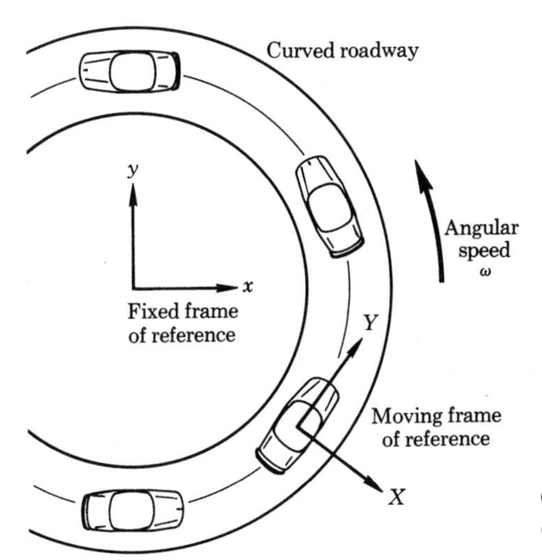

FIGURE 5-7

Centrifugal force on automobile rounding a curve

The left side of this equation is still in the nature of a force, and thus the form of Newton's second law is preserved.

To illustrate the idea of centrifugal force further, let us calculate the effect of the Earth's rotation on the acceleration of gravity (see Figure 5-8). The force of gravity as experienced at or above the Earth's surface is weaker than expected because of the Earth's angular speed $\omega = \dfrac{2\pi \text{ rad/day}}{86{,}400 \text{ sec/day}} = 7.27 \times 10^{-5}$ rad/sec. The actual force toward the center is $F = mg$, where $g = 9.84$ m/sec^2, obtained from the law of gravitation, which acts whether there is rotation or not. The centrifugal acceleration at the Equator, where $R = 3960$ mi $= 6.37 \times 10^6$ m, is

$$a_c = \omega^2 R = (7.27 \times 10^{-5})^2(6.37 \times 10^6) = 0.03 \text{ m/sec}^2$$

Thus the acceleration of gravity *observed from the Earth* is

$$a = 9.84 - 0.03 = 9.81 \text{ m/sec}^2$$

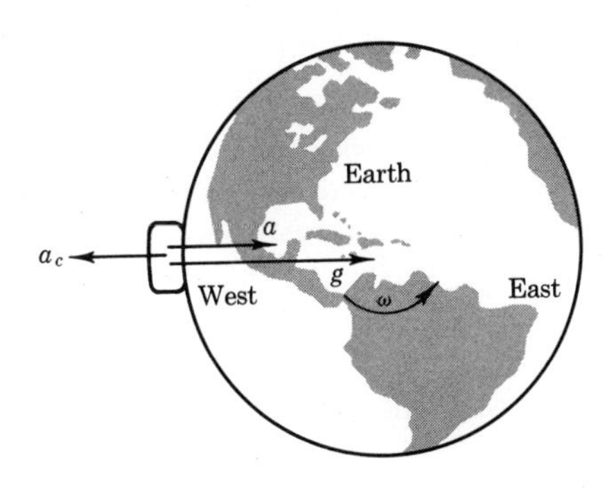

FIGURE 5-8

Effect of Earth's rotation on the acceleration of gravity

From an external view, a point on the Earth's surface is "dropping away" from any object falling toward it or resting upon it, which results in less acceleration toward the surface. Note that the effect vanishes at the poles of the Earth, where there is no linear velocity. Two interesting questions are:

1. What would be the effect on life on Earth if the planet were to speed up its rotation so that the "day" was only a minute long?

2. What would be the nature of existence if our Earth were only a hollow shell with gravity acceleration of only $\frac{1}{1000}$ of its actual value?

The phenomenon of "weightlessness," as experienced by astronauts in orbit about the Earth (Figure 5-9), can now be explained from either of the two viewpoints. First let us examine, from the spacecraft, the motion

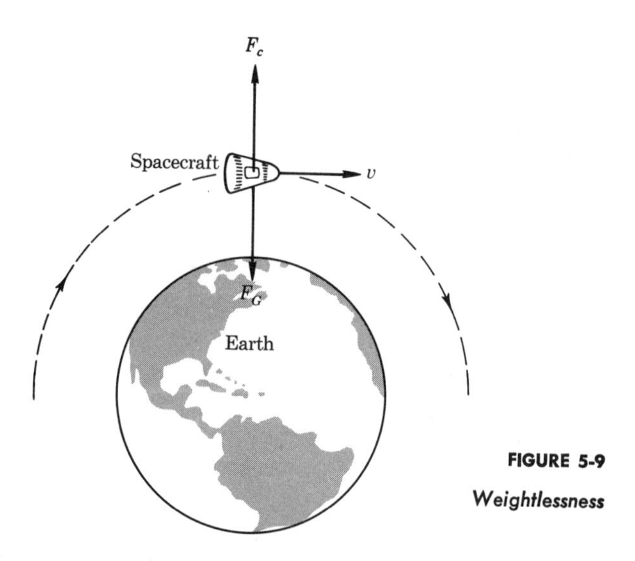

FIGURE 5-9
Weightlessness

of any mass m. There is a gravitational attraction toward the Earth, of mass M_E, of amount $F_G = k_G m\, M_E/R^2$. Because of orbital motion, there is a centrifugal force $F_c = mv^2/R$ outward (assuming the orbit to be circular). There is no net radial force, and the mass m say, the astronaut, experiences no net force either, and hence he "floats" in his cabin. Alternately, from the Earth's viewpoint, the spacecraft has a velocity tangential to the orbit at any instant, but is continually *falling* toward the Earth because of the unbalanced force of gravity. The amount of drop is exactly that required to keep the ship at constant radius. Both the ship and the astronaut are in "free fall," just as if the cable on an elevator had broken.

We can answer the question, when does a projectile become a satellite? A low-speed missile will travel in a parabola over a limited plane of

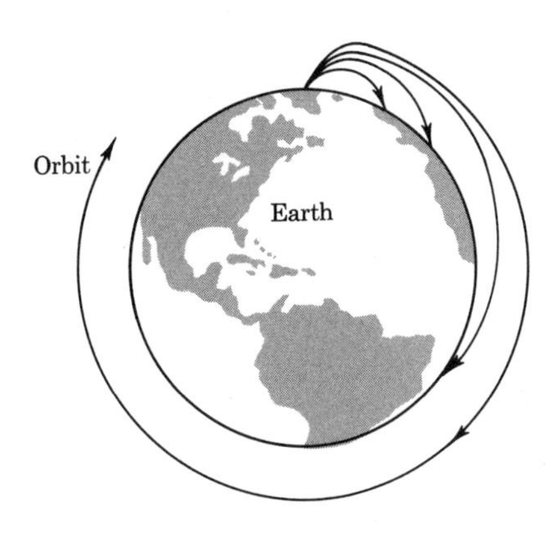

FIGURE 5-10

Paths of missiles at various speeds of launching

the Earth, as shown later (Section 10-6). If its initial speed were great enough (Figure 5-10), the object would never land, but would go into orbit.

Although astronauts seem to experience little trouble because of weightlessness, it is probable that spacecraft of the future will be rotated so as to provide an "artificial gravity." The resulting local centrifugal force on the occupants will not be distinguishable from a real gravitational force acting toward the wall. In fact, we may estimate the rate of rotation needed to give the full strength of 1 g, 9.8 m/sec^2. First we set $a_c = \omega^2 R = g$. For a vessel of radius 5 m, the angular speed would have to be $\sqrt{9.8/5} =$ 1.4 rad/sec, which means $1.4/2\pi = 0.22$ revolutions/sec. It is likely that one would compromise on a roll that gives a lower acceleration than 9.8 m/sec^2, in that it would be easier to achieve and also more comfortable for the passengers of the spaceship.

The effects just discussed all refer to *radial acceleration*. Another force related to angular motion may also be observed. Imagine a person standing on a rotating circular platform. He attempts to walk outward along a radial line drawn on the surface, but he experiences a sideways force that is proportional to his speed and to the rate of rotation of the table. His path will be a slight curve that deviates from the straight-line course he had planned. From the viewpoint of an observer on the ground, the person's path is a much larger sweeping curve caused by the combination of the motion relative to the platform and the rotation of the platform itself. If Newton's laws hold in the fixed frame, they cannot hold in the moving frame because of its rotation. The walker's departure from the line is interpreted as the effect of an additional *Coriolis force*, which is of course quite real from his viewpoint.

Since the Earth rotates about an axis, observations from its surface are always subject to the Coriolis force. Another simple situation can be used to illustrate this. A helicopter hovering above the ground drops a package of supplies to some explorers who are stranded directly below, at a point on the Equator in South America. The parachute fails to open and the package falls by gravity. Our first inclination would be to predict that the surface of the Earth, as it moves from west to east, would move out

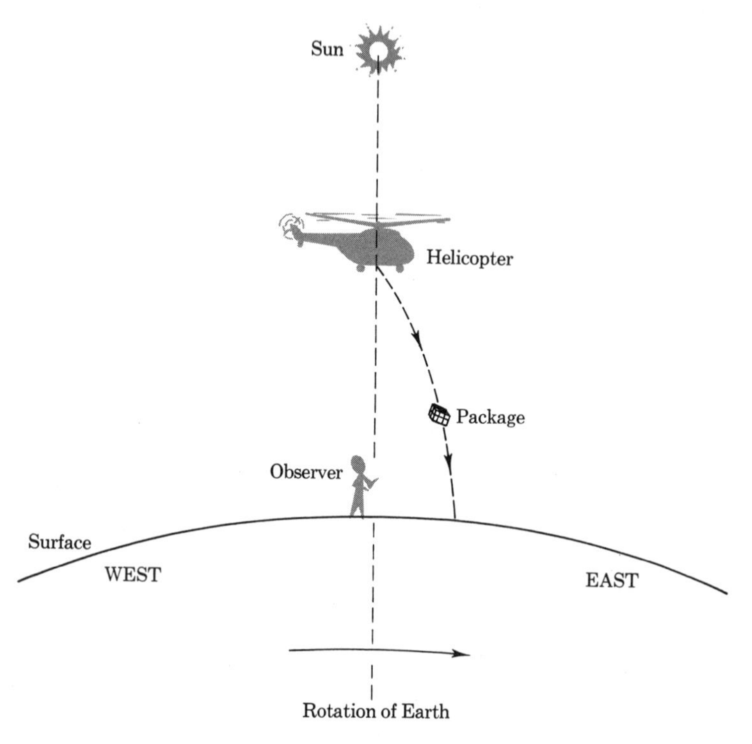

FIGURE 5-11

Illustration of Coriolis force

from under the package, so it should land to the west of the explorers. Then, realizing that the atmosphere accompanies the Earth in its rotation, we might predict that the package would land directly below the helicopter. In actuality, it strikes to the east of the vertical because of the Coriolis force, as can be seen in Figure 5-11. The motion can be viewed from two different observation points:

1. From a frame of reference that is located at the center of the Earth, but which does not rotate with it. If an observer could watch from such a point, he would first see the helicopter and package rotating with high speed about him at a height h above the ground. When the package is released, the force of gravity would accelerate it toward the center of the Earth in a curved path, according to Newton's second law.

2. From a frame of reference on the Earth's surface, which is rotating. The observer would see the package fall in a path that is slightly off from the vertical, and which in his frame he could describe by Newton's laws only by introducing a special force. The amount of sideways displacement is small—only 0.7 m in a drop of 1000 m—but observable.

For a long-range missile, a significant deflection would occur. The Coriolis force is an important factor in our climate and weather in that it plays a part in the characteristic counterclockwise motion of air masses in the Earth's atmosphere in the northern hemisphere.

5-7 COMPOSITE FORCES IN MATERIALS

Having identified the forces that can be viewed as primary, we are in a position to give at least qualitative descriptions of those due to the addition or superposition of many particle interactions. The relation between the atomic (or microscopic) forces and the composite (or macroscopic) forces will be revealed in a conceptual sense.

Interatomic Forces

The force between pairs or groups of atomic particles can be found by the addition of electrical interactions. In what follows, several examples are given to illustrate the effect.

The case of two single charges—for example, protons, H^+—is the simplest. The electrostatic force of repulsion between them is proportional to $1/R^2$, as discussed in Section 5-3.

Next, consider two ions, such as sodium (Na^+) and chlorine (Cl^-). When their distance of separation is large compared with their dimensions, they interact as point charges, with a $1/R^2$ force of electrostatic attraction. If, however, the ions are close enough for the electron orbits to overlap, there is a repulsion force between them. At some point the forces are just balanced, permitting the sodium and chlorine ions to form a stable compound, NaCl. The *ionic bond* in crystals forms the basis of the strength with which certain bulk materials are held together.

Now let us examine the interaction of molecules, each composed of two or more atoms, and each electrically neutral. The force between widely separated molecules is negligible, because there is no net electrical charge on either. If they approach each other closely enough, the way the positive and negative charge is distributed in the molecules becomes important. For example, in the water molecule there is a permanent average separation of positive and negative charges—(see Figure 5-12(a)). Such an arrangement is called a *dipole*. Pairs of molecules such as dipoles—(see Figure 5-12(b))—interact by virtue of the several forces: $++$, $+-$, $--$, and $-+$. The net force is proportional to $1/R^4$.

Finally, consider the interaction of atoms such as argon that have closed electron shells and whose internal charge is symmetrically located. Only in very close proximity will two atoms of such elements interact, and then by reason of *induced dipoles*—that is, mutually distorted charge distributions. In this situation, the force varies as $1/R^6$. It is easy to see that the larger the power of R, the weaker the force at large distances of separation. Thus the three types of force—ionic, dipole, and induced dipole—are in order of decreasing strength.

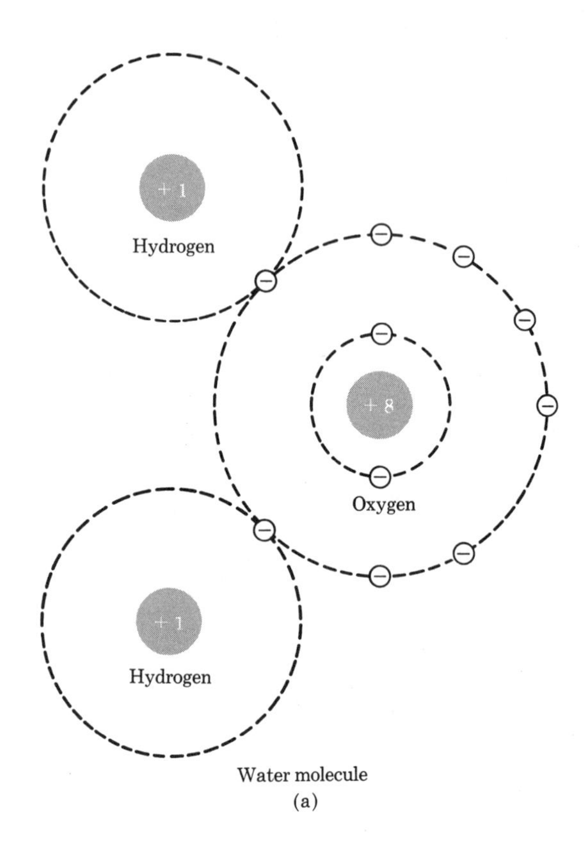

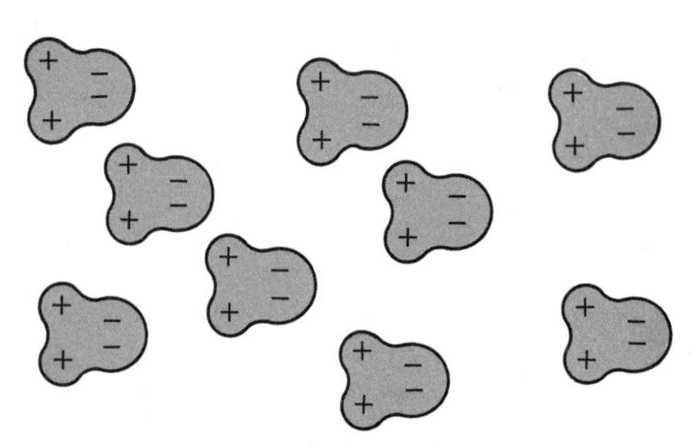
(b) Alignment of dipoles in water

FIGURE 5-12

Interaction of water molecules

Forces Between Magnetic Poles

We are familiar with the ability of a bar magnet to pick up a nail or iron filings, and with the magnetic compass that helps us to find the direction north. The detailed atomic explanation of such manifestations of magnetism is rather complicated, and it will be sufficient now to note that the effects are ascribed to circulating electron flows within the metal.

The ends of a permanent magnet in the form of a long thin bar are called its *poles*, one of which is labeled "north" since it would tend in many places on the Earth to point in that direction; while the opposite end of the bar is the magnet's "south" pole. It is easy to verify by an experiment that like poles repel each other and unlike poles attract each other (exactly the same as in the rule for electrical charges). Thus a north pole of one magnet attracts a south pole of another, but repels a north pole, and so on. The amount of force between magnet poles depends on the "strength" of their poles, symbolized by p. Also, the force diminishes with separation according to our now-familiar inverse-square law. For the force between the ends of two bar magnets at a distance R from each other, there is a formula (first stated by Coulomb)

$$F = k_M \frac{p_1 p_2}{R^2}$$

where the magnetic proportionality constant is taken to be $k_M = 10^{-7}$ as used in Section 5-4. This equation thus serves to define pole strength. The force between two poles of unit strength separated by 1 meter is

$$F = \frac{10^{-7}(1)(1)}{(1)^2} = 10^{-7}\ \text{N}$$

Elastic Forces

One often encounters substances that can easily be stretched or compressed, and which return to their original size and shape after being distorted. Examples are a rubber band or a steel string on a guitar. Such materials are classed as *elastic*. The restoring force in many substances is found to be proportional to the displacement—an expression of *Hooke's law*, which holds so long as the displacements are not so great as to permanently change or break the bonds holding the substance together. A simple model will help us to appreciate such behavior qualitatively.

We assume that two atoms attract at large distances of separation, repel at short distances, and do neither at a certain separation that might be regarded as "natural." Let us consider, then, a material that obeys Hooke's law, composed of many atoms in a chain (Figure 5-13). For very small displacements we can think of the object as behaving much like a rubber band. If we fix one end and stretch the elastic material, all of its atoms are forced slightly farther apart, and tend to go back to their original position. When the material is compressed, the atoms are closer to each

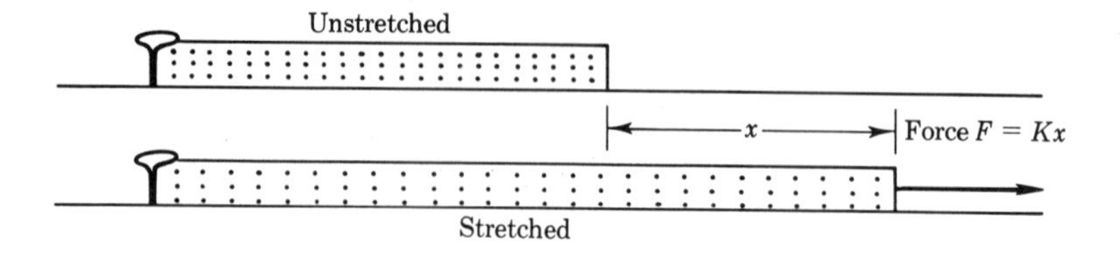

FIGURE 5-13

Stretching of rubber band

other and again seek to return to their "natural" position. On an atomic scale and in the chain of atoms as a whole, we may state that the force is proportional to the displacement x; i.e.,

$$F = Kx$$

where the *force constant* K measures the strength of the bond. It is readily verified by experiment that a coil spring exhibits this very same behavior; and so for ease in picturing the atomic interactions, we often sketch tiny springs connecting the particles.

A simple experiment can yield an estimate of the constant K. We attach a known weight to a long rubber band and measure the amount of stretching. Suppose the weight is found to be 0.40 lb and the elongation is 0.8 in.; then $K = F/x = 0.40/0.8 = 0.5$ lb/in. For measurements in the metric system, if a force of 2 N gives a displacement of 0.05 m, then $K = F/x = 2/0.05 = 40$ N/m. Similar effects are noted with coil springs used in automobiles, in scales for weighing, on doors, or in furniture. Even though the restoration of original length is due to complicated twisting action, the expression $F = Kx$ still holds.

Many materials that we normally think of as rigid exhibit some elasticity—for instance, the concrete and steel of a building are flexible. The top of the Empire State Building is said to swing several inches back and forth under the influence of winds.

5-8 FRICTION

When the surfaces of large-scale objects move past each other in contact or when a body moves through the atmosphere, forces are developed that tend to retard the motion. These forces of *friction* are much more difficult to explain than are individual-particle forces, since the effects involve many atoms and molecules—the amount depending on the particular surfaces, the type of contact, and the specific materials. Friction generally involves variables with uncertain effect: motion that can be predicted most elegantly neglecting friction is often far from resembling real behavior. The use of greases and oils between bearing surfaces reduces friction by providing a layer of molecules that are free to slide past each other.

Friction can be studied from either a large-scale or small-scale viewpoint. One of the simplest cases to visualize is the process of sanding a board of wood. To a considerable degree, the resistance to motion is provided by the interference of the grains of sand affixed to the sandpaper with the indentations in the wood. The grains are, of course, bonded to the paper by forces that exceed those associated with the close contact with the atoms of the wood. The mechanical impacts of the peaks and valleys in the two surfaces provide the resistance to motion, and eventually pieces of both surfaces break off—the familiar process of wear that takes place as well in tires, bearings, and so on.

This model fails us when we discover that very smooth and highly polished surfaces can exhibit far more friction than rough ones when the air between surfaces is withdrawn, as in the moving parts of spacecraft operating in a vacuum. The explanation must be based on interatomic forces that act more strongly, the smaller the average distance between particles. The same cohesive mechanism that holds matter itself together operates here to provide a high degree of adhesion of separate surfaces.

A frictional force that is dependent on the speed is developed when a material object such as an atomic particle, a projectile, a car, a boat, or a plane moves through a fluid (such as air or water). Since the air or water molecules are far apart, the process of adhesion is no longer relevant, and we must consider some alternate mechanism. First, suppose the object is moving *slowly*. The fluid is not greatly disturbed, but merely flows past the object with a speed that depends on the distance of the stream from the surface. Figure 5-14 shows these velocities for a sphere. In addition to the flow velocity, there is a chaotic motion that transfers molecules from regions of lower speed to those of higher speed. This has the effect of causing one layer to exert a frictional force on an adjacent layer. The net result is to retard the progress of the object with a force *proportional to the speed*:

$$F = cv$$

where c depends on the nature of the fluid and the radius of the sphere. If, instead, an object such as a bullet moves very *rapidly* through the medium,

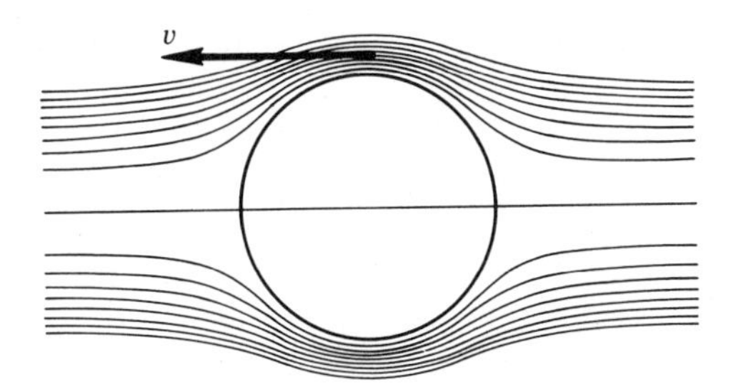

FIGURE 5-14

Low-speed flow of air past a sphere

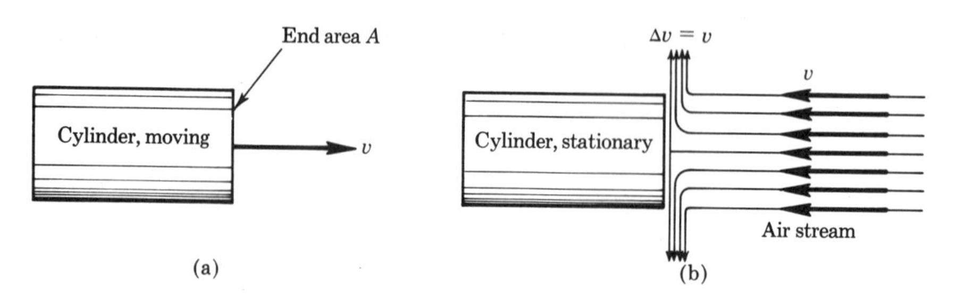

FIGURE 5-15

Friction on high-speed object

say air, the fluid is affected greatly. The resistive force depends mutually on the amount of air that strikes the body each second, which is proportional to v, as well as on how much change in v is experienced by the molecules. The change in momentum corresponds to a force, according to Newton's laws. Picture a cylinder of end area A, moving parallel to its length, as in Figure 5-15(a). If its speed is v, the situation is equivalent to air of speed v striking a stationary cylinder, as in Figure 5-15(b). The airstream is stopped, with a change in v that is equal to v. The result is that the force is *proportional to the square of the speed*:

$$F = Cv^2$$

where C depends on the fluid density and the area of the projectile. Part of the reason that jet planes are flown as high as 30,000 feet above the Earth is that the air density is significantly lower at such heights, and friction is correspondingly reduced. The above relations hold only for the limits of low and high speeds, and the dependence on v is in general to some power between 1 and 2. For speeds of objects near and exceeding the speed of speed of sound (Mach 1), much more complicated expressions are required.

Friction for the case of objects such as wheels that roll on road surfaces has a somewhat different explanation. In this case, a deformation of the surface of contact takes place. As seen in Figure 5-16, a hard-metal wheel causes a depression in the softer road surface, and there is a continual horizontal component of force on the wheel provided by the wall of the

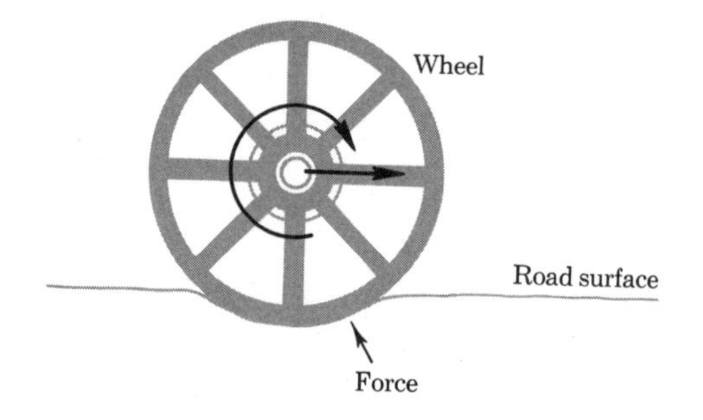

FIGURE 5-16

The origin of rolling friction

depression. On the other hand, a soft tire on a concrete road is flattened, which again distorts the contact from an ideal line to a broad surface. To minimize road friction for purposes of gasoline economy, one would inflate tires to the highest practical pressure—at the expense of comfort and safety, of course.

Since there are many particles involved, and since friction depends on materials, shape, and size, along with the nature of any lubricating agent that may be present, it is not possible to express frictional forces in detailed mathematical form. A gross description that holds for many surfaces is based on the general experience that the sliding frictional force increases with the force pressing the surfaces together, all other things being equal.

Let the force that acts perpendicular to the two surfaces be labeled F_N, the normal force. Then the force of friction F_F acting to prevent motion of sliding is less than or equal to μF_N. Since the quantity μ applies to surfaces at rest with respect to each other, it is called the *coefficient of static friction.* The meaning of the phrase "less than or equal to" requires some comment. Suppose there is a heavy box to be moved along the floor. The normal force is the reaction of the floor and is thus equal to the weight. Before we start to push the box, the applied force is obviously zero, and the force of friction is also zero. We give a small push, but no motion occurs, meaning that the net force is zero, and F_F is correspondingly small. As we increase the applied force through the value μF_N, the box starts to slide, meaning that the applied force has exceeded the *static* friction force. Once the box is in motion, a coefficient of *kinetic* friction must be used in the relation $F_F = \mu F_N$. Table 5-1 shows the range of measured values for different interacting materials. Let us now consider a few examples of motion involving friction.

TABLE 5-1

COEFFICIENTS OF FRICTION

MATERIALS	STATIC	KINETIC
Steel on steel	0.58	
Steel on steel, with motor oil	0.20	
Ski wax on dry snow	0.04	
Rubber on concrete		1.02
Ice on ice	0.1	0.02

First, suppose a car is parked on a hill with its brakes locked. At what angle will it just start to slide? As seen in Figure 5-17, the component of force of gravity down the incline $F_W = W \sin\theta$ is balanced by the force of friction F_F. The normal force pressing the surfaces together is $F_N = W\cos\theta$. Then, the friction force is up to its maximum value

$$F_F = \mu F_N = \mu W \cos\theta = F_W = W \sin\theta$$

Eliminating W, we find

$$\tan\theta = \mu$$

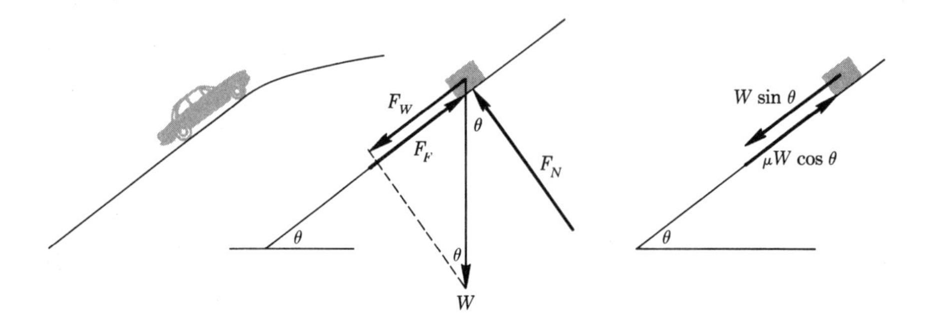

FIGURE 5-17

Friction on an incline

If, for example, μ were approximately 1, as for rubber on a dry concrete street, the maximum safe angle would be $\pi/4$, or 45 deg.

Next, let us find the correct angle θ for banking a highway (Figure 5-18). We want to be sure that cars moving at speed v do not slide toward the outside of a curve of radius R when the coefficient of static friction is μ.

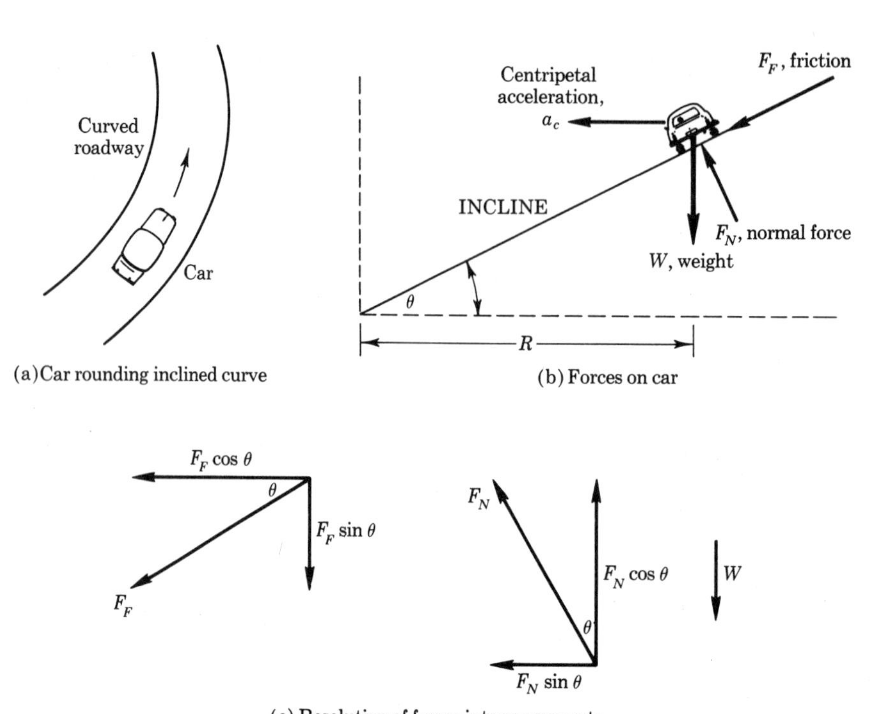

FIGURE 5-18

Motion on a banked highway curve

The forces acting on a car are

W = weight, vertically
F_F = friction, down the incline
F_N = normal, perpendicular to the incline

We have two choices—to find force components relative to the incline or to find them relative to the usual x- and y-axes. For variety, let us use the latter approach. The horizontal components of F_F and F_N provide the necessary centripetal acceleration $a_c = v^2/R$,

$$F_N \sin\theta + F_F \cos\theta = ma_c$$

The vertical forces are balanced,

$$F_F \sin\theta + W = F_N \cos\theta$$

However, $F_F = \mu F_N$ and $W = mg$. Substituting and dividing equations,

$$\frac{\sin\theta + \mu\cos\theta}{\cos\theta - \mu\sin\theta} = \frac{a_c}{g}$$

Rearranging,

$$\tan\theta = \frac{a_c - \mu g}{\mu a_c + g}$$

For example, let the curve have radius $R = 50$ m and suppose that μ is normally 0.8. A car with speed $v = 60$ mi/hr (26.8 m/sec) requires $a_c = (26.8)^2/50 = 14.4$ m/sec². Then,

$$\tan\theta = \frac{14.4 - 0.8(9.8)}{(0.8)(14.4) + 9.8} = 0.31$$

or

$$\theta \cong 17 \text{ deg}$$

If the roadway were covered with ice, reducing the coefficient of friction to nearly zero, it would obviously be necessary to reduce speed. Then $\tan\theta = a_c/g$, $v^2 = gR\tan\theta$, or

$$v = \sqrt{(9.8)(50)(0.308)} = 12.3 \text{ m/sec} \quad \text{or} \quad 27.5 \text{ mi/hr.}$$

Motion Involving Friction

Let us examine the effect of friction on the motion of an object. Suppose, for example, that a returning Apollo spacecraft enters the Earth's atmosphere and experiences frictional force in the air proportional to the vehicle's speed v. If we ignore other forces, such as gravity, Newton's law $\mathbf{F} = m\mathbf{a}$ may be written

$$m\frac{\Delta v}{\Delta t} = -cv$$

where the negative sign is included to indicate that the frictional force is opposite to the motion and that Δv is always negative—that is, that v is decreasing. The acceleration, as the rate of change of speed, is seen to be proportional to the speed itself; so the more slowly the spacecraft moves, the smaller the acceleration. We encounter many physical situations to which such a mathematical form applies. Let the ratio c/m be abbreviated as λ, so that

$$\frac{\Delta v}{\Delta t} = -\lambda v$$

If we start counting time when the speed is v_0,* then the speed at any time t is the exponential

$$v = v_0 e^{-\lambda t}$$

with e as the natural logarithm base, 2.718. . . . Figure 5-19 shows the way in which the speed decreases with time. Note that each equal interval of time results in the same fractional loss in speed.

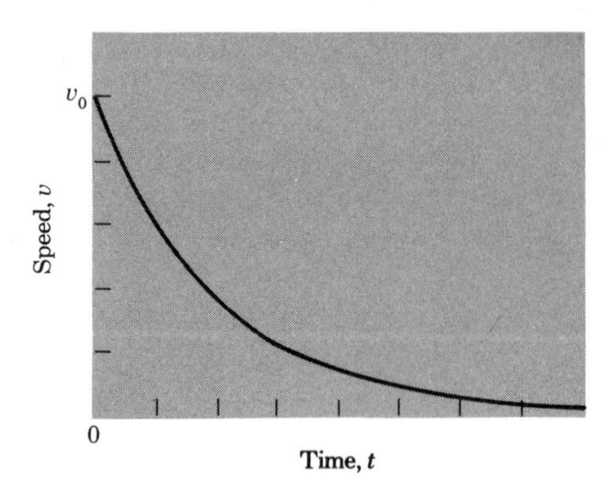

FIGURE 5-19

Slowing with friction proportional to speed

As a second example, consider the effect of "drag" on a plane or rocket if a constant force F_A is applied by a jet engine or a rocket motor. With a frictional force proportional to speed but acting in the opposite direction, the net force is $F_A - cv$, and Newton's second law becomes

$$F_A - cv = m\frac{\Delta v}{\Delta t}$$

If the object starts at rest ($v = 0$ at $t = 0$), the initial acceleration is $a_0 = F_A/m$. As speed increases, the acceleration drops off, and the speed at any instant is lower than without friction. Over long time intervals, the speed

* By use of the calculus, $dv/v = -\lambda\, dt$. Integrating, $\ln v = -\lambda t + C$; and if $v = v_0$ at $t = 0$, then $C = \ln v_0$. Thus $\ln v/v_0 = -\lambda t$. Taking the antilog of both sides, $v/v_0 = e^{-\lambda t}$.

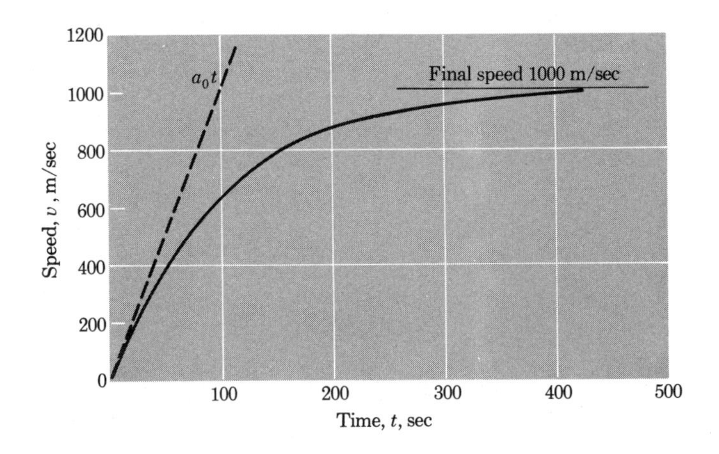

FIGURE 5-20

Motion with air resistance proportional to speed
$m = 10^3$ kg, $F_A = 10^4$ N, $c = 10$ kg/sec
By use of the calculus, $v = F_A/c(1 - e^{-ct/m})$

approaches the value F_A/c for which a is zero—that is, when the frictional force balances the applied force. Figure 5-20 shows the trend for a particular example.

Force as a Unifying Concept

The array of forces discussed in this chapter may be a source of bewilderment to the student of physics. Although the single idea of force has been displayed, there remain at least half a dozen different types of force, distinguished according to their origin. It is interesting, in this connection, to look both backward and forward in time, as man expands his understanding of his surroundings. To early man, a demon or god lay behind (or sometimes *in*) each physical phenomenon or entity. He surely identified separate spirits in the sea, in the wind, in lightning, in the Sun, Moon, and stars, and so on. Our progress has been great, therefore, in arriving at no more than half a dozen distinct types of force. On the other hand—we might well ask—if there is but one nature, why should we not be able to define a single type of force, of which electricity, magnetism, gravity, and the others are simply manifestations or variations? Scientists continually ask that very question, and are actively striving to achieve further unity in the explanation of the still-diverse features of physical phenomena. As we shall see, success in relating electricity and light has been achieved, and the concepts of gravitational and mechanical force brought into some harmony. However, no acceptable connection has yet been established between electrical forces and gravitational forces, even though we suspect they have a common basis. This situation merely points up the fact that physics is by no means a closed subject, and that there yet remains many challenges to man's intellectual talents.

PROBLEMS

5-1 Find the gravitational force between two people each of mass 80 kg, with a distance of 2 m between their effective centers. What fraction of a "g" force toward each other is there?

5-2* At what fraction of the distance between centers of the Earth and the Moon (2.39×10^5 mi) is there exactly zero gravity? How many miles is that above the Moon's surface?

5-3 How high above the Earth's surface would a person of weight 250 lb have to go to "lose" 50 lb of weight?

5-4* As a spacecraft such as Apollo 11 orbits the Moon near the surface, there is a difference in gravitational force depending on whether the ship is on the far or the near side from the Earth. Estimate the percentage difference due to this effect, noting that $M_M/M_E = 0.0123$, $R_M/R_E = 0.272$, the distance between centers of Earth and Moon is 239,000 mi, and the radius of the Earth is 3960 mi. State clearly what approximations are made.

5-5 How many coulombs of positive (and negative) electricity does the isotope ${}_{92}U^{238}$ have?

5-6 Find the coulomb force between the helium nucleus, charge $+2e$, and the electron charge $-e$, if they are separated by 2.6×10^{-11} m.

5-7 Show that gravity has little to do with keeping the hydrogen atom stable.

5-8 How high could an athlete jump on a planetoid 40 mi in radius if he could jump 6 ft on Earth? Assume the density of the planetoid to be the same as the Earth's and the Earth's radius to be approximately 4000 mi.

5-9 At what altitude will the acceleration of gravity be 16.1 ft/sec^2? 8.05 ft/sec^2?

5-10 Suppose that a dwarf star had the same radius as that of the Earth, but a density 10^4 times as great. How much would a nickel coin (5 grams) weigh on such a star?

5-11 If enough charge could be transported from the Earth to the Moon, the attraction between the two could be doubled. Calculate (a) the amount of charge, and (b) the mass of hydrogen that would have to be ionized to get this charge.

5-12 Two small, light spheres, each of mass 0.2 g, are charged equally with 4×10^{-9} C and hung by threads of the same length. If the centers of the spheres are 3 cm apart, and gravitational attraction between spheres is neglected, what angle will the threads make with the vertical?

5-13 An alpha particle $({}_2He^4)^{++}$ is fixed at the Earth's surface. How high above it would a second alpha particle float?

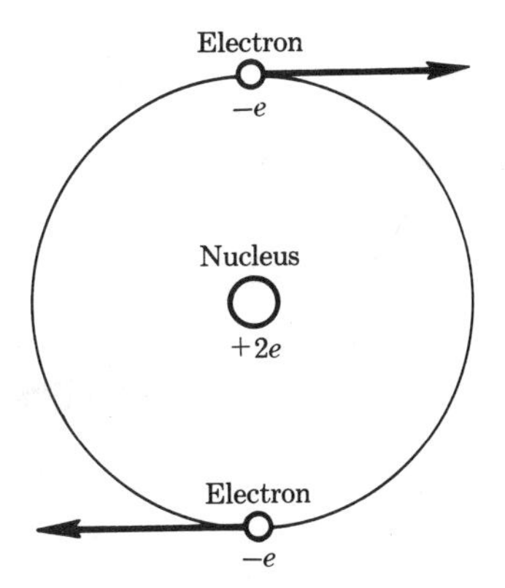

FIGURE P5-14

5-14 Find the magnitude and direction of magnetic force between two electrons in the helium atom if we visualized them as opposite to each other in one orbit of radius 2.6×10^{-11} m, and of speeds 4.5×10^6 m/sec (see Figure P5-14). How would this force compare in size with the electrostatic attraction of each to the nucleus? (See Problem 5-6.)

5-15 What would one predict that the speeds of two protons would be if the electrostatic repulsion equaled the electromagnetic attraction? What are the possible meanings of this result in terms of relativity?

5-16 If we try to find the relation between the angle made by electroscope leaves and the charge contained, by adding forces, what appears to happen at the vertex? What effect has the fact that gold is a good conductor of electricity?

5-17 Investigate the possibility of measuring the gravitational constant k_G in the laboratory by hanging masses by cables from the ceiling (see Figure P5-17). For equal masses of 1 kg, composed of lead, what is the closest distance of centers? Estimate the angle through which a cable would be deflected. Even if such angles could be measured, could k_G be obtained? What is actually being measured?

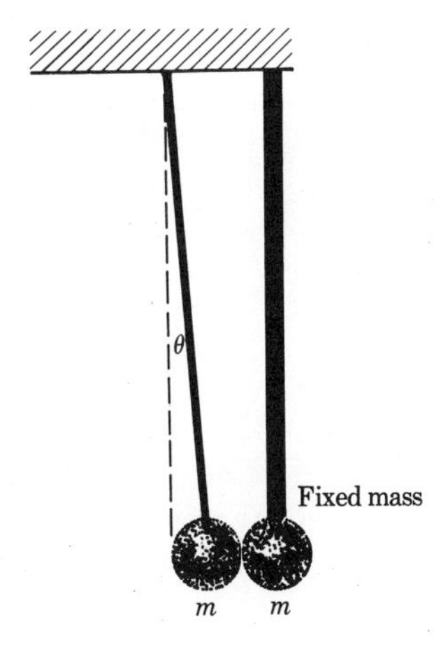

FIGURE P5-17

5-18 Calculate the strength of each of two identical bar magnets if the forces between the poles is 4×10^{-3} N when separated by 0.03 m. How much charge would be required on each of two objects to obtain the same force?

5-19 On what radius loop must a 1200-ft/sec jet plane fly in order to create the condition of weightlessness within the cabin at the top of the trajectory?

5-20 The acceleration of gravity is 25 m/sec² on the surface of Jupiter, whose mass and radius are respectively 318 and 11.2 times that of the Earth. With Jupiter's short day of 10 hours, by what fraction is this acceleration weakened by centrifugal force?

5-21* A marble is rolled onto and aimed toward the center of a phonograph record which turns clockwise. Discuss the factors that enter into predicting the shape of the path we see, and the path as seen from the frame of the record.

5-22 A collision with a strange planet from outer space sets our Earth into a higher speed of rotation, and everything not fastened down at the Equator "rises" slowly from the surface. How many hours long is our new "day"?

5-23* An intercontinental ballistic missile is fired toward the east with an initial speed of 200 m/sec at an angle of 30 deg from the ground. Ignoring the rotation of the Earth, variation of gravity with height, and air friction, how high, for how long a time, and how far will it go?

5-24 A space traveler has a mass of 75 kg. (a) What is his weight on the Earth? (b) What is his weight when his ship is in orbit at height 800 km? (c) If his ship is accelerating outward on a radial line at that orbit with $a = 19.6$ m/sec², what is his apparent weight?

5-25 Data on the orbital flight of the first American astronaut John Glenn are as follows: launch date, Feb. 20, 1962; weight of spacecraft, 2500 lb; time per orbit, 88.3 min; average height above the Earth, 131 mi; number of orbits, 3 (see Figure P5-25). Calculate the speed of the spacecraft in m/sec and mi/hr and the acceleration toward the Earth. How does the latter

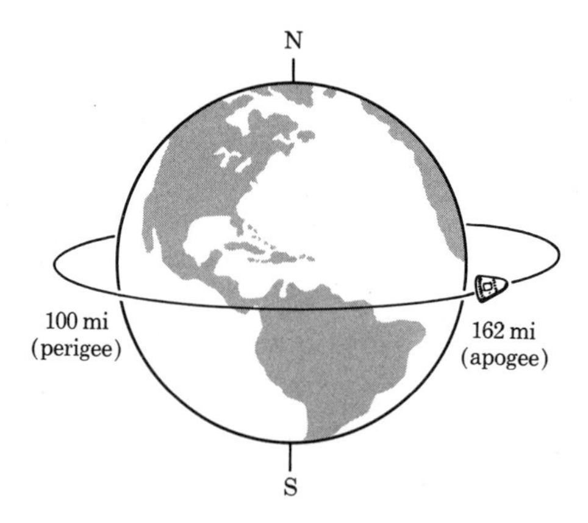

FIGURE P5-25

compare with gravitational acceleration at that height? Explain how accurate information on the orbit of a space flight permits a good measurement of the acceleration of gravity.

5-26 A particle of mass m is whirled in a circle with angular speed ω by means of an elastic cord of initial length R_0 that has a force constant K. (a) Devise an expression for the amount of elongation x of the cord. (b) Compute x if $m = 0.05$ kg, $R_0 = 0.4$ m, and $K = 0.2$ N/m, when the mass makes 10 revolutions per minute.

5-27 The oxygen molecule O_2 acts as if it had an elastic-force constant K of 1200 N/m. If this bond is stretched in length by an amount 2×10^{-11} m from the normal 1.2×10^{-10} m, how large is the restoring force? Would you judge the force to be electrostatic, electromagnetic, or gravitational? Explain.

5-28 A car starts to coast with initial speed 15 m/sec down a hill of height 120 m and length 2000 m. If the coefficient of friction is 0.05, what speed does the car have at the bottom?

5-29 A curve in the highway is banked to allow cars to turn safely at high speeds. Show that the tangent of the proper angle, for a turn without need of frictional forces, is given by the ratio of the centripetal and gravitational accelerations.

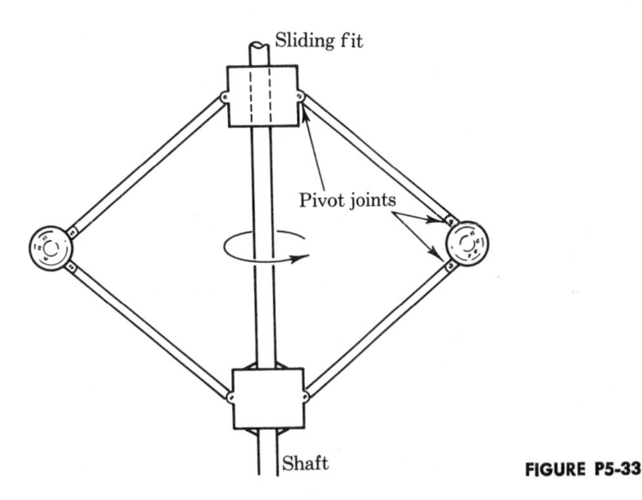

FIGURE P5-33

5-30 The speed limit is posted at 25 mi/hr (11.2 m/sec) for a freeway exit of radius 80 m. If the angle the roadway makes with the horizontal is 5 deg, what coefficient of friction was assumed in the design?

5-31 A coin of mass 5 grams is placed at a radius of 4 cm on a 78-rpm phonograph record while it is rotating at normal speed. The coin will start to slide if located at a larger radius. Find the coefficient of friction between the surfaces.

5-32* We try to climb out of a hemispherical depression with radius r and a coefficient of friction μ with the walls. Show that the height to which we can climb is

$$r[1 - (1 + \mu^2)^{-1/2}]$$

measured from the bottom.

5-33 A simple design of a governor for an engine appears in Figure P5-33. If the speed of rotation becomes excessive, the spheres pull farther apart and a valve is closed. Analyze the forces in this device.

THE FIELD CONCEPT

6

One of the functions of a science is to develop broad generalizations that include many particular phenomena. The concept of the *field,* which is a *property of space* containing interacting objects, goes beyond the evaluation of individual forces. In some ways it resembles the idea of *number,* as a generalization of the process of counting.

6-1 TEST BODIES AND FIELDS

The use of a device or "tester" to measure effects at a point is very familiar to us. For instance, the light meter indicates whether the illumination is proper for taking a photograph; the magnetic compass shows us the direction in which we are going; the pressure gauge tells us when our tires need more air; the thermometer gives us the temperature of an object or of the air. Other devices are the electrician's plug-in lamp that tells him if the line is "hot"; the accelerometer, the strain gauge, and the seismograph that measure degree of acceleration, mechanical disturbance, and earthquake activity, respectively; and the survey meter used to check the presence of nuclear radiation.

What do all of these have in common? First, each is a particular device chosen because it responds to the right conditions (a magnet would be of no value in measuring wind speed, for example). Second, the testing process involved in the use of each of these devices does not appreciably disturb the situation (a voltmeter that caused a battery to run down would be absurd). We are now going behind these practical devices to more basic *test bodies* used to detect and measure conditions at a *point*. Three steps are involved:

1. We insert, in a region of space, a mass m_1 to sense gravity, a charge q_1 at rest to see if electrical forces are present, or a moving charge characterized by $(qv)_1$ to detect magnetic effects. (The subscript "1" is attached in each case to remind us that these are test bodies.)

In each case the amount of material in the test body is assumed to be vanishingly small, so as not to disturb what is being measured. If we observe a tendency for motion to occur, we say qualitatively that there is a *field* present that exerts an influence on objects placed in that region of space.

2. We measure the force, including magnitude and direction, that the test body experiences.

3. We compute the force *per unit* of mass, charge, or charge-speed by dividing the forces by m_1, q_1, or $(qv)_1$, respectively. (In the last case, directional aspects of the velocity need to be considered.) Suppose that we measured a force of 3.2×10^{-16} N on a test charge of 1.6×10^{-19} C. Then the electrical force per unit charge would be

$$\frac{(F_E)_1}{q_1} = \frac{3.2 \times 10^{-16}}{1.6 \times 10^{-19}} = 2000 \text{ N/C}$$

The virtue of such a quotient is that it is independent of the charge. If we doubled the test charge, the force would be doubled, but $(F_E)/q$ would be the same as before. We give the name *field intensity* or simply *field* to such ratios. Thus the electric-field intensity is

$$\mathcal{E} = \frac{(F_E)_1}{q_1}$$

Similar ratios are $\mathcal{G} = (F_G)_1/m_1$, the gravitational field, and $\mathcal{B} = (F_M)_1/qv_1$, the magnetic field. Having found the field at a point, we can dispense with the test bodies and are in a position to predict the force on *any* object we wish to place at the point. The relations are simply written

$$F_G = m\mathcal{G}, \qquad F_E = q\mathcal{E}$$

The expression $F_E = q\mathcal{E}$ says in words, "If we know the electric force per unit charge at a point (the electric field $\mathcal{E}$), then the force F_E on a charge q is simply q times as large."

Fields are vector quantities, just as forces are. In the case of electrical and gravitational fields, we write $\mathbf{F}_E = q\boldsymbol{\mathcal{E}}$ and $\mathbf{F}_G = m\boldsymbol{\mathcal{G}}$. Because of special directional aspects, the magnetic force is somewhat more complicated. For charges moving perpendicular to the magnetic field, the force is at right angles to both $\mathbf{v}$ and $\boldsymbol{\mathcal{B}}$ and is of magnitude

$$F_M = qv\mathcal{B}$$

This magnetic force will be discussed in detail in Chapter 7. The units of the three fields are readily verified to be N/kg, N/C, and N-sec/C-m.

In order to dispel the idea that this procedure is unduly artificial and cumbersome, let us consider an example. Exploratory spacecraft of the Mariner class, as they travel to the planet Mars, experience forces from the Earth, the Moon, Mars, the Sun, and—to a lesser degree—from the other planets. Using the gravitational-*force* laws, it would be necessary to calculate all forces for each spacecraft, inserting its proper mass. By the use of the *field*, which does not depend on which vehicle is involved or the source of the effect, the gravitational behavior of the whole set can be found at once. In the case of magnetic forces, the formula for forces between charges in motion is complicated, and the use of the magnetic field helps provide an intermediate simplifying step. Finally, the concept of field provides a generalization that is characteristic of the science of physics; for instead of describing the force on a particular body, we now can determine the force on any body.

6-2 INVERSE-SQUARE FIELDS AND LINES OF FORCE

The similarity in mathematical form of the forces of gravity and electrostatics makes it convenient to describe their fields together, even though we realize that they differ completely in origin and strength. Simple numerical illustrations of each are given below.

Gravity Field at the Earth's Surface

The procedure for finding the gravitational field at any point is particularly easy for a single mass M such as that of the Earth. By our definition, the field has magnitude

$$\mathcal{G} = \frac{F_G}{m}$$

or

$$\mathcal{G} = \frac{k_G M}{R^2} \qquad \textit{(gravitational field)}$$

which is the force per unit mass "created" at a distance R from the center of the planet. Its value at the Earth's surface is readily computed, using the figures $k_G = 6.67 \times 10^{-11}$ N-m²/kg², $M = 6.0 \times 10^{24}$ kg, and $R = 6.4 \times 10^6$ m. The result is

$$\mathcal{G} = 9.8 \text{ N/kg}$$

This number is the same as that for the acceleration of gravity at the surface, $g = 9.8\ \mathrm{m/sec^2}$, as one can see by referring to previous calculations, or by noting that both $\mathcal{G}$ and g are forces per unit mass. Their identity is a consequence of the fact that the mass of inertia and that of gravity are the same; that is, the property of matter that produces resistance to change of motion is also the property that gives rise to gravitational attraction.

Once we know this field value, we can find the force on *any* object located on the Earth. For example, on a man of mass 75 kg the gravity force would be

$$F_G = m\mathcal{G} = (75\ \mathrm{kg})(9.8\ \mathrm{N/kg}) = 735\ \mathrm{N}$$

On an automobile of mass 1000 kg, the force would be

$$F_G = m\mathcal{G} = (1000\ \mathrm{kg})(9.8\ \mathrm{N/kg}) = 9.8 \times 10^3\ \mathrm{N}$$

Electric Field in Vicinity of a Nucleus

Let us find the magnitude of the electric field at any distance R from a nucleus of gold $_{79}\mathrm{Au}^{197}$, with 79 protons in the nucleus. The charge is 1.6×10^{-19} C on each proton, and the total charge on the nucleus is

$$Q = (79)(1.6 \times 10^{-19}\ \mathrm{C}) = 1.26 \times 10^{-17}\ \mathrm{C}$$

The electric field (force per coulomb) is thus $\mathcal{E} = F/q$ or

$$\mathcal{E} = \frac{k_E Q}{R^2} = \frac{[9 \times 10^9\ \mathrm{N\text{-}m^2/C^2}][1.26 \times 10^{-17}\ \mathrm{C}]}{[R(\mathrm{m})]^2}$$

$$= \frac{1.13 \times 10^{-7}}{R^2}\ \mathrm{N/C}$$

Now let us proceed to find the force on an alpha particle (the nucleus of $_2\mathrm{He}^4$) at a distance $R = 10^{-11}$ m. Now the atomic number of helium is $Z = 2$, and thus the charge on the alpha particle is $q = (2)(1.60 \times 10^{-19}\ \mathrm{C}) = 3.20 \times 10^{-19}$ C. Then,

$$F = q\mathcal{E} = (3.20 \times 10^{-19})\,\frac{1.13 \times 10^{-7}}{(10^{-11})^2} = 3.6 \times 10^{-4}\ \mathrm{N}$$

A picturesque graphical description of the field was devised by Faraday.* One draws continuous lines or curves that are in the direction of the field at every point. Remembering how the field is measured, we see that such paths would be traced out by a small test body if allowed to move slowly in that region of space. For convenience, the number of lines that one draws through each unit of area perpendicular to the field at a point,

* Michael Faraday (1791–1867) was an imaginative, practical, and productive English investigator. Self-educated, he volunteered his services as assistant in the Royal Institution, and eventually became its Director. Among his many contributions were the discovery of the principles of electrolysis, the motor, and the generator.

is equal to the numerical strength of the field. Although Faraday attributed more significance to the lines than is necessary, the idea is helpful for visualizing both the direction and the strength of fields.

Electric-field and Gravitational-field Lines

The field picture of a point negative charge or a mass is very easily drawn according to the above specifications (Figure 6-1). We know that the field is everywhere directed inward, thus the lines must be straight. The lines of a positive charge are also straight, but directed outward. We might reasonably ask if there is such a thing as gravitational *repulsion*. Although no "negative mass" has been observed, the possibility of its existence is not ruled out.

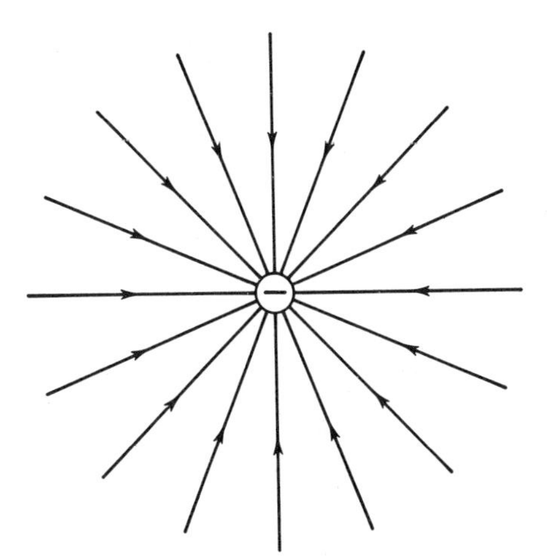

FIGURE 6-1

Field lines of negative charge or spherical mass

By repeated use of Coulomb's law of electrostatic force, we can find the field due to two like point charges, drawn in Figure 6-2(a). We note that the field lines from each charge never connect or intersect. In contrast, those due to two unlike charges, Figure 6-2(b), do join the two particles. This is an illustration of the correct statement that *electric-field lines start on positive charges and end on negative charges.*

The technical term *flux*, symbolized by Φ, is a designation of the product of field strength and area.* Thus if the electric field $\mathcal{E}$ is the same everywhere and is directed perpendicularly through an area A, the electric flux is $\Phi_E = \mathcal{E}A$. Similarly, one can write $\Phi_G = \mathcal{G}A$ for gravity and $\Phi_M = \mathcal{B}A$ for magnetism. The units of each of the fluxes are different. For

* The use of the word "flux" in this context apparently originated with Faraday, who imagined that there was a "flow" of lines emanating from and entering gravitational, electrical, or magnetic bodies. The product of field strength and area, then, was the total number of lines through the area.

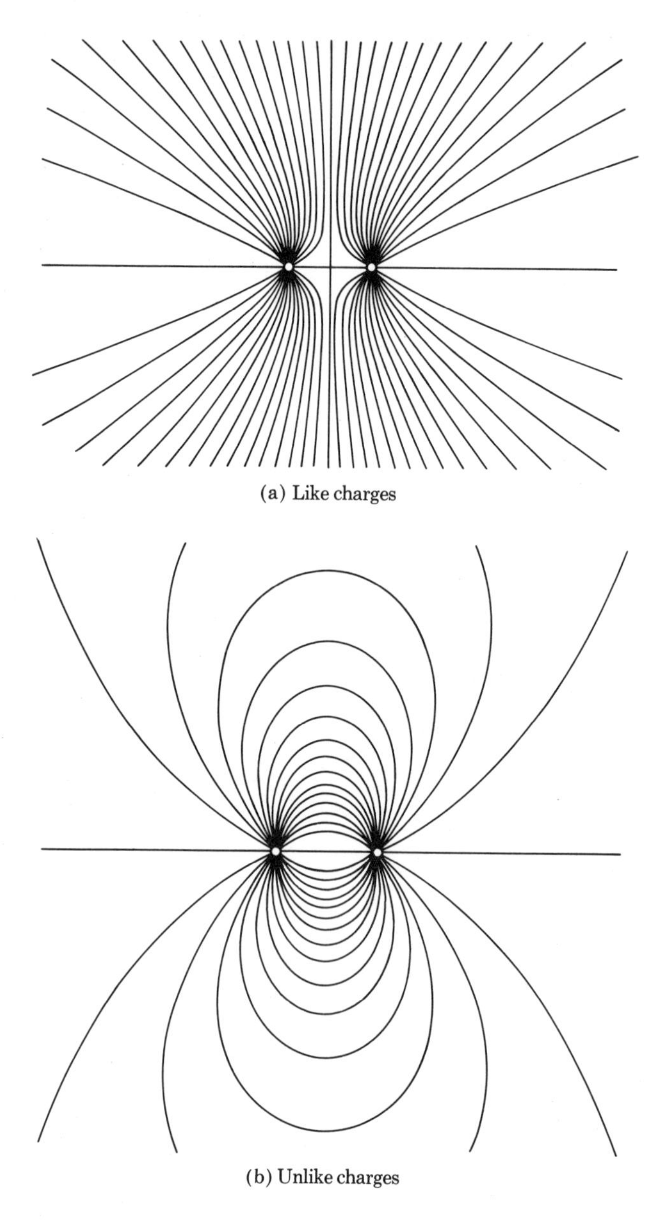

FIGURE 6-2

Electric field lines due to two charges

instance, those of electric flux are (N/C)(m^2) or N-m^2/C, while those of gravitational flux are (N/kg)m^2 or N-m^2/kg. The flux must be interpreted as the product of the area of the surface and the component of the field that is perpendicular to the surface. If the field varies with position and the surface is curved, it will be necessary to add up fluxes through separate small areas.

Gauss's Law

A very useful method for finding the field of a collection of charges or masses was developed by Gauss, a German physicist and mathematician. We can deduce his generalization by studying field lines from a particle. Because of the close similarity between the mathematical forms of electric and gravitational fields, it will be necessary to examine only one type of field. Let us draw a sphere of radius R about a point positive charge Q. The electric-field lines radiate outward and pierce this surface, as shown in Figure 6-3. Now, the field strength is the same everywhere on the sphere, is perpendicular to the surface, and is of value $\mathcal{E} = k_E(Q/R^2)$. The area

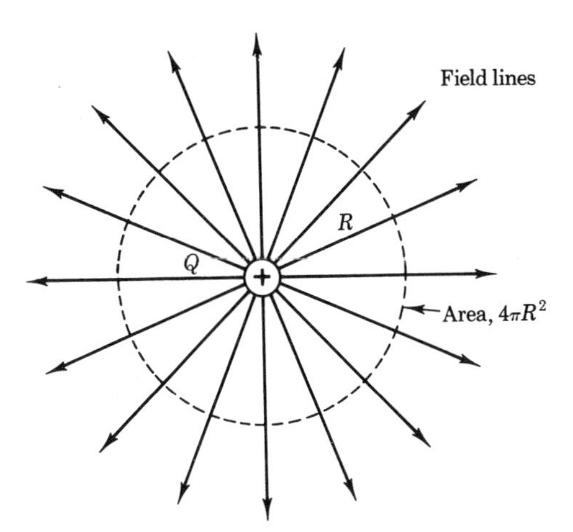

FIGURE 6-3
Electric flux and Gauss's law

through which the lines pass is $A = 4\pi R^2$. Then the product of field and area, the flux, is

$$\Phi_E = 4\pi k_E Q \qquad (\textit{Gauss's law})$$

In words, *the electric flux over a closed surface is proportional to the total enclosed charge*. We have used a special geometry to "derive" Gauss's law, but the theorem is quite general.

If a region is free of charges, the flux over the surface is zero, according to Gauss's law. To verify this, let us consider a region of space in the vicinity of the point charge (but not enclosing it). We draw a special box, as in Figure 6-4, with four sides parallel to radial lines from the origin, and two sides as portions of spherical surfaces. No flux lines cross the sides of the box. The flux entering the inner face of the box is $\Phi_1 = \mathcal{E}_1 A_1$ while that leaving the outer face is $\Phi_2 = \mathcal{E}_2 A_2$. The ratio of areas is $A_2/A_1 = R_2^2/R_1^2$ and the ratio of electric fields is $\mathcal{E}_2/\mathcal{E}_1 = R_1^2/R_2^2$. Substituting, $\Phi_1/\Phi_2 = 1$. This means that the flux *in* is equal to the flux *out*, and the *net* flux is zero over the whole surfaces, proving Gauss's law once more.

The law gives us a procedure for finding various electric and gravitational fields without the necessity of adding the effects of charges or masses individually. The method is most convenient when the object has

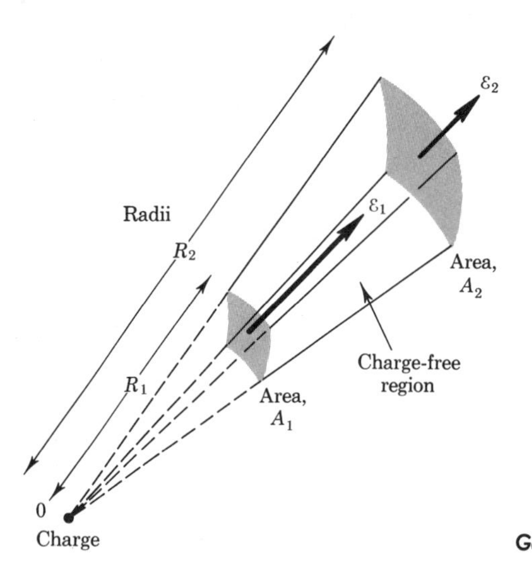

FIGURE 6-4

Gauss's law for a region without charge

a certain symmetry of form. Let us further illustrate Gauss's law by other examples.

In the Geiger counter used for uranium prospecting and for detecting radioactivity in general, a length of wire along the axis of a cylindrical tube is given an electric charge—Figure 6-5(a). We should like to know the electric field in the vicinity of the wire. The physical situation is simplified if we consider a very long wire of charge q per unit length, and look for $\mathcal{E}$ at a small distance y from the center of the wire, as in Figure 6-5(b). If the point of observation is equidistant from the ends, we can deduce that the field is pointed radially, since all axial components of $\mathcal{E}$ due to charges at

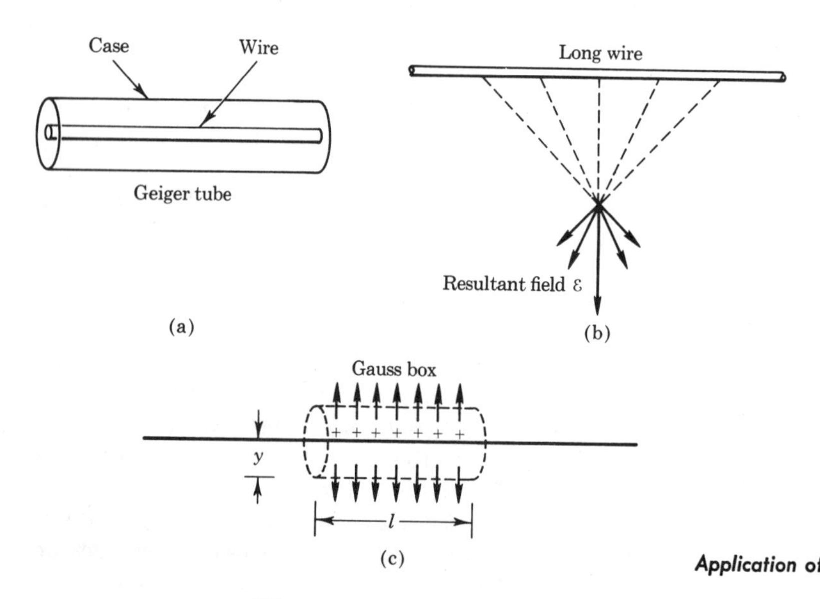

FIGURE 6-5

Application of Gauss's law to charged wire

equal distances from 0 will cancel. The "Gauss box" is chosen to be a short cylinder of length l and radius y, enclosing a section of the wire—see Figure 6-5(c). Over its sides of area $A = 2\pi yl$, the field is the value to be found, $\mathcal{E}$. The field component perpendicular to the end faces is zero. Now, the total charge enclosed by the surface is $Q = lq$, and by Gauss's law $\Phi_E = 4\pi k_E Q$, or

$$\mathcal{E}A = \mathcal{E}2\pi yl = 4\pi k_E lq$$

Thus

$$\mathcal{E} = \frac{k_E 2q}{y} \qquad \textit{(electric field near a long wire)}$$

We note that the field varies inversely with the distance—not with the inverse square of distance, as for a point charge.

Drilling into the Earth shows that the density of soil and rock varies with depth. If viewed from a great distance, however, the Earth could be considered as composed of a set of uniform concentric shells. The total field at some distance from the center of one of these could be found by

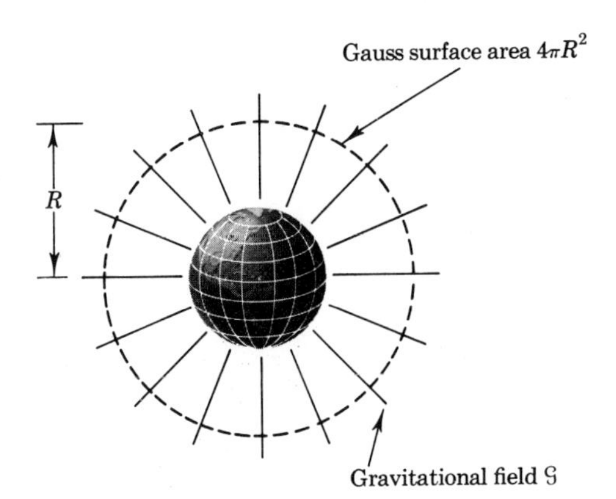

FIGURE 6-6
Application of Gauss's law to a planet

adding the effects of point masses. It is much easier to employ Gauss's law to obtain the gravitational field for the whole planet of mass M at a distance R from its center. This time we draw a spherical surface of radius R around the planet, as in Figure 6-6. Its area is $4\pi R^2$, and the total enclosed mass is M, so Gauss's law tells us that

$$\Phi_G = \mathcal{G}4\pi R^2 = 4\pi k_G M$$

or

$$\mathcal{G} = \frac{k_G M}{R^2} \qquad \textit{(gravitational field outside a spherical body)}$$

This expression looks very familiar, but its new meaning must be carefully considered. It says that *the field outside a sphere is the same as if all the mass were concentrated at its center*. We come to the general conclusion that the field outside a solid sphere that has similar layers or is uniform is the same as that of a point mass at the center. This obviously simplifies calculations, and allows us to immediately find $\mathcal{G}$ above a planet, or near a sun, and the electric field $\mathcal{E}$ outside a nucleus as well, assuming that the protons that go to make it up are uniformly distributed.

By careful application of Gauss's method, we can find the field *inside* the Earth, of radius R_E. We draw the surface as a spherical boundary of some smaller radius R measured from the center. There will be a shell of material of thickness $R_E - R$ *outside us*, which gives no field since the mass is not contained by the spherical surface. However, there will be a sphere of radius R that *we are outside*. It has a volume (and mass) that is smaller than that of the Earth by a factor $(R/R_E)^3$. By Gauss's law

$$\Phi_G = \mathcal{G}4\pi R^2 = 4\pi k_G \left(\frac{R}{R_E}\right)^3 M_E$$

or

$$\mathcal{G} = \frac{R}{R_E}\,\mathcal{G}_E$$

where $\mathcal{G}_E$ is the gravitational field at R_E. Thus we see that outside a planet the field varies inversely as the square of the distance from the center, but inside varies *linearly* with the distance. At the center of the planet the field is zero.

As another example, let us find the electric field near a sheet of charge, with amount σ per unit area. For this case, we draw a cylindrical box that goes through the sheet, as in Figure 6-7(a). Again if the plane area is very large, we know that the field is perpendicular to the surface, the sideways components having canceled. Over the sides of the box, there is zero perpendicular component of field, but over the two ends of total area $2A$, the field is $\mathcal{E}$. The total charge inside the closed surface is that on the sheet, of amount σA. Applying Gauss's law, $\Phi_E = \mathcal{E}2A = 4\pi k_E A\sigma$, or

$$\mathcal{E} = k_E 2\pi\sigma \qquad \textit{(electric field of a plane sheet)}$$

It is interesting to note that the field does not depend on how far one is away from the plane sheet of charge. Our derivation applies strictly to an infinite uniformly charged sheet, and for a finite disc of charge would be smaller because of "edge effects." On the other hand, if we got very close to such a disc, the above result would be quite accurate. We find a different field near a large plane *conductor* on which we place charge. Then, repulsion causes the charges to move to the outside surface and effectively to isolate those on each side. Our "Gauss box" lies only on one side, as shown in Figure 6-7(b). With end area A and charge density σ, Gauss's law yields $\Phi_E = \mathcal{E}A = 4\pi k_E A\sigma$, or

$$\mathcal{E} = k_E 4\pi\sigma \qquad \textit{(electric field of a charged plane conductor)}$$

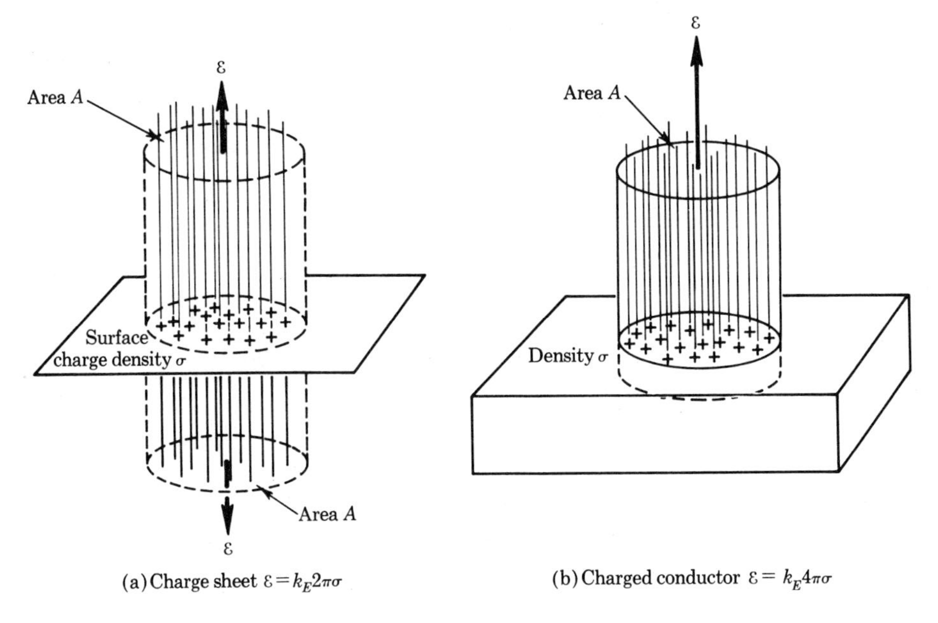

FIGURE 6-7

Electric fields near planes of charge

Of particular note is the factor of 2 in comparison with the field due to a sheet of charge.

We are now in a position to find the electric field between two charged conducting plates whose surface dimensions are large compared with their separation. An example is a capacitor (or "condenser") used in radio circuits for tuning. When connected to a battery or other electrical supply, as shown in Figure 6-8, the plates assume charge densities of equal but opposite sign. Because of mutual electrostatic forces, electrons are drawn to the lower side of the top plate and driven to the lower side of the bottom

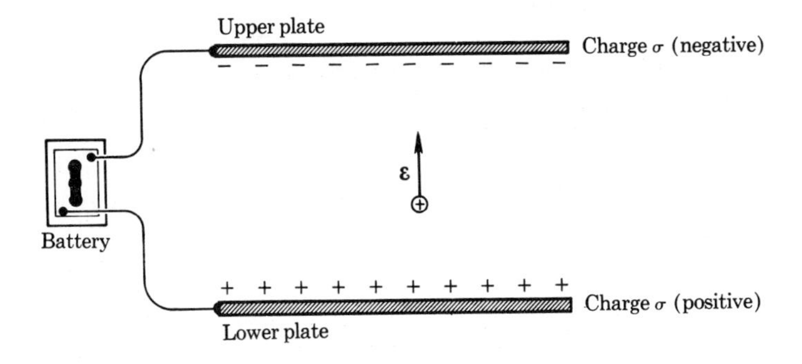

FIGURE 6-8

Field within capacitor

plate, leaving the upper side positive. Each plate now appears as a one-sided sheet of charge. Let us suppose that the plates are large enough that we can use the previous conclusions on the field from an infinite sheet. The fields of the two plates are of the same magnitude, $\mathcal{E} = k_E 2\pi\sigma$, and in the same direction—upward; and thus the total is

$$\mathcal{E} = k_E 4\pi\sigma \qquad \textit{(field between parallel plates, sign of charge opposite)}$$

If the plates were instead given equal charges of the *same* sign—say positive—the fields would be oppositely directed and cancel each other, giving *zero* field. In the space between plates of finite size this field is zero only at the exact center of the gap.

We may reason out another important conclusion from this experiment. Picture the two plates as extending out indefinitely, and in effect completely enclosing the point of observation (we are reminded of the geo-

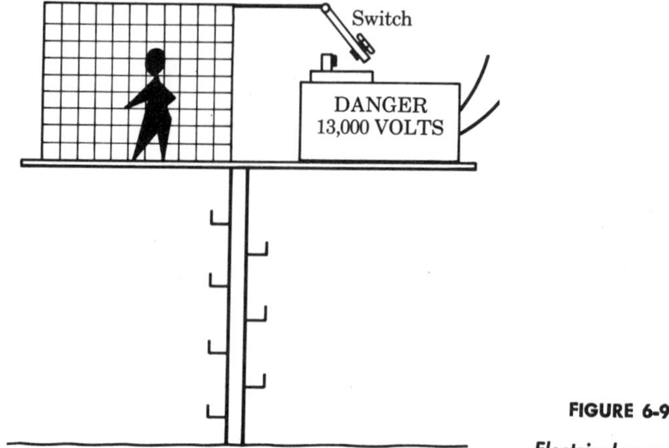

FIGURE 6-9
Electrical cage

metric theorem that parallel lines *meet* at infinity). The field is zero inside, as shown above. Thus we can state that the electric field inside a charged closed surface is zero, a theorem that is generally true, although we took a special case to prove it. In order to appreciate this result better, imagine the following "experiment." A large metal cage, solid or of metal screen, is constructed on a platform atop an electrical power pole (Figure 6-9). With the switch open, we climb the pole and get in the cage. The switch is closed and we remain inside, completely unharmed, since the electric field is zero. Naturally, we should be sure the outside of the cage has been discharged before stepping out.

The final application of Gauss's law will be to find how the field varies with position in a region where the charge density is not uniform, as in a vacuum tube found in electrical circuits. Figure 6-10(a) shows a simple case

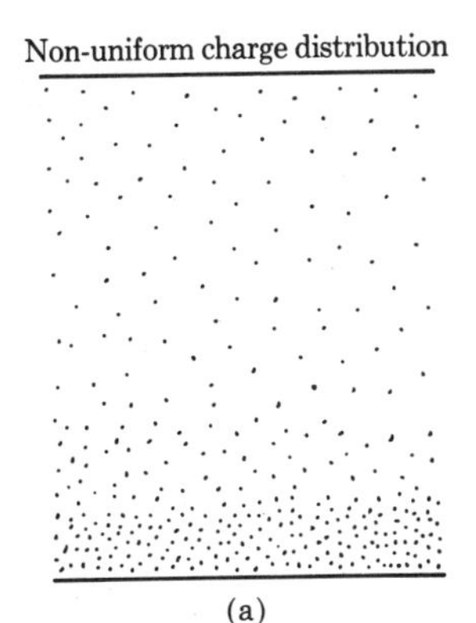

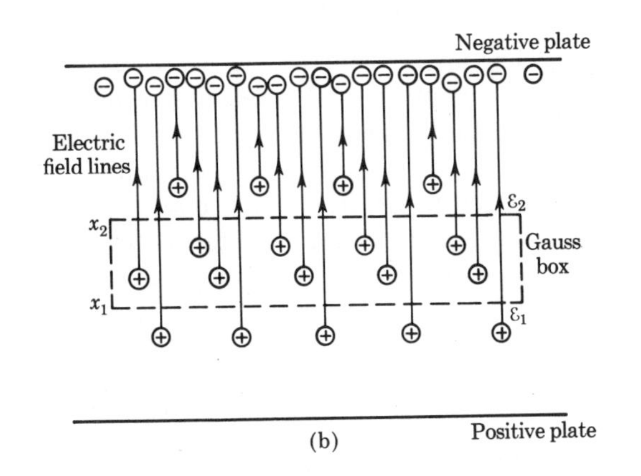

FIGURE 6-10

Field in region with varying charge density

in which the charge density decreases as we go up from one plate to another. The lines of electric field start on the positive charges in the space and end on negative charges on the upper plate, as shown in Figure 6-10(b). We draw a flat "Gauss box" with surfaces parallel to the plates. The fields at x_2 and x_1 are taken as $\mathcal{E}_2$ and $\mathcal{E}_1$, and the net flux through the closed surface is $\mathcal{E}_2 A - \mathcal{E}_1 A$ or $A\Delta\mathcal{E}$. The total charge inside, assuming a particle density n and charge per particle q, is $nqA\Delta x$. Gauss's law then yields

$$A\Delta\mathcal{E} = 4\pi k_E nqA\Delta x$$

or

$$\frac{\Delta\mathcal{E}}{\Delta x} = 4\pi k_E nq$$

This tells us the rate of change of field with position for any dependence of charge density on position.

6-3 FIELDS DUE TO EXTENDED OBJECTS

The field that results from the combined action of more than one particle may be deduced by the process of addition of effects, or *superposition*. This is feasible because the individual forces are vectors, obeying standard rules for combination. Gauss's law was seen to provide a quick and easy way of finding the total field from some simple geometric arrangements of mass and charge. Unfortunately, its application becomes prohibitively difficult if the shape is not regular and symmetric. We must then turn to a process of direct summations, which will be illustrated for systems often encountered in the physical world.

The Electric Dipole

The idea of superposing electric fields is applied to the *electric dipole,* which is simply a pair of charges of opposite signs separated by a small distance (see Figure 6-11). The electron and the proton of the hydrogen atom may be visualized as an example. The combined field varies in strength both with the distance away and direction from the dipole. Let us find $\mathcal{E}$ along the axis at a distance x from the midpoint of the two particles as if they were fixed in position. Now

$$\mathcal{E} = \mathcal{E}_+ - \mathcal{E}_-$$

where the distance R appearing in the expressions for the two electric fields of magnitude $k_E q/R^2$ is $x - s/2$ for the $(+)$ charge and $x + s/2$ for the $(-)$ charge. Inserting and doing the algebra,

$$\mathcal{E} = k_E q \frac{2sx}{[x^2 - (s/2)^2]^2}$$

At great distances from the dipole, such that $s/2$ is much smaller than x and can be neglected, the denominator is approximately x^4 and

$$\mathcal{E} \cong \frac{2k_E qs}{x^3}$$

Notice that if we let s be zero, there would be no field, since the effects of the two charges would exactly cancel.

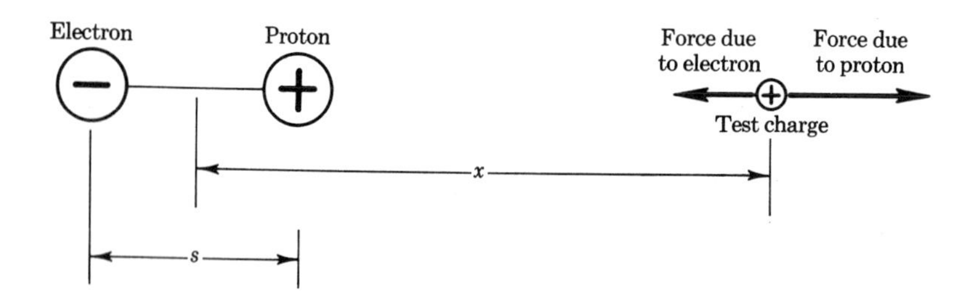

FIGURE 6-11

Electric dipole

Let us compare the field of the hydrogen atom with that of the hydrogen ion (the proton), at say 100 atomic radii from the positive charge. The atom's field $\mathcal{E}_a$ from the approximate dipole formula is $2k_E er/(100.5r)^3$, while the singly charged proton's field $\mathcal{E}_p$ is $k_E e/(100r)^2$. Their ratio $\mathcal{E}_a/\mathcal{E}_p$ is very close to 0.02, which checks our earlier observation that widely separated atoms interact far more weakly than do widely separated ions.

The product qs is called the *electric moment* M of the dipole. It is often

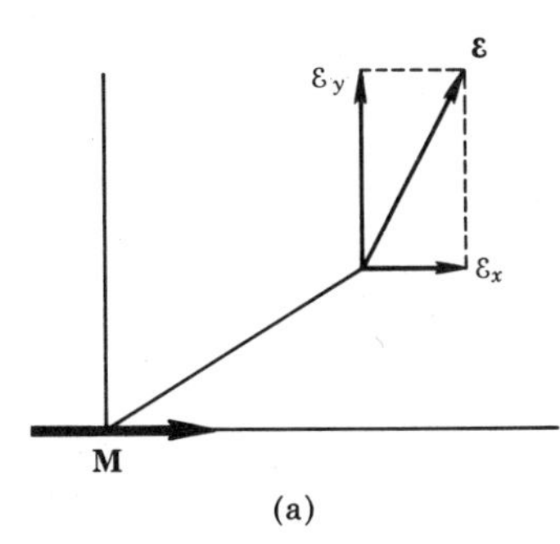

(a)

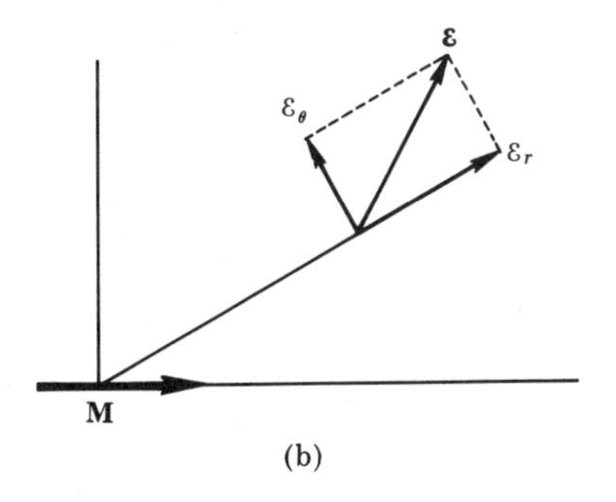

(b)

FIGURE 6-12

Field lines due to electric dipole

useful to form the vector **M**, of magnitude qs and directed from the negative charge to the positive charge. Figures 6-12(a) and (b) show the electric-field vector at a point due to a dipole, and how it can be resolved into components in the x, y or in the r, θ directions.

Combined Gravitational Effects

For an astronautical example of superposition, let us calculate the gravitational field at a spaceship that is 0.8 the distance from Earth to the Moon, when the relative positions of Earth, Moon, and Sun are as shown in Figure 6-13(a). First we find the magnitudes of each of the three fields $\mathcal{G}_E$, $\mathcal{G}_M$, and $\mathcal{G}_S$. At the Earth's surface at radius 3960 mi, the field is 9.8 N/kg—i.e., 1 g. We shall make use of ratios to find $\mathcal{G}$ values elsewhere. At a distance $0.8R_M$, where $R_M = 239{,}000$ mi (Table 10-1), the field due to the Earth is

$$\mathcal{G}_E = 9.8\left(\frac{R_E}{0.8R_M}\right)^2 = 9.8\left(\frac{3.96 \times 10^3}{1.91 \times 10^5}\right)^2 = 0.00421 \text{ N/kg}$$

The field due to the Moon, of mass 0.0123 times that of the Earth, is

$$\mathcal{G}_M = 9.8\left(\frac{R_E}{0.2R_M}\right)^2\left(\frac{M_M}{M_E}\right) = 9.8\left(\frac{3.96 \times 10^3}{0.48 \times 10^5}\right)^2 (0.0123)$$
$$= 0.00082 \text{ N/kg}$$

Finally, the field caused by the Sun, of mass 3.33×10^5 times that of the Earth—at a distance that is very close to the radius a_E of the Earth's orbit, 92.9×10^6 miles—is

$$\mathcal{G}_S = 9.8\left(\frac{R_E}{a_E}\right)^2\left(\frac{M_S}{M_E}\right) = 9.8\left(\frac{3.96 \times 10^3}{92.9 \times 10^6}\right)^2 (3.33 \times 10^5)$$
$$= 0.00589 \text{ N/kg}$$

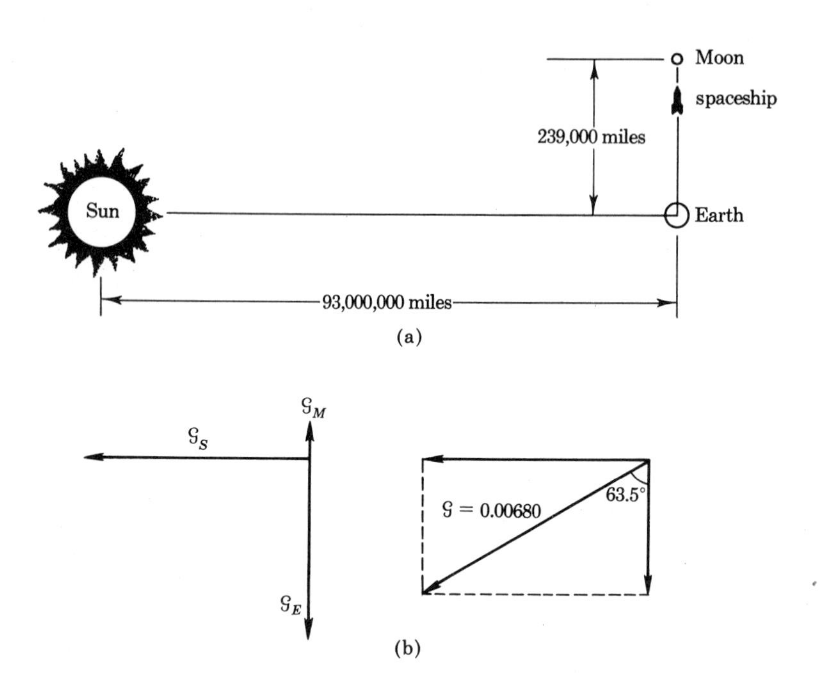

FIGURE 6-13

Gravitational field at spaceship 0.8 of the way from Earth to Moon

The net field in the "vertical" direction is 0.00421 − 0.00082 = 0.00339, while that in the "horizontal" is close to 0.00589. The vector resultant is—Figure 6-12(b)—

$$\mathcal{G} = \sqrt{(0.00339)^2 + (0.00589)^2} = 0.00680 \text{ N/kg}$$

The angle between the field and the line joining Earth and Moon is found from $\tan\theta = \dfrac{0.00680}{0.00339} = 2.01$, or $\theta = 63.5$ deg. With such information continuously available to a computer on the spaceship, correction forces that ensure safe landing can be applied.

6-4 MAGNETIC FIELDS OF MACROSCOPIC BODIES

The existence of forces between objects composed of iron had been known from early times, and the magnetic compass made possible the wide explorations in the sixteenth and seventeenth centuries. The ultimate recognition that such ferromagnetism was due to electronic currents was a relatively recent development. The idea that there were physical poles of magnetism at the ends of bar magnets—analogous to the positive and negative electric charges—was reasonable because of the superficial resemblance of the observed laws of force.

Field Lines of Bar Magnets

The force between two "poles" of a long thin magnet is given by the experimental relation

$$F = \frac{k_M p_1 p_2}{R^2}$$

Two like poles—each labeled N (for "north")—will repel each other, as will two S (for "south") poles; two unlike poles—an N and an S—will attract each other. A field of magnetism can be defined just as for the gravitational or electrostatic cases, letting the ratio $\mathbf{F}/p$ be the field, where p is the strength of a test body. Then, for any pole of strength p, the magnetic field is

$$\mathcal{B} = \frac{k_M p}{R^2}$$

Vector addition of the contributions of several poles is completely analogous to that for several charges or masses because of the inverse-square dependence. Figure 6-14 shows how small iron filings line up with the direction of the field of a bar. Note the similarity to the gravitational and electric fields.

A *magnetic moment* can be defined for a bar magnet of length s and pole strength p, by direct analogy to the electric moment of the charge

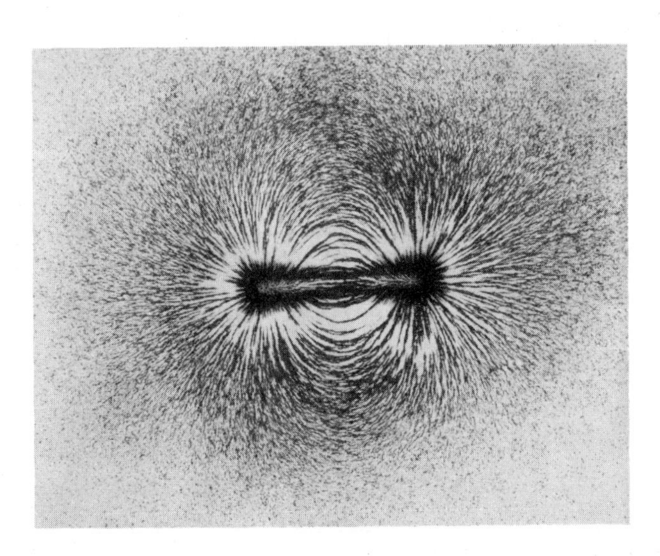

FIGURE 6-14

Iron filings in field of bar magnet

dipole: we merely replace q by p, and the moment is thus

$$M = ps$$

Gauss's Law for Magnetism

The similarity between force formulas for electrostatic charges and bar-magnet poles would suggest that there is a complete analogy between charges and poles. This is incorrect, according to the following arguments. First, let us attempt to isolate one pole. We know that single charges exist—for example, electrons or protons. However, if we cut a bar in two, or four, or any number of pieces, there is always found to be a north and a south pole on the ends of each piece, as illustrated in Figure 6-15. Thus *there is no magnetic substance analogous to charge that can be isolated.* Accepting this, we settle for a magnet with two poles, and compare the lines of force with those from a dipole. The field pattern due to an electric dipole and a magnetic dipole are the same so long as we are very far away. Close scrutiny of the lines near two charges or the magnet shows quite a difference, as shown in Figure 6-16. Whereas the lines of an electric field begin on the (+) charge and end on the (−) charge, the lines for magnetic field are *continuous*. Again, we find that there are really no poles on which lines begin or end. We try to apply Gauss's law formally by constructing a box

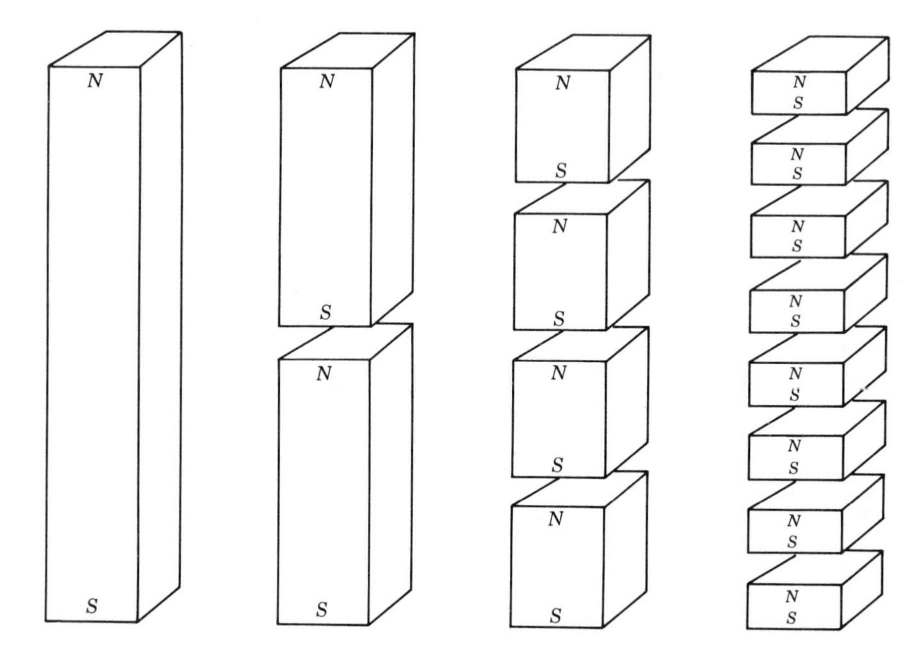

FIGURE 6-15

Attempt to isolate a single magnetic pole

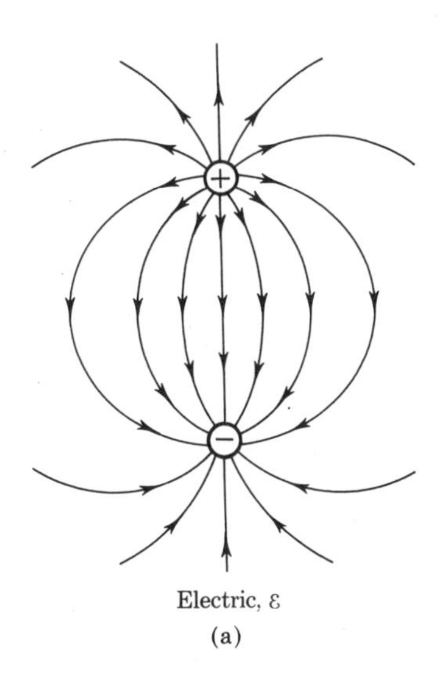

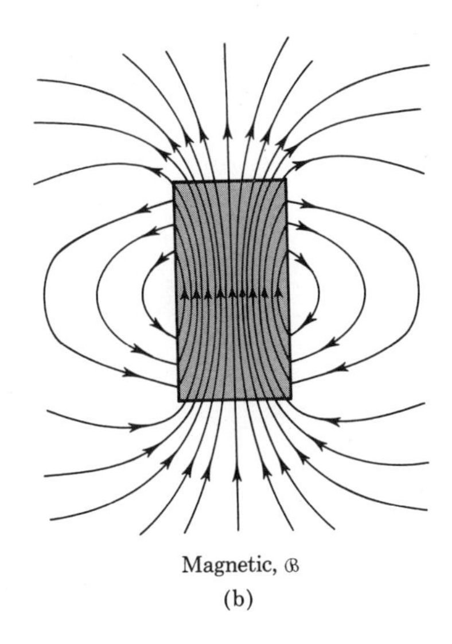

FIGURE 6-16

Comparison of field lines of unlike charges and a bar magnet

with sides parallel to the field and ends perpendicular, enclosing a bundle of lines within the bar magnet, as in Figure 6-16(b). If we make measurements inside the magnet using a minute detector in an appropriate cavity, we find the magnetic field is everywhere directed upward. Thus through the ends of our "Gauss box," there is an outward flux of exactly the same magnitude as that of the inward flux, and they cancel to give zero; and thus the total flux over the surface surrounding the presumed poles is zero. We are led to the important conclusion that *there is no magnetic material corresponding to electric charge.* This deduction was made for bar magnetism, but it can be shown to be completely general; and although it is a negative statement, it is as powerful a concept as the inverse-square law of electrostatics.

The Earth's Magnetic Field

At the surface of the Earth and surrounding it is a weak magnetic field, the cause of which is not precisely known. One popular explanation is based on the motion of molten iron deep in the interior of the planet. The Earth has a field that very roughly resembles that of a bar magnet (see Figure 6-17). Since there are no fixed magnetic poles in the Earth and the field lines vary so widely in direction over the surface, the comparison is useful only for a qualitative view. The needle of a compass will point, at any location selected, at an *angle of declination* d from geographic north—Figure 6-18(a). Charts and tables of this angle assist in navigation. The needle will also assume an *angle of inclination* i with the horizontal—Figure

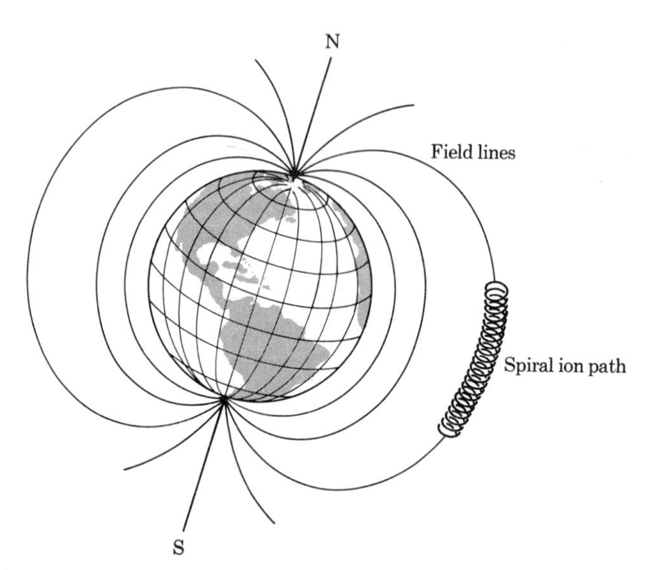

FIGURE 6-17

Earth's magnetic field

6-18(b). The magnetic poles of the Earth are defined as those places where the needle would be vertical. There is a slow but erratic change with time of these spots. Finally, the strength of the field is low, but it varies widely over the globe and depends as well on local topography. Although the field is weak, it has a profound effect on charged particles coming from the Sun and from outer space. Ions can move along magnetic-field lines freely, but are forced into circular paths about the lines. Charges can oscillate back and forth, as shown in Figure 6-17, creating ionization along their

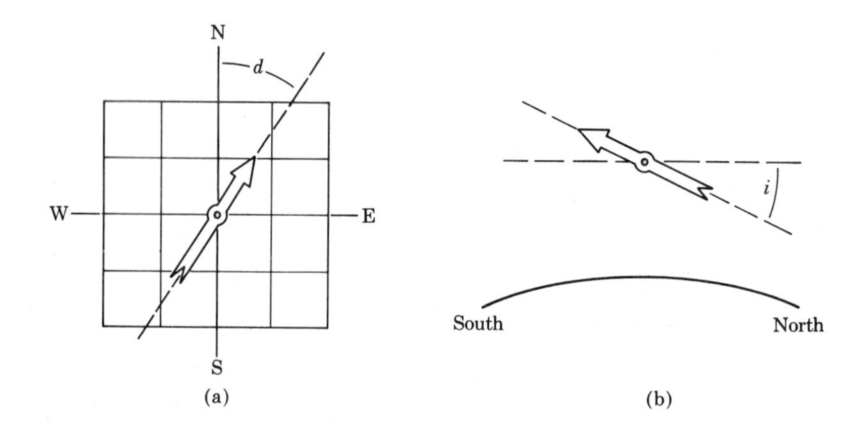

FIGURE 6-18

Declination and inclination of a compass

paths. The northern lights (aurora borealis) are explained in terms of this process, as are the radiation belts several thousand miles above the Earth's surface.

The Deeper Meaning of Fields

The field provides us with a convenient way of describing and adding forces when there are many particles involved. There is, however, a fundamental meaning attached to the field that goes far beyond this mere usefulness. The modern view is that an object such as a charge or mass modifies or "disturbs" the space around it. The field is that disturbance. When another particle is placed at a point, it is said to be immersed in that field, and experiences a force. Similarly the particle creates *its* field, which the original object feels. The field thus serves as a connecting "medium" between bodies. This is in contrast to imagining that there are forces acting directly between them, which is the older "action-at-a-distance" idea, now discarded. As an example, picture a radio station and a receiver many miles apart. Under the action-at-a-distance view, the sudden shift of one electron at the station would cause an *instantaneous* change in the electric forces at the receiving set. It is well verified that a signal with the large but *finite speed of light* $c = 3 \times 10^8$ m/sec spans the distance and creates the effect. These signals, which are electric and magnetic fields varying in direction and time, are the electromagnetic waves. Now look at the situation in gravity forces and fields. Presumably a sudden shift in position of a planet would generate gravitational signals or waves that would spread out through the universe with speed c, to announce the change and to cause field adjustments. Such gravitational waves, if they exist, are very feeble, being smaller than those of electricity by factors such as the 10^{42} that we derived in Section 5-3. Experimental verification of the gravitational-wave theory is not available as yet.

Fields are now believed to be the only entities in nature, and that all matter merely consists of "condensed" quantum fields. To quote physicist Freeman Dyson, "Even to a hardened theoretical physicist it remains perpetually astonishing that our solid world of trees and stones can be built of quantum fields and nothing else. The quantum field seems far too fluid and insubstantial to be the basic stuff of the universe."*

PROBLEMS

6-1 Find the gravitational field at a radius 1075 mi from the Earth's surface, which is 3960 mi from the center, using ratios rather than direct calculation.

6-2 The radii of the orbits of the Earth and Mars about the Sun are 93 million miles and 142 million miles, respectively. Calculate the gravi-

* "Field Theory," *Scientific American*, p. 57 (April, 1953).

tational field of the Sun at the orbit of Mars, and compare it with the field there due to the Earth, using ratios as in Section 6-3.

6-3 In order to extract and use the heat of the interior of the Earth, it is proposed to drill a hole 100 mi deep. The question is asked—What is the gravitational field at the bottom?

6-4 How far from the Earth toward the Sun must a body be so that the Sun's gravitational field just balances the Earth's? (Ignore the field due to the Moon.) The Sun is 9.3×10^7 mi away and its mass is 3.3×10^5 Earth masses.

6-5 A force of 0.01 N is required to hold a charge of $12\mu C$ at a point. What is the electric-field strength at this point?

6-6 The Echo balloon in orbit for communication tests is a sphere of aluminum-coated 0.005-in.-thick plastic, 100 ft in diameter, weight 137 lb. If it were given a charge of 50 μC, what surface-charge density would it have? What would be the strength of the electric field right at its surface, viewed as an infinite plane? Compare with the field from a point charge at the center of the sphere.

6-7 A deflector for positive ions of hydrogen in an accelerator consists of a flat circular disc of large radius uniformly charged by an amount 0.6 $\mu C/m^2$. As an ion passes the plate, how much force does it experience?

6-8 Calculate the electric field at 5 cm from a long wire with charge 60 $\mu C/m$.

6-9 (a) Find the electric field at 1.59×10^{-10} m from the hydrogen atom, with $s = 5.3 \times 10^{-11}$ m, $e = 1.60 \times 10^{-19}$ C, using the approximate dipole relation for a point on the axis formed by the proton and electron. (b)* How much percentage error is there in using the approximate formula?

6-10 Show that if the cartesian components of electric field of a dipole are given as

$$\mathcal{E}_x = \frac{k_E M}{r^3}(3\cos^2\theta - 1), \qquad \mathcal{E}_y = \frac{k_E M}{r^3}(3\cos\theta\sin\theta)$$

then the polar components are

$$\mathcal{E}_r = \frac{k_E M}{r^3}2\cos\theta, \qquad \mathcal{E}_\theta = \frac{k_E M}{r^3}\sin\theta$$

6-11 (a) Find the direction and magnitude of the electric field from a dipole of strength $M = 0.3$ m-C, at position $r = 0.2$ m, $\theta = 30$ deg, using cartesian relations from Problem 6-10. (b) Check results by use of polar coordinates.

6-12 A radio antenna behaves as a dipole with alternating polarity. Sketch the directions and relative magnitudes (let $k_E M/r^3 = 1$) of fields at a certain distance out for $\theta = 0$ deg, 45 deg, 90 deg, 135 deg, and 180 deg, with the electric moment directed to the right. What happens to all vectors if the moment is reversed?

6-13 Find the electric field as a function of distance from the completely ionized carbon atom $({}_6C^{12})^{(+6)}$. What is the magnitude of the field at 4×10^{-10} m?

6-14 What is the electric field and flux at the surface of a sphere of radius 2 cm if a positive charge of 3×10^{-6} C is located at the center? What if the sphere radius is changed to 2 m? What happens if the charge is not located at the center?

6-15 What is the electric-field strength midway between two charges of $+100\ \mu$C and $-30\ \mu$C located 40 cm apart?

6-16 If a drop of water of mass 25 μg were allowed to fall under gravity force with air resistance, it would reach a maximum speed of 400 m/sec. If the drop is then given 20 electronic charges, what electric field must be applied to maintain a uniform speed of fall of 300 m/sec? (Assume that frictional force is proportional to speed.)

6-17 Two discs of radius R are charged equally but of opposite sign charge σ, and fixed with their planes $2a$ apart (see Figure P6-17). The electric-field component along the axis through the centers is given by

$$\mathcal{E}_y = 2\pi k_E \sigma (2 - f - g)$$

where

$$f = \left[1 + \left(\frac{R}{a+y}\right)^2\right]^{-1/2} \quad \text{and} \quad g = \left[1 + \left(\frac{R}{a-y}\right)^2\right]^{-1/2}$$

Find the ratio of the field to that between infinite plates with the same σ, for y/a equal to 0, 0.5, 1.0, in the case $R = a$. Discuss the trends.

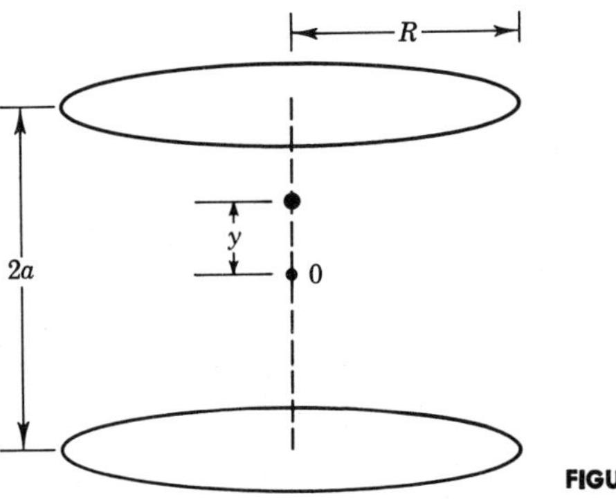

FIGURE P6-17

6-18 Spheres charged with 1.5, 2.0, and 2.5μC of positive electricity are placed on the x-y plane at points (0, 2), $(-\sqrt{3}, -1)$, and $(\sqrt{3}, -1)$, where the units are meters. Find the magnitude (in units of $10^{-6}\,k_E$) and direction of the electric field at the origin.

6-19 You are asked to design a simple model of a *gravimeter*, a "tester" for gravitational field. Its purpose is to detect differences in g that might be due to oil deposits in the Earth. As in Figure P6-19, a mass m is attached to a spring of force constant K, a pointer, and a scale. If the mass selected is 1.5 kg and we wish to detect variations of 0.01 m/sec² by deflections of 1 mm, what must K be? How far does the spring stretch when the mass is first attached, assuming $g = 9.80$ m/sec² at that point?

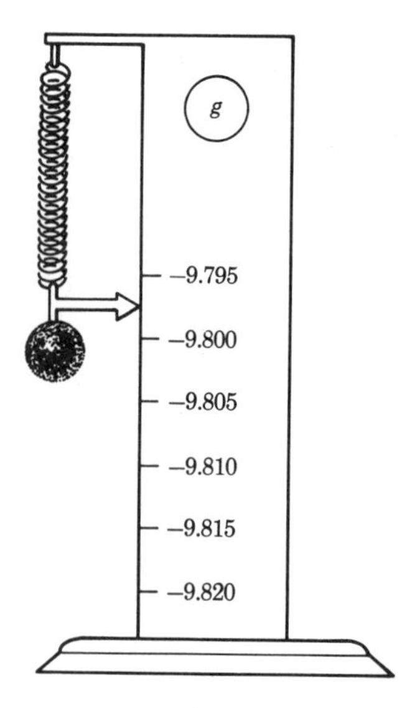

FIGURE P6-19

6-20 A spaceship leaving the Earth travels in a circular orbit that will allow a rendevous with Mars (see Problem 2-20). Calculate the components of gravitational field $\mathcal{G}$ along and perpendicular to the path, when the ship is farthest from the line between Earth and Mars. Why can you neglect the forces due to the planets?

6-21 A metallic sphere of radius a is given a uniform surface charge of σ C/m². (a) Using Gauss's law, find the field at a distance x along a radius from the surface. (b) If the distance x is fixed and the sphere radius a is allowed to approach infinity, keeping the same σ, what is the field intensity? (c) Compare this result with that from an infinite-plane charged conductor and that from an infinite sheet of charge, and discuss.

6-22* The behavior of ocean tides is governed in part by the difference $\Delta\mathcal{G}$ in the gravitational field of the Moon on opposite sides of the Earth and the relative effects on the water. Show that $\Delta\mathcal{G}/\mathcal{G}$ (where $\mathcal{G}$ is the field due to the Moon at the center of the Earth) is $\frac{4x}{(1 - x^2)^2}$, where $x = R_E/R_{ME}$, with $R_{ME} = 239{,}000$ mi (the distance between centers of the Moon and the Earth) and $R_E = 3960$ mi (the Earth's radius). Evaluate $\Delta\mathcal{G}/\mathcal{G}$ to show that the variation is about 1 part in 15.

6-23* At the moment of an eclipse of the Sun, the gravitational field at a point on the Earth's surface nearest the Sun is weaker than if the Moon were on the opposite side. Verify this with vector diagrams. By use of ratios, as in Problem 6-22, compute the fractional difference in $\mathcal{G}$ (and g). Has the Sun a significant role in this phenomenon?

ELECTROMAGNETISM

7

Electric charges that are in motion with respect to each other exert mutual forces that are designated *electromagnetic*. These forces act in addition to the electrostatic forces governed by Coulomb's law. The field concept, we shall see, has great benefit in describing such new interactions. Still more forces come into play whenever changes in electric and magnetic fields with time are produced. The delineation of this complex subject in terms of the motion of electrically charged particles is our objective in this chapter. We shall seek to describe the connection between electricity and magnetism in terms of experiments and mathematical formalism.

7-1 MAGNETIC FORCES AND FIELDS

We have seen in the previous chapter how the concept of a *field* was a convenience in describing gravitational and electrostatic phenomena. The magnetic field is a similarly important concept. The experiment involving charged-particle accelerators (Section 5-4) showed that magnetic force obeys an inverse-square law, and that the product of charge and speed determines the strength of the force. One can imagine the difficulty of finding all the mutual forces in a maze of wires with currents going in several directions, if all the combinations of qv values and distances have to be dealt with. Nevertheless, we should like to be able to determine the force on a certain moving charge at a point, whatever other moving charges causing the force are present. The field concept greatly helps to simplify the problem. The *magnetic field** is defined as the force per unit of charge-speed qv; however, it is important to note that the magnetic force depends on both the *direction* of the field and that of the moving charge.

A few simple experiments will help us appreciate this fact. Let us set up our laboratory at a place on the Earth where a magnetic-compass needle is aligned horizontally along the north–south direction. We have good evidence that there exists a magnetic field, and shall take north as the direction of the field $\mathcal{B}$. We observe the effect of the field on moving positive ions of charge q and speed v, aimed at various angles from the north. When the velocity $\mathbf{v}$ is parallel to $\mathcal{B}$, as in Figure 7-1(a), no deflection is observed, implying that the force is zero. If, instead, $\mathbf{v}$ is directed perpendicularly to $\mathcal{B}$, say toward the east—as in Figure 7-1(b)—the particle is seen to be deflected upward, demonstrating the existence of a force. Now we aim the charge with $\mathbf{v}$ parallel to the ground, but at some

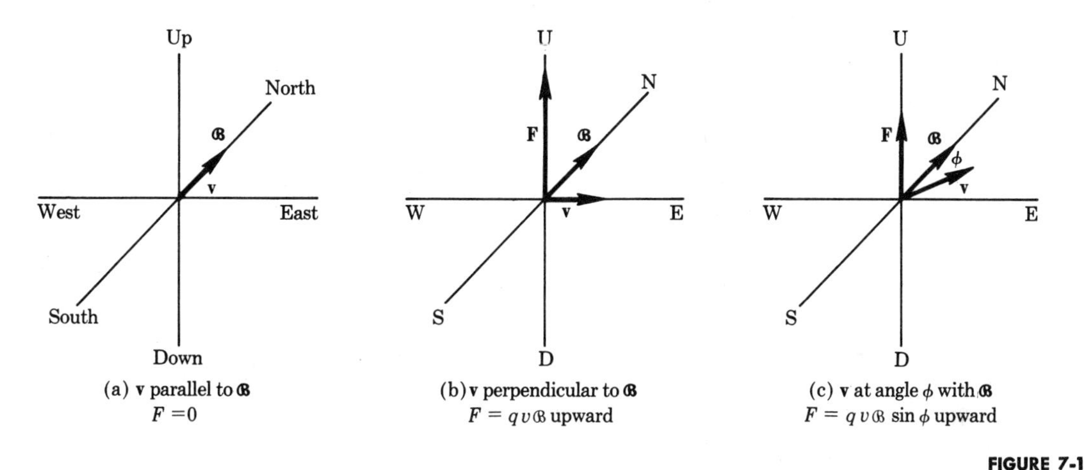

(a) $\mathbf{v}$ parallel to $\mathcal{B}$
$F = 0$

(b) $\mathbf{v}$ perpendicular to $\mathcal{B}$
$F = qv\mathcal{B}$ upward

(c) $\mathbf{v}$ at angle ϕ with $\mathcal{B}$
$F = qv\mathcal{B} \sin\phi$ upward

FIGURE 7-1

Forces on a moving charge in a magnetic field

angle ϕ with the north, as shown in Figure 7-1(c). We find the force to be $\sin\phi$ times as large as before. The conclusions reached from these and similar tests are that the force on a moving charge in a magnetic field is

1. proportional to the charge q, the speed v, and the field $\mathcal{B}$;

2. perpendicular to the velocity and field vectors (a screwdriver points in the direction of $\mathbf{F}$ if we turn it as if $\mathbf{v}$ were rotated to line up with $\mathcal{B}$);

3. proportional to the sine of the angle between $\mathbf{v}$ and $\mathcal{B}$.

* The vector is given the specific name *magnetic induction,* but in common usage the term "magnetic field" is frequently used.

We thus write

$$F = qv\mathcal{B} \sin \phi \qquad \text{(force on a moving charge in a magnetic field)}$$

The shorthand of vectors can be used to express all of our observations. If **A** and **B** are two vectors with angle ϕ between them, the "vector cross-product" $\mathbf{A} \times \mathbf{B}$ is a third vector **C** of magnitude $AB \sin \phi$ and direction perpendicular to the plane formed by **A** and **B**. Thus the compact statement of the force is

$$\mathbf{F} = q\mathbf{v} \times \boldsymbol{\mathcal{B}}$$

The unit of the magnetic field can be deduced from the relation $\mathcal{B} = F/qv$ to be N-sec/C-m, which combination is given the name *weber* per square meter, Wb/m^2, after the nineteenth-century electrical researcher W. E. Weber.

The above experiments demonstrate the important fact that a charge moving along or opposite to the direction of a uniform magnetic field is unaffected by it. Note that $\sin \phi = 0$ for either $\phi = 0$ or $\phi = 180$ deg. We also find that a charge moving in a vacuum perpendicularly to the magnetic field will execute a *circle.* Recall from Section 2-4 that motion at constant speed v on a circle of radius R involves a constant centripetal acceleration toward the center. The magnetic force, always directed perpendicularly to $\mathbf{v}$ so that $\sin \phi = 1$, provides that acceleration. Thus $F_M = qv\mathcal{B} = ma_c = m(v^2/R)$, or the radius of the circle is

$$R = \frac{mv}{q\mathcal{B}} \qquad \text{(radius of motion of a charge in a uniform magnetic field)}$$

The circular motion of electrons, protons, and other ions in the plane perpendicular to the magnetic field is encountered in many natural phenomena and applications. This relation serves as a partial basis for interpreting observations of cosmic particles that reach the Earth. If the field is known, measurement of the radius of curvature provides the ratio of momentum mv to the charge q. It also provides a description of the circulating motion of charges in the cyclotron, in which a uniform field is maintained. Recalling that the angular speed in a circle is $\omega = v/R$, we find

$$\omega = \frac{q\mathcal{B}}{m}$$

which is called the *cyclotron (angular) frequency.* The value of ω depends only on the field intensity and the ratio of charge to mass of the particle. Consider an electron of speed 2×10^7 m/sec in a magnetic field of 3 Wb/m^2. With mass 9.11×10^{-31} kg, the radius of motion is

$$R = \frac{(9.11 \times 10^{-31}\ \text{kg})(2 \times 10^7\ \text{m/sec})}{(1.60 \times 10^{-19}\ \text{C})(3\ \text{Wb/m}^2)} = 3.8 \times 10^{-5}\ \text{m}$$

This is a surprisingly small radius, being only about $\frac{1}{25}$ mm.

7-2 CHARGE FLOW AND CURRENT

Although individual moving charges can be isolated in the laboratory, the operation of most electrical devices is based on the flow of large numbers of electrons and ions. Such a flow is called a *current*. The relation between individual particle numbers, speed, and charge can easily be deduced. Suppose that we project a continuous stream of positively charged particles—say protons, each with charge q and speed v, as illustrated on a highly magnified scale in Figure 7-2(a). Let the number of charges per cubic meter of the beam be labeled n. Since each has a speed v, in the time of one second it will travel v meters—Figure 7-2(b). Let us observe any square meter area perpendicular to the flow. In 1 sec a column of charge will sweep

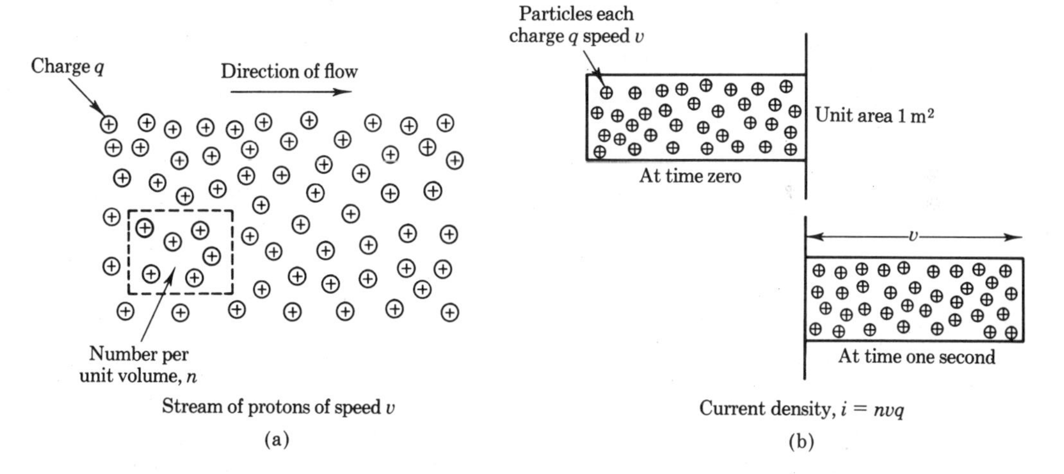

FIGURE 7-2

Relation of particle flow and current

past, having length v m and thus containing nv particles. Each bears a charge q, and the total quantity of electrical charge that passes per second, called the *current density*, is

$$i = nvq \frac{\text{C}}{\text{sec-m}^2}$$

The term *ampere*, abbreviated A, is given to the unit *coulomb per second*, in honor of the French physicist A. M. Ampère. The units of current density are thus amperes per square meter A/m^2. Fractions of the ampere that are frequently used are the milliampere (mA) $= 10^{-3}$ A, and the microampere (μA) $= 10^{-6}$ A.

Let us calculate the current density and current in a stream of cross-sectional area $A = 2 \times 10^{-5}$ m^2, when a proton particle number density $n = 10^{15}$/m^3 flows along with a speed $v = 5 \times 10^6$ m/sec. Each proton has a charge $q = 1.60 \times 10^{-19}$ C. The current density is

$$i = nvq = (10^{15}/\text{m}^3)(5 \times 10^6 \text{ m/sec})(1.60 \times 10^{-19} \text{ C}) = 800 \text{ A/m}^2$$

Then the current as the rate of flow of charge through the area A is

$$I = iA = (800 \text{ A/m}^2)(2 \times 10^{-5} \text{ m}^2) = 0.016 \text{ A} \quad \text{or} \quad 16 \text{ mA}$$

The starting current for an automobile may be as high as 100 amperes. We can estimate the speed of electrons of charge 1.60×10^{-19} C in a copper wire, on the assumption that there is a particle density of $8 \times 10^{28}/\text{m}^3$ (corresponding to 1 free electron per copper atom). In a wire of cross-sectional area 4×10^{-6} m², the speed is

$$v = \frac{I}{nqA} = \frac{100}{(8 \times 10^{28})(1.60 \times 10^{-19})(4 \times 10^{-6})}$$
$$= 2 \times 10^{-3} \text{ m/sec}$$

Note that this is a surprisingly low speed.

Let us find the current that is equivalent to a single charge q moving at speed v in a circle of radius r. Observing at a point on the orbit, we find the time for each revolution to be $t = 2\pi r/v$. The amount of charge passing in that time is q; hence the current is

$$I = \frac{q}{t} = \frac{q}{2\pi r/v} = \frac{qv}{2\pi r}$$

For example, compute the current corresponding to an electron charge $q = 1.60 \times 10^{-19}$ C, of speed 2.2×10^6 m/sec, circulating around a proton at a distance of 5.3×10^{-11} m:

$$I = \frac{(1.60 \times 10^{-19} \text{ C})(2.2 \times 10^6 \text{ m/sec})}{2\pi(5.3 \times 10^{-11} \text{ m})}$$
$$= 1.07 \times 10^{-3} \text{ A} \quad \text{or about} \quad 1 \text{ milliampere.}$$

The current can be related also to the charge per unit length of a stream or wire. Picture a section of stream of length l and cross-sectional area A. Its volume is $V = lA$. Now, if the charge population is n per m³, and each charge is q coulombs, then the total charge is nVq. The charge per unit length, which we shall label Q, is

$$Q = \frac{nVq}{l} = nqA$$

Comparing this with our expression for current, $I = nvqA$, we find

$$I = Qv$$

Thus the current is the product of charge per unit length and charge speed.

We are now able to describe the force on a current-carrying wire placed in a magnetic field. In our special laboratory at a place where the Earth's magnetic field is due north, we run a long wire in an east–west direction and send a current I through it, say by a battery. Each moving charge experiences a force, and since the charges are guided by the wire, any given cross-section of the wire of length Δl feels a force F. Suppose that the wire has cross-sectional area A, and the number of charges per unit volume is n. The total charge in this region is thus $nqA\Delta l$, which we insert where q appears in our force formula $F = qv\mathcal{B} \sin \phi$. Recall also that $I = nAqv$, to obtain

$$F = I\mathcal{B} \sin \phi \, \Delta l \qquad \textit{(magnetic force on a current segment)}$$

If we let the current be a vector **I**, the force per unit length is easily seen to be

$$\mathbf{F}_1 = \mathbf{I} \times \boldsymbol{\mathcal{B}}$$

which states that the force is perpendicular both to the direction of flow and to the field.

For example, let us find the force on a 15-cm length of our wire, assuming a 2-A current and the Earth's magnetic-field value of 5×10^{-5} Wb/m^2. Now $\sin \phi = \sin 90° = 1$, and

$$F = I\mathcal{B}\Delta l = (2 \text{ A})(5 \times 10^{-5} \text{ Wb/m}^2)(0.15 \text{ m})$$
$$= 1.5 \times 10^{-5} \text{ N}$$

The forces on wires through which currents flow make possible the electric motor, meters to measure current and voltage, and many other electrical systems. We shall discuss these applications of the fundamental principles of electromagnetism in Chapter 18.

7-3 FIELDS AND CURRENTS

The force on either individual charges or on currents can be found readily by the foregoing methods, assuming that the direction and magnitude of the field $\boldsymbol{\mathcal{B}}$ is known. We now turn to the question: What is the nature of the field? The discovery of the magnetic field due to a current was made in 1819 by Oersted. After attending a lecture on electricity, out of curiosity he brought a magnetic compass near a wire connected to a battery and was amazed to find that the needle was strongly affected and lined up perpendicularly to the current. Although the discovery was accidental, he was able to recognize its importance. His revelation of the effect immediately stimulated investigations by others, including Ampère who made the first thorough study of the magnetic field. Ampère found that a set of magnetic-compass needles placed equidistant from a long

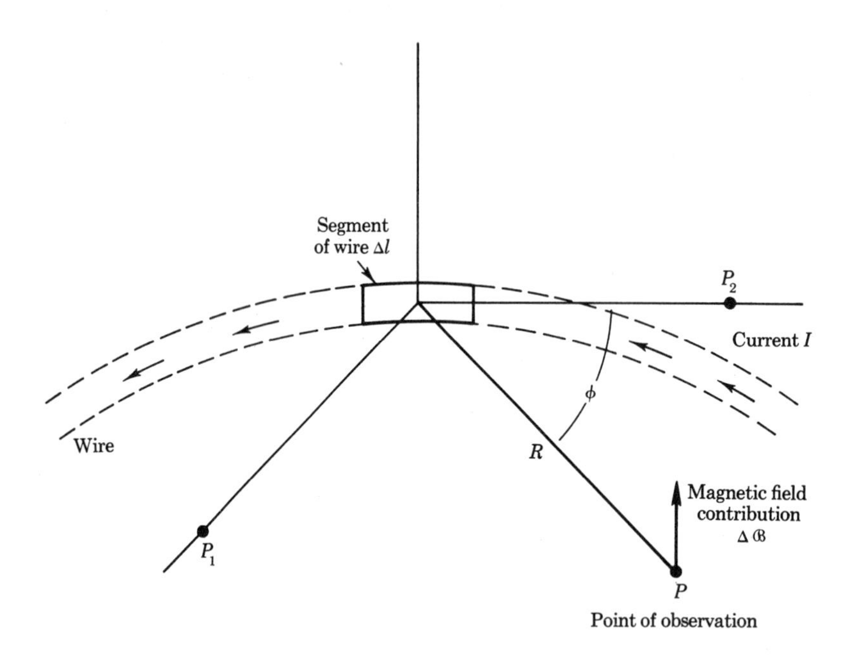

FIGURE 7-3

Magnetic field according to Ampère's law

section of current-carrying wire would line up (neglecting the Earth's field) to form a circle around the wire; also, the larger the current, or the nearer the compass to the wire, the stronger the deflection of the compass needle.

Let us put Ampère's conclusions in mathematical form. Consider a wire of any shape through which a current I flows. Each small segment of the wire contributes to the magnetic field by virtue of the charges moving within it. The amount of this field is proportional to the segment length, and inversely proportional to the square of distance to the wire R. Also if the point of observation P is on a line that makes an angle ϕ with the segment of the wire (see Figure 7-3), a factor $\sin \phi$ appears. The contribution to the total field at P resulting from the current through a small segment of wire of length Δl is

$$\Delta \mathcal{B} = \frac{k_M I \sin \phi \, \Delta l}{R^2} \qquad (\textit{Ampère's law})$$

(Here the symbol Δ implies a small amount, not a change.) Note that if the observation is made at a point such as P_1, perpendicular to the current, then $\phi = \pi/2$, $\sin \phi = 1$, and the factor is not needed; if at P_2, parallel to the current, $\phi = 0$, $\sin \phi = 0$, and the field is zero. The total field due to all segments of the wire can be obtained by adding up the separate contributions, taking proper account of any changes in R or ϕ. We can illustrate this superposition process by finding the field at the center of a *circular loop* of current. This would be encountered in the circulation of electrons in the atom, in a single turn of wire in an electromagnet, or in the beams within a particle accelerator. For this case, all lines from the point of observation are perpendicular to the segments of current, and $\sin \phi = 1$. Then, with a field per unit length of $k_M I/R^2$ and a total length of the circular loop $2\pi R$, the total magnetic field is simply

$$\mathcal{B} = k_M \frac{2\pi I}{R} \qquad \textit{(magnetic field at the center of a circular current loop)}$$

As expected, the larger the current, the stronger the field; and the larger the loop, the weaker the field, since the charges are farther away from the point of observation. We note in passing that the field varies as $1/R$, not $1/R^2$. The next case of interest is the *long straight wire*, through which passes a current I. We would expect the field to be the same at every point that is the same distance y from the wire, to fall off with increasing distance from the charges, and to increase with current. Applying the process of addition or superposition, we view the wire through which the current passes as made up of a large number of small segments of length Δx, as in Figure 7-4(a). The total magnetic field at a point of observation is the sum of the contributions $\Delta\mathcal{B}$ from each element of current. The way in which these

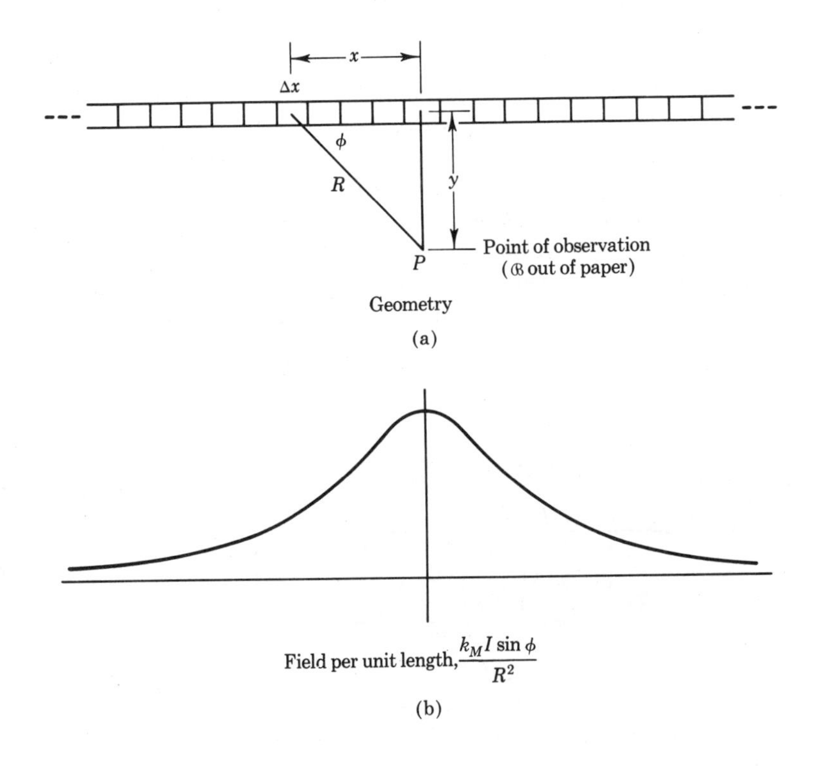

FIGURE 7-4

Magnetic field due to long straight wire through which current flows

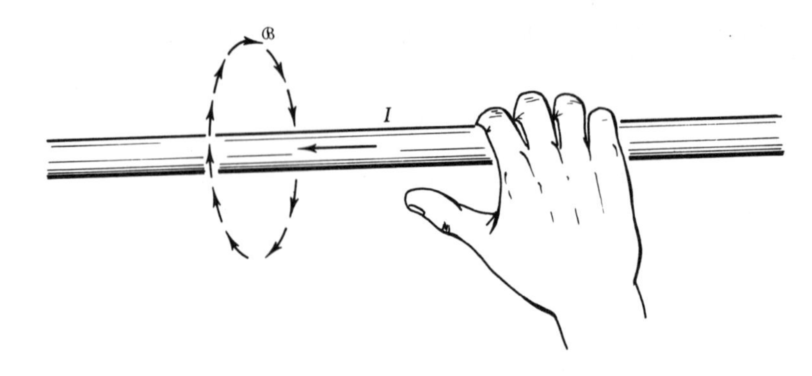

FIGURE 7-5

Right-hand rule for finding field

vary with position along the wire can be seen in Figure 7-4(b), a plot of the field per unit length of wire. The total field is found to be of the form

$$\mathcal{B} = k_M \frac{2I}{y} \qquad \textit{(magnetic field of a line current)}$$

which reminds us of the expression for the electric field of a long charged wire. There is an important difference, however. The magnetic field is *tangential* to a circle around the wire, whereas the electric field was found to be *radial*. An aid to memory in this respect is the "right-hand rule": *the magnetic field is in the direction of the fingers if the right hand is wrapped around the stream, with the thumb pointing in the direction of the positive current.* The field vectors at various points on a circle about the wire are shown in Figure 7-5, giving meaning to Ampère's experiments with magnetic needles. The magnitude of $\mathcal{B}$ is constant, but the direction is chang-

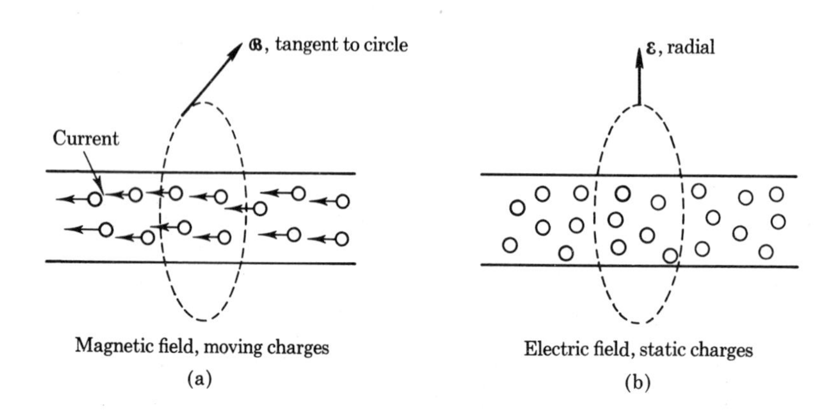

FIGURE 7-6

Comparison of field directions

ing. The distinction between the electrically charged wire and the current in a wire is shown in Figure 7-6.

Let us calculate the magnetic field at a distance 5 cm from a current of 40 mA:

$$\mathcal{B} = k_M \frac{2I}{y} = 10^{-7} \frac{2(0.04)}{0.05} = 1.6 \times 10^{-7}\ \mathrm{Wb/m^2}$$

To obtain a field of the order of 3 $\mathrm{Wb/m^2}$ as used in a previous example, it is seen that either a much higher current than 40 mA is needed or that the magnet must be constructed of many wires.

Circulation Theorem

A powerful and general theorem that is equivalent to Ampère's law can now be demonstrated. Let us draw a circle of radius y and circumference $C = 2\pi y$ about the long straight wire carrying a current I. Then we form a product called the "circulation." Using the $\mathcal{B}$ just derived, the circulation is

$$\mathcal{B}C = k_M \frac{2I}{y} 2\pi y$$

or

$$\mathcal{B}C = 4\pi k_M I \qquad \textit{(alternate form of Ampère's law)}$$

This is a formulation of the *circulation theorem,* which states that the product of field strength and the circumference of a loop drawn around the wire that produced the field is proportional to the current through the loop.

We have chosen a special, simple case to demonstrate the theorem, but it can be shown to be correct for a closed path of any shape and for any arrangement of currents.* The theorem can be used to *derive* fields for certain other cases—for instance, in the *solenoid* electromagnet, which is found in control switches, in coils to lift steel, in generators, and in motors. A solenoid can be pictured as a set of N coils connected in series and piled closely on each other, then stretched out like a coil spring to length l—see Figure 7-7(a). A simple form of a solenoid is the coil in a doughnut shape or *toroid,* as shown in Figure 7-7(b). Suppose there are N turns of wire, each carrying a current I. Let us take as the loop the circle formed by the central axis of the solenoid of length l. The total current enclosed is NI, and the product of the field and the circumference of the loop is $\mathcal{B}l$. Applying the circulation theorem, $\mathcal{B}l = 4\pi k_M NI$; and since the number of turns per unit length is $n = N/l$, we have the field on the axis of the toroid

$$\mathcal{B} = 4\pi k_M nI \qquad \textit{(field in a toroid and in the center of a long solenoid)}$$

* The general statement is: the sum of the products of the lengths of segments and the components of the field along each is proportional to the total current through the closed path.

(a) Section of a solenoid

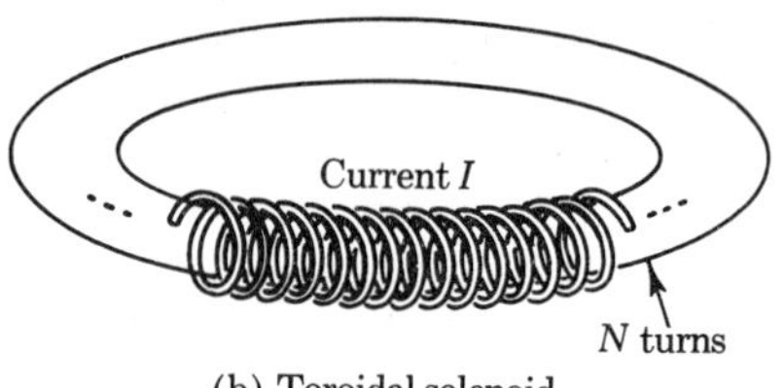

(b) Toroidal solenoid

FIGURE 7-7

Solenoid electromagnet

This relation also holds at the center for a long straight solenoid, which can be visualized as a section of toroid of very great diameter. If the straight solenoid is long in comparison with its width, it is found that the magnetic field at the center is twice the field at the ends. This may be verified by use

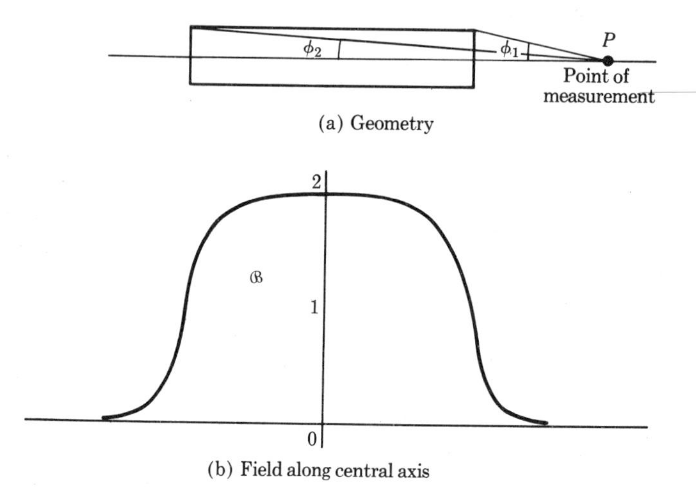

(a) Geometry

(b) Field along central axis

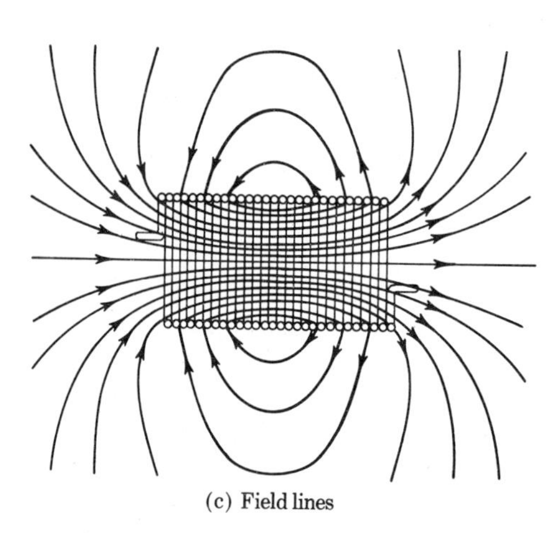
(c) Field lines

FIGURE 7-8

Variation of magnetic field in solenoid

of the general formula for field on the axis of a solenoid,* which is

$$\mathcal{B} = 2\pi k_M n I(\cos\phi_2 - \cos\phi_1)$$

where the angles are as shown in Figure 7-8(a). For a long thin solenoid, note that at the ends, $\phi_1 \cong 90$ deg, $\phi_2 \cong 0$ deg; at the center, $\phi_1 \cong 180$ deg, $\phi_2 \cong 0$ deg. The variation of field along the axis is shown in Figure 7-8(b).

Proof that Magnetic Poles Do Not Exist

The nature of the magnetic field of a long current-bearing wire provides an excellent opportunity to test the conclusion drawn earlier that magnetic poles analogous to electric charge poles do not exist. We again apply Gauss's law by choosing a surface in the form of a shell concentric with the wire, as in Figure 7-9. Recall from Section 6-2 that Gauss's law for the relation of electric flux over a surface and contained charges was of the form $\Phi_E = 4\pi k_E Q$. If distinct magnetic poles existed, we would expect

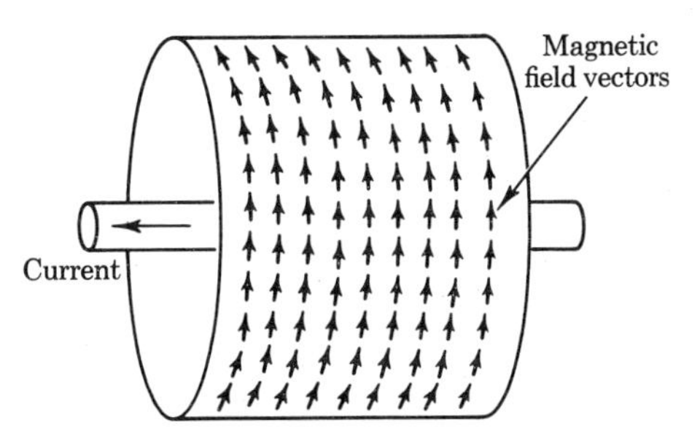

FIGURE 7-9

Gauss's law for magnetism

by analogy to find the magnetic flux over a closed surface containing magnetic material to be governed by a formula such as $\Phi_M = 4\pi k_M P$, where P would be the total contained-pole strength (see Section 5-7). However, if we plaster the surface with the field vectors, all tangential to the curved portion, we see that nowhere is there a component of $\mathcal{B}$ that crosses the surface. The ends of the box have no lines through them either, since the field is perpendicular to those faces. Thus Φ_M is zero, and hence P is zero. The conclusion, stated formally, is that through a closed surface surrounding any arrangement of steady currents, the total magnetic flux is zero:

$$\Phi_M = 0 \qquad \textit{(no magnetic poles)}$$

We arrived at the above deduction for a particular type of current; but it can be shown to be completely general, on the basis that magnetism in solid materials can be traced back to currents in atoms. The conclusion

* Derived by adding up the field contributions of the separate current loops of which it is composed.

that isolated poles do not exist, although deceptively simple, forms one of the cornerstones of electromagnetic theory.

In order to put the idea of poles of a bar magnet in a proper framework, it is necessary to say that the gross shape of the field from a magnet resembles superficially that of an electric dipole, but that the details of the origin of the field preclude the existence of isolated poles.

7-4 ELECTRIC FIELDS INDUCED BY CHANGING MAGNETIC FIELDS

The motion of electric charges has been shown to create steady magnetic fields. A very important phenomenon, discovered at about the same time by Faraday in England and Henry in America, is that a magnetic field, changing in time, can cause electric charges to move.* We can demonstrate this process of *induction* of electric forces with a very simple apparatus—a loop of wire attached to a meter, and a bar magnet—Figure 7-10(a). As we bring the magnet near the loop, the current, as measured by the meter, rises from zero. When we stop, the current falls to zero. We observe a similar response if a loop is moved with respect to a fixed magnet, or if we make the loop area larger or smaller by bending the wire. A

* Joseph Henry of Princeton had been the first to develop a strong electromagnet by insulating the separate wires from each other, and had given a great deal of information to Morse to help perfect the telegraph. Michael Faraday had previously observed that a current-carrying wire experienced a force in a magnetic field. In 1830 and 1831, respectively, they found the new effect independently, but Faraday was the first to publish his results.

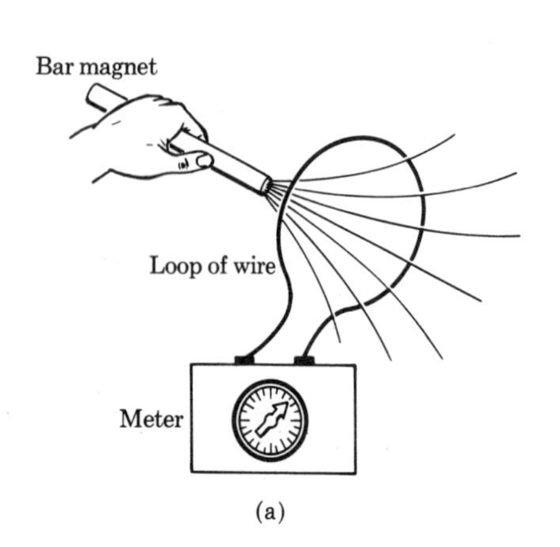

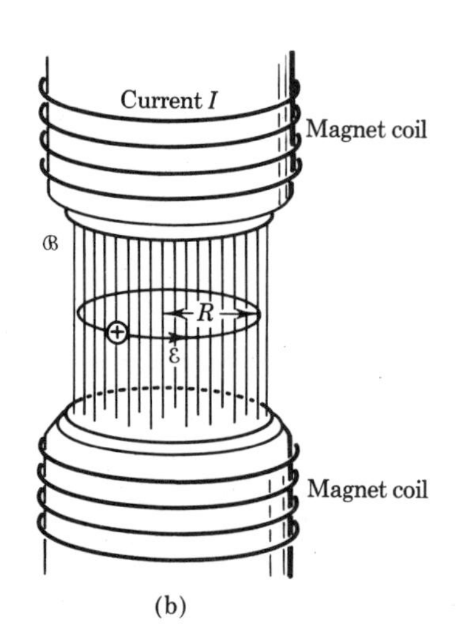

FIGURE 7-10

Electric fields induced by changing magnetic fields

qualitative explanation of this phenomenon is advanced. First, the existence of currents indicates that electric fields are generated in the wire, creating charge flow. Second, the observations of the several ways in which a current can be generated suggests that the effect depends on the strength of the magnetic field, on the area of the loop, and on the rate at which either is changed.

We may now illustrate Faraday's discovery mathematically and numerically. Visualize, as in Figure 7-10(b), an electromagnet that consists of iron wrapped with coils, as in a solenoid. Assume the device to be large enough to provide a uniform field. A current I is passed through the coils, giving rise to a field $\mathcal{B}$ that is proportional to I, as was shown for a loop or solenoid in Section 7-3. An electric charge is introduced in the space between the electromagnet poles, and travels in a circle of radius R (Section 7-1). Now, the current I is increased by a small amount, so that in a time Δt the magnetic field is increased by $\Delta\mathcal{B}$. Experimentally, the charge experiences an *electric field* that is perpendicular to $\boldsymbol{\mathcal{B}}$ and tangential to the circle. The magnitude of the field is

$$\mathcal{E} = \frac{R}{2}\frac{\Delta\mathcal{B}}{\Delta t}$$

that is, proportional to the rate at which $\mathcal{B}$ changes with time. Let us compute this induced field for an example in which $R = 1.5$ m and the rate of change of magnetic field is 1 Wb/m²-sec. Then $\mathcal{E} = (1.5/2)(1) =$ 0.75 Wb/m²-sec. These units are the same as N/C. *Faraday's law* is a general statement of the electric field due to changing magnetic fields. We can derive it from our special case by multiplying both sides of the equation by the circumference of the circle $C = 2\pi R$. Then $\mathcal{E}C = \pi R^2\,\Delta\mathcal{B}/\Delta t$; but the area of the loop is $A = \pi R^2$, and the magnetic flux is $\Phi_M = \mathcal{B}A$. Combining,

$$\mathcal{E}C = -\frac{\Delta\Phi_M}{\Delta t} \qquad (\textit{Faraday's law})$$

This says in words that the *product of the induced field and the circumference of the loop is equal to the rate of change of magnetic flux with time.* The negative sign has a very special significance, related to the direction of the field. The presence of $\mathcal{E}$ causes the charge to gain speed and hence to cause an increase in current. The latter creates a magnetic field, in accord with Section 7-3, directed in just such a way as to oppose the increase in the magnetic field of the coils. This is an example of *Lenz's law*, named after a Russian scientist working in the same period as Faraday and Henry. It states that an induced current produces effects opposing the original action. Thus, if we thrust a magnet through a loop of wire and generate a current thereby, the current will produce a magnetic field that makes it more difficult to move the magnet.

Electric fields are induced when a conductor is moved across a field in such a way as to "cut the magnetic lines." This can be also interpreted as an area effect, as follows. Let us move a conductor of length l through a uniform field $\mathcal{B}$ at a constant speed v (Figure 7-11). The number of lines cut in a time Δt is the product of $\mathcal{B}$ and the area swept through by the wire: $l\Delta x = lv\Delta t$. Thus the change in flux (or the lines cut) is $\Delta\Phi = \mathcal{B}lv\Delta t$. The product of the electric field induced in the wire and the length over which the action takes place is $\mathcal{E}l$, which according to Faraday's law is $\Delta\Phi/\Delta t$. Thus $\mathcal{E}l = \mathcal{B}lv$, or

$$\mathcal{E} = \mathcal{B}v \qquad \textit{(electric field induced by cutting magnetic lines)}$$

The induced electric field is in general perpendicular to both $\mathbf{v}$ and $\boldsymbol{\mathcal{B}}$. Let us calculate the amount of electric field produced in a 1-meter length of

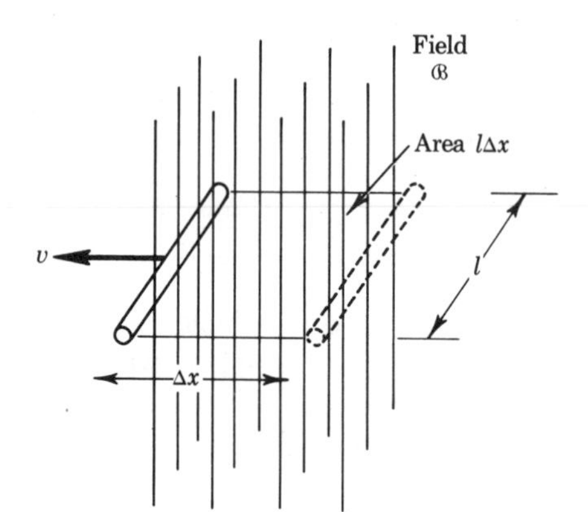

FIGURE 7-11

Electric field induced by cutting magnetic lines

wire when it is moved at a speed of 20 m/sec through the Earth's magnetic field of around 5×10^{-5} Wb/m². Then,

$$\mathcal{E} = \mathcal{B}v = (5 \times 10^{-5})(20) = 10^{-3} \text{ N/C}$$

As we shall discuss later (in Chapters 17 and 18), Faraday's law describes the electric currents produced in a transformer and in a generator such as in the automobile or the larger-scale units used for commercial electric power. In a generator, the wires of a turning armature cut lines of magnetic force to generate electric fields, which in turn drive electric charge.

7-5 MAGNETIC FIELDS INDUCED BY CHANGING ELECTRIC FIELDS

Conventionally, most discoveries in science result from the study of phenomena and the interpretation of measurements, using mathematics. More rarely, a discovery is made by pure reasoning, based on a belief or

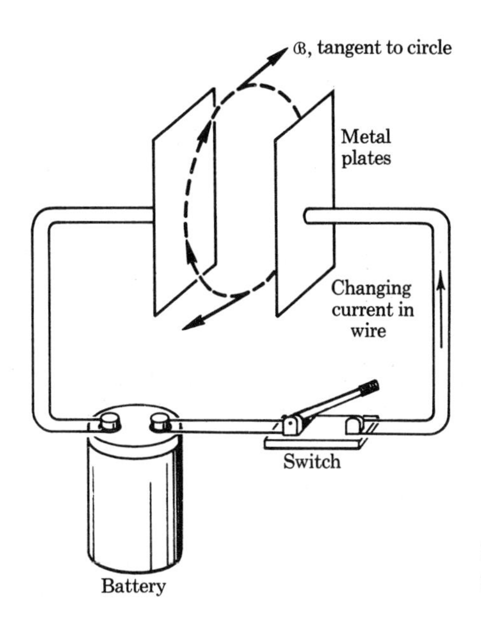

FIGURE 7-12

Experiment to show induced magnetic field and displacement current

hope that nature should behave in a certain manner. The completion of the picture of electricity and magnetism by Maxwell* is considered one of the remarkable examples of the latter process. On the grounds that nature can be described in a symmetric way, Maxwell reasoned that if changing magnetic fields create electric fields, then changing electric fields should create magnetic fields. Let us perform an experiment that verifies Maxwell's assumption. As in Figure 7-12, we close a switch that connects a battery to two parallel plates with a vacuum between them. A current flows in the conducting wires for a short time as the charge on the plates builds up, and an electric field is established. As expected, there will be a readily measured magnetic field near the wire. However, if a sensitive detector is placed *between* the plates, a very weak magnetic field is observed, which is surprising since no electricity is conducted across the vacuum. The field is tangent to a circle of radius R and of circumference $C = 2\pi R$ drawn parallel to the plates. Recall that the circulation formula for magnetic fields due to steady currents (Section 7-3) is $\mathfrak{B}C = 4\pi k_M I$. For the region between plates, we find instead that $\mathfrak{B}C$ is proportional to the rate

* James Clerk Maxwell (1831–1879) was a brilliant Scot who was presenting papers on geometry before he was fifteen years old. He did both theoretical and experimental work in several areas, but is best known for his theory of gases as particles in motion and his elegant theory of electromagnetism and waves. The latter is generally regarded now as one of the major developments of the nineteenth century, although acceptance was delayed because few could understand his work at the time. As an admirer of Faraday's research on electricity, Maxwell used many analogies to other phenomena, for example those in fluids. We shall describe his contribution and how it fits in with the rest of electromagnetism.

of change of the electric flux. The constant of proportionality, according to Maxwell, is (in our notation) the ratio k_M/k_E, so that

$$\mathcal{B}C = \frac{k_M}{k_E}\frac{\Delta\Phi_E}{\Delta t} \qquad \text{(induced magnetic field)}$$

The above states that the *product of the induced magnetic field and the circumference of a closed loop is equal to the time rate of change of the electric flux through the loop* (except for constants of proportionality). The similarity to Faraday's formula, $\mathcal{E}C = -(\Delta\Phi_M/\Delta t)$, is evident. We note, though, that there is no negative sign, and the factor k_M/k_E is present.

Let us deduce and compute the amount of the magnetic field produced in the space between plates. At a point on a circle of radius R and of circumference $C = 2\pi R$, electric flux $\Phi_E = \pi R^2 \mathcal{E}$, we easily find

$$\mathcal{B} = \frac{k_M}{k_E}\frac{R}{2}\frac{\Delta\mathcal{E}}{\Delta t}$$

with units webers per square meter. Suppose we control the supply of charges to the plates in such a way as to cause the electric field to rise linearly in time. Let it go from zero to 3.6×10^3 N/C in 10^{-6} sec. At a radius $R = 0.01$ m,

$$\mathcal{B} = \left(\frac{10^{-7}}{9 \times 10^9}\right)\left(\frac{0.01}{2}\right)\left(\frac{3.6 \times 10^3}{10^{-6}}\right) = 2.0 \times 10^{-10}\ \text{Wb/m}^2$$

We see that this is weaker than the Earth's magnetic field by a factor of about 10^5. If the electric plates in our example happened to be in a recorder using magnetic tape, we could be assured that the induced field would have no appreciable effect on the quality of reproduction.

Maxwell proposed that, in general, the magnetic field was produced by the combination of an ordinary current as a flow of charge and a "displacement current" I_d associated with the changing electric field. We can see that if we insert $I_d = \dfrac{1}{4\pi k_E}\dfrac{\Delta\Phi_E}{\Delta t}$ into Maxwell's formula, it will yield $\mathcal{B}C = 4\pi k_M I_d$. An important deduction to be made is that the *total* current —of charge plus displacement—is continuous through a circuit loop having a gap that charge cannot cross.

Maxwell's Equations

The descriptions given in this chapter of electrostatic and electromagnetic phenomena are known collectively as *Maxwell's equations,* in honor of his important contributions to the subject. For completeness, we shall summarize them at this point.

1. Coulomb's inverse-square law for the electric field from a point charge may be generalized using Gauss's law, for the total electric flux over

a surface due to any set of enclosed charges Q, as

$$\Phi_E = 4\pi k_E Q$$

2. The total magnetic flux through a closed surface, whether or not it contains magnetic material, is

$$\Phi_M = 0$$

3. Electric fields are induced by changing magnetic fluxes, according to Faraday's law, and around a closed path the circulation is

$$\mathcal{E}C = -\frac{\Delta\Phi_M}{\Delta t}$$

4. Magnetic fields are produced by steady charge flow in accord with Ampère's formula as generalized by Maxwell, so that around a closed path the circulation is

$$\mathcal{B}C = 4\pi k_M I + \frac{k_M}{k_E}\frac{\Delta\Phi_E}{\Delta t}$$

That the essence of all of electromagnetic phenomena can be expressed so compactly and elegantly is a tribute to Maxwell's brilliance, aside from the fact that many dedicated experimentalists and theorists had paved the way with prior discoveries. The progress of science depends mutually on the effort of many and on the flashes of inspiration of a few. Its pattern is often erratic, there are many false starts, and authoritative but erroneous concepts remain in vogue for surprisingly long periods. Scientists realize that the historical course of scientific investigation can have alternate paths. They also believe there are many discoveries yet to be made, and that it is only a matter of time before a given discovery will be achieved by someone.

PROBLEMS

7-1 A proton and deuteron of the same speed move in a uniform magnetic field of certain strength. Show that the proton will move on a radius that is one-half that of the deuteron, but that the forces on the two are the same.

7-2 A proton and an alpha particle each move at speed 10^7 m/sec perpendicular to the Earth's magnetic field of strength 0.5×10^{-4} Wb/m^2. Noting that the proton's and alpha particle's charges are 1.6×10^{-19} C and twice that, respectively, and that their masses are 1.67×10^{-27} kg and 4 times that, respectively, compute their radii of motion.

7-3* What magnetic force would the Echo balloon of Problem 6-6 experience in the Earth's magnetic field of 4×10^{-5} Wb/m^2 at 1000 mi in a 118-min orbit? How does this compare with the force of gravity on the balloon?

7-4 Electrons in the great radiation belts that lie outside the Earth by a distance of about 3 Earth radii may have speeds of 10^7 m/sec. If the magnetic field of the Earth at that distance is 10^{-8} Wb/m^2 and is perpendicular to the electron velocity, what will be the radii of the electron orbits about the field lines?

7-5 A positive ion is shot along the positive x-axis into a uniform magnetic field that is directed into this page. What is the direction of the force? Is its path clockwise or counterclockwise as we look along the field direction? Repeat for an electron injected in the same way. Comment on the possibility of causing collisions between positive and negative charges.

7-6 In the *mass spectrograph,* ions of two or more isotopes travel semicircular paths in a magnetic field between a source and a collecting plate (see Figure P7-6). If the field is 0.08 Wb/m^2, compute the radius of an ion of oxygen $({}_8O^{16})^{(-2)}$ with speed 5×10^4 m/sec.

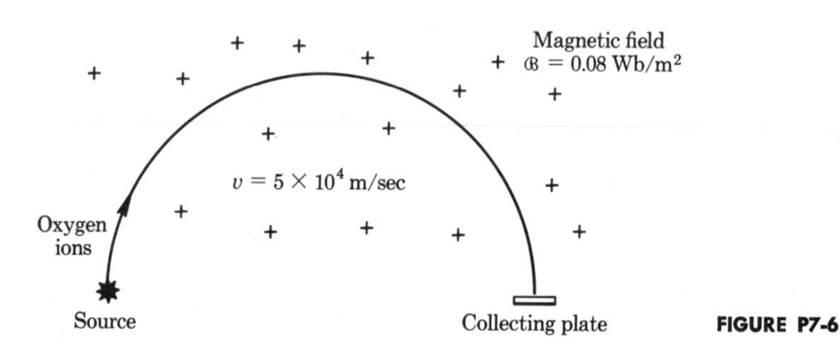

FIGURE P7-6

7-7 Show that the collector plates in a mass spectrograph (see Problem 7-6) for distinguishing protons and deuterons must be separated by a distance $x = 2m_pv/q\mathcal{B}$. Find the separation for charges of speed 10^5 m/sec in a field of 0.02 Wb/m^2.

7-8 An electron is introduced as a test body into a region where the fields are unknown. The observed motion is as plotted in Figure P7-8. Deduce what the type and direction of fields might be.

7-9 What vertical displacement would be experienced by electrons of

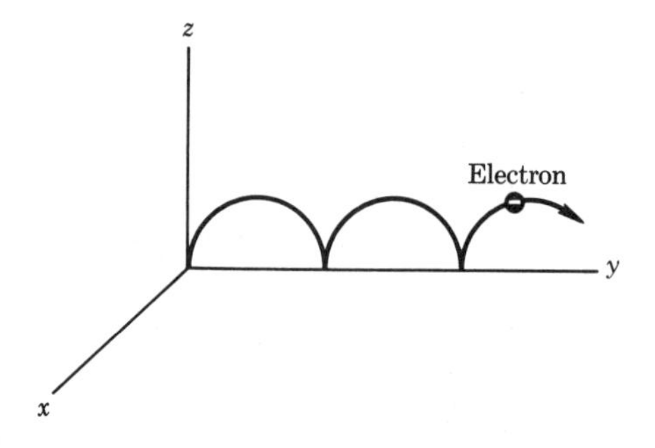

FIGURE P7-8

speed 6×10^7 m/sec as they pass through a television tube of length 0.24 m located at a point in California where the magnetic field strength is 5.2×10^{-5} Wb/m² and the angle of inclination is 60 deg? Is this an important effect?

7-10 An auto battery develops 100 A to the starter for 5 sec in a wire of radius 0.2 cm. Calculate the current density and the total charge that has passed. How many electrons is this?

7-11 Find the force on a wire 2 m long carrying a 3-A current perpendicular to the Earth's magnetic field, where $\mathcal{B} = 6 \times 10^{-5}$ Wb/m².

7-12 A wire carrying a current can be made to float without apparent support between the poles of an electromagnet. (a) What must be the relative directions of the field, the wire, and the Earth's surface? (b) Calculate the current required in a 0.25-m long section of copper wire, gauge No. 0000 (diameter 11.68 mm, density 8.90×10^3 kg/m³) to allow suspension in a field of 0.04 Wb/m².

7-13* Using the expression for the magnetic field opposite a straight section of wire of length L,

$$\mathcal{B} = \frac{2k_M I}{y\sqrt{1 + (2y/L)^2}}$$

(a) find the magnetic field at the center of a square loop of wire with current I, length of sides L. Compare this result with the field due to several circular loops of (b) the same area, (c) the same wire length, (d) the same diameter. Discuss the results.

7-14* Two limiting cases of the magnetic field on the axis *outside* a straight solenoid of length L and radius a, area of loops A, are found to be (a) $\mathcal{B} = \frac{k_M n I A}{X^2}$ for a distance X from the *end* of a very *long* solenoid, with $X \gg a$, (b) $\mathcal{B} = \frac{2k_M n I A L}{x^3}$ for a distance x from the *center* of a very *short* solenoid, with $x \gg a$. Prove these relations. (*Suggestion*: Note that for small ϕ that $\cos \phi \cong 1 - \phi^2/2$.) What is the value of the equivalent pole strength of a long solenoid? What is the magnetic dipole moment (Section 6-3) of a short solenoid?

7-15 After walking on a rug on a dry day, a man touches a metal doorknob and draws a spark. If the current was 5 microamperes for $\frac{1}{4}$ sec, how much charge was drawn and how many electrons flowed?

7-16 The insulated rim of a wheel of radius 0.5 m is given a total charge of 0.80 coulombs, and rotated with angular speed 15 rad/sec. What amount of electric current is produced?

7-17 Find the drift speed of electrons in a copper wire of specific gravity 8.9. Assume on the average 1 free electron per atom of mass about 10^{-25} kg, when the current density is 4×10^5 A/m². Discuss the low magnitude of this result.

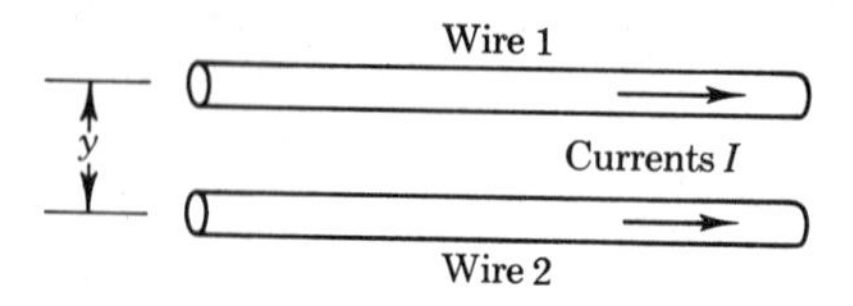

FIGURE P7-19

7-18 A magnetic field is produced by a stream of electrons in a wire. If an additional electron is sent along the length of wire *outside* it, in what direction will the charge tend to move? This is closely related to the "pinch effect" in electrical discharges.

7-19 Each of two long parallel wires—labeled wire 1 and wire 2—carries a current I in the same direction (see Figure P7-19). Show that the resulting force on a unit length of wire 2 is $2k_M I^2/y$. What is the direction of the force?

7-20 Find the proper separation of two electrical-power transmission lines of mass 0.5 kg/m so that the magnetic force (see Problem 7-19) between them due to 1000 amperes in each line does not exceed that of gravity.

7-21 In a classic apparatus devised by J. J. Thomson in 1897, an electron of speed v_0 enters a region with uniform electric and magnetic fields at right angles to each other and to the initial velocity (see Figure P7-21). (a) Show that the ratio of field strengths must be adjusted to $\mathcal{E}/\mathcal{B} = v_0$ to maintain a straight path. (b) Show that if the magnetic field is turned off, the electron moves in a parabola $y = cx^2$. What is the value of c? Show that the ratio of charge to mass (e/m) can be obtained from the knowledge of the two fields and the deflection y of the electron on moving a distance x.

7-22 The Thomson apparatus (Problem 7-21) is used to measure the ratio of charge to mass for an electron. Compute v_0 and e/m using the following data:

$$\mathcal{E} = 3000 \text{ N/C}$$
$$\mathcal{B} = 1.2 \times 10^{-4} \text{ Wb/m}^2$$
$$\text{path length } x = 0.15 \text{ m}$$
$$\text{deflection } y = 0.0096 \text{ m}$$

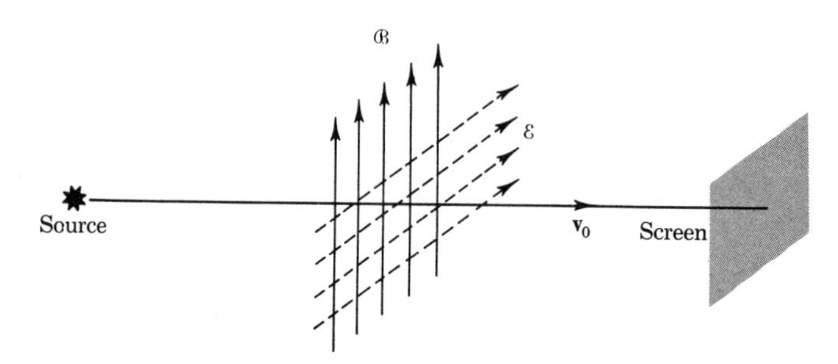

FIGURE P7-21

How great is the error of this result, using the present accurate e/m value of 1.759×10^{11} C/kg?

7-23 We need a long straight solenoid that will give the same value of magnetic field at *all* interior points along its axis. How would its construction differ from that of an ordinary straight solenoid?

MATTER AND ENERGY

8

We are already familiar in a qualitative way with the role of energy and the conversion between its forms in our modern technological civilization. We know that heat is released by the burning of fuels such as oil, coal, gas, and wood. Electrical energy is produced by chemical action, as in the storage battery. In a hydroelectric plant, the force of gravity on water produces mechanical energy that is converted into electrical energy. All processes in nature involve the exchange or transfer of energy in its variety of forms—mechanical, thermal, electrical, gravitational, and nuclear.

The concept of *energy* as a physical quantity will now be developed as it relates to motion or the possibility of motion. Although the concept is simply a consequence of the laws of motion, it serves as a generalization that is of central importance in understanding, analyzing, and predicting the behavior of systems ranging from atoms to galaxies. The fact that energy is transferred but not lost in physical processes places it in the category of *conserved* quantities that provide unity in the subject of physics. The discovery that matter and energy are equivalent, and hence can be transferred, one to the other, provides an even broader generalization of nature.

8-1 THE MEANING OF WORK

We use the terms *work* and *energy* qualitatively in our daily activities. As we exert our muscles to lift an object, we say we are doing "work," of amount depending on the weight of the object and how far we lift it. We go on to say that we have used up our "energy" and must eat to replenish it. While these are correct notions, the science of physics sharpens the ideas by providing mathematical definitions of the quantities, units for their measurements, and laws describing their relation.

Work done on a body is defined as *the product of force and displacement* (in the direction the force acts). Thus, if a constant force F is applied so as to cause an object to move along a *straight line* through a distance x, then the work W done on the object is

$$W = Fx \qquad \textit{(work, constant force)}$$

The unit of work is the newton-meter which product is given the name *joule* after the scientist Joule who contributed to physics an understanding of the relation of heat and mechanics. The unit is abbreviated J. Let us compute the work done as a man of weight 750 N (about 168 pounds) climbs a vertical ladder through a height of 4 meters—(Figure 8-1(a). The work he does to lift himself against gravity is $W = Fx = (750\text{ N})(4\text{ m}) = 3000\text{ J}$. The importance, in our definition of work, of the phrase "in the direction the force acts" can be emphasized by changing the example slightly. Suppose the ladder is placed against the wall at an angle θ with the vertical, as in Figure 8-1(b). The component of the force of gravity that the man acts against is now $mg \cos \theta$. If the distance along the ladder that he moves is x, the work is $W = mgx \cos \theta$.

Some of the other important features of work are listed below:

1. When the force is zero, no work is done, even if the object is in motion and thus experiences displacements. For example, a locomotive that coasts with its cars at a constant speed down a grade that just overcomes friction does no work on the rest of the train.

2. When there is no displacement, no work is done, even though a force is applied. We may push hard against a brick wall, but since it does

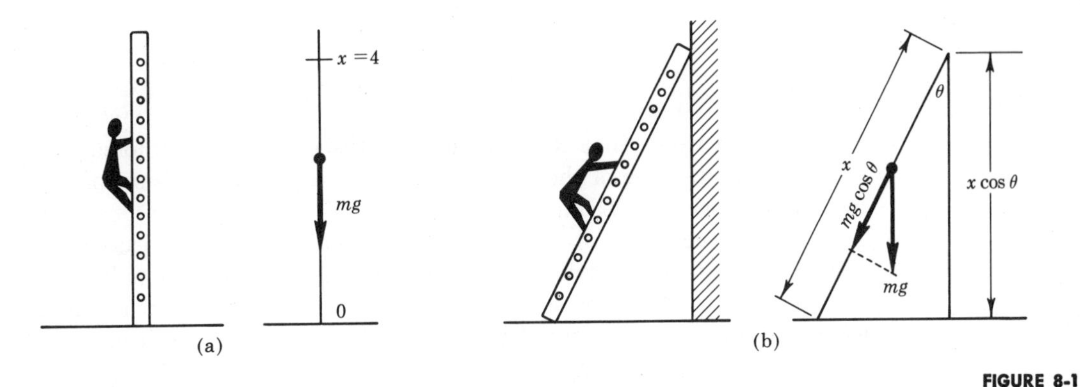

FIGURE 8-1
Work in climbing a ladder

not move, the work done on it is zero. (The fact that we get tired is of course due to work done within our body.)

3. When the force and displacement are oppositely directed, we say that the work done by the agent is negative—that is, the object does work on the agent. If we lower a box from a closet shelf to the floor, we do negative work.

4. When the force and displacement are at right angles, no work is done, since there is no displacement in the direction the force acts, as required by our definition. Examples in this category are the force of gravity on an object moving horizontally, and centripetal forces such as that provided by a magnetic field on a charged particle.

5. When there is more than one force acting, the total work is either the sum of the amounts of work done by each force, $F_1x + F_2x + \ldots$, or the sum of the forces times the displacement, $(F_1 + F_2 + \ldots)x$. This result is obtained by factoring out the common x.

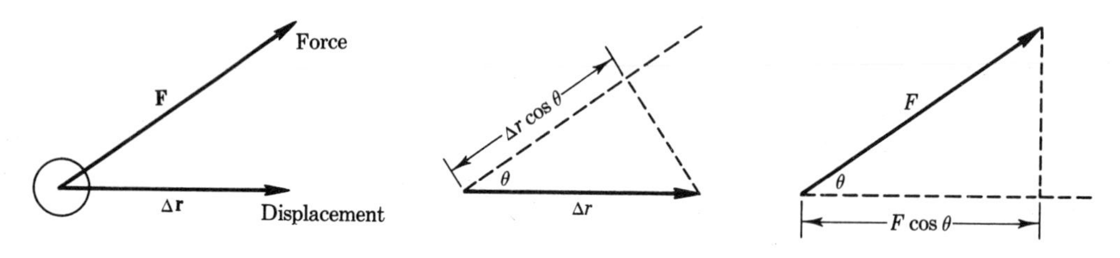

FIGURE 8-2

Work as a product of force and displacement

We may find the amount of work either by multiplying the component of **F** along **Δr** by the displacement Δr to get $\Delta W = (F \cos \theta)\Delta r$, or by multiplying F by the component of **Δr** along **F** to get $\Delta W = F(\Delta r \cos \theta)$—Figure 8-2 shows these alternate views. Since the results are identical, we omit parentheses, and write

$$\Delta W = F \Delta r \cos \theta$$

which says in words that the work done is the product of the force, displacement, and cosine of the angle between the directions. The set of five features of work discussed above conforms to this expression.

The concept of work applies to all types of forces. Let us examine a case involving *friction*, as discussed in Section 5-7. A locomotive pulls with a force F on a train of cars such that there is a constant speed. Since there is no acceleration, the force of friction F_F acting on the cars must be equal and oppositely directed to F, as illustrated in Figure 8-3. Now, $F_F = \mu_k F_N$, where μ_k is the coefficient of kinetic friction, and the normal force is equal to the weight mg. Let us find the work done by the locomotive on

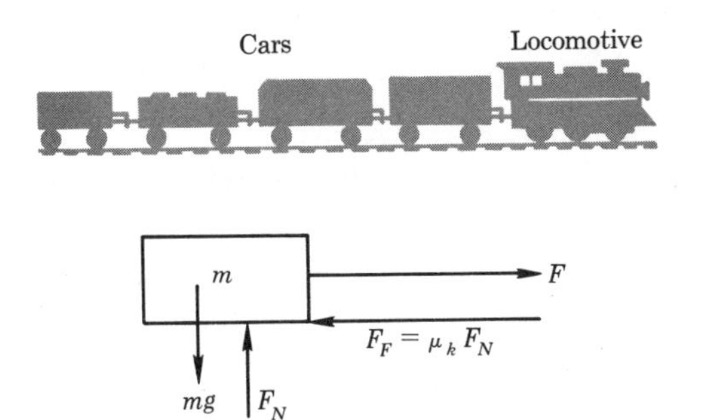

FIGURE 8-3

Work done against friction

the cars of weight 3×10^5 lb in moving them 1 mile down the track, assuming $\mu_k = 0.2$. $F = F_F = 0.2\,(3 \times 10^5) = 6 \times 10^4$ lb. Then, $W = Fx = (6 \times 10^4 \text{ lb})(5280 \text{ ft}) = 3.17 \times 10^8$ ft-lb. Here the British unit of work, the foot-pound (ft-lb) is used instead of the metric newton-meter or joule. Note that conversion factors between metric and British energy units are

$$1 \text{ ft-lb} = 1.36 \text{ J}$$
$$1 \text{ J} = 0.738 \text{ ft-lb}$$

In the previous illustration the forces were constant. Now let us turn to a situation where the force varies with distance along the path. As in Figure 8-4(a), we wish to stretch a spring in such a way that the end moves from 0 to x. The elastic force, as discussed in Section 5-7, varies as $F = Kx$. We can no longer merely multiply force by distance. Instead, we divide the distance into small equal intervals Δx over which F is constant—as in Figure

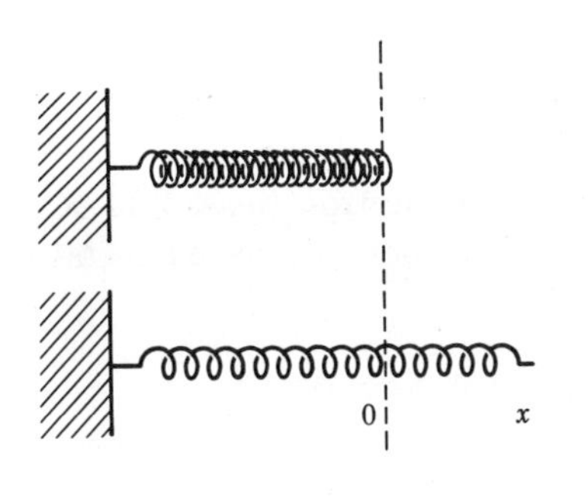

(a) Stretching a spring

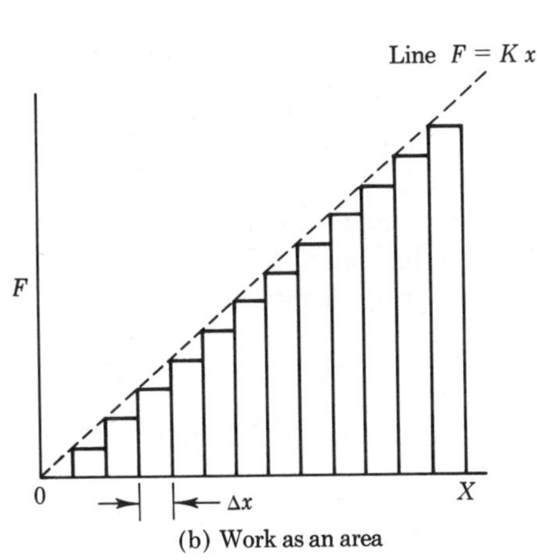

(b) Work as an area

FIGURE 8-4

Work with a variable force

8-4(b)—then compute the amount of work $\Delta W = F\Delta x$ for each, and add the results. This procedure is exactly the same as for the accumulation of speed in Section 2-3. Thus we arrive at the conclusion that *work is equal to the area under the curve of force as it depends on distance.* Applying this to our elastic-force curve, which forms a triangle, the area is $W = \frac{1}{2}x(Kx)$ or, more compactly, $W = \frac{1}{2}Kx^2$.

Work done on a body can set it into motion, or can place it in a position such that future motion is possible. In either case, the object is said to possess the property that we call *energy.*

8-2 KINETIC ENERGY

The work done in setting a body into motion is given the special name *kinetic energy.** The simplest example of motion (other than the unique case of a body at rest) is that in a straight line, with constant force F and acceleration a related by $F = ma$. As in Section 2-3, we know that the speed, starting from rest, is obtained from $v^2 = 2ax$. Let us calculate the work for this situation:

$$W = Fx = max = \frac{mv^2}{2}$$

This new combination of mass and speed is symbolized by E_k, and called the kinetic energy

$$E_k = \tfrac{1}{2}mv^2 \qquad (\textit{kinetic energy})$$

We say that an object *possesses* kinetic energy by reason of its motion. Although we derived the formula for the case of uniform acceleration, it holds generally for low-speed objects. We note that the kinetic energy depends on the *square* of the speed, and that it is a scalar quantity, just as work is. Regardless of the direction of a particle, if its speed is v, then E_k is $\frac{1}{2}mv^2$.

Let us find E_k for a 1000-kg automobile moving at a speed of 60 km/hr (16.7 m/sec). Then, $E_k = \frac{1}{2}(10^3 \text{ kg})(16.7 \text{ m/sec})^2 = 1.39 \times 10^5$ J. The units are verified by forming kg-m²/sec² = N-m = J.

Now let us calculate the kinetic energy of an electron (mass 9.11×10^{-31} kg) that has a speed of 10^5 m/sec, a likely value for electrons in a metal at ordinary temperatures:

$$\begin{aligned} E_k &= \tfrac{1}{2}mv^2 = \tfrac{1}{2}(9.11 \times 10^{-31} \text{ kg})(10^5 \text{ m/sec})^2 \\ &= 4.55 \times 10^{-21} \text{ J} \end{aligned}$$

The formula we have derived turns out to be correct, even if the force applied varies with position and hence with time. To show this, we picture the work done in small enough steps so that the force is constant over

* From the Greek words *kinetikos* ("motion") and *ergon* ("work").

each. Then, for a displacement Δx, the work is $\Delta W = F\Delta x$. Thus we may write Newton's law as $F = m(\Delta v/\Delta t)$, so that $\Delta W = m(\Delta v/\Delta t)\Delta x = m\Delta v(\Delta x/\Delta t) = mv\Delta v$. However, mv is the momentum p and, identifying this work with the change in kinetic energy,

$$\Delta E_k = p\Delta v$$

This type of expression is similar to the one for our previous treatment of work. By comparison, we see that the total kinetic energy gained over many steps is the *area under the curve of momentum as it depends on speed.* Noting that $p = mv$ is a linear expression in v, we see that the triangular area under the line is $\frac{1}{2}(v)(mv) = \frac{1}{2}mv^2$. Thus we again find $E_k = \frac{1}{2}mv^2$, a result which is independent of the character of the applied force.

8-3 FORMS OF ENERGY

Any energy that is *stored* and which can be released under proper conditions to yield energy of motion is classified as *potential energy.* Many objects have this possibility—that is, possess potential energy because of position or condition. Examples are the following:

A metal spring that can unwind and drive the mechanism of an alarm clock.

A storage battery that can turn over the motor in a car.

Water in a reservoir that can fall and turn an electric generator beneath a dam.

Two chemicals, such as H_2 and O_2, that can cause a vigorous reaction when mixed.

The uranium fuel that can undergo fission and propel a nuclear submarine.

The collection of hydrogen or uranium in a nuclear weapon that can explode with great destructive force.

Potential energy, symbolized by E_p, is defined as *the measure of capacity of a system to do work.* Because some of the situations above are too complex to treat at this point, we shall start with some simple examples. Let us calculate E_p for a mass m that has been lifted against gravity to some height. For the time being, we shall assume that there is no friction, as if the object were in a perfect vacuum. This example, although idealized, will give us a start toward establishing principles applicable to real problems where falling is possible—e.g., a situation involving water power. The force we apply in this uniform field is $F = m\mathcal{g}$ and the work we do to raise the mass to any height x from the ground (Figure 8-5) is Fx. By the definition of potential energy,

$$E_p = Fx$$

If we carry it on further to height h, the total work done (and potential energy "stored" in the object) is Fh. We know from experience that if we

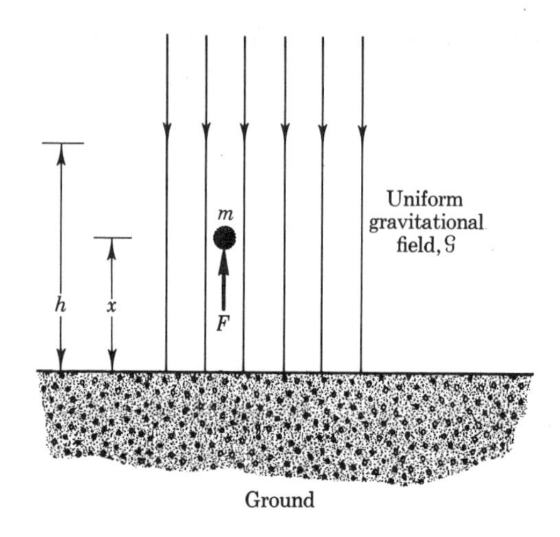

FIGURE 8-5

Work and potential energy of a mass in a gravitational field

release the object it will fall and, since it picks up speed, will gain kinetic energy E_k. Next, let us deduce the relation between E_p and E_k. On falling through a distance $h - x$, its speed is $v = \sqrt{2a(h - x)}$, where the acceleration is F/m. Then,

$$E_k = \tfrac{1}{2}mv^2 = F(h - x)$$

However, since $E_p = Fx$, we find that $E_k + E_p = Fh$. We can make a chart, Table 8-1, that shows that the sum of kinetic and potential energies

TABLE 8-1

DEMONSTRATION OF CONSERVATION OF MECHANICAL ENERGY

Position	Kinetic Energy E_k	Potential Energy E_p	Mechanical Energy E
Top, h	0	Fh	Fh
Any point, x	$F(h - x)$	Fx	Fh
Bottom, 0	Fh	0	Fh

is always the constant Fh, which we call the *mechanical energy* E. We say that E is *conserved*, for E_p may be transformed into E_k or vice versa. A very general and powerful principle has been demonstrated, that of *conservation of mechanical energy:*

> THE MECHANICAL ENERGY OF A SYSTEM—THE SUM OF THE POTENTIAL AND KINETIC ENERGIES—IS CONSTANT IF THERE IS NO FRICTION.

Let us now examine the nature of electrical potential energy, as in the storage battery or any other electrical device. A simple illustration is the process of carrying a positive charge from one plate to another, as in Figure 8-6. Recall from Section 6-2 that the electric field between parallel plates is uniform, as in the gravity problem. The force we apply in the uniform electric field is $F = q\mathcal{E}$, and the work we do to raise the charge to any height x is Fx. Again, the potential energy is

$$E_p = Fx$$

and we could proceed as before to develop the conservation principle. We thus see that the principle is of broader meaning than is implied simply by the word "mechanical."

Since energy is conserved when a particle moves without friction in a gravitational or electric field, we say that the field is *conservative*. In general,

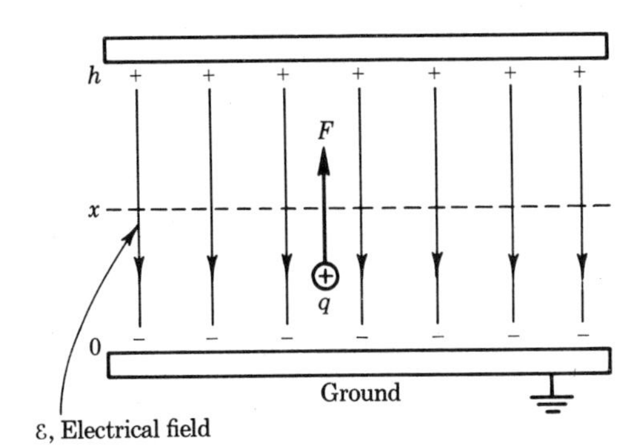

FIGURE 8-6

Work and potential energy of a charge in an electric field

electromagnetic fields are not conservative, however. Also, we shall show later that friction, such as that due to air through which the particle moves, causes the conservation principle to be violated.

The principle of energy conservation permits us to obtain certain information about the motion of objects that would require much more effort to secure by use of Newton's laws. A good example is the flight of a projectile through a curved path above the surface of the Earth. The object might be a baseball thrown from the outfield toward home plate, or a shell fired from a gun, or an intercontinental ballistic missile. The body is given an initial velocity v_0 at some angle with the surface (Figure 8-7). We should like to know how high above the ground it will go, neglecting friction. Newton's laws could be applied to find the actual path, but it is much easier to use the conservation principle. Its initial kinetic energy is $E_k = \frac{1}{2}mv_0^2$, while its potential energy is $E_p = 0$. The total (mechanical) energy is thus $E = E_k + E_p = \frac{1}{2}mv_0^2$. When it reaches the top of its path, $y = H$, its vertical component of velocity is reduced to zero, but its original horizontal component is unchanged, there being no forces along the x-axis. Then, $E_k = \frac{1}{2}mv_{0x}^2$; and the potential energy at that point is $E_p = mgH$.

Since the total energy E is constant, we merely write

$$E = \tfrac{1}{2}mv_0^2 = \tfrac{1}{2}mv_{0x}^2 + mgH$$

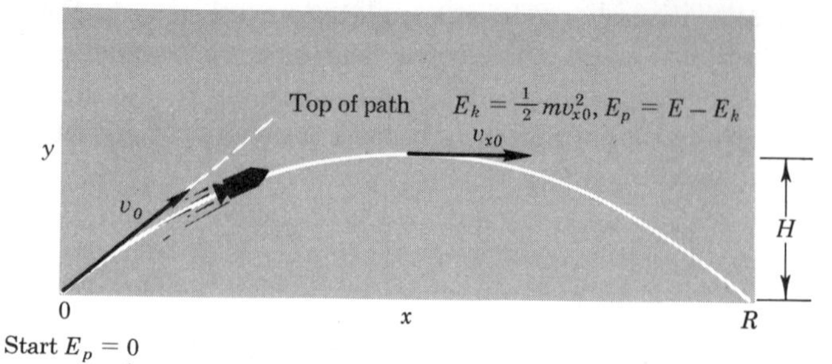

FIGURE 8-7

Application of conservation of energy to the flight of a projectile

However, $v_0^2 = v_{0x}^2 + v_{0y}^2$ by the Pythagorean theorem, and we solve for the maximum height of trajectory:

$$H = \frac{v_{0y}^2}{2g}$$

Certain difficult motion problems can be solved almost by inspection using the principle of energy conservation. For example, let us find the speed at any point of an object sliding due to gravity down a frictionless curve of any shape, as shown in Figure 8-8. The gravitational and normal forces are continuously varying with time. The problem is two-dimensional, requiring the component forms of Newton's second law. Application of the conservation principle, however, gives us the potential energy at the top, $E_p = mgh$, and the kinetic energy there, $E_k = 0$, so that the total is $E = mgh$. At the bottom, $E_p = 0$, $E_k = \frac{1}{2}mv^2$, and thus $v = \sqrt{2gh}$. We may find the speed at any other point in a similar manner—without needing to know anything about the curve.

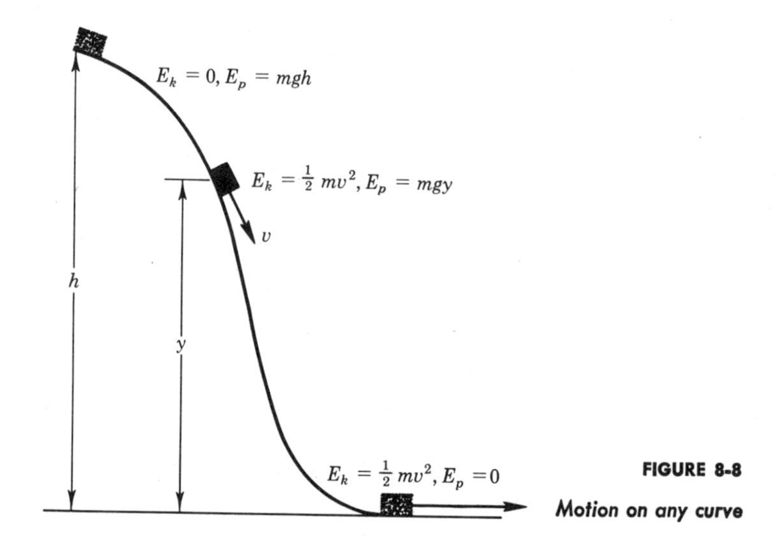

FIGURE 8-8

Motion on any curve

Labor-saving Devices

The law of conservation of energy provides a useful approach to the understanding of *machines*, defined as devices that extend man's limited muscular capacity. Basic examples are the lever, the pulley, and the incline. In each of these devices, a heavy weight or force is moved a small distance by application of a small force over a long distance. The work done on the object is never greater than the energy supplied, but the use of such devices offers the advantage of convenience and ease.

The lever. A large force F can be produced by the *lever*,* which consists of two unequal arms pivoted at a point called a *fulcrum* (Figure 8-9). A small force f is applied over a large distance X, allowing a large force F to act over a small distance x. The amounts of work done are

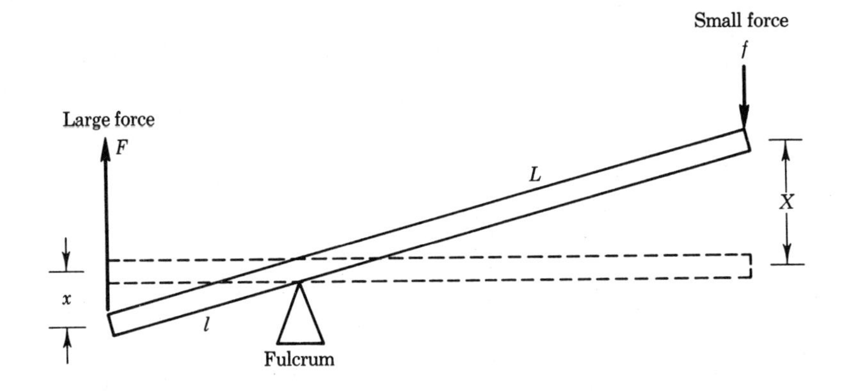

FIGURE 8-9

The lever as a simple machine

exactly the same—if there is no friction—and $fX = Fx$. However, since $X/L = x/l$ by similar triangles, we can find the amount by which force is magnified,

$$\frac{F}{f} = \frac{L}{l}$$

The ratio of output and input forces is called the *mechanical advantage* (MA). Let us find the mechanical advantage of the hammer when used for pulling nails. The fact that the handle is not on a line with the fulcrum and nail (Figure 8-10) is immaterial, and the equivalent lever is shown. Let l = 0.5 in. and L = 8 in. Then MA = 8/0.5 = 16, and a force of 10 lb can extract a nail with frictional force 160 lb.

* According to legend, the ancient Greek mathematician and inventor Archimedes said, referring to the lever, "Give me a place to stand and I will move the world."

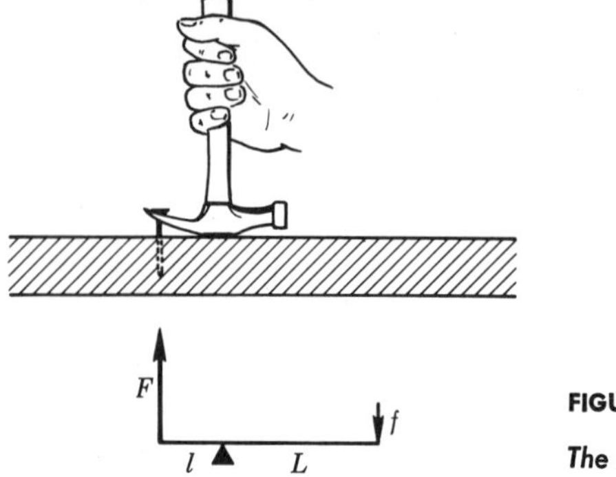

FIGURE 8-10

The hammer

The pulley. The single wheel with rope, as shown in Figure 8-11, has a mechanical advantage of no more than unity, since the distances through which F and f act are the same. The virtue of this simple *pulley* is merely one of convenience in lifting or pulling. However, in the more

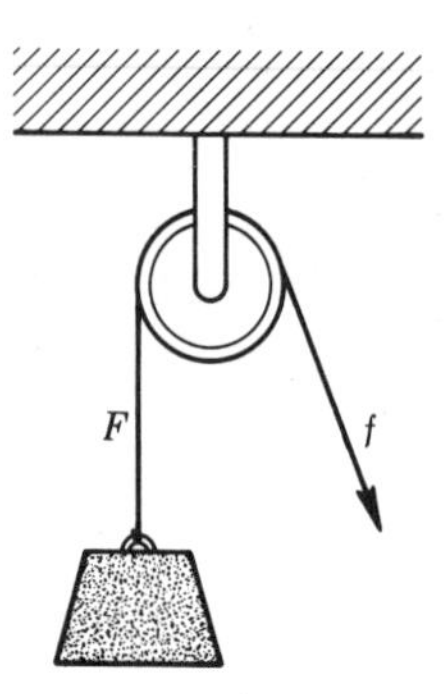

FIGURE 8-11

Simple pulley

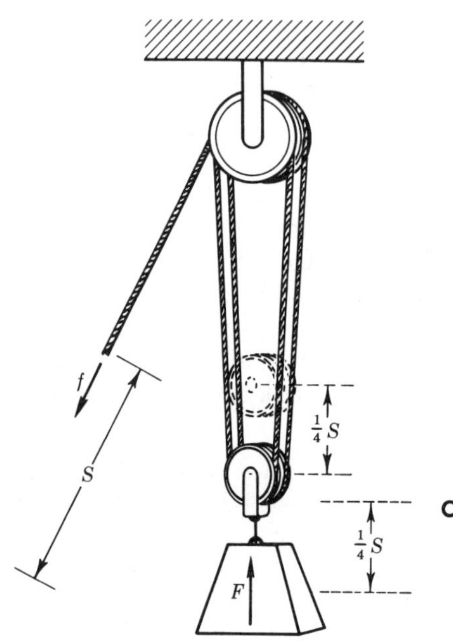

FIGURE 8-12

Complex pulley

complicated pulley of Figure 8-12, if we pull the end of the rope through a distance S (and if there is no stretching), each of the four supporting lengths of rope shortens by $s = \frac{1}{4}S$, and the weight is lifted the same distance. The applied force is f, and the force produced is F. Neglecting friction, the amounts of work done are the same:

$$fS = Fs$$

and

$$\text{MA} = \frac{F}{f} = \frac{S}{s} = 4$$

The incline. A weight F may be raised from one level to another, as from the ground to a loading platform, by sliding it along an *inclined plane* as shown in Figure 8-13. The work input is fS, equal to the output Fs if friction is ignored. However, from the triangle in Figure 8-13 we see that $s/S = \sin\theta$ and thus

$$\text{MA} = \frac{F}{f} = \frac{S}{s} = \frac{1}{\sin\theta}$$

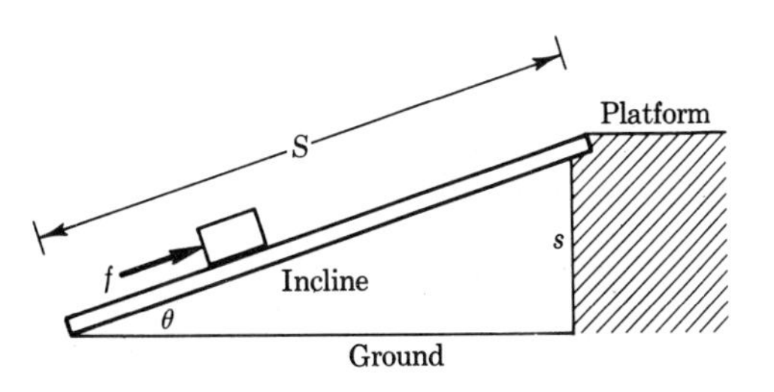

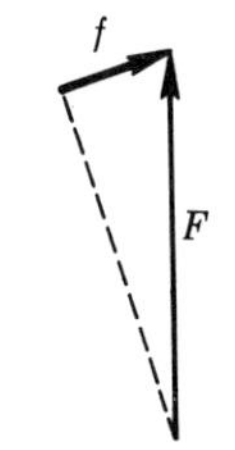

FIGURE 8-13
Inclined plane

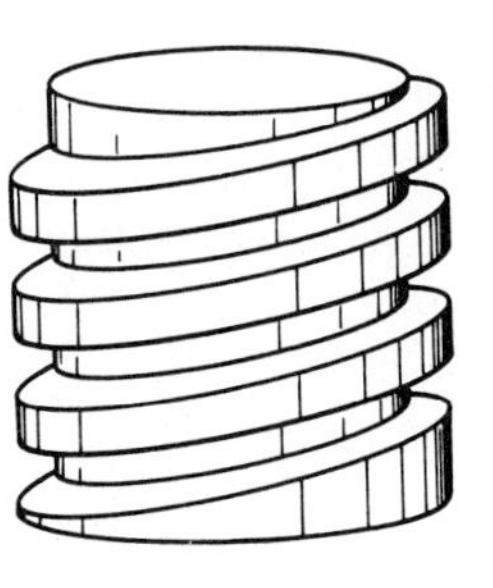

FIGURE 8-14
Inclined ramp

For a 30-deg angle, $\sin\theta = 0.5$, and the mechanical advantage is 2. This "machine" originated very early in history, having been used to build the pyramids of Egypt, around 2500 B.C. The inclined ramp sometimes used instead of stairs in buildings merely consists of an inclined plane wrapped around a cylinder (Figure 8-14). The screw and the worm gear are relatives of this machine.

Properties of Potential Energy

A more general mathematical statement of the conservation law can now be given. Suppose that we permit an object to move from one point to another. Kinetic and potential energy values can be assigned: if their sum is constant,

$$E_{k1} + E_{p1} = E_{k2} + E_{p2} = \text{a constant}$$

By transposition,

$$E_{k2} - E_{k1} = -(E_{p2} - E_{p1})$$

In other words, the change in kinetic energy is the negative of the change in potential energy, or

$$\Delta E_k = -\Delta E_p$$

This suggests that it is the *changes* in the energies that are important rather than the *quantities* themselves. The relation of potential energy to the coordinate system in which it is given requires special discussion. First, we recall that a system of coordinates is an artificial scale that we impose on a space, and has nothing to do with any motion that occurs in that space. For instance, we could set up a scale in a building with the basement level labeled "0" ("zero"), the first floor "1," etc. Nothing prevents us, however, from instead marking the basement "-1," the first-floor level "0," the second floor "$+1$," etc. An object falling down the stairway in the building would behave the same no matter where our zero of coördinates was located. It is only a matter of convenience where we place the origin, but once we do set it, our mathematical formulas to describe the motion must stay consistent with that choice. Since changes in potential energy are more meaningful than potential energy itself, we are free to say where E_p is zero, and calculate its changes as we move to other places. The criterion of convenience implies that we seek simple expressions. In the case of a uniform field, such as that of gravity near the Earth's surface, it is logical to say that the potential energy is zero at the ground, $x = 0$. The gravitational field is $\mathcal{G}$, we apply a force $m\mathcal{G}$ to lift a mass m to a height x, and do work $m\mathcal{G}x$. Thus the potential-energy function is

$$E_p = m\mathcal{G}x \qquad \textit{(potential energy in a uniform gravitational field)}$$

The analogous electrical case involves two charged plates. As in Figure 8-6 we carry a positive charge q from the lower plate (assumed negative) toward the upper plate. It is handy to let the potential energy be zero at the lower plate, $x = 0$—at "ground" level. The electric field is $\mathcal{E}$, we apply a force $q\mathcal{E}$, do work $q\mathcal{E}x$, and the potential-energy function is

$$E_p = q\mathcal{E}x \qquad \textit{(potential energy in a uniform electrical field)}$$

Potential Energy with a Variable Force

The discussion of potential energy so far has been limited to problems involving constant force, and we now turn to the more general case of forces that vary with position. For example, when a rocket is sent out into space it experiences the force of gravity varying as $1/R^2$, where R is measured from the center of the Earth. Similarly, the field at distance R from the center of an electrically charged metal ball is of the inverse-square type. Such forces are called *central* forces, being directed toward or away from a center. Figure 8-15 shows some of these situations.

Our first consideration will be to decide where to locate the "ground" or point of zero potential energy. We know, from our previous discussion about floors in a building, that the choice is arbitrary. The aspect of con-

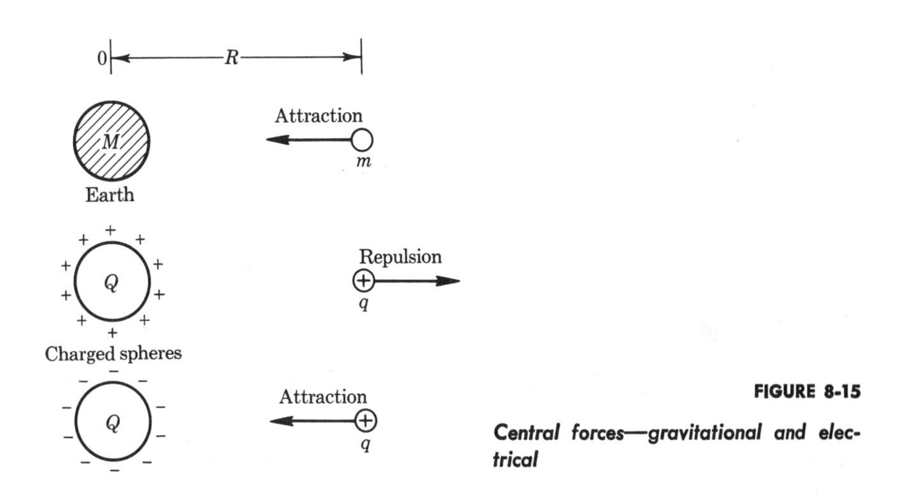

FIGURE 8-15

Central forces—gravitational and electrical

venience dictates our choice. Using the surface of the body as the point of zero potential energy, as before, requires the formula for potential energy to depend on the size of the body generating the force: every planet or charged sphere would have its own special potential-energy expression. We then ask, is there any point in space that is equally distant from all centers of force, such that the potential energy would have the same form for all objects? A little thought shows that the answer is *at infinity*. As we start moving inward from infinity toward any object, the force is zero, and the size, shape, mass, or charge will not matter until we get closer to the source of field. Let us develop a formula for E_p by calculating the work done to bring a positive charge q from infinity up to a distance R from another positive charge Q.

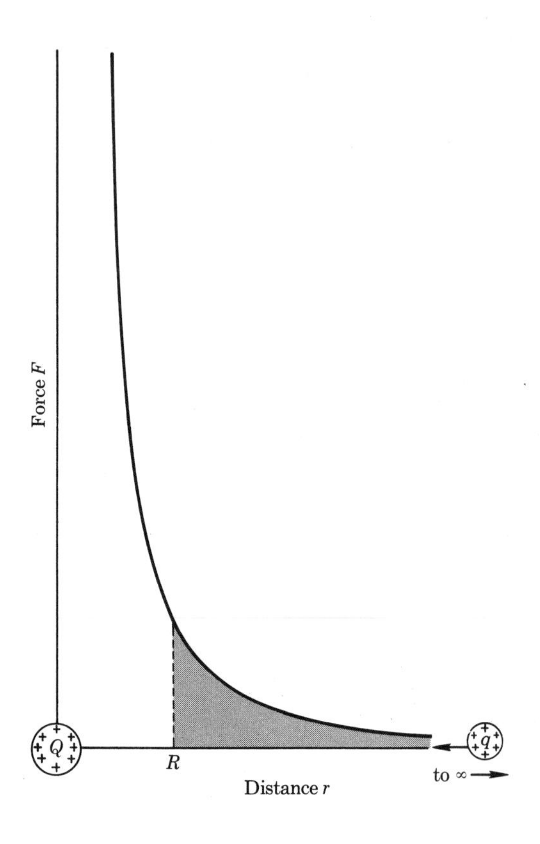

FIGURE 8-16

Potential energy as the area under a force curve

The force against which q moves is $F = k_E Qq/r^2$, where r is variable. Applying our principle that the work is the area under the force curve, shaded in Figure 8-16, we find* that

$$E_p(R) = \frac{k_E Qq}{R} \qquad \textit{(electric potential energy, positive charge distribution)}$$

Note that the potential energy varies inversely as R, not as R^2. A plot is shown in Figure 8-17(a).

A positive charge released anywhere will move outward from the region of high potential energy to the region of low. On the other hand, if we bombard a positively charged nucleus from the outside with a proton or alpha particle, there is a potential-energy *barrier*, just as found on climbing a hill against gravity.

* The calculus shows us that the desired integral is

$$\int_R^\infty \frac{1}{r^2}\,dr = -\frac{1}{r}\bigg|_R^\infty = \frac{1}{R}$$

Now suppose the charge Q is *negative*. The situation would be reversed, with attraction for the positive charge q. Then, we must write

$$E_p(R) = -\frac{k_E Qq}{R} \qquad \textit{(electric potential energy, negative charge distribution)}$$

if Q is the magnitude of the source charge. For convenience, we can use the previous formula and merely insert $+Q$ or $-Q$ as needed. This potential-energy *well*, corresponding to a negative charge, is plotted in Figure 8-17(b). What physical meaning can be attached to a negative potential energy? The charge located at R seeks to move inward, and we must supply it with energy by doing work to get it out to infinity. Think of a positive nucleus, with electrons in their neighborhood. This is also an attractive force situation. In a classical model, the electrical force provides the necessary centripetal force to keep the electrons on circular paths, and the atom is stable. The electrons are trapped in a potential-energy well of the same shape as that sketched in Figure 8-17(b). Since the potential energy is negative, we must inject energy to send the electrons far away—that is, to ionize the atom.

Finally, we come back to gravity. Now the force is always attractive, which means that the potential energy of a mass m influenced by a mass M is mathematically exactly the same as that of the negative Q and positive q. We can immediately write down

$$E_p(R) = -\frac{k_G Mm}{R} \qquad \textit{(gravitational potential energy)}$$

It is interesting to compare the potential energy formulas for the inverse-square field and for the uniform field, especially since the points of zero potential energy were chosen infinitely far apart. Using the gravity

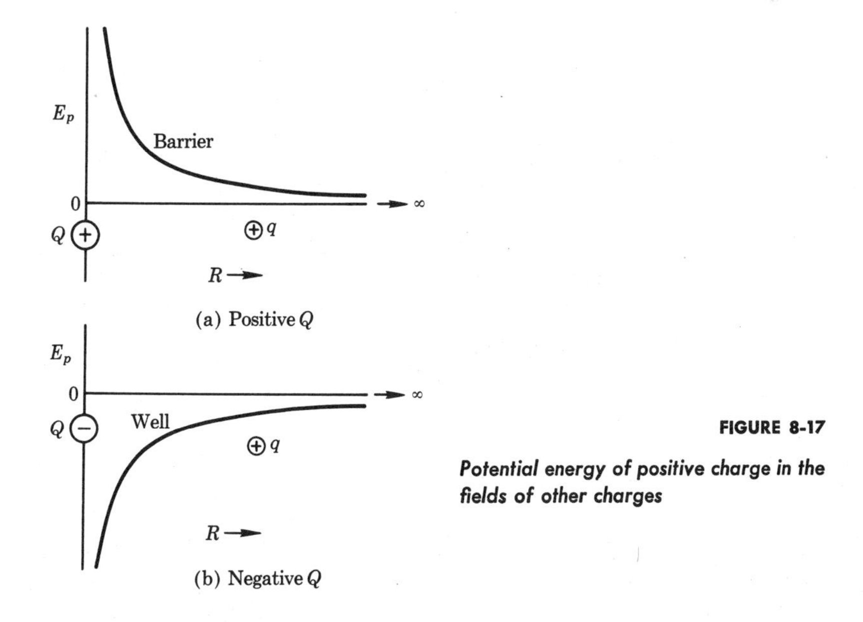

FIGURE 8-17

Potential energy of positive charge in the fields of other charges

expression, let us calculate the potential-energy difference between two points. One will be the Earth's surface R_E and the other any higher level R, both measured from an origin at the center of the Earth. Now,

$$\Delta E_p = E_p(R) - E_p(R_E) = -k_G Mm\left(\frac{1}{R} - \frac{1}{R_E}\right)$$

Rearranging,

$$\Delta E_p = k_G Mm \frac{R - R_E}{RR_E}$$

Let the height above the Earth be $x = R - R_E$. If x is small compared with the Earth's radius, R is very close to R_E in the denominator. Thus

$$\Delta E_p \cong \left(\frac{k_G Mm}{R_E^2}\right)x$$

The quantity in parentheses is seen to be the force on a mass m at the Earth's surface, which is also $m\mathcal{G}$. Hence

$$\Delta E_p \cong m\mathcal{G}x$$

which is exactly what we found before for the uniform field. This shows that the place at which we set the potential energy equal to zero does not matter, since we calculate potential-energy *differences*. For electrical problems, where charges are carried from point to point in a circuit, it is convenient to use some particular place in the circuit, or the connection to the ground. In the study of atomic collisions or of the motion of astronomical bodies, infinity is more convenient.

8-4 POTENTIAL AND POTENTIAL DIFFERENCE

We have seen that the potential energy depends on the mass m or on the charge q that is being moved, as well as on the source and the position. We should like to find a quantity that describes the *region of space*, regardless of what object is present. The quantity that gives this more general view is called the *potential* V, and is defined as the potential energy of a unit mass or positive charge—or what is the same, the work required to place a unit mass or unit positive charge at a point. The potential plays a similar role in referring to energy as field did with respect to force. To show the relation between potential and potential energy, suppose that we lift a mass m from the Earth's surface to a height x against the approximately uniform field $\mathcal{G}$. The potential energy it now has is $E_p = m\mathcal{G}x$. The potential, however, refers to a *unit* mass; hence

$$V = \mathcal{G}x$$

To illustrate using numbers and units, let $\mathcal{G} = 9.8$ N/kg (which is also $g = 9.8$ m/sec²), and suppose $x = 50$ m. Then

$$V = \mathcal{G}x = (9.8 \text{ N/kg})(50 \text{ m}) = 490 \text{ N-m/kg} = 490 \text{ J/kg}$$

The units of V are found to be those of energy per unit mass, as they should be.

The same procedure holds for electric fields. We carry a charge q from a plate at ground or zero potential energy to a distance x toward another plate. It has a potential energy $E_p = q\mathcal{E}x$. The potential, however, refers to a unit charge; hence

$$V = \mathcal{E}x$$

To illustrate, suppose $\mathcal{E} = 400$ N/C and $x = 0.005$ m. Then $V = \mathcal{E}x = (400 \text{ N/C})(0.005 \text{ m}) = 2 \text{ N-m/C} = 2 \text{ J/C}$. The units are energy per unit charge. This particular combination of units is given the name *volt*:*

1 volt = 1 joule/coulomb

The word for the unit is abbreviated V, not to be confused with the symbol for potential itself. The connection of the word "volt" with electrical usage is familiar to us. When we see a sign "DANGER, HIGH VOLTAGE," it means that the equipment is at high potential above ground, and charges released there would accelerate to ground (through our bodies if we were not careful and got too close!).

The potential has the property of having an arbitrary reference level, just as does the potential energy. Thus it is the *potential difference* that is significant. For example, consider an electrical wire at 200-volt potential, connected to a battery at 50 volts, while the "ground" is taken as zero potential. The potential difference between wire and battery is 150 volts. An important distinction must be made between potential energy and potential, in spite of their apparent similarity in form. Potential energy refers to the *system* of two or more particles, whereas potential refers to the *medium* into which we *may* place a test particle or body. In referring to electrical circuits, "potential" is more widely used; in dynamic-motion problems "potential energy" is more common. Thus the potential can be formally added to the collection of quantities that describe a region, along with field.

Equipotential Surfaces

There is an intimate relationship between field and potential. Consider the geometric meaning of the expression for a uniform electric field $V = \mathcal{E}x$ or a uniform gravitational field $V = \mathcal{G}x$. These are equations of

* After Alessandro Volta, an electrical experimenter of the seventeenth century, credited with the first laboratory demonstration of electricity.

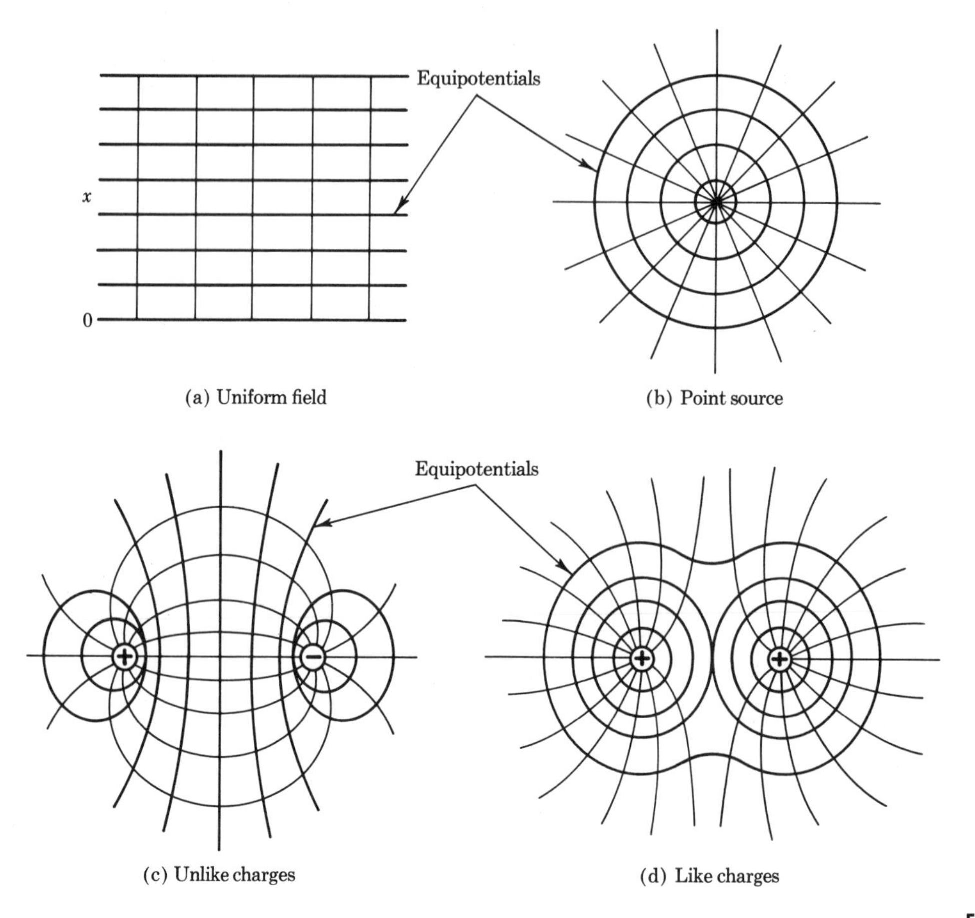

FIGURE 8-18

Equipotentials and fields for several charge arrangements

planes parallel to the ground (the plane $x = 0$)—Figure 8-18(a). Everywhere on the surface at x the potential is the same, hence it may be called an equal-potential surface or, better, *equipotential surface*. Now let us draw in the field lines, which run vertically. Obviously, the lines are everywhere perpendicular to the equipotentials. On the other hand, look at the inverse-square field of a point source—Figure 8-18(b). There, $V = C/R$, where C depends on whether the field is gravitational or electrical. This equation is that for a sphere of radius R. Everywhere on a sphere the potential has that value, so the surface is equipotential. If we sketch some of the curves, along with the known field lines, again they are found to be perpendicular, which is always the case. Then, for two unlike charges, shown in Figure 8-18(c), we connect all the points that have the same V to obtain the set of equipotential surfaces, along with the field lines that we developed in Section 5-5. Finally, we show the equipotentials of a pair of like charges—Figure 8-18(d). The important property of an equipotential line or surface is that charge or mass can be moved along them without work being required.

The potential at a point due to several masses or charges can be obtained by simple addition, since energy is a scalar quantity. Let a set of objects be given numbers 1, 2, 3, . . . , N. The total gravitational and electrical potentials are

$$V_G = -k_G\left(\frac{M_1}{r_1} + \frac{M_2}{r_2} + \ldots + \frac{M_N}{r_N}\right), \quad V_E = k_E\left(\frac{Q_1}{r_1} + \frac{Q_2}{r_2} + \ldots + \frac{Q_N}{r_N}\right)$$

where the correct sign (+ or −) is given to each Q, depending on whether the charge is + or −. To see the advantage of the potential, let us find the energy needed to project a spacecraft from the Earth's surface to the Moon's surface.

One way to find the answer would be to add up the amounts of work done in each step of the motion between the Earth and the Moon. It is

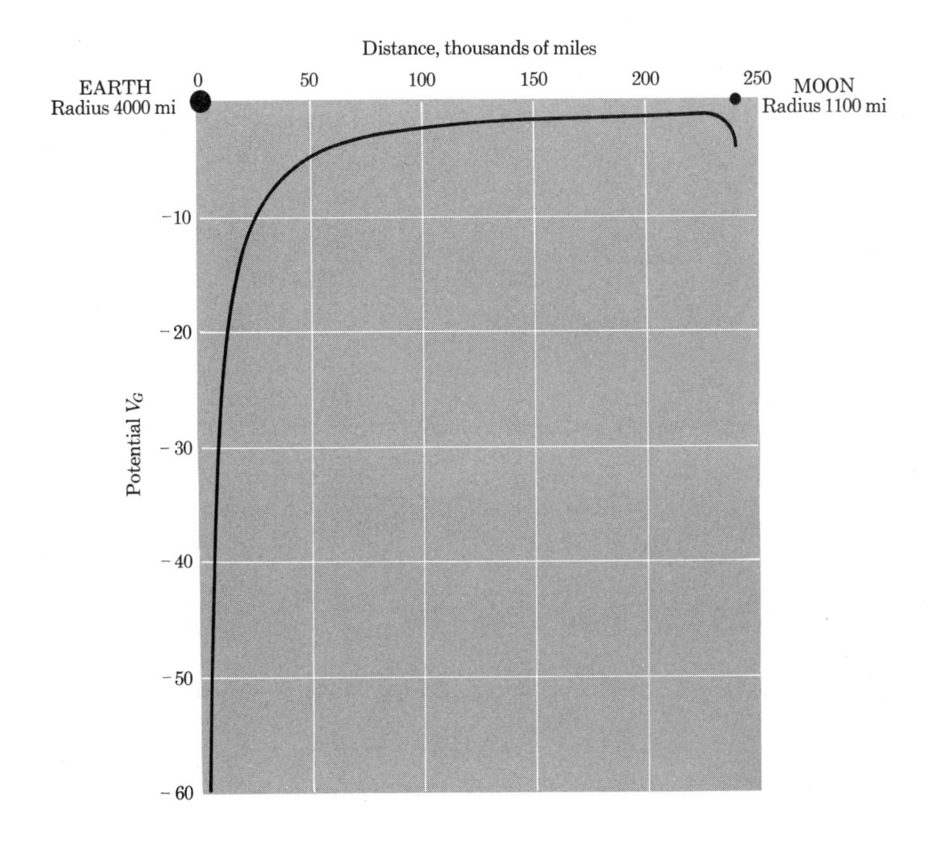

FIGURE 8-19

Gravitational potential along a line between the Earth and the Moon

much easier, however, to find the difference in potential, which is the work per kilogram. For any point in space, the gravitational potential due to the two bodies is

$$V_G = -k_G\left(\frac{M_E}{r_1} + \frac{M_M}{r_2}\right)$$

where r_1 and r_2 are measured to the centers of the Earth and Moon respectively—see Figure 8-19(a). The work required for the trip is simply the difference between potentials at the two surfaces, where the proper values of r_1 and r_2 are inserted in the general expression. At the surface of the Earth, we use $r_1 = R_E$, $r_2 = R - R_E$; at the surface of the Moon, we use $r_1 = R - R_M$, $r_2 = R_M$. Figure 8-19(b) shows a plot of the potential at all points on a line between the two masses. To get from the surface of the Earth to the surface of the Moon in a direct rocket flight, it is necessary to climb a "hill" and then "coast" down to the landing spot. A soft landing is not possible unless we provide energy of reverse thrust, since the rocket picks up kinetic energy in the last part of its flight.

We encountered the electrical dipole, in Section 6-3, as a pair of opposite charges of the same magnitude q, separated by a distance s. The potential due to a dipole is readily seen to be

$$V_E = k_E q\left(\frac{1}{r_1} - \frac{1}{r_2}\right)$$

if subscripts 1 and 2 refer to the + and − charges. Let us find V at the point halfway between the charges. Then, $r_1 = s/2$, $r_2 = s/2$ and $V = 0$. This is surprising until we realize that the equipotential through the origin extends to infinity. A charge can be brought in along that plane without doing any work on it.

The Electron-volt as a Unit of Energy

From the potential concept comes a unit that is widely used in atomic and nuclear physics: the *electron-volt*. To see how it arises, let us compute *the work required to carry one electron through a potential difference of one volt*. Now

$$W = (\Delta V)q = (1\text{ volt})(1.6 \times 10^{-19}\text{ coulomb}) = 1.6 \times 10^{-19}\text{ joule}$$

This is a very small amount of energy in terms of large-scale objects. For comparison, the work to lift 1 kg through a height of 1 m is (9.8 N) (1 m) = 9.8 J. On an atomic scale, however, 1.6×10^{-19} J is a very convenient amount, since one deals with changes in energy and with potential differences of this order of magnitude. This quantity of energy is called the *electron-volt*, and is abbreviated eV:

$$1\text{ eV} = 1.6 \times 10^{-19}\text{ J}$$

Once defined, the unit may be used for energy in any form, electrical or otherwise. For example, let us find the energy in electron-volts of the electron in a metal, Section 8-2. The kinetic energy was 4.55×10^{-21} J, or

$$E_k = \frac{4.55 \times 10^{-21}\ \text{J}}{1.6 \times 10^{-19}\ \text{J/eV}} = 0.028\ \text{eV}$$

In nuclear reactions, processes involve very much higher energies, and the *kiloelectron-volt, million-electron-volt,* and *billion-electron-volt* units are common; these are abbreviated keV, MeV, and BeV* respectively.

Van de Graaff Accelerator

The electric potential of a metal conductor can be raised to millions of volts by the continued addition of charge. This forms the basis of the *Van de Graaff accelerator,* named after the man who developed it. Charge is supplied to the interior of a metal shell by a discharge from a moving belt, as illustrated in Figure 8-20. The charges repel each other, and migrate to the outside of the shell, leaving the inside free of electric fields. A large electric field exists between the sphere and ground, through which protons or deuterons are accelerated in a vacuum. At energies of the order of 10 million electron-volts, many nuclear reactions can be created by the bombarding ions, making the accelerator a valuable physics research apparatus.

* The "gigavolt" (GeV) is also often used instead of BeV.

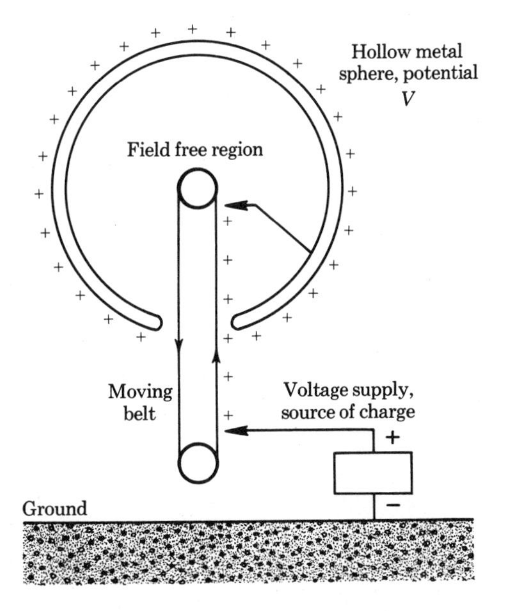

FIGURE 8-20
Van de Graaff accelerator

Escape Speed from the Earth

Jules Verne, in his book *From the Earth to the Moon*, written in 1865, anticipated our present space program. His "cannon" consisted of a deep hole dug in the Earth, and his projectile was given an initial velocity toward the Moon by a mammoth explosion of gunpowder. The modern technique for space travel is preferable—a two-step process, first from Earth to orbit, and then from orbit to some great distance, where the Earth's pull is negligible. The concept of potential is perfectly designed, however, to find the *escape speed*—that is, the speed an object must have as it leaves the surface in order to escape from the planet completely. At the Earth's surface the gravitational potential is $V_G = -(k_G M_E/R_E)$, which is the work done in bringing a unit mass from infinity. To send any object out to infinity again, we must supply a positive energy of just this magnitude for each kilogram of mass. If a rocket of mass m were given kinetic energy $\frac{1}{2}mv_e^2$ at the Earth and just makes it away with no energy left over, then

$$\tfrac{1}{2}mv_e^2 = \frac{k_G M_E m}{R_E}$$

The mass cancels out, which means that the *escape speed* v_e *is the same for all objects*, neglecting air friction. Then,

$$v_e = \sqrt{\frac{2k_G M_E}{R_E}}$$

We could substitute numbers, but it is more convenient to rearrange the formula using the surface gravitational acceleration $g = k_G M_E/R_E^2$. Then,

$$v_e = \sqrt{2gR_E} \qquad \textit{(escape speed from Earth)}$$

This checks dimensionally, has the right form, and is unusual in that it is also the speed reached in free fall from a height R_E in a uniform field. Inserting $g = 9.81$ m/sec², $R_E = 6.37 \times 10^6$ m, we find the escape speed to be $v_e = 11.2$ km/sec, which is also about 25,000 mi/hr or 7.0 mi/sec. The kinetic energy for escape is $E_k = \frac{1}{2}mv_e^2 = 2mgR_E$. For a 1-kg mass this is

$$2(9.81)(6.37 \times 10^6) = 1.25 \times 10^8 \text{ J}$$

A more exact treatment should take account of the initial speed of the projectile on the Earth's surface, about 0.46 km/sec, but this is small and air resistance has been neglected anyway.

Energy Relations in Planetary or Atomic Orbits

The concepts of potential and potential energy are very revealing with regard to the behavior of objects in satellite orbit, such as the Moon about the Earth, or a spacecraft circulating about the Earth or Moon, or an electron about a nucleus (in the classical model). The potential energy of a particle under an attractive force is

$$E_p = -\frac{C}{R}$$

where C is $k_E Qq$ for charges and $k_G Mm$ for masses. The kinetic energy is, of course, $E_k = \frac{1}{2}mv^2$. If the satellite is to be in circular orbit of radius R, the attraction must provide the centripetal force

$$\frac{mv^2}{R} = \frac{C}{R^2}$$

which leads to the conclusion that $\frac{1}{2}mv^2 = C/2R$, or

$$E_k = -\frac{E_p}{2}$$

The total mechanical energy is thus

$$E = E_k + E_p = \frac{E_p}{2} \quad \text{or} \quad -E_k$$

To arrive at a *negative* total energy is at first startling. Think, however, of the process by which the two particles can be completely separated—called *escape from orbit* in the astronautical sense and *ionization* in the atomic sense. We must add energy in amount ΔE from some external source to remove the body to infinity, leaving zero kinetic and potential energies.

For the case of ionization of hydrogen, let us calculate ΔE, noting that $C = k_E e^2$ and thus

$$\Delta E = \frac{k_E e^2}{2R} = \frac{(9 \times 10^9)(1.60 \times 10^{-19})^2}{2(5.3 \times 10^{-11})} = 2.18 \times 10^{-18}\ \text{J}$$

Thus the energy required to ionize one hydrogen atom is

$$\Delta E = \frac{2.18 \times 10^{-18}\ \text{J}}{1.60 \times 10^{-19}\ \text{J/eV}} = 13.6\ \text{eV}$$

Measurements verify this result to be rather accurate, in spite of the fact that the classical model is used instead of the correct quantum-mechanical description.

A second example is the escape of a spacecraft from orbit around the Earth. The amount of energy that must be produced by the rocket engine and the change in speed may be calculated. As shown above, $\Delta E = E_k$, which says very simply that the energy addition must exactly match the original energy in orbit. Let the speed in orbit be v_0 and the speed after sudden acceleration be v_1. Then,

$$\Delta E = \tfrac{1}{2}mv_1^2 - \tfrac{1}{2}mv_0^2 = E_k = \tfrac{1}{2}mv_0^2$$

or

$$v_1^2 = 2v_0^2$$

$$\Delta v = v_1 - v_0 = (\sqrt{2} - 1)v_0 = 0.414\ v_0$$

The change in speed is thus 0.414 times the original speed. For instance, it can easily be shown that an 84-min orbit demands a speed v_0 close to 8 km/sec. The change in speed, which we shall call the *escape speed from orbit*, is thus only $\Delta v = (0.414)(8)$ or 3.3 km/sec. Compare this result with the 11.2 km/sec change needed for direct escape from the Earth, as calculated earlier: although the total energy that must be expended in order to reach a point in deep space is the same for both modes, it is more convenient to put a spacecraft in orbit first, using two or more rocket stages.

8-5 POWER—CONSERVATIVE FORCES

The definition of work as a product of force and distance says nothing about the time required. For example, we could walk slowly up a flight of stairs or run up, and the work done is the same.

The quantity called *power* is the rate at which a certain force does work. Suppose that W is the work done in a time t. The power is then

$$P = \frac{W}{t} \qquad \textit{(power, constant rate of work)}$$

The unit of power in the mks system is the joule per second, which is called the *watt* (W) after James Watt, the inventor of the steam engine. This power unit is familiar to us since household electric lightbulbs are rated as 25W, 60W, 100W, etc, which indicates the rate at which they consume electrical energy. When the power W varies with time, the above relation can be considered as the *average* power over a time t. For a numerical example, look again at the situation of a locomotive that pulls a train of cars at a uniform speed, doing work mainly against the frictional force between wheels and track. Suppose the applied force is 4×10^5 N and that it requires 1 hr to go a distance of 50 km. The work done is

$$W = (4 \times 10^5\ \text{N})(5 \times 10^4\ \text{m}) = 2 \times 10^{10}\ \text{J}$$

The time is 3600 sec; thus the power is

$$P = \frac{W}{t} = \frac{2 \times 10^{10}\ \text{J}}{3.6 \times 10^3\ \text{sec}} = 5.56 \times 10^6\ \text{W}$$

Large units of measure for power are needed for processes involving vehicles or generating plants for electricity:

$$1 \text{ kilowatt (kW)} = 10^3 \text{ W}$$
$$1 \text{ megawatt (MW)} = 10^6 \text{ W}$$

The British system of power units, as expected, has the ft-lb/sec but also the horsepower (hp) which is 550 ft-lb/sec or 33,000 ft-lb/hr. Conversion between the mks and British systems are

$$1 \text{ horsepower} = 550 \text{ ft-lb/sec} = 746 \text{ watt}$$
$$1 \text{ watt} = 1.34 \times 10^{-3} \text{ horsepower}$$

For our train problem, the horsepower developed by the train would be

$$(5.56 \times 10^6 \text{ W})(1.34 \times 10^{-3} \text{ hp/W}) = 7500 \text{ hp}$$

The term "horsepower" is also familiar as being the common rating for automobile engines. A compact car might have a 100-hp rating, a larger one a 300-hp rating. It is interesting to note that Watt also provided the calculations leading to the unit called horsepower. He observed that an average horse walking at 2.5 mi/hr exerted a force of 150 lb.

Having defined the watt, we have the option of using the *watt-second* as a unit of energy equal to the joule. Derived from the W-sec is the *kilowatt-hour* (kWh): the total amount of energy resulting from a power of 1 kW acting for 1 hr. Electrical bills for a house are based on the usage in kilowatt-hours, as we may recall. A rough cost figure for electricity is 1 cent per kWh. Physically, a high power rating for any device means that much work can be done in a short time. Thus a high-hp automobile can accelerate quickly, overcoming inertia and frictional forces readily. An electrical-power-station rating of 1000 megawatts can provide continuous light and work energy for a city of hundreds of thousands of people.

Just as we encounter forces and potential energies that vary with position and time, we find that in general W is a function of time. Power must then be defined in terms of small intervals Δt, with work done ΔW, and

$$P = \frac{\Delta W}{\Delta t}$$

This general relation enables us to find the connection between power and speed. The power developed by an applied force F on an object moving with a speed v along the line of the force is simply

$$P = Fv$$

To demonstrate this, write $P = \Delta W/\Delta t$, but recall that $\Delta W = F\Delta x$. Then $P = F\Delta x/\Delta t = Fv$. Returning to our train problem of a previous paragraph, we find the speed to be

$$v = \frac{50{,}000 \text{ m}}{3600 \text{ sec}} = 13.9 \text{ m/sec}$$

With a force of 4×10^5 N, the power is calculated to be

$$P = (4 \times 10^5 \text{ N})(13.9 \text{ m/sec}) = 5.56 \times 10^6 \text{ W}$$

as we found before.

The total work output may be found from a known dependence of power on time. For the case of a constant power, we merely write

$$W = Pt$$

If the power varies with time, however, we must return to the form $\Delta W = P\Delta t$, and add up small contributions. We then observe that the work, by analogy with previous situations, is the area under the curve of power as it depends on time.

The power expression $P = Fv$ is completely general, being correct for motion with either force or speed either varying with time or constant. Since power is defined from work, this P refers to the power due to only one specific force on a body. We must examine carefully the usual situation where several forces act, and where there is a *net force*. The distinction will be brought out by studying motion with and without friction.

The power rating of an automobile must be understood as the maximum that can be developed by the engine. The operating power may be any value lower than this maximum, depending on driving conditions. Let us examine the power, energy, and force aspects of a car starting from a standstill. As we well know, the burning of gasoline in the engine provides the energy of propulsion. Much of this energy is wasted in the exhaust gas heat or in frictional heating of the moving parts, but we can assume that the engine provides an effective force F_A that can move the car forward.

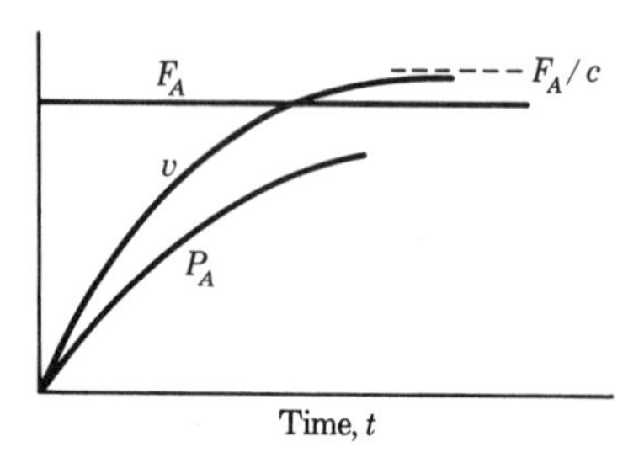

(a) Start up with constant force

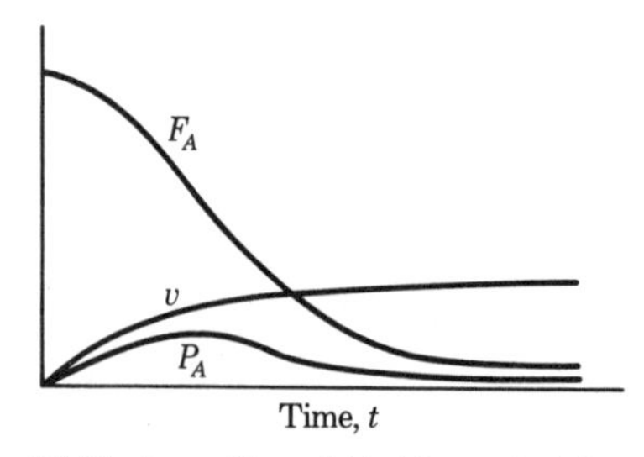

(b) Start up with variable (decreasing) force

FIGURE 8-21

Relation of force, speed, and power

However, there is also a frictional force of air resistance F_F, which opposes the car's progress. Each force does work, thus each force develops power. We are usually interested in the power $P_A = F_A v$ provided by the engine, since that demands the use of fuel. The speed of the vehicle, however, depends on the *net* force, $F_A - F_F = ma$. If we were to apply a constant force F_A and friction were proportional to speed ($F_F = cv$), we would obtain the same type of motion as described in Section 5-8: the speed would rise from zero at the start and approach a maximum corresponding to zero acceleration—i.e., $F_A - F_F = 0$ or $v = F_A/c$. Such a mode of operation would result in an excessive final speed—Figure 8-21(a). In practice we instead start with a large force F_A to get an initial large acceleration, but reduce F_A as the car picks up speed. We arrive gradually at a final steady safe speed, as illustrated in Figure 8-21(b). Thus the power rises from zero to a maximum and drops to a low constant value.

Conservative and Nonconservative Fields

We introduced the concepts of conservation of mechanical energy through the examples of gravitational and electrostatic forces, being careful to exclude forces of friction. Now we should like to examine the meaning of nonconservative fields and extend the conservation law to more nearly real systems.

A *conservative field* is defined as one for which mechanical energy of an isolated body is conserved. This is equivalent to saying that a field is *conservative* if (1) the work done in moving a test body between two points is independent of the route taken, or (2) if the work done in carrying a test body around a closed path is zero. The truth of (1) for the gravity field is easily recognized. If we climb steps to get to the door of a building, or walk up a ramp, the work done against the force is the same: mgh. A ball thrown vertically with a certain kinetic energy will return to our glove with the same energy, neglecting any slowing by air resistance.

A more careful analysis for the case of the electric field of a spherical charge Q will also verify these definitions. Let us start at point A on an equipotential curve at radius R_1—see Figure 8-22(a)—and move a unit positive charge directly inward to point B, at radius R_2. The work done is the potential energy difference

$$k_E Q \left(\frac{1}{R_1} - \frac{1}{R_2}\right)$$

Instead, we go a roundabout route, as in Figure 8-22(b), from A on a circle to C, then in to D, and around to B—a much longer path. While we are on the equipotential surfaces, no work is done, since the component of field along the displacement is zero. The only work is done in going from C to D; but since the potential difference is the same as from A to B, the total work is identical. One might ask, what if we go in some wandering path,

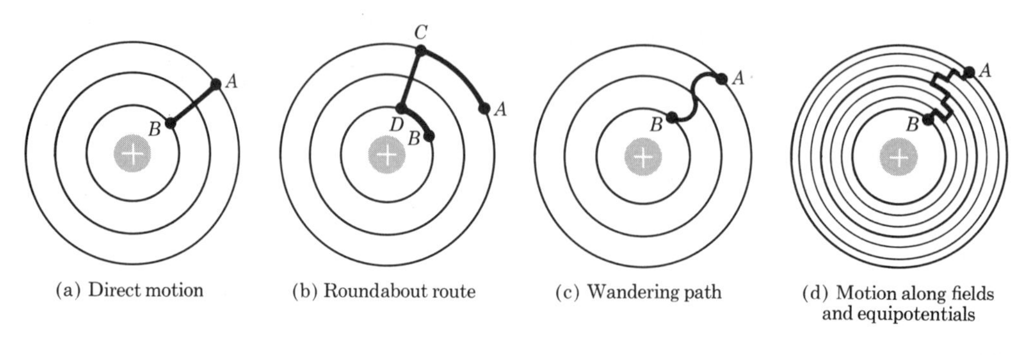

FIGURE 8-22

Properties of conservative field

as illustrated in Figure 8-22(c)? Such a route can be replaced with a set of alternate movements along field lines and along equipotentials—Figure 8-22(d). The total radial distance remains $R_1 - R_2$. Definition (2) above is a natural consequence of (1), since work that we do on a charge in moving it inward can be followed by an equal negative work done as it returns, and the total is zero.

Several of the fields that are important in physics satisfy the requirements for conservation. Examples are the uniform field, already discussed in detail; the inverse-square field of gravity, electrostatics, or bar magnets (Chapter 6); and the linear field that produces vibration, to be discussed in Chapter 11.

An elementary illustration will demonstrate that motion involving friction is *nonconservative*, even though the field is conservative. If we push a box along a very smooth incline and then let it slide freely back down again, its final kinetic energy will be equal to the potential energy we originally gave it. However, if the incline is rough, work is done against friction in both directions, so there will be less mechanical energy at the end of the motion than in the case without friction.

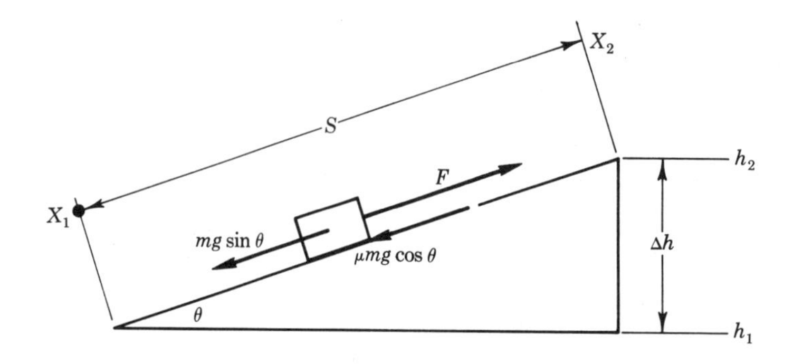

FIGURE 8-23

Work against nonconservative friction force

The idea of conservation of energy is so appealing that we should like to extend it to include nonconservative problems—for example, those including friction. A straightforward logic permits us to do so. For purposes of illustration, let us push a box up a rough ramp from point 1 to point 2, as in Figure 8-23. We obviously do more work in moving it than if the surface were smooth. However, we provide the same amount of potential energy in both cases. The total (mechanical) energy of the box at the end of the motion is reduced because of frictional work; i.e.,

$$E_2 = E_1 - W_F$$

The frictional "loss" went into the medium through which the body passed or slid over, giving its molecules a greater motion. The medium is heated, as our experience tells us. Suppose we let the original *energy of the medium* be U_1 and later the higher value $U_2 = U_1 + W_F$. Combining the two equations, we obtain

$$U_1 + E_1 = U_2 + E_2$$

which says that thc *grand total energy*, $E + U$ (mechanical plus heat), is constant. Thus we see that conservation of energy holds if we look at a larger system.

8-6 HEAT AS A FORM OF ENERGY

The notions of heat and cold are very familiar to us, including the need for warmer or cooler clothing or for heating and air conditioning, as conditions change; and we use the thermometer to tell us the temperature. Less well appreciated, however, is the fact that heat is a form of kinetic energy due to molecular motion. Such heat energy can be stored, transferred, or transformed from and converted into other forms of energy. The terms "heat" and "thermal energy" may be used interchangeably.

A qualitative molecular picture of this relation is easily visualized. In a solid or liquid, the atoms are in continuous motion, but held together by attractive bonds. Mathematically, these bear a similarity to sets of connecting springs. We shall find temperature to be a measure of the kinetic energy contained in a substance. The relation is linear and direct, in that the energy is proportional to the temperature.

Temperature Scales

We shall treat temperature for the present as a quantity that can be measured by a suitable device called a *thermometer*. The most familiar type is the graduated glass tube with a colored liquid, the level of which rises as the liquid expands and falls as it contracts. The temperature scale in most

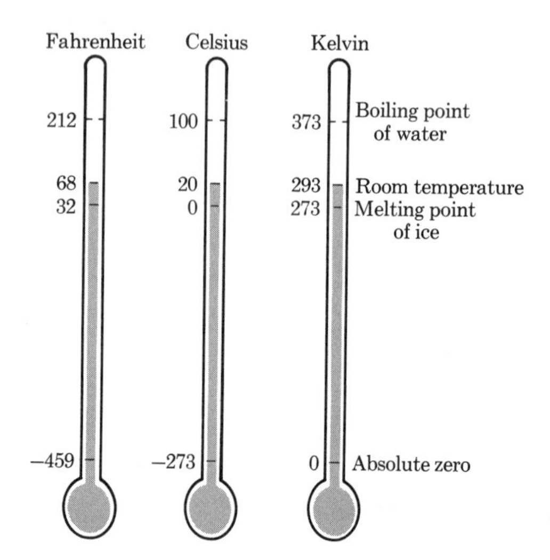

FIGURE 8-24

Temperature scales

common usage is the *Fahrenheit* (F). On this scale, the melting point of ice is 32°F and the boiling point of water is 212°F.

The *Celsius** scale (C) is used in scientific measurements and is standard in most European countries. The melting point of ice is 0°C; the boiling point of water is 100°C. The *Kelvin* (K) or *absolute* scale uses the same spacing of degrees as the Celsius, but its zero is at −273.15°C ("absolute zero"), where particle motion is at a minimum. Figure 8-24 shows the relation of the scales. Since the respective spans between the boiling and freezing points of water are 180°F and 100°C, the Fahrenheit degree is $\frac{100}{180} = \frac{5}{9}$ as large as the Celsius degree. Conversions between the two systems are

$$T\ (°\mathrm{F}) = \tfrac{9}{5}T\ (°\mathrm{C}) + 32$$

$$T\ (°\mathrm{C}) = \tfrac{5}{9}[T\ (°\mathrm{F}) - 32]$$

As a simple example, let us find the Celsius temperature corresponding to healthful room conditions, 68°F: $T\ (°\mathrm{C}) = \frac{5}{9}(68 - 32) = 20°\mathrm{C}$.

Relation of Heat and Temperature

One of man's earliest discoveries was that water in a container could be heated by placing it over a fire. When scientists sought the explanation centuries later, it was natural to imagine that there was a flow of some substance from the source of heat to the water. It is therefore understandable why heat was long considered as something quite different from kinetic energy in the mechanical sense. Large-scale motion can be seen and the impact of a moving object can be felt, and the surface of the skin senses

* Formerly *centigrade.*

temperature changes. The difference, however, lies only wherein the kinetic energy resides. Mechanically, the particles of an object are moving along together; thermally, there is an internal energy of chaotic motion of atoms and molecules. Further consideration will verify for us that heat and mechanical effects are related. Because of friction, one burns his hands as he slides down a rope; a piece of metal gets hot if we pound it with a hammer or drill a hole in it. Energy in the form of heat is converted into motion in all types of engines—the steam locomotive, the gasoline engine, the jet airplane, and the rocket. Not until the nineteenth century was this relation between mechanical and thermal energy well understood,* and as an unfortunate consequence of earlier ignorance an entirely different set of units came to be used for the two forms of energy.

Heat Units

The quantity of heat required to raise 1 gram of water 1 degree Celsius was defined as a *calorie* (cal). Experiments demonstrating the equivalence of heat and mechanical energy were originally conducted by Joule. By turning a paddle wheel in water and measuring the work done and the temperature rise of the liquid, he established that 1 joule of work was the same as 4.185 calories of heat. Thus if a sample of water of 5 kg mass was brought from its melting point (0°C) to room temperature (20°C), the number of calories of heat supplied was (5000)(20) = 100,000 cal. This corresponds to an energy

$$\frac{100{,}000 \text{ cal}}{4.185 \text{ cal/J}} = 2.39 \times 10^3 \text{ J}$$

The *kilocalorie* (kcal) as 1000 calories or 1 *Calorie* (Cal) is used in the metric system. This is also the unit of energy referred to in food and diets. The conversions between heat and energy units and mechanical units are

$$1 \text{ kcal} = 4185 \text{ J}$$
$$1 \text{ J} = 2.39 \times 10^{-4} \text{ kcal}$$

The *British thermal unit* (Btu), used in engineering, is an amount of heat required to raise a mass of water corresponding to 1 pound by 1 degree Fahrenheit. Conversion factors between the Btu and metric units are

$$1 \text{ Btu} = 1.05 \times 10^3 \text{ J}$$
$$1 \text{ kWh} = 3.41 \times 10^3 \text{ Btu}$$

Usually the idea of temperature is applied to something with many atoms or molecules—the air, a floor, an icicle, a radiator or stove, an engine,

* It is recorded that Count Rumford deduced the connection by noting that the mechanical effort of boring the cylinders of cannon caused a great rise in temperature.

the Sun, etc. The atoms have a variety of energies, and we relate their average energy to temperature.

Conservation of Thermal Energy

When two objects at different temperatures are brought together, either by contact or by mixing, the energy of particle motion tends to be shared between them. For example, let the original internal energies of two bodies be U_1 and U_2, having temperatures T_1 and T_2. Assuming that there is no change in motion or level of the system, the final internal energy will be

$$U = U_1 + U_2$$

with a common temperature T. This is merely an example of conservation of energy.

8-7 THE EQUIVALENCE OF MASS AND ENERGY

Einstein's critical re-examination of the meaning of motion led him, as discussed in Section 3-4, to a drastic modification of the ideas of distance and time. One of the consequences (Section 4-7) was that he found the mass of an object to be dependent on its speed v:

$$m = \frac{m_0}{\sqrt{1 - (v/c)^2}}$$

where the mass at rest is m_0, and c is the speed of light 3×10^8 m/sec. In order to describe motion of atomic particles, which can have speeds approaching c, we must repeat our derivation of kinetic energy from the viewpoint of special relativity. Newton's second law of motion is correct relativistically if formulated in terms of the rate of change of momentum:

$$F = \frac{\Delta p}{\Delta t} = \frac{\Delta(mv)}{\Delta t}$$

Since the mass depends on speed, which in turn changes with time, the kinetic energy must be computed step by step. For a displacement Δx, the work done is $\Delta W = F\Delta x$; but since $F = \Delta p/\Delta t$, $v = \Delta x/\Delta t$, and $\Delta W = \Delta E_k$, then the change in kinetic energy is

$$\Delta E_k = v\Delta p$$

The kinetic energy is now the area under the curve of speed as it depends on momentum. We can easily find v as it depends on p by combining $p = mv$

and the relativistic mass–speed relation. As listed in Appendix D,

$$v = \frac{pc}{\sqrt{p^2 + (m_0c)^2}}$$

(A plot is shown in Figure 8-25.) We see that no matter how large the momentum becomes, the speed never exceeds c. At low speeds, of course, v is merely p/m_0; but by the time v gets as large as half the speed of light $c/2$, significant deviation from the straight line will have taken place. The area can be estimated graphically, but by use of the methods of calculus an exact result is obtained, as

$$E_k = mc^2 - m_0c^2 \qquad \textit{(relativistic kinetic energy)}$$

Far-reaching interpretations may be made for this very simple formula. Let us rewrite it as

$$mc^2 = m_0c^2 + E_k$$

We see that mc^2 has the units of energy, which suggests that *mass and energy are equivalent physical quantities,* with c^2 the constant of proportionality between them. Recalling that m_0 was taken to be the mass of a body at rest, we may consider m_0c^2 as a static or *rest energy,* inherent in matter and not dependent on motion. Since E_k is the energy of motion, we must conclude

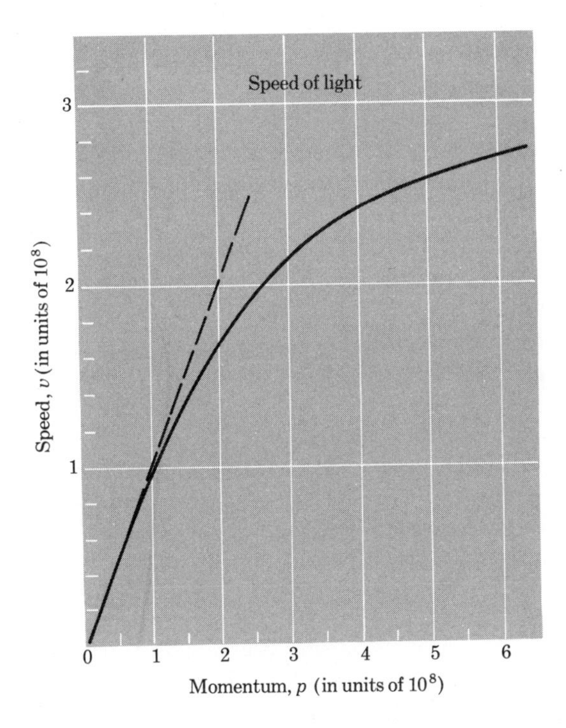

FIGURE 8-25

Relation of speed and momentum in special relativity (mass 1 kg at rest)

that mc^2 is the *total energy* E. In words, then,

$$\text{total energy} = \text{rest energy} + \text{kinetic energy}$$

Einstein's famous formula is the most compact statement of this equivalence of mass and energy:

$$E = mc^2 \qquad \textit{(Einstein's formula)}$$

This relation implies that the supply of energy to a body through work "creates" mass or, conversely, that mass can be converted into useful energy. This reminds us of the ability of kinetic energy to be transformed into potential energy and vice versa, with the total mechanical energy conserved. According to the theory of relativity, the sum of mass and energy, which we can call "*mass-energy*," is conserved.

The equivalence of mass and energy was predicted by Einstein long before any experimental evidence had been obtained. The possibility of transforming mass into kinetic energy, with its world-shaking implications, was hardly noticed at first, although a few science-minded feature writers of the time suggested that a transatlantic voyage of an ocean liner might require only a cupful of water as a source of propulsion energy.

Let us calculate the rest energy for an electron:

$$m_0c^2 = (9.11 \times 10^{-31}\ \text{kg})(3 \times 10^8\ \text{m/sec})^2 = 8.20 \times 10^{-14}\ \text{J}$$

This is also

$$\frac{8.20 \times 10^{-14}\ \text{J}}{1.60 \times 10^{-19}\ \text{J/eV}} = 0.51 \times 10^6\ \text{eV} \quad \text{or} \quad 0.51\ \text{MeV}$$

Every electron has this inherent amount of energy, to which kinetic energy is added by any acceleration. Similarly, every proton and neutron in the elements of the periodic table have rest energies in proportion to their masses. A set of relativistic formulas that allow one to work back and forth between the physical variables v, p, E_k, E, and m is given in Appendix D.

Let us compute the speed a particle will attain when the amount of kinetic energy added just equals its rest energy. Here, E_k has to be also m_0c^2, the total mass is $2m_0c^2$, or $m/m_0 = 2$. Thus

$$\frac{1}{\sqrt{1 - (v/c)^2}} = 2$$

Solving for v/c we obtain 0.866—that is, the speed is almost 90 per cent that of light. The actual value is

$$v = (0.866)(3 \times 10^8) = 2.6 \times 10^8\ \text{m/sec}$$

To help us visualize the magnitude of the energy release from matter, we make a few comparisons. The possible energy from the conversion of only 1 kilogram could accomplish any of the following:

1. melt, heat, and boil an ice cube 0.2 mi on a side;
2. replace 3 million tons of the best coal or 670 million gallons of gasoline as fuel;
3. provide all the heat energy used by the 3 billion people of the Earth for a period of 8 hr;
4. lift 10 million men of weight 160 lb to the Moon and on out into space;
5. cause as much physical damage as 22 megatons of explosive such as TNT.

We may express E_k in terms of speed and rest mass:

$$E_k = (m - m_0)c^2 = \frac{m_0c^2}{\sqrt{1-(v/c)^2}} - m_0c^2 = m_0c^2\left[\frac{1}{\sqrt{1-(v/c)^2}} - 1\right]$$

This formidable expression must be used if we want to describe atomic particles whenever their speed is close to c, or whenever their total energy appreciably exceeds the rest energy. It must not be viewed as *different* from the familiar $\frac{1}{2}mv^2$ but rather as the *correct* form, which goes into

$$E_k \cong \tfrac{1}{2}m_0v^2$$

as an *approximation*,* good for low speeds—where v is much less than c.

We have seen that relativistic formulas are complicated and not easy to think about. Thus most people try to avoid using them if they can, even when working with electrons, atoms, and nuclei. Whether this practice is safe or not depends on the accuracy required by the problem. Table 8-2 shows the error in using classical formulas as a function of v/c. A fair rule for slide-rule accuracy is to use classical formulas only if v is less than $0.1c$. One common error should be avoided—that of substituting the relativistic mass into the nonrelativistic kinetic-energy formula. Such a hybrid is good only if $v \ll c$, in which case there is no need to use the relativistic mass anyway.

* We may perform a binomial expansion to obtain

$$\left[1 - \left(\frac{v}{c}\right)^2\right]^{-1/2} = 1 + \left(\frac{1}{2}\right)\left(\frac{v}{c}\right)^2 + \left(\frac{3}{8}\right)\left(\frac{v}{c}\right)^4 + \cdots$$

Thus

$$E_k = \left(\frac{1}{2}\right) m_0v^2\left[1 + \left(\frac{3}{4}\right)\left(\frac{v}{c}\right)^2 + \cdots\right]$$

If $v/c \ll 1$, then $\frac{3}{4}(v/c)^2$ and higher-order terms are very much less than 1, and can be neglected. For example, if $v = 10^5$ and $c = 3 \times 10^8$, then $v/c = 0.33 \times 10^{-3}$, $\frac{3}{4}(v/c)^2 = 8 \times 10^{-8}$, less than 1 part in 1,000,000.

TABLE 8-2

PERCENTAGE ERROR IN CLASSICAL FORMULAS*

$\frac{v}{c}$	Mass or Momentum	Kinetic Energy
0	0 %	0 %
0.01	0.005	0.0075
0.1	0.5	0.75
0.5	13	20
0.9	56	69
0.99	86	92
1.00	100	100

* The magnitude of the fractional error is $\frac{C - R}{R}$, where C is classical (approximate) and R is relativistic (exact).

8-8 LIGHT AS A FORM OF ENERGY

We are well aware of the fact that light from the Sun warms the Earth, the air, and our bodies. This permits us to accept readily the idea that radiation carries or consists of energy. It emanates from a heated object, and upon being absorbed by a substance (e.g., air, ground, tissue) stimulates its molecules into increased motion. Thus light energy is converted into thermal energy. If a substance—e.g., a lightbulb filament—is heated sufficiently, it will release energy as light, a form of electromagnetic energy.

There are two simplified descriptions of light—the *wave* model and the *particle* model. Which view is more convenient to use depends on the situation. According to the wave version, light involves electric and magnetic fields that vary periodically with frequency ν. According to the particle view, light consists of *photons,* whose energy is proportional to ν. The speed of motion of light in free space, in either case, is c. It is more convenient to use wave theory, for example, to describe the electromagnetic disturbances involved in radio and television with frequencies of vibration in the vicinity of 10^7 cps (cycles per second); while the particle theory is often applied to *gamma* rays and X rays with frequencies around 10^{20} cps.

The amount of energy carried by an individual photon is extremely small. For the typical visible radiation from the Sun or a lightbulb, the photon energy is around 1 eV, 1.6×10^{-19} J. Let us compute the number of such particles that would have to be absorbed to raise 5 kg of water by 20°C. In Section 8-6 we calculated the total energy required to be 2.39×10^3 J; the number of photons this represents is

$$\frac{2.39 \times 10^3}{1.6 \times 10^{-19}} = 1.5 \times 10^{22}$$

which is a remarkably large number.

The photon is regarded as a particle in the sense that it may collide with electrons and nuclei and impart energy to them. However, since the photon moves only at the speed of light, it does not exist in a state of rest. It has no rest mass, as does a true particle, and thus consists of pure energy. The Einsteinian equivalence of mass and energy $E = mc^2$ (Section 8-7) permits us to define an *effective mass* m_P of the photon:

$$m_P = \frac{E}{c^2}$$

It is interesting to compare this effective mass for a 1-eV photon with the rest mass of an electron, 9.11×10^{-31} kg. The photon's mass is

$$m_P = \frac{E}{c^2} = \frac{1.6 \times 10^{-19}}{9 \times 10^{18}} = 1.78 \times 10^{-38} \text{ kg}$$

which is only 2×10^{-8} times that of the electron.

We may now proceed to find the momentum p as the product of mass E/c^2 and speed c. Thus the momentum of the photon is

$$p = \frac{E}{c}$$

Photons are capable of ejecting electrons from metals, in what is called the *photoelectric effect*. Visualize a simple vacuum tube with one electrode composed of tungsten (Figure 8-26). The glass envelope is coated with an opaque substance overall except for a small area through which light can pass. When a pencil of light impinges on the plate, a current flows. The explanation of this phenomenon was originally provided by Einstein in 1905. Electrons normally experience some binding force that holds them in the metal. However, a photon can impart all of its energy

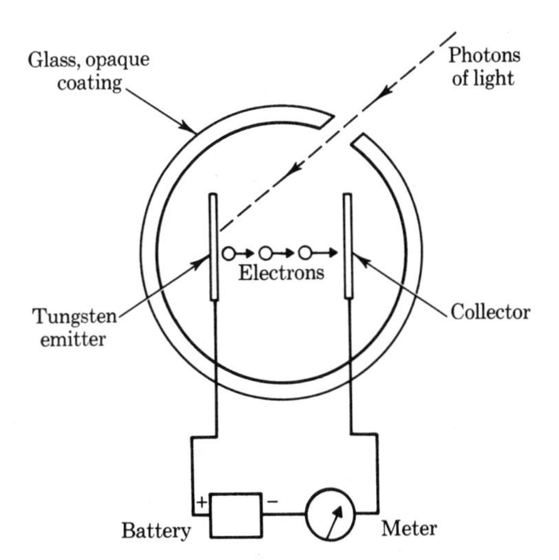

FIGURE 8-26
Photoelectric tube

E_P to an electron that can then escape to the metal surface, be accelerated by the electric field across the gap between electrodes, and contribute to a measurable current. Automatic door controls make use of such devices: as a person approaches and interrupts a light beam, the current in the photocell changes, and a signal is sent to activate the motor that opens the door.

The amount of kinetic energy borne by various electrons on escape from the surface will differ, depending on their original internal energy, the number of collisions that occur on the way out, and the character of attraction forces in the metal. Suppose that the photon energy E_P is converted completely to electron energy. If a potential V_s is applied to stop the collection of electrons completely, we can deduce the *work function* W as the energy required to overcome internal forces. The principle of conservation of energy yields

$$W = E_P - V_s e$$

The magnitude of W depends on the element of which the emitter is composed or on its surface coating; Table 8-3 gives the appropriate value for a few elements. Note that the work function is small for elements Cs, K, and Na; it is relatively easy to dislodge the one valence electron in these substances.

TABLE 8-3
PHOTOELECTRIC WORK FUNCTIONS

MATERIAL	W (in electron-volts)
Al	4.1
Cu	4.5
Cs	1.9
K	2.2
Na	2.3
Mo	4.2
W	4.5

As discovered by Planck, the relation between photon energy and the frequency of light is $E = h\nu$, where h is called *Planck's constant*, 6.63×10^{-34} J-sec. For example, our 1-eV photon has a frequency

$$\nu = \frac{E}{h} = \frac{1.60 \times 10^{-19}}{6.63 \times 10^{-34}} = 2.4 \times 10^{14} \text{ cps}$$

We can see that the removal of electrons from a metal of a certain work function W by the photoelectric effect requires light of a certain minimum frequency; this *threshold frequency* ν_0 for the effect in tungsten, work function $W = 4.5$ eV, is

$$\nu_0 = \frac{W}{h} = \frac{(4.5)(1.60 \times 10^{-19})}{6.63 \times 10^{-34}} = 1.1 \times 10^{15} \text{ cps}$$

Energy and Existence

Through our study of energy and its conservation, we have begun to realize how universal is the concept of energy. It embraces all basic processes ranging from atoms to astronomy, and serves as an intellectual abstraction for the understanding of interactions of all types—whether they be electromagnetic or gravitational or nuclear. We are also in a better position to appreciate the role of energy (or mass-energy) as a physical entity in our world and life. At some point in time, the Sun was formed and began its nuclear process of conversion of mass into energy. The Earth appeared, with its energy-containing minerals. Plants and animals, as systems that accept and transform energy, then populated the planet. The remains of the prehistoric plants and animals became deposits of coal and oil, potential energy sources ultimately to be used by man. Light energy from the Sun continually bombards the atmosphere and surface of our globe, and is converted into thermal energy that appears as the energy in the weather. It also maintains the life processes in plants, which in turn supply the animals and man with the chemical energy of food. We tap various resources to provide energy for the control of our thermal environment. We take the gravitational energy of falling water or the stored chemical energy of coal or the mass-energy of uranium, and develop electrical energy, which may then be transformed into mechanical energy for manufacturing processes or chemical energy for producing materials. The satisfaction of all of man's needs for survival and comfort thus hinges upon the availability of energy and upon his ability to transform energy.

PROBLEMS

8-1 A man climbs a flight of 20 steps with treads and risers each of length 8 in. If his weight is 160 lb, how much work does he do?

8-2 An electron is given a constant force of 2×10^{-16} N in an electric field. How much work is done in carrying the charge through 0.008 m in the direction of the force?

8-3 An automobile of mass 1500 kg moving on a level highway experiences a frictional force corresponding to $\mu_k = 0.4$. How much work against friction is done in traveling each kilometer?

8-4 To put the top block of weight 20 tons on the pyramid of Cheops at Gizeh, height 481 ft, base 756 ft (see Figure P8-4), how many workers each capable of 100-lb force were required, assuming the ramp lay along the flat side of the pyramid and neglecting friction?

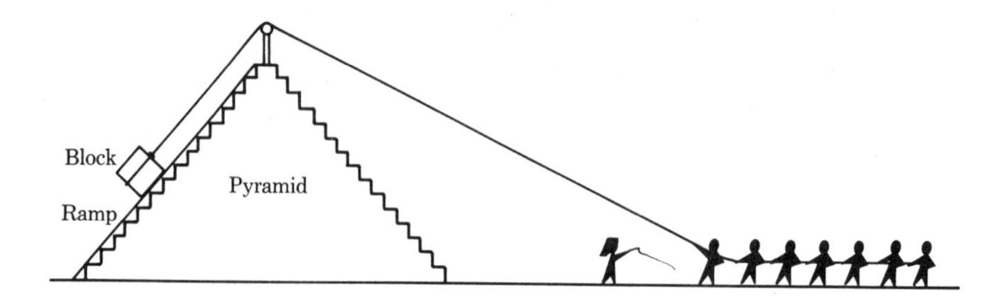

FIGURE P8-4

8-5 One wire of length 0.04 m in an electric motor carries a current of 0.25 A and experiences a magnetic field of 0.3 Wb/m^2, causing it to move on a radius of 0.01 m (see Section 7-1). How much work is done per revolution? How many such wires would be needed to develop 0.1 joule per turn?

8-6 Two people push with equal force of 25 lb at right angles on a steamer trunk, and move it a distance of 6 feet. Find the work done by each and the sum. Also calculate the work they do with their combined (vector) forces. Discuss the results.

8-7 A proton experiences an electric field of 10^4 N/C in the 0.05-m gap of a linear accelerator. Find the electric force on the proton and the work done in carrying it across the gap.

8-8 Compute the kinetic energy of a proton, $m = 1.67 \times 10^{-27}$ kg, that has a speed 3×10^4 m/sec.

8-9 The potential energy of a positive ion $(Cs^{133})^{(+)}$ released at the surface of a negatively charged plate of a spaceship accelerator is 7.5×10^{-15} joule. Find its speed on reaching a positive plate at "ground."

8-10 Calculate the potential energy in joules and electron-volts of an electron at distance 5.3×10^{-11} m from a proton. The charge of each particle is 1.60×10^{-19} coulomb.

8-11* Investigate the strategy of driving a car on a highway in the hills. Is it more efficient to take it out of gear and coast down each incline, then accelerate to reach the next peak, or is it more efficient to maintain a constant speed? (Consider friction and the matter of time elapsed.)

8-12 A car of mass 1500 kg is moving on level ground at speed 60 mi/hr. It encounters a 5-deg hill, and just enough force is applied to balance friction. Find how far it will go along the incline before stopping, using conservation laws.

8-13 A driver's examination has the following question: "A car with speed 30 mi/hr hits head-on with another moving at 40 mi/hr. The damage will be the same as a car hitting a solid wall at (check one) ___ 70 mi/hr, ___ 10 mi/hr, ___ 50 mi/hr, ___ 35 mi/hr." Assuming that all kinetic energy goes into physical damage, what is the best answer?

8-14 Calculate the work done in sliding a 50-kg barrel of radius 0.4 m up to a platform 2 m high using a frictionless incline of 30 deg (see Figure P8-14). Assume a force parallel to the incline and through the center of the barrel. What is the mechanical advantage of this "machine"?

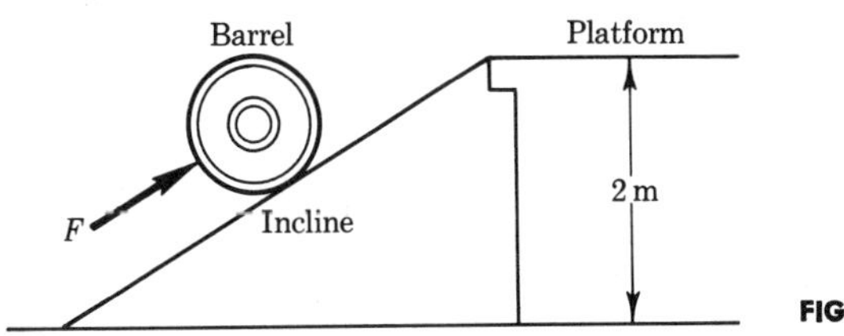

FIGURE P8-14

8-15* Two men are in a car that suddenly has a flat tire, and they must lift the car 4 in. off the ground to change wheels as shown in Figure P8-15. They have access to a board 16 ft long and 2 in. thick; a log of 3 in. diameter and 14 in. length; and a block and tackle with six cables. Assuming that the car has a total weight of 2200 lb, that its frame is 12 in. off the ground, and that the two men weigh 150 lb each, what appears to be the best method? Are there any others?

FIGURE P8-15

8-16* Find the motion of a ballistic missile, using Newton's laws. If the projectile is released at an angle θ with the ground, the horizontal component of velocity is $v_0 \cos \theta$. This remains unchanged, since the force of gravity acts vertically. The vertical component of velocity $v_0 \sin \theta$ is reduced to zero, then increases to its initial value by the end of the range. Find the maximum height and the range as they depend on angle θ. Discuss the results and compare with a solution based on conservation.

8-17 (a) Show that the potential energy as a function of position above the Earth's surface—if the reference level is chosen as any definite x_0 rather than zero—is $E_p(x) = mg(x - x_0)$. (b) Verify that, if a particle is dropped from height x_1, the mechanical energy is then $E = mg(x_1 - x_0)$. (c) Prove that $E_k = -\Delta E_p$ for motion between any two points x_2 and x_3.

8-18 Find the speed of an electron accelerated through a potential difference of 1 volt. Is this speed in the relativistic region? If the potential difference is 1 kV, what percentage mass increase occurs?

8-19 A charge of 10^{-8} C can be produced by friction. What would be the potential of a 1-cm sphere so charged?

8-20 Find the electric potential at the center of a ring of radius 0.5 m, with total charge 7.5×10^{-10} C.

8-21 Find the electric potential at a distance of 4×10^{-11} m from the nucleus of the helium atom $_2He^4$. Assume that the nucleus, the two electrons, and the point of observation all lie on a line, and that the radii of electron orbits are 2.6×10^{-11} m.

8-22 Find the electric potential V at a distance x from the plane of a ring of radius a, total charge Q, along an axis through an origin at the center of the ring.

8-23 How much energy in joules and kilowatt-hours would be required to ionize 1 kg of free hydrogen atoms? How many Btu is this?

8-24 In a nuclear experiment, a proton of kinetic energy E_k moves in a circular orbit of radius R due to a uniform magnetic field $\mathcal{B}$. What energy must—(a) an alpha particle, and (b) a deuteron—have to circulate in the same orbit?

8-25* A projectile is fired vertically from the Earth's surface with an initial velocity of 10 km/sec. Neglecting air friction and rotation of the globe, how high will it go? (Assume the Earth's radius to be 6400 km.)

8-26 Find the gravitational acceleration and the escape speed for the planet Jupiter. Note that $M_J/M_E = 318$ and $R_J/R_E = 11.2$.

8-27 What is the escape speed from the Moon, if $M_M/M_E = 0.0123$ and $R_M/R_E = 0.272$? From a surface-level orbit around the Moon?

8-28 How heavy a load can a 40-hp hoist lift at a steady speed of 126 ft/min?

8-29 A volume of 15 m^3 of water goes over a waterfall 30-m high every second. Find the power involved in this process. Where does the energy go?

8-30 A runner does the 26-mi–385-yd marathon in a time of 2 hr 30 min, exerting an average force on the ground (to propel himself) of 20 lb. What horsepower does he develop? Noting that 1 calorie = 4.185 joules, how many calories of energy are expended? How does this compare with the average total daily need of about 2000 Calories of food energy?

8-31 Electricity costs about 1.2 cents per kilowatt-hour. What is this in cents per kilocalorie? How much would it cost to drive a 300-hp electric car from New York to Chicago, a distance of 800 miles, at an average speed of 40 mi/hr? How would this compare with an ordinary car that gets 20 mi/gal at 35 cents/gal?

8-32 A rocket is propelled with a thrust of 3×10^5 lb. When moving at a speed of 4000 mi/hr, what power does it develop in ft-lb/sec? In horsepower? In kilowatts?

8-33 The amount of power developed by a car of weight 3000 lb against air friction when moving with uniform speed of 60 mi/hr on a level highway is 150 hp. How much force is applied? If the friction remains constant as the car goes onto a 3-deg grade, estimate the new horsepower required to maintain speed.

8-34 (a) Find the temperature at which the reading is the same on the Celsius and Fahrenheit scales. (b) At what point is the numerical value of temperature as much above zero on one scale as it is below zero on the other?

8-35 The Rankine (R) temperature scale has the same zero point as the Kelvin scale, but is graduated in Fahrenheit degrees. What is room temperature, 20°C, in degrees Rankine?

8-36 How great a temperature rise would result from the fall of 1 kg of water through 200 m, if all kinetic energy is converted into thermal energy?

8-37 How much heat is developed (in J and cal) in the brake linings of a 1000-kg car stopped from 30 m/sec (about 70 mph)?

8-38 Show that 1 cal/gm-°C = 1 Btu/lb-°F, and that 1 kcal/mole-°C = 0.0435 eV/molecule-°C.

8-39 We put a kettle containing 5 kg of water on the stove. How many kcal, joules, eV, and ft-lb are required to bring it from room temperature (20°C) to the boiling point (100°C)?

8-40 A container of volume 22.4 liters holds 32 grams of oxygen gas at 0°C. If the total heat content is 0.810 kcal, what is the energy of each molecule of O_2 in joules? In electron-volts?

8-41 Sketch a graph of the gravitational potential in the region from the Sun to the Earth to the Moon, at an instant when they are lined up. *Note:* $M_S/M_E = 3.3 \times 10^5$, $M_M/M_E = 0.0123$, with orbits $a_E = 93 \times 10^6$ mi, $a_M = 2.39 \times 10^5$ mi.

8-42 The energy required to dissociate 1 molecule of H_2O into H_2 and $\frac{1}{2}O_2$ by electrical means is 69 kcal per mole of H_2O. What amount of energy per molecule dissociated is that, in electron-volts?

8-43 A rocket of mass 7×10^4 kg reaches the height of 150 km with a speed of 2000 m/sec, having been provided with 4×10^{11} joules of energy through the burning of propellant. How much energy was lost because of frictional resistance of the air through which the rocket passed?

8-44 The United States consumes about 10^{16} watt-hours of electrical energy a year. How many kilograms of matter would have to be converted into energy to yield this amount?

8-45 How many kilograms of mass must be converted into energy to ionize 1 kg of hydrogen? (See Problem 8-23.)

8-46 A constant force is applied to a particle of rest mass m_0 for a time t. The momentum is then $p = Ft$. Find a formula for the speed v as it depends on time. Will the relation reduce to classical results if v is much less than c?

8-47* A constant force F is applied to a spaceship, which eventually approaches the speed of light. The distance traveled as a function of time may be found, by forming the area under the $v(t)$ curve of Prob. 8-46, to be

$$s = \frac{c^2}{a_0}\left[\sqrt{1 + \left(\frac{a_0 t}{c}\right)^2} - 1\right]$$

where a_0 is the initial acceleration F/m_0. How long would it take for a spaceship with constant excess thrust corresponding to 1 "g" to reach the nearest star, 4.3 light-years away?

8-48 Find the amount of mass-energy in joules and MeV of a proton at rest and at speed 0.999 times that of light.

8-49 What speed must a particle have in order to increase its rest mass by 10%? By what fraction then is the low-speed formula $E_k = \frac{1}{2}m_0v^2$ in error?

8-50 What is the lowest frequency of light that can dislodge an electron from cesium?

8-51 How many photons of frequency 2×10^{13} cps must fall on a surface each second to give a total incident power of 1 watt?

8-52 Yellow-orange light has a frequency of around 5×10^{14} cps. What energy in joules and eV is this? How many of such photons of light would have to be absorbed by 1 gram of water to raise its temperature 1°C?

MOTION OF MANY-PARTICLE SYSTEMS

9

The concept of fields has enabled us to find the forces due to many particles by the process of addition or superposition. This is especially helpful in providing a description of magnetic effects caused by moving charges. We shall now study the motion of connected or interacting particles—as in large-scale bodies—that are held together by strong atomic forces.

9-1 TYPES OF MOTION

Up to now, we have applied Newton's laws of motion only to objects that could be treated as single particles or point masses. The types of motion that they experienced were of several simple classes (Figure 9-1): the case of *no motion* is obvious; a falling body moves in a *straight line* with respect to the surface of the Earth, while a projectile moves in a *plane curve;* the spiraling of charge in a uniform magnetic field is an example of motion in a *space curve*.

For objects that can be viewed as moving like point particles, size and shape have no meaning. We know, however, that all real bodies of matter have dimensions, whether they be molecules or galaxies. Even a molecule of H_2O is a rather complex structure—three interconnected nuclei plus many electrons. The wheel of an automobile with its tire, rim, hub, and other parts is similarly complicated. Such objects rotate as well as

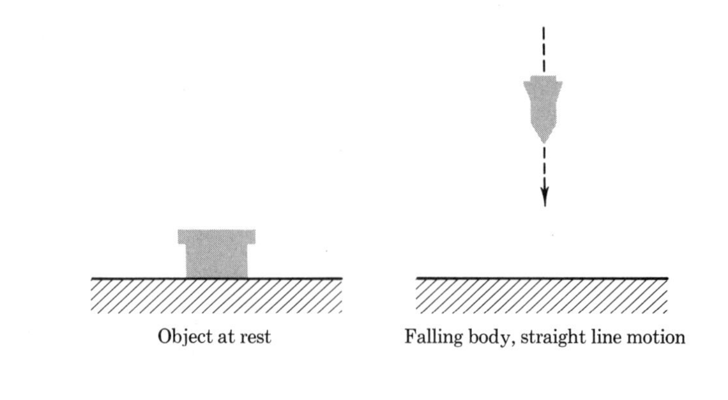

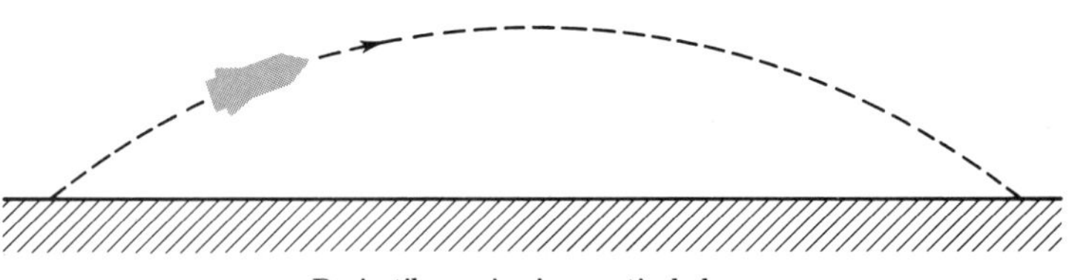

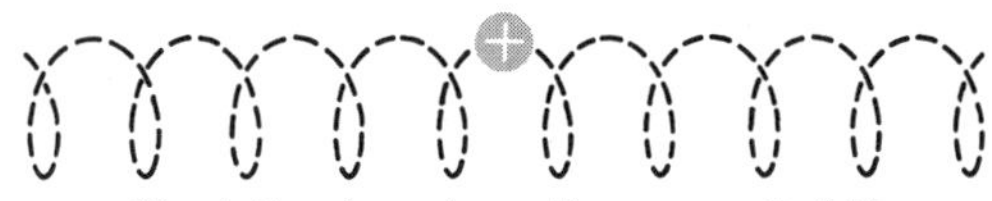

FIGURE 9-1

Types of particle motions

move along some path when they experience forces, which clearly taxes our powers of physical description. We shall find that it is possible to continue to use Newton's laws to determine the motion.

While we analyze this more involved motion, it will be convenient to have the picture of an object in mind. Visualize, then, a simple two-particle structure with rigid connection; examples are the barbell used by weight-lifters, the diatomic molecules of hydrogen H_2 (two equal masses), or nitric oxide NO (two unequal masses). We ignore any mass in the connecting rod or electrons, and also assume that there is no vibration along the line connecting the particles, now viewed as point masses.

We can visualize many types of motion, but shall restrict our attention to four possibilities:

no motion at all;
translation only;
rotation only;
translation and rotation.

(These are sketched in Figure 9-2.) We may also classify motions in terms of actions on the body. If no forces are present or if they are in balance, an object will remain at rest, or in constant motion in a straight line (uniform translation). This state is called *translational equilibrium.* As we shall see shortly, if there are no torques or they are in balance, an object will remain at rest or in motion in a circle (uniform rotation). This state is called *rotational equilibrium.*

The term *static equilibrium* is reserved for the case where the system undergoes neither translation nor rotation. Static equilibrium can in turn

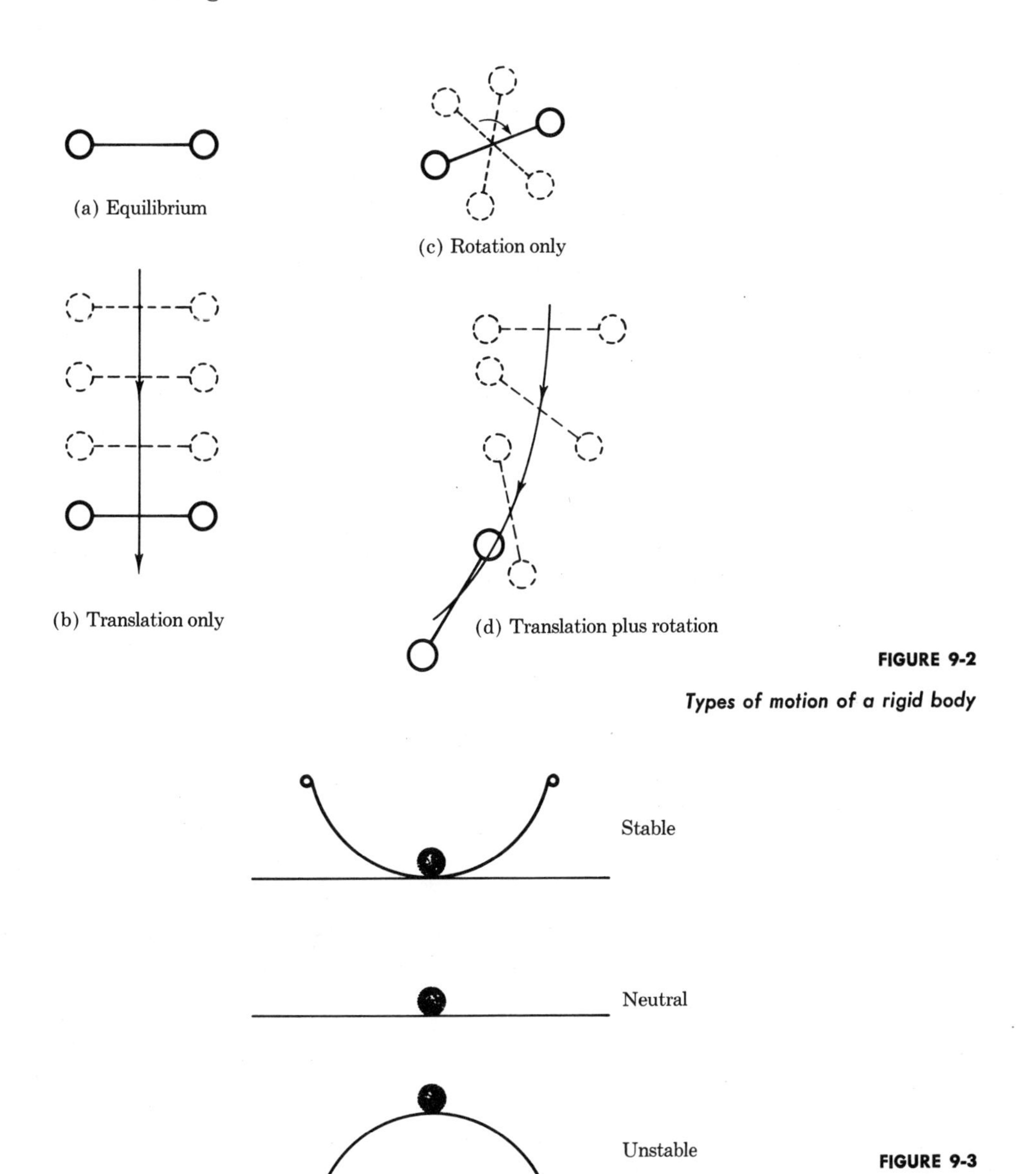

FIGURE 9-2

Types of motion of a rigid body

FIGURE 9-3

Classes of static equilibrium

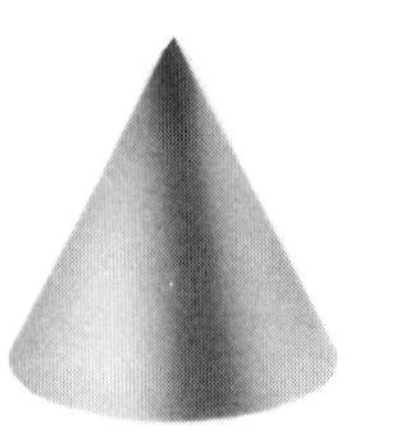

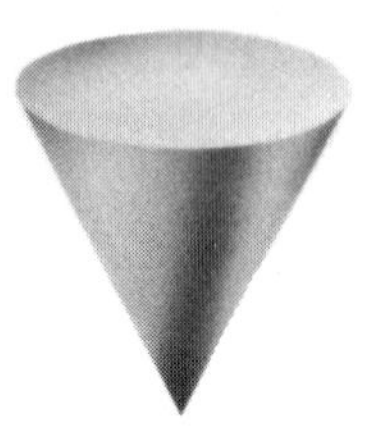

FIGURE 9-4

Balancing a cone

be further classified, according to the result of disturbing the system, as stable, neutral, or unstable. If, as in Figure 9-3, a marble is put in the bottom of a hemispherical bowl, then moved slightly, it will return to its former position—that is, it was originally in a state of *stable equilibrium*. If it were on a level surface, it would remain in a new position, so it was originally in *neutral equilibrium*. If it were originally resting on an inverted bowl, the marble would move away on being displaced, thus it was originally in *unstable equilibrium*.

The reader is invited to decide to which class of equilibrium each of the cones shown in Figure 9-4 belongs.

9-2 EQUILIBRIUM

We first examine the simplest case of static equilibrium, in which there is no motion at all as the result of a balance of influences.

Let us look at the H_2 molecule or the weight-lifter's dumbbell in a uniform field—say, the Earth's gravity—assuming them to be rigid. Each mass will experience a force $F = mg$ downward (Figure 9-5). If we insist that there be no acceleration—i.e., net force zero—we must support the structure by an upward force equal to $2F$. This is a special case of Newton's first law: *if the net force is zero, there will be no change in translational velocity.* We can easily guess that the point of application of a single supporting

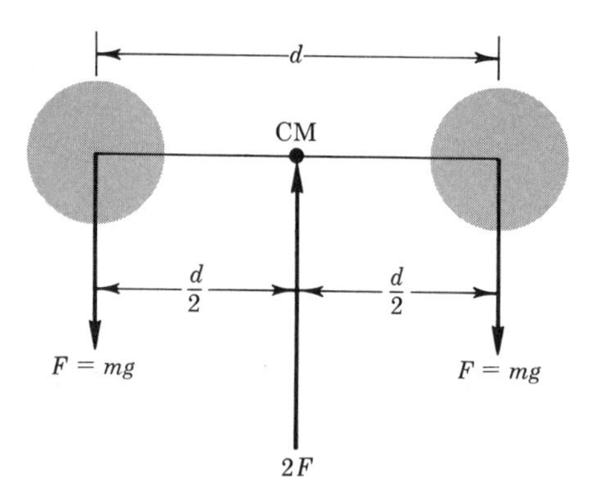

FIGURE 9-5

Support at the center of mass

force should be midway between the two particles, for only then will there be no tendency to rotate: the two particles will balance on the supporting point. This point is the simplest example of the *center of mass*, a concept that plays an important role in motion problems.

It is necessary to examine carefully the tendency to rotate. Experience tells us that the amount of twisting action depends not only on the magnitude of the force applied, but also on its distance from the point about which the rotation takes place. A measure of the ability to cause rotation is *torque*: the torque N is defined as a product of two factors—the distance from the pivot to the point of action of the force, called the *lever arm;* and the component of force perpendicular to that line. Looking again at the hydrogen molecule in Figure 9-5, the lever arm is $d/2$ and the force is F. There is a torque $(d/2)F$ on one atom, tending to rotate the molecule in the clockwise direction about the midpoint, and another equal torque tending to rotate it counterclockwise. In order to ensure that no rotation takes place, we have to locate the support in such a way that opposing

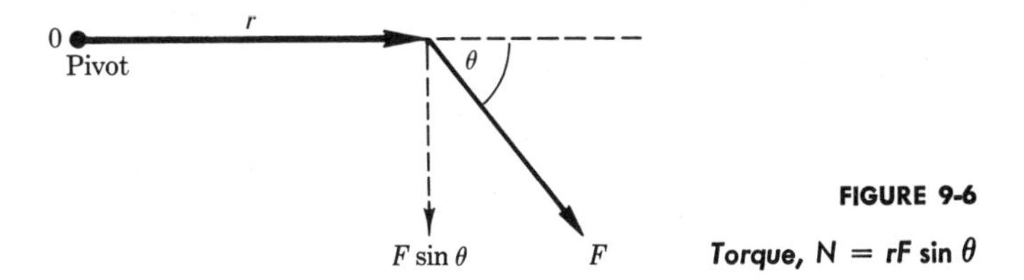

FIGURE 9-6
Torque, $N = rF \sin\theta$

torques are in balance. Suppose instead, however, that a force is directed at an angle θ from the horizontal, as in Figure 9-6. The force component tending to cause rotation is $F \sin\theta$ and the torque is

$$N = rF \sin\theta$$

For example, if we pull with a force of 30 N at right angles ($\theta = \pi/2$, $\sin\theta = 1$) to the handle of a hammer of length 0.2 m, the torque we apply is $N = (0.2 \text{ m})\ (30 \text{ N})\ (1) = 6$ N-m. The units of torque are seen to be the same as those for work.

As another example, suppose we tighten a nut by use of a wrench of lever arm 6 in. by applying a force of 20 lb at an angle θ of 30 deg to the handle of the wrench. The torque is

$$N = (0.5 \text{ ft})(20 \text{ lb})(\sin 30°) = 5.0 \text{ lb-ft}$$

Notice that we also could regard the torque as the product of the force F and the lever arm $r \sin\theta$, which is the distance from 0 out to the extension of the line of force.

Just as we consider forces to the right as positive and forces to the left as negative, we label counterclockwise torques as positive (tending to increase the angle) and clockwise torques as negative (tending to decrease the angle). A balance of torques means the same as a *net torque* of zero. Clearly, if the hydrogen molecule is initially at rest and the net torque on it is zero, there will be no tendency for the structure to rotate. Such a state is an example of *static rotational equilibrium.*

Now let us turn to a two-particle structure with unequal masses—for example, the molecule nitric oxide NO, with relative masses (and weights) 14 for nitrogen and 16 for oxygen, and d the separation of the particles (around 10^{-10} m). Where should we locate the upward force of 30 units to prevent both translation and rotation? The origin will be placed for convenience at the pivot point—Figure 9-7(a)—an unknown distance $\bar{r}$ from the midpoint. Let us form the two torques, whose magnitude must be equal:

$$N_1 = r_1F_1 = \left(\frac{d}{2} + \bar{r}\right) 14$$

$$N_2 = r_2F_2 = \left(\frac{d}{2} - \bar{r}\right) 16$$

Equating and solving for the unknown $\bar{r}$ gives

$$\bar{r} = \frac{16\left(\frac{d}{2}\right) + 14\left(-\frac{d}{2}\right)}{16 + 14} = \frac{d}{30}$$

The point at which we apply the force is again called the *center of mass.* It is slightly nearer the heavier atom, as we might expect.

The location of the origin is immaterial, as the reader may verify by repeating the calculation with origin at the center of either atom, or even

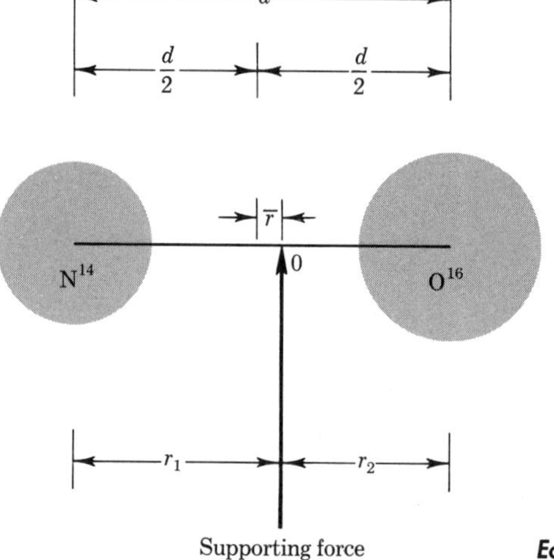

FIGURE 9-7

Equilibrium of torques on a rigid NO *molecule*

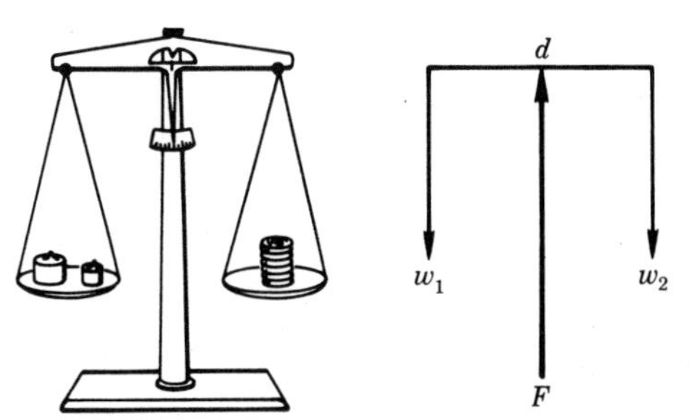

FIGURE 9-8

Simple balance

outside the range of distance d. Once the coordinate system is specified, the location of the center of mass is given by the condition on equality of torques

$$(F_1 + F_2)\bar{r} = F_1 r_1 + F_2 r_2$$

and, letting $F_1 = m_1 g$, $F_2 = m_2 g$ as for gravity,

$$\bar{r} = \frac{m_1 r_1 + m_2 r_2}{m_1 + m_2} \qquad \textit{(center of mass of two particles)}$$

Let us find the center of mass of the hydrogen atom, consisting of a proton of mass $m_1 = 1.67 \times 10^{-27}$ kg, which is 1836 times that of the electron of mass $m_2 = 9.11 \times 10^{-31}$ kg. The "centers" of the particles are separated by an average distance $d = 5.3 \times 10^{-11}$ m. We place the origin at the proton, so that $r_1 = 0$. Then,

$$\bar{r} = \frac{m_2 r_2}{m_1 + m_2} = \frac{d}{m_1/m_2 + 1} = \frac{5.3 \times 10^{-11}\ \text{m}}{1836 + 1} = 2.7 \times 10^{-14}\ \text{m}$$

This is such a very small distance that we can say that the center of mass of the hydrogen atom is practically *at* the proton.

We are familiar with the classical scale or balance used for comparing weights, as in Figure 9-8. The supporting force F is $w_1 + w_2$, and the equal torques are $(d/2)w_1$ and $(d/2)w_2$. A little algebra shows that $w_1 = w_2$ and $F = 2w_1$, as expected.

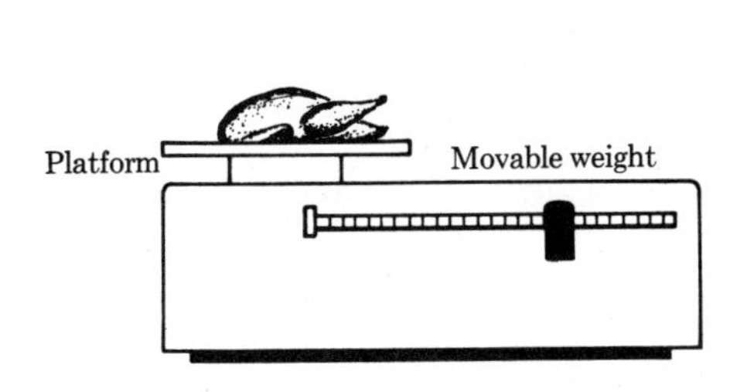

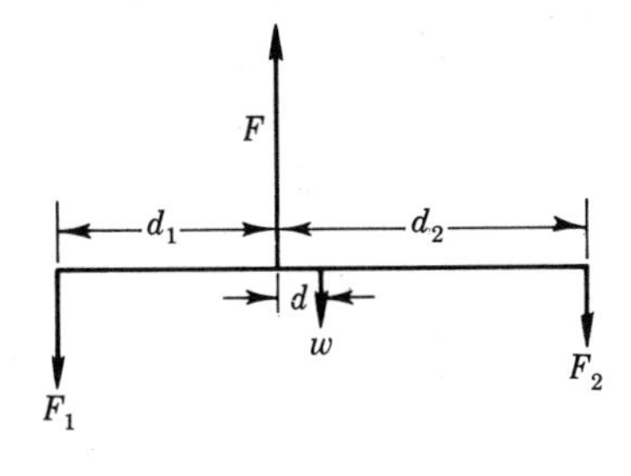

FIGURE 9-9

Scale with unequal arms

In a simple grocer's scale (Figure 9-9), the arms are unequal and the connecting bar has a definite weight. If the bar is uniform, for purposes of computing torques we may imagine its weight concentrated at its center of mass. The force balance is then

$$F = F_1 + F_2 + w$$

and the torque balance is

$$F_1 d_1 = F_2 d_2 + wd$$

These equations can easily be solved to obtain one of the distances:

$$d_2 = \frac{F_1 d_1 - wd}{F_2}$$

Let us investigate the problem of equilibrium for a uniform ladder leaning against a smooth wall, as in Figure 9-10. We know that there must be friction between the ladder and the ground, or else the ladder would fall

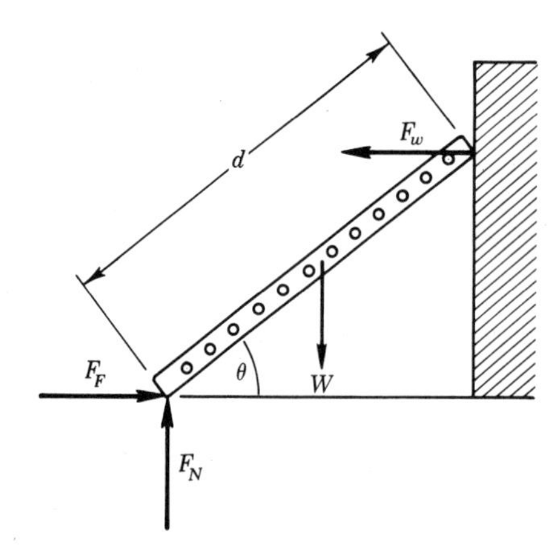

FIGURE 9-10
The ladder as a problem in equilibrium

at once. Let us find the necessary coefficient of friction. First we must identify the forces and torques on the body, and then apply the conditions for equilibrium—that is, forces up = forces down; forces right = forces left; torques clockwise = torques counterclockwise. We may then solve the resulting equations in general for the unknowns. Consider the following quantities:

W = weight of ladder, acting downward at the center of mass
F_w = reaction of the wall
F_F = frictional force between ladder and ground
F_N = reaction of the ground

It is clear that

$$F_w = F_F$$
$$W = F_N$$

and we use the relation $F_F = \mu F_N$, as in Section 5-7. Using the middle of the ladder as the pivot, we may draw up the following table of torques:

Counterclockwise		*Clockwise*	
Force	*Lever Arm*	*Force*	*Lever Arm*
F_w	$\left(\frac{d}{2}\right) \sin \theta$	F_N	$\left(\frac{d}{2}\right) \cos \theta$
F_F	$\left(\frac{d}{2}\right) \sin \theta$		

Equating torques in the opposite senses and simplifying,

$$F_N \cot \theta = F_w + F_F = 2F_F = 2\mu F_N$$

Then

$$\mu = \frac{\cot \theta}{2}$$

For example, if the angle is 60 deg then $\cot \theta = 1/\sqrt{3}$, and the smallest coefficient of friction that will prevent sliding is

$$\mu = \frac{1}{2\sqrt{3}} = 0.289$$

9-3 TRANSLATION AND THE CENTER OF MASS

We have seen that a structure consisting of two atoms could be supported in a uniform gravitational field by the application of a single force at the center of mass. Experience tells us that if the support were suddenly removed, the object would fall without turning; or, if we exerted a stronger force upward, the object could be accelerated upward, again without rotating. These observations can be stated generally, using Newton's laws of motion.

For example, when the "dumbbell" molecule NO falls freely, the forces on the two masses are F_1 and F_2, and the accelerations are a_1 and a_2, here both equal to $a = g$, the acceleration of gravity. Applying Newton's second law to each atom,

$$F_1 = m_1 a$$

$$F_2 = m_2 a$$

Now, the total force on the structure, which undergoes an acceleration a, is

$$F = F_1 + F_2 = (m_1 + m_2)a$$

Had there been no gravity, we could have produced the very same motion by applying a force at the center of mass (thus preventing rotation) of

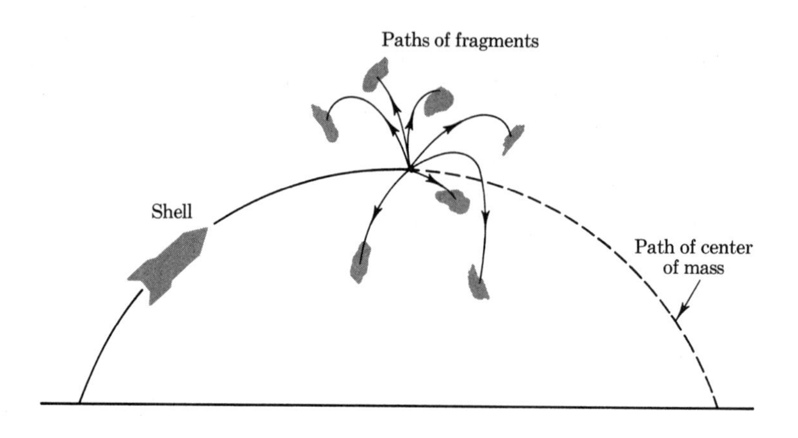

FIGURE 9-11

The bursting shell

amount F. We could have imagined the two masses concentrated there, giving a total mass $m = m_1 + m_2$. Our formula tells us that $F = ma$. In other words, *the translational motion of a system is given by Newton's second law of motion using the resultant external force applied to the total mass as if located at the center of mass.* The derivation using vectors leads to the same conclusion, in the form

$$\mathbf{F} = m\bar{\mathbf{a}}$$

where $\bar{\mathbf{a}}$ is the acceleration of the center of mass.

In the case that there are no external forces, we would then deduce that there is no acceleration of the center of mass. A graphic example, in case there is an external (gravity) force, is the behavior of the fragments of a shell that bursts in flight (Figure 9-11). Even though fragments fly in various directions, the center of mass proceeds undisturbed in the original trajectory. This situation suggests to us that a new type of conservation is involved. Just as mechanical energy is conserved, the momentum of a system of particles is conserved—i.e., remains constant—if there are no external forces. In the following chapter we shall elaborate on this concept.

Evaluation of Center of Mass for Different Objects

If there are many particles of masses m_i, where $i = 1, 2, 3$, etc., the general form of the expression for the location of the center of mass is

$$\bar{r} = \frac{\sum_i m_i r_i}{\sum_i m_i} \qquad \textit{(center of mass)}$$

where the symbol $\sum_i$ means the sum over all particles—e.g., $\sum_i m_i = m_1 +$

$m_2 + \ldots$. We can easily verify that this expression reduces to the expression for two particles. A material object such as a metal rod or a wheel or bowling pin consists of billions of particles, and we cannot hope to add up the individual effects of each one to obtain the center of mass. In general, the process of integration of the calculus is required. However, we can often guess the location of the center of mass of regular objects composed of a large number of particles. For example, the center of mass of a long thin uniform rod is at the center of the rod, and at the center of a uniform sphere or cube as well. Starting with such knowledge, we can deduce where the center of mass of more complicated objects is located.

For instance, let us find the center of mass of a figure formed by bending a rod at right angles (Figure 9-12). We can regard this rod as being

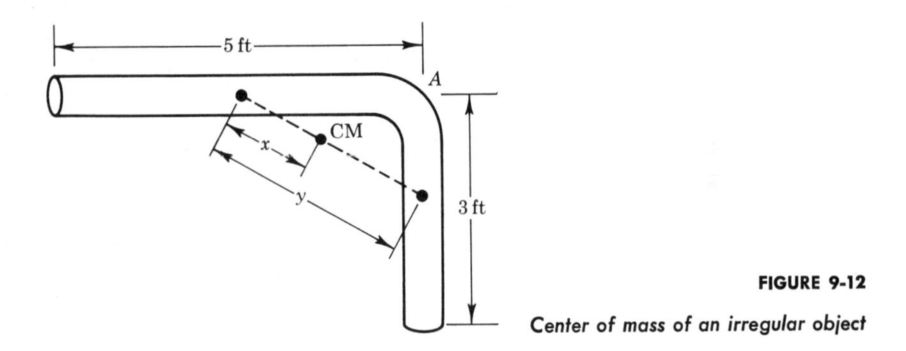

FIGURE 9-12

Center of mass of an irregular object

composed of two pieces by mentally cutting it at point A. Then we have two rods of lengths 5 ft and 3 ft. The center of mass of each would be as indicated at points $2\frac{1}{2}$ ft and $1\frac{1}{2}$ ft from the respective ends. It is clear that the center of mass of the combination would be on a line connecting these two points. The exact distance may be determined by imagining the entire mass of each side to be located at its center of mass. From the relation

$$5x = 3(y - x)$$

where

$$y = \sqrt{(1.5)^2 + (2.5)^2} = 2.92 \text{ cm}$$

we obtain

$$x = \tfrac{3}{8}y = 1.10 \text{ cm}$$

The center of mass (or *center of gravity*) of a flat object may be determined experimentally, as shown in Figure 9-13, by making use of the fact that since each particle of matter in a body is attracted toward the Earth, the *weight* of a body is simply the resultant of all these forces of attraction. If the body is suspended by some arbitrary point A, then the center of gravity must lie on a vertical line below A. Choosing a new point

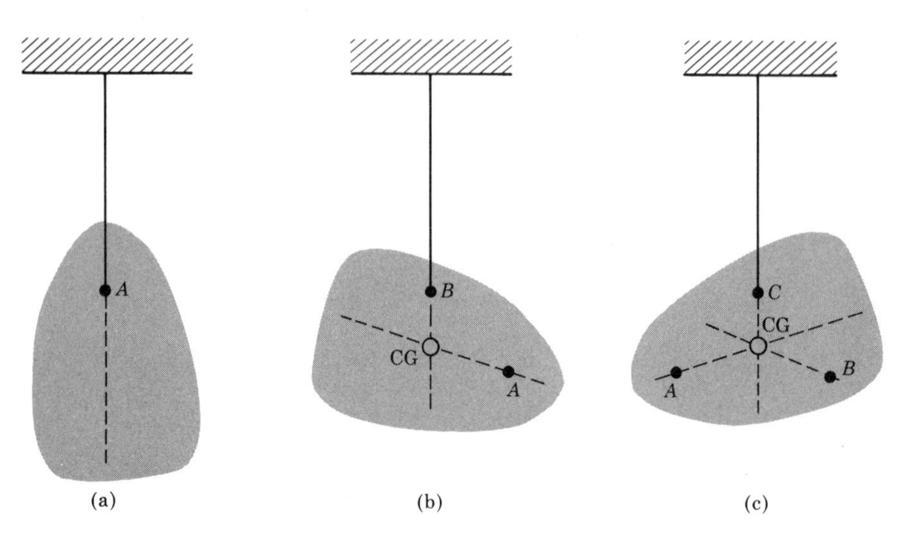

FIGURE 9-13

Experimental measurement of the center of mass of a flat plate

of suspension, B, we then find that the same argument is true for the new orientation. The center of gravity is thus located at the intersection of the vertical lines that were drawn while the body was suspended from two different points. If the body is suspended by a third point C, a vertical line through this point will be found to pass through the point of intersection of the first two lines. The center of mass of an isosceles triangle can be found by such methods, or by calculus, to be located $\frac{2}{3}$ of the way from the apex to the base—of course, along the altitude.

9-4 ROTATION AND MOMENT OF INERTIA

We shall now examine the features of a rigid object that determine its resistance to being turned. Just as mass is a measure of translational inertia, we find that the *moment of inertia* is a measure of rotational inertia. Let us examine the situation where only rotation is allowed.

Our analysis will be simpler if we consider only one particle, located at a point as in Figure 9-14. The motion possible on application of a force F perpendicular to the radial line is circular, which suggests the use of polar coordinates r and θ to locate the particle. We recall Newton's second law in the form

$$F = \frac{\Delta p}{\Delta t}$$

in words, force equals the ratio of the change in momentum to the elapsed time. Multiplying both sides of the equation by r, letting $N = rF$ (which

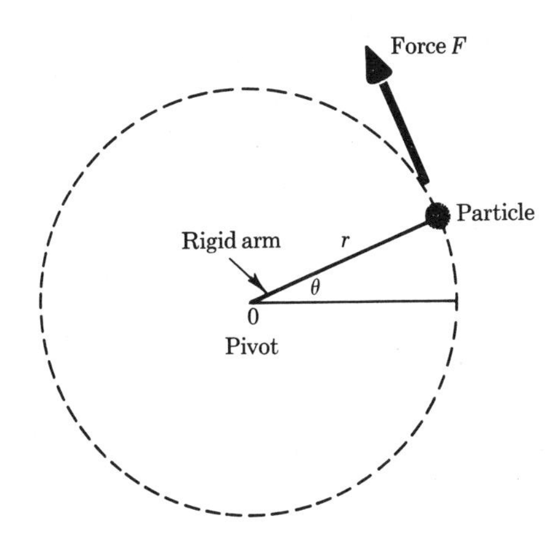

FIGURE 9-14

Rotation and angular momentum for a particle

is the torque), and letting $L = rp$ be a new quantity called the *angular momentum*, we find

$$N = \frac{\Delta L}{\Delta t}$$

This is *Newton's law for angular motion*, which states: *the externally applied torque is equal to the time rate of change of the angular momentum.*

We saw earlier, in the study of the hydrogen and NO molecules, that if the net torque on a structure at rest is zero, the system will not tend to rotate. We can now generalize that result by use of Newton's law for angular motion, by noting that if N is zero, the rate of change of the angular momentum L is zero. In other words, if the net external torque is zero, the angular momentum is constant. A new conservation concept has emerged, which we shall discuss more fully in the next chapter.

Moment of Inertia

We now introduce a quantity that is analogous to mass, but which refers to rotational motion. Consider once more the single particle in Figure 9-14. For a fixed arm of length R, the motion is in a circle. Then $L = pr = mvR$ and, since $v = \omega R$, $L = mR^2\omega$. Newton's law then becomes

$$N = mR^2 \frac{\Delta\omega}{\Delta t}$$

The rate of change of angular speed is the angular acceleration $\Delta\omega/\Delta t = \alpha$. Also, letting the product mR^2 be abbreviated as I—called the *moment of inertia*—we have

$$N = I\alpha \qquad \text{(Newton's law for rotation of a rigid body)}$$

The analogy of this expression to $F = ma$ has been forced, so that corresponding factors are N and F as applied influences, α and a as accelerations, and I and m as measures of inertia.

A rigid two-particle structure with pivot at the center of mass obeys the same form of Newton's law for angular motion if we take N to be the *net* torque and I to be the *total* moment of inertia, such that $I = I_1 + I_2 = m_1r_1^2 + m_2r_2^2$. For example, for the diatomic molecule NO, it is easy to see that $r_1 = \frac{8}{15}d$, $r_2 = \frac{7}{15}d$, and, with $m_1 = 14$ and $m_2 = 16$,

$$I = 14(\tfrac{8}{15}d)^2 + 16(\tfrac{7}{15}d)^2 = 7.5d^2$$

For the study of translation, we found earlier that we could assume the entire mass of an object to be located at the coordinate of the center of mass. For purposes of studying rotation, a convenient quantity to work with is the *radius of gyration:* the distance k from the axis of rotation where all the mass should be concentrated to give the same moment of inertia I as the extended body. Such a single particle would have a value of I equal to mk^2, so that

$$k = \sqrt{\frac{I}{m}}$$

TABLE 9-1
MOMENTS OF INERTIA

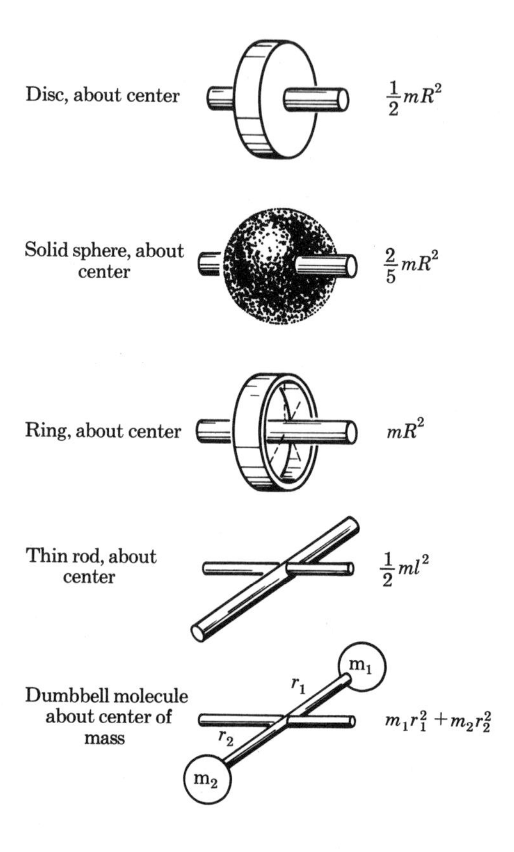

For example, the NO molecule has a total mass $14 + 16 = 30$. Its moment of inertia is $I = 7.5d^2$ about the axis through the center of mass. Thus its radius of gyration is

$$k = \sqrt{\frac{7.5}{30} d^2} = \frac{d}{2}$$

Calculations of moments of inertia for common-shaped objects can be made by superposition, or in general by integration. All of the particles in a thin ring of radius R, rotating in its own plane about its center, are at the same distance from the axis. The total I is thus the total mass times the square of the radius—i.e., $I = mR^2$. We would expect that I for a disc of radius R containing the same mass as the ring to be smaller than this, since its particles are nearer the axis, on the average. It turns out that $I = \frac{1}{2}mR^2$ for the disc. Since the moment of inertia is a measure of resistance to changes in rotation, it is easier to set a disc into motion (or to stop its turning) than to do the same with a ring of the same total mass. Table 9-1 shows I values for a variety of rigid bodies.

9-5 ANALOGY BETWEEN TRANSLATION AND ROTATION

In our study of motion, we emphasized similarities between translation and rotation. For example, translational speed is a linear displacement divided by the time interval, $v = \Delta x/\Delta t$, and rotational speed is an angular displacement divided by the time interval, $\omega = \Delta\theta/\Delta t$. We can take advantage of this analogy both in the process of thinking about various motions and in writing formulas to describe motion, especially when we restrict our attention to the rotation of rigid bodies.

Additional analogies between translational and rotational momentum have been indicated: for instance, linear momentum is the product of mass, as a measure of translational inertia, and speed, as a rate of change of position along a line. Also, angular momentum is the product of moment of inertia, as a measure of rotational inertia, and angular speed. Thus $p = mv$ has its analog $L = I\omega$.

Let us turn now to the work done by forces acting in a straight line in comparison with those involving twisting action. In the first case, work is the product of force and the distance through which the force acts, and we would anticipate rotational work to be a product of torque and angle of displacement. To verify this, suppose we apply a force F tangential to a circle on which a particle moves. Now, $W = Fs$, where s is the arc length. However, the angle of rotation is $\theta = s/R$, and the torque is $N = FR$. The product of these is $N\theta = (FR)(s/R)$, which is again Fs, or work. Thus rotational work is $W = N\theta$. Note in passing that the units are the same as those for torque, since angles measured in radians are not represented by absolute units. The analogy carries over further to kinetic energy, which is of course $E_k = \frac{1}{2}mv^2$ in the translational sense. We can readily predict that rotational kinetic energy is $E_k = \frac{1}{2}I\omega^2$, and check the relation by considering a point mass moving on a circle of radius R with constant speed v. Now, the moment of inertia is $I = mR^2$ and $\omega = v/R$. Combining, $E_k = \frac{1}{2}mR^2(v/R)^2$. Although we have used the point mass as our example, it is a straightforward matter to demonstrate the relations for a rigid rotating object. Table 9-2 lists a number of analogous transational and rotational expressions for physical quantities.

TABLE 9-2

ANALOGIES BETWEEN TRANSLATIONAL AND ROTATIONAL QUANTITIES

TRANSLATIONAL		ROTATIONAL
x	Coordinate	θ
$v = \dfrac{\Delta x}{\Delta t}$	Velocity	$\omega = \dfrac{\Delta\theta}{\Delta t}$
$a = \dfrac{\Delta v}{\Delta t}$	Acceleration	$\alpha = \dfrac{\Delta\omega}{\Delta t}$
$p = mv$	Momentum	$L = I\omega$
F	Force, torque	N
m	Inertia	I
$F = ma$	Newton's second law	$N = I\alpha$
$\frac{1}{2}mv^2$	Kinetic energy	$\frac{1}{2}I\omega^2$
Fx	Work and potential energy	$N\theta$
Fv	Power	$N\omega$

9-6 COMBINED MOTION

We shall now examine the movement of objects that have freedom of both translation and rotation. Simple examples are the baton thrown in the air, a diatomic molecule that has just collided with another particle, or a wheel on a moving automobile.

Reviewing our findings for separate motions, we have the forms of Newton's second law $F = ma$ and $N = I\alpha$. If no change in motion occurs, both F and N must be zero. For a change in translation only, N is zero; and for a change in rotation only, F is zero.

For combined motion, we can continue using the two forms of Newton's second law; but we should keep in mind that forces and torques are related, as are linear and angular distances, speeds, and accelerations. Also, the total energy is the sum of the translational and rotational parts. It is important to note that we may view the complete motion of a rigid object as composed of translation of the center of mass plus rotation about the center of mass. Let us apply these ideas to some familiar examples. We flip a coin of radius R, as in Figure 9-15(a). The force F is momentarily applied to the edge of the coin. There are two effects—the

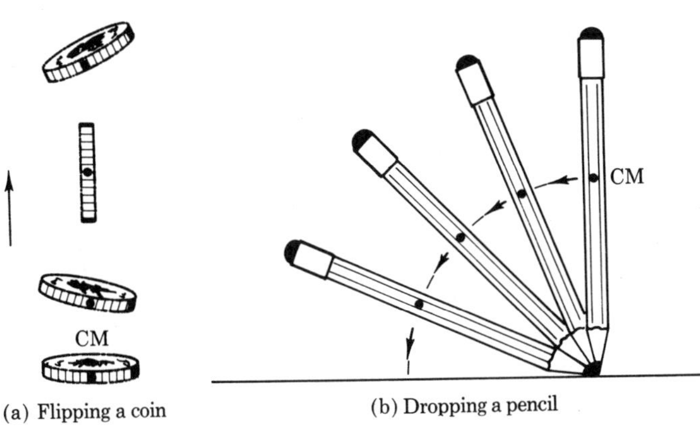

FIGURE 9-15

Illustrations of translation of CM plus rotation about CM

whole mass is accelerated vertically, and the coin is set into rotation. Ignoring gravity, we find the two equations that describe the motion to be

$$a = \frac{F}{m} \quad \text{and} \quad \alpha = \frac{FR}{I}$$

Another example is the pencil falling to the table in Figure 9-15(b). The center of mass clearly traces out a 90-deg arc of a circle. At the same time, the pencil has itself rotated through 90 deg. These are separate motions governed by $a = F/m$ for the acceleration of the center of mass, and $\alpha = N/I$ for the angular acceleration of the whole object.

Rolling on a Level Surface

Suppose that we are asked "What feature of a ball and wheel distinguishes their motion from that of other shapes?" The obvious answer would be that they *roll.* As we shall see, the *way* that they roll depends on several things—mass, moment of inertia, initial motion, and (especially) friction. The following discussion applies to the ball, wheel, disc, cylinder, or ring.

If we toss a ball on the floor in just the right manner, with a special v and ω, it will move off with no slipping or sliding, a case that we shall call "natural rolling." The velocity of a point P on the ball that touches the floor is exactly zero at that instant (Figure 9-16). In this situation there is no velocity of the surface of the sphere relative to the floor, and there is no sliding friction: the ball keeps moving forward and turning. (The normal force and the weight are perpendicular to the plane and through the center of mass, and do nothing.) It is easy to see, from Figure 9-16, that the angular and linear speeds are "matched." If in a time t the

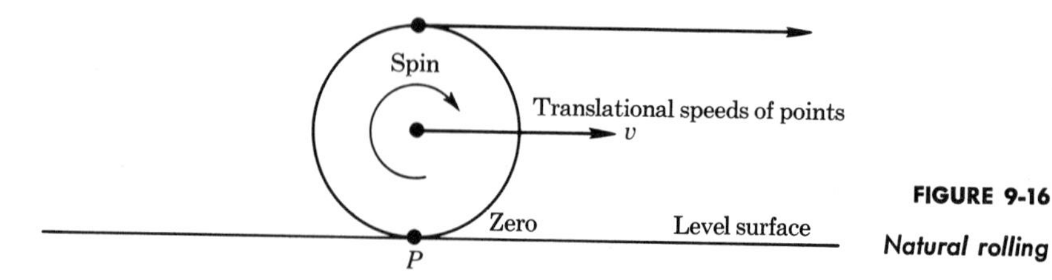

FIGURE 9-16

Natural rolling

ball makes 1 revolution, a point on its edge moves a distance $s = 2\pi R$ *as viewed from the center of mass* (Figure 9-17). However, the center of mass moves a distance $x = 2\pi R$. (We can imagine unrolling the circumference and laying it flat with P always in contact with the surface.) The angular speed is $\omega = 2\pi/t$, the linear speed $v = 2\pi R/t$. Dividing,

$$v = \omega R \quad \text{or} \quad \omega = \frac{v}{R}$$

Such motion would continue indefinitely if it were not for air resistance and distortion of the surfaces, as discussed in Section 4-6.

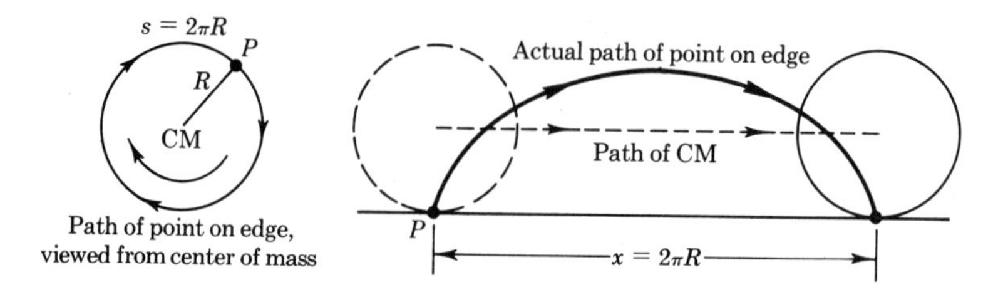

FIGURE 9-17

One revolution of natural rolling

Energy in Combined Motion

When a system includes both translation and rotation, there is a kinetic energy associated with each type of motion. The total kinetic energy is the sum of these:

$$(E_k)_t + (E_k)_r = E_k$$

To illustrate, let us compute the energies for an automobile wheel. Suppose that $m = 8$ kg and $v = 22.4$ mi/hr, which is also 10 m/sec. A wheel of radius $R = 0.4$ m is assumed to be a uniform disc. Its moment of inertia is

$$I = \tfrac{1}{2}mR^2 = \tfrac{1}{2}(8)(0.16) = 0.64 \text{ kg-m}^2$$

and its angular speed is

$$\omega = \frac{v}{R} = \frac{10 \text{ m/sec}}{0.4 \text{ m}} = 25 \text{ rad/sec}$$

The rotational kinetic energy is

$$(E_k)_r = \tfrac{1}{2}I\omega^2 = \tfrac{1}{2}(0.64)(25)^2 = 200 \text{ J}$$

while the translational kinetic energy is

$$(E_k)_t = \tfrac{1}{2}mv^2 = \tfrac{1}{2}(8)(10)^2 = 400 \text{ J}$$

The total is then 600 J. Note in passing that $(E_k)_r$ is exactly one-half of $(E_k)_t$, which holds for any disc that is rolling without slipping.

Rolling on an Incline Without Slipping

Let us use our new ideas on combined motion to find the acceleration of a ball or wheel along an incline (see Figure 9-18). As in Section 5-7, where we discussed the motion of an object on an incline, there is a component of weight $W_{\parallel}$ down the incline and a force of friction F_F up the incline. These provide two effects—translation, since there is a net force down the

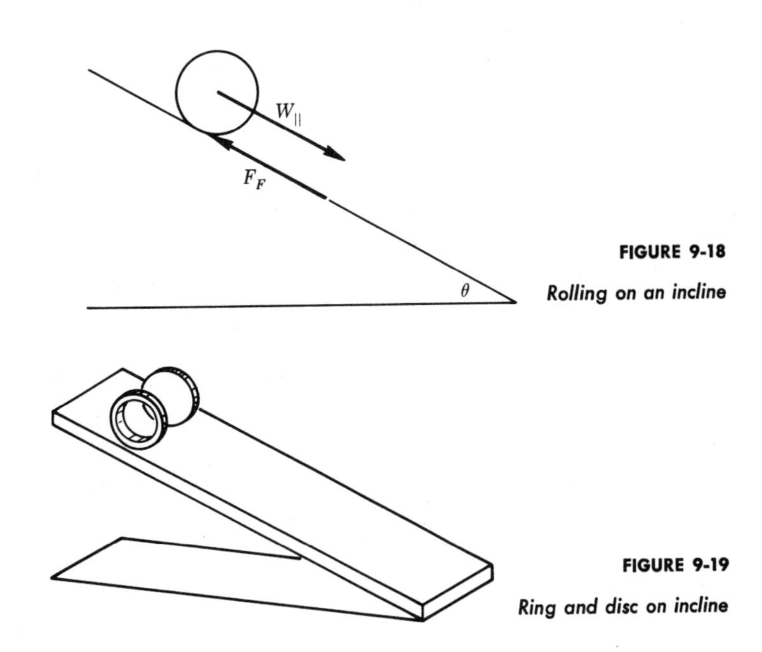

FIGURE 9-18
Rolling on an incline

FIGURE 9-19
Ring and disc on incline

slope; and rotation, since there is a clockwise torque about the center of mass. By applying Newton's linear and angular laws, we can find the acceleration. The interesting result is that, for objects of equal mass, the larger the moment of inertia, the smaller the acceleration. For example, a ring with $I = mR^2$ will take longer to reach the bottom of the incline than a solid disc with $I = \frac{1}{2}mR^2$ (Figure 9-19). The same conclusion is reached from the viewpoint of energy.

Rolling with Sliding

Interesting effects come into play when the requirement for natural rolling is not met, as when the angular speed is very large and there is slippage against a surface. For example, if we toss a ball on the floor with a large initial *forward spin*, the point on the ball in contact with the floor is moving backward. This results in a frictional force forward which slows the rotation but accelerates the ball, which scoots away along the floor. If instead we give the ball a large *backward spin*, there is a large backward friction, which stops the ball quickly; but since there still remains some angular motion, the ball will roll backward. This effect, commonly known as "backspin," is familiar to golfers, bowlers, and table-tennis players.

Mechanical Devices

Many motions involving acceleration of machinery by weights, pulleys, inclines, and cables, where both translation and rotation take place, can be found by a straightforward pattern of analysis. For example, let us consider a block pulled up an incline—Figure 9-20(a)—with a hoist consisting of a disc of mass M with a light shaft; a cable is attached to the weights. We should like to find the direction and amount of the acceleration of the system when released, neglecting friction.

One approach is to apply Newton's laws separately to each of the masses and the rotating disc, assuming that there are different tension forces in the connecting cables—Figure 9.20(b). Abbreviating these as T_1 and T_2, the balance equations—assuming acceleration of m_1 up the incline—are

$$T_1 - m_1 g \sin\theta = m_1 a$$

$$T_2 R - T_1 R = I\alpha$$

$$m_2 g - T_2 = m_2 a$$

To solve, we divide the middle equation by R, and let $\alpha = a/R$, $I = \frac{1}{2}MR^2$. Adding all three resulting equations eliminates the tensions, and we solve for the acceleration:

$$a = g\,\frac{m_2 - m_1 \sin\theta}{m_1 + m_2 + M/2}$$

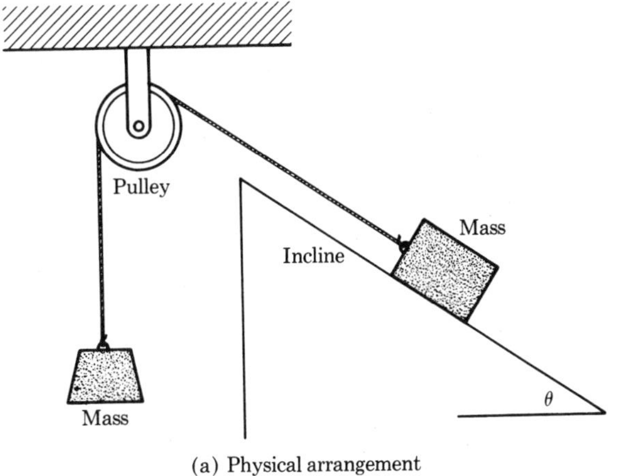

(a) Physical arrangement

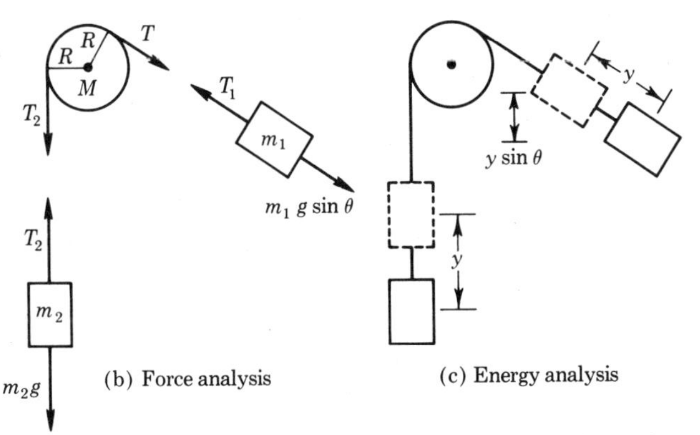

(b) Force analysis

(c) Energy analysis

FIGURE 9-20

Analysis of motion of connected systems

The same result can be obtained by application of the principle of conservation of mechanical energy—Figure 9.20(c). Let the potential energy be zero at the start of the motion, when the kinetic energy is of course also zero. Then $E = E_k + E_p = 0$. As the system moves, changes in E_k are the negative of changes in E_p:

$$\Delta E_k = -\Delta E_p$$

Suppose the mass m_2 drops a distance y, with a loss of potential energy $-m_2gy$. At the same time, m_1 gains potential energy $m_1gy \sin \theta$, and

$$\Delta E_p = m_1gy \sin \theta - m_2gy$$

The change in kinetic energy is

$$\Delta E_k = \tfrac{1}{2}m_1v^2 + \tfrac{1}{2}m_2v^2 + \tfrac{1}{2}I\omega^2$$

Again, $I = \frac{1}{2}MR^2$ and $\omega = v/R$. Combining, we can solve for v^2 and form $v^2/2y$, which is also the acceleration a. The reader can easily verify that this agrees with the previous expression.

Gyroscopic Motion

We cannot leave the subject of rotational motion without some mention of the stability of systems such as the gyroscope. The mathematical complexity of the subject is great, but we can gain some appreciation of the processes by examining some familiar examples. A simple experiment that anyone can do will demonstrate gyroscopic action. We first take a coin and try to balance it on its edge. Unless we are extremely careful, it will immediately fall over. However, if we roll the coin along the surface of a smooth table or floor, it will remain upright for long distances. We also find that the faster the coin revolves, the longer time it takes to fall over. The same effect is found with a free automobile tire or wheel, or with a bicycle. The unusual stability in these cases is due to the rotational inertia that the object possesses.

We have seen the toy top—Figure 9-21(a)—consisting simply of a disc and an axle, that will spin with its point on a table or floor for long times before falling over. The toy gyroscope is similar, only it includes a surrounding metal ring. Newton's law for rotation in vector form provides us with a basis for explaining gyroscopic motion:

$$\mathbf{N} = \frac{\Delta \mathbf{L}}{\Delta t}$$

The directions of the vectors $\mathbf{N}$ and $\mathbf{L}$ are taken to be the same as those in which right-handed screws would move, with clockwise torque or clockwise rotation, respectively.

Let us perform some basic experiments. We first place a top with its point on the table, as in Figure 9-21(a), and give the shaft a twist. The torque is clearly directed downward, and in a time Δt we give a change in angular momentum ΔL, which is also downward according to Newton's law, and the top starts spinning. Next, we set the top in motion with its tip resting on a support, as in Figure 9-21(b). The top does not fall, but its axis starts to rotate in the horizontal plane, a phenomenon called *precession*. Newton's law again provides the explanation. The initial angular-momentum vector $\mathbf{L}_0$ is pointed to the right. The force of gravity W acting through the center of mass of the system causes a clockwise torque that is a vector directed into this page. In a time Δt, a small change in angular momentum $\Delta \mathbf{L} = \mathbf{N}\Delta t$ takes place. Viewed from above, as in Figure 9-21(c), $\mathbf{L} = \mathbf{L}_0 + \Delta \mathbf{L}$ is at an angle $\Delta\phi$ from the original direction, which corresponds to rotation of the top's axis. If the spin velocity is large, the force of the support on the end of the top is equal and opposite to the weight, and the axis remains horizontal. The relation of angular

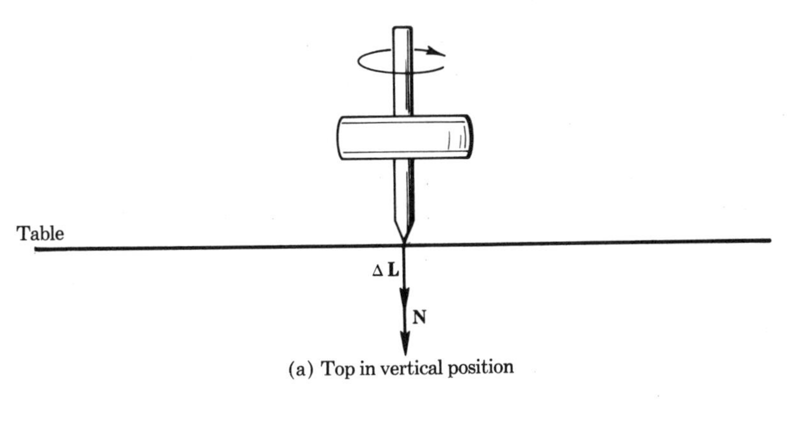

(a) Top in vertical position

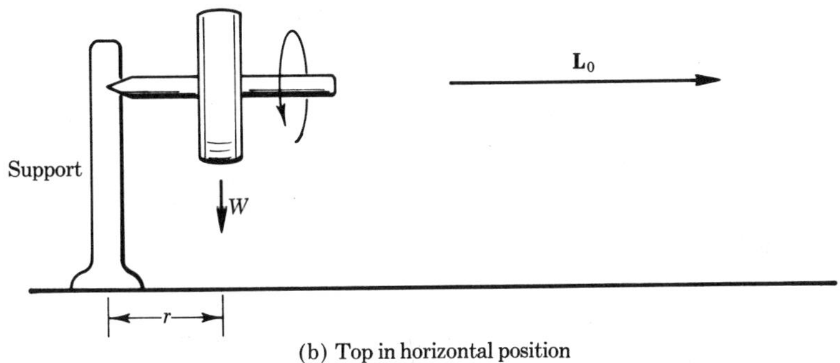

(b) Top in horizontal position

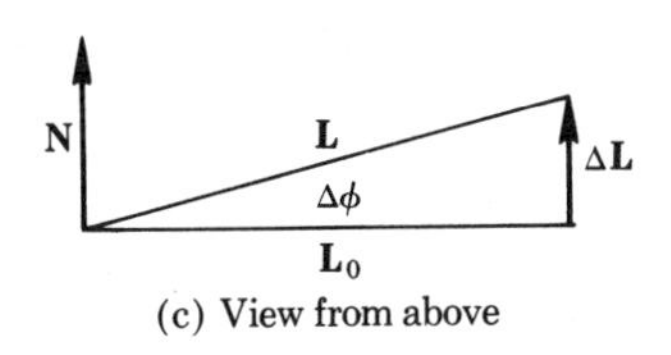

(c) View from above

FIGURE 9-21

Gyroscopic action

speeds of precession ω_p and spin ω is readily deduced. If the moment of inertia is I, then the initial angular momentum is $L_0 = I\omega$. Now, from Figure 9-21(c), $\Delta L = L_0\Delta\phi$, and also $\Delta L = N\Delta t$. Equating, we find $\omega_p = \Delta\phi/\Delta t = N/I\omega$. Thus the higher the spin, the slower the precession, and vice versa. As the spin of the top slows due to friction, the rate of precession is found to increase.

The gyroscope plays an important part in the guidance system of ships, airplanes, missiles, and satellites. The basic arrangement is a wheel rotating at high speed on a shaft, with the structure effectively supported at its center of mass. Gravity and forces of the support produce no torque, since the lever arms are always zero, and thus the initial rotational motion is preserved, regardless of the behavior of the vehicle. Even if a ship that is guided by the device curves or tips over, the axis of the gyroscope continues to point in its original direction.

9-7 TORQUES ON DIPOLES AND CURRENTS

Electrical and magnetic fields, as we saw earlier, exert forces on individual charges at rest or in motion. When several charges are involved, torques are produced.

First let us examine the electric dipole from Section 6-3. When placed in a uniform electric field $\mathcal{E}$, as in Figure 9-22, the positive charge will tend to move downward, the negative charge upward. With lever

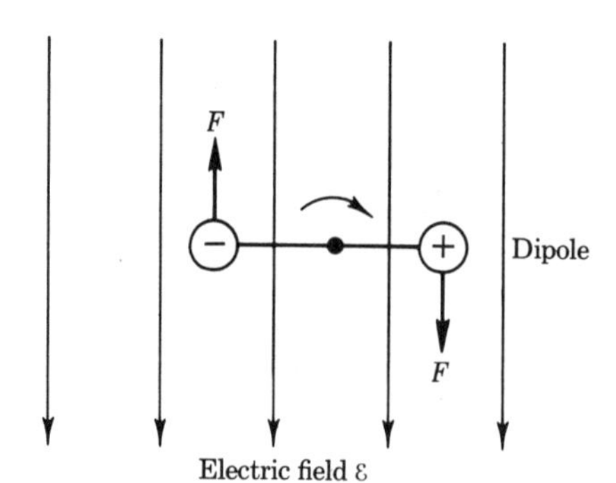

FIGURE 9-22

Torque on electric dipole

arms $d/2$ for rotation about the midpoint, and forces $F = \mathcal{E}q$ on each particle, the total torque is

$$N = \mathcal{E}qs$$

Or letting the electric moment be $M = qs$,

$$N = M\mathcal{E} \qquad \textit{(torque on an electric dipole)}$$

The torque on a bar magnet in a magnetic field $\mathcal{B}$ is easily found by analogy. If the pole strength is p and the magnetic moment is $M = ps$,

$$N = M\mathcal{B} \qquad \textit{(torque on a magnetic dipole)}$$

When either the electric or magnetic dipole is aligned at some angle θ with the field, the torque is smaller than the above values by $\sin\theta$. For example, if $\theta = 0$, $\sin\theta = 0$, then the forces on the electric charges are along the axis of the dipole and no tendency to rotate is experienced. When $\theta = 90°$, $\sin\theta = 1$, and the torque is again $N = M\mathcal{E}$. When a freely moving compass needle or bar magnet is placed in a magnetic field, the torque brings the moment $\mathbf{M}$ into line with $\mathcal{B}$ so that θ and the torque are zero.

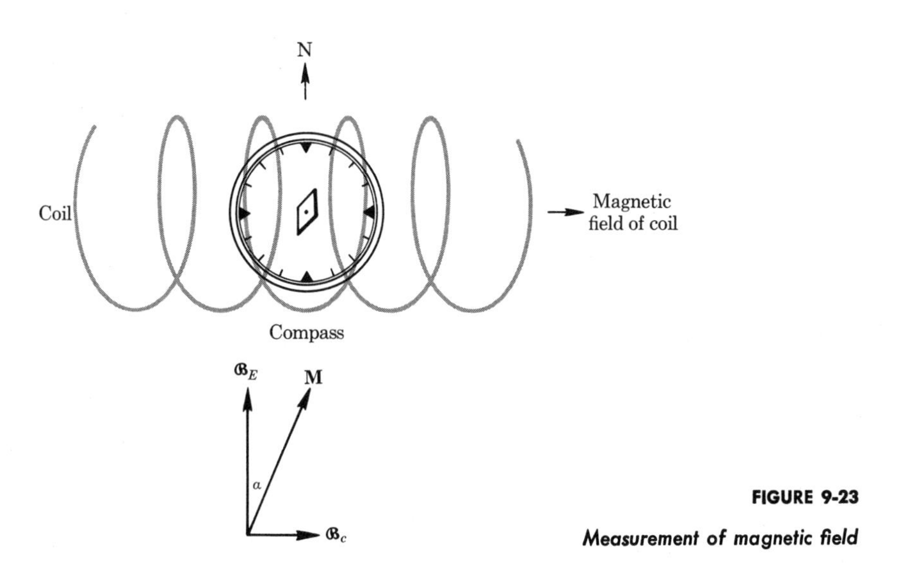

FIGURE 9-23

Measurement of magnetic field

An experiment to test this effect is easily set up as shown in Figure 9-23. A current is sent through a solenoid to produce a magnetic field $\mathcal{B}_c$ inside it, parallel to the coil axis. Then, the needle of a magnetic compass placed within the coil will be deflected from north through an angle α, such that the torques due to the coil and to the Earth's field $\mathcal{B}_E$ are equal:

$$N_E = M\mathcal{B}_E \sin \alpha = M\mathcal{B}_c \sin (90° - \alpha) = N_c$$

or

$$\tan \alpha = \frac{\mathcal{B}_c}{\mathcal{B}_E}$$

For example, if $\mathcal{B}_E = 5 \times 10^{-5}$ Wb/m^2 and $\mathcal{B}_c = 2 \times 10^{-5}$ Wb/m^2, then $\tan \alpha = 0.4$ and $\alpha = 22$ deg. This apparatus is an elementary version of a magnetometer for measuring magnetic fields. If the field due to the coil is known, the measured angle yields the strength of the unknown field.

Torques on current loops produce motion in electric motors, and are closely related to the whole subject of magnetism in solids. Picture, as in Figure 9-24, a rectangular loop of wire of dimensions L and W, mounted on an axis and free to rotate. A current I flows in each side of the loop. There are no forces on the vertical legs, but there is a force of amount $F = \mathcal{B}IL$ on each of the horizontal legs, recalling Section 7-2. The lever arms are each $W/2$; hence the total torque is

$$N = 2(\mathcal{B}IL)\frac{W}{2} = \mathcal{B}ILW$$

However, the *area* of the loop is $A = LW$; hence

$$N = IA\mathcal{B}$$

The similarity of dependence of torque on magnetic field suggests that we let $IA = M$, so that

$$N = M\mathcal{B}$$

Thus a current loop of area A, magnetic moment $M = IA$, is equivalent to a bar magnet of moment $M = ps$. We derived this expression for a rectangular loop, but it can be shown that the magnetic moment of a loop depends on its *area*, not on its shape.

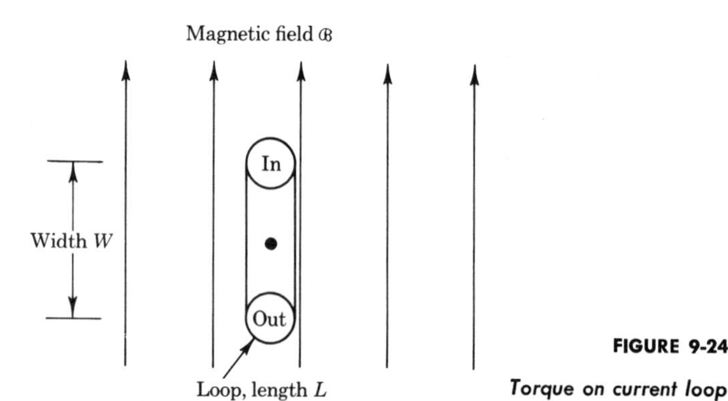

FIGURE 9-24
Torque on current loop

An interesting and important relation exists between two rotational aspects—mechanical and electrical—of a charge moving in a circular path. The magnetic moment is $M = IA$, but as shown in Section 7-2, the current of a charge q moving on radius R is $I = \dfrac{q}{2\pi R/v}\cdot$ Thus

$$M = \pi R^2 \left(\frac{qv}{2\pi R}\right) = qvR/2$$

This is the charge's electrical characteristic. As a rotating mass, it has an angular momentum $L = mvR$. Now let us form the ratio of the two properties:

$$\frac{M}{L} = \frac{qvR/2}{mvR} = \frac{q}{2m} \qquad \textit{(ratio of magnetic moment to angular momentum)}$$

Notice that the quotient of magnetic moment and angular momentum (called the *gyromagnetic ratio* in atomic physics) is directly proportional only to the ratio of the basic quantities of charge and mass. Magnetic properties depend on this quantity, as will be shown later. The numerical magnitude for an *electron in an orbit* is

$$\frac{M}{L} = \frac{e}{2m} = \frac{1.60 \times 10^{-19}\ \text{C}}{(2)(9.11 \times 10^{-31}\ \text{kg})} = 8.79 \times 10^{10}\ \text{C/kg}$$

The Subject of Mechanics

When we hear the words "mechanical" or "mechanics," we tend naturally to think of machines or mechanisms involving wheels, gears, shafts, levers, and so forth. In the vocabulary of physics, however, *mechanics* refers to the area of study involving forces, torques, and motion, regardless of the nature of the forces. We thus employ its principles, including Newton's laws and the energy concept, to describe and analyze motions of electric charge or of gravitational masses or of atomic nuclei. The word "mechanics" was first used when the science of physics consisted principally of the study of devices that involved rigid structures and "mechanical" forces. This era was long before electricity was discovered, before magnetism was other than a curiosity, before gravitation was fully understood, and surely before the existence of nuclei was suspected. It is rather remarkable to contemplate that the laws developed by Galileo and Newton as early as the 1600's to describe such unsophisticated systems as falling bodies have stood the test of time and remain so widely applicable to the varied and complex phenomena of nature. This may be regarded either as a demonstration of the genius of these early scientists or perhaps as a reminder that physical laws remain while men and their artifacts appear and disappear.

PROBLEMS

9-1 In tightening the nuts that hold a wheel on an auto, forces of 24 lb are applied to each of the two 8-in. arms of a wrench. How much torque is applied?

9-2 How far from the center of the Earth is the center of mass of the Earth–Moon system? *Note:* $M_M = 0.272\ M_E$, and the separation of centers is 239,000 mi.

9-3 One die of an otherwise identical pair of dice has the "1-spot" loaded with lead of mass 0.3 gram. If the "honest" die has sides of length 1 cm and a mass of 10 g, by how much is the center of mass of the "loaded" die shifted?

9-4 A neutron (mass $m = 1$) approaches a deuteron at rest (mass $M = 2$). Show that the center of mass is at a point $\frac{2}{3}$ of the way between the particles. To which nucleus is the center of mass nearer?

9-5 Find the center of mass of the water molecule H_2O, assuming average distances between atoms, in nanometers (10^{-9} m), as follows: O–H, 0.0958; H–H, 0.1518.

9-6 Find the center of mass of a thin rod that has a density that varies linearly from zero to ρ_L along its length L, by drawing analogy to a triangular plate (see Section 9-3).

9-7 One of the two arms of an analytical balance was manufactured 1 mm longer than the proper 12-cm length. How much error (specify high or low) will this introduce in the determination of the mass of a 200-gram chemical sample placed on the pan attached to the longer arm?

9-8* The angle between a ladder and the ground must not be too large, or a person will tip over backwards as he climbs. Find the largest safe angle, assuming that the wall exerts no force on the uniform ladder (length d, weight w). The center of mass of his body (weight W) is located at a perpendicular distance l from the center of the ladder. *Suggestion:* Take torques about the point of contact of the ladder and the ground.

9-9 A truck is weighed at a highway checking station by putting only the front wheels on the scale platform (w_1), then only the back wheels (w_2). Derive a formula for the correct weight w.

9-10 How high can a man of mass 75 kg climb on a uniform ladder of length 4 m, mass 12 kg at 60 deg from the horizontal? Let the coefficient of friction with the ground be 0.6 and with the wall be zero. If he had a longer or shorter ladder could he climb a greater fraction of the way up?

9-11* The correct angle of highway banking to prevent cars from sliding was investigated in Section 5-7. Find the angle to prevent tipping over, first when the car is not moving, then when it has a speed v on radius R. Assume that the car's center of mass is at a distance y from the ground and that it is halfway between the wheels, a distance $2x$. Show how a low center of mass is desirable, especially for racing cars.

9-12 A motorcycle with high enough speed can travel on the inside of a circular vertical wall. (a) Derive an expression for the necessary angle that the plane of the cycle must have with the vertical, assuming that friction prevents sliding down. (b) Find the angle if the speed is 20 m/sec and the radius of motion is 15 m.

9-13 A phonograph motor is to be designed to accelerate a stack of eight 45 rpm records, each of mass 0.040 kg and diameter 0.175 m, to full speed in 2 sec. What is the minimum torque? What force does that imply on a drive shaft of 0.4 cm radius?

9-14 Find the moment of inertia of the H_2O molecule about an axis that is through the center of mass and perpendicular to the plane of the atoms. (See problem 9-5 for interatomic distances.)

9-15 Show that the moment of inertia of any two particles of mass m_1, m_2, and distances to the center of mass r_1, r_2, is m_1r_1s, where $s = r_1 + r_2$, if the axis is through the center of mass.

9-16 What is the moment of inertia of a semicircular plate about the center of the straight side?

9-17 Find (a) the center of mass of the mixed molecule of hydrogen and deuterium with the HD atom separation of 7.4×10^{-11} m, and (b) its moment of inertia about an axis through the center of mass.

9-18 The brakes on a car essentially operate as shown in Figure P9-18. The spreading of the lined brake "shoes" increases their contact with the interior of the drum. Find the amount of work, the value of the torque, and

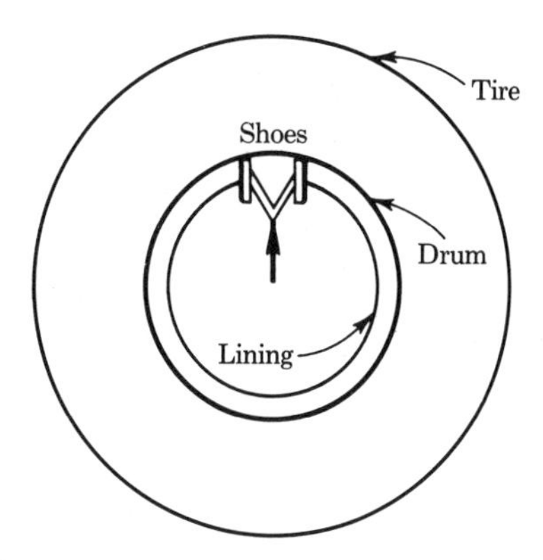

FIGURE P9-18

the constant force that must be applied on each of four 12-cm drums to stop a car of mass 2000 kg with speed 30 m/sec and tire diameter 0.4 m, within a distance of 100 m.

9-19 An automobile engine develops 100 hp at a speed of 2700 rpm. What torque is developed?

9-20 A flywheel comes off its shaft and rolls away on the floor. Assuming it to be a ring of weight 300 lb, and center-of-mass speed 1.5 ft/sec, how much kinetic energy does it have?

9-21 (a) Show that the power developed in a system by a torque N with angular speed ω is $P = N\omega$. (b) Compute the torque on an airplane propeller produced by a rotary engine developing 3000 hp, while turning over at 2000 rpm.

9-22 A wheel of a car moving at 60 ft/sec comes off and goes rolling down a level highway. A frictional force of 2 lb acting tangentially to the tire of radius 14 in. tends to slow the motion. The wheel weight is 48.3 lb and its radius of gyration is 9 in. Find the distance the wheel travels before it stops, using the principle of energy conservation.

9-23 Describe the following in terms of rotational motion, forces, and torques: (a) a sailing platter, (b) the well-passed football, and (c) the yo-yo.

9-24* A meter stick is held vertically with one end on the floor. It is then allowed to fall. Find the speed of the center of mass as it depends on angle, assuming that the end touching the floor does not slip. *Note:* For a thin rod rotating about one end, the moment of inertia is $\frac{1}{3}ml^2$.

9-25 Show that in natural rolling the fraction of the total kinetic energy of a solid sphere that is rotational is $\frac{2}{7}$. If I for a hollow sphere is $\frac{2}{3}mR^2$, what is this fraction?

9-26 A bowling ball with a diameter of 10 in. and a weight of 12 lb rolls along an alley with a velocity of 14 ft/sec. Calculate its total kinetic energy.

9-27 A flywheel with moment of inertia of 25 kg-m^2 has a rotational kinetic energy of 1500 joules. What is its speed of rotation in revolutions per minute?

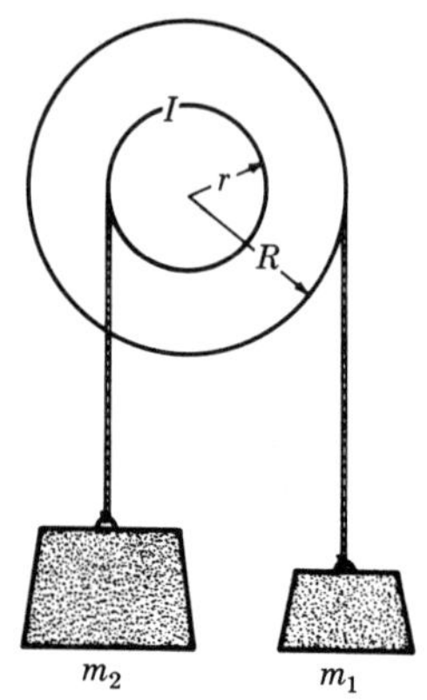

FIGURE P9-28

9-28* Find the angular acceleration of a wheel driven by two weights (see Figure P9-28) using Newton's laws of motion, but neglecting friction. Let $m_1 = 5$ kg, $m_2 = 15$ kg, $r = 0.10$ m, $R = 0.25$ m, and $I = 2$ kg-m^2 for the wheel.

9-29* What speed will the system in Problem 9-28 have after mass m_1 has moved through 0.5 m, according to conservation of mechanical energy?

9-30 What is the torque in the Earth's magnetic field of 5×10^{-5} Wb/m^2 on a compass magnet of length 2 cm, pole strength 10^{-6} placed perpendicular to the field line?

9-31 How much torque would be required to hold an iron rod of 5-cm length and magnetic moment 8.5 A-m^2 perpendicular to a strong field of 10 Wb/m^2? What amount of force at each end of the rod is required? Express also in pounds.

9-32 A magnetic compass is placed at the center of a circular loop of current-carrying wire, the plane of which is parallel to the Earth's magnetic field. Show that the angle of deflection of the needle when the current is turned on is found from

$$\tan \alpha = \frac{2\pi k_M I}{a \mathcal{B}_E}$$

9-33 Referring to Problem 9-32, what is the strength of the Earth's field at a point where a current of 3 A in a loop of 8-cm radius gives a deflection from North of 30 deg?

9-34 Find the magnetic moment of a ring of wire of radius R that has a total charge Q, and rotates about the axis through the center of the plane with angular speed ω.

9-35 The magnetic field at the Earth's Equator due to a dipole of moment M at the center and lined up with the poles of the Earth (see Section 6-4) is

$$\mathcal{B} = \frac{2k_M M}{R_E^3}$$

If the field is 5×10^{-5} Wb/m^3, estimate the Earth's magnetic moment.

9-36* What is the angular momentum of the Earth as a uniform sphere? Assuming that the Earth is a dipole of magnetic moment 1.55×10^{22} A-m^2, what uniform charge per kilogram of mass would be required to give this strength? How many iron atoms per free electron does this correspond to? Discuss the meaning of this result.

INTERACTIONS AND CONSERVATION LAWS

10

The physical universe can be described either in terms of interactions of basic particles—electrons, photons, protons, neutrons, and so on—or in terms of interactions between structures ranging from molecular to astronomical size. The laws of motion provide a means for prediction in physical problems at either level, using several quantities such as mass, charge, velocity, acceleration, force, momentum, and energy, as we have seen in preceding chapters. Now, we shall be introduced to some simplifying assumptions or methods to help describe and analyze complex situations. First we shall examine the concept of *conservation* which is a powerful device for this purpose, and serves as a unifying concept in physics.

10-1 CONSERVED QUANTITIES

If some physical quantity does not change with time during a process, we call that quantity a "constant of the motion," and say that the quantity is *conserved*. We are already familiar with some of these. For example, when two chemicals react, the total mass of materials is constant (neglecting mass–energy conversion); or again, a falling body has both potential and kinetic energy, but the total mechanical energy is constant. Clearly, our ability to describe processes is improved whenever it is possible to discover such constant entities, since we can use their initial values at all later times and not be concerned with changes.

In addition to the familiar examples of mass and energy, there are other fundamentally conserved quantities. For instance the total amount of electric charge in the universe is believed to be constant. Although we can "create" electrons from energy, they arise in pairs of opposite sign; and if we think of electrons as positive matter and antielectrons as negative matter, the resultant net charge is zero, and thus no change in total charge takes place. (Similar statements can be made about the total number of nucleons in the universe, which we shall discuss in Chapter 24.)

For an example of how conservation is useful in solving problems, recall the balanced chemical equation

$$2H_2 + O_2 \rightarrow 2H_2O$$

We use the principle of chemical valence to balance the equation with the factors of 2, but this may be traced to more fundamental quantities such as charge or electron spin, to be studied in Chapter 23.

Some of the quantities in nature which are conserved are very nebulous and can only be described mathematically: "strangeness" is a property of fundamental particles related to their lifetimes; "parity" is a mathematical property conserved in some interactions and not in others; "isotopic spin" is a number used to distinguish between a proton and a neutron.

We have already shown the universality of conservation of energy, and now shall develop the necessary mathematical description for two other conserved quantities: *linear momentum* and *angular momentum.* The motion of assemblies of particles has been shown to be governed by the general forms of Newton's laws $\mathbf{F} = \Delta\mathbf{p}/\Delta t$ and $\mathbf{N} = \Delta\mathbf{L}/\Delta t$. In each case, the motion is determined by *external* forces and torques. A large class of physical situations can be analyzed, however, by choosing the system of interacting particles large enough so that *no external forces* come into play. As an example, a charge in the field of another obviously experiences a force external to it, but the two particles taken together with an imaginary boundary around them are *isolated* from the outside.

10-2 INTERACTIONS AND LINEAR MOMENTUM

We may now ask: What laws govern systems with no external forces? The total momentum of a group of *isolated* particles, labeled 1, 2, 3, etc., is

$$\mathbf{p} = \mathbf{p}_1 + \mathbf{p}_2 + \mathbf{p}_3 + \ldots = \sum_i \mathbf{p}_i$$

The mutual internal forces between pairs of particles are equal and opposite in direction, and their vector sums add to zero. There being no external forces, $\mathbf{F} = 0$, and thus

$$\frac{\Delta \mathbf{p}}{\Delta t} = 0$$

This can be true *only if* **p** *does not change with time*—a statement that is deceptively simple in view of its power to analyze physical problems. It deserves the formality of calling it the *principle of conservation of linear momentum:*

> IF NO EXTERNAL FORCES ACT ON A SYSTEM OF PARTICLES, THE TOTAL LINEAR MOMENTUM REMAINS CONSTANT IN TIME.

Let us look at some examples with objects that have mass and velocity, and thus momentum, to show that the principle encompasses various types of two-body interactions. In each, m and v will refer to the smaller body and M and V to the larger.

An amateur sailor attempts to jump from a rowboat to the shore and falls on his face in the water—Figure 10-1(a). This mishap is easily explained by conservation of momentum. Before the man leaps, the momentum of the system (including him and the boat) is zero, which is of course a constant. By giving himself a forward momentum toward land, there must be imparted to the boat an equal and opposite momentum. Mathematically, the momenta of man and boat are respectively $m\mathbf{v}$ and $M\mathbf{V}$ on separation, such that

$$m\mathbf{v} + M\mathbf{V} = 0$$

Thus $\mathbf{V} = -(m/M)\mathbf{v}$, in the opposite direction from the man's motion.

When a rifle is fired—Figure 10-1(b)—there is always a recoil that jars the shoulder. The bullet of mass m is sent forward with velocity $\mathbf{v}$, and the gun of mass M goes backward with velocity $\mathbf{V}$. The recoil speed is easily found. Before firing, the total momentum was zero, and must remain so. Again $m\mathbf{v} + M\mathbf{V} = 0$ and $\mathbf{V} = -(m/M)\mathbf{v}$. Suppose v is 150 m/sec and m is 2 grams, while M is 6 kg. The gun's speed is found to be $V = 0.05$ m/sec.

Notice that the two energies are *not* equal but vastly different. In the above example, where the rifle's mass was 6 kg and the bullet's mass 2 grams, the kinetic energies are

Bullet: $E_k = \frac{1}{2}(0.002)(150)^2 = 22.5$ joules

Rifle: $E_k = \frac{1}{2}(6)(0.05)^2 = 0.0075$ joule

Thus the energy of the bullet is 3000 times that of the rifle.

Suppose that there is some momentum to start with, as when a golf club strikes a ball resting on a tee—Figure 10-1(c). Originally the total momentum of the system is $M\mathbf{V}$, afterward $M(\mathbf{V} + \Delta\mathbf{V}) = m\mathbf{v}$. Equating, we find the club is slowed by the amount $\Delta\mathbf{V} = -(m/M)\mathbf{v}$. For a typical speed of the ball of 68 m/sec (about 152 mi/hr) and a ratio of masses of $\frac{1}{8}$, the loss is 8.5 m/sec.

In other instances both objects may have momentum, as when a space

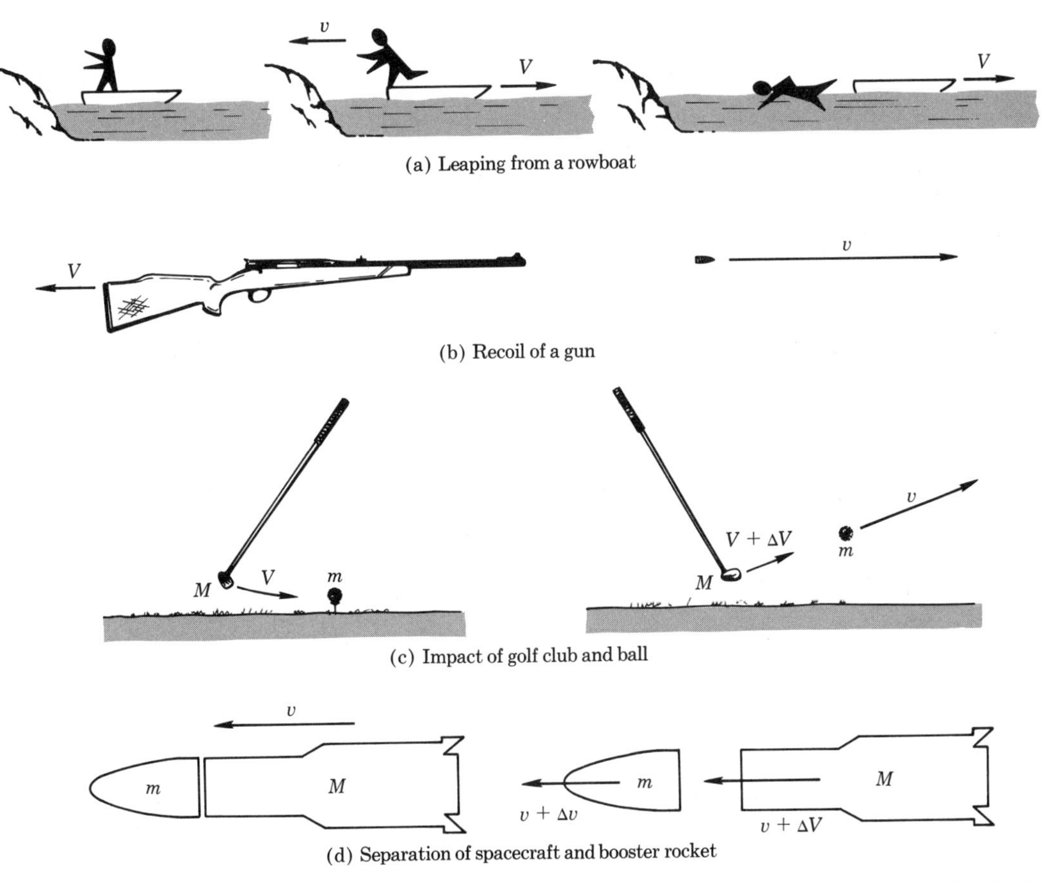

(a) Leaping from a rowboat

(b) Recoil of a gun

(c) Impact of golf club and ball

(d) Separation of spacecraft and booster rocket

FIGURE 10-1

Examples of conservation of linear momentum

capsule of mass m is suddenly disengaged from its booster rocket of mass M—Figure 10-1(d). They moved along together with speed v originally, with total momentum $(m + M)\mathbf{v}$. Each experiences changes in velocity, and afterward the total momentum is $m(\mathbf{v} + \Delta\mathbf{v}) + M(\mathbf{V} + \Delta\mathbf{V})$. Equating, we find that $\Delta\mathbf{v} = -(m/M)\Delta\mathbf{V}$, meaning that the capsule gains speed, while the booster loses speed. The reader can check that similar relations can be developed for the collision of two atomic or nuclear particles coming together from opposite directions. In each of these examples, note that the conservation principle is actually being violated if gravity is present, since then the objects are *not* free from external forces. The error is very small, however, because the times for the collisions or interactions are extremely short, and the forces thus acting are large in comparison with external forces. For example, one can estimate that there is a force of hundreds of pounds between a colliding clubhead and a golfball. Gravity as an external force clearly can have no sensible effect on the process.

Such reactions as the one between golf club and ball are called

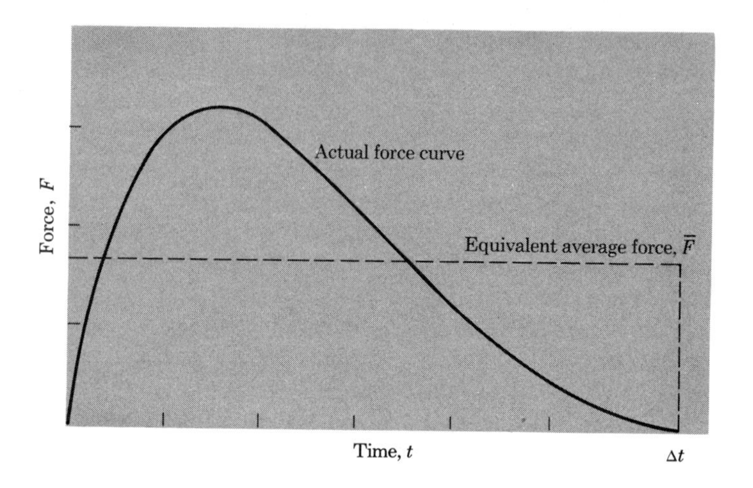

FIGURE 10-2

Force varying with time and impulse produced

impulsive, the term *impulse* that describes them being defined as the product of the average force $\overline{\mathbf{F}}$ acting on impact and the time Δt it lasts. The impulse is equal to the change in momentum:

$$\overline{\mathbf{F}}\Delta t = \Delta \mathbf{p}$$

as is easily seen from Newton's second law.

The above equation can be used just as it appears if the force is constant over the time of action Δt. For most problems, however, the force is *not* a constant in time but rises rapidly and then falls off, as in Figure 10-2. In this situation, we find the average force $\overline{F}$ by making the area under the curve of the actual F the same as that of the dotted rectangle. For example, let us find the average force between a golf club and a ball of mass 0.05 kg, if the ball picks up a speed of 100 m/sec in a time of 0.0025 sec:

$$\Delta p = m\Delta v = (0.05 \text{ kg})(100 \text{ m/sec}) = 5.0 \text{ kg-m/sec}$$

$$\overline{F} = \frac{\Delta p}{\Delta t} = \frac{5.0 \text{ kg-m/sec}}{0.0025 \text{ sec}} = 2000 \text{ N} \quad \text{or} \quad \cong 450 \text{ lb}$$

Inelastic collisions. If some of the kinetic energy of motion of colliding objects goes into heat, the reaction is said to be *inelastic.* For example, a bullet fired into a thick block of wood will slow down because of friction and remain imbedded in the wood. *Total* energy is conserved, but kinetic energy is not. Suppose that initially the bullet has mass m and speed u, while the block M is at rest. The momentum of the system is mu. After the collision, the two objects move on together with momentum $(m + M)v$. Equating,

$$mu = (M + m)v$$

or

$$v = \frac{m}{M + m} u$$

To illustrate, let $m = 0.05$ kg, $M = 10$ kg, and $u = 400$ m/sec. Then $v = (0.05/10.05)(400) = 2$ m/sec. The only information that can be gained from consideration of energy conservation is the temperature rise of the system. Few processes in nature are perfectly elastic because friction or its equivalent are inevitable. In order to analyze reactions fully, some knowledge of the details of heating effects is required. As a second example, picture a neutron of mass $m = 1.67 \times 10^{-27}$ kg, speed $u = 2.2 \times 10^5$ m/sec, being absorbed by a U^{235} atom, mass $M = 392 \times 10^{-27}$ kg. The resultant atom, U^{236}, has a speed

$$v = \left(\frac{1.67}{392 + 1.67}\right)(2.2 \times 10^5) = 930 \text{ m/sec}$$

10-3 CONSERVATION IN ROCKET MOTION

The behavior of a rocket engine—as in missiles, aircraft, and spacecraft involves a particularly interesting application of the principle of conservation of momentum. A chemical fuel is burned and the high-speed gases are continuously discharged from the rocket nozzle (Figure 10-3). Just as the firing of a bullet causes a gun to recoil in the opposite direction, the ejection of gas molecules propels the rocket. Suppose that the mass of propellant is discharged from the rocket at a speed v_p and at a rate r (in kg/sec). In a time Δt, a mass $\Delta m = r\Delta t$ escapes, implying a momentum loss of $(r\Delta t)v_p$. The rocket of mass m experiences a momentum gain of magnitude $m\Delta v$. Equating and dividing by Δt,

$$rv_p = m\frac{\Delta v}{\Delta t}$$

From this form of Newton's second law, we can identify the force of the propellant on the rocket—the "thrust"—as

$$F_p = rv_p$$

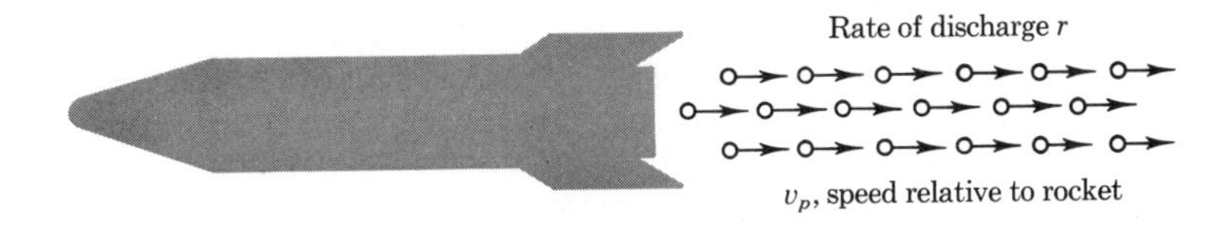

FIGURE 10-3
Rocket propulsion

For example, suppose that for the Atlas missile, $r = 100$ kg/sec and $v_p = 4500$ m/sec. Then, $F_p = 4.5 \times 10^5$ N. This amounts to about 100,000 lb, meaning that such a thrust could *hold up* a weight of 50 tons, or impart an acceleration of 1 "g" upward from the Earth's surface to a rocket of 25 tons.

It is a common misconception that the exhaust gases "push on the air" and that this reaction forces the rocket to move. Clearly this is wrong, since rocket propulsion can take place in a complete vacuum.* The reaction force is in fact due to the discharge process itself and is based on the conservation of linear momentum.

Newton's formula holds accurately even if the system is acted on by external forces, because they do not affect the violent process of expulsion of burning gases to any degree. Thus we may generalize the equation to include gravity $\mathbf{F}_G$ and friction of air $\mathbf{F}_F$:

$$\mathbf{F}_{\text{total}} = \mathbf{F}_p + \mathbf{F}_G + \mathbf{F}_F = m\frac{\Delta \mathbf{v}}{\Delta t}$$

The friction will always be negative, acting to slow the rocket, while the sign of gravity will depend on whether launching or landing is the maneuver under consideration.

As the fuel burns and is discharged, the mass of the rocket continually decreases, and thus its acceleration increases even if the thrust is constant. Let us examine the relationship of this variable mass and the speed at any time, neglecting gravity and friction, as if the rocket were starting from rest far out in space. If the mass of the whole system—rocket plus fuel—were m_0 at time zero, at any later time t it would be

$$m = m_0 - rt$$

Then, the ratio of mass to initial mass is found to be†

$$\frac{m}{m_0} = e^{-v/v_p}$$

where v is the speed of the rocket relative to the point it started and v_p is the speed of the propellant relative to the *rocket*. Let us calculate what ultimate speed the Atlas-type vehicle would reach in gravity-free vacuum. Assume the mass of fuel to be 25,000 kg and that of the total structure and payload 10,000 kg, giving a mass m_0 of 35,000 kg. Take the thrust as 4.5×10^5 N, with $v_p = 4.5 \times 10^3$ m/sec and $r = 100$ kg/sec. At "burnout," when all fuel is gone, $m = 10{,}000$ kg and m_0/m is 3.5. Taking the logarithm of the

* As late as 1920, Dr. Robert Goddard, a pioneer in rocket studies, was taken to task by *The New York Times* for asserting that travel in the vacuum of space was possible. In 1969, after the landing on the Moon, the *Times* apologized.

† Note that $\Delta m = -r\Delta t$ (a reduction), and that $m\Delta v = rv_p\Delta t$. Eliminating Δt, we obtain $\Delta m/\Delta v = -(m/v_p)$. This equation is of the same form as that for friction slowing in Section 5-7; by comparison, $m = m_0e^{-v/v_p}$.

rocket formula and solving for v, we obtain

$$v = v_p \ln\left(\frac{m_0}{m}\right) = (4.5 \times 10^3)(\ln 3.5) = 5.6 \times 10^3 \text{ m/sec}$$
$$= 5.6 \text{ km/sec}$$

This is less than the speed needed for a minimum-radius orbit around the Earth, which is 8.0 km/sec, and only one-half the Earth escape speed of 11.2 km/sec (see Section 8-4). Recall also that gravity and air resistance are still not accounted for. For satellite and interplanetary travel, however, two or more rocket stages are required; and if the second stage is appreciably lighter than the first stage, one can show that the final speed is roughly *doubled.* This would yield 11.2 km/sec with the specifications given. The length of time of the rocket blast is easily found to be

$$t_b = \frac{(m_0 - m)}{r} = \frac{25{,}000 \text{ kg}}{100 \text{ kg/sec}} = 250 \text{ sec}$$

The basic mass–speed equation can be written in terms of the ratio of fuel mass used to original total mass:

$$\frac{m_f}{m_0} = 1 - e^{-v/v_p}$$

A typical question is: What fraction of fuel will give a certain final speed—say, twice v_p? Immediately, we find that

$$\frac{m_f}{m_0} = 1 - e^{-2} = 0.865$$

or 86.5% of the system mass must be fuel. As we may note, the higher the speed demanded, the larger the fuel fraction and the smaller the payload that can be transported. From another viewpoint, very high exhaust speeds allow a low fuel load. This suggests that a stream of charged ions, accelerated by electric forces, would serve as an extremely effective propellant.

10-4 PARTICLE COLLISIONS

Molecules, atoms, and nuclei in general involve two types of motion: *translational,* that of the body as a whole, and *internal,* that of the component parts. Thus corresponding translational and internal energies are changed or shared when two particles collide. As a start in studying the complex nature of such collisions, let us consider interactions that are *elastic,* such that internal energy is not involved. We shall also suppose, in the following examples, that mass is not converted into energy. On the atomic scale, the collision between a neutron and certain nuclei is a good example, while on a larger scale, we may visualize the collision of two billiard balls. Also, we shall apply the laws of conservation of mechanical energy and linear momentum. If there is no potential energy, we may state these simply as

$$E_k = \text{constant}$$
$$\mathbf{p} = \text{constant}$$

For any two bodies of masses m_1 and m_2 that collide with initial velocities $\mathbf{u}_1$, $\mathbf{u}_2$ and final velocities $\mathbf{v}_1$, $\mathbf{v}_2$, the conservation formulas state* that

$$\tfrac{1}{2}m_1u_1^2 + \tfrac{1}{2}m_2u_2^2 = \tfrac{1}{2}m_1v_1^2 + \tfrac{1}{2}m_2v_2^2 \qquad \textit{(kinetic energy)}$$
$$m_1\mathbf{u}_1 + m_2\mathbf{u}_2 = m_1\mathbf{v}_1 + m_2\mathbf{v}_2 \qquad \textit{(momentum)}$$

A convenient alternative equation for elastic collisions connects the relative velocities of the two particles before and after collision:

$$\mathbf{u}_1 - \mathbf{u}_2 = \mathbf{v}_2 - \mathbf{v}_1 \qquad \textit{(relative velocity)}$$

We can readily demonstrate and explain this general expression for one type of interaction—the "head-on collision," in which particles approach and recede from each other on the same line, as shown in Figure 10-4. This is but one of the many ways in which two objects can collide. In general, their paths make some angle with each other.

Let us rearrange the energy and momentum equations for this case as

$$m_1(u_1^2 - v_1^2) = m_2(v_2^2 - u_2^2)$$
$$m_1(u_1 - v_1) = m_2(v_2 - u_2)$$

and divide equations, noting that $x^2 - y^2 = (x + y)(x - y)$, to yield

$$u_1 - u_2 = v_2 - v_1$$

The speed of approach of particle 2 to particle 1 before collision is the same as the speed of separation of the two after collision.†

Let us illustrate the use of the last two equations involving momentum and relative speed. An object of mass $m_1 = 1.2$ kg with speed $u_1 = 30$ m/sec overtakes another object of mass $m_2 = 2.0$ kg and speed $u_2 = 8$ m/sec and collides elastically. Then,

$$1.2(30 - v_1) = 2.0(v_2 - 8)$$
$$30 - 8 = v_2 - v_1$$

* Particle 1, usually of primary interest, bombards particle 2, of secondary interest. Conditions "before" are symbolized by u, which precedes v in the alphabet.

† If one combines algebraically the relative-velocity equation $\mathbf{u}_1 - \mathbf{u}_2 = \mathbf{v}_2 - \mathbf{v}_1$ with the momentum equation $m_1\mathbf{u}_1 + m_2\mathbf{u}_2 = m_1\mathbf{v}_1 + m_2\mathbf{v}_2$, the final velocities for any head-on collision are

$$\mathbf{v}_1 = \frac{2m_2\mathbf{u}_2 + (m_1 - m_2)\mathbf{u}_1}{m_1 + m_2} \quad \text{and} \quad \mathbf{v}_2 = \frac{2m_1\mathbf{u}_1 + (m_2 - m_1)\mathbf{u}_2}{m_1 + m_2}$$

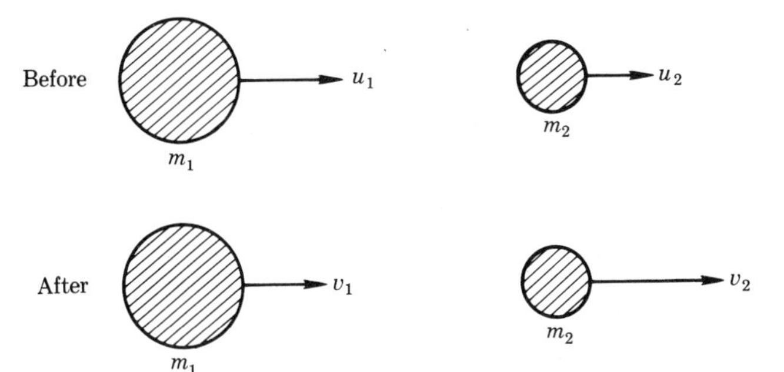

FIGURE 10-4

Head-on collision of two particles

Solving, $v_1 = 2.5$ m/sec, $v_2 = 24.5$ m/sec.

The conservation laws permit the analysis of all types of collisions. Some special cases now can be examined rather easily.

1. *Collision involving a very large mass.* The elastic collision between very light objects (as projectiles) and very massive ones (as targets) is especially simple to analyze. Good examples are a bullet striking a boulder or a stray golfball hitting the highway; the collisions between an electron and an atom or between a neutron and the nucleus of uranium may also be considered.

Let m_1 be much smaller than m_2, which is initially at rest, $u_2 = 0$. Dividing the momentum equation through by m_2 gives

$$\frac{m_1}{m_2}\mathbf{u}_1 = \frac{m_1}{m_2}\mathbf{v}_1 + \mathbf{v}_2$$

The quotients of the masses are very small and thus may be neglected. This tells us that

$$\mathbf{v}_2 \cong 0$$

The more massive body has a small speed after collision. We easily recognize the truth of this—one does not knock elephants over with BB shot. Let us examine the vector equation for relative motion with $\mathbf{u}_2 = 0$, as set originally, and $\mathbf{v}_2 \cong 0$, as just found. This leaves only

$$\mathbf{v}_1 \cong -\mathbf{u}_1$$

The velocity of the light object is approximately reversed, with the speed remaining the same. It bounces back, a fact we know from practicing tennis against a wall. Approximately, the change in momentum of the light object is $2m_1u_1$, equal to that of the massive object, m_2v_2; hence $v_2 \cong 2m_1u_1/m_2$.

Let us now reverse the situation, allowing a massive object m_1 to be the projectile striking a light object m_2 at rest, $u_2 = 0$. Dividing the momentum equation by m_1 yields

$$\mathbf{u}_1 = \mathbf{v}_1 + \frac{m_2}{m_1}\mathbf{v}_2$$

Neglecting the second term on the right, we see that the velocity of the heavy object is practically unchanged:

$$\mathbf{v}_1 \cong \mathbf{u}_1$$

To verify, think of the difficulty a football line composed of small players has in stopping a 250-lb fullback. From the relative-velocity equation,

$$\mathbf{v}_2 \cong 2\mathbf{u}_1$$

meaning that the target is given about twice the speed of the projectile.

2. *Collisions of a neutron and a stationary nucleus.* As an illustration of interest from the study of nuclear chain reactors or the evaluation of neutron radiation effects on tissue in the human body, let us consider a neutron of mass m_1, speed u_1 impinging on a nucleus of mass m_2, speed $u_2 = 0$. Let the ratio m_2/m_1 be labeled A, which is close to the mass number of the target. Our equations become

$$u_1 = v_1 + Av_2$$

$$u_1 = v_2 - v_1$$

Eliminating v_2, we obtain the ratio of neutron speeds after and before collision:

$$r = \frac{v_1}{u_1} = \frac{A - 1}{A + 1}$$

We may deduce from the above that the collision of a neutron with a very heavy nucleus ($A \gg 1$) leaves the speed almost unchanged ($v_1 \cong u_1$), while the collision with the nucleus of hydrogen ($A = 1$) causes v_1 to be zero. (For this latter special case, the proton is projected forward with the same speed as the neutron had, leaving the neutron at rest.) To illustrate, let us find the ratio of neutron speeds and energies after and before a head-on collision, when the target nucleus is C^{12}. Now $\frac{v_1}{u_1} = \frac{12 - 1}{12 + 1} = \frac{11}{13}$ and the energy ratio is $\left(\frac{11}{13}\right)^2 = 0.716$. If instead the target is U^{235}, $\frac{v_1}{u_1} = \frac{234}{236}$ and the energy ratio is $\left(\frac{234}{236}\right)^2 = 0.983$. This indicates that the neutron's energy *loss* is considerably greater on impact with the lighter nucleus, which explains why light elements H, C, Be, etc. are preferred as "moderators" for neutrons in nuclear reactors, where slow neutrons produce fission of uranium.

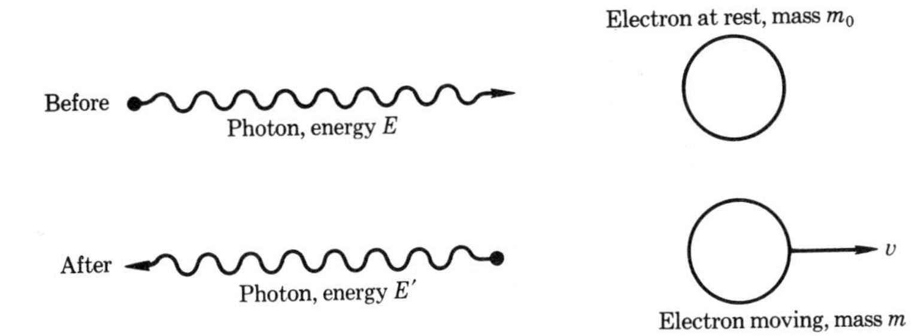

FIGURE 10-5

Collision of photon and electron, the Compton effect

3. *Collisions of a photon and an electron at rest.* Examples 1 and 2 above did not require the use of relativistic concepts—the dependence of mass on speed and the equivalence of mass and energy. Here we turn to a more advanced type of collision, that of a photon and an electron, in which the electron's speed after collision is great enough to require the use of relativistic relations. Also, we shall note the fact that the momentum of a photon (Section 8-8) is the quotient of its energy and the speed of light. Keeping these two special factors in mind, we may now apply the conservation laws to the "Compton effect," resulting from the bombardment of an electron at rest, mass m_0, by a photon of energy E. In the case of a head-on collision (Figure 10-5), the photon is reflected backward with a new energy E', and the electron is projected forward with a new relativistic mass m. Conservation of mass-energy gives

$$E + m_0c^2 = E' + mc^2$$

while conservation of momentum gives

$$\frac{E}{c} = -\frac{E'}{c} + mv$$

there being no original electron momentum. The two equations may now be solved for the new photon energy in terms of the original value. The algebra required* to eliminate the speed v is left to the reader, and only the result is given:

$$E' = \frac{1}{1/E + 2/E_0}$$

where E_0 is the rest energy m_0c^2.

Thus we see that the photon energy after collision depends on how

* The best procedure is to form $mv = c\sqrt{m^2 - m_0^2}$ and eliminate m between the equations.

energetic the photon was initially, compared to the mass-energy (Section 8-7) inherent in the electron (0.51 MeV). If E is low, say 0.01 MeV, we can ignore the second term in the denominator, noting that the energy is almost unchanged—analogous to a tennis ball bouncing off a wall. If, however, E is very large, say 10 MeV, we can neglect the $1/E$ in the denominator and obtain a final energy of $\frac{1}{2}m_0c^2$ or 0.25 MeV. Except for this small amount, all of the initial energy has gone into projecting the electron forward with a high relativistic mass. There is obviously no classical analogue to this, since a large mass striking a small one would be expected to lose practically no energy. The first measurement of this collision process was carried out by the American physicist Arthur H. Compton, after whom the effect is named.

General Elastic Collision, Center-of-mass View

The head-on collision, as we have noted, is a special situation. In general objects come together from different directions, just as two cars might collide at an intersection—say, one from the north and one from the west—leaving each other at angles that depend on the laws of mechanics and probability. The final energies of the particles depend on these angles, and many different values would be found if the collision experiment were performed repetitively with identical initial conditions.

We shall now outline the treatment of the general elastic collision. First let us look at the "before" and "after" views of the event as seen in the laboratory where the experiment is done. We call this the "lab" system of coordinates—Figure 10-6(a). The vector lengths represent velocities, and the circles show where the particles are located just before and after the collision. Often the only items of interest are the angle of scattering θ and the magnitude of v_1 for one particle. As before, the equations of conservation apply, but there are many combinations of final directions θ and ϕ that the particles can have, depending on the details of the impact. These details can best be seen if one is "at the scene"—that is, at the *center of mass* of the two particles, which lies somewhere between them at all times. We label this the CM coordinate system. Consider what an observer would see from this viewpoint. The particles would appear to approach on a line, as seen in Figure 10-6(b), and after contact to recede on another line. Each particle has a velocity relative to the center of mass. As it turns out, the two *speeds* remain constant, providing some simplification of the algebra, which at best is involved. Note that the collision as seen from CM amounts only to a rotation of the "dumbbell" formed by the particles, through some angle α. (In one type of collision found frequently, *every* final direction is equally likely—a phenomenon called *isotropic scattering*.)

The procedure for finding the final velocities is to travel mentally from the lab to the CM, observe the collision there, then "come out" to see what really happened in the lab. This requires connecting relations for

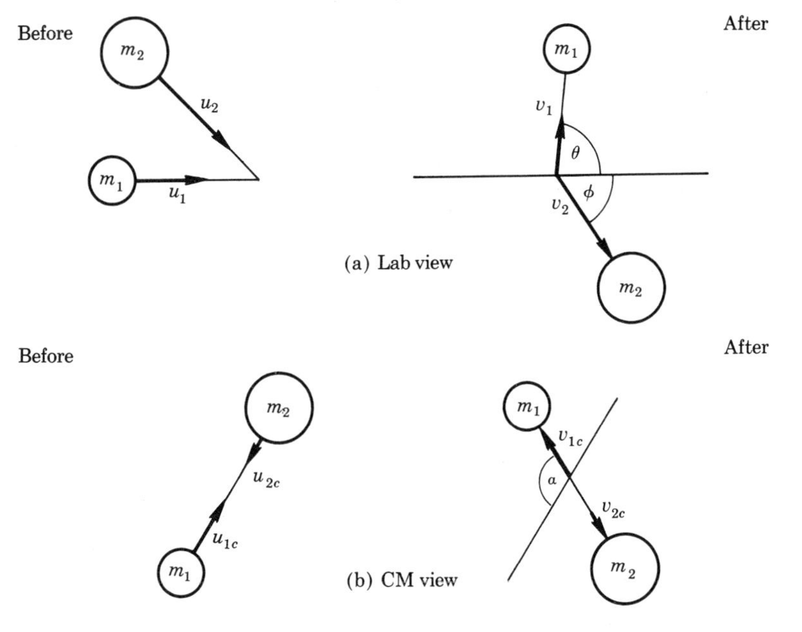

FIGURE 10-6

Collision in laboratory and center-of-mass coordinate systems

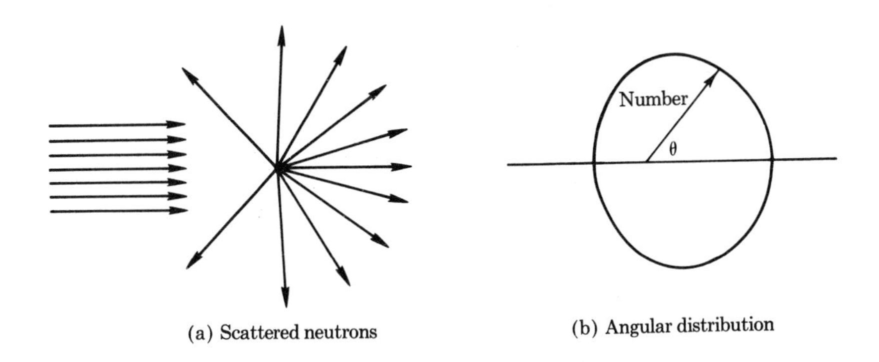

FIGURE 10-7

Collisions of neutrons with carbon nuclei

the momentum of the system, which is unaffected by the collision of its component particles, which are treated as isolated objects. We shall note only one phenomenon arising from such an analysis, one that is important in the collisions of gas particles or neutrons with nuclei. Although an incoming particle can be scattered through many different possible angles, it tends to "remember" the fact that it carried initial momentum with it; that is, the scattered particle tends to preserve a component of velocity *forward*. This effect is given various names, such as "persistence of velocity" or "average forward scattering." Figures 10-7(a) and 10-7(b) show the emergent directions of particles when the target is bombarded again and again—first a pictorial description, then a polar diagram of the *angular distribution*. In the polar diagram, the number of particles scattered at each angle θ is indicated by the length of radial line to the curve. This particular example is of neutrons of low speed striking free carbon nuclei for which A is 12.

10-5 CONSERVATION OF ANGULAR MOMENTUM

The analogy between translation and rotation suggests the possibility of a conservation law for *angular momentum* that parallels the one for *linear* momentum. This comparison leads us to a very fundamental principle of physics, applicable to all motions of isolated systems. As we saw in Section 9-4, the rotation of a system of particles is governed by Newton's second law in the angular form:

$$N = \frac{\Delta L}{\Delta t}$$

where N is the total external torque and L is the angular momentum, equal to mvr for a point or $I\omega$ for a body. If the external torque is zero, regardless of internal torques, $\Delta L/\Delta t$ is zero, and L *must be constant in time.* This *principle of conservation of angular momentum* is, in words:

> IF THE NET TORQUE ON A SYSTEM OF PARTICLES IS ZERO, THE TOTAL ANGULAR MOMENTUM REMAINS CONSTANT.

Following are some simple examples in which the law is applicable:

1. An ice skater (or ballet dancer) spins with arms extended, giving the body its maximum moment of inertia. When the arms are pulled close in to the body, the moment of inertia I is reduced. Since $L = I\omega$ is constant, ω must increase, and the skater (or dancer) revolves more rapidly.

The effect can be seen in a system less complicated than the human body. Picture two spheres of mass m, connected by a spring. We stretch the spring, and then set the system in rotation, as in Figure 10-8. The force due to the spring pulls the masses together; and as the distance of separation decreases, the angular speed increases, while the angular momentum remains constant.

2. An artificial satellite of mass m is in orbit about the Earth at a radius R, speed v. By the application of outward rocket thrust, perpendicular to the orbit path, the satellite enters a larger orbit of radius R'. We find that the vehicle *slows down,* since

$$mvR = mv'R'$$

or

$$v' = \frac{R}{R'}v$$

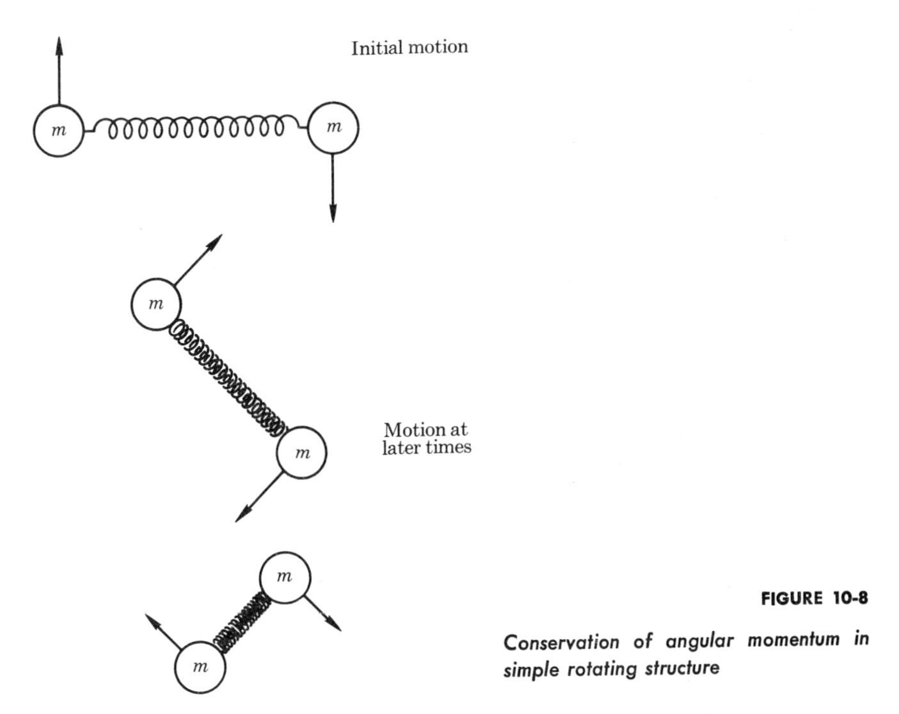

FIGURE 10-8

Conservation of angular momentum in simple rotating structure

In both cases, there was no need to know the detailed nature of the force that created the effect.

One may ask why we do not consider the Earth in this example—after all the satellite is "connected" to the Earth by the force of gravity in just as real a fashion as the skater's arms are to his body. Strictly speaking, the *system* to be considered should have included the Earth, but the effect of the change in conditions is negligible because of the Earth's tremendous mass and moment of inertia. According to the laws of conservation of linear and angular momentum, every time each of us takes a step we impart a tiny momentum to the Earth, but the effect is entirely unnoticeable. Even a battery of atomic bombs fired off at the surface would do little to change the Earth's orbit or rotation. Thus in many situations we are able to select a limited portion of the whole system.

3. Collisions of molecules are also governed by the angular-momentum conservation principle. As a simple example, suppose that a hydrogen atom strikes another hydrogen atom that is part of an H_2 molecule at rest (Figure 10-9). The atom in the molecule recoils, setting the molecule into rotation about its center of mass, which is moving through space. As before, let u_1 and v_1 be the projectile speeds before and after collision. The only initial angular momentum, measured with reference to thc CM of the H_2 molecule, is that of the H atom, mu_1r, where r is one-half the separation of the atoms. Afterward it is $-mv_1r$ for the atom (assuming backward reflection), plus an amount $I\omega$ for the molecule. Conservation of angular momentum then states that

$$mu_1r = -mv_1r + I\omega$$

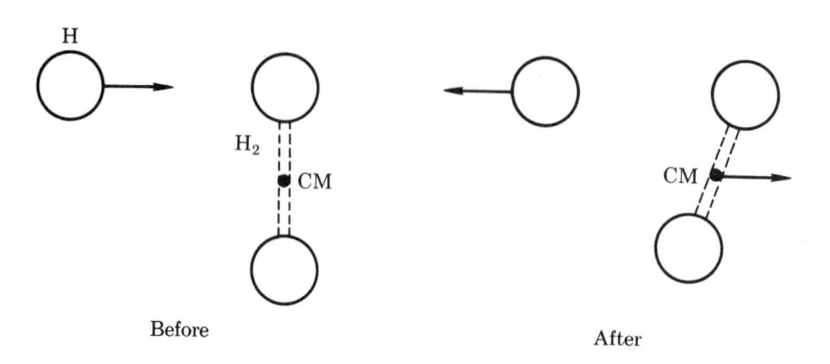

FIGURE 10-9

Collision of H atom and H_2 molecule

Conservation of kinetic energy, both translational and rotational, tells us that

$$\tfrac{1}{2}mu_1^2 = \tfrac{1}{2}mv_1^2 + \tfrac{1}{2}(2m)V^2 + \tfrac{1}{2}I\omega^2$$

while conservation of linear momentum is

$$mu_1 = -mv_1 + 2mV$$

where V is the speed of the center of mass. The three equations are easily solved if we insert $I = 2mr^2$ and $\omega = v/r$, where v is the speed of each of the H atoms in the molecule about the CM. As in Problem 10-30, we find that $v_1 = 0$ and $v = V = u_1/2$, which means that the projectile stops, giving the molecule all its momentum and energy. Thus the center of mass of the H_2 molecule ends up with the same speed as that of atom rotation about the CM.

10-6 INTERACTION OF ASTRONOMICAL BODIES OR ATOMIC PARTICLES

We have seen that both the gravitational and the electrostatic forces between particles vary inversely as the square of the distance of separation. Thus there is a close analogy between the interaction of two celestial objects such as the Sun and the Earth and the classical elementary quantum-mechanical view of the interaction of two atomic particles such as the proton and the electron (Figure 10-10). One major difference between the two cases is that charges may be like or unlike, corresponding to attraction or repulsion, whereas gravitational force is only attractive.*

The mathematical details of the general motion of objects are somewhat more involved than we wish to discuss here, but there are several

* This difference can conceivably be eliminated if the hypothesis were correct that the expansion of the universe is due to the force of gravitational repulsion of galaxies composed of matter and antimatter.

concepts that may be illustrated by simple cases. We shall invoke the conservation laws of energy and angular momentum in the study of what is called the *central-force* problem, in which the forces between two particles depend only on the radial separation of centers. The two bodies are viewed as being isolated from all others, or at least influenced little by external forces. Then the principle of conservation of angular momentum is applicable:

$$\mathbf{L} = \text{constant}$$

Nothing will be said yet about the nature of the force acting except that it is along a line between the objects. The fact that **L** is constant with so little information specified suggests that angular momentum is very basic to nature. In the following examples, one of the bodies will be assumed to be much more massive than the other. The typical relative sizes favor this assumption—for example, the Earth's mass is only 3×10^{-6} times that of the Sun. The lighter object of mass m is considered to be a point mass or point charge.

Let us list several phenomena that we should like to investigate and explain using central-force dynamics:

1. the Earth's motion about the Sun, and that of the other planets in our solar system (the Moon's orbit about the Earth has similar interest);
2. the motion of a spacecraft in orbit around the Earth or Moon;
3. the launching of spacecraft into satellite path;
4. the motion of comets that return periodically or of those that appear only once, and the behavior of meteorites;
5. the deflection of high-speed charged particles by nuclei of the atom.

This is clearly a large order; nevertheless, we shall attempt a single formulation that covers all such problems. The first feature common to all of these situations is the applicability of the force law of the inverse-square type:

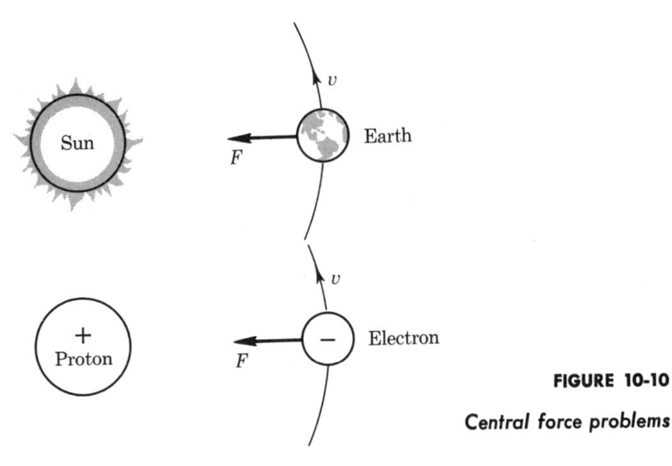

FIGURE 10-10

Central force problems

Force: $F = \frac{-C}{r^2}$; *Potential energy*: $E_p = \frac{-C}{r}$

The form and sign of C depends on the two interacting objects. For gravity, $C = k_G Mm$; for electrical charges of magnitude Q and q, $C = k_E Qq$ for attraction, $C = -k_E Qq$ for repulsion. Note that these conventions guarantee that the sign on F is $+$ for outward force, in the direction of increasing r; and $-$ for the reverse case.

The second common feature is that the laws of conservation of energy and angular momentum are applicable. Now,

$$E_k + E_p = E$$

or

$$\tfrac{1}{2}mv^2 - \frac{C}{r} = E$$

where E is the total (mechanical) energy, a constant. Also, the angular momentum, which is constant, is

$$L = mr^2 \frac{\Delta\theta}{\Delta t}$$

Let us think for a moment what information is needed. The motion as it proceeds in time will be known if the two coordinates of an object r and θ are available as functions of t. The alternate is the relation between r and θ, called an *equation of the orbit*. For example, if r is the same for all θ, the orbit is a circle. In general, the paths of particles that experience central forces turn out to be the curves that we recognize as the circle, the ellipse, the parabola, and the hyperbola. Which of these paths the body follows depends on E, L, and the sign of C in the force law.

For our investigation, we may picture the body that provides the central force as a sun, and consider various projectiles coming near it. Suppose a planet with very high velocity comes from outer space aimed in such a way as to pass far from this sun. We would guess that its path will be very nearly a straight line—Figure 10-11(a). If, however, the planet has less energy or comes closer, the sun will attract the planet, deflect it, and the path will be a curve, which turns out to be a hyperbola—Figure 10-11(b). If, now, the energy and angular momentum are just enough to permit the planet to escape the sun's gravitational field, the path is a parabola—Figure 10-11(c). A planet that collided with another body before approaching the sun might have still lower kinetic energy. Referring back to our formula for conservation of energy, we see that E could be negative. In this case, the planet is trapped in an orbit, generally in the shape of an ellipse—Figure 10-11(d). Finally, with a particular initial energy and direction of motion, the orbit may be a circle—Figure 10-11(e)—which is a special case of the ellipse.

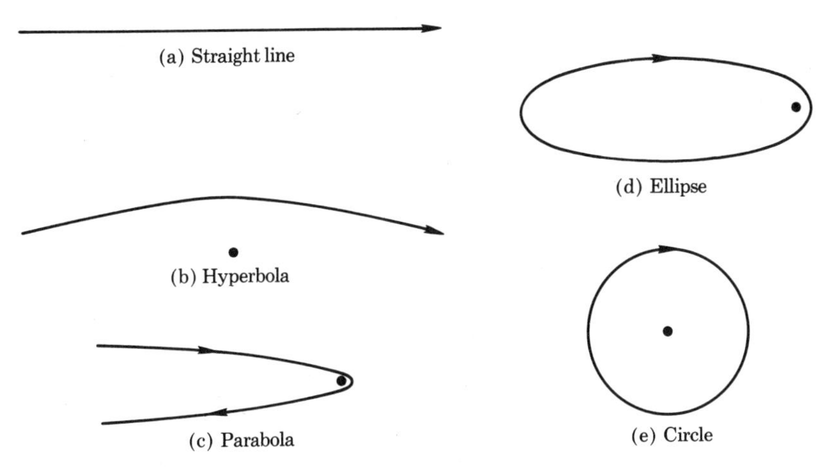

FIGURE 10-11

Orbits of astronomical bodies

We can compare each of these various paths with the others in terms of its *eccentricity* ϵ, which is a measure of the deviation from the circular shape. Figure 10-12 shows the range of values of ϵ for the curves discussed above. The mathematical meaning of eccentricity can be appreciated by studying the ellipse, which exhibits the smallest distortion from the circle of all the curves shown. Also, since the ellipse is the path traced out in orbit by each of the planets in our solar system, it takes on special importance. First, recall that the sum of the distances from a point on the curve to fixed points (foci 0 and 0′) is a constant, which permits the ellipse to be

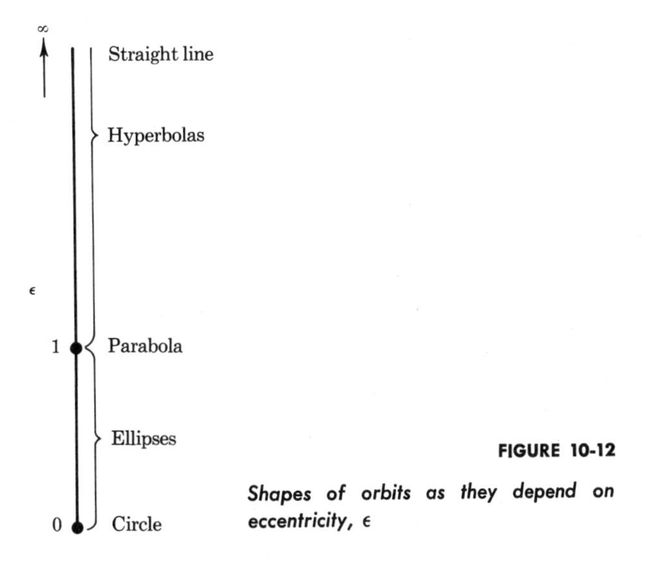

FIGURE 10-12

Shapes of orbits as they depend on eccentricity, ϵ

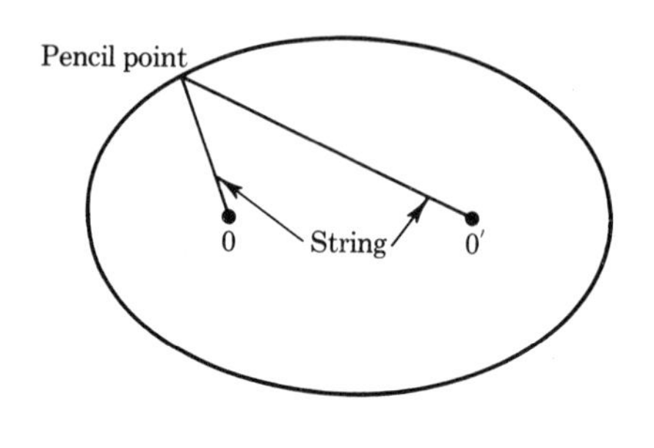

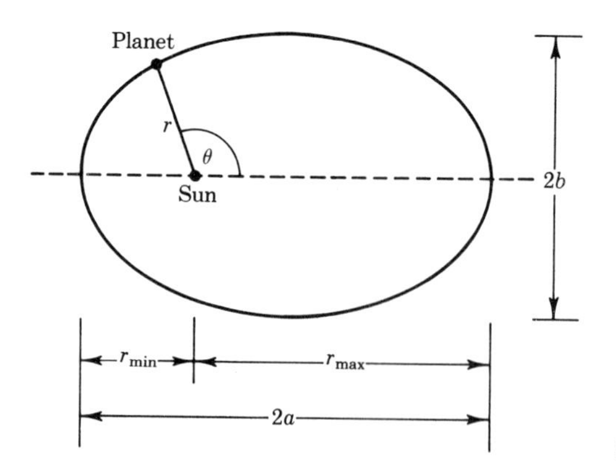

FIGURE 10-13

Properties of the ellipse

drawn merely by use of pencil and string, as in Figure 10-13(a). Its longest and shortest dimensions are $2a$ and $2b$ (the major and minor axes) as shown in Figure 10-13(b). The eccentricity of the ellipse is now defined by

$$\epsilon = \sqrt{1 - (b/a)^2}$$

Thus, when $b = a$ (the circle), $\epsilon = 0$; and when b/a becomes small (a long thin ellipse), ϵ approaches 1.

By application of the laws of conservation of energy and angular momentum for any of the paths shown in Figure 10-12, the eccentricity is found to be generally related to E, L, and other constants by

$$\epsilon = \sqrt{1 + \frac{2E}{m}\left(\frac{L}{C}\right)^2}$$

Comparison of the geometric and mechanical expressions for ϵ shows that a particle moves in a circular path, $\epsilon = 0$, if its total energy has a particular value $E = -\frac{m}{2}\left(\frac{C}{L}\right)^2$. Its kinetic energy is then $E_k = C/2R$, which can be checked by equating centripetal force mv^2/R with the force of attraction

C/R^2, and forming $E_k = \frac{1}{2}mv^2$. If the kinetic energy of the particle were either higher or lower, the orbit would be an ellipse instead, with the center of the attracting body at one of the foci of the ellipse. We see that if the total energy is exactly zero, then $\epsilon = 1$, which corresponds to a parabola. If E is positive, it means that ϵ is greater than 1, a condition for hyperbolas. Finally, as ϵ becomes very large, the hyperbola becomes a straight line.

Astronomical observations of our solar system have been collected over centuries and continue to be improved. Data on our system are given in Table 10-1. The mass (M), semi-major axis (a), period of revolution (T),

TABLE 10-1

ASTRONOMICAL DATA

	M/M_E	a/a_E	T/T_E	ϵ	R/R_E
Mercury	0.055	0.388	0.241	0.206	0.38
Venus	0.817	0.722	0.615	0.0068	0.96
Earth	1.000	1.000	1.000	0.0167	1.00
Mars	0.108	1.525	1.88	0.0934	0.53
Jupiter	318	5.20	11.9	0.0484	11.2
Saturn	95.2	9.54	29.5	0.0557	9.5
Uranus	14.6	19.2	84.0	0.0472	3.9
Neptune	17.3	30.1	165	0.0086	4.2
Pluto	0.93	39.4	248	0.249	0.4
Sun	333,432	—	—	—	109
Moon	0.0123	0.00257	0.0748	—	0.273

a_E = average distance from Earth to Sun, 92.9×10^6 mi (149.5×10^6 km)
M_E = mass of Earth, 5.98×10^{24} kg
T_E = period of Earth about Sun, 365.2 days
R_E = radius of Earth, 3.96×10^3 mi (6.37×10^6 m)

and radius (R) for each celestial body are expressed compactly in terms of ratios to the corresponding values for the Earth, and values of the eccentricity (ϵ) of elliptical orbit are listed. From the table we see that the eccentricities of the planets range from the smallest 0.007 for Venus, to 0.017 for the Earth, to the largest 0.249 for Pluto. The maximum and minimum distances from the planets to the Sun, as they move in their elliptical paths, can be calculated by use of our geometric and mechanical relations. Thus it is found that*

$$r_{\min} = a(1 - \epsilon) = \frac{\mathcal{R}}{1 + \epsilon}, \qquad r_{\max} = a(1 + \epsilon) = \frac{\mathcal{R}}{1 - \epsilon}$$

* These expressions may also be used for any of the paths about a central attractive body: for the circle, $\epsilon = 0$, $r_{\min} = r_{\max} = \mathcal{R}$; for the ellipse, $0 < \epsilon < 1$, $(\mathcal{R}/2) < r_{\min} < \mathcal{R}$; for the parabola, $\epsilon = 1$, $r_{\min} = (\mathcal{R}/2)$; and for the hyperbola, $\epsilon > 1$, $r_{\min} < (\mathcal{R}/2)$.

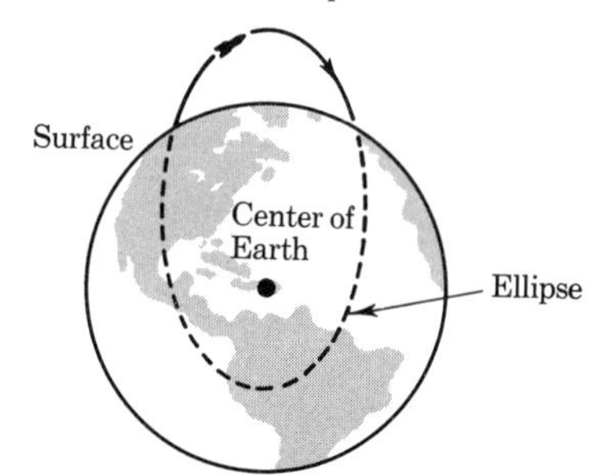

FIGURE 10-14

Path of projectile launched at low speed from Earth

where $\mathcal{R} = L^2/mC$ is approximately equal to a, which is the average distance from the Sun.

The distance of nearest approach of the Earth to the Sun, using $\mathcal{R} \cong a = 92.9 \times 10^6$ mi, $\epsilon = 0.017$, is easily found to be $r_{\min} = 91.3 \times 10^6$ mi, while the greatest distance is $r_{\max} = 94.5 \times 10^6$ mi. The differences from the average are small, and have far less to do with seasonal changes in weather than does the change in the axis of the Earth relative to the incoming sunlight.

One of the first artificial satellites, Explorer X, has an ϵ of 0.946 and ranges between 100 mi and 145,000 mi from the Earth's surface. Its orbit is thus a very elongated ellipse. A still more exaggerated ellipse is that of Halley's comet, which has appeared every 76 years at least since the third century B.C., and which is expected to return in the year 1986.

We may now ask: What happens to any projectile fired with a low initial speed *from* the Earth? Its path will be an ellipse, since the kinetic energy is low; but the path is highly eccentric and intersects with the Earth's surface, which stops the object, as shown in Figure 10-14. Only if its minimum radius is larger than that of the globe will it go into free orbit.

What is the chance of the Earth being struck by an errant planet or star? There are two ways in which this can happen. The first is the extremely unlikely case of a body aimed directly toward the center of the Earth. The second is the case of an object with a small initial angular momentum, along with a low kinetic energy. Such a body would seek to go into an orbit of radius *smaller* than that of the Earth and would thus intercept the Earth's surface. One can show, however, that the "target area" is very small. For instance, if a particle emerged from the Sun with a speed of 10 km/sec and with random original direction, the chance is only about 1 in 10^{12} that it would hit the Earth.

Prior to Newton's discovery of the law of gravitation, some important conclusions about planetary motion were developed by the German astronomer Johannes Kepler. He had studied the very accurate astronomical data of the Danish astronomer Tycho Brahe, and without the benefit of our present mathematics had stated what have become known as *Kepler's laws:*

KEPLER'S FIRST LAW: THE PLANETS MOVE IN ELLIPSES WITH THE SUN AS FOCUS.
KEPLER'S SECOND LAW: THE LINE BETWEEN THE SUN AND A PLANET SWEEPS OUT EQUAL AREAS IN EQUAL TIMES.
KEPLER'S THIRD LAW: THE SQUARE OF THE PERIOD OF REVOLUTION OF A PLANET VARIES AS THE CUBE OF ITS DISTANCE FROM THE SUN.

We can prove Kepler's second law by direct application of the law of conservation of angular momentum. As in Figure 10-15, the triangular area A swept out in a short time t is one-half the base times the altitude, $A = \frac{1}{2}r(r\theta)$. Then,

$$\frac{A}{t} = \frac{1}{2}r^2\frac{\theta}{t}$$

However, $\theta/t = \omega$, and since $L = mr^2\omega$,

$$\frac{A}{t} = \frac{L}{2m}$$

which is a constant.

Kepler's third law relating a planet's period of revolution to its distance from the Sun can easily be proved for a circular orbit. The force of gravitational attraction between the Sun and a planet provides the centripetal force needed to maintain a circular orbit of radius R:

$$\frac{k_G M m}{R^2} = \frac{mv^2}{R}$$

The *period* (the time for one cycle), however, is $T = 2\pi R/v$. Combining to eliminate v, we obtain

$$T = 2\pi\left(\frac{R^3}{k_G M}\right)^{1/2}$$

Squaring both sides gives Kepler's result that the square of the period is proportional to the cube of the orbit radius. The relation holds also for ellipses if R is replaced by the distance a (see Problem 10-43).

From measurements of the length of year of any planet and a knowledge of its path, the mass M of the Sun can be calculated. For example,

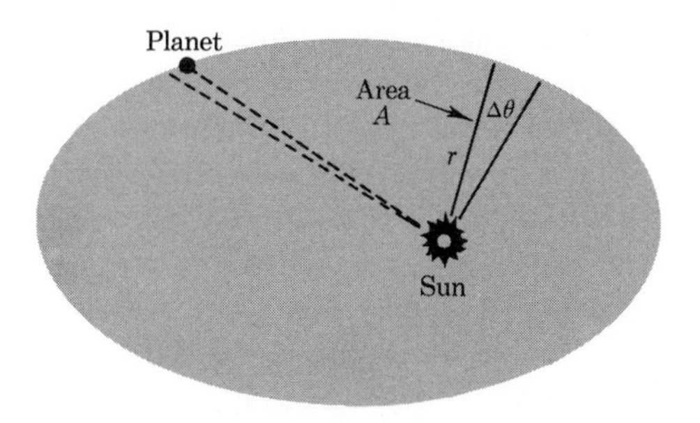

FIGURE 10-15

Kepler's second law that equal areas are swept out in equal times

using data for the Earth (see Table 10-1), $a = 149.5 \times 10^9$ m and $T = 365.2$ days or 3.16×10^7 sec. Then, using $k_G = 6.67 \times 10^{-11}$ in the rearranged expression $M = (2\pi/T)^2 a^3/k_G$ we deduce that the Sun has a mass of about 2×10^{30} kg.

Launching Spacecraft

The analysis of the process of placing a spacecraft, such as one of the Apollo series, into orbit is a most complex operation. In determining the actual flight from the ground to a point above the air layer, we must include many factors in the application of Newton's laws or in use of the energy equation:

1. The thrust may vary with time, according to the plan of launching. Its direction with respect to the rocket coordinates can be changed somewhat during flight, as well. With reference to coordinates at the center of the Earth, there will be both radial and angular components of thrust.
2. The mass of the system will vary with time, according to the rate of discharge of propellant, which is related to the thrust, as discussed in Section 10-3.
3. The acceleration of gravity will vary with position somewhat during the initial vertical motion, as discussed in Section 5-2.
4. The force of friction will depend on the shape of the vehicle, on its speed and probably its acceleration, and on the air density.
5. The air density will decrease with height in a generally exponential manner.

A glance at these qualitative considerations tells us why there is a need for instantaneous computation of position and motion, tracking stations, and the option for automatic and manual control of the spacecraft. Furthermore, the Earth's rotation, with an angular speed of 7.3×10^{-5} rad/sec, gives the rocket a "built-in" speed if the launching is from west to east. This corresponds to a surface speed ranging from 0.46 km/sec at the Equator to zero at the pole.

Collision of Charged Particles of Like Sign

It is one of the remarkable facts of nature that the same laws of motion can be applied to bodies of astronomical dimensions and to those of atomic size. With respect to the latter, the interaction of two electrically charged nuclei was first studied in detail by Ernest Rutherford in England in the early part of this century. Rutherford led research laboratories at Manchester and Cambridge, where an unusual number of discoveries were made. He named alpha rays coming from the radioactivity of uranium, developed a general theory of radioactivity, and identified many of the elements involved in this phenomenon. He and his students studied the

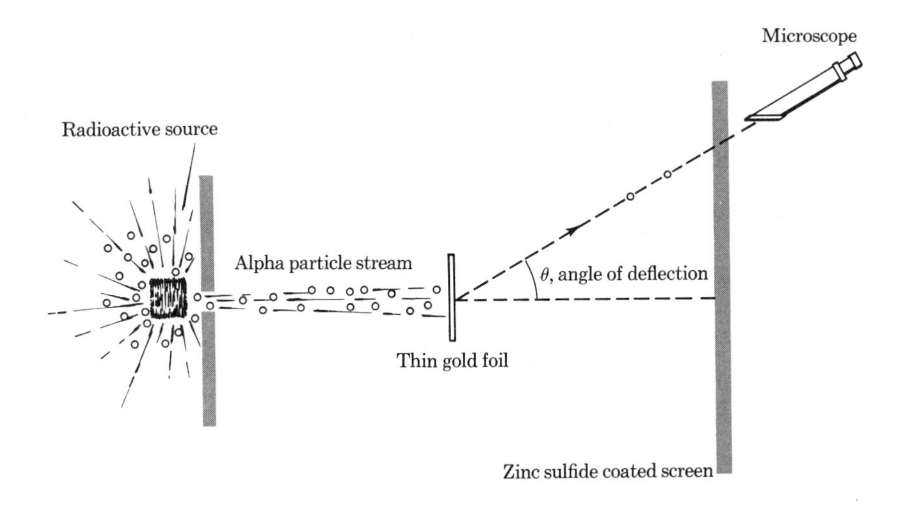

FIGURE 10-16

Rutherford's alpha particle-scattering experiment

scattering of alpha particles, known to be helium ions, as they passed through very thin films of gold and platinum. Their apparatus consisted of a radioactive source, which provided a stream of alpha particles (Figure 10-16). On the opposite side of the plates was a screen coated with zinc sulfide, which emitted light where a particle struck, viewed by a microscope. They found that the number of ions deflected through large angles was far greater than was expected on the basis of the thickness of the target and the then-current model of the atom, as a uniform distribution of positive

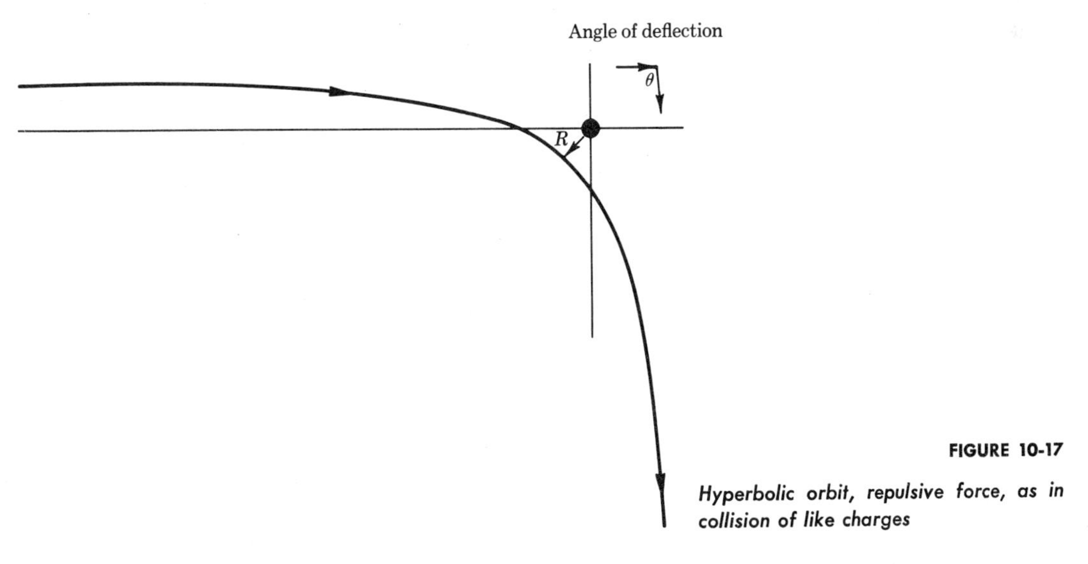

FIGURE 10-17

Hyperbolic orbit, repulsive force, as in collision of like charges

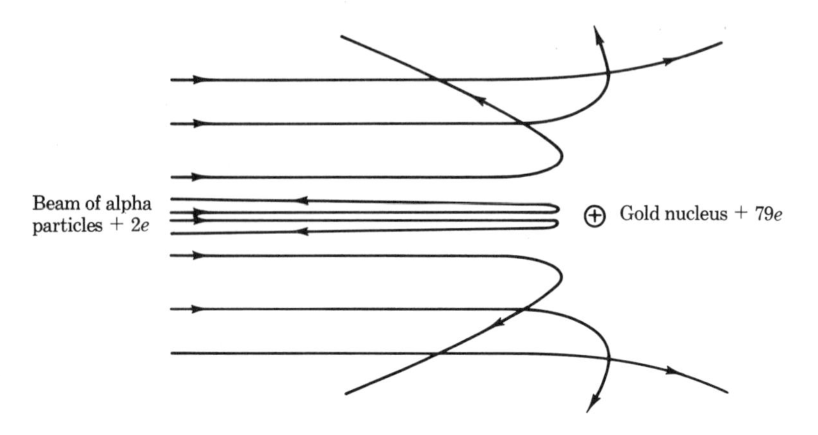

FIGURE 10-18

Deflection of alpha particles by gold nuclei

and negative charge. In fact, some were scattered sufficiently to return toward the source. Rutherford made the brilliant deduction in 1911 that the alpha particles passed through a very strong electric field of a highly concentrated "nucleus" of charge, of dimension thousands of times smaller than the atom. He proposed that the nucleus was responsible for gravitational and radioactive effects, and that it contained all of the atom's positive electricity. The mathematical understanding of his experiment follows from the planetary laws, with the important difference that there is a repulsive force between the alpha particle and the nucleus. The energy of this system can never be negative or zero, hence the eccentricity ϵ is greater than 1. The motion of an alpha particle, as a doubly charged ${}_2He^4$ ion, in the electrostatic field of the gold nucleus ${}_{79}Au^{197}$ with its 79 positive charges, describes a hyperbola. In contrast to the path of a high-speed projectile traveling past a planet, the alpha particle moves on the branch of the hyperbola that lies on the other side of the central body, as illustrated in Figure 10-17. The closest distance of approach again depends on energy and angular momentum. Figure 10-18 shows a series of scattering curves for such a collision of positive charges.

The Use of Words and Symbols for Physical Quantities

We have used symbols to represent physical quantities in our studies ever since we learned formulas such as $A = LW$ for the area of a rectangle. Later, the unknown x became familiar, even if it was not fully understood. As we enter the subject of physics, we are deluged with symbols—a different one for each quantity such as distance, time, mass, force, and so on. After a while we run out of letters in the Roman alphabet, since there are far more than twenty-six distinguishable physical quantities. Even if we employ capitals, lower-case, and script letters, and part of the Greek alphabet, we are eventually forced to use the same letter for two different things. It would be perfect if all symbols could remind us of the full name of the object being considered—e.g., t for *t*ime, m for *m*ass, F for *f*orce, etc. Unfortunately, we encounter many other quantities—such as *t*emperature and *t*orque, *m*olecular weight and *m*omentum, *f*requency and *f*lux—that doom this effort to defeat.

Symbols in physics are similar to words in a language. They are invented by someone and become popular, but meanings change and new terms appear. In earlier literature on mechanics, we find T and V for kinetic and potential energy. Encountering conflicts with *t*emperature and *v*olume, some writers have adopted KE and PE, or E_k and E_p, for these energies. The reader is cautioned to check on usage in a given reference. The list of symbols and units that we use in this text is given in Appendix B.

Just as in our language as a whole, there are many different sources of technical words in physics. The ancient Greek is the origin of such words as "kilogram" (*chilioi*, a thousand; *gramma*, a small weight), "mass" (*maza*, a barley cake). On the other hand, the Anglo-Saxon language gives us "time" (*tima*), and the Arabic provides "algebra" (*al-jabr*, reunion of broken parts). Elements were occasionally named according to a characteristic property —e.g., iridium (Latin *iris, iridis*, rainbow) after the colorful nature of solutions formed by compounds of the element. More recently, noted persons are honored—e.g., the elements lawrencium and einsteinium.

Words must even now be coined to meet the needs of new discoveries. Attempts are often made to select words that have some association with the old languages. For example the "neutron," identified in 1932, was named from the Latin (*neuter*, neither) implying neither positive nor negative charge. Some new words are formed by abbreviation of longer words or phrases—e.g., "quasar" ("quasi-stellar"). Others are in the class of acronyms, in which the initial letters of a phrase are combined. For instance, the word "laser" stands for "*l*ight-*a*mplification by *s*timulated *e*mission of *r*adiation."

The subject of semantics tells us that we must not confuse the word with the object itself. Stuart Chase in *The Tyranny of Words* comments on the misuse of abstract words: "We are continually confusing the label [the word "dog" in the sentence 'This is a dog'] with the non-verbal object, and so giving a spurious validity to the word, as something alive and barking in its own right. When this tendency to identify expands from dogs to higher abstractions such as 'liberty,' 'justice,' 'the eternal,' and imputes living, breathing entity to them, almost nobody knows what anybody else means."*

There is a similar hazard in science. We become used to working with the formulas $F = ma$ or $E_k = \frac{1}{2}mv^2$, and tend to expect physical measurements to conform. They seldom do, for many reasons. The formula may

* New York: Harcourt, Brace & World, Inc., 1938, p. 9.

be an approximation, in this example, if the speed were very high. More important, we must remember that the real objects are influenced by many forces and effects that our simple idealized formulas cannot describe. It is our task to interpret nature by physical and mathematical generalization, but we cannot expect nature necessarily to conform with our concepts.

PROBLEMS

10-1 A shell of mass 5 kg is fired with speed 350 m/sec from an artillery weapon of mass 600 kg. What is the speed of recoil?

10-2 How much speed does a rabbit of mass 2 kg gain on being struck with a 5-gram bullet of speed 600 m/sec? Assume that the bullet does not go through.

10-3 A 75-kg man standing on a frozen lake of negligible friction throws a 4-kg stone with a velocity of 6 m/sec. What velocity does the man acquire?

10-4 A machine gun fires 50-gram bullets at 330 m/sec. The gunner can exert a force of only 450 N against the gun. What is the maximum number of bullets he can fire per minute?

10-5 A radioactive nucleus of mass 350×10^{-27} kg at rest decays by emitting an alpha particle (helium nucleus) of mass 6.7×10^{-27} kg with a momentum of 8.2×10^{-21} kg-m/sec. What are the velocities of recoil of the residual nucleus and of the emitted alpha particle?

10-6 A 2-kg block is dropped from a height of 2.5 m onto a spring of constant 2800 N/m. What is the distance of compression of the spring?

10-7 A bat is in contact with a baseball of mass 0.15 kg for 0.4 sec, changing its velocity from 30 m/sec toward the plate to 40 m/sec back toward the pitcher. What impulse was it given and what average force during the blow?

10-8 A lead sphere of mass 2 kg and speed 6 m/sec overtakes a second lead sphere of mass 1.5 kg and speed 4 m/sec. They stick together and move on. What is the final speed? What amount of the initial energy is dissipated as heat?

10-9 A golfball of mass 0.05 kg dropped to a concrete sidewalk from a height of 2 m will bounce to a height of about 1.2 m. How much energy is dissipated as heat? How much momentum is imparted to the sidewalk?

10-10 Two pendulums each of length l are initially as indicated in Figure P10-10. The raised pendulum is released and strikes the second *inelastically*. How high does the center of mass rise after the collision? Show that energy is lost in the collision.

10-11 To obtain a thrust of 1,000,000 lb, as in the F-1 Moon rocket, what rate of discharge of propellant is needed, assuming 4 km/sec exhaust speed?

10-12 What is the fractional fuel weight needed in a rocket to reach the same final speed as that of the exhaust gas relative to the rocket?

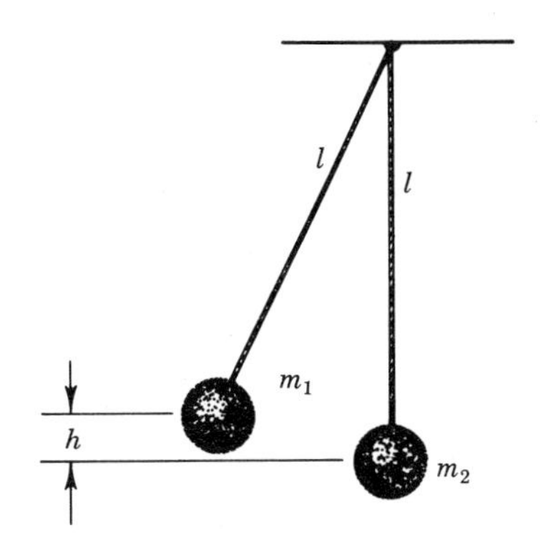

FIGURE P10-10

10-13 (a) Escape velocity from the Earth is approximately 11 km/sec. If a rocket has an exhaust-gas speed of 8 km/sec, what fraction of the initial rocket mass is left on escape? (b) In a vertical firing in a uniform gravitational field, the variation of rocket mass is

$$m/m_0 = e^{-(v+gt)/v_p}$$

With propellant speed 4 km/sec and burning time 1 min, how much fuel is used to reach 2000-m/sec speed?

10-14 Write an expression for the acceleration of a rocket perpendicularly from the Earth's surface: constant thrust, but variable gravity force, frictional force proportional to the square of the speed and to the density of air, with air density falling off exponentially with height above the ground.

10-15 Suppose that equal masses moving in the same direction collide elastically and head-on. Show that conservation of momentum and energy requires that the masses exchange velocities.

10-16 Two automobiles of equal mass m and speed u approach an intersection, one from the south, one from the east. After an elastic collision, the first car is going north at one-half its original speed. At what direction from north and how fast is the other car going? What fraction of the initial total energy does it now have?

10-17 A heavily loaded truck of weight 20,000 lb, moving at 55 mi/hr, strikes a parked Volkswagen car of weight 1600 lb. (a) What speed would the VW gain and how much would the truck lose if the collision were perfectly elastic? (b) Repeat the calculation for a perfectly inelastic collision.

10-18* (a) A very light particle such as an electron collides with a massive particle such as the hydrogen atom, at rest. Assuming an elastic collision in which the direction of the light particle is *reversed*, derive from fundamentals an expression for its final speed. (b) What fraction of the electron's initial kinetic energy is lost in such a collision, assuming $m_{\mathrm{H}}/m_e \cong 1840$?

10-19* A deuteron of mass 2 collides head-on with a triton of mass 3, such that the ratio of initial speeds is $v_d/v_t = 4$. Find the final ratio of speeds for the two possible relative velocities of approach. *Suggestion:* use footnote, p. 276.

10-20 When the radioactive atom of radium Ra^{226}, mass 3.75×10^{-25} kg, disintegrates, it emits an alpha particle ${}_2He^4$, mass 6.65×10^{-27} kg, with an energy 4.78 MeV. What is the speed of the alpha particle? What is the speed of recoil of the product nucleus of radon Rn^{222}, mass 3.68×10^{-25} kg?

10-21 Calculate the kinetic energies of the products of the nuclear reaction ${}_9F^{19} + {}_1H^1 \rightarrow {}_8O^{16} + {}_2He^4 + Q$—where Q is the energy released in the process, 8.1 MeV—if the initial kinetic energies of the reacting particles are negligible.

10-22 A ping-pong ball of mass 2.5 gram moving with speed 5 m/sec collides head-on with a bowling ball of mass 6 kg at rest. Approximately what speed is given to the latter?

10-23 Compare the qualities of BeO and D_2O as moderators for a nuclear reactor.

10-24 Show that the momentum $p = mv$ of a particle with speed near that of light is related to the kinetic energy by

$$p = \sqrt{2m_0E_k + (E_k/c)^2}$$

10-25 A radioactive nucleus carbon-14, mass 2.34×10^{-26} kg, emits an electron of 0.155 MeV kinetic energy. What momentum does the electron have? What is the recoil speed of the heavy particle? *Note:* Relativity need not be considered in calculating the momentum of the *carbon atom.*

10-26 Show that the kinetic energy imparted to an *electron* struck by a photon of energy E is

$$E_k = \frac{E}{1 + E_0/2E}$$

If E is 1 MeV, what fractions are imparted to the electron and the photon?

10-27 A particle of water of mass m is discharged at speed v from the end and perpendicular to one arm of a sprinkler of mass M, length l, in the form of a uniform pipe pivoted at its center. Find the resulting angular velocity of the sprinkler.

10-28 An artificial satellite of mass M is coasting about the Earth at speed v. A burst of exhaust gas of mass m, and speed v_p relative to the rocket motor is released. What new speed does the satellite immediately have, according to conservation of linear momentum?

10-29 A playground merry-go-round, moment of inertia I, is rotating at angular speed ω. (a) If a boy of mass m suddenly sits down at radius R and rides along, what angular speed does the system assume? (b) How would this result differ if instead he was running on the ground tangentially at speed v in the direction the merry-go-round was moving?

10-30* Verify the conclusions reached about particle speeds in molecular collisions in example 3, Section 10-5.

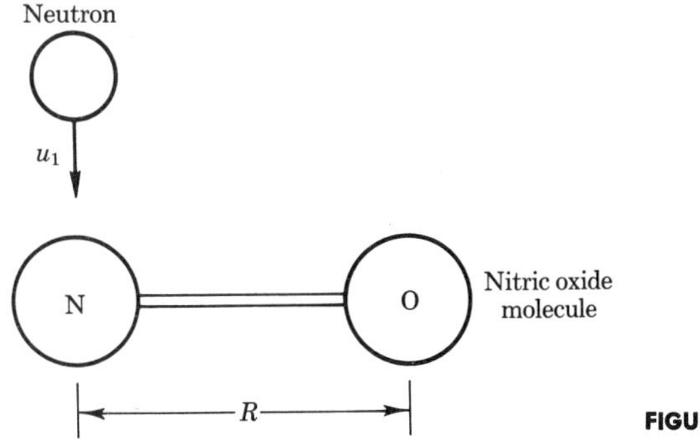

FIGURE P10-31

10-31 A neutron of mass $m_1 = 1$ amu and speed u_1 strikes the nitrogen atom in the molecule of nitric oxide, NO (see Figure P10-31) masses $m_2 = 14$ amu, $m_3 = 16$ amu, and with fixed separation R. Write the set of equations required to determine the motion of the particles after collision.

10-32 Compute how long it takes from the start to the end of an eclipse of the Moon.

10-33 At what angle must a rocket of speed 4000 mi/hr be fired from the surface of the Earth at the Equator if it is to leave perpendicularly to the surface, as viewed by an observer far away from the Earth? What velocity relative to the Earth does it have?

10-34 Find the distance of nearest approach of Mars to the Sun, and then the closest distance between Mars and the Earth.

10-35 Show that Pluto sometimes comes closer to the Sun than the orbit of Neptune, which is generally considered nearer.

10-36* Verify that if the Explorer X, $\epsilon = 0.947$, reaches a minimum height of 100 mi, that its maximum height is around 1.45×10^5 mi. Show that the angular momentum for a very elongated ellipse is approximately

$$L = m\sqrt{2gr_{\min}}R_E$$

and calculate its value for Explorer X of mass 35 kg.

10-37 A meteorite with eccentricity 1.05 and mass 50 kg comes within 1 Earth radius R_E of the surface of the Earth. What is the shape of its path? Calculate its angular momentum, assuming $r_{\min} = \mathcal{R}/(1 + \epsilon)$.

10-38 Calculate the period of revolution T of an electron moving about the proton in a circle if Kepler's third law were applicable, and $R = 5.3 \times 10^{-11}$ m.

10-39 From the area of the ellipse πab, find the area in square miles swept out each second by the line between the Earth and the Sun.

10-40 A communications satellite is launched with a 24-hour period so that it remains above a certain spot on the Equator at all times. In which direction must it rotate? What radius and speed must it have, according to Kepler's third law?

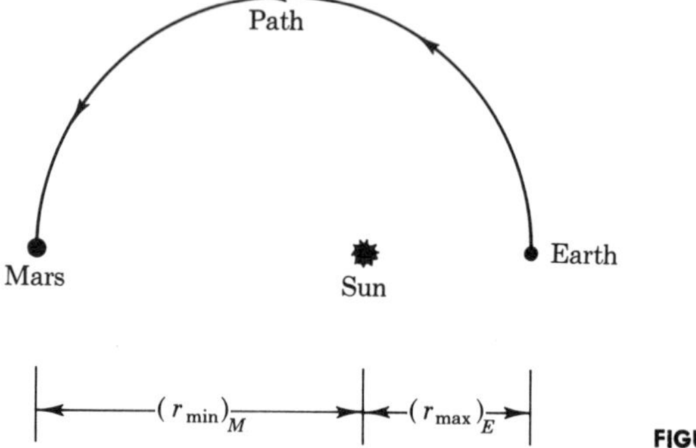

FIGURE P10-42

10-41 A less serious proposal than that in Problem 10-40 is to launch a satellite eastward in a $\frac{24}{25}$-hour orbit (or westward in a $\frac{24}{23}$-hour orbit) so that it crosses overhead exactly on the hour, and serves as a standard clock. What is wrong with this idea?

10-42* Show by use of Kepler's third law that the time for a semielliptical path of a spacecraft about the Sun, going from Earth orbit to Mars orbit (see Figure P10-42) is 240 days, and that the path is 350,000,000 mi. *Note:* The perimeter of an ellipse is approximately $2\pi\sqrt{\dfrac{a^2 + b^2}{2}}$.

10-43 Using Kepler's second law, $T = 2mA/L$, noting that the area of an ellipse is $A = \pi ab$, verify Kepler's third law: $T = 2\pi(a^3/k_G M)^{1/2}$.

10-44* The "three-body problem" of astronomy consists of the description of the motion of three gravitationally interacting masses such as the Sun, the Earth, and the Moon. It cannot be solved "exactly"—that is, in terms of simple mathematical functions and geometric figures such as the ellipse —as can problems involving the motion of only one planet about the Sun, a "two-body problem." Investigate, using diagrams and central-force principles, the motion of the Earth and the Moon about the Sun with emphasis on (a) the difference between the time between successive "full Moons" (29.5 days) and the period of the Moon about the Earth (27.3 days); (b) the progress in time of an eclipse of the Sun (by the Moon) and an eclipse of the Moon (by the Earth).

VIBRATION AND WAVE MOTION

11

Many phenomena of nature are found to be periodic or oscillatory. Some of these, such as the orbits of atomic or planetary masses about each other, are periodic because of the essentially *circular* nature of their paths. Others, such as the motions of springs and pendulums, are naturally capable of *vibration* or *oscillation*. The study of such motion in this chapter will serve as a preface to the general subject of waves.

11-1 PERIODIC MOTION

Let us list some well-known examples of periodic or vibrational motion:

1. The motion of the Moon about the Earth (or that of the Earth about the Sun) brings the orbiting mass back to its original position every month (or year).

2. The electrons in the atom may be imagined to execute circular paths about the nucleus.

3. A pendulum swings back and forth, always regaining its initial displacement after a definite period of time.

4. A piano string, a bell, a drum, and a woodwind reed all vibrate in a manner to give us the sounds we know as music.

5. The voltage on electrical circuits in the home alternate in polarity, implying that charge is moving in a periodic way.

6. An automobile will bounce up and down on its coil springs after the wheels go over a bump in the road.

These are but a few of the innumerable illustrations found in nature or among man's devices. Although diverse in character, they may be treated mathematically in the same way. In this section we shall review some of the properties of oscillation or vibration—such as period and frequency—and then develop the mathematics for the description of periodic motion.

A *cycle* is defined as motion that brings the system back to its original state, regardless of where one starts. For example, arrival at the same point on a circle after one revolution completes a cycle. The length of time to execute a cycle is called the *period* T, usually measured in seconds. The number of cycles each second is the *frequency* ν, related to the period by

$$\nu = \frac{1}{T} \qquad T = \frac{1}{\nu}$$

For instance, the voltage mentioned above (example 5) has a frequency of 60 cps (cycles per second), and the period is $\frac{1}{60}$ sec.

Motion in a circular path, discussed first in Chapter 2, serves as a good example of periodic motion. First, recall that the location of a point on a circle can be expressed in either cartesian or polar coordinates. If A is the radius (instead of r), then

$$x = A \cos \theta, \qquad y = A \sin \theta$$

as in Figure 11-1. Next, suppose that we wish to represent the location of

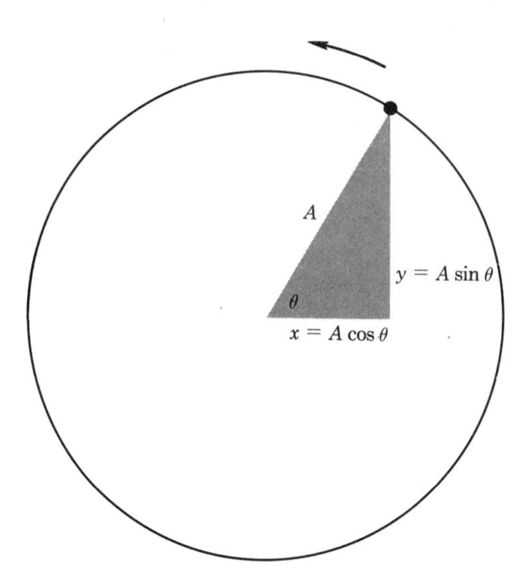

FIGURE 11-1

Periodic circular motion

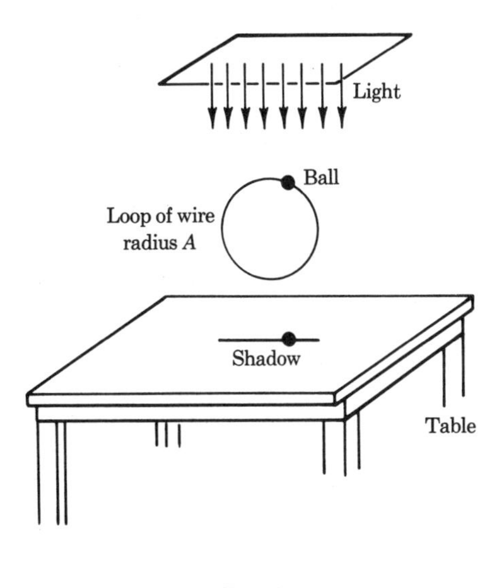

(a) Experiment

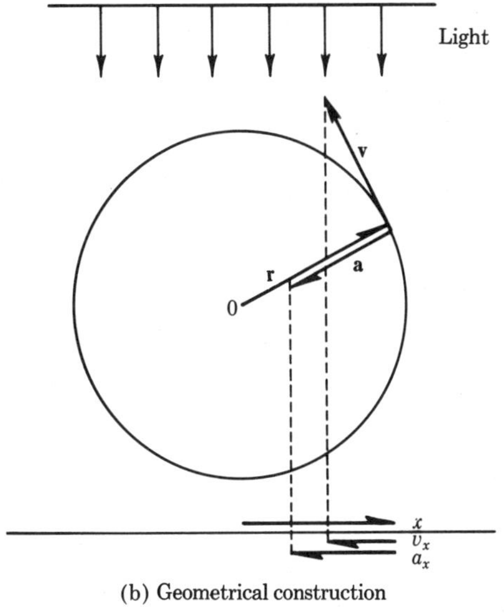

(b) Geometrical construction

FIGURE 11-2

Relation of circular and simple harmonic motion

an object moving in a circular path—for example, a spacecraft in orbit of radius A about the Earth. If the speed v is constant, the angular speed $\omega = v/A$ will also be constant, and the angle turned in time t will be $\theta = \omega t$. Then,

$$x = A \cos \omega t, \qquad y = A \sin \omega t$$

Let us express the motion in terms of period or frequency. The time for one complete revolution corresponds to an angle of 2π. At any time t, by proportions, $t/T = \theta/2\pi$, and $x = A \cos (2\pi t/T)$. Alternately, since $\nu = 1/T$, $x = A \cos 2\pi\nu t$.

Many of the motions that we listed earlier are not circular in nature. Their common feature, however, is that in each case the acceleration of the particle is proportional to the displacement. Let us verify this statement by performing an experiment. Above a table, as shown in Figure 11-2(a), we mount a circular loop of wire with a ball that can slide freely, and illuminate it from above, as by a skylight window. We then cause the ball to move on the radius A with angular speed ω, and note that its shadow on the table moves horizontally. The vectors representing position **r**, velocity **v**, and acceleration **a** are shown in Figure 11-2(b). Their magnitudes are A, ωA, and $\omega^2 A$, respectively.* The x-components of the vectors can be viewed also as shadows on the table. From the diagram we see that

* Recall that for radius R, the tangential speed is $v = \omega R$, and the centripetal acceleration is $a = v^2/R$ or $\omega^2 R$.

$$x = A \cos \omega t$$
$$v_x = -\omega A \sin \omega t$$
$$a_x = -\omega^2 A \cos \omega t$$

The important feature to note is that a_x, the acceleration in the x-direction, is proportional to the displacement x at all times, but is in the opposite direction. The phrase *simple harmonic motion* is used to identify such behavior. Note that the speed is a maximum when the displacement is a minimum, and vice versa—that is, these two quantities are *out of phase* with each other. In this case the phase angle is $\pi/2$ or 90 deg.

11-2 EXAMPLES OF SIMPLE HARMONIC MOTION

Since the acceleration of a body is proportional to the force applied to it, we can generalize by stating that simple harmonic motion takes place when the force that tends to restore the object to its original position is proportional to the displacement. Following are several examples of this situation.

Oscillating Mass on a Spring

Let us first perform some experiments with an elastic spring. Starting with the spring lying on a smooth table in unstretched condition, as in Figure 11-3, we pull on the end with a force F. As indicated earlier, the larger the force, the larger the displacement, and $F = Kx$. By Newton's third law, the spring exerts an equal and opposite force on our hand or any object we attach to the end. This force is written

$$F = -Kx \qquad \textit{(restoring force)}$$

The negative sign denotes that the force is in the negative x-direction if the displacement is in the positive x-direction.

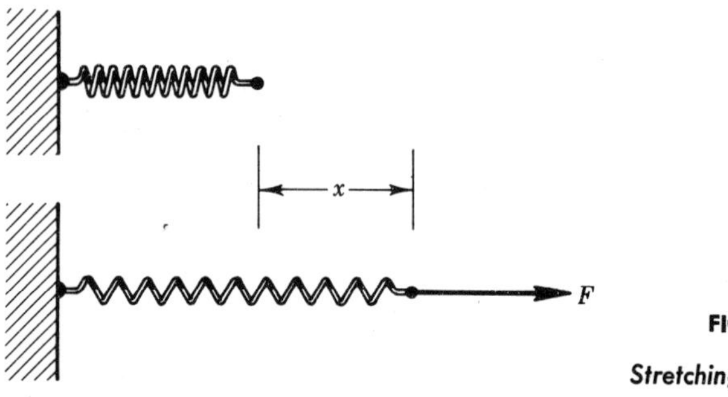

FIGURE 11-3
Stretching a spring

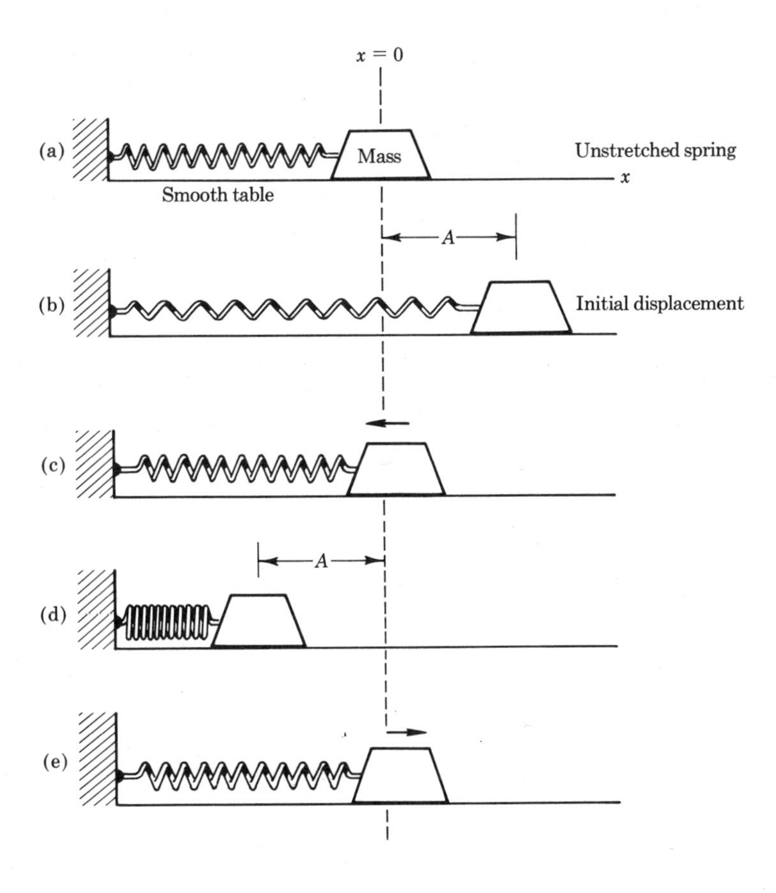

FIGURE 11-4

Oscillations of mass on spring

Now let us attach a mass m to the end of the unstretched spring—Figure 11-4(a). Then we pull the mass to a distance $x = A$—Figure 11-4(b)—and release it at time $t = 0$. By Newton's second law, the mass is accelerated by an initial value

$$a_0 = -\frac{KA}{m}$$

The motion is toward the equilibrium position $x = 0$—Figure 11-4(c). The acceleration decreases as x gets smaller. When the mass reaches the neutral position, its displacement, the force, and the acceleration are all zero, but it has a maximum speed. After it moves past $x = 0$, the force is in the $+x$ direction, and the object attains a maximum displacement $x = -A$—Figure 11-4(d)—when the acceleration is a maximum:

$$a_{\max} = \frac{KA}{m}$$

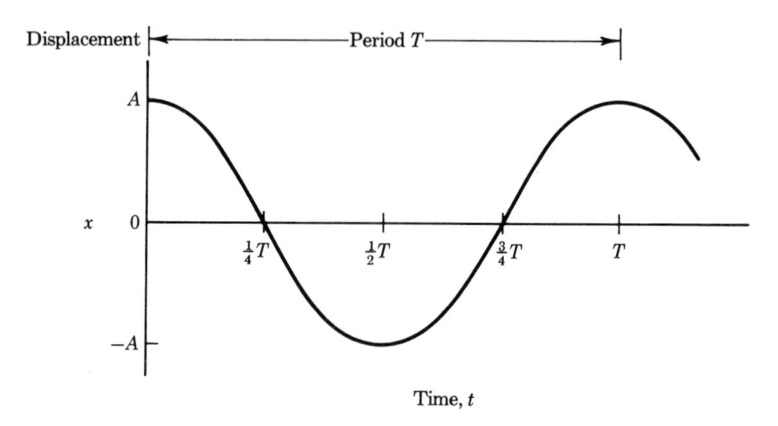

FIGURE 11-5

Graph of vibration cycle

Now the speed is zero, and the direction of motion reverses. A graph of these results is shown in Figure 11-5. The curve is of the form

$$x = A \cos \sqrt{\frac{K}{m}}\, t$$

The motion is periodic, as we can see from the cosine dependence on time. The period T is the time elapsed to bring x from position A through the cycle back to A again, such that the cosine becomes $\cos 2\pi$ or

$$\sqrt{\frac{K}{m}}\, T = 2\pi$$

Thus,

$$T = 2\pi \sqrt{\frac{m}{K}} \qquad \textit{(period of oscillating mass on spring)}$$

The frequency for this oscillating system is the number of such complete cycles each second,

$$\nu = \frac{1}{T} = \frac{1}{2\pi}\sqrt{\frac{K}{m}} \qquad \textit{(frequency of oscillating mass on spring)}$$

and the angular frequency is

$$\omega = 2\pi\nu = \sqrt{\frac{K}{m}}$$

Other simple forms for the displacement are then

$$x = A \cos \frac{2\pi t}{T} = A \cos \omega t$$

Steps in the vibration cycle that we readily recognize are listed in Table 11-1.

TABLE 11-1

SELECTED POINTS IN A CYCLE OF SIMPLE HARMONIC MOTION

t	$2\pi t/T$	$\cos(2\pi t/T)$	x
0	0	1	A
$\frac{1}{4}T$	$\pi/2$	0	0
$\frac{1}{2}T$	π	-1	$-A$
$\frac{3}{4}T$	$3\pi/2$	0	0
T	2π	1	A

Let us find T, ν, and ω for a mass of 0.4 kg attached to a spring that elongates 0.125 meters for each N of applied force—i.e., $K = 8$ N/m. The period is

$$T = 2\pi\sqrt{\frac{m}{K}} = 2\pi\sqrt{\frac{0.4\text{ kg}}{8\text{ kg/sec}^2}}$$
$$= 2\pi\sqrt{0.05} = 1.4\text{ sec}$$

The frequency of vibration is $\nu = 1/T = 0.71$ cps, and the angular frequency is $\omega = 2\pi\nu = 4.5$ rad/sec.

An important fact to note is that the oscillation of the mass on the spring, expressed as $x = A\cos\omega t$, is identical to the x-coordinate of the point moving on a circle. The circular motion is essentially a geometric effect, while the vibrational motion depends on the nature of the restoring force. If the restoring force were somehow dependent on the square of the displacement—not on the first power of x—the motion of the mass would be entirely different.

Energy in the harmonic oscillator alternates between the potential and kinetic forms, but with constant *total*. To demonstrate this, we recall from Section 8-1 that the work done in stretching an elastic substance by an amount x is $\frac{1}{2}Kx^2$. This is also the stored potential energy $E_p = \frac{1}{2}Kx^2$. At any instant, $x = A\cos\omega t$, where $\omega = \sqrt{K/m}$. Then $E_p = \frac{1}{2}KA^2\cos^2\omega t$. The kinetic energy at that time is $E_k = \frac{1}{2}mv^2$, but $v = -\omega A\sin\omega t$ and thus $E_k = \frac{1}{2}mA^2\omega^2\sin^2\omega t$. The sum is $E = E_k + E_p = \frac{1}{2}KA^2$ or $\frac{1}{2}m\omega^2A^2$, where we have used the fact that $\omega^2 = K/m$ and $\cos^2\omega t + \sin^2\omega t = 1$. The total energy depends on the squares of the original amplitude and the frequency of vibration. Figure 11-6 shows the variation with time of the two types of energy.

For the previous mass–spring example, if the maximum amplitude is $A = 0.03$ m, the total energy is

$$E = \tfrac{1}{2}m\omega^2A^2 = \tfrac{1}{2}(0.4\text{ kg})(4.5\text{ sec}^{-1})^2(0.03\text{ m})^2 = 3.6 \times 10^{-3}\text{ J}$$

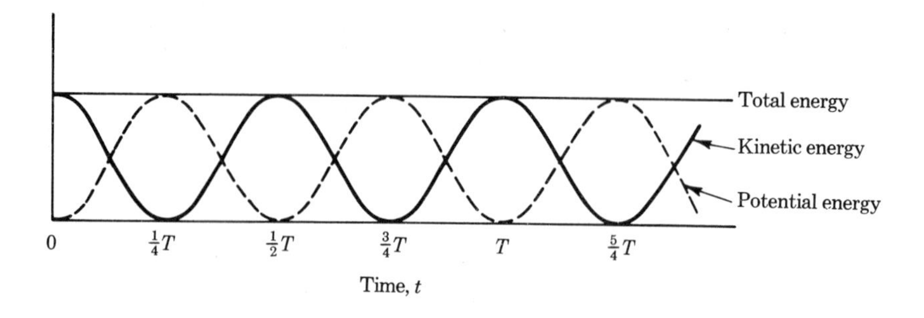

FIGURE 11-6

Energy in the harmonic oscillator

As a check, we compute the potential energy at full displacement:

$$E_p = \tfrac{1}{2}KA^2 = \tfrac{1}{2}(8\ \text{N/m})(0.03\ \text{m})^2 = 3.6 \times 10^{-3}\ \text{J}$$

The maximum speed occurs when the mass passes through the origin. The total energy is all kinetic, hence

$$v_{\text{max}} = \sqrt{\frac{2E}{m}} = \sqrt{\frac{7.2 \times 10^{-3}\ \text{J}}{0.4\ \text{kg}}} = 0.134\ \text{m/sec}$$

Torsional Oscillator

Another illustration of periodic motion is the torsional oscillator, which consists of a disc that can rotate and twist a thin rod, as shown in Figure 11-7. When turned through any angle θ, an elastic restoring *torque* $N = -c\theta$ results, where c is called the *torsion constant* of the rod. If the disc has a moment of inertia I, then by Newton's angular law, an angular acceleration occurs of amount

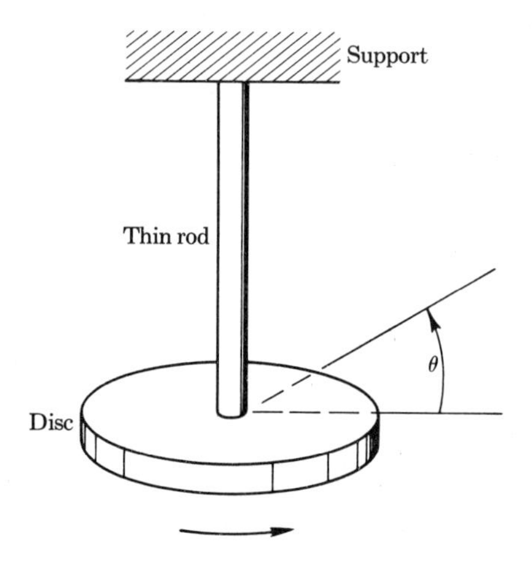

FIGURE 11-7

Torsional oscillator

$$\alpha = \frac{N}{I}$$

Analogy with the equation for the mass and spring tells us that

$$\theta = \theta_0 \cos \sqrt{\frac{c}{I}}\, t$$

where θ_0 is the maximum amplitude of oscillation. By further comparison, the period is $T = 2\pi\sqrt{I/c}$, from which $\nu = 1/T$ and $\omega = 2\pi\nu$ are easily found.

Let us use the torsional oscillator for the measurement of the moment of inertia of an object of complicated shape. A wheel is hung on a thin flexible rod and the torque measured as it depends on angle. If $N = 15$ N-m for an angle 30 deg, $\theta = \pi/6$, then

$$c = \frac{N}{\theta} = \frac{15 \text{ N-m}}{\pi/6 \text{ rad}} = 28.6 \text{ N-m/rad}$$

Now, suppose the period T is found to be 5 seconds; then,

$$I = c\left(\frac{T}{2\pi}\right)^2 = (28.6 \text{ N-m/rad})\left(\frac{5 \text{ sec}}{2\pi}\right)^2 = 18.2 \text{ kg-m}^2$$

Torsional vibrations are very useful in watches for keeping accurate time, since the period depends only on the fixed physical quantities I and c.

Oscillations of an Electrical Discharge

We are familiar with electrical sparks and the discharges that produce light in fluorescent bulbs. These mixtures of positive and negative ions are examples of a medium called a *plasma*. If the charges of opposite sign are not distributed uniformly, electrical fields will be set up and oscillations of charge will occur. Suppose that the electrons all happen to be displaced a distance x, as illustrated in Figure 11-8. The electric field $\mathcal{E}$ experienced by an individual electron in this ionic medium is proportional to displacement x, and since the force on an electron of mass m is $-\mathcal{E}e$, the conditions for harmonic oscillation are met. Since the mass of an electron is very small, plasma oscillations can have frequencies in the millions of cycles per second.

Oscillations in Magnetic Fields

As another illustration of harmonic motion, picture a small bar magnet of moment M and moment of inertia I, suspended at its center by a very fine thread and placed in a magnetic field $\mathcal{B}$. When the magnet is displaced slightly from its equilibrium direction, the restoring torque due to the field is $-M\mathcal{B} \sin \theta$—or, for small angles, $-M\mathcal{B}\theta$. By direct analogy to

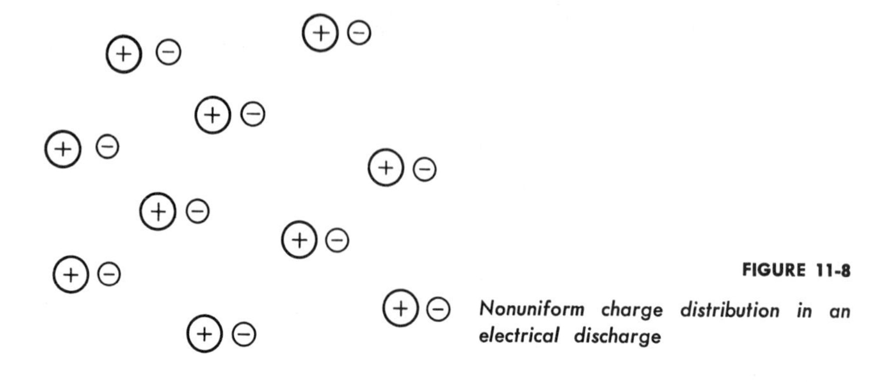

FIGURE 11-8

Nonuniform charge distribution in an electrical discharge

the torsional oscillator, we can identify the torsion constant as $M\mathcal{B}$, and immediately write the period as

$$T = 2\pi \sqrt{\frac{I}{M\mathcal{B}}}$$

The stronger the field, the shorter will be the period. Such a simple device can be used to compare the strengths of various magnetic fields. Measurements of the periods T_1 and T_2—in two different locations—yield the ratio of the two fields $\mathcal{B}_1/\mathcal{B}_2 = T_2^2/T_1^2$.

The Pendulum

Simple harmonic motion in an angular sense is exhibited by the familiar pendulum (Figure 11-9), at least if the angles are small. Let us analyze its action. The vertical force of gravity on the pendulum bob,

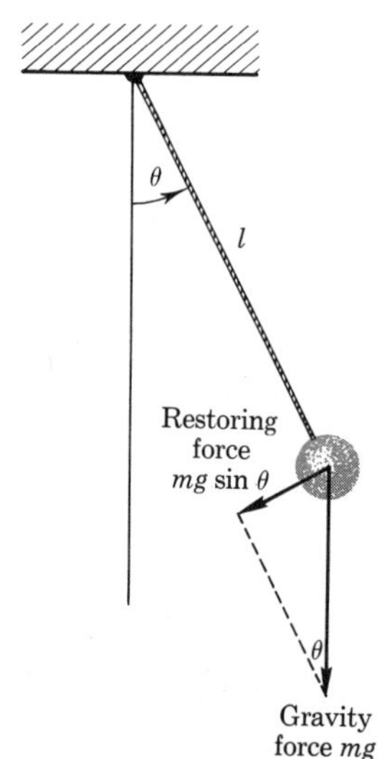

FIGURE 11-9

Forces on the simple pendulum

treated as a point particle of material, is mg. The component of restoring force that is perpendicular to the line l and thus in the direction opposite to increasing θ is $-mg \sin \theta$. For small angles, $\sin \theta \cong \theta$, and since the arc on the circle made by the displacement is $s = l\theta$, the force along the arc becomes $-mg(s/l)$. Equating this, by Newton's second law, to ma gives the equation

$$a = -\frac{g}{l} s$$

where the mass has canceled out. By direct comparison with the previous spring problem, we see that the arc of swing is

$$s = A \cos \sqrt{\frac{g}{l}}\, t$$

where A is the maximum arc. Then the period is*

$$T = 2\pi \sqrt{\frac{l}{g}} \qquad \textit{(period of a pendulum)}$$

We recognize the agreement between this formula and our experience that, the longer the arm of the pendulum, the greater the time per swing. Seconds can be measured by adjusting the length l to give a period of 2 sec—that is, 1 sec for each direction of swing. The numerical value of the length is

$$l = g\left(\frac{T}{2\pi}\right)^2 = (9.81 \text{ m/sec}^2)\left(\frac{2 \text{ sec}}{2\pi}\right)^2$$

$$= 0.994 \text{ m} \quad \text{or} \quad 39.1 \text{ in.}$$

If we look at the relation in the opposite way, a measurement of the period of motion of a pendulum with accurately known length l enables us to compute the acceleration of gravity at any location.

11-3 DAMPED, FORCED, AND RESONANCE VIBRATIONS

If any object capable of vibration is set into motion, it will oscillate with decreasing amplitude and finally come to rest because of friction. For examples, think of a plucked string on a musical instrument, or a child's swing moving in the air. The oscillation can be made to continue, however, if we prompt it by giving a push at just the right time. By bowing the string on a violin we maintain a constant note; by "pumping" the swing we can keep the amplitude constant or increase it; by jumping up and down

* A more correct value is $T \cong T_0(1 + \frac{1}{4}x^2 + \frac{9}{64}x^4)$, where T_0 is $2\pi\sqrt{l/g}$, x is $\sin(\alpha/2)$, and α is the angle of swing from the vertical.

on a diving board or trampoline, we keep these objects in motion. The mathematical treatment of such behavior is a straightforward extension of the concept of simple harmonic motion. In addition to the restoring force F_R, there is an applied force F_A that depends on time, and a friction force F_F that depends on speed. We shall assume low-speed motion such that the friction is proportional to speed, as in Section 5-8. Let us use as an example the mass and spring in order to be specific; thus the three forces are as follows:

Applied: $F_A(t)$
Restoring: $F_R = -Kx$
Friction: $F_F = -cv$

By Newton's law,

$$F_A(t) - Kx - cv = ma$$

This formula contains displacement, speed, and acceleration, all of which are related. There are several special cases of interest, however, which are now discussed.

1. *No applied force and no friction.* A freely oscillating object is characterized by $F_A = 0$, $c = 0$, and thus we revert to the simple harmonic motion, $ma = -Kx$, for which

$$x = A \cos \omega_0 t$$

where $\omega_0 = \sqrt{K/m}$ is the natural angular frequency. The subscript zero is attached to distinguish ω_0 from other frequencies to be found shortly.

2. *No applied force but with friction.* Again $F_A = 0$, leaving

$$ma = -Kx - cv$$

The object oscillates, but the amplitude decreases with time because of friction; that is, *damped vibrations* take place. This response is illustrated by springs or pendulums or tuning forks in air. The amplitude varies as $A(t) = e^{-\gamma t}$, where $\gamma = c/2m$. We see that the exponent γt increases with time, and $A(t)$ decreases exponentially. The larger the friction factor c or the smaller the mass m, the more rapidly the vibrations die away. The *damping time* $1/\gamma$ is that required for the amplitude to fall off by a factor $1/e$. Since there is friction, the frequency is changed from ω_0 to a new value $\omega = \sqrt{\omega_0^2 - \gamma^2}$, so long as γ is less than ω_0. Note what happens if the damping factor γ gets as large as ω_0. Then $\omega = 0$, denoting an infinite period—that is, no oscillation at all. A pendulum released in cold oil exhibits this behavior.

3. *Applied force with friction.* Let us now consider *forced vibrations*, with an external force present. Suppose the mass on a spring is subjected to a force that is of the form of a cosine function $F_A(t) = f \cos (\omega t - \phi)$,

where a phase angle ϕ accounts for the possibility that the force and vibration are not synchronized. Here ω may be any of a variety of possible frequencies. For an analogy, imagine bouncing on a diving board in various ways. We know in advance what will happen. If the frequency and timing are wrong, we tend to stop any vibrations that have been started. If, however, the impulses are right, we can set up a large amplitude vibration that can give us an impetus into the air. For forced vibrations in general, the amplitude A depends on the amplitude f and frequency ω of the applied force F_A, the mass m, the force constant K, and the friction constant c. At a special value of ω, called the resonance angular frequency, *resonance vibrations* of very large amplitude can be set up. For example, the possibility of dangerous resonance vibrations in bridges over which soldiers march has long been known; the order to break step results in randomly applied forces instead of a periodic one. Also the tuning of a radio or TV set involves an adjustment of electrical components to obtain resonance; many phenomena of light absorption and radiation may be described in terms of resonances; when the energy of bombarding nuclear particles coincides with natural internal motion of the target nuclei, the chance of reaction is large.

It seems remarkable that so many diverse processes display simple harmonic motion, or damped vibration, or resonance. The reason is not hard to understand: the *linear* restoring force is the simplest approximation of the correct force; friction is inevitable and its linear dependence on speed is the simplest expression that includes motion; applied forces arise from *other* vibrating sources, and hence are of a sine or cosine form.

11-4 MOLECULAR FORCES AND VIBRATIONS

Vibrations also take place on a molecular scale inasmuch as the forces between atoms are roughly proportional to the displacement. To appreciate this fact, let us review qualitatively the nature of the bonds.

We described (Section 1-2) the atomic structure of some of the elements at the low end of the periodic table in terms of nuclei and surrounding electron clouds. The atom as a whole is neutral and will affect other atoms negligibly—so long as the separation of particles is much greater than atomic dimensions. This qualifying phrase is needed because net electrical forces will arise between atoms when they approach each other more closely (Figure 11-10). Nonspherical charge distributions will be mutually induced. When the atoms come very close together, there will be a strong repulsion of the two nuclei (Section 5-7). In addition to the electrical force, there is an "electron-exchange" effect that may be fully explained only in terms of quantum mechanics. Roughly, it consists of an attraction that favors two adjacent atoms sharing or exchanging their outermost electrons. The very existence of diatomic molecules such as H_2 and O_2 is a demonstration of this force. The combined forces of several kinds may be repre-

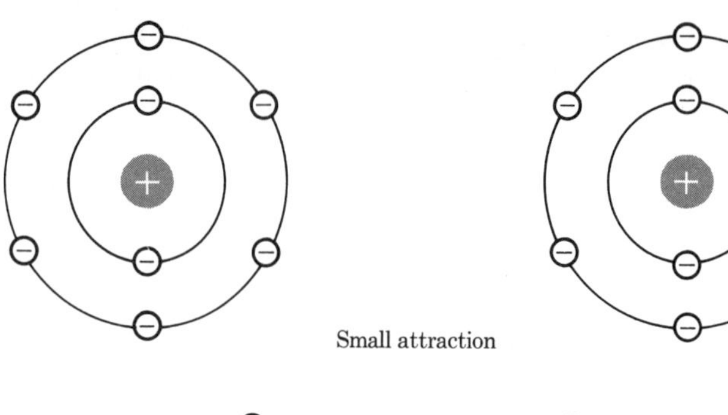

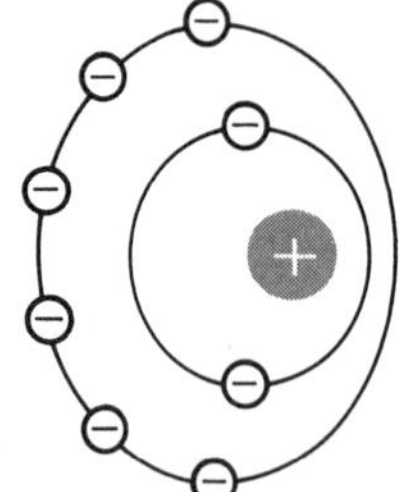

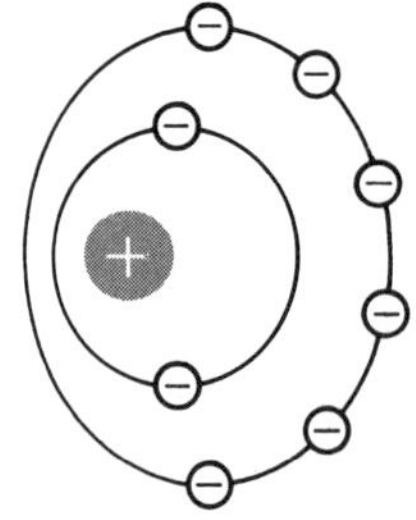

Strong repulsion

FIGURE 11-10

Encounter of two oxygen atoms—a qualitative picture

sented in a single diagram of force as it varies with the distance of separation of two particles. A more useful method of displaying interactions, as in Figure 11-11, uses a plot of *potential energy* as it depends on d, the distance of approach of the center of one atom to that of the other. The curve in Figure 11-11, which was drawn for oxygen, is not symmetric. The left side goes up rapidly, indicating that repulsion of the nuclei prevents the atoms from coming close to each other. The right side goes gradually to zero, denoting that the addition of an amount of energy E_D (here, 5.1 eV) will cause the particles to separate—that is, it will cause the molecule to *dissociate* into its component atoms.

We may draw a picturesque analogy of the behavior of this system to a familiar problem of a golfball rolling into a ditch next to a steep bank. The ball will oscillate back and forth about the center of the ditch (indefinitely, if there were no friction). If we could impart energy to the ball by a blow from the golf club, the oscillations would be more vigorous; with luck, we could give it enough energy to cause it to escape to level ground. So it is with a molecule under the influence of another. When the temperature of the gas is raised, the collisions that occur are more energetic. If the temperature and energy are high enough, dissociation of the molecule takes place. The amount of this energy is the depth of the "potential well."

A simple formula that fits the potential-energy curve for many diatomic molecules is the one devised by Morse:

$$E_p = E_D(e^{-2ax} - 2e^{-ax})$$

where x is the displacement from the normal position d_e—i.e., $x = d - d_e$—and a is a constant. Approximate values for some molecules are listed in Table 11-2.

TABLE 11-2

MOLECULAR CONSTANTS

Molecule	E_D (electron-volts)	d_e (meters)	a (m^{-1})
H_2	4.45	0.75×10^{-10}	1.94×10^{10}
O_2	5.09	1.20×10^{-10}	2.68×10^{10}
NO	5.30	1.15×10^{-10}	3.06×10^{10}

Note that the curve is roughly parabolic over a small region of the potential-energy graph near d_e in Figure 11-11. We can fit the graph by such a function over a small region by choosing the zero of potential energy at the bottom of the curve. Then the parabola is of the form

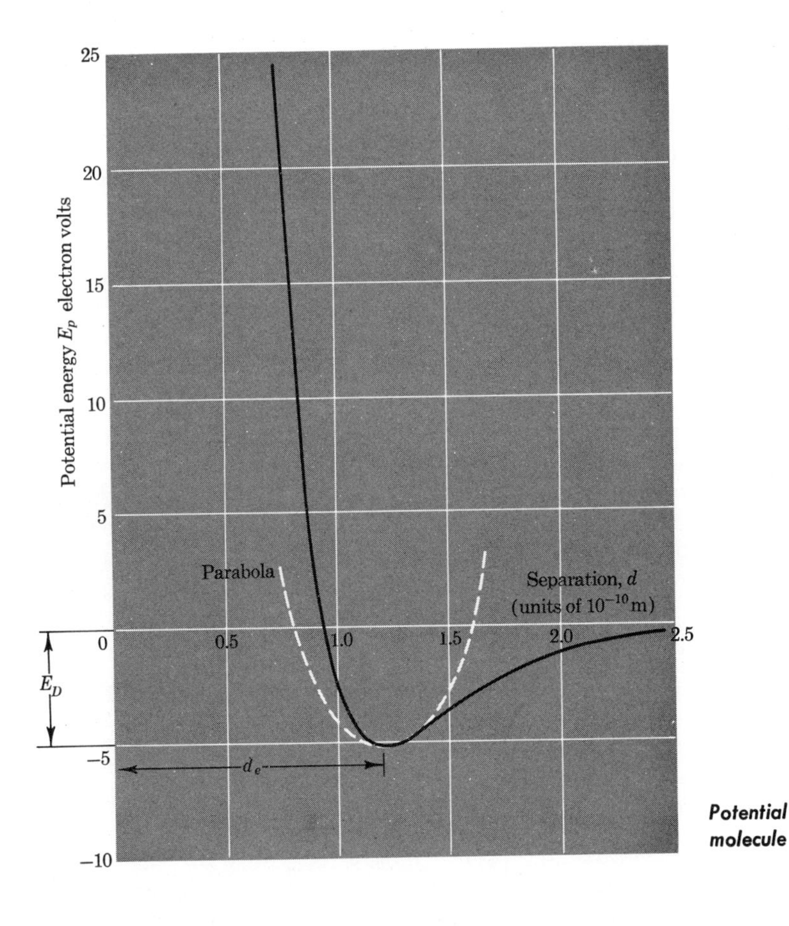

FIGURE 11-11

Potential energy diagram for oxygen molecule

$$E_p = \frac{1}{2}K_1x^2$$

where the subscript 1 reminds us that single molecules are involved.

The formula for potential energy is similar to that for a mechanical system, the mass on a spring, which executes simple harmonic motion (Section 11-2). We can thus visualize the two atoms, as in Figure 11-12, connected by an *equivalent* spring, and vibrating about a neutral position.

The classical motion of two masses m_1 and m_2 thus connected can be found by application of Newton's laws. Suppose the masses are pulled farther apart than d_e by an amount x. Mutual restoring forces of magnitude K_1x are produced. Thus we write the two equations of motion for the particles, whose separate displacements are x_1 and x_2, with $x = x_1 + x_2$, as seen in Figure 11-12:

$$m_1a_1 = -K_1x$$

$$m_2a_2 = -K_1x$$

Now, if we multiply the equations by m_2 and m_1, respectively, and add, then

$$m_1m_2a = -K_1(m_1 + m_2)x$$

where

$$a = a_1 + a_2$$

This becomes identical in form to the simple harmonic equation if we introduce a quantity called the *reduced mass:*

$$M = \frac{m_1m_2}{m_1 + m_2}$$

Then,

$$F = Ma = -K_1x$$

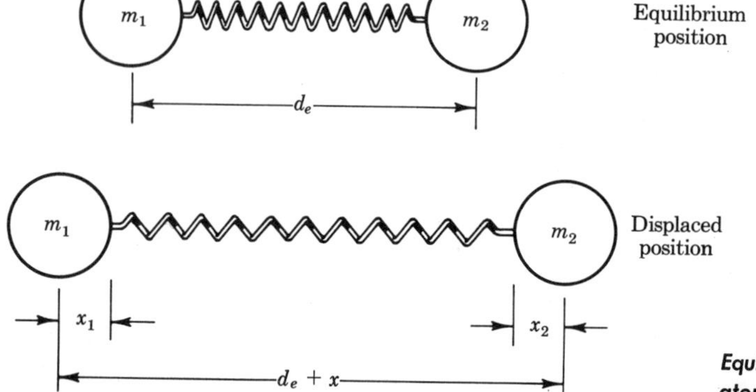

FIGURE 11-12

Equivalent spring connection between atoms

whose solution we know to be

$$x = A \cos \omega t$$

where A is the initial stretching and $\omega = \sqrt{K_1/M}$. We have thus verified that the nature of interatomic forces allows vibratory motion of the simple harmonic type, at least for small amplitudes near the neutral position.

Numerical values of the spring constant can be obtained from Table 11-2 by noting that the parabola fits the exponential Morse curve if the slopes of the two functions near their lowest points are set equal. Comparing, we find that

$$K_1 = 2a^2 E_D$$

For example, for oxygen

$$K_1 = 2(2.68 \times 10^{10}/\text{m})^2(5.09 \times 1.6 \times 10^{-19}\ \text{J})$$

$$K_1 = 1170\ \text{N/m}$$

Let us find the *classical* frequency of vibration of the O_2 molecule. It is easy to see that $M = m/2$ for a diatomic molecule with equal masses m. Thus $M = (8)(1.67 \times 10^{-27}) = 1.34 \times 10^{-26}$ kg. Then, the frequency is

$$\nu = \frac{\omega}{2\pi} = \frac{1}{2\pi}\sqrt{\frac{K_1}{M}} = \frac{1}{2\pi}\sqrt{\frac{1170\ \text{N/m}}{1.34 \times 10^{-26}\ \text{kg}}} = 4.7 \times 10^{13}\ \text{cps}$$

We shall have occasion shortly to use the spring-like nature of atomic forces for the appreciation of a variety of large-scale phenomena.

11-5 THE PRODUCTION AND PROPAGATION OF WAVES

A disturbance that is produced at a point in a medium that has elastic properties can be transmitted to other points. Two simple experiments will illustrate this statement. Let us attach a long coil spring to a far wall and hold it taut with our hand. We then compress, as in Figure 11-13, a small section of the spring and let go. The compressed region moves along the spring with a certain speed. Such a pulse is in the class of *longitudinal* disturbances, being along the length of the spring. Next, let us tightly stretch a long wire or rope between two widely separated walls. Near one end, we create a small kink or pulse in the line and then let go. The kink "moves" along with constant speed (see Figure 11-14). Physical reasoning will tell us what causes the observed motion: by displacing the wire from its normal position, we have put it under tension; and on release, the elastic restoring force gives the section involved an acceleration toward the normal position. This motion carries along the immediately adjacent

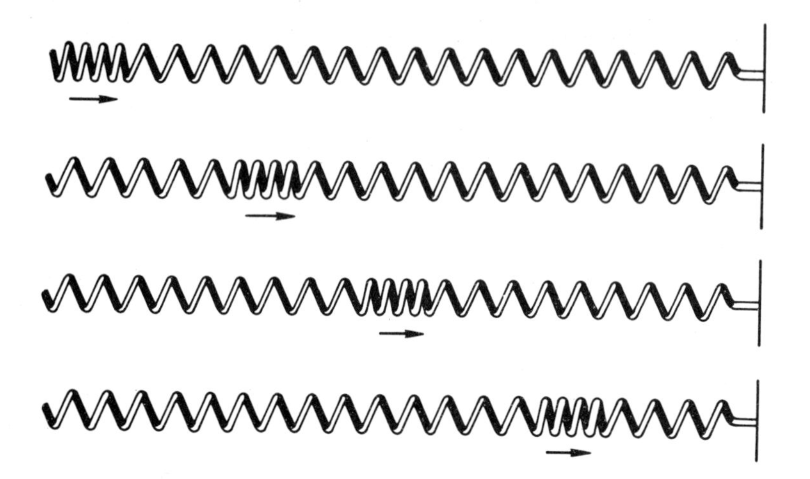

FIGURE 11-13

Motion of a compressed section of a coil spring

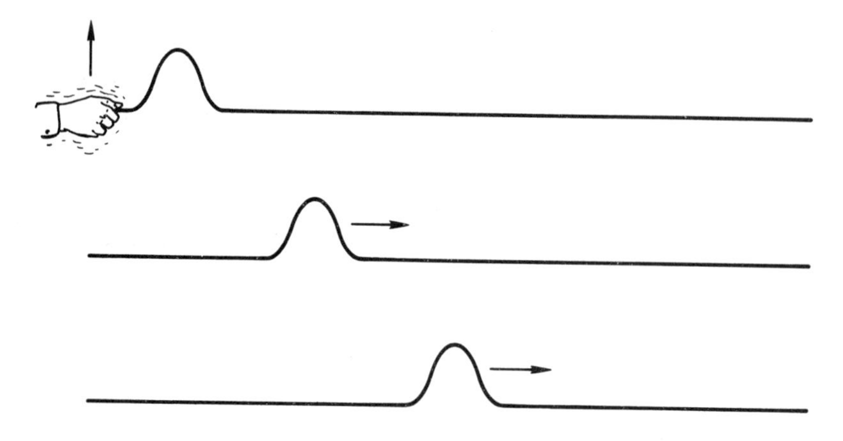

FIGURE 11-14

Motion of a pulse along a wire

sections. They in turn impart motion and energy to their neighbors. By this means the impulse is transmitted along the wire, and if there were no friction and the wire were infinitely long, the disturbance would go on indefinitely. This disturbance is of the *transverse* variety, being at right angles to the length and to the direction in which the impulse is propagated.

We now examine some of the properties of wave propagation. If we observe such a pulse over an interval of time, the disturbance moves along with speed V, and without changing shape, as sketched in Figure 11-15. Let us compare what two different observers see. According to an observer riding the crest, the point on the curve is given by the function

$$y = f(X)$$

where X is measured from the peak. However, an observer fixed at the origin sees the point at coordinate x, where, from Figure 11-15,

$$X = x - Vt$$

and thus he sees the curve as the function

$$y = f(x - Vt)$$

that is, for a disturbance moving in the $+x$-direction, the displacement depends on the combination $x - Vt$. The functional form of f is not specified until we say how the end of the wire is moved. The positive pulse was formed by a single quick displacement upward. Had we shifted the end downward, a *depression* would have been sent along the wire, as in Figure

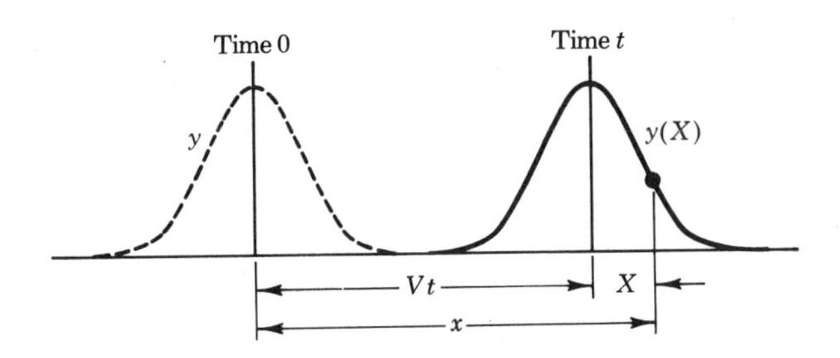

FIGURE 11-15

Observation of shape from moving peak and fixed frame

11-16. It is easy to see that execution of two or more motions in succession will create wave "trains" of several loops that also travel along the rope. The last sketch—of continuous vibration—shows the result of a repetitive agitation of the end. The wave pattern is seen to be dominated by the motion at the origin, the *initial conditions*. Let us now use some mechanical device to generate simple harmonic vibrations. As long as the end of the wire is agitated, periodic disturbances proceed along the x-axis.

Several possible mechanisms will serve the purpose—a weight on a spring, a flexible bar, or a slotted wheel (Figure 11-17). Electromagnetic devices could also be used. Each of these provides simple harmonic motion in the y-direction of some amplitude A:

$$y = -A \sin \frac{2\pi t}{T}$$

We assume that the vibration has been going on for some time, but we start counting time $t = 0$, when the end is at the neutral position and the motion is downward (this is the significance of the $-$ sign). The displacement for

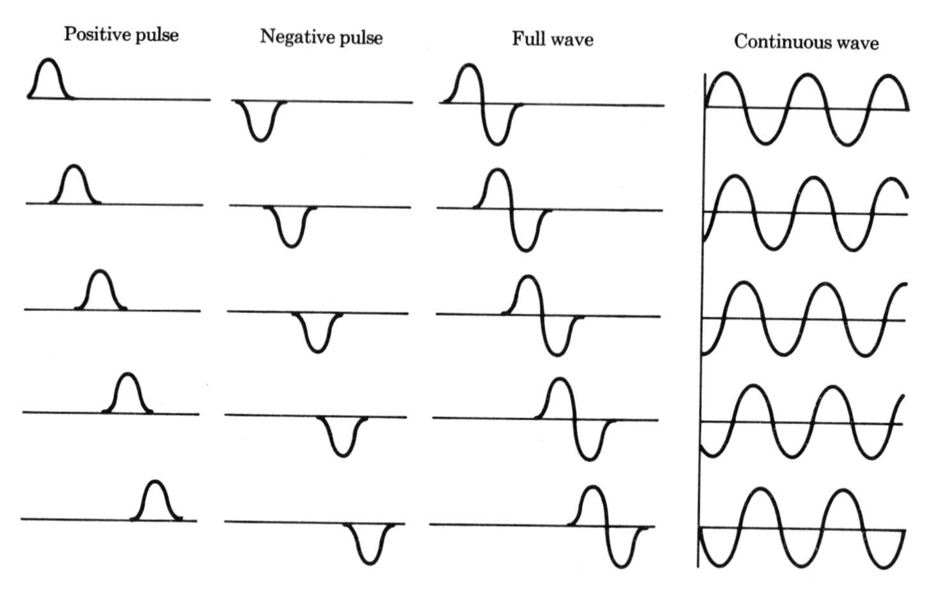

FIGURE 11-16

Construction of waves from pulses

any point in the wire and at any time $y(x, t)$ must simplify to this form at $x = 0$ and must also be some function of $x - Vt$. A little trial and error shows that the proper expression is

$$y(x, t) = A \sin \left[\frac{2\pi}{VT} (x - Vt) \right]$$

As a check, let us simply set $x = 0$. Then,

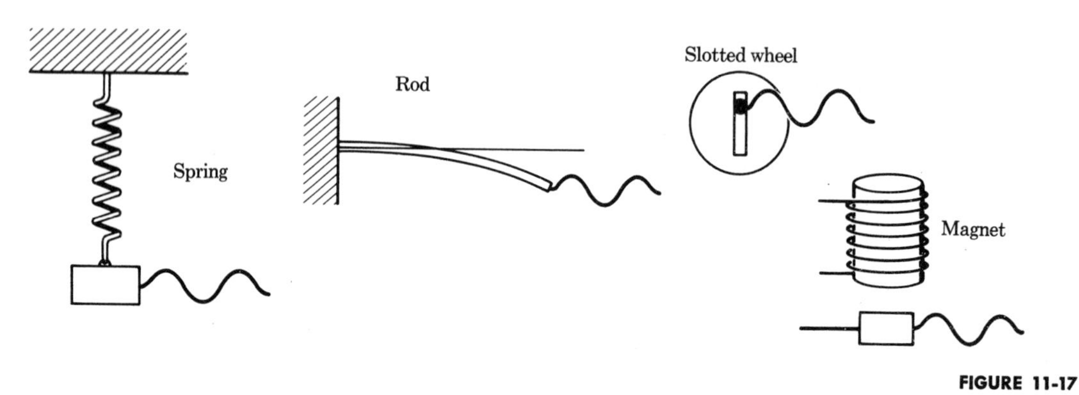

FIGURE 11-17

Methods of vibrating the end of a wire or rope

$$y(0, t) = A \sin \frac{2\pi}{VT}(-Vt) = -A \frac{\sin 2\pi t}{T}$$

which is in agreement with our assumption for the driving motion.

The displacement may be written in several equivalent forms, depending on which property we wish to emphasize. First, let us view the wave at time zero. Then,

$$y\,(x, 0) = A \sin \frac{2\pi x}{VT}$$

which is plotted in Figure 11-18. If $y = 0$ at $x = 0$, then the distance out to the point x_1, where y is *again* zero is such that

$$A \sin \frac{2\pi x_1}{VT} = 0$$

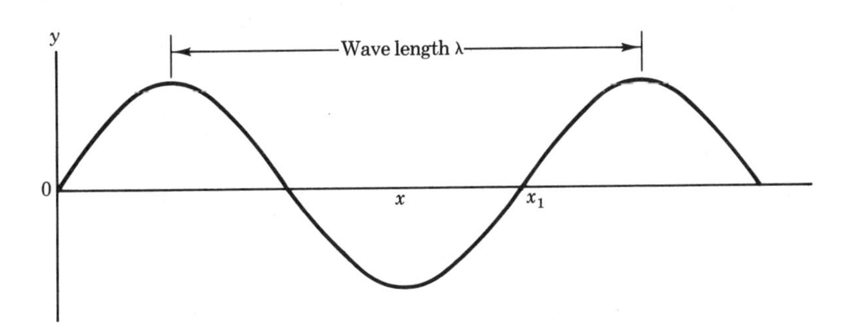

FIGURE 11-18

Sine wave at $t = 0$, $y = A \sin \dfrac{2\pi x}{VT}$

The sine is zero and the trend in the curve is repeated when the angle is 2π; hence $2\pi x_1/VT = 2\pi$ or $x_1 = VT$. Let us call this distance the *wavelength* λ, noted to be the same for any two like points. Thus we obtain the following important relation:

$$\lambda = VT$$

The wavelength is the product of the speed of the wave V and the period T of the driving force or vibration. The first alternate form is thus

$$= A \sin 2\pi \left(\frac{x}{\lambda} - \frac{t}{T}\right)$$

Letting $T = 1/\nu$, then $\lambda = V/\nu$ or

$$V = \lambda\nu \qquad \textit{(relation of wave speed, wavelength, and frequency)}$$

This is the most useful connection between the three quantities to remember. One should note the following facts: (1) the frequency ν or period T of the wave depends only on the source of vibrations; (2) the wave speed V depends only on the medium in which the wave is propagated; (3) the wavelength λ depends mutually on V and ν, hence on both medium and source. If we introduce a new quantity called the propagation constant, of magnitude

$$k = \frac{2\pi}{\lambda}$$

and use the angular frequency

$$\omega = 2\pi\nu = \frac{2\pi}{T}$$

we obtain a very compact form of the displacement:

$$y = A \sin (kx - \omega t)$$

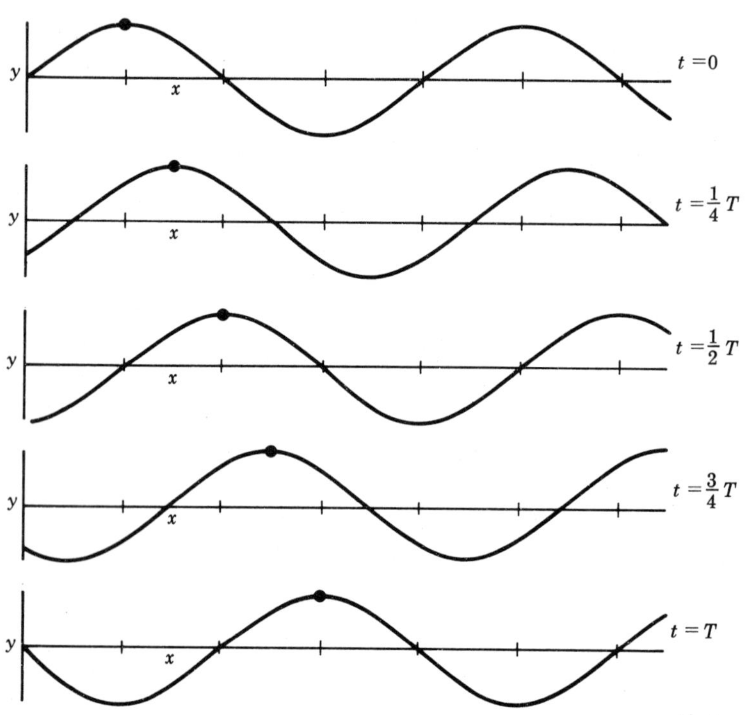

FIGURE 11-19

Photograph of a wave at various times

It is essential that we understand what this mathematical form of the wave means graphically and physically. For this purpose we select three different views:

1. First let us verify that the formula represents a disturbance that moves in the positive x-direction. Imagine that we have a moving-picture camera that can take a series of photographs of the wave. From the formula for the wave, let us construct a set of such pictures. At $t = 0$, $y = A \sin kx$, which is a sine curve in space, as plotted in Figure 11-19. As time goes on, we shall watch the first peak, which is marked with a heavy dot. Suppose our next movie frame is at $t = T/4$, $\omega t = 2\pi\nu(T/4) = \pi/2$. Then, $y = A \sin (kx - \pi/2)$, which is also $y = -A \cos kx$. The peak has moved to the right. One more frame will be enough to check the trend: $t = T/2$, $\omega t = \pi$, $y = A \sin (kx - \pi)$ or $y = -A \sin kx$. How fast is the wave moving? The peak shifted to the right a distance $\lambda/2$ in a time $T/2$, so its speed is $V = \dfrac{(\lambda/2)}{(T/2)} = \dfrac{\lambda}{T} = \lambda\nu$, as already stated.

2. If instead we were to *stand* at a particular point, such as $x = \lambda/2$, then at $t = 0$ we would see the peak coming toward us; at $t = T/4$ it would be alongside, and at $t = T/2$ it would be going away.

3. Suppose, however, that we were able only to observe what was happening at our location, say $x = \lambda$. Then $kx = k\lambda$, which is 2π; hence $y = A \sin (2\pi - \omega t) = -A \sin \omega t$. The displacement would oscillate in time as a sine function, going from $-A$ to 0 to A and back again.

Wave Speed

Let us now calculate the speed of the transverse waves on a stretched string. To do this we shall use a "device" to start the string moving and see how this disturbance propagates down the length of stretched string. In order to keep our problem simple, we shall restrict the motion to linear displacement and make the string very long so that disturbances will keep moving without being interrupted by boundaries such as tied end points.

Consider the end point of an infinite string of mass per unit length ρ and tension F_T, attached to a transverse driving mechanism that is turned on at time $t = 0$. Let the mechanism move with a constant speed u. After a short time τ the string will assume the position shown in Figure 11-20.

A "kink" has thus been generated in the string that will move with the wave speed V along the string. Then, after a time τ it will be a distance $V\tau$ along the string, while the end will have moved up a distance $u\tau$, as indicated. The mass of string of length $V\tau$, mass $m = \rho V\tau$, has been given a momentum $p = \rho V\tau u$ in a time τ. By Newton's second law, $F = \Delta p/\Delta t$ or $F = \rho V\tau u/\tau$, and the force is equal to the y-component of tension $F_T \sin \theta$, where $\sin \theta = u\tau/V\tau = u/V$. Then, $F_T(u/V) = \rho Vu$, or

$$V = \sqrt{\frac{F_T}{\rho}}$$

which is the formula for the wave speed in a wire or string. The speed is high if the tension is great or if the string is very light.

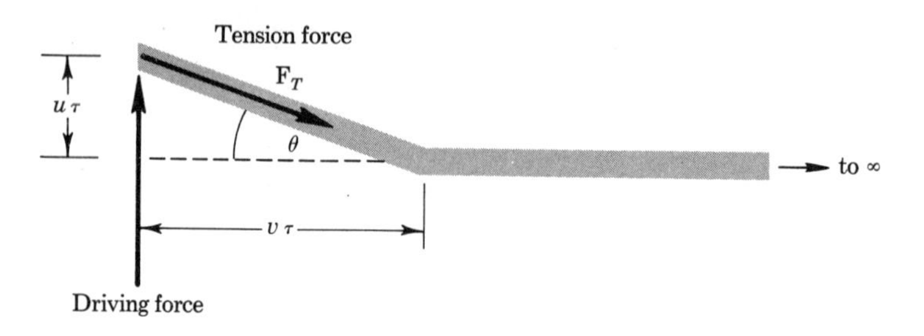

FIGURE 11-20

Derivation of wave speed in a string or wire

For example, let us find V for a wire of length 50 cm and mass 10 grams if stretched by a weight of 30 lb (134 N). Now, $\rho = \dfrac{0.01\ \text{kg}}{0.5\ \text{m}} = 0.02$ kg/m. Then

$$V = \sqrt{\frac{134}{0.02}} = 82\ \text{m/sec}$$

Momentum, energy, and power in a wave. We have seen how a pulse or a wave progresses along the length of a stretched wire. It should be emphasized that the momentum of the displaced particles is in the y-direction, not in the direction of the waves. The total mechanical energy, as usual, is made up of the kinetic part due to motion, plus the potential part due to the slight stretching of the wire. As time goes on, and a pulse proceeds along the x-direction, the energy is contained in different segments of the material; hence energy is communicated from one end to the other. No significant motion along that axis takes place, however.

In the next chapter, we shall examine further the traveling wave in several different physical phenomena, including sound, light, electrical vibrations, and "matter waves."

PROBLEMS

11-1 The frequency of alternating potential used commercially and in households is 60 cps. Calculate the period and the angular frequency.

11-2 Find the displacement at one-eighth of the complete cycle if $x = 4 \cos 2\pi t/T$.

11-3 The general form of the displacement in simple harmonic motion is $x = A \cos (\omega t - \phi)$. What is the value of ϕ if the displacement at time zero is $+A$? If it is $-A$?

11-4 The position of a bead moving on a circular loop of wire is $y = A \cos (\omega t - \phi)$ and its vertical component of velocity is $v_y = A\omega \sin (\omega t - \phi)$.

Show where the bead is located at time zero, indicate the phase angle ϕ, and verify the velocity relation.

11-5 If the displacement of a particle in harmonic motion is given (in meters) by $x = 5 \cos 1.5t$, where x is measured positive upward, find the particle velocity at $t = 0.8$ sec. When will that same *speed* be reached again?

11-6 The motion of a harmonically oscillating mass of 0.15 kg on a spring is governed by the equation $y = 0.2 \cos 3t$, with y in meters and t in seconds. Find (a) the frequency of vibration; (b) the force constant K; (c) the displacement at $t = 2.5$ sec.

11-7 A coil spring for an automobile has a value of K of 7.5×10^5 N/m. What is the natural frequency of vibration if the mass supported is 300 kg?

11-8 Estimate the value of K for the coil springs of an automobile if the application of one-quarter of the weight of 3000 lb causes a compression of 2 in. Express in both pounds per feet and newtons per meter.

11-9 Coil springs on a rolling vehicle absorb the shock of bumps. Prepare a description of the motion of an automobile as it goes over a depression in a road—for example the ditch left after a new water line has been installed.

11-10 At what time in a cycle of vibration of a mass on a spring are the kinetic and potential energies equal in magnitude?

11-11 Calculate the maximum kinetic energy for a vibration characterized by $K = 3 \times 10^5$ N/m, $m = 1500$ kg, $A = 0.05$ m. Verify that the maximum potential energy has the same numerical value.

11-12 If the restoring force for some system varied according to $F = -kx^n$, where n was some positive number greater than 1, how would you go about finding the form of the potential energy? Would the cosine function properly describe the motion? Would conservation of energy still hold or not?

11-13 The balance wheel of a watch has a spiral spring that provides a torsional restoring force. If the moment of inertia of the wheel is 5×10^{-7} kg-m^2, what value of the torsion constant is needed to give a period of 2 sec?

11-14 The center of a thin metal rod of mass 4 kg and length 0.8 m is attached to a wire with torsion constant $c = 12$ N-m/rad (see Figure P11-14). What is its small-angle period?

11-15 Imagine a hole drilled through the center of the Earth from the United States to the Indian Ocean. An object is dropped down the hole. Recall from Section 6-2 that the interior gravitational field varies linearly with distance from the center of the Earth. Which of the following do you expect to happen to the object? (a) It goes on through and out into space; (b) it stops at the center; (c) it oscillates indefinitely; or (d) it oscillates with decreasing swings and eventually stops at the center. Explain.

11-16 (a) Using the results of problem 11-15, find the numerical-equivalent force constant K for motion of a 1-kg mass through the Earth. (b) What is the period of oscillation of a mass of 1 kg if air resistance is neg-

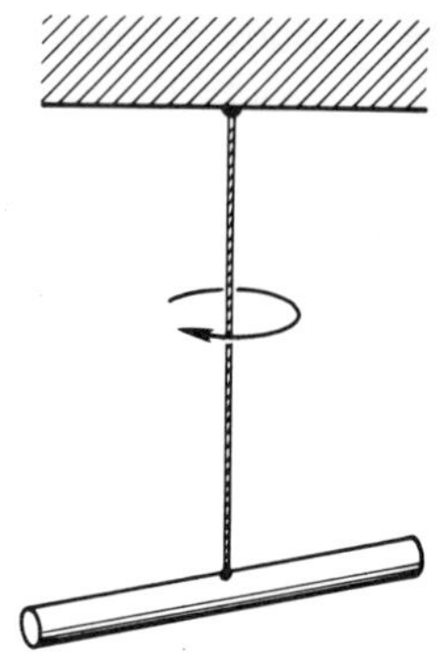
FIGURE P11-14

lected? (Assume the Earth is a uniform sphere of radius 6.37×10^6 m, with surface field 9.8 N/kg.) (c) The time to traverse a chord through the Earth is the same as to go completely through a diameter. It has been proposed that transcontinental tunnels be built using this fact. Discuss the feasibility of such a transportation system.

11-17 A small bar magnet with moment of inertia $I = 8 \times 10^{-7}$ kg-m^2 and magnetic moment $M = 3 \times 10^{-3}$ A-m^2 (see Section 9-7) is set into small-amplitude vibrations of frequency 2.5 cps. What is the strength of the field where the magnet is located?

11-18* Show that an electron located near the midpoint of a line joining two equal negative charges would execute simple harmonic motion, approximately. Evaluate the force constant K in terms of the magnitudes of charges and the distance of separation.

11-19 A baseball bat is suspended at its smaller end and allowed to swing freely (see Figure P11-19). By use of Newton's second law in angular form and analogy with simple pendulums, show that for small amplitudes, the period of oscillation is $T = 2\pi\sqrt{\dfrac{I}{mgl}}$.

11-20 A piston of an automobile engine executes nearly simple harmonic motion. If the crank shaft turns at 3000 rpm, and the piston has a mass of

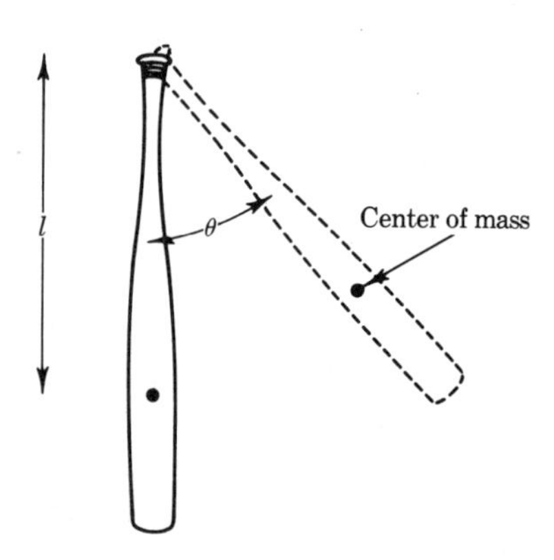

FIGURE P11-19

2 kg, moving with a 15-cm stroke, what force is needed to reverse the piston at the end of its path?

11-21 Find the period and frequency of a simple pendulum of length 4.5 m at a point where the acceleration of gravity is 9.8 m/sec^2.

11-22 How long would a pendulum arm have to be in order to give a period of 1 min?

11-23 Calculate the acceleration of gravity from the following data:

Pendulum Length (meters)	*Period* (seconds)
0.5	1.4
0.8	1.8
1.0	2.0

What is the average of absolute errors?

11-24 A measurement of g is made in the laboratory, with the following data on period as dependent on length of the pendulum:

l (meters)	T (seconds)
0.18	0.85
0.32	1.14
0.74	1.73

Deduce the acceleration of gravity.

11-25 A cuckoo clock has a pendulum that is supposed to have a period of 0.5 sec but the clock loses 3 min a day. (a) Find the length of the pendulum. (b) How much adjustment, in millimeters up or down, is needed?

11-26 A steel ball of mass 400 kg on a cable of 5-m length is used to wreck walls of old buildings, by allowing the ball to swing through an arc of 60 deg. What amount of kinetic energy does this pendulum have just before impact?

11-27* Galileo said in the sixteenth century that a pendulum of length 100 cubits (1 cubit = 18 inches) would have the same period for a 90-deg swing from the vertical as for a 1-deg or smaller swing. What are the actual periods and errors that are made in this statement, assuming $g = 9.806$ m/sec^2 in Italy?

11-28 The French astronomer Jean Richer discovered in 1671 that his pendulum clock ran slow in French Guiana by 2.5 minutes per day compared with time in Paris ($g = 9.81$ m/sec^2). Find the value of g he experienced.

11-29 The ballistic pendulum, first described by Robins in 1742, is used to measure the speeds of bullets. A bullet fired with speed v into a suspended wooden block will set up oscillations of certain period and maximum angle. Derive a formula from momentum and energy considerations to permit calculation of v from measurements.

11-30 The resonance angular frequency for oscillations of a mass m attached to a spring of constant K and natural angular frequency $\omega_0 = \sqrt{K/m}$ can be shown to be $\omega_R = \sqrt{\omega_0^2 - 2\gamma^2}$. Compute ω_R using $K = 8$ N/m, $m = 0.4$ kg, and $\gamma = 0.25$ sec^{-1}. How does the result compare with the natural frequency?

11-31 The time for a tuning fork to decrease in amplitude by a factor $e = 2.718$ is 8 sec, when immersed in oil. How long does it take for the amplitude to drop to *one-half* its initial value?

11-32 Two skaters of mass 50 kg and 70 kg take a muscle-building spring (force constant 12,000 N/m) out onto the ice to experiment with two-body vibration. What is their natural frequency?

11-33 Calculate the classical frequency of vibration of the diatomic molecule HCl, atomic masses 1 and 35 (1 amu $= 1.67 \times 10^{-27}$ kg), $E_D = 4.40$ eV, $a = 1.91 \times 10^{10}$/m.

11-34 Using the potential-energy constants for the diatomic molecule H_2 from Table 11-2, compute the force constant K_1 and the classical frequency of vibration ν.

11-35* Expand the Morse potential-energy formula in a series using $e^{-y} = 1 - y + y^2/2 - y^3/6 + \ldots$, to verify that the function is parabolic near zero displacement. Show that $K_1 = 2a^2E_D$ is the approximate force constant.

11-36 For a wave with speed $V = 5000$ m/sec and frequency $\nu = 60$ cps, calculate the wavelength, the angular frequency, and the propagation constant k.

11-37 If thinking beings exist on other planets, they would probably signal using light (speed 3×10^8 m/sec) from hydrogen atoms of 21-cm wavelength. What frequency is this?

11-38 Calculate the wavelength of radio signals from station WOR, New York City, frequency 710 kc/sec (kilocycles per second).

11-39 The wavelength of the prominent yellow light from sodium lamps that are sometimes used for street illumination is 0.589 μ. Compute the frequency of this light.

11-40 A steel wire 4-m long has a mass of 50 grams and is stretched by a hanging weight of 2 kgf. What is the speed of waves on this wire?

11-41 A metal wire is attached to one prong of a tuning fork, whose frequency of vibration is 276 cps. The other end is passed over a pulley and supports a weight of 10 lb. The linear density of the wire is 0.012 lb/ft. (a) What is the speed of waves on this wire? (b) What is the wavelength?

11-42 A wire has a linear density of 0.25 kg/m. What is the wave speed when the tension force is 80 N? What frequency of vibration results in a wavelength of 0.9 m?

WAVE PHENOMENA

12

Many familiar effects may be described as wave motion. The surface of water on a lake into which a stone is dropped can be observed to be rising and falling at a particular point, while the disturbance moves outward in ever-widening circles. Energy is being transferred along the direction in which the wave progresses. That waves involve energy is readily understood by anyone who has been on a boat in rough weather. A disturbance at one end of a rope or wire is transmitted along its length. The vibration of wires in all stringed musical instruments is of a similar nature. Although it is not as obvious to us as the above examples, the transmission of sound in air or of light through space may be similarly viewed as a wave motion. As the Earth's surface is exposed periodically to sunlight and darkness, one finds that waves of heat energy penetrate into the ground and rocks. Finally, the theory of wave mechanics tells us that the motion of matter can be viewed in terms of wave propagation. In this chapter we shall study the character of wave motion further, and show the way in which many phenomena of interest may be described by waves.

12-1 THE ADDITION OF WAVES

The relations derived in the previous chapter refer *only* to waves that are sent out into a medium of indefinite extent. If the wire or bar or tube is

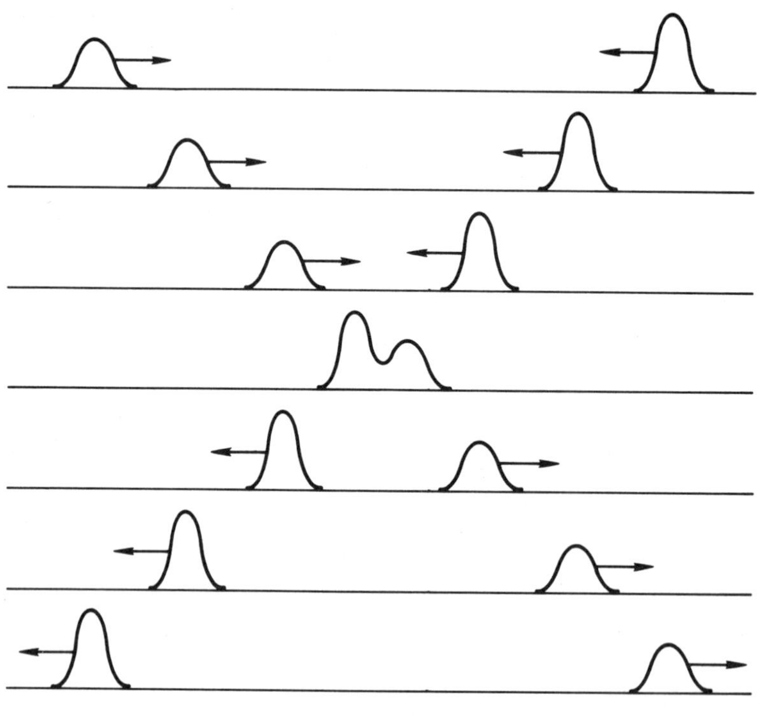

FIGURE 12-1

Superposition of pulses moving in opposite directions

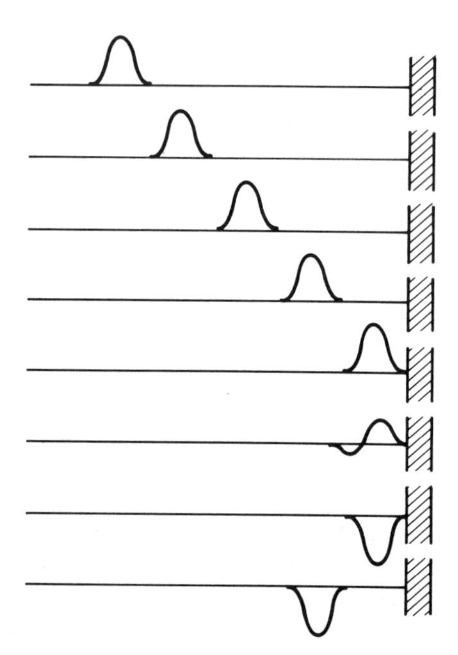

FIGURE 12-2

Reflection of a single pulse

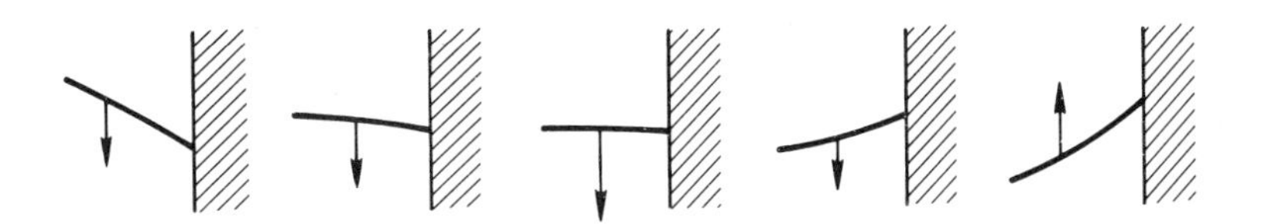

FIGURE 12-3

Motion of wave at fixed end

of definite length L, a reflection effect takes place at the end. Waves come back toward the source of vibrations and are added or *superposed* on the original wave. In order to appreciate this effect, let us first return to the pulse picture and imagine that two people sent out single disturbances at each end of a wire. Let the pulses be of slightly different amplitude. Each moves along independently until they cross (Figure 12-1), when the two add together to from a peak that is momentarily larger than either. Instead of involving two persons, let us merely attach the rope to a wall and send out a pulse. It will be reflected as seen in the sequence of Figure 12-2. Notice especially that the returning signal is a depression, or a negative pulse.

How the reflection comes about can be seen if we view the region near the far end of the wire at various times, as in Figure 12-3. As the wave approaches the fixed point, the last section of the wire moves down through $y = 0$. As it comes up again, it sends a signal back along the length. In general, if a continuous wave $y_R = A \sin (kx - \omega t)$ is sent to the right and reflected waves are returned, the two waves may "interfere" with each other constructively or destructively—that is, the combination may be greater or smaller than the individual waves, depending on the phase relations of the component waves.

Standing Waves

In one special combination of frequency, speed, and length of the line, the wave to the left is of the form $y_L = A \sin (kx + \omega t)$. Then, the total displacement is the sum

$$y_T = y_R + y_L = A \sin (kx - \omega t) + A \sin (kx + \omega t)$$

Using the trigonometric identity for the sum of sines,*

$$y_T = 2A \sin kx \cos \omega t$$

The curve representing the wire has an amplitude $2A \sin kx$ and a cosine time dependence as in simple harmonic motion. It is called a *standing wave*. We have not yet taken account of the condition that the far end of the wire is fixed. As we shall see, this requirement limits the type of waves that can be sustained in the region from $x = 0$ to $x = L$. Now, at any time t, for

* That is, $\sin x \pm \sin y = 2 \sin \frac{1}{2}(x \pm y) \cos \frac{1}{2}(x \mp y)$.

$x = L$, the amplitude y has to be zero. This will occur when $\sin kL = 0$, or when $kL = \pi$. Since $k = 2\pi/\lambda$, we find

$$\frac{\lambda}{2} = L$$

This says that one loop of the sine wave fits between the ends of the wire, as shown in Figure 12-4(a). As a check, put $k = 2\pi/\lambda = \pi/L$ back in $\sin kx$, to obtain the space form $\sin(\pi x/L)$. This is seen to be zero at $x = L$. This particular vibration, called the *fundamental mode*, is shown for many times during the cycle. The required frequency is $\nu = V/\lambda = V/2L$, which we shall label ν_1. This wave is not the only one that is zero at $x = L$. We see that $\sin kL = 0$ also if

$$\lambda = L$$

since $k = 2\pi/\lambda$, $kL = 2\pi$, and $\sin 2\pi = 0$. Such a wave, as sketched in Figure 12-4(b), has two loops. The amplitude is zero at $x = 0$, $L/2$, and L, which are points called *nodes*. The points of maximum amplitude called *antinodes* are at $x = L/4$ and $x = 3L/4$. The frequency of this second wave, V/L, is higher than that of the fundamental $V/2L$; in fact, it is twice as large, $\nu = 2\nu_1$. This vibration is called the *first overtone*, the frequency being "over" that of the fundamental. The pattern by which we construct the different modes of standing waves is now seen. As in Figure 12-4(c), $\frac{3}{2}\lambda = L$ or $\nu = 3\nu_1$.

If there is an integer number of loops labeled $n = 1, 2, 3$, etc, the space part of the wave is given by $\sin n\pi x/L$. The fundamental is only the

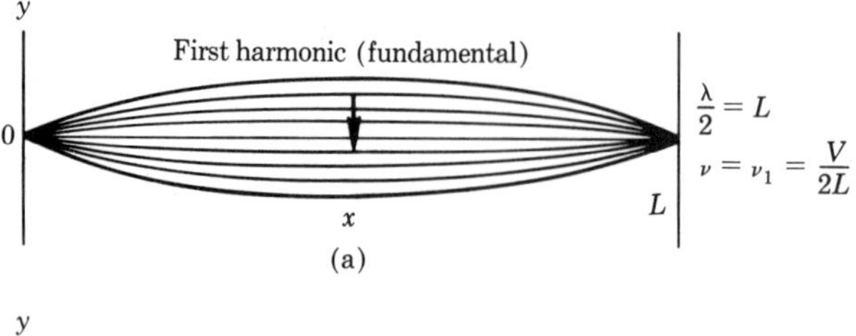

(a)

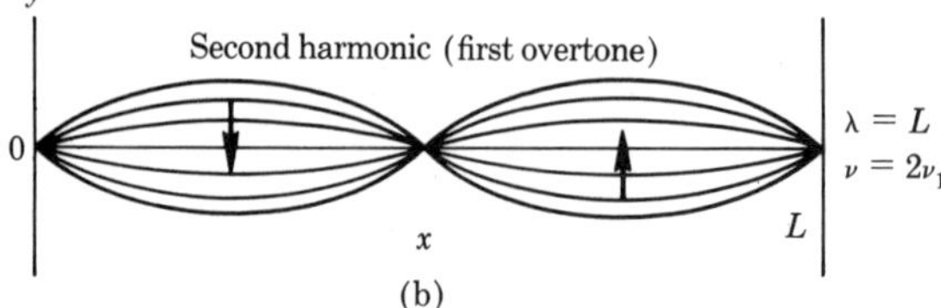

(b)

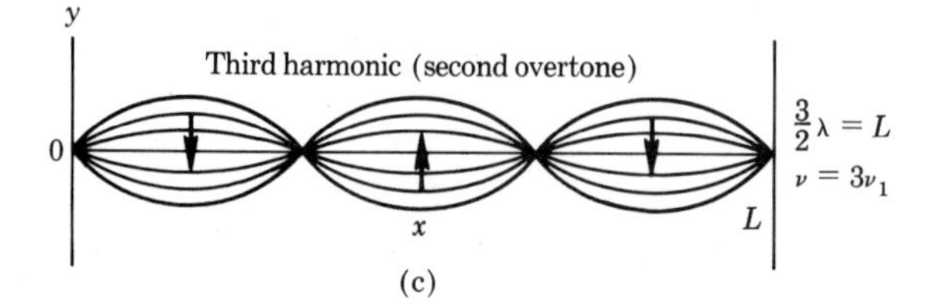

(c)

FIGURE 12-4

Standing waves (arrows indicate the instantaneous directions of motion of the wire)

first of many possible modes; thus it is called the *first harmonic* $n = 1$, followed by the second harmonic $n = 2$, the third, $n = 3$, etc. Relations that are easily proved are $\nu_n = n\nu_1$, $k_n = nk_1$, $\omega_n = n\omega_1$. Each of the possible standing waves may be written as a single formula:

$$y_n = 2A_n \sin n(k_1 x - \omega_1 t)$$

Let us calculate how to obtain standing waves in a wire stretched between two points, with one end fixed and the other vibrated at frequency ν. The amount of tension can be adjusted. Let the length of vibrating wire be $L = 0.8$ m, and let us assume that the driving frequency is $\nu = 60$ cycles per second. Suppose the wire has a linear density of 2 grams per meter. The amount of tension needed to obtain the fundamental can be found as follows. Now, $\lambda = 2L = (2)(0.8) = 1.6$ m, and $V = \lambda\nu = (1.6)(60) = 96$ m/sec. Then, using the relation at the end of Section 11-5, $F_T = \rho V^2 = (0.002)(96)^2 = 18.4$ N. To obtain the second harmonic $\lambda = L$, the speed must be increased by a factor of 2, meaning that the tension must go up by a factor of 4.

Let us now ask "How high a frequency can be produced?" From what we have seen so far, there would appear to be no limit to the number n that identifies a harmonic. However, when we take account of the fact that matter is not a continuous substance, but is composed of a large but finite number of atoms, we realize that there must be some highest frequency. Picture, as in Figure 12-5(a), a wire that consists microscopically of a chain of atoms connected together by elastic-force "springs." The fundamental mode corresponds to a standing wave in which the chain swings back and forth, as in Figure 12-5(b). The first overtone has two separate loops, as in Figure 12-5(c). We can continue drawing higher harmonics, until Figure 12-5(e), where the number of loops that fit between the ends of the wire is equal to the number of atoms in the chain, N. It is not possible for a wave to be shorter than the distance between atoms, a. The maximum value of the harmonic index n is $n_m = N$, corresponding to the highest frequency

$$\nu_m = n_m \nu_1$$

However, since $\nu_1 = V/2L$, and $L = Na$, we find that

$$\nu_m = \frac{V}{2a}$$

It is interesting to note that the shortest period of vibration $T_m = 1/\nu_m$ is $2a/V$, which happens to be the time for the wave to go between adjacent atoms such as A and B in Figure 12-5. The highest frequency is exceedingly large. For instance, in steel the speed of waves is about 3000 m/sec, while the atoms are separated by only 2.3×10^{-10} m. Thus

$$\nu_m = \frac{3000}{2 \times 2.3 \times 10^{-10}} = 6.5 \times 10^{12} \text{ cps}$$

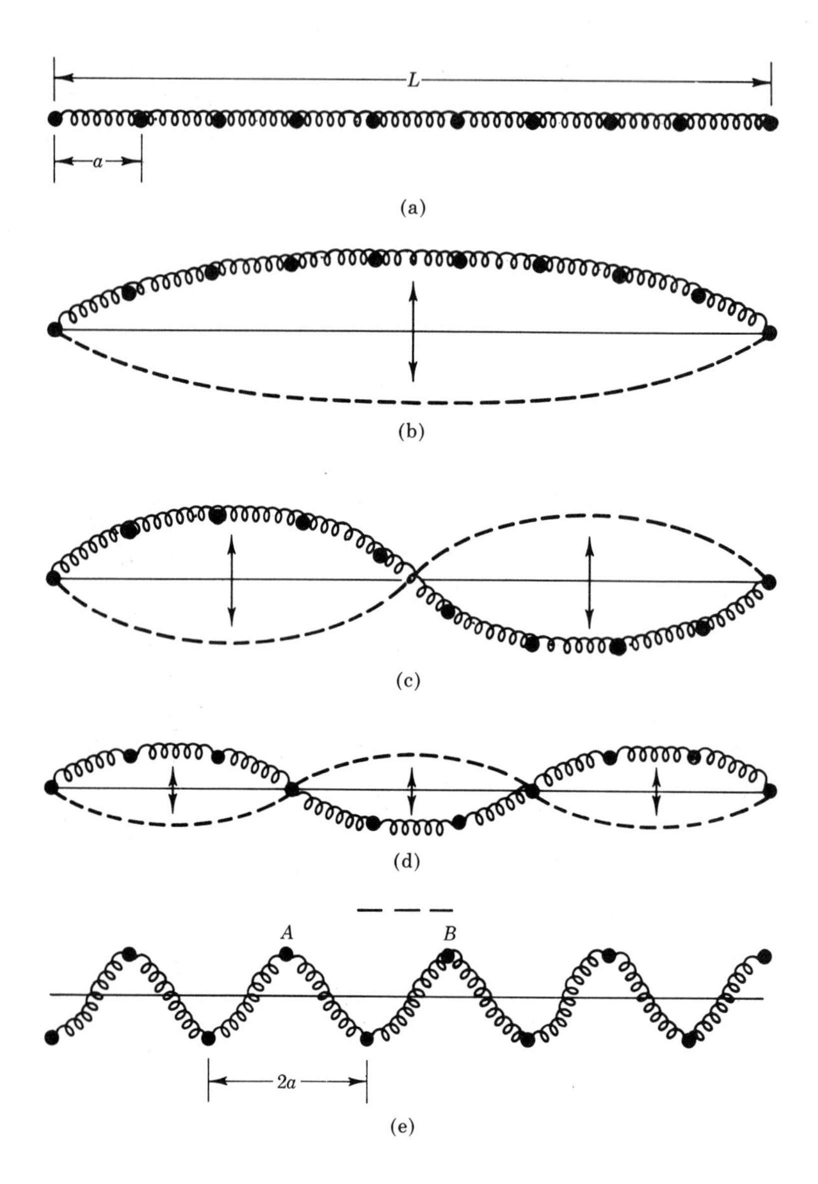

FIGURE 12-5

Largest frequency of vibration

We find that in any frequency range there can be many different waves. The actual number can be calculated by use of our relation $\nu_n = n\nu_1$ or $n = \nu_n/\nu_1$. Thus for a range $\Delta\nu$, the number of harmonics is $\Delta n = \Delta\nu_n/\nu_1$. For example, in a wire of length 0.8 m and for speed 704 m/sec, the fundamental frequency is $\nu_1 = V/2L = 440$ cps. In a frequency range, say, from 4.4×10^5 to 4.5×10^5, $\Delta\nu_n = 10^4$, we find $\Delta n = 10^4/440 = 23$ as the total number of harmonics.

We used the mechanical vibration of a stretched wire to help understand the mathematical form of waves. We now consider the wave nature of sound. "If a tree falls in a forest with no one there to hear it, is there any sound?" is an old question. The answer is "yes" or "no" depending on whether we are thinking in physical or physiological terms. We shall now devote our attention to the physical production and transmission of sound, although the mechanism and functions of the ear and the nervous system are very interesting subjects, and are taken up in Chapter 24. We shall not be concerned here with speech, although there are aspects of sound that involve recording and reproducing words by electrical and mechanical means. We are most familiar with the passage of sound through air, although we are aware of the fact that solids and liquids can also carry vibrations from place to place: steel, as in a railroad track, or wood, as in the floor of a house, or water in a swimming pool are examples that come to mind. Naturally, we are more concerned with those frequencies that register in our ears, but we know also that animals can hear higher notes, such as that of "dog whistles." Insects are known to communicate frequencies that are not audible to humans. Most generally, the subject of sound includes the production and transmission of disturbances in all kinds of elastic media. Sounds include not only those of a definite frequency, as in pure musical tones, but also those that have no recognizable pattern, such as noises; they include disturbances that are too low in intensity for us to hear, and even those that are so loud that the ear cannot handle them.

A simple device that produces sound is a tuning fork, consisting of a U-shaped metal bar with a wooden handle. Tapping the fork with a rubber mallet sets up vibrations of a definite frequency. If we could actually view the molecules of air near the surface of the fork, as in Figure 12-6, we would see alternate regions of high particle density (called *condensations*) and low

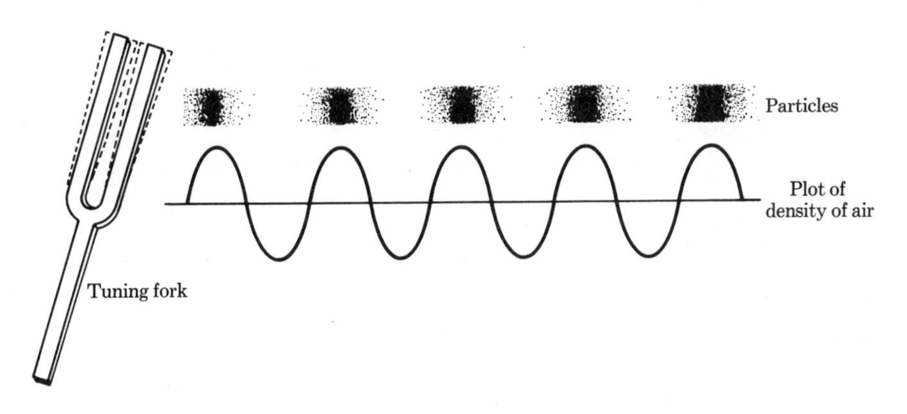

FIGURE 12-6

Generation of sound waves by vibrating tuning fork

particle density (called *rarefactions*). A plot of the change in density from average value against position is found to be the familiar sine curve:

$$\rho = \rho_0 \sin (kx - \omega t)$$

where ρ_0 is the maximum change in density. These are *longitudinal* waves, where the term refers to the fact that the displacements are *along* the direction of motion of the wave, in contrast to the *transverse* waves in a vibrating wire.

The speed of sound in a gas, liquid, or solid may be derived in terms of the properties of the medium, which in turn are related to intermolecular forces.

The speed of sound in air is measured to be 331.4 m/sec or 1087 ft/sec at 0°C (32°F). For every 1°C increase, the speed increases by 0.6 m/sec. To find the speed at room temperature (20°C), we must add $(20)(0.6) =$ 12 m/sec to the value at 0°C. Thus the total is 343.4 m/sec, which is also 1127 ft/sec. The delay noticed between the lightning flash and the thunderclap is explained by this rather low speed. (To tell how far away lightning strikes, simply assume that each 5-sec lapse before hearing the sound corresponds to 1 mi.) If the timers for the 100-meter run used as a starting signal the crack of the pistol rather than the flash of smoke, their timings would be in error by $\frac{100}{343} = 0.3$ sec. When a jet plane "breaks the sound barrier," it means that its speed relative to the air is 1127 ft/sec. For very high speeds, the terminology *Mach number* is used. Thus a speed 3 times that of sound—i.e., 3381 ft/sec or 2305 mi/hr—would be called "Mach 3." At high altitudes, the density of the air is low, and since the speed of sound varies as $1/\sqrt{\rho}$, we find V to be considerably larger there than at sea level. For example, at a height of 10 km (6.4 mi), the air density is only 28 per cent of its sea-level value. The sound speed would then be $1127/\sqrt{0.28} =$ 2130 ft/sec = 1452 mi/hr. From the definition of Mach number, a plane flying at Mach 2 at this altitude would be going at 2904 mi/hr.

Another type of motion that can be called "sound" consists of the molecular vibrations in a solid. Some of the newer theories of solids are based on the sound-like motion of the molecules making up the structure. The atoms may vibrate in both longitudinal and transverse modes simultaneously.

In order to appreciate the behavior of sound it is necessary to understand the instruments used to produce the vibrations. We shall look first at musical instruments because of their facility in illustrating certain physical principles. In our experiment that disclosed the existence of standing waves (Section 12-1), we caused the wire to vibrate in a very special way. When a string is plucked, as on a guitar, or bowed, as on a violin, the resulting motion includes many harmonics. The relative amplitudes (and thus the energy output) of these many harmonics depend on the way the forces are applied. The difference between a pleasing note played on a violin by a professional musician and the screech produced by an amateur is, from a physics viewpoint, only a matter of the distribution of harmonics that are excited.

Stringed Musical Instruments

The construction and operation of stringed musical instruments can be explained in terms of standing waves. In the back of a piano, for instance, we see a great array of strings of graduated lengths and diameters, all under tension. To achieve the correct notes, which are merely the frequencies of the standing waves, the appropriate length L and the density ρ are selected and the tension F_T is adjusted by the piano tuner. We notice that the bass strings are wrapped with coils, which provide extra density without loss of needed flexibility.

To illustrate the magnitudes of the numbers involved, let us look at one of the wires in a piano. Suppose that the linear density of the wire is 0.025 kg/m and that it is under a tension of $F_T = 1000$ N. The speed of waves in this wire is

$$V = \sqrt{\frac{F_T}{\rho}} = \sqrt{\frac{1000\text{ N}}{0.025\text{ kg/m}}} = \sqrt{4 \times 10^4(\text{m/sec})^2} = 200\text{ m/sec}$$

Let the wire be vibrated at its basic frequency, $\nu = 250$ cps. The period is $T = 1/\nu = 0.004$ sec, and the wavelength is

$$\lambda = \frac{V}{\nu} = \frac{200\text{ m/sec}}{250\text{ cps}} = 0.8\text{ m}$$

Wind Instruments

When we blow across the opening of a narrow bottle or closed metal tube, a whistling note is heard. Standing waves of sound are created as the result of reflection at the closed end, where there is a node, and at the open end, where there is an antinode. The lowest frequency (longest wavelength) is obtained when the wave is as shown in Figure 12-7. The length L contains one-quarter of a wave

$$L = \frac{\lambda}{4}$$

The fundamental frequency is thus

$$\nu = \frac{V}{\lambda}$$

or

$$\nu = \frac{V}{4L}$$

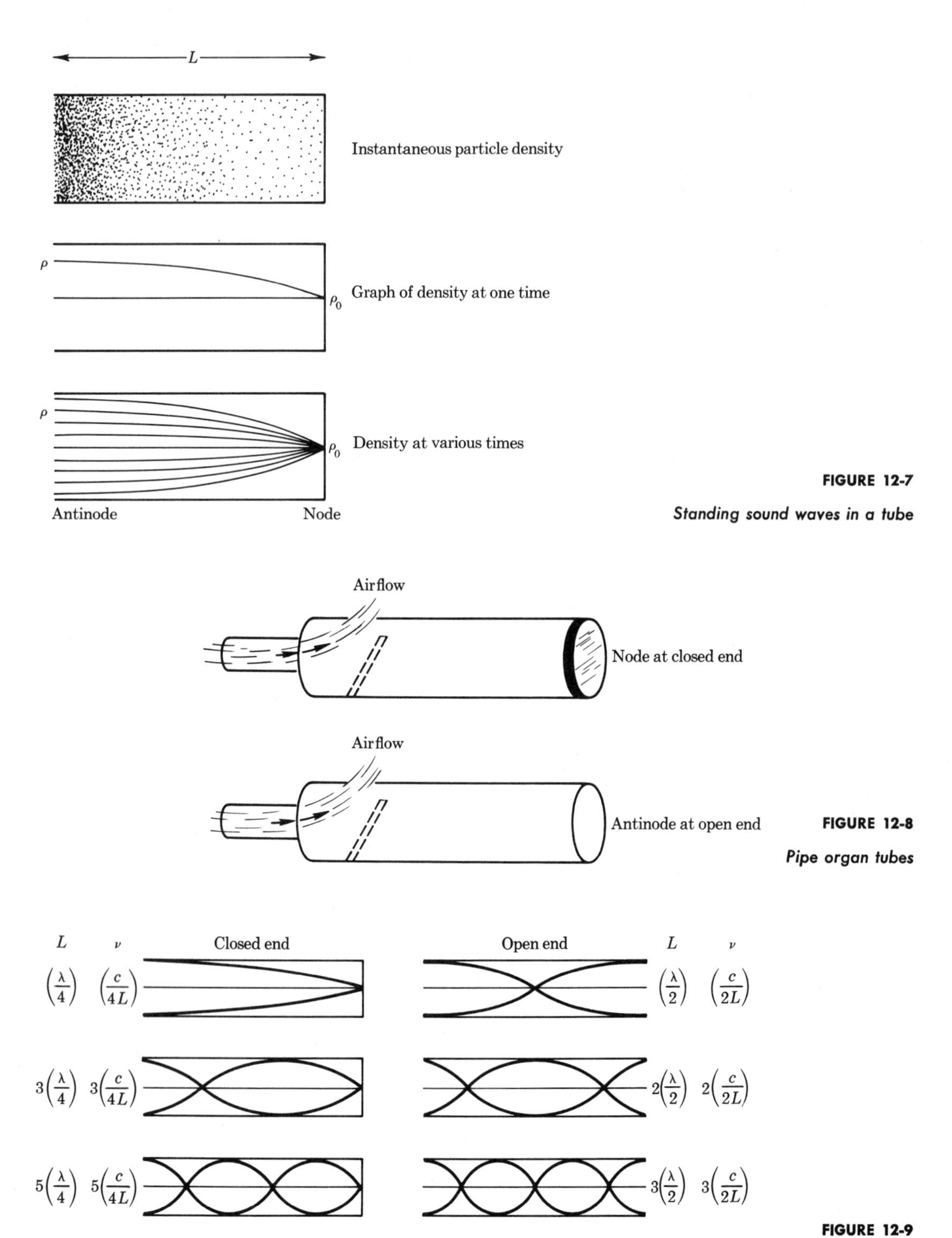

FIGURE 12-7

Standing sound waves in a tube

FIGURE 12-8

Pipe organ tubes

FIGURE 12-9

Frequencies in closed and open pipes

Higher frequencies are also present, with $\nu = 3V/4L$, $5V/4L$, etc., as can be verified by requiring different wave forms to fit within the length of the bottle or tube.

A pipe organ is constructed of hollow tubes of various lengths; each length allows a fundamental frequency of sound vibration when air passes over its open end. The other end may be either closed or open, as in Figure 12-8. For the two cases, the standing sound waves have, respectively, a node and an antinode at the ends. By drawing graphs that represent the density of molecules as a function of position (Figure 12-9), the possible frequencies can be deduced. Notice that the fundamental frequency for the open pipe is twice that of the closed pipe of the same length. Frequencies in the open-pipe case consist of *all* multiples 1, 2, 3, etc. of the fundamental, while frequencies in the closed tube are *odd* multiples 1, 3, 5, etc.

Let us compute the length of an open organ pipe that gives a frequency $\nu = 440$ cps, at room temperature, where $V = 343$ m/sec. Now, $\lambda = V/\nu = \frac{334 \text{ m/sec}}{440 \text{ cps}} = 0.78$ m. To produce the fundamental note, the length must be one-half a wave in length, $L = \lambda/2 = 0.78/2 = 0.39$ m, or about 15 in.

Relatively few string or wind instruments give notes of a single frequency, and we would find music rather dull if they did. The quality of notes from a musical instrument is determined by which harmonics are encouraged and which are suppressed. It is the special combination of the fundamental note with its overtones that gives the pleasing sensations. Instruments themselves are complicated structures, for example, the violin, which was developed over years of experimenting and is not at all easy to analyze.

Beats

When two people close together whistle the same tune at the same time, unusual pulsating sound effects are heard. Notes are heard of lower frequency than those being produced by either person. This is a simple example of the phenomenon of "beats," involving the superposition of sound waves of slightly different frequency. Suppose the waveforms that combine at a fixed point of observation—say, the origin $x = 0$—are pure notes of the same amplitude

$$y_1 = A \sin \omega_1 t$$

$$y_2 = A \sin \omega_2 t$$

The sum may be formed by trigonometry to be

$$y = y_1 + y_2 = 2A\left[\sin \frac{(\omega_1 + \omega_2)\,t}{2} \cos \frac{(\omega_1 - \omega_2)\,t}{2}\right]$$

Let us abbreviate this expression by letting the difference in angular frequencies be $\Delta\omega = \omega_1 - \omega_2$, and the average be $\bar{\omega} = \dfrac{\omega_1 + \omega_2}{2}$. Then,

$$y = \left(2A \cos \frac{\Delta\omega t}{2}\right) \sin \bar{\omega} t$$

Suppose ω_1 and ω_2 are rather close to each other—as, for instance, if $\nu_1 = 256$ cps and $\nu_2 = 250$ cps. A plot of y, as in Figure 12-10, shows the rapidly varying oscillation ($\sin \bar{\omega} t$) with amplitude $2A \cos (\Delta\omega t/2)$ that reaches a maximum only infrequently. Whenever the amplitude reaches a peak—i.e., when $\cos \frac{1}{2}\Delta\omega t = \pm 1$ or $\frac{1}{2}\Delta\omega t = \pi, 2\pi, \ldots$—a pulse of sound called a *beat* is found. The period of beats is $T_b = 2\pi/\Delta\omega$, and the frequency is $\nu_b = \Delta\omega/2\pi$, which means that *the frequency of beats between two notes is the difference in their frequencies.* Thus ν_b is 6 when $\nu_1 = 256$ and $\nu_2 = 250$.

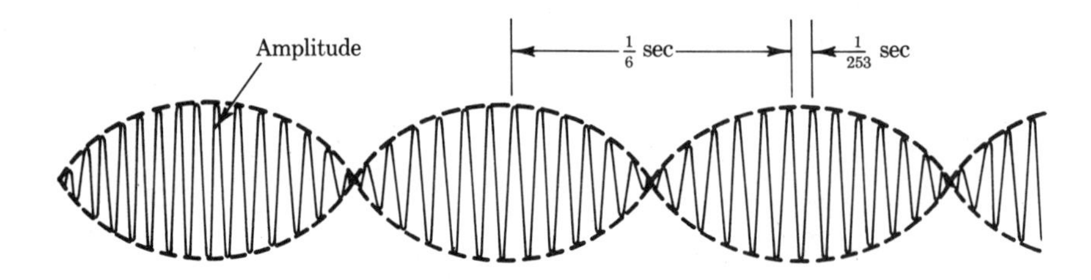

FIGURE 12-10

Beats of frequencies 250 and 256 cps

The presence or absence of beats determines whether music is harmonious or not. Psychologically, we find a combination of notes pleasant if they have very near the same frequency, for example, as sung by two people or played by two instruments. If the two frequencies are such that strong beats of certain pitch occur, the music is unmelodious. Similarly, if the fundamentals and harmonics of several notes played simultaneously as on a piano tend to augment each other without extraneous beats, the resultant sound is considered harmonious. To illustrate, let us consider the effect of playing notes whose fundamentals are 100 cycles per second and 200 cycles per second, an octave higher. The second harmonic of the first note will match the fundamental of the second. Or, if the fundamentals are 100 and 150, in ratio 3:2, the resultant sound will have pure frequencies of 100, 150, 200, 300, 450, 500, 600, 700, 750, etc.

In general, if simultaneous notes have a frequency ratio of two small numbers, 1:1, 2:1, 3:1, 3:2, etc., the music is pleasing. Now, however, let the two notes be 100 and 205. The second harmonic of the lower note, with frequency 200, produces undesirable beats of 5 cps with the higher note. A 1-cps beat also occurs between the fourth and second harmonics of the two notes. The phenomenon allows the piano tuner to adjust the instrument by sensing frequencies that are very low—i.e., the beats.

12-3 ELECTROMAGNETIC VIBRATIONS AND WAVES

The light by which we see and radio and television signals are examples of a wave motion in space that is called *electromagnetic* radiation. The colors of the visible spectrum correspond to different light frequencies, while the dial or channel settings on a radio or TV set correspond to frequencies used by the stations. Let us now look into ways of producing electromagnetic waves.

The electric field between two large parallel plates coated with charges of opposite sign was found in Section 6-2 to be uniform, of magnitude $\mathcal{E} = 4\pi k_E \sigma$ and directed from the + plate to the − plate. Clearly, if we interchange the charges, the field will point in the opposite direction (Figure 12-11). Suppose it were possible to alternate the polarity suddenly and periodically. A plot of the amount of charge on one of the plates, say the lower, would be as shown in Figure 12-12(a). Now let us plot the resulting field, letting + be up, − down—Figure 12-12b. A test body such as an electron placed in the space would be pulled up and down alternately. It is much easier to alternate the polarity as a sine curve rather than as the "square" functions just described. The alternating voltage of commercial electrical outlets is the form $\cos 2\pi\nu t$ where the frequency ν is 60 cps. Thus, if the charge density σ is varied according to $\sigma = \sigma_0 \cos 2\pi\nu t$, the electric field will vary in time between the plate as $\mathcal{E} = \mathcal{E}_0 \cos 2\pi\nu t$, where $\mathcal{E}_0 = 4\pi k_E \sigma_0$. It will go positive and negative in direction in accord with Figure 12-13. The similarity between these electric-field "vibrations" and the displacement in time of a mass on a spring is obvious.

Just as periodic motion of the end of a rope can give rise to waves traveling along it, periodic motion of charges result in electromagnetic waves. The physical mechanisms are entirely different, of course, but there is a mathematical analogy. As we shall see in greater detail in Section 19-1, the periodic reversal of polarity of the two charges of an electric dipole at frequency ω causes changes in electric and magnetic fields in space. At a large distance z away, the fields vary in a pure sine form, represented by the expressions

$$\mathcal{E}_x = \mathcal{E}_0 \cos (kz - \omega t)$$

$$\mathcal{B}_y = \mathcal{B}_0 \cos (kz - \omega t)$$

where z is in the direction of the wave progression. The two fields point

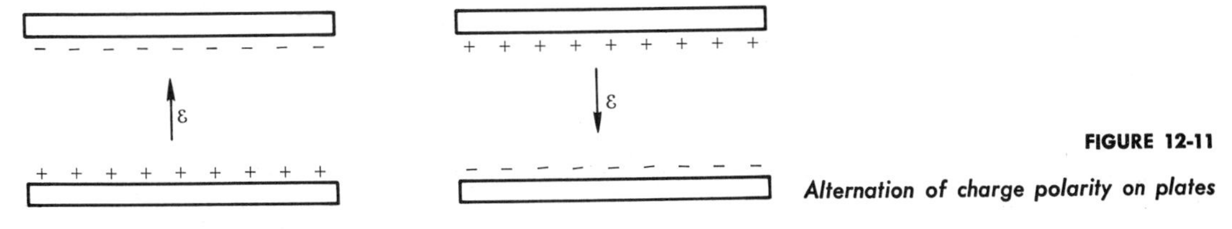

FIGURE 12-11

Alternation of charge polarity on plates

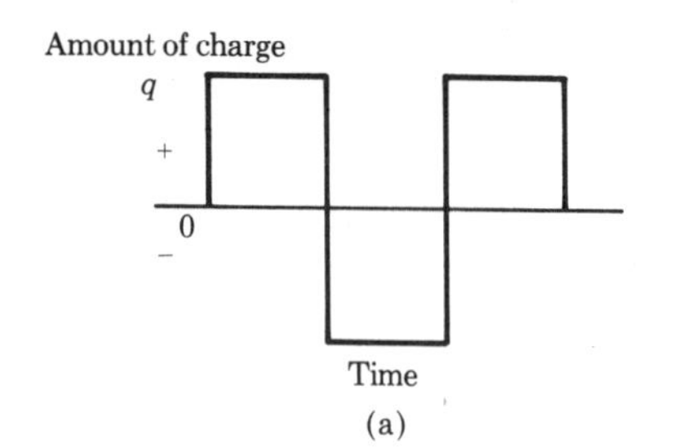

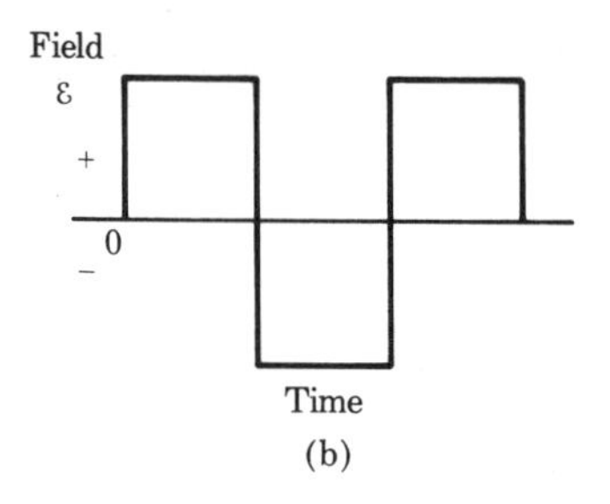

FIGURE 12-12

Variation of charge on lower plate and strength of electric field with time

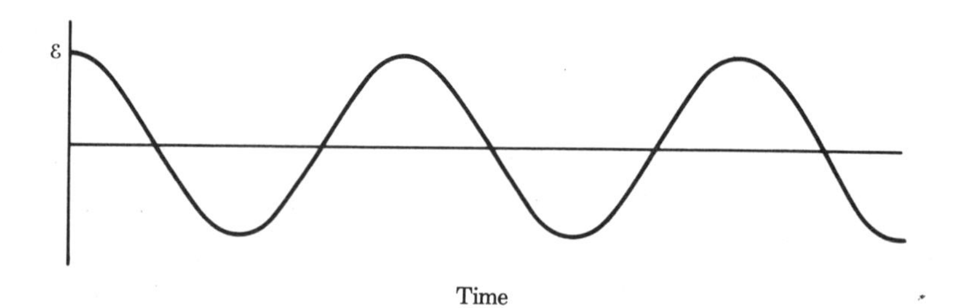

FIGURE 12-13

Electric field varying as $E = E_0 \cos 2\pi\nu t$

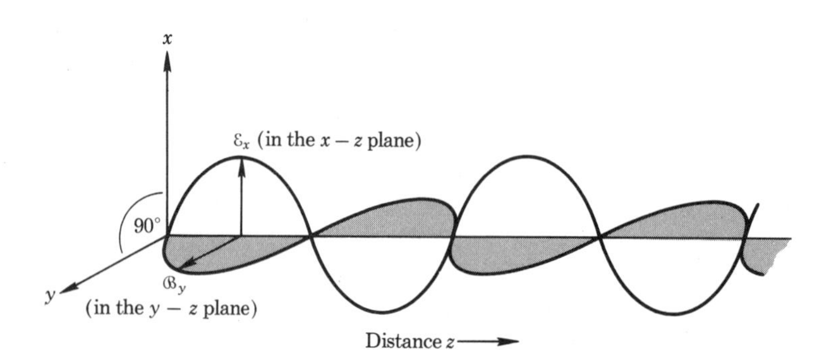

FIGURE 12-14

Graph of electric and magnetic fields in electromagnetic wave at an instant

only in the x- and y-directions, respectively. Figure 12-14 shows what an instantaneous "photograph" would reveal about the fields along a section of the z-axis. If one remains at a point, he sees the alternation of $\mathcal{E}_x$. There is a maximum upward, then zero, then a maximum downward.

12-4 THE DOPPLER EFFECT

As a whistling train or an ambulance with siren approaches us, we hear a higher note than after the source of sound goes past. Also, if we move relative to a source of sound, the frequency appears higher or lower depending on whether we approach or recede from it. The term *Doppler effect** is given to this phenomenon. Our knowledge of wave motion can be applied to find the amount of frequency change. First consider the simplest case, the *moving observer*. Suppose we approach, with uniform speed v_o, a fixed sound source which emits waves of definite length λ, as in Figure 12-15(a). At rest, we would receive each second ν peaks, where $\nu = V/\lambda$. Moving, we encounter $\nu' = \frac{V + v_o}{\lambda}$, a greater number. The *ratio* of frequencies is seen to be

$$\frac{\nu'}{\nu} = 1 + \frac{v_o}{V}$$

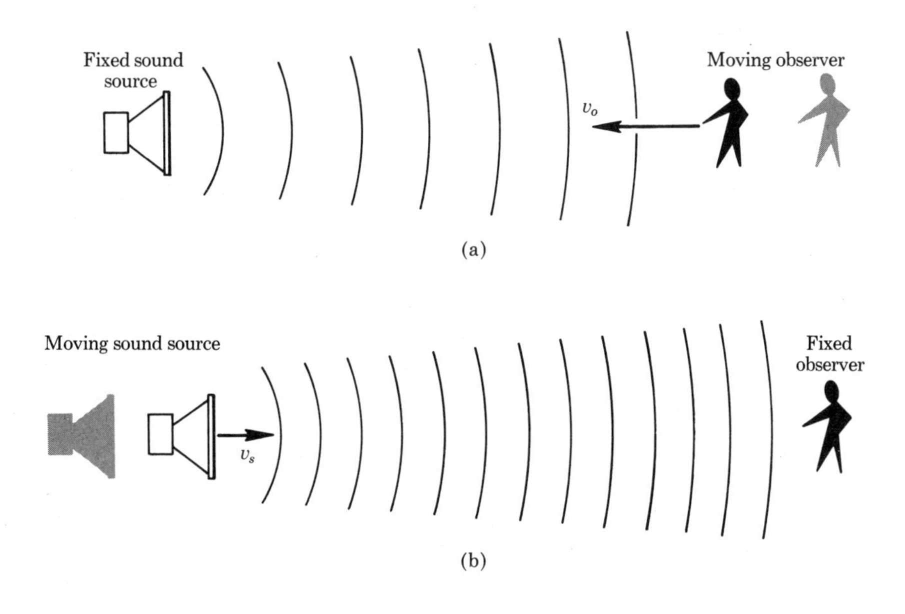

FIGURE 12-15

Doppler effect. In case (a), the observer meets more peaks; in case (b), the waves are closer together

Note that ν' is larger than ν, as we know from experience. To handle the case in which we recede from the source instead, merely change the sign on v_o. The *moving source* that approaches us with speed v_s emits waves of a new length—Figure 12-15(b). Between times for maximum amplitude, the

* After Christian Doppler, who described the process in 1842.

source has moved by a fraction v_s/V of a wavelength. It tends to catch up with the waves just emitted. The waves experienced at a distance are shorter—i.e., $\lambda' = \lambda - \frac{v_s}{V}\lambda$. Now, the two frequencies are $\nu' = V/\lambda'$ and $\nu = V/\lambda$. Their ratio is λ/λ' or

$$\frac{\nu'}{\nu} = \frac{1}{1 - v_s/V}$$

Again, the frequency is higher than when the source and observer are at rest. If the source were moving away instead, we would change the minus sign to a plus.

A general relation for any motion of source and observer is

$$\frac{\nu'}{\nu} = \frac{1 \pm v_o/V}{1 \pm v_s/V} \qquad \textit{(Doppler effect for sound)}$$

where the correct pair of signs to use is readily selected by the knowledge of the nature of the effect.

As an example, find the observed frequency when a train approaches with speed 50 m/sec, whistling at 256 cps. Take the speed of sound as in air at 0°C, 331 m/sec. Now $\frac{\nu'}{\nu} = \frac{1}{[1 - (50/331)]} = 1.18$ and $\nu' = 302$ cps. Next, suppose that while we are riding in a train, we hear the whistle of a following train of the same speed. The speeds of the source and observer *relative* to each other will be zero and $\nu' = \nu$, and no change in frequency will be heard.

The Doppler effect also occurs when sources of light approach or recede from us. Measured changes in the frequency of light allow us to estimate the speeds of stars or galaxies. If the frequency of the light is lower than normal—that is, shifts toward the red end of the spectrum—one concludes that the star is receding from us. Much discussion of the possibility of an expanding universe has resulted from such measurements. Although the effect is in the same sense, it is not proper to use the above formula developed for sound, since the speed of light c is always the same (Section 3-4). Einstein's theory of relativity states that the new frequency is the same whether we approach the source or the source approaches us. Instead of two formulas, which by analogy to sound would be

$$\frac{\nu'}{\nu} = 1 + \frac{v_o}{c} \qquad \text{and} \qquad \frac{\nu'}{\nu} = \frac{1}{1 - v_s/c}$$

there is but one formula, as shown in Appendix D:

$$\frac{\nu'}{\nu} = \sqrt{\frac{1 + v/c}{1 - v/c}} \qquad \textit{(Doppler effect for light)}$$

where v is the relative speed of approach. For separating objects, we change both signs. For $v \ll c$, the ratio is approximately $1 \pm v/c$, as easily proved by use of the binomial expansion.

Other examples of the Doppler effect are: (1) the variety of frequencies observed from a heated gas in which moving atoms emit light; (2) the nuclear reaction between moving neutrons and nuclei; (3) the reflection of radar waves from aircraft or ships or moving cars, or from artificial Earth satellites, to determine velocities.

12-5 MATTER WAVES AND ENERGY LEVELS

There is another unusual and important concept of wave motion— that waves are associated with every moving particle. DeBroglie, in 1924, first postulated that every moving particle with momentum p has associated with it a wave whose wavelength λ is given by

$$\lambda = \frac{h}{p}$$

where h is Planck's constant, 6.63×10^{-34} J-sec.

This validity of this hypothesis has been proven by its predictive capability. For example, the discrete energy levels of the electrons in atoms can be shown to exist if one assumes that the energy levels represent standing waves of the matter wave associated with the particles. Although calculations in general involve some higher mathematics, we can illustrate the concept.

Assume that an electron with mass m and momentum $p = mv$ (nonrelativistic) is located in a region of the x-axis between $x = 0$ and $x = a$. Whenever the electron reaches either of these end points, it is perfectly reflected. Then, we can say that it is confined in a one-dimensional square "well," as shown in Figure 12-16. The lines going upward at $x = 0$ and $x = a$ represent barriers that cannot be penetrated for any energy which the electron might have, so we say they are of "infinite" potential.

Now, if a standing wave is to exist in this region, the wave must go to zero at $x = 0$ and $x = a$ at all times, just as in the string with fixed ends; or

$$y_T = 2A \sin kx \cos \omega t = 0$$

which occurs only if $ka = n\pi$, where n is an integer. Recall, however, $k = 2\pi/\lambda$ and $\lambda = h/p$, so

$$p = n \frac{h}{2a}$$

Thus, the energy of the electron (which is all kinetic energy) in this region is given by

$$E = \frac{p^2}{2m} = \frac{n^2h^2}{8ma^2}$$

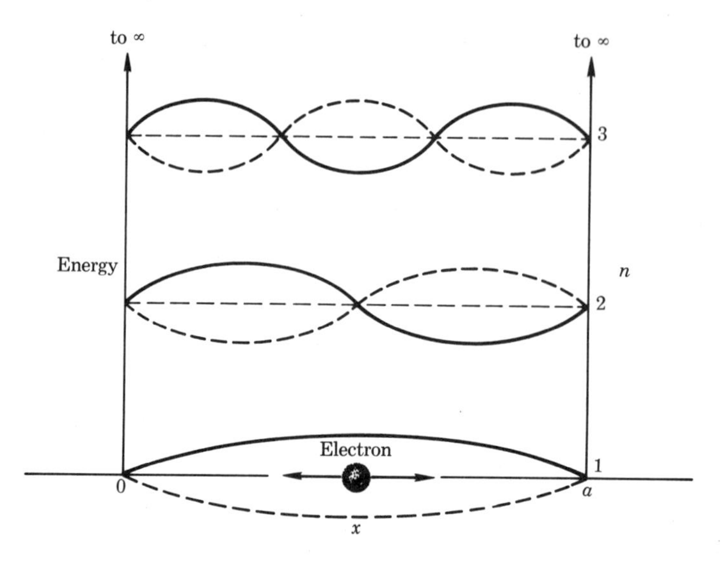

FIGURE 12-16

Electron in a square well, as a matter wave

and can only be those values given by $n = 1, 2, 3$, etc. This concept is called *quantization* and is a direct result of the assumption that a moving particle has a *wave* associated with it.

We shall reserve further discussion of the physical significance of this concept until Section 22-2. It will suffice here to note that de Broglie reasoned that since electromagnetic radiation and light could be viewed as either waves or particles, then matter might be regarded either as a particle or as a wave.

The Variety of Waves in Nature

We have cited many examples of physical situations that exhibit wave character, including motions of a string, rope, or wire, of a water surface, sound in air or solid materials, of electromagnetic disturbances, and even matter. It is natural to ask the question "Why should these diverse phenomena be governed by the same mathematical description?" The answer, for many situations, goes back to the fact that the medium being disturbed involves restoring forces between atoms that are proportional to the displacement from their normal position. The application of the laws of mechanics then yields the harmonic motion or waves in the form of sine or cosine functions. In other cases, the physical processes may be different, but the mathematical description turns out also to require wave functions. We should keep in mind, however, that the motion of every system cannot be expressed in such an elementary way. Obviously when the displacements are large—as when water crashes on the beach in a storm—simple sine- and cosine-wave forms are not applicable at all.

Problems

12-1 Compute the wavelength in air at room temperature corresponding to the lower and upper limits of audible frequencies, about 30 cps and 16,000 cps.

12-2 Sound travels through two different paths as illustrated in Figure P12-2. If the path difference is one-half wavelength, what is the amplitude heard? If it is one wavelength?

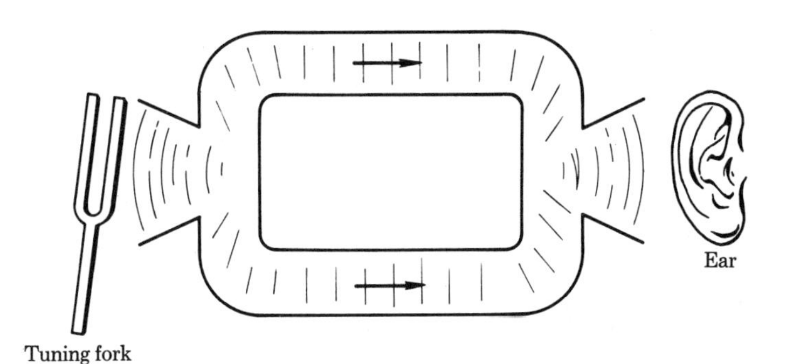

FIGURE P12-2

12-3 Two waves of the same amplitude, speed, and frequency (but out of phase) are added. Letting

$$y_1 = A \sin(\omega t - kz)$$
$$y_2 = A \sin(\omega t - kz - \phi)$$

find the sum $y = y_1 + y_2$ by use of trigonometric formulas—see, for example, the footnote on page 331. Discuss the result.

12-4 Find the wave velocity and the frequency of the third harmonic (second overtone) of a wire with length 1.4 m, linear density 0.006 kg/m, and under 4 N tension.

12-5 At what frequency will a wire 2-m long vibrate in the fifth harmonic mode, if the wave speed is 3000 m/sec?

12-6 How many harmonics are there between frequencies 2.7×10^4 and 2.8×10^4 if the fundamental is 50 cps? What harmonics are the two end values?

12-7 How many harmonics are there between 5400 cps and 6900 cps for a wire with length 0.3 m, wave speed 180 m/sec?

12-8 A violin string with length 0.3 m, atom number density $7 \times 10^{28}/\text{m}^3$, has a fundamental frequency of 440 cps. What is the highest frequency possible? Is it in the audible range?

12-9 An atomic bomb explodes at a distance of 100 miles away. Find the times elapsed for the flash to be seen, heard, and felt through the ground (seismic waves have a speed of about 5.5 km/sec), assuming room temperature.

12-10 Find the frequency ν, the angular frequency ω, and the magnitude of the propagation constant k for a wave of sound with the following properties: $\lambda = 5$ m, $V = 343$ m/sec.

12-11 A rocket plane has a speed of 4000 mi/hr at an altitude where the density of air is $\frac{1}{2}$ that at sea level. At what Mach number is the plane traveling?

12-12 Find the wavelength of sound with frequency 7100 cps in air at temperature 40°C (104°F).

12-13 How accurate is the rule of thumb on sound speed "5 sec = 1 mi," when tested at room temperature (20°C)?

12-14 An apparatus to demonstrate transverse waves in a wire is shown in Figure P12-14. If the density of the wire is 0.015 kg/m, and its length is 0.75 m, what mass will be required on the platform to give the second-harmonic standing waves?

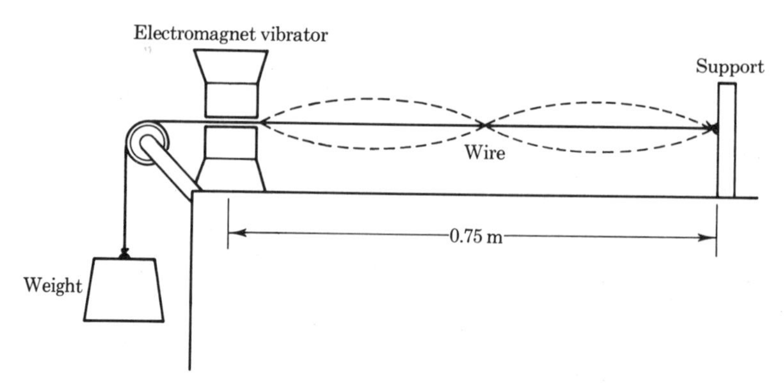

FIGURE P12-14

12-15 In cold weather, would an organ note sound higher or lower than in hot weather? Explain.

12-16 What fundamental note is sounded by an open organ pipe of length 4 ft, operated at room temperature?

12-17 Find the fundamental frequency at room temperature of an open organ pipe that is 2.5 m long.

12-18 How much difference is there between the fundamental frequency of a closed organ pipe of length 1.8 m in hot weather (30°C), and cold weather (18°C)? What are the two frequencies?

12-19* (a) What are the beat frequencies possible between various harmonics of the frequencies 100 and 205? (b) Show that no distinguishable beats are possible if the ratio of two frequencies is $\frac{1}{3}$.

12-20 Two closed organ pipes of length 0.75 m and 0.91 m are sounded simultaneously. What beat frequencies below 50 cps are there between various harmonics, assuming a sound speed of 344 m/sec?

12-21 Suppose that a piano tuner uses a 440-cps note as standard, and finds that the note 1 octave higher (supposed to be double the frequency) produces beats at 2 cps. What is the actual frequency of the higher note?

12-22 On a piano, the frequencies of "middle C" and the G above it are 261.33 and 391.99 cps, respectively. Calculate the first three beat frequencies between the appropriate harmonics of these notes.

12-23* The "ideal temperament" of the C scale on a piano has frequencies that bear the following ratios:

C	D	E	F	G	A	B	C
1	9:8	5:4	4:3	3:2	5:3	15:8	2

Taking the lower C to be 260 cps, if the notes C, E, and G are played together, what harmonic frequencies augment each other?

12-24 The sound from a police car's siren with frequency 900 cps seems to be 920 cps to a stationary observer. Is the car approaching or receding from the observer? How fast?

12-25 A whistle of frequency 600 cps is mounted on the end of an arm of length 0.5 m, and rotated at 60 revolutions per second. What maximum and minimum frequencies are heard?

12-26 A radioactive source Na^{22} that emits gamma rays with frequency 3.0×10^{20} cps is mounted on a loudspeaker that vibrates with frequency 120 cps and amplitude 2 mm. What is the largest *shift* in the frequency of the gamma rays that can be measured by an observer? (See Fig. P12-26.)

12-27 A train whistle has a definite frequency of 2000 cps. The engineer sounds his whistle when he is 3 km from the station, and the stationmaster notes the time and measures the frequency to be 2050 cps. If it maintains constant speed, how much later will the train go through the station?

12-28 The frequency of light from a certain lamp is observed to be shifted from its normal value ν to a lower frequency ν'. Show that this implies a speed of recession of the source of

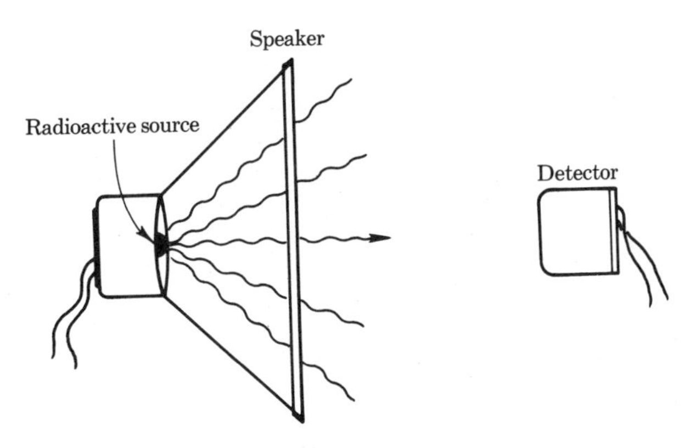

FIGURE P12-26

$$v = \frac{c[1 - (\nu'/\nu)^2]}{1 + (\nu'/\nu)^2}$$

12-29 A shift of the frequency of light from a distant galaxy is observed. Referring to Problem 12-28, if the change is from 1.43×10^9 to 1.40×10^9 cps, what is the apparent speed of recession of the galaxy?

12-30 An inventor proposes a device for highway patrolmen to measure the speed of oncoming cars at night by observing a Doppler shift. Investigate the feasibility of this idea.

12-31 Find the first two energy levels, in electron-volts, of an electron of mass 9.1×10^{-31} kg confined to a box of width 10^{-8} cm (about the size of an atom).

MECHANICAL AND THERMAL PROPERTIES OF MATTER

13

We now turn to the properties of matter in bulk form, consisting of many atoms or molecules that interact with each other. Some of the phenomena to be explained in this chapter are the action of elastic forces, vibrations, expansion caused by heating, changes of state, buoyancy, and the flow of fluids. We shall seek an appreciation of the relation between small-scale processes, often called *microscopic*, and large-scale or *macroscopic*, behavior.

13-1 THE ROLE OF THERMAL ENERGY AND TEMPERATURE

Much of the macroscopic behavior of matter may be explained in terms of basic atomic forces, the concept of temperature, and the energy of motion of atomic particles that we call *thermal energy*. Let us consider a particularly familiar substance—water—as it goes through several changes in state when we give it more and more energy, keeping in mind that many substances behave in a similar way.

We shall start with a sample of water at the lowest possible temperature, absolute zero, defined as the temperature at which molecular motion is a minimum. Thus the sample will be in the form of ice, a relatively hard and rigid solid. As thermal energy is supplied by some external source, motion of the molecules increases in magnitude as the temperature rises.

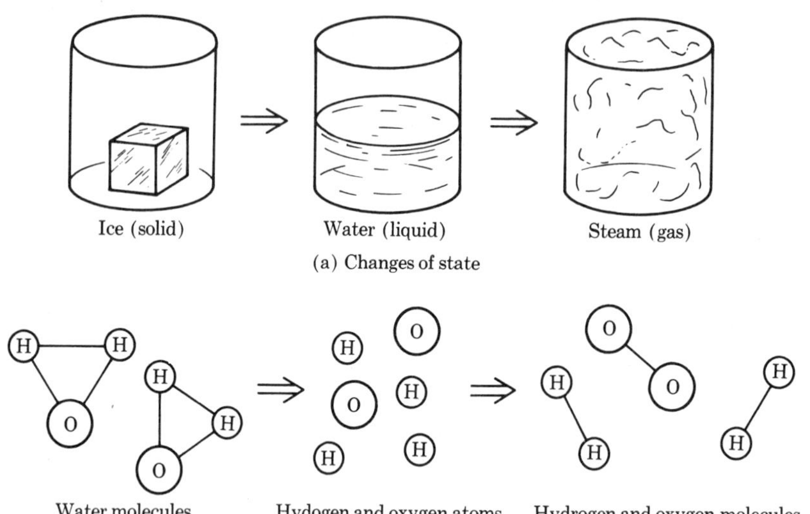

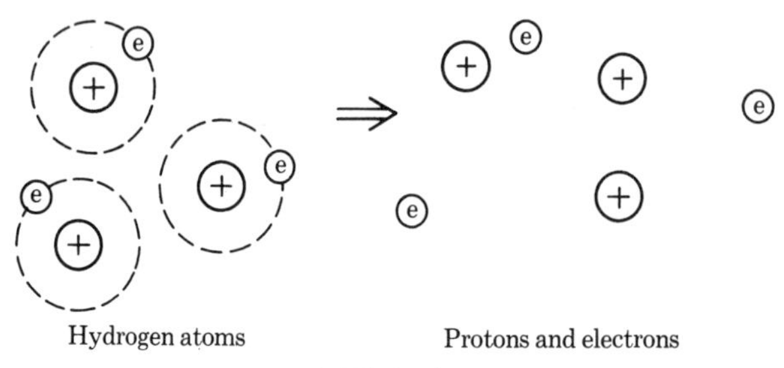

FIGURE 13-1

Effects of heating water

The ice expands somewhat as a result. Eventually, the vibrations become sufficiently vigorous to change the ice to water, a liquid—Figure 13-1(a). In crystalline materials such as ice, salt, and many pure metals, there is a sudden change from the solid form to the liquid form. This change is accompanied by a large absorption of thermal energy at a constant temperature, called the *melting point* (0°C for H_2O). One can visualize that this energy has in some way relaxed or otherwise altered the bonds between the molecules. Other noncrystalline substances, called *amorphous* ("no form"), soften gradually as they are warmed. Familiar examples are plastics and wax. A *liquid*, on the other hand, is characterized by its ability to flow. Thus, under the influence of gravity, its surface forms a level and the liquid fits the shape of the vessel containing it. It should be realized, however, that under sufficient pressure, solids can also be made to flow between plates or through holes, as in rolling mills for sheets or dies for drawing wire. The distinction between solids and liquids is not sharp if the range of pressure is extended. Similarly, liquids and gases, especially at low speed, behave the same—as *fluids*.

In a liquid, there is little resistance to sliding of layers of molecules past each other, even though there are forces between particles causing them to cluster into a relatively unique volume. The raindrop is a good example. It is spherical when first formed, with negligible external forces. On passing through the air, the larger drops are distorted into the familiar tear-shape, but the molecules of water remain closely packed. On striking the ground, the raindrop flattens out and is broken up by the impact.

When a liquid is in contact with the air or a vacuum, there is a tendency for some energetic molecules to overcome the surface forces of attraction and escape from the surface, a process of *evaporation*. Thus there is a cooling of the liquid as each molecule escapes. We are personally familiar with this phenomenon—the evaporation of perspiration from the skin is a mechanism for regulating body temperature. The term *vaporization* refers to a large-scale transition of liquid into the gaseous state. Usually a quantity of heat is absorbed at a particular temperature called the *boiling point* (100°C for H_2O at standard atmospheric pressure).

An experiment will reveal the relation between a liquid and a vapor or gas. We put some water in a container equipped with a pressure gauge. At a given temperature, molecules of water continuously leave the surface of the liquid, while others return and are caught. There will be a constant average number of molecules in the space, giving rise to a measurable pressure, called the *vapor pressure* of the liquid. Raising the temperature of the liquid clearly makes it possible for more molecules to escape from the surface, and the vapor pressure increases. When this vapor pressure reaches the pressure of the surrounding atmosphere, the liquid boils (as discussed further in Section 14-3). This happens at the "normal" boiling point of the liquid if the surrounding gas is air at standard pressure. However, the boiling point decreases as the atmospheric gas pressure decreases. For example, when the pressure is lowered by a pump or by an altitude change, water may "boil" at any temperature down to nearly 0°C.

With the absorption of enough heat energy, the intermolecular forces in a liquid can be overcome almost completely. Then the substance goes into the state called a *gas*, where the molecules are free to move about, bouncing off each other or against the walls of the vessel. A gas fully occupies whatever volume of space is available to it, because of the relative lack of forces between molecules. Except for collisions of neutral molecules with each other, the particles remain independent. This can be explained if it is assumed that the range of the forces between particles is small compared with the average distances between them. Since there is much space between particles, the gas is much more readily compressed than was the same material in the liquid state. It exerts a pressure (a force per unit area) on the walls of any confining vessel because of incessant bombardment by particles. Further heating gives the molecules a higher average kinetic energy, and the gas is raised to a correspondingly higher temperature. We may view the motion of the molecules as consisting of translation, rotation, and vibration, each molecule consisting of 3 atoms in the case of H_2O. On colliding, molecules exchange energy, with that of rotation and vibration giving rise subsequently to the emission of light radiation of a low frequency, below the visible range.

Continued heating causes the molecules of H_2O to dissociate into their component atoms of hydrogen and oxygen, and these readily recombine to form molecules of H_2 and O_2. We may now trace the fate of these components, which are simply different compounds still in gaseous state, and which appear to exhibit the same various types of motion described for the case of water molecules. Addition of even more energy causes the H_2 and O_2 molecules to dissociate into free atoms of H and O—Figure

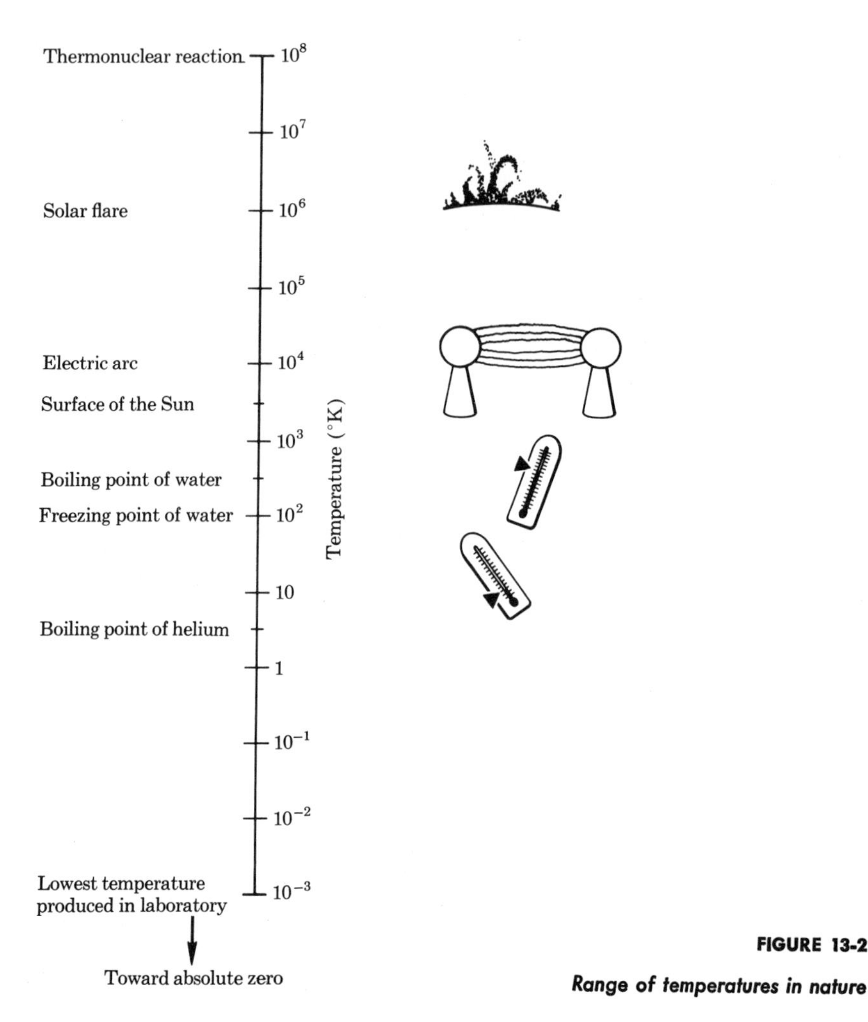

FIGURE 13-2
Range of temperatures in nature

13-1(b). Each is a neutral, stable structure, although each is highly reactive chemically. The electron and proton which make up the hydrogen atom are normally separated by an average value of about 0.5×10^{-10} m. The hydrogen atom can, however, absorb energy and this average separation can increase. The energy of excitation may be re-emitted in the form of light radiation, bringing the atom back to a normal energy state. A similar process may occur in the oxygen atom. This absorption and emission of energy will be discussed in greater detail in Section 22-1.

When even more energy is imparted to it, the H atom dissociates into a free electron and a free proton—Figure 13-1(c). As ions these can recombine with each other or with other charges. The protons can then be given still more energy of translation by the application of an electric field. At high enough speeds they may experience nuclear reactions with the other nuclei present, those of oxygen and hydrogen. In the case of oxygen, excitations of the motions of the nuclear components—neutrons and protons—can result in high-frequency light emission. This phenomenon will be discussed further in Section 23-2.

Finally, at extremely high temperatures, the nuclei of deuterium, the rare heavy isotope of hydrogen, can combine to form heavier nuclei. In this fusion or *thermonuclear* process, large quantities of kinetic and radiation energy are released—in the reaction, mass is converted into energy in accord with Einstein's relation, $E = mc^2$. As we can see in Figure 13-2, the range of temperatures with which we must deal in physics is very large. (The lowest value, absolute zero on the Kelvin scale, $-273.1°C$, does not appear because of the use of logarithms in the figure.)

With the conclusion of this qualitative discussion, we now turn to the problem of analyzing the mechanical and thermal behavior of matter consisting of a large number of basic particles. We shall try to relate the simple molecular structure to many-particle substances.

13-2 ELASTICITY

Using our analogy between atomic forces and mechanical-spring forces (Section 10-1), let us try to explain the behavior of elastic objects, such as a coil spring or a length of wire, which resume their initial length after a stretching force has been released. The experimental observation that the *force tending to restore* the original shape and size of elastic materials *is proportional to the displacement* is called *Hooke's law*. Such behavior is limited to a range of forces and displacements that do not permanently deform the material.

Let us imagine first a thin metal wire or rod of original length L and cross-sectional area A. One end is fixed, and to the other end we apply a force F (Figure 13-3). The displacement l of the end of the wire is proportional to F:

$$F = Kl$$

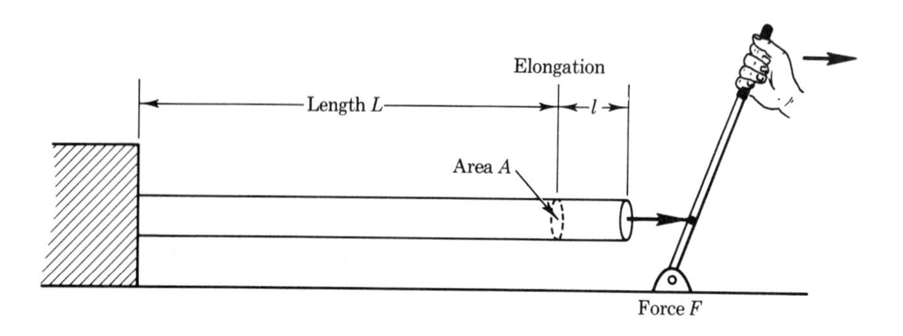

FIGURE 13-3

Stretching an elastic wire

where K is the *force constant* of this particular wire, a macroscopic quantity. Suppose the same type of experiment is done with several samples of the same material having different lengths and areas. Then we find that the force per unit area F/A, which we call the *stress*, is proportional to the fraction the wire elongates l/L, called the *strain*. Thus, for an elastic material,

$$\text{stress} \sim \text{strain}$$

or

$$\frac{F}{A} = E_y\left(\frac{l}{L}\right) \qquad (\textit{Hooke's law})$$

The constant of proportionality, E_y, called *Young's modulus of elasticity*, depends on the *material* of which the wire is made and not on its length and area. It is related to the force constant for a given sample by

$$E_y = \frac{KL}{A}$$

Values of Young's modulus are given for several materials in the first column of Table 13-1. As an example of its use, let us find the change in length of a steel pillar 1.4 m in diameter and 4 m long, that supports a weight 9×10^8 N—for example, as part of a building. Young's modulus for steel is $22 \times 10^{10}\ \text{N/m}^2$. The column cross-sectional area is $\pi d^2/4 = \pi(1.4)^2/4 = 1.54\ \text{m}^2$. The change in length is then

$$l = \frac{F}{A}\frac{L}{E_y} = \frac{(9 \times 10^8\ \text{N})(4\ \text{m})}{(1.54\ \text{m}^2)(22 \times 10^{10}\ \text{N/m}^2)}$$
$$= 0.0106\ \text{m}$$

which is about a centimeter.

Let us now look at the forces and displacements from the microscopic

TABLE 13-1

ELASTIC PROPERTIES OF MATERIALS

MATERIAL	MODULUS (in units of 10^{10} newtons/m²) E_y	E_b	E_s	ν
Aluminum	7	3	3	0.13
Copper	11	10	4	0.32
Glass	5.4	3.6	2.2	0.25
Iron	20	15	8	0.27
Mercury	0	2.4	0	0.50
Rubber	1.0×10^{-4}	1.9×10^{-3}	3×10^{-5}	0.49
Steel	22	18	8.5	0.30
Tungsten	35	18	15	0.17
Water	0	0.22	0	0.50

or atomic viewpoint. Typical metals have a lattice structure, with atoms arranged in regular patterns. The three most common arrangements, shown in Figure 13-4(a), are face-centered cubic, body-centered cubic, and hexagonal. The relation of structure to elastic properties can be more readily explained if we examine an idealized simple cubic arrangement of atoms, as illustrated in Figure 13-4(b). We imagine there to be force bonds that act as springs between the atoms. Along each side of a unit volume there are N particles giving a total of N^3 per m³. The spacing of particles is $a = 1/N$. Since the stretching is along one direction only, we can view the wire as a bundle of N^2A *chains* of connected atoms. Let us propose that the force f between individual atoms is proportional to the amount x that their normal separation is increased. Thus we write

$$f = K_1 x$$

where the atomic force constant K_1, as here defined, clearly does not depend on the large-scale form of the wire. By Newton's law of action and reaction, applied successively, f is also the force applied at the end of one chain. The total force for the N^2A chains is then $F = N^2Af = N^2AK_1x$. If each atomic bond stretches an amount x, the total elongation of a chain of length L, containing NL atoms, is $l = NLx$; combining the two equations and rearranging, we obtain

$$\frac{F}{A} = NK_1\left(\frac{l}{L}\right)$$

Comparison with the stress–strain relation shows that Young's modulus is merely N times the force constant of one atom bond,

$$E_y = NK_1$$

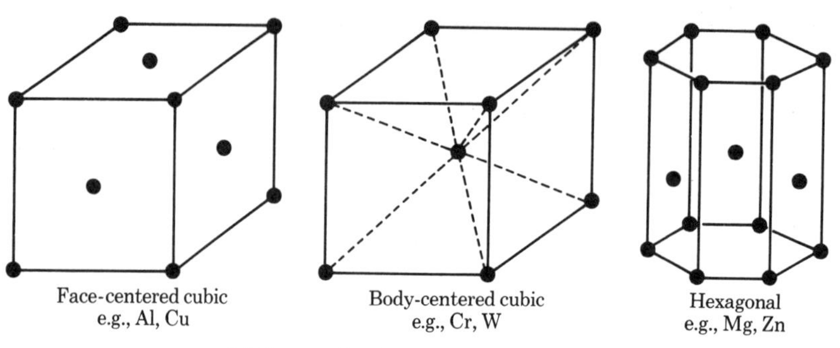

(a) Common lattice arrangements in metals

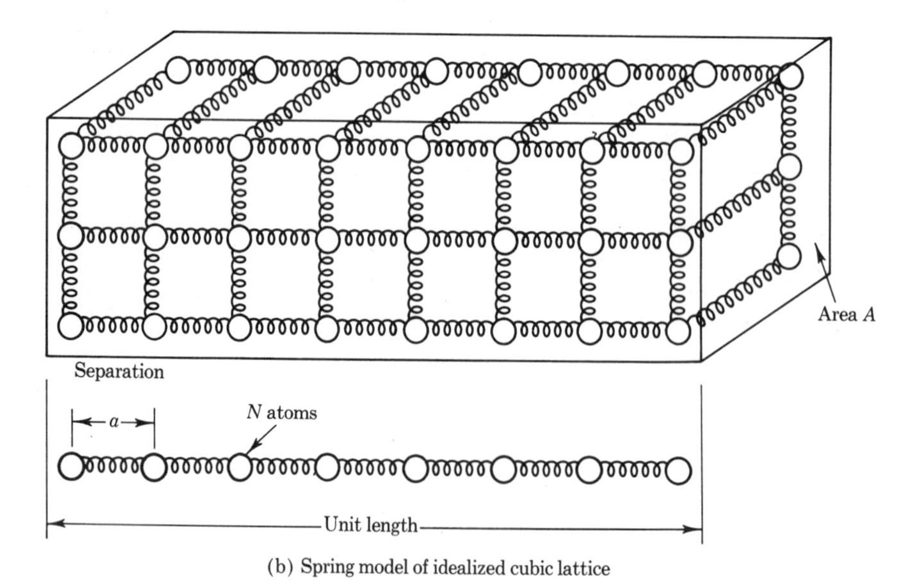

(b) Spring model of idealized cubic lattice

FIGURE 13-4

Lattice structure of metals

and thus must be *dependent only on the material*, just as noted before. Also, we see that the force constant of the wire is related to the atomic-force constant by

$$K = \frac{NAK_1}{L}$$

The small-scale properties of iron can be estimated on the basis of our special model, keeping in mind that iron is actually a more complicated structure involving both face- and body-centered arrangements. In one kilomole of iron, of mass 55.85 kg, there are 6.02×10^{26} atoms (Avogadro's number, Section 4-4). By proportions, in 1 m^3 containing 7.85×10^3 kg, the number of atoms is

$$n = \frac{7.85 \times 10^3}{55.85}(6.02 \times 10^{26}) = 8.5 \times 10^{28}$$

Hence the number per unit length of chain is

$$N = \sqrt[3]{n} = \sqrt[3]{85 \times 10^{27}} = 4.4 \times 10^9/\text{m}$$

The spacing of atoms is $a = 1/N = 2.3 \times 10^{-10}$ m. Young's modulus is 20×10^{10} N/m^2, and the force constant per atom is

$$K_1 = \frac{E_y}{N} = \frac{20 \times 10^{10}\ \text{N/m}^2}{4.4 \times 10^9/\text{m}} = 45\ \text{N/m}$$

As a check on this relation between small-scale and large-scale views, let us find the potential energies due to stretching. For *one* bond only,

$$E_{p1} = \tfrac{1}{2}K_1x^2$$

The number of participating atoms (and bonds) in the particular sample of wire of length L and area A is N^2AL. The total potential energy is

$$E_p = (N^2AL)E_{p1} = (N^2AL)(\tfrac{1}{2}K_1x^2)$$

On introducing the expressions for K_1 and x, we find this to be exactly the same as that for the whole wire as it elongates by an amount l:

$$E_p = \tfrac{1}{2}Kl^2$$

Our macroscopic and microscopic models are thus consistent with respect to energy.

Suppose that an iron wire is stretched by 1%. What is the potential energy stored in each atom? Now,

$$x = \frac{l}{NL} = \frac{l}{L}a = (0.01)(2.3 \times 10^{-10}) = 2.3 \times 10^{-12}\ \text{m}$$

$$E_{p1} = \tfrac{1}{2}K_1x^2 = \tfrac{1}{2}(45\ \text{N/m})(2.3 \times 10^{-12}\ \text{m})^2 = 1.2 \times 10^{-22}\ \text{J}$$

This is also

$$\frac{1.2 \times 10^{-22}\ \text{J}}{1.6 \times 10^{-19}\ \text{J/eV}} = 7.5 \times 10^{-4}\ \text{eV}$$

a very small energy compared with that of electrons in atoms.

Volume Changes

The linear-chain model of stretching is oversimplified, since changes also take place in other dimensions. When we stretch a rubber band or a wire we know that it becomes thinner. Let us deduce the amount of this contraction. Let the diameter change from D to $D - d$, as the length changes from L to $L + l$ (Figure 13-5). Suppose that the total volume is unchanged, a reasonable assumption—at least for rubber and plastic materials. Then,

$$\frac{\pi}{4} D^2 L = \frac{\pi}{4} (D - d)^2 (L + l)$$

Multiplying this out and neglecting all but the largest terms in the equation, we find $d/D \cong 0.5l/L$. In general, one can write $d/D = \nu(l/L)$ where ν is called *Poisson's ratio*. Experimental measurements give $\nu = 0.49$

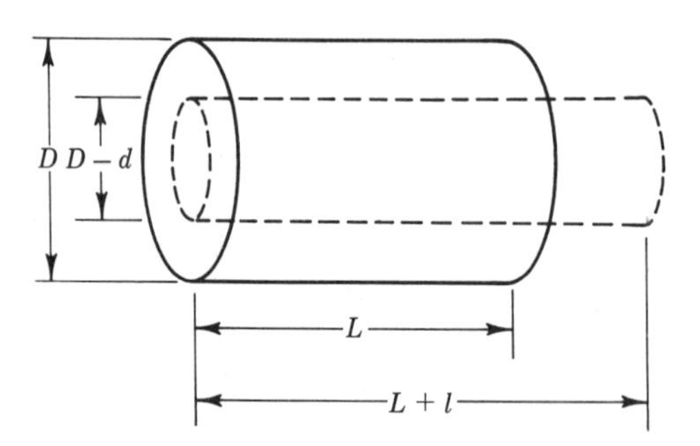

FIGURE 13-5

Dimension changes in stretching

for rubber, but 0.32 for copper, 0.30 for steel, and as low as 0.13 for aluminum; Table 13-1 gives additional values. If ν is less than 0.5 it means that the volume does not remain constant as stretching takes place.

The application of a uniform external pressure to a solid results in a compression. Such a situation seldom is found in structures, because forces will usually be different in one direction than another. However, we could imagine a three-dimensional vise that could provide equal forces on all sides of a cube. A block placed in deep water would experience a uniform hydrostatic pressure. The pressure $P = F/A$ will be the stress. The volume changes from V by an amount v, and v/V will be the strain. Then

$$\frac{F}{A} = E_b \frac{v}{V}$$

where E_b is called the *bulk modulus*. We would expect it to be related to Young's modulus E_y. Let us try to find the dependence. Visualize forces applied to three of the faces, with the remaining three (opposite) faces held fixed. Each side will compress a fraction l/L that is the result of the force in that direction, $(1/E_y)(F/A)$, less the expansion of $(\nu/E_y)(F/A)$ due to each of the other forces. Thus

$$\frac{l}{L} = \frac{1 - 2\nu}{E_y} \frac{F}{A}$$

The volume change is $v = (L + l)^3 - L^3$, which is approximately $3L^2l$, and thus $v/V = 3l/L$. Combining,

$$E_b = \frac{E_y}{3(1 - 2\nu)}$$

For a material with $\nu \cong \frac{1}{3}$, such as copper, $E_b \cong E_y$. If ν is close to 0.5, as for rubber, E_b is very much larger than E_y, which means that stretching or compression is easy in one direction but very difficult in all three directions at once.

We find that the bulk moduli of solids and liquids are very high, which implies that a very large force is required to give any compression. For example, if a piece of iron, with $E_b = 15 \times 10^{10}$ N/m^2, is to undergo a 1% change in volume, $v/V = 0.01$, the force per unit area or pressure must be

$$\frac{F}{A} = P = (15 \times 10^{10})(0.01) = 1.5 \times 10^9 \text{ N/m}^2$$

which is about 15,000 atm or 220,000 lb/in.^{2}. The term *compressibility* χ is used to describe the ability of a substance to be compressed. It is defined by

$$\chi = -\frac{\Delta V/V}{\Delta P}$$

which is the fractional change in volume per unit of pressure change. If there were no force initially, then the change in pressure is $\Delta P = F/A$; the change in volume is $\Delta V = v$, and we see that the magnitude of compressibility χ is the reciprocal of the bulk modulus E_b. Thus, for iron, $\chi = 6.7 \times 10^{-12}$ m^2/N.

Shear Stress and Strain

The elongation of a solid is only one of the possible distortions it can experience. Picture, as in Figure 13-6, a cube of material to which we apply a force tending to slide planes of atoms past each other. The bonds *between atoms in the layers* must be stretched, and tend to restore the original position.

Such a shifting of atoms is called *shear*, involving stresses and resulting strains which are different in character than those previously discussed. Actually, we use the term in a similar sense when we say the head of a bolt "shears off," or that we will "shear" a piece of metal. Whenever a rod or wire is twisted, shearing stresses and strains are involved.

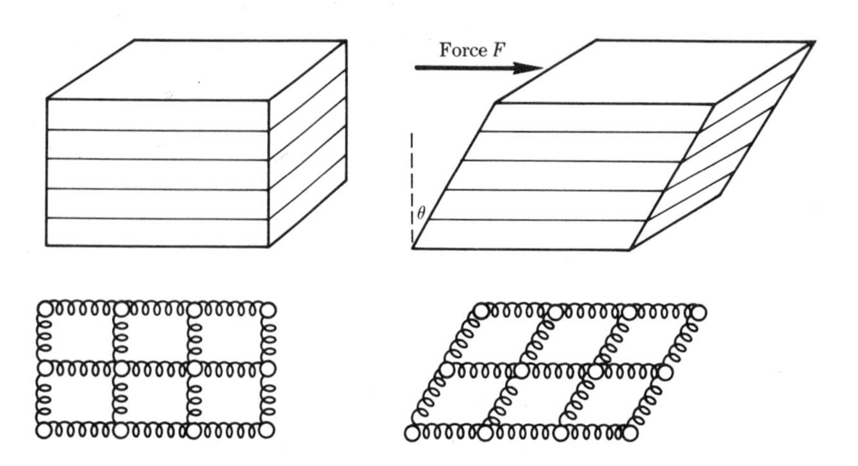

FIGURE 13-6

Shearing effects in solids

The torsion constant c, discussed in Section 11-2, is related to shear as the force constant K is related to stretching. A *shear modulus* E_s indicates by how much the layers shift. The ratio of applied tangential force F to the area A is called the *shearing stress*; the ratio of arc s to side S, the angular shift, is called the *shearing strain*. The proportionality of shear stress to shear strain is expressed as

$$\frac{F}{A} = E_s \frac{s}{S}$$

It can be shown by a derivation, similar to that for the relation of bulk modulus to Young's modulus, that

$$E_s = \frac{E_y}{2(1 + \nu)}$$

Thus, for a material with $\nu \cong \frac{1}{3}$, we find that $E_s \cong \frac{3}{8}E_y$. For example, for copper, $\nu = 0.32$ and the ratio of $E_s = 4 \times 10^{10}$ N/m^2 to $E_y = 11 \times 10^{10}$ N/m^2 is 0.36.

By eliminating ν, a single relation between the three moduli E_y, E_b, and E_s is obtained. Most compactly,

$$\frac{3}{E_y} = \frac{1}{E_s} + \frac{1}{3E_b}$$

The general description of elasticity requires the use of tensor analysis for both the large- and small-scale views, and the role of imperfections in the regular lattice must be accounted for. Our treatment thus provides a basis for qualitative discussion only.

13-3 THERMAL EXPANSION

Most substances increase in volume when heated, a process called *thermal expansion*. Our atomic picture of matter provides a qualitative understanding of this effect. The addition of heat energy to any object, composed as it is of bound atoms and molecules, tends to increase the thermal agitation. From the microscopic viewpoint, this energy of motion "competes" with the energy of binding. On the average, the component particles are farther apart than at a lower temperature. This corresponds to an overall expansion of the object. Let us re-examine Morse's potential-energy curve (Figure 11-11) to see qualitatively how expansion arises. We assume that such a graph as in Figure 13-7 can be used to describe roughly the forces between atoms in a lattice of a solid. If the potential-energy curve were a perfect parabola as for a spring connection, there would be an equal chance that the displacement would be as often positive as negative. The average x would be zero. However, the actual curve is flatter in the direction of $+x$, which favors the positive displacements.

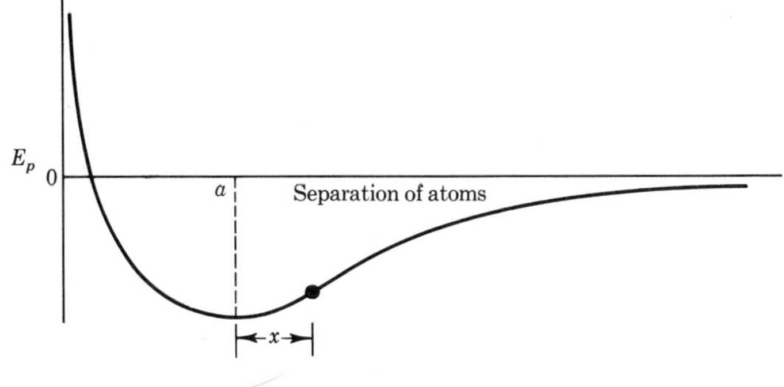

FIGURE 13-7

Interatomic potential energy

Thus subatomic vibrations are said to be *anharmonic*. However, materials expand about the same amount in each direction, except in cases where the crystal structure is *not* the same along different axes.

Let us look at a long thin bar of metal (Figure 13-8) that is heated throughout to cause a temperature increase from T to $T + \Delta T$. Measurements of length before and after give l and $l + \Delta l$; thus the length has increased by a fraction $\Delta l/l$. The *linear coefficient of thermal expansion* is defined by the ratio

$$\alpha = \frac{\Delta l/l}{\Delta T}$$

If α is known, we can compute either the change in length $\Delta l = l\alpha\Delta T$ or the new length $l' = l(1 + \alpha\Delta T)$. For example, suppose a thin wire of iron, length 0.5 m, is heated from 20°C to 100°C. Its coefficient α is about 11×10^{-6}/°C. The increase in length is thus

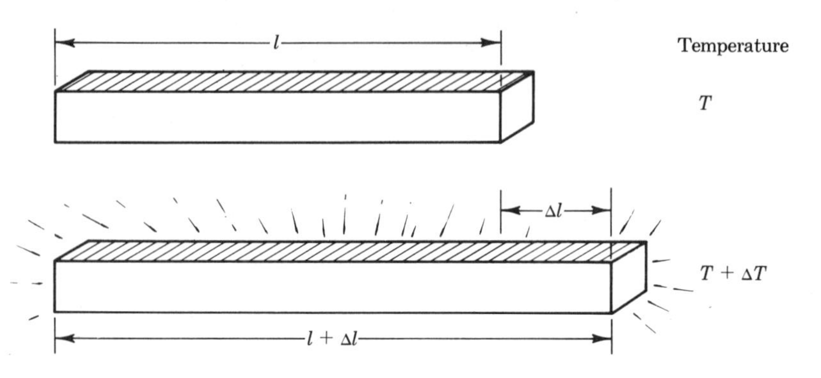

FIGURE 13-8

Thermal expansion of a solid bar

$$\Delta l = l\alpha\Delta T = (0.5\text{ m})(11 \times 10^{-6}/°\text{C})(80°\text{C}) = 0.00044\text{ m}$$

which is only about $\frac{1}{20}$ millimeter. Amounts of expansion depend on the type of material; numerical values of α are listed in Table 13-2.

TABLE 13-2

COEFFICIENTS OF LINEAR THERMAL EXPANSION (measured at 0°C)

MATERIAL	α (in units of $10^{-6}/°\text{C}$)
Aluminum	24
Copper	17
Steel	12
Silver	18
Pyrex glass	3.6
Ice	51
Diamond	3.5
Paraffin	588

COEFFICIENTS OF VOLUME THERMAL EXPANSION (measured at 20°C)

MATERIAL	β (in units of $10^{-3}/°\text{C}$)
Alcohol, ethyl	1.2
Water	0.207
Mercury	0.182

A well-known application of thermal expansion is the creation of a tight bond of a ring to a rod (Figure 13-9). The inside diameter of the ring is deliberately made a little smaller than the rod diameter. On heating, the ring expands; it is slipped over the rod and allowed to cool, giving a high-pressure fit. We would find it very complicated to describe the ring expansion if we viewed it as a three-dimensional figure. The problem is easy if we look at the *circumference*, which expands almost linearly. For example, suppose a copper ring of inside diameter 2.0 cm is to go around a rod of 2.005-cm diameter. How much increase in temperature of the ring will allow it to be slipped on? The fraction that the diameter and circumference must change is $0.005/2.0 = 0.0025 = \Delta l/l$. The linear coefficient of copper is $\alpha = 17 \times 10^{-6}/°C$. Thus the temperature increase needed is

$$\Delta T = \frac{\Delta l/l}{\alpha} = \frac{0.0025}{17 \times 10^{-6}} = 150°C$$

Mechanical thermometers or thermostats for controlling temperature are based on differences in expansion coefficients of metals. A bimetallic

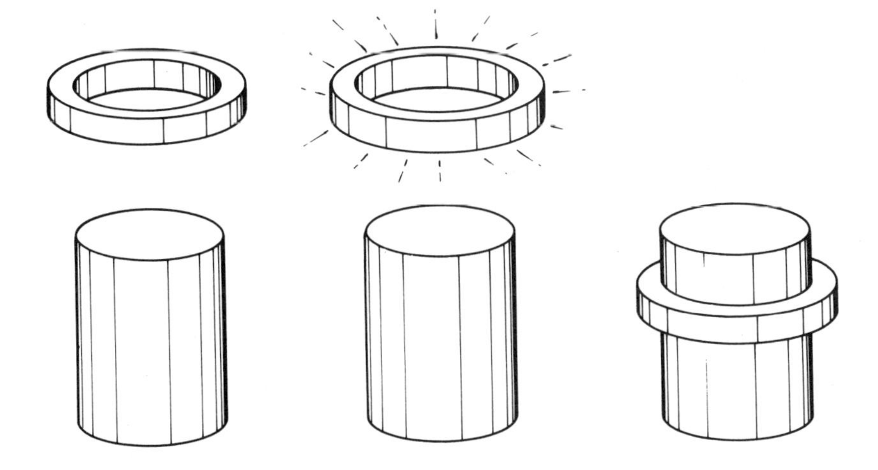

FIGURE 13-9

Fitting a ring on a rod

strip—composed of Al and Cu, for example—will bend as illustrated in Figure 13-10 when the temperature is increased. If the rise is ΔT, the new lengths are the following:

$$L_A = L_0(1 + \alpha_A \Delta T)$$
$$L_C = L_0(1 + \alpha_C \Delta T)$$

Notice that the bimetallic strip will now constitute the arc of a circle. The strips have radii $R + t/2$ and $R - t/2$, respectively, and lengths

$$L_A = \left(R + \frac{t}{2}\right)\theta \quad \text{and} \quad L_C = \left(R - \frac{t}{2}\right)\theta$$

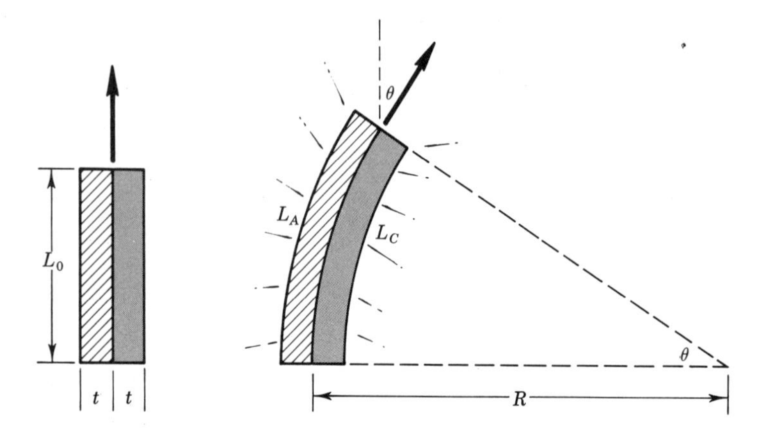

FIGURE 13-10

Differential expansion of bimetallic strip

and on combining equations we obtain

$$L_0(\alpha_A - \alpha_C)\Delta T = t\theta$$

Therefore, the angle is directly proportional to the temperature change ΔT, and the angular deflection of a needle attached to one end is directly proportional to temperature changes.

Although we examined only the effect of temperature on the length of a bar, it should be understood that expansion is taking place in the other two directions as well. This will be brought out by examining the change in *volume* of a rectangular solid of dimensions l, w, and h. The result of a temperature rise ΔT is a new set of dimensions l', w', h' (Figure 13-11), where l'/l, w'/w, and h'/h are all equal to $1 + \alpha\Delta T$ if the coefficient of expansion is the same in all directions. The ratio of volume after heating to that before is

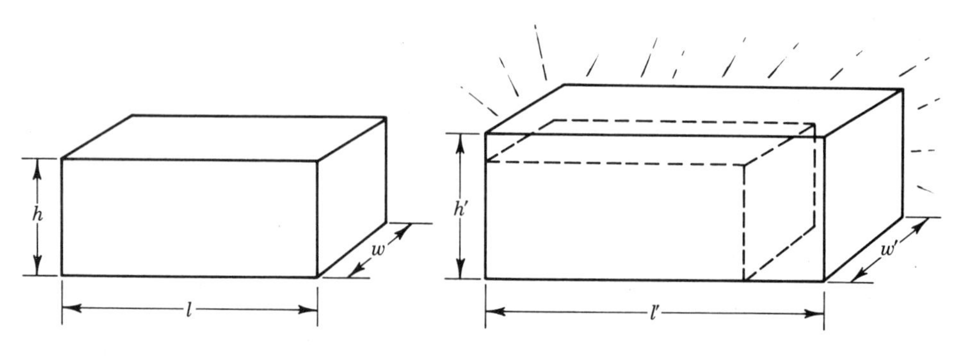

FIGURE 13-11

Volume expansion of heated solid

$$\frac{V'}{V} = \frac{l'w'h'}{lwh} = (1 + \alpha\Delta T)^3$$

Multiplying this out gives a complicated expression; but if we realize that the changes in dimensions are quite small, we can neglect all terms involving $(\alpha\Delta T)^2$ and $(\alpha\Delta T)^3$ and find

$$\frac{V'}{V} \cong 1 + 3\alpha\Delta T = 1 + \beta\Delta T$$

where the quantity 3α is labeled β, the *volume coefficient of expansion.*

Other properties of interest change along with volume. For instance, although the greater separation of the particles on heating does not change

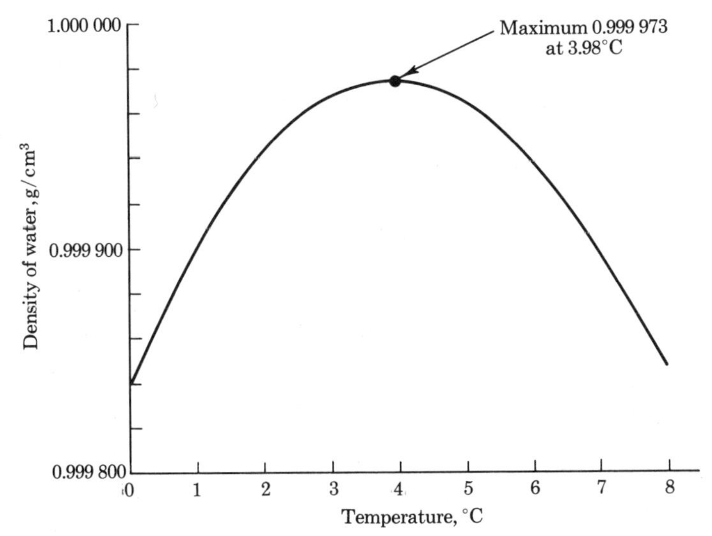

FIGURE 13-12

Density of water near its freezing point

the mass of the sample M, it does reduce the density or mass per unit volume. Initially $\rho = M/V$, and finally $\rho' = M/V'$. Thus the density ratio is*

$$\frac{\rho'}{\rho} = \frac{V}{V'} = \frac{1}{1 + \beta\Delta T} \cong 1 - \beta\Delta T$$

Next, let us see how n, the number of atoms per unit volume, changes with temperature. For any substance, n is directly proportional to density ρ, and we may form the ratio

$$\frac{n'}{n} \cong 1 - \beta\Delta T$$

* Recall that $(1 + x)^{-1} = 1 - x + x^2 - \ldots \cong 1 - x$ for small x.

The coefficient β is thus seen to be either the fractional change in volume or in atom density per degree Celsius. (Values for a few liquids are given in Table 13-2.) For example, at 20°C water has a coefficient $\beta = 2.07 \times 10^{-4}$/°C, which means that 1°C rise in temperature gives only 0.02% expansion. Water exhibits unusual behavior with temperature near its freezing point. As it is cooled from room temperature, its density increases, as do typical substances, but a maximum is reached close to 4°C. Below this point and into the ice phase it *expands* on cooling. The amount of the change is very slight, as shown on the expanded scale in Figure 13-12. This is a most fortunate property of nature. Life on Earth would be quite different or would not exist otherwise. Cold water in winter would not be buoyed to the surface as it approached freezing, but would sink to the bottom of bodies of water, which would soon be filled with ice.

13-4 THE GAS LAW

The three properties that characterize a gas, as we learn in chemistry, are the pressure P (N/m²), the volume V (m³), and the temperature T (°K). Experiments by Boyle and by Charles, with interpretation by Dalton, led to the following compact relation between these variables:

$$\frac{PV}{T} = Nk \qquad \textit{(general gas law)}$$

where N is the number of particles in the sample and k is *Boltzmann's constant* 1.38×10^{-23} J/°K. The product Nk is constant, regardless of changes in the variables in this *equation of state*. The formula becomes plausible when we associate pressure with wall bombardment by molecules, and temperature with particle speed and kinetic energy. Let us consider a cylinder of gas having an adjustable piston and a means for adding or removing heat (Figure 13-13) and perform several experiments.

1. We fix the volume V by holding the piston, and increase the temperature T by heating. Now, since

$$P \sim T$$

the pressure increases. The number of molecules per unit volume remains the same, but their speeds are higher; hence there is a greater number of collisions with the wall, and the pressure is increased.

2. We fix the temperature T by providing the necessary heat transfer, and decrease V by pushing the piston down. Since

$$P \sim \frac{1}{V}$$

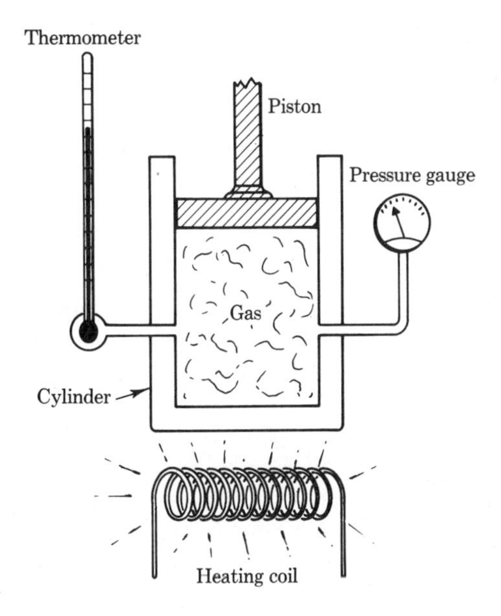

FIGURE 13-13

Apparatus for tests of gas law

the pressure increases. Although molecular speeds are unchanged because T is fixed, the number of molecules per unit volume increases; hence the momentum transfer increases, and the pressure is increased.

3. We fix the pressure P by allowing the piston to move freely to seek its natural balance against some external force, and increase the value of T by heating the gas. Since

$$V \sim T$$

the volume increases. The atom density is reduced by expansion, but the greater molecular agitation of the gas is such that pressure remains constant. Other combinations can be explained in a similar manner.

To illustrate the use of the gas law, let us consider a weather balloon containing helium gas, launched from the ground, with a volume of 4 m^3. We wish to compute the final volume and the number of atoms per cubic meter (the atom number density) at its final altitude, say 18 mi. Assume that the gas cools from room temperature 25°C (298°K) to 0°C (273°K) and experiences a pressure change, from 1 atm to 0.25 atm, in rising to that height. Now,

$$\frac{P_1 V_1}{T_1} = \frac{P_2 V_2}{T_2} = Nk$$

and

$$V_2 = \frac{P_1}{P_2}\frac{T_2}{T_1} V_1 = \left(\frac{1}{0.25}\right)\left(\frac{273}{298}\right)(4) = 14.7 \text{ m}^3$$

The atom density is $n = N/V_2 = P_2/(kT_2)$. The pressuie is $P_2 = (0.25)(1.01 \times 10^5) = 0.25 \times 10^5$ N/m²; hence

$$n = \frac{0.25 \times 10^5}{(1.38 \times 10^{-23})(273)} = 6.6 \times 10^{24}/\text{m}^3$$

Variation of Air Pressure with Height

We note the convenience of using ratios of pressure, volume, or temperature in the solution of gas problems. Also, we often make comparison with a reference state called *normal temperature and pressure* (NTP), which is 0°C (273°K) and 1 atmosphere (14.7 lb/in.² or 1.01×10^5 N/m²).

We are aware that the pressure and density of air becomes lower as one goes far above the Earth's surface. It is more difficult to breathe at high altitudes, which explains why supplies of oxygen are sometimes carried by expert mountain climbers. Rocket probes reach heights where the particle density is very low compared to that on the Earth's surface. The change with height may be deduced from fluid statics and the general gas law. If one moves upward a small distance Δh, the pressure changes by $\Delta P = -\rho g \Delta h$ (a reduction in P). However, $P = \rho kT/m$. Forming a quotient and rearranging,

$$\frac{\Delta P}{\Delta h} = -\left(\frac{mg}{kT}\right)P$$

an equation analogous to that describing air friction in Section 5-8. The solution is thus

$$P = P_0 e^{-mgh/kT}$$

where the pressure is P_0 at the Earth's surface, $h = 0$. This formula says that the pressure falls off exponentially with height above the Earth's surface in the ideal condition of uniform temperature and gravitational field. For air of approximate molecular weight 29, the molecule mass is $m = (29)(1.67 \times 10^{-27}\text{ kg}) = 4.84 \times 10^{-26}$ kg. Now, at 0°C, $T = 273$°K, $kT = (1.38 \times 10^{-23}\text{ J/°K})(273\text{°K}) = 3.77 \times 10^{-21}$ J. Then, using $g = 9.81$ m/sec²,

$$\frac{mg}{kT} = \frac{(4.84 \times 10^{-26}\text{ kg})(9.81\text{ m/sec}^2)}{3.77 \times 10^{-21}\text{ J}} = 1.26 \times 10^{-4}/\text{m}$$

which is also 0.126/km. Thus $P/P_0 = e^{-0.126h}$, where h is in kilometers. For example, at the top of Mount Everest, height 29,028 ft or 8.85 km, we would predict that the pressure is only a fraction $e^{-1.12} = 0.330$ of the value at sea level.

13-5 BUOYANCY OF FLUIDS

An object that is less dense than the fluid in which it is immersed will float—the swimmer, the boat, and the balloon are familiar examples—while one that is more dense will sink. The explanation of this effect in terms of forces and density was given by Archimedes,* whose principle is:

A BODY IS BUOYED UP BY A FORCE EQUAL TO THE WEIGHT OF THE FLUID DISPLACED.

This statement may be demonstrated for a block of material of volume V submerged in a fluid of density ρ (Figure 13-14). The forces on the top and bottom surfaces due to fluid pressures are $F_1 = \rho g h_1 A$ and

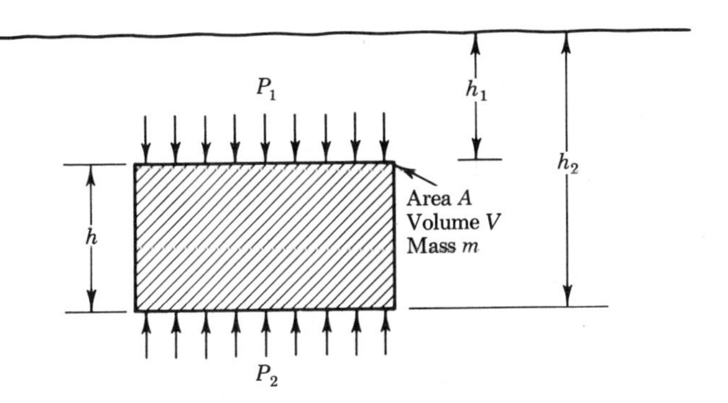

FIGURE 13-14

Archimedes's principle of buoyancy

$F_2 = \rho g h_2 A$. The magnitude of the net upward force is $F_2 - F_1 = \rho g A h = \rho g V = mg$, which is the weight of fluid displaced. The block will sink, rise, or remain in the same position depending on whether the weight of the block W is greater than, less than, or the same as the buoyant force mg. To prove this, let us examine the forces acting when the object is submerged (Figure 13-15). There is a downward weight W and a buoyant force upward F_B. If F_B is greater than W, there will be a net force $F_B - W$ upward, accelerating the body to the surface, where it will float with just the correct amount under the surface to make $F_B = W$. If, however, W is greater than F_B, the object will be accelerated downward, and thus sink. We are most often interested in the case where the object floats without moving. Although we have derived Archimedes's principle using a block of material, it applies to an object of any shape. A few examples will bring out the meaning of the principle.

1. A balloon of volume V is filled with helium gas and weighted down by the basket and passengers of total weight w so that the balloon hovers

* As the story goes, Archimedes (287–212 B.C.) leaped from his bathtub on discovering his principle and ran naked into the street shouting "*Eureka!*" ("I have found it!")

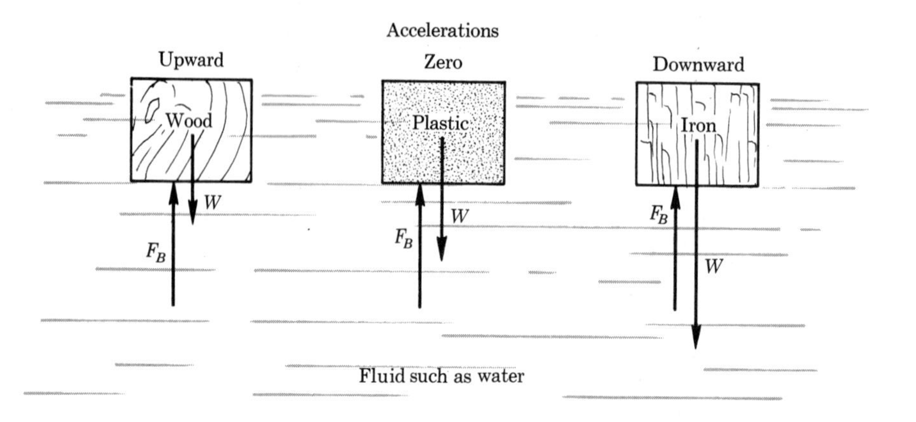

FIGURE 13-15

Motion of objects with gravity and buoyancy

over the Earth. The fluid that the balloon displaces is air, of weight $W_A = \rho_A g V$; this is also the buoyant force on the balloon. Now, the helium in the bag weighs $W_{He} = \rho_{He} g V$. The force balance drawn in Figure 13-16 shows that

$$W_A = W_{He} + w$$

Let us compute the allowed value of w. For a bag of volume $V = 200\ m^3$, using $\rho_A = 1.293\ kg/m^3$ and $\rho_{He} = 0.178\ kg/m^3$, we find that

$$(\rho_A - \rho_{He})gV = (1.293 - 0.178)(9.8)(200) = 2190\ N$$

which is 492 lb.

2. An iceberg of density $\rho_I = 917\ kg/m^3$ floats in sea water of density $\rho_W = 1025\ kg/m^3$, with a large fraction of the iceberg beneath the surface (Figure 13-17). We should like to find this fraction f. If the total ice volume is V, then the volume of water displaced is fV. The weight of that much water is $\rho_W g f V$, which is also equal to the buoyant force. The total downward force is the weight of all the ice $\rho_I g V$. These forces are in balance:

$$\rho_W g f V = \rho_I g V$$

Then,

$$f = \frac{\rho_I}{\rho_W} = \frac{917}{1025} = 0.895$$

With so little showing above the surface, one can easily see why icebergs are a menace to shipping in northern waters.

3. We wish to find the volume of a very irregular metal object by use of Archimedes's principle. To do so we first find its weight in air, W, then its weight when fully immersed in water, W'. Now, the difference in

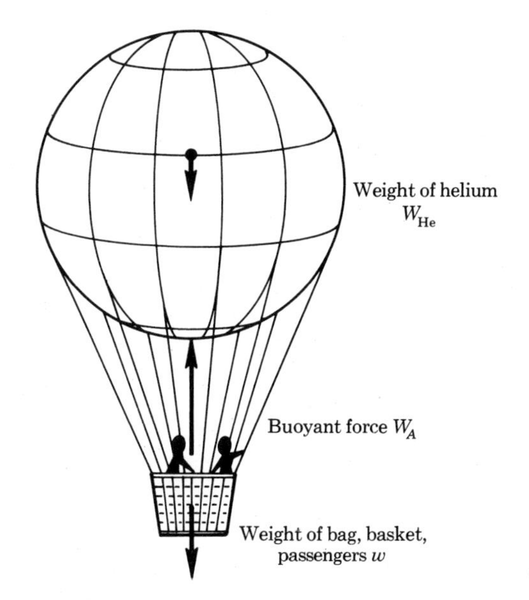

FIGURE 13-16

Forces on a balloon

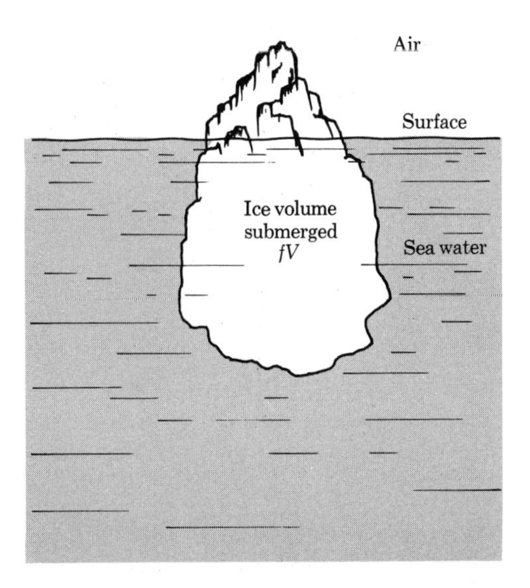

FIGURE 13-17

Iceberg in sea water

weights, $W - W'$, is the buoyant force, which is the weight of water displaced, $\rho_W gV$. Thus

$$V = \frac{W - W'}{\rho_W g}$$

but this is also the volume of the object. Archimedes is credited with having so tested the purity of a gold crown for King Hiero of Alexandria. Once the volume is known, the density of an object is computed from

$\rho = W/V$ and compared with the standard value, in the case of gold 19.3×10^3 kg/m^3.

13-6 FLUID FLOW

The law of conservation of mechanical energy provides a simple basis for the analysis of fluid flow so long as certain requirements are met. Let us observe the pattern of the flow of water in a glass funnel as shown in Figure 13-18(a). The level drops slowly as the water pours out the bottom at relatively high speed. The amount of liquid that crosses any horizontal plane in the funnel each second is the same, since water is practically *incompressible.* If the speed of liquid is low enough, the particles will follow smooth lines of flow, or streamlines, as shown in Figure 11-16(b). The paths do not cross over each other, and the flow is said to be of the *streamline* type, in contrast with eddying or turbulent flow. The motion of any fluid involves a combination of the flow of the bulk material and the random motion of individual molecules. As we shall see in Chapter 16, the molecular agitation tends to slow the forward motion, and the fluid experiences viscous forces. When this frictional effect is small, the fluid is said to be *nonviscous.*

The flow of water in the pipes of a metropolitan water-supply system fits these special requirements, as does the flow of gases such as air over the surface of an airplane wing when the speed of the plane is low. Let us now trace a single particle of fluid throughout a circuit that involves changes in speed, height, and area of passage through which it flows. At each point we may determine its energy and apply the law of conservation of energy. This straightforward procedure leads to what is called *Bernoulli's principle.*

The simple section of pipe shown in Figure 13-19 exhibits all of the variables of interest. Consider flow into and out of the shaded region bounded by planes 1 and 2, where the pressures are P_1 and P_2. In one second, a volume v_1A_1 (as indicated by the shading) will enter at 1 and an

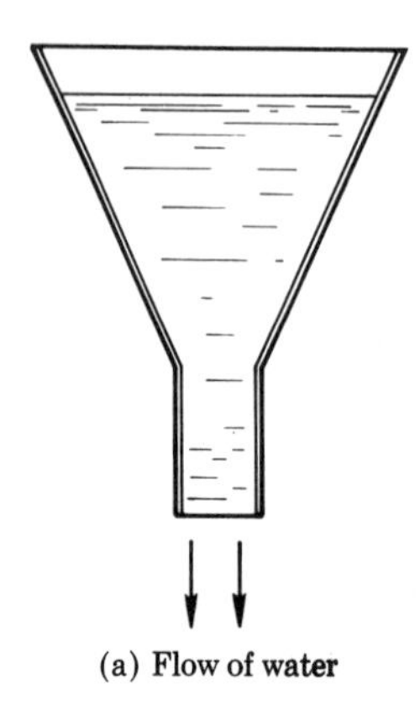

(a) Flow of water

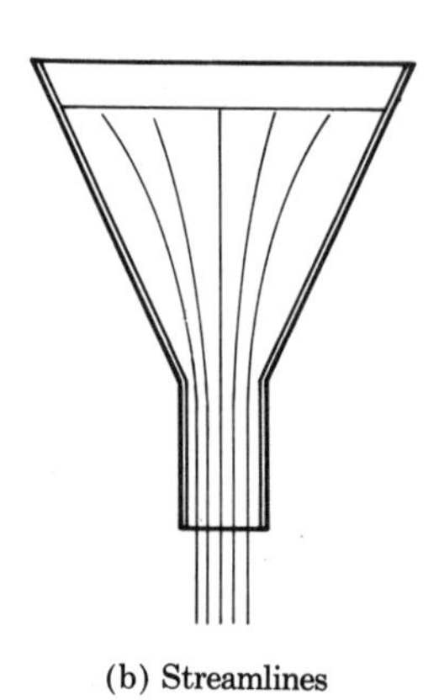

(b) Streamlines

FIGURE 13-18
Incompressible and streamline flow

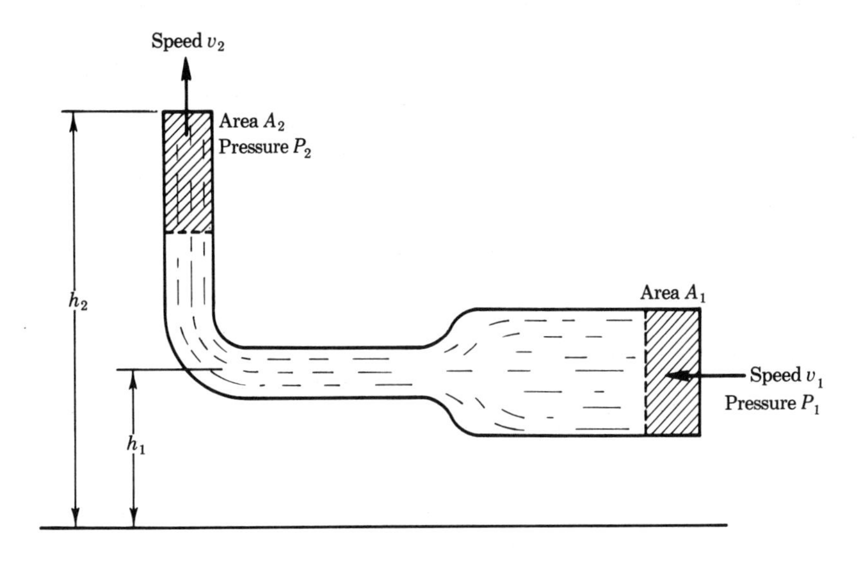

FIGURE 13-19

Derivation of Bernoulli's principle

equal volume v_2A_2 will leave at 2. The mechanical power supplied at plane 1 is $F_1v_1 = P_1A_1v_1$, while that removed at plane 2 is $F_2v_2 = P_2A_2v_2$. The energies (kinetic plus potential) in a unit volume are $\frac{1}{2}\rho v_1^2A_1 + \rho gh_1$ and $\frac{1}{2}\rho v_2^2A_2 + \rho gh_2$. The principle of conservation of energy states that the net power supplied is equal to the change in total energy occurring per second. Thus

$$P_1A_1v_1 - P_2A_2v_2 = (\tfrac{1}{2}\rho v_1^2 + \rho gh_1)v_1A_1 - (\tfrac{1}{2}\rho v_2^2 + \rho gh_2)v_2A_2$$

Using the fact that $A_1v_1 = A_2v_2$ and rearranging,

$$P_1 + \tfrac{1}{2}\rho v_1^2 + \rho gh_1 = P_2 + \tfrac{1}{2}\rho v_2^2 + \rho gh_2 \qquad \textit{(Bernoulli's principle)}$$

This theorem states that *the sum of the pressure and the total energy per unit volume is constant from point to point in a flowing fluid.*

Many special cases can be cited as illustrations of Bernoulli's principle.

1. *Pressure of a column.* For a fluid at rest, $v = 0$ everywhere, and the pressure difference between levels is equal to the weight of the column between them, as previously shown. To see this, we write

$$P_1 + \rho gh_1 = P_2 + \rho gh_2$$

and transpose to get

$$P_1 - P_2 = \rho g(h_2 - h_1)$$

or

$$\Delta P = \rho g\Delta h$$

which is the column weight, as derived earlier in Section 4-6 by a process of piling up cubes of matter. For example, the difference in pressure from sea level to the lowest point in the ocean off Hawaii, 35,630 ft or 10,860 m, is

$$\Delta P = \rho g \Delta h = (1025 \text{ kg/m}^3)(9.81 \text{ m/sec}^2)(10{,}860 \text{ m}) = 1.09 \times 10^8 \text{ N/m}^2$$

which is about 1000 times atmospheric pressure.

2. *Pressure in a moving column.* For a fluid flowing in a vertical pipe of constant cross-sectional area, the pressure difference is still $\rho g \Delta h$. To see this, note that the volume of fluid passing every level each second is constant; hence the speeds v_1 and v_2 are the same everywhere. The kinetic-energy terms cancel out in Bernoulli's equation.

3. *Pascal's principle.* For fluids at rest, $v = 0$, and at the same level (or nearly so) $h_1 = h_2$. The pressure P is the same at all points. This is *Pascal's principle,* dating back to 1647. A simple example of its use is in the hydraulic press, as shown in Figure 13-20. The application to the fluid of a small force f by a piston of area a gives a pressure $P = f/a$. This same pressure is transmitted by the fluid to another piston of area A, which then exerts a force $F = PA$ or

$$F = \frac{A}{a} f$$

By choosing suitable areas for the pistons, large weights can be lifted by relatively small forces. This is the principle of operation of the hydraulic jack and similar devices. For example, let $A = 0.5 \text{ m}^2$ and $a = 0.01 \text{ m}^2$. With a force as small as 20 lb, a weight of 1000 lb can be raised. In this use of Bernoulli's principle, the speed of the moving fluid is taken to be practically zero. Note that although we multiply forces, we do not get work for nothing. If the first piston moves a distance l, it does work $W = fl$ and moves a column of fluid of volume $V = la$. This same volume is swept out by the second piston, $V = LA$. The second piston moves a distance

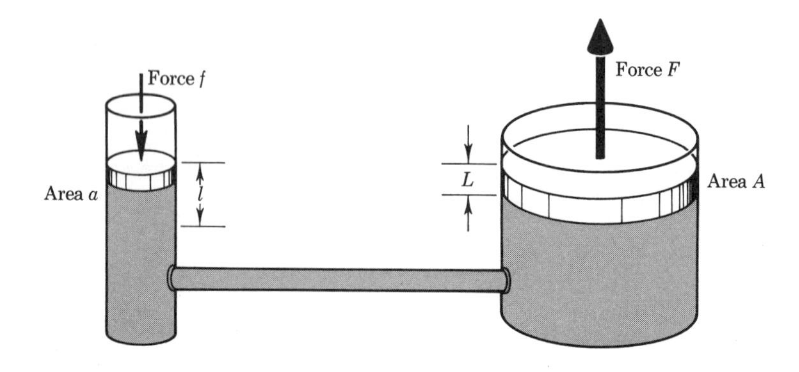

FIGURE 13-20

Hydraulic press

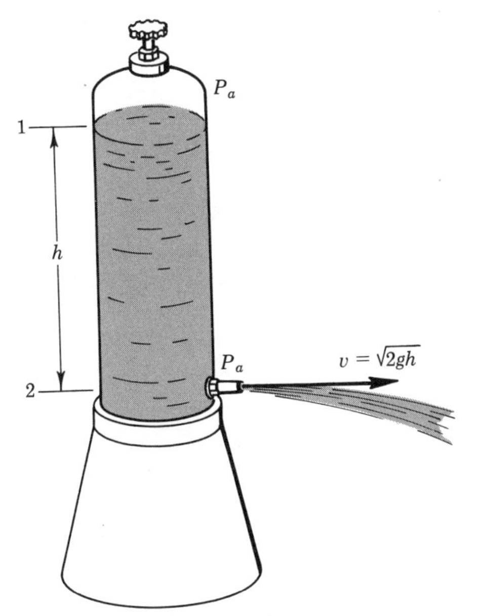

FIGURE 13-21

Flow from opening in bottom of tank—Torricelli's theorem

$$L = \frac{a}{A}l$$

The work output is $W = FL$, the same as the input fl, neglecting friction.

4. *Torricelli's theorem.* A liquid flowing from a hole in the bottom of a large tank of height h (Figure 13-21) has a speed $v = \sqrt{2gh}$. To prove this, consider point 1 at the top of the column, where the pressure is the atmospheric value $P_1 = P_a$ and $h_1 = h$. Let point 2 be at the opening where the fluid again experiences atmospheric pressure $P_2 = P_a$. We choose our coordinates so that $h_2 = 0$ at the bottom. Then, by Bernoulli's principle,

$$P_a + 0 + \rho gh = P_a + \tfrac{1}{2}\rho v^2 + 0$$

Simplifying,

$$v = \sqrt{2gh}$$

This is called *Torricelli's theorem* (1643). This speed is seen to be the same as if each particle of the fluid dropped freely under the acceleration of gravity; here, however, the velocity is horizontal rather than vertical. Conservation of energy as applied to a unit volume of fluid also gives the result directly. For example, the speed of flow of water out of the bottom of a column of height 2.5 m is $v = \sqrt{2gh} = \sqrt{2(9.8)(2.5)} = 7.0$ m/sec.

5. *Pressure–speed relation.* For flow along a single level, h constant, but with variable area of fluid along the path of flow, Bernoulli's principle gives

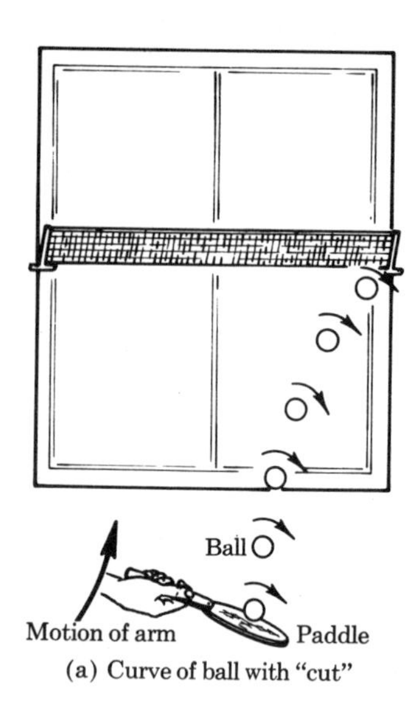

(a) Curve of ball with "cut"

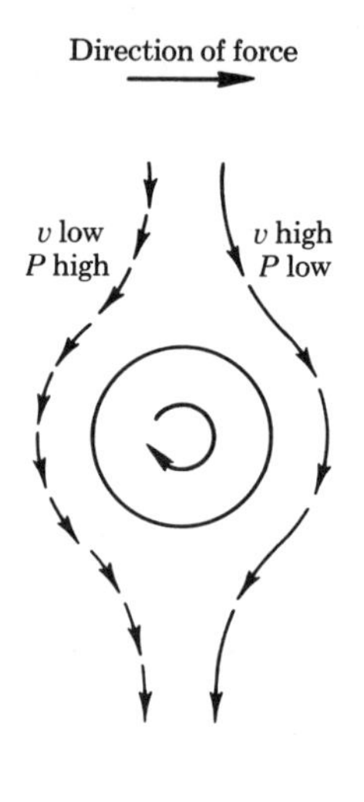

(b) Effect of rotation on air speeds

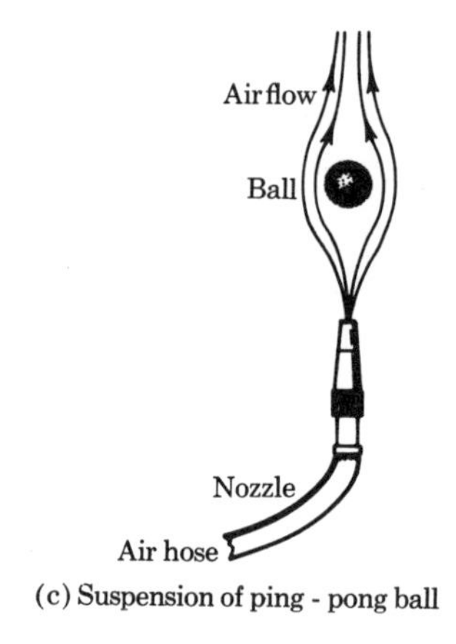

(c) Suspension of ping - pong ball

FIGURE 13-22

Illustrations of Bernoulli's principle

$$P_1 + \tfrac{1}{2}\rho v_1^2 = P_2 + \tfrac{1}{2}\rho v_2^2$$

which states that *where the speed is low, the pressure is high, and where the speed is high, the pressure is low.* The curving of a ping-pong ball if we "cut" the ball is explained by Bernoulli's principle. If a right-handed person strikes the ball with the paddle handle slightly forward, as in Figure 13-22(a), a clockwise rotation (as viewed from above) is set up. The air streaming past the sphere is slowed down by the surface motion of the left side. Thus there is a higher pressure on that side and the ball curves right—Figure 13-22(b). This conclusion may also be demonstrated by a simple laboratory experiment. If a fine stream of air is directed upward from a nozzle, as shown in Figure 13-22(c), a ping-pong ball will remain suspended by the stream. If the ball starts to fall off the jet, say to the right, the stream causes the ball to spin, which slows the stream on the right side. The pressure is higher there and the ball is forced back toward its original position.

Other illustrations of Bernoulli's principle are as follows:

6. *Flight of heavier-than-air machines,* first successfully achieved by the Wright brothers in 1903, is an example of an application of Bernoulli's principle. An airplane wing's cross-section or *airfoil* (shown in Figure 13-23), has a shape that favors a higher air speed over the top, hence a lower pressure. The pressure difference thus provides a *lift* to the wing. This supplements the vertical component of force due to the "angle of attack"—that is, the angle the wing is tipped from the horizontal. The passage of the air over all surfaces causes a *drag,* tending to prevent forward

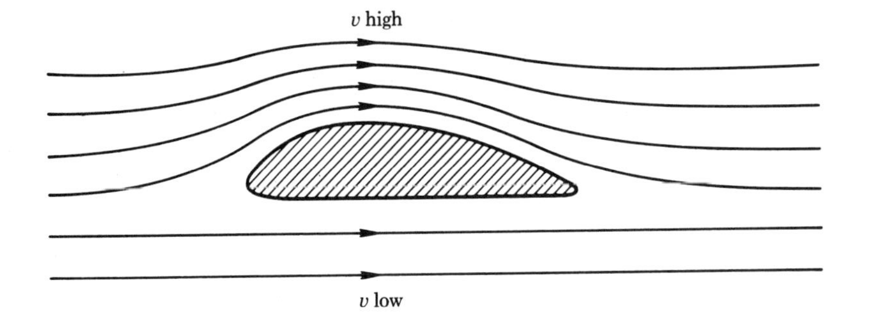

FIGURE 13-23

Airfoil in streamline flow

motion, and the relative efficiency of the plane depends on the ratio of the lift to drag forces.

7. *The air speed of a plane* can be measured continually by a Pitot tube, as shown in Figure 13-24. Air of speed v_1 passes over the holes in the outer tube but comes to rest $v_2 = 0$ when it enters the inner tube. The manometer measures the pressure difference $P = P_2 - P_1$, from which is computed the speed:

$$v_1 = \sqrt{\frac{2\Delta P}{\rho}}$$

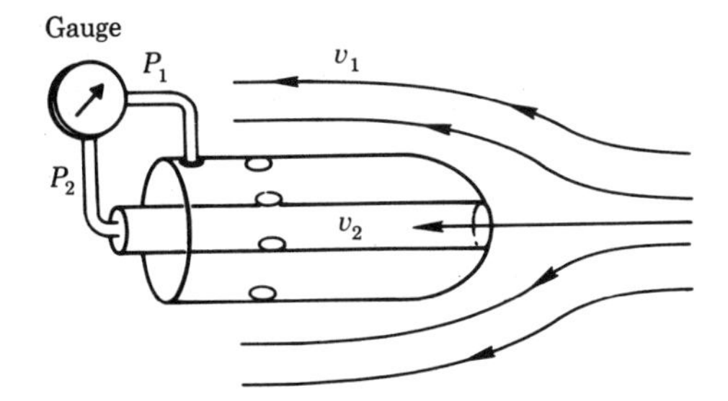

FIGURE 13-24

Pitot tube for measurement of air speed

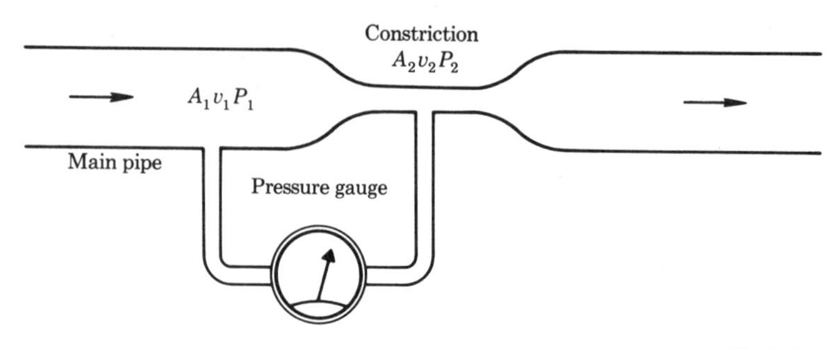

FIGURE 13-25

Venturi meter for measuring fluid flow rates

8. *The volume flow rate of a liquid,* $J = vA$, can be measured by the *Venturi meter*. A constriction is made in a pipe, and the difference in pressure between that point and the main stream is measured (Figure 13-25). Now, the volume flow rate vA is constant because the fluid is assumed to be incompressible. Thus $v_2 = v_1(A_1/A_2)$, and the application of Bernoulli's formula gives an expression for J (see Problem 13-32).

9. *The thrust of escaping rocket gases* (see Section 4-3), may be deduced by a simple argument. Gas at high pressure P and nearly zero speed escapes at high speed v through a nozzle into a near-vacuum, pressure essentially zero, as in Figure 13-26. Bernoulli's principle simplifies to $P = \frac{1}{2}\rho v^2$ for this situation. Now, the volume flow rate of gas through the nozzle area A is

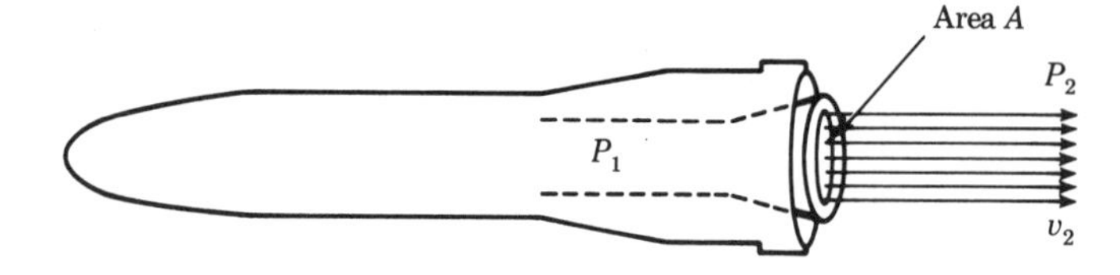

FIGURE 13-26

Rocket nozzle

vA, the mass flow rate is ρvA, and thus the momentum carried away from the rocket each second by the exhaust gases is $(\rho vA)v$. By Newton's law, this rate of change of momentum is identified with a force F on the rocket. Combining expressions, we find the thrust to be

$$F = 2AP$$

Let us find the pressure difference required in a rocket to get a thrust of 4×10^5 N with a nozzle area of 0.008 m², as the gas escapes to a vacuum:

$$\Delta P = \frac{F}{2A} = \frac{4 \times 10^5 \text{ N}}{2(0.008)} = 2.5 \times 10^7 \text{ N/m}^2$$

Since 1 atmosphere is 1.01×10^5 N/m², this corresponds to about 250 atm or 3700 lb/in.² of internal pressure.

10. *Relations between pressure and flow in water systems* such as shown in Figure 13-27 are simply calculated. Let the pipe diameters be 0.1 m and 0.04 m and the pressure P_1 be 6.8×10^5 N/m² (100 lb/in.²). Suppose that the total flow rate is $J = 0.02$ m³/sec, the same in both pipes since the water is almost incompressible. Let us find the speeds and P_2. Now, $A_1 = \pi r_1^2 = \pi(0.05)^2 = 0.00785$ m², $A_2 = \pi r_2^2 = \pi(0.02)^2 = 0.00126$ m². Thus $v_1 = J/A_1 = 0.02/0.00785 = 2.5$ m/sec and $v_2 = J/A_2 = 0.02/0.00126 =$

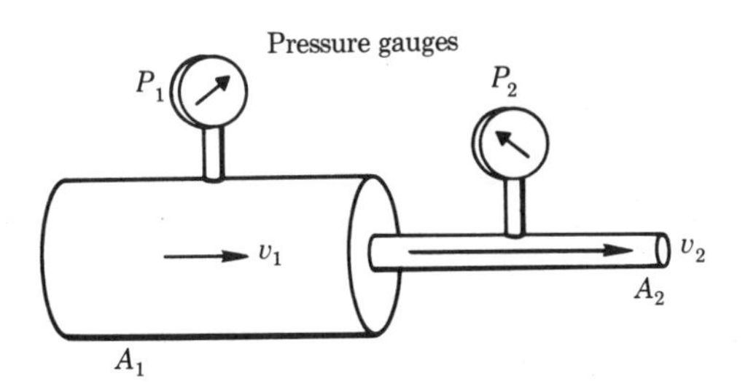

FIGURE 13-27

Pressure-speed relation in water pipes

15.9 m/sec. The density of water is 1000 kg/m³. Thus, by Bernoulli's formula,

$$\begin{aligned} P_2 &= P_1 + \tfrac{1}{2}\rho(v_1^2 - v_2^2) = 6.8 \times 10^5 + \tfrac{1}{2}(1000)[(2.5)^2 - (15.9)^2] \\ &= 5.6 \times 10^5 \text{ N/m}^2 \end{aligned}$$

or about 81 lb/in.².

13-7 THERMAL PROCESSES

In the several transformations of water described in Section 13-1, it was presumed that there was a source of thermal energy. As discussed in Section 8-8, the primary source of such energy is the Sun. However, wood, coal, oil, and natural gases are used for most of the energy needs of the world. The process of combustion to release thermal energy consists of burning, an oxidation process. We shall not examine these chemical reactions, but merely note some values of heat of combustion of various materials in Table 13-3. Natural gases contain varying amounts of ethane, methane, and propane, depending in part on the location of the well. We can readily see from Table 13-3 the advantage of oil or coal over wood, and understand why hydrogen is an excellent fuel for propulsion of a space rocket.

Let us now consider in more detail the process of heating a substance. As thermal energy is added, the temperature of the substance rises, as discussed in Sections 8-6 and 13-1. This implies an increase in the kinetic energy of the molecules. The energy absorbed goes into exciting vibrations of the system, not only of individual molecules but also of the structure as a whole. Because of differences in molecular structure, substances have

TABLE 13-3

HEAT OF COMBUSTION

MATERIAL	kcal/kg
Acetylene, C_2H_2	12,000
Coal	8,000
Ethane, C_2H_6	11,000
Gasoline	12,000
Hydrogen, H_2	60,000
Methane, CH_4	3,400
Propane, C_3H_8	23,000
Sugar, $C_{12}H_{22}O_{11}$	4,000
TNT, $C_7H_5O_6N_3$	1,800

different responses to the addition of heat energy. The term *specific heat c* is given to the amount of energy required to raise 1 kilogram of any material by 1°C. The value of c for water, by definition, is simply 1 kcal/kg-°C in the mks system.

For many engineering and industrial purposes, use is made of the *British thermal unit* (Btu), which is defined as the heat needed to raise the temperature of 1 pound of water 1°F. The following conversion factors are useful:

$$1 \text{ Btu} = 0.252 \text{ kcal} = 1055 \text{ J}$$

$$1 \text{ Btu/lb-°F} = 1 \text{ cal/g-°C} = 1 \text{ kcal/kg-°C}$$

Values of the specific heat of several well-known materials are listed in Table 13-4. If an amount of energy ΔU is added to an object of mass M and specific heat c, the temperature rise ΔT is obtained from

$$\Delta U = Mc\Delta T$$

TABLE 13-4

SPECIFIC HEAT OF SOLIDS AND LIQUIDS (0°C, 1 atm)

MATERIAL	c (kcal/kg-°C)
Water	1.00
Alcohol	0.6
Aluminum	0.21
Copper	0.094
Glass	≅0.12
Ice	0.5
Iron	0.11
Lead	0.031
Mercury	0.033
Sea water	0.93

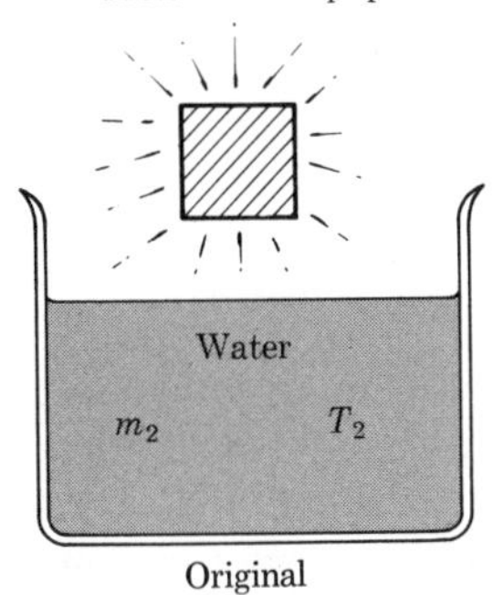

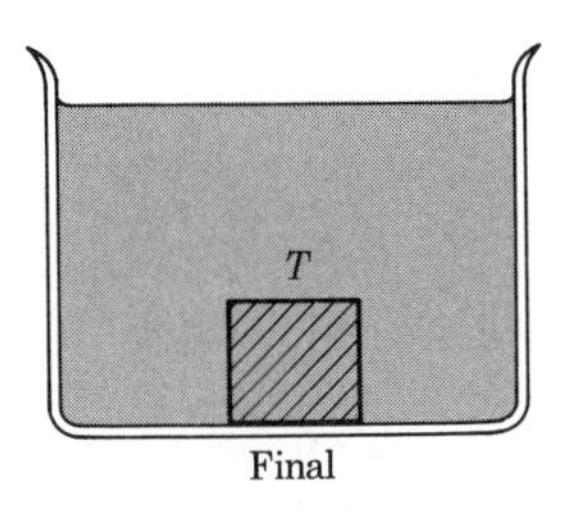

FIGURE 13-28

Experiment for measuring specific heat

The larger the mass of the sample, or the larger its specific heat, the greater the energy required to give a certain temperature increase. The effect of mixing two substances at different temperatures is easily predicted, since we can treat heat energy as a special manifestation of mechanical energy and require conservation in a closed system. Let us perform a simple experiment (Figure 13-28). We drop a piece of hot metal into an insulated container of cold water, and allow the combination to exchange heat, share energy, and come to equilibrium. The metal loses energy while the water gains energy, but the *total* remains the same and the system comes to a common temperature T. Let the mass, specific heat, and initial temperature be respectively m_1, c_1, and T_1 for the metal, and m_2, c_2, and T_2 for the water. The heat lost by the metal is $\Delta U_1 = M_1c_1\Delta T_1$, equal to that gained by the water $\Delta U_2 = M_2c_2\Delta T_2$. However, $\Delta T_1 = T_1 - T$ and $\Delta T_2 = T - T_2$. Thus

$$M_1c_1(T_1 - T) = M_2c_2(T_2 - T)$$

The final temperature is easily found to be

$$T = \frac{M_1c_1T_1 + M_2c_2T_2}{M_1c_1 + M_2c_2}$$

The quantity $C = Mc$, called the *heat capacity*, is useful as a measure of the ability of a certain *object* to absorb or deliver heat. For the metal–water mixture, the formula simplifies to

$$T = \frac{C_1T_1 + C_2T_2}{C_1 + C_2}$$

Suppose 1.5 kg of iron at temperature 80°C, specific heat 0.11 kcal/kg-°C, is placed in a beaker of 2 kg of water, 1 kcal/kg-°C, at 20°C. The heat capacities are $C_1 = (1.5)(0.11) = 0.165$ kcal/°C and $C_2 = (2)(1) = 2$ kcal/°C. Thus the final temperature is

$$T = \frac{(0.165)(80) + (2)(20)}{0.165 + 2} = 24.6°\text{C}$$

The *calorimeter* is a device for measuring specific heat by using this relation in reverse. The three temperatures are measured; two masses and the specific heat of H_2O are known; the unknown c is computed.

Heat Balances in Change of State

When a solid melts, the change into a fluid form is a process labeled *fusion*. A large amount of heat energy is absorbed at the single temperature, the melting point. No temperature change occurs, because all of the energy goes into breaking the bonds that hold the solid structure together. For each kilogram of material that melts, an amount of heat L_f, called the *latent heat of fusion*, is absorbed. Similarly, when a liquid boils, the transition from liquid to gas is a process called *vaporization*. At the single temperature, the boiling point, *a latent heat of vaporization* L_v is absorbed by each kilogram. Table 13-5 shows data on changes of state for such materials. For these transitions the heat absorbed by a mass M is $U = ML$.

The principle of conservation of energy may be used to calculate the effect of mixing materials where changes of state as well as temperature occur. A familiar experiment will illustrate. We drop an ice cube into a glass of water and allow it to melt. If the system is insulated, no heat will be gained or lost by the system. Let us calculate the final temperature when the ice originally is somewhat below the freezing point of water, and the water is at room temperature. We assign and tabulate the various numbers needed as follows:

	Ice (i)	*Water* (w)
Mass (kg)	0.04	0.25
Specific heat (kcal/kg-°C)	0.5	1.0
Temperature (°C)	−10	20

The ice absorbs heat in three steps—it warms to the temperature of fusion T_f (0°C), it melts at that point, and as water it warms (with specific heat c_w) to a final temperature T. Its total energy change is

$$\Delta U_i = M_i c_i (T_f - T_i) + M_i L_f + M_i c_w (T - T_f)$$

The water in the glass cools, and gives energy to the ice of amount

$$\Delta U_w = M_w c_w (T_w - T)$$

Conservation of energy states that $\Delta U_i = \Delta U_w$, from which we can find T. For our example,

$$T = \frac{M_w c_w T_w + M_i[c_w T_f - c_i(T_f - T_i) - L_f]}{(M_w + M_i)c_w}$$

Substitution of numbers including $L_f = 79.7$ kcal/kg, gives a final temperature of 5.5°C.

TABLE 13-5
DATA ON CHANGES OF STATE
(at 1 atm)

Material	Melting Point (°C)	Heat of Fusion (kcal/kg)	Boiling Point (°C)	Heat of Vaporization (kcal/kg)
Water	0	79.7	100	539
Helium	*	*	−269	6.0
Hydrogen	−259	15	−253	107
Nitrogen	−210	6.2	−196	48
Oxygen	−219	3.3	−183	51
Acetone	−95	22	56	125
Mercury	−39	2.7	357	71
Sodium	98	32	892	10200
Lead	327	6.3	1700	220
Aluminum	660	93	2467	2510
Tungsten	3370	44	5900	1200

* Not solid at 1 atmosphere pressure

We now have an appreciation of the numerical description of thermal processes in materials as of the middle of the nineteenth century. At the time, although there was evidence that matter was made up of particles, the full connection between thermal properties of matter in bulk form with atomic and molecular processes was yet to be established. In the next chapter we shall make the connection between the previously well-known science of mechanics, the molecular concept of matter, and observations on thermal processes.

PROBLEMS

13-1 A steel guitar wire of diameter 0.8 mm is stretched from length 0.2 m to 0.21 m. How much tension force results?

13-2 Determine Young's modulus E_y for the element lead (Pb) from the following experimental data: wire diameter 2 mm, original wire length 0.8 m, elongation with 4-kg mass attached 0.67 mm. How does the elastic behavior of Pb compare with that of steel?

13-3 Calculate the atomic force constant K_1 for copper from the observed stretching of 0.2% in length with a force per unit area of 2.2×10^8 N/m² and atomic spacing 3.6×10^{-10} m.

13-4 Find the force constant K, the elongation, and potential energy stored in a steel telegraph wire of diameter 2 mm and length 150 m if it is in tension at 6×10^7 N/m².

13-5* (a) A cube of material of original side L and Poisson's ratio ν is placed in a vise and a force F is applied. Show that the fractional change in volume is $\frac{1}{3}$ of that if pressure were applied to all faces. (b) On applying a vise force of 400 kgf to a cube of glass 1 cm on a side, how much volume change occurs?

13-6 Copper transatlantic communication cables are laid on the bottom of the ocean (specific gravity 1.025) at an average depth of 2 miles. By what fraction would the volume of the wire be reduced because of the water pressure?

13-7 How large a gap between adjacent 25-ft lengths of steel railroad track would be needed if the temperature ranges from $-30°F$ to 150°F?

13-8* The marks on a steel tape are correct at 20°C. The distance between two points in the desert where the temperature is 42°C appears to be 150 m. What is the true distance?

13-9 An aluminum collar of inside diameter 3 cm fits very tightly on a steel pipe. The joint is heated to permit the collar to be removed easily. How much increase in temperature is needed to give a clearance of 0.1 mm?

13-10 A tight fit between a disc of radius r and a metal rim of inner radius R can be obtained by first *cooling* the disc. Show that the temperature change needed is $T = \dfrac{(R/r)^2 - 1}{2\alpha}$. *Hint:* Consider expansion of the area.

13-11 Find the angle of deflection in degrees of a thermostat composed of 0.001-m-thick strips of Al and Fe, length 0.02 m, when heated by 100°C. Note the relation in Section 13-3.

13-12 Will a swimmer float better in very cold water or very warm water the instant he dives in? Explain. What will happen as he cools off or heats up?

13-13* An aluminum container of volume 5×10^{-4} m^3 is completely filled with water and heated from 20°C to 24°C. (a) How much expansion in the volume of the container is there? (b) How much *actual* expansion of the water takes place? (c) What amount of pressure on the wall of the vessel is developed?

13-14 A mercury thermometer is constructed with a spherical Pyrex-glass bulb of radius 0.3 cm and connected to a column of inside radius 0.10 mm. When the bulb at 20°C is inserted in boiling water at 100°C, what rise in the mercury level of the tube will occur?

13-15 A fixed volume of argon gas at atmospheric pressure (atomic weight 39.94) is heated from 0°C to 100°C. How many atmospheres pressure on the walls of the container does it exert? How many atoms per unit volume are there?

13-16 A 3-liter volume of air at atmospheric pressure is heated from 0°C to 400°C. If the container does not expand, by what factor does the pressure increase, and what is its actual value in N/m^2?

13-17 Calculate the actual reading and percentage change for a mercury barometer at atmospheric pressure 760 mm, if the temperature changes from 20°C to 35°C. Assume no change in the size of the tube.

13-18 A balloon of mass 60 kg used for cosmic-ray observations in the upper atmosphere has a possible volume of 1000 m^3, but has on launching only 200 m^3 of helium at NTP (see Section 13-2). Assuming that there is no change in gas temperature, but that the air density varies with height in kilometers as $\rho_0 e^{-0.126h}$, how high will the balloon go?

13-19 Suppose that Archimedes's measurements on a certain crown were 0.60 kgf in water and 0.64 kgf in air. Was the crown pure gold or not?

13-20 An ice cube is placed in a drinking glass which is then filled to the top with water. When the cube melts, how much water overflows? Explain.

13-21* A "hard-hat" diver plus gear appears to have a weight of 90 kgf in fresh water, but a weight of 85 kgf in sea water. (a) What is the actual weight and mass? (b) A crank-and-wheel arrangement is used to pull him up (see Figure P13-21). What force must be applied to the cable to bring him up to the surface, from 10 m down in salt water, in 9 sec? (c) Discuss any effects that have been ignored.

13-22 We can construct a crude hydrometer (such as used to test the specific gravity of storage-battery solutions) by driving a nail into the end of a piece of wooden dowel rod giving a total mass of 1 gram. If the rod has 0.5-cm diameter, how far apart should marks be placed to indicate "oil" and "water"?

FIGURE P13-21

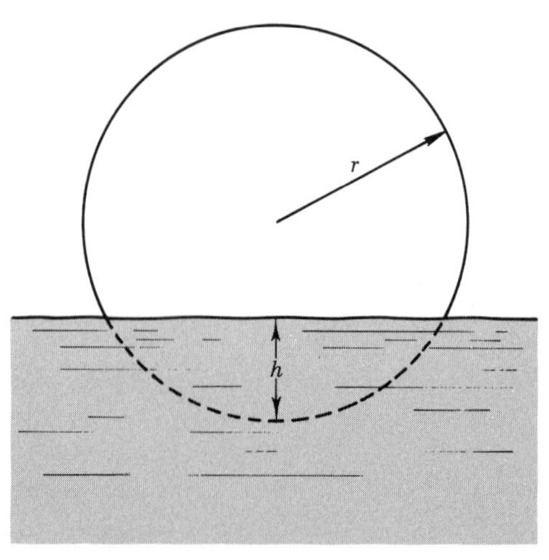

FIGURE P13-25

13-23 What percentage error is introduced by neglecting the buoyant force of air in weighing a sample of gold, using balancing brass weights? If we bought 10 troy ounces (12 per pound) at \$36 per ounce, as weighed by our scales, how much value would we lose or gain?

13-24 We weigh an evacuated container, then admit air, and weigh again, noting an increase in weight. Using Bernoulli's principle, what relation is there between internal pressures of the gas at the top and bottom of the vessel?

13-25 To weigh a ping-pong ball, we find how much it depresses the surface of water (see Figure P13-25). If the ball diameter is 3.7 cm and the depth it sinks is 0.7 cm, what is its mass? *Note:* the volume of a spherical segment is $(\pi/3)h^2(3r - h)$.

13-26 A hollow metal sphere of mass m and volume V is released at a depth h in a body of water. How soon does it reach the surface, neglecting friction?

13-27 On diving into a swimming pool, buoyant forces ensure that we pop up to the surface immediately. Calculate the force and acceleration on a 50-kg boy, specific gravity 0.95 with breath held, and the time that it takes to come up from a depth of 3 m.

13-28* Water of density 62.4 lb/ft^3 flows in a "main" of cross-sectional area 25 in.^{2} at pressure 60 lb/in.^{2} and flow rate 30 lb/sec into a pipe of area 2 in.^{2} inside a house. (a) What is the speed of the water in the main? (b) In the pipe? (c) What is the pressure in the house, assuming no friction? *Note:* Density must be in slugs/ft^3 for kinetic-energy calculations.

13-29 A hydraulic jack has piston diameters of 2 cm and 0.15 cm. How much force must be applied to lift the front end of a car of mass 750 kg? How many strokes in a cylinder of length 0.3 m will be needed to lift the car a distance of 6 cm?

13-30 Find the absolute pressure at the bottom of a water reservoir of height 50 m. With what speed will water flow into an open pipe at the ground level?

13-31* Hydrogen gas at 100 times atmospheric pressure and at temperature 2000°C is used to propel a nuclear rocket. (a) How does the density compare with the standard value of 0.09 kg/m³? (b) If the pressure drops to the vacuum of space, and the average density in the nozzle is one-half the initial value, what is the exhaust speed? (c) What will be the amount of thrust for a nozzle of area 0.008 m²?

13-32 Show that the volume flow rate, as measured by a Venturi meter, is $J = C\sqrt{\Delta p}$ where the constant is

$$C = \left[\frac{\rho}{2}\left(\frac{1}{A_2^2} - \frac{1}{A_2^2}\right)\right]^{-1/2}$$

13-33 An idea is advanced for the storage of radioactive wastes at the bottom of the sea. Water containing small amounts of highly active elements is to be mixed with a liquid slightly denser than water, sealed in flexible plastic bags, and dropped overboard from a ship. Investigate this idea, taking into account the variation in pressure in water with depth, the compressibility of water and the liquid, and buoyancy effects.

13-34 A metal sample of mass 500 grams is heated to 100°C by immersion in boiling water, then put in a calorimeter containing 200 grams of water, $c = 1.0$, at its freezing point. They eventually come to a common temperature of 20°C. What is the specific heat of the metal and of what element does it appear to be composed?

13-35 Samples of three phases of water are mixed: 1 kg of steam at 100°C, 1 kg of ice at 0°C, and 50 kg of water at 20°C. What is the final temperature? Which phase contributed most to the result?

13-36 In a steam plant for producing electricity, if all the heat of combustion of a metric ton (1000 kg) of coal is used to bring water from 20°C to boiling and to vaporize it, how many kilograms of steam will be produced?

MOLECULAR THEORY OF GASES

14

The behavior of a single particle under the influence of a force, we have seen, is rather easy to describe using the methods of mechanics. A collection of rigidly connected particles admits study through the concepts of center of mass and rotational inertia. When we turn to bulk matter, with its fantastic number of particles, a detailed description of the history of each particle would become hopelessly complex, and we are forced to look at the *average* behavior of the large assembly. Phenomena taking place in systems of particles are understandable only in terms of the laws of probability—that is, from a statistical point of view.

In this chapter we shall first demonstrate how difficult is a detailed mechanical description because of the number of particles involved, and introduce the *kinetic theory* as it refers to a gas. This simple statistical picture permits us to understand the particle origin of pressure, the relation between pressure, volume, and temperature, and the connection between heat and temperature. In the next chapter, we shall take up the science of *thermodynamics*, which helps specify what natural processes are possible, and the still more general *statistical mechanics*, which embraces both previous subjects.

14-1 PARTICLES, INTERACTIONS, AND THE KINETIC THEORY

There is a very *large number of particles* in a volume as small as a cubic centimeter of water. In this mass of 1 gram, there are about 3.3 ×

10^{22} molecules of H_2O, each of which contains 3 atoms. The hydrogen atoms are relatively simple, each with only the proton and the electron; but the oxygen atom has 8 electrons, and in the nucleus, 8 protons and 8 neutrons. If we add up all these numbers, in 1 cm³ of water we have roughly 8.6×10^{23} particles to deal with. The number of *pairs of interactions* of particles is then enormous. Figure 14-1 shows how pairs increase as the number N of particles increases. The arithmetic sum of the numbers up to $N - 1$ is seen to apply:

$$\text{pairs} = 1 + 2 + 3 + \ldots + (N - 1) = \frac{N(N-1)}{2}$$

Thus for 1 cm³ of water there are around 7.4×10^{47} pairs of interactions.

There are *several types of force* acting between the particles, as we have already discussed. The electrostatic interactions of nuclei and electrons are dominant forces, although gravitational forces are present. The steady and time-dependent currents of various charges give rise to additional electrical and magnetic effects. Forces that have no classical analogy are present because of the wave-mechanical nature of matter. Nuclear forces are involved in the structural stability of the groups of neutrons and protons.

Each of the particles has some degree of possible *motion in space*. In addition to the x-, y-, and z-coordinates, there are the components of velocity in the three directions. Translational, rotational, and vibrational degrees of freedom exist. As will be shown in Chapter 22, one can define the coordinate positions only within certain limits set by wave mechanics.

Externally imposed forces on the surface and body of a sample of material may cause internal changes. Involved here again are electrical and magnetic fields, steady or varying with time (including electromagnetic waves), gravity, mechanical forces, and heat energy.

To achieve any order at all in the physical description of nature under such circumstances is admittedly difficult. We must select those processes that are more important than others, and develop simplified models with which experimental measurements can be compared. There are two opposite approaches to the problem of describing the action of matter composed of very many particles. One, as we used in Chapter 13 for forces in solids,

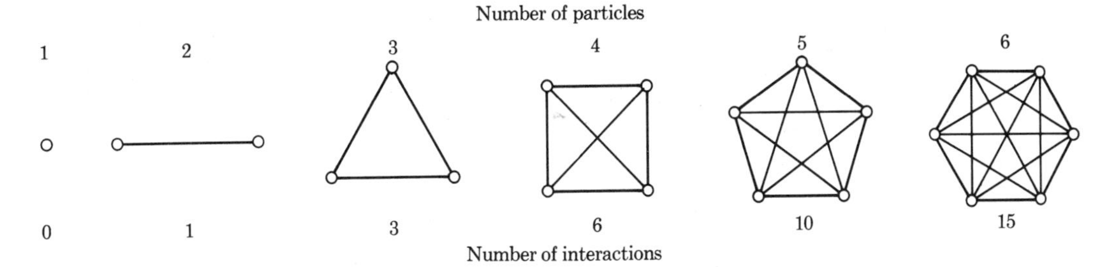

FIGURE 14-1

Interactions of particles

is to attempt to analyze the behavior of successively more complex structures—the atom, the simple molecule, the group of molecules. Considerable success has been met through this avenue, but the mathematics becomes unwieldy even when serious simplifying assumptions are made.

The alternate approach is to accept the fact at the beginning that there is such a very large number of particles that the medium can be viewed as almost uniform and continuous. Then we may take account of the known particle interactions, but use large-scale (macroscopic) statistical averages rather than small-scale (microscopic) details.

Kinetic Theory of Gases

Around the middle of the nineteenth century, the theory of gases, based on the idea of individual particle motion, was developed by the work of Maxwell, Boltzmann, Clausius, and others. The model they used was very simple. Gas atoms or molecules were considered as small hard spheres that bounced off each other and off the walls of a containing vessel. The total amount of energy within the gas was taken to be the sum of the kinetic energies of individual particles. Observation of trends and accurate measurements on gases made by many investigators had led to several definitions of gas properties and relations between them. The three principal quantities were: volume, pressure, and temperature. The meaning of *volume* is obvious, as the space within the retaining boundaries of the sample of gas. *Pressure* is defined as the force on a unit area of the wall. Physically, this force is attributed to the continual bombardment of the wall by individual particles, and the resultant transfer of momentum upon reflection. We shall consider temperature for the moment as *the quantity observed by a device such as a thermometer*. Such a definition cannot be expressed in terms of fundamental quantities length, mass, and time, but must remain an *operational* definition. Shortly, we shall be able to associate increases in temperature with increases in the average motion of the molecules of a substance and we can specify absolute zero as a point at which molecule motion is negligible. The individual particles are in continual motion and exchange energy by colliding with each other. As discussed in Section 10-2, a high-speed particle can impart some of its energy to a low-speed particle by elastic collision. At any instant there will be many speeds represented, but few of either very high or very low values.

We shall concentrate our attention on a state of *thermal equilibrium*, which means that there are no changes with time in the large-scale view of the system. At all times, there is assumed to be the same number of particles and the same average energy in each element of volume. For example, if we introduce hot coffee into a *perfectly* insulated Thermos bottle and wait until all motion resulting from pouring settles out, we could consider the liquid to be in equilibrium. There must be no exchange of material or energy with the surroundings. The last qualification is necessary to distinguish equilibrium from the *steady state*, where there are flows in and out of the region under consideration and constant conditions inside.

14-2 PRESSURE OF A GAS

We have already discussed (Section 8-3) the concept that heat energy is associated with molecular motion, and that the higher the temperature, the greater the kinetic energy of the molecules. Also, we recall (from Section 13-4) that the temperature, pressure, and volume of a gas are related experimentally by the gas law $PV = NkT$. We should now like to examine these ideas on a particle basis by analyzing pressure, a large-scale effect, in terms of motions of individual molecules.

The kinetic theory gives us a means of predicting the pressure on the wall of a container of gas (Figure 14-2). Every time a particle rebounds from the surface, it experiences a change of momentum, which by the laws of mechanics provides a force. The amount of such force is easy to compute for a particle striking the surface perpendicularly with speed v. The change in particle momentum is

$$mv - (-mv) = 2mv$$

If the particle strikes at an angle, only the x-component of momentum is changed, and v_x would appear instead of v. To find the pressure, it is necessary to add up the effects of all the particles with their various directions and speeds. This complicated process can be avoided if we imagine a special gas in which all particles have the same speed v, and are merely bouncing back and forth in the three directions x, y, and z, with $\frac{1}{6}$ of them going in, say, the positive x-direction. Suppose that N is the total number of molecules in the container of volume V. Then the number per unit volume is

$$n = \frac{N}{V}$$

The particle-current density associated with each direction is $(n/6)v$, which is also the number striking the wall per second. The pressure is the total change in momentum of all the particles in the stream; i.e.,

$$P = \frac{nv}{6} 2mv = \tfrac{1}{3} nmv^2$$

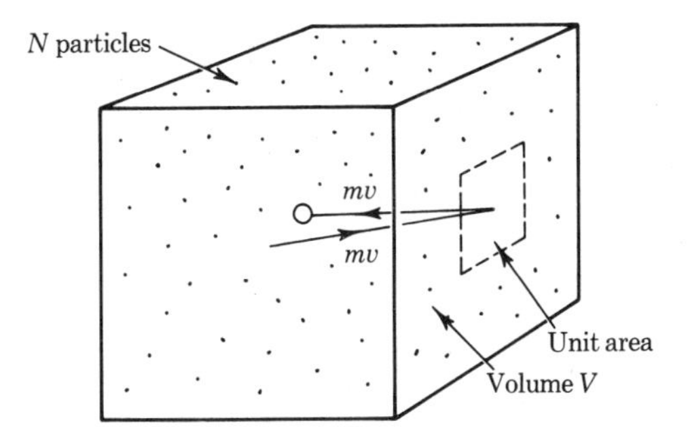

FIGURE 14-2

Container of gas under pressure

However, the kinetic energy of each of the particles of common speed is

$$E_k = \tfrac{1}{2}mv^2$$

leading to the expression

$$P = \tfrac{2}{3}nE_k$$

Comparing with the experimental law in the form

$$P = nkT$$

we see at once that kinetic energy of translational motion and temperature are directly related:

$$E_k = \tfrac{3}{2}kT \qquad \textit{(relation of kinetic energy and temperature)}$$

Since this expression does not contain the mass of the particle, it holds for all gases at the same temperature. Also, if $E_k = \frac{1}{2}mv^2 = \frac{3}{2}kT$, then $v = \sqrt{3kT/m}$.

Let us calculate the energy for a particle at room temperature, 20°C or 293°K. Then,

$$E_k = \tfrac{3}{2}(1.38 \times 10^{-23}\ \text{J/°K})(293\text{°K}) = 6.07 \times 10^{-21}\ \text{J}$$

This is, of course, an extremely small number. For 1 m^3 of a gas at NTP, containing 2.7×10^{25} molecules, the total internal energy of translation is

$$(2.7 \times 10^{25})(6.07 \times 10^{-21}) = 1.64 \times 10^5\ \text{J}$$

In order to appreciate the magnitudes of particle speeds, let us compute v for air, composed of about 20 per cent oxygen and 80 per cent nitrogen. For convenience, we regard the mixture as consisting of a single type of molecule of molecular weight $(0.2)(32) + (0.8)(28) = 28.8$. Then, the particle mass would be

$$m = (28.8)(1.67 \times 10^{-27}\ \text{kg}) = 4.81 \times 10^{-26}\ \text{kg}$$

At room temperature, $T = 293$°K,

$$v = \sqrt{\frac{3kT}{m}} = \sqrt{\frac{3(1.38 \times 10^{-23})(293)}{4.81 \times 10^{-26}}} = 502\ \text{m/sec}$$

We note that this is comparable to the speed of sound in air, found to be 343 m/sec at room temperature (Section 12-2). This general agreement as to order of magnitude might be expected, since it is by individual particle motion that signals of compression and rarefaction are sent. From such numbers we can also acquire a feeling for the size of microscopic effects. For example, the number of particles that strike a square meter of the container wall each second is

$$\frac{nv}{6} = \frac{(2.7 \times 10^{25})(502)}{6} = 2.26 \times 10^{27}/\text{m}^2\text{-sec}$$

The force of each particle is

$$2mv = 2(4.81 \times 10^{-26})(502) = 4.83 \times 10^{-23}\ \text{N}$$

It is clear why we cannot feel the individual air molecules bombarding our skin—the force of each is extremely small.

14-3 IDEAL AND REAL GASES

The general gas law is often called the "ideal" or "perfect" gas law, since it is mathematically rigorous for hypothetical molecules that occupy negligible volume and interact elastically only at the exact instant of col-

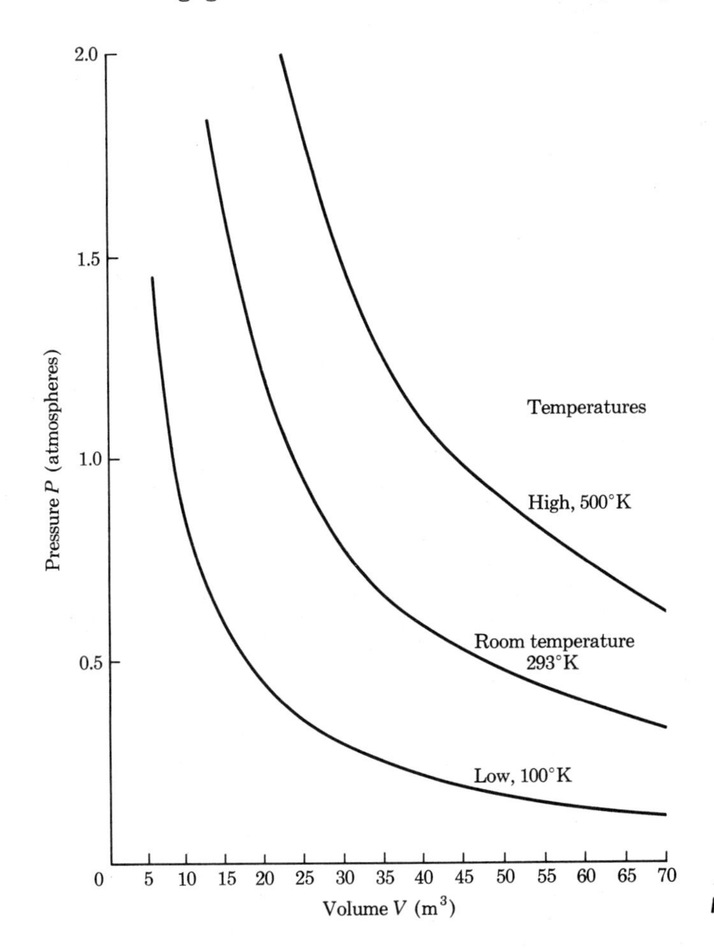

FIGURE 14-3

Plot of the gas law $PV = NkT$

lision. In spite of such restrictions, the general gas law is well verified in the laboratory for gases at low density and high temperatures. The dependence of pressure on volume is plotted in Figure 14-3 for several temperatures, using the general-gas-law formula.

For many problems it is desirable to use the *kilomole* (kmole) as the basic quantity of material in working with the gas law. One kilomole contains Avogadro's number $N_a = 6.02 \times 10^{26}$ of particles, regardless of the type of atom or molecule. The number of particles in n kmoles is nN_a; hence the gas law reads

$$PV = nN_a kT$$

or

$$PV = nRT$$

where $R = N_a k = (6.02 \times 10^{26})(1.38 \times 10^{-23}) = 8314$ J/°K-kmole is the *general gas constant*. The constant R is often expressed in calories. If we recall that 1 kilocalorie is 4185 J, then

$$R = \frac{8314 \text{ J/°K-kmole}}{4185 \text{ J/kcal}} = 1.986 \text{ kcal/°K-kmole}$$

(For rough calculations, the number 2 is good enough.)

We can readily find the pressure for our example of air at room temperature. In 1 kmole $N = N_a = 6.02 \times 10^{26}$, volume $V = 22.4 \times 10^3$ liters $= 22.4$ m³, we then have

$$P = \frac{(6.02 \times 10^{26})(1.38 \times 10^{-23} \text{ J/°K})(293\text{°K})}{22.4 \text{ m}^3}$$
$$= 1.09 \times 10^5 \text{ N/m}^2$$

The kinetic view of pressure allows us to understand *Dalton's law of partial pressures,* which states that the total pressure of a mixture of gases is the sum of the pressures that would be exerted by each component gas if it alone were present. The simplest case involves two gases—say, nitrogen and oxygen. If nitrogen alone at temperature T fills a container of volume V, then $P_N V = N_N kT$; if only oxygen were present instead, $P_O V = N_O kT$. Now, if N_N and N_O molecules at temperature T are present in a container, the momentum transfer of each type on collision with the walls gives rise to pressures P_N and P_O given by $N_N kT/V$ and $N_O kT/V$, using the argument for one gas only. The total pressure is

$$P = P_N + P_O = (N_N + N_O)\frac{kT}{V}$$

The law is as accurate as the perfect gas law, and can be applied to any number of gases, so long as there is no chemical action that results in combination to form new molecules.

Relation of Vapor Pressure and Temperature

We now examine the connection of pressure and temperature in a system that consists of both a liquid and a vapor "phase." Picture a closed container with water in the bottom and, at the start, a vacuum in the space above it. Molecules in the interior of the liquid experience no net force since they are surrounded by others, but those near the surface have a net component of force downward. Some molecules have enough kinetic energy to escape from the liquid. The higher the temperature, the more particles have adequate energy, and thus the greater the rate of this vaporization. On the other hand, molecules of water vapor above the surface move about freely but on striking the liquid are readily caught.

Thus the rate of this condensation is dependent on the number of particles present—i.e., the vapor pressure. A balance of the two processes is soon established, such that *the vapor pressure in the space is determined by the temperature of the liquid.* Table 14-1 shows this relation for water. The

TABLE 14-1

VAPORIZATION OF WATER

Liquid Temperature (°C)	Vapor Pressure (mm Hg)
−10	2.0
0	4.6
10	9.2
20	17.5
70	234
90	526
100	760
120	1490
200	11600

table helps us understand some familiar liquid–vapor effects. Suppose there are means for keeping the vapor pressure constant, say at 234 mm Hg (millimeters of mercury). If heat is continually supplied to the liquid, its temperature will rise to the 70°C value shown in the table and eventually all of the liquid will vaporize.

If instead we heat water in an open container with air above the liquid, at atmospheric pressure 760 mm, and the temperature below 100°C, relatively slow evaporation *from the surface* will occur. On heating to 100°C, vapor bubbles with 760 mm inside pressure can develop *within the liquid,* and the boiling process occurs. No further temperature increase is needed. If the atmospheric pressure is lower, as on a mountain top—e.g., 526 mm—boiling will occur at 90°C, according to the table. Conversely, if we constrain the vapor, as in a pressure-cooker, letting the pressure go to 1490 mm, boiling would occur at 120°C. This higher temperature of course permits more rapid cooking of food.

Real Gases

The perfect gas law $PV = NkT$ is accurate for real gases at low density, but fails at high density, where the volume occupied by molecules is not negligible. The finite size of molecules and their mutual forces acting over large distances become important. A formula suggested by van der Waals provides semiempirical correction terms to P and V. First, the space available for particles to move about is less than V by the effective volume b occupied by the molecules*; hence one replaces V by $V - b$. Second, intermolecular attractions tend to hold particles together and thus reduce the pressure they can exert on the container walls. If the volume of gas is

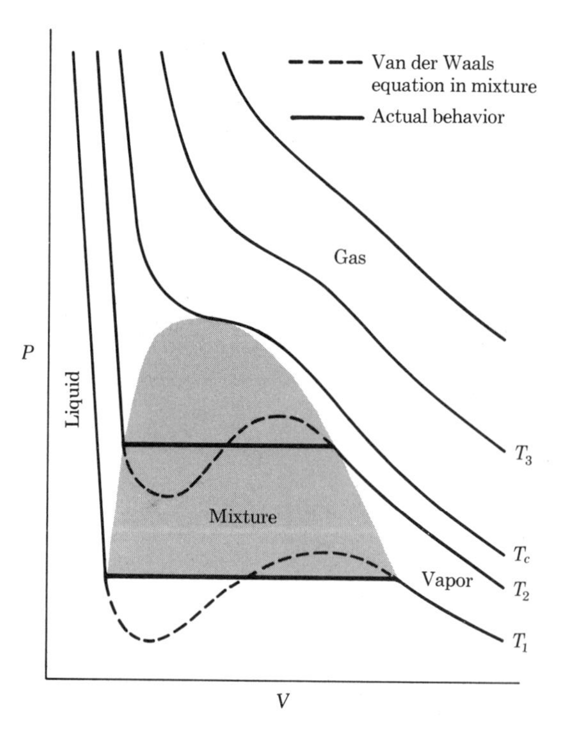

FIGURE 14-4
Isothermals for a real gas

halved, the number of such force bonds across any unit area is increased by a factor of four (since there are twice as many particles on each side). Thus van der Waals added a correction a/V^2 to the pressure P. His formula was then

$$\left(P + \frac{a}{V^2}\right)(V - b) = RT$$

* When a particle of radius r collides with another, their centers can come no closer than $2r$, which corresponds to a volume $\frac{4}{3}\pi(2r)^3$. Since two particles are involved, the excluded volume per molecule is one-half this, and $b = N_a 16\pi r^3/3$ for a kilomole containing N_a molecules.

where a, b, and R are to be regarded as empirical constants. Typical PV plots at various fixed temperatures T_1, T_2, etc. are shown in Figure 14-4. The curves are called isothermals ("equal temperature").

Let us trace the curve labeled T_1, starting at the far right with a large volume and low pressure. As the pressure is increased, the volume drops until it reaches the edge of the shaded region. Liquid and vapor exist together in equilibrium in this zone. With care, the substance can be made to follow a part of the van der Waals curve, but the situation is unstable. Normally, there is a large reduction of volume (and a corresponding rise in density) with little change in pressure as the gas condenses into a liquid. Then a very sharp rise in pressure is required to compress the liquid into a smaller volume.

For a higher temperature such as T_2, the mixture zone is narrower. At a critical temperature T_c the gas is transformed into a liquid without sudden change of density. The peak of the mixture zone is the *critical point* characterized by P_c and T_c.

By comparing the properties of the experimental gas curves and the van der Waals formula, the constants a, b, and R are found, leading to information on molecular size and forces. The formula is only approximate, since no single set of constants will fit all gases over a wide range of conditions.

At ordinary temperatures and pressures, the corrections to the perfect gas law are small. For instance, in N_2 (critical-state values $P_c = 33$ atm, $V_c = 0.12$ m^3 per kmole, $T_c = 126$°K), the correction to pressure at NTP is found to be about 0.2% while that to volume is 0.5%.

14-4 THE STATISTICAL BASIS OF SPECIFIC HEAT

The relation between particle energy and temperature enables us to find the theoretical value of the *specific heat*, which we recall from Section 8-6 to be the amount of energy that must be added to raise the temperature of 1 kilogram by 1°C. If the kinetic energy of translation of 1 molecule is $E_k = \frac{3}{2}kT$, then the total internal energy of N molecules is

$$U = NE_k = N\tfrac{3}{2}kT$$

To effect a temperature rise ΔT, the amount of energy that must be added is $\Delta U = N\frac{3}{2}k\Delta T$, but this must also be the same as the energy change on a large-scale basis, $\Delta U = cM\Delta T$. Comparing and letting $M = Nm$, where m is the mass of one particle, we see that

$$c = \frac{3}{2}\frac{k}{m} \qquad \text{(specific heat of a monatomic gas)}$$

This provides a means of determining specific heat on an atomic basis. Let us compute its value for argon, with molecular weight 39.4, and mass per atom

$$m = (39.4)(1.67 \times 10^{-27}\ \text{kg}) = 6.58 \times 10^{-26}\ \text{kg}$$

Then,

$$c = \frac{3}{2}\frac{k}{m} = \frac{3}{2}\frac{1.38 \times 10^{-23}\ \text{J/°C}}{6.58 \times 10^{-26}\ \text{kg}} = 315\ \text{J/kg-°C}$$

Now we shall investigate the specific heat of a more complicated molecule, as in a diatomic gas such as H_2, O_2, or NO. As in Chapter 9, we may visualize such a molecule as a rigid dumbbell with two point masses separated by a fixed distance, which is effectively the case for low temperatures. In addition to motion of translation, there will be rotation. We should like to know how any energy added to the gas will be shared by these types of motion. We recall that for pure translational motion, particles are free to move in three directions—for which we use the coordinates x, y, and z—and that the kinetic energy is $E_k = \frac{3}{2}kT$. This suggests that to each of the three directions an energy $\frac{1}{2}kT$ is ascribed. This idea is dignified by the phrase *law of equipartition of energy*, which states that the energy is equally shared among the degrees of freedom of motion. Although we merely guessed at the law, it can be shown by statistical methods to be quite general.

Let us apply the concept of equipartition to the diatomic molecule. Since it can rotate about each of the two axes perpendicular to the connecting bond, we say that it has two additional degrees of freedom. (Motion about the bond itself has no rotational inertia, since we have taken the particles to be points.) Now, the energy of rotation per molecule is

$$E_R = 2(\tfrac{1}{2}kT)$$

and the sum of translational and rotational energies is

$$E = \tfrac{3}{2}kT + kT = \tfrac{5}{2}kT$$

The specific heat may be derived as before, and we see that we merely use $\frac{5}{2}$ wherever $\frac{3}{2}$ appeared. Thus, for a molecule of mass m,

$$c = \frac{5}{2}\frac{k}{m} \qquad \textit{(specific heat of a diatomic gas)}$$

These relations are accurately verified by experimental measurements on gases. The media must be sufficiently rarefied so that there is no appreciable interaction between molecules, and must be at low enough temperatures so that there is little vibration or other types of excitation.

The specific heat has reference to a unit mass of material, while the *heat capacity* refers to a certain mass M of a substance. Now, 1 kmole of gas has a mass of $M = N_a m$, and its heat capacity at constant volume will be

$$C = Mc$$

where a capital letter is now used for heat capacity per kmole. For monatomic materials, $c = \frac{3}{2}k/m$; hence $C = (N_a m)(\frac{3}{2}k/m) = \frac{3}{2}R$, and similarly for a diatomic gas $C = \frac{5}{2}R$. For example, the heat capacity per kmole of argon (and of any other monatomic gas) is

$$C = \tfrac{3}{2}R = \tfrac{3}{2}(8314) \cong 12{,}500\ \text{J/°K}$$

Specific Heat of Solids

Generally for systems containing many atoms N, the number of vibrational degrees of freedom is $3N$. We are now in a position to predict the specific heat of a solid, in which the spring-like connections prevent rotation of the molecules. From the principle of equipartition of energy, the energy per degree of freedom is $\frac{1}{2}kT$. Since vibration involves both kinetic and potential energy, the energy per particle is $2(\frac{1}{2}kT) = kT$. Thus the total internal energy of a solid is $3N(kT)$ or

$$U = 3NkT$$

The specific heat—at least at high temperature, where all degrees of freedom are possible—is

$$c = \frac{1}{Nm}\frac{\Delta U}{\Delta T} = \frac{3Nk}{m}$$

For 1 kmole, containing N_a particles,

$$C = 3N_a k = 3R$$

The figure 6 kcal/kmole thus is frequently used for specific heat of most materials at high temperature.

14-5 MOLECULAR DISTRIBUTIONS

Basic to the kinetic theory is the demonstration by statistical logic that the speeds of particles—say, in the air of the room—range from very low to very high values, with a most likely speed that depends on temperature.

As our conservation methods showed us (Section 10-4), a collision between particles of high and low speeds results in a sharing of energy; hence we would expect there to be very few particles of either large or small energy. It is too much to expect that all come to the *same* energy, rather, they will be distributed over the energy range. The term *distribution function* is given to the formula that relates number to speed. It can be also considered as a mathematical expression for a graph.

The important idea of a distribution function can be brought out by experiments in games of chance. Suppose a pair of dice is rolled on a table.

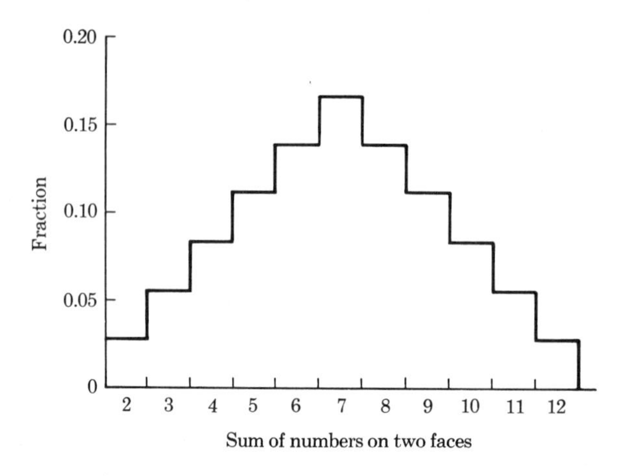

FIGURE 14-5

Probability graph for dice

The sum of the numbers on the faces that come up can be anything from 2 to 12, but the most likely number is 7, because it can be obtained in more ways than any other. Since there are thirty-six possible combinations, and since the "most probable" sum 7 can be obtained by six of these combinations, the chance of obtaining a 7 is $6/36 = \frac{1}{6}$. If one rolls the dice hundreds of times, this fraction is verified. The sums 2 and 12 are least probable, each occurring only $\frac{1}{36}$ of the time. Let us plot a bar graph of the results, as in Figure 14-5. This is a rough *probability-distribution* graph that allows us to find the fraction of all the throws that will yield any given sum. The total area under the "curve" is 1.000, which says the probability of *something* coming up is 1, as it must. Suppose we wanted to find the chance of a sum above 7—i.e., the numbers 8 through 12. The area under that part of the curve can be found to be 0.417.

The distribution of the population of the United States can be plotted on a graph from census data, as in Figure 14-6. A smooth curve can be drawn because there are so many people involved—over 200 million. We see at once that the most probable age of adults is 37.0 years. The fraction

FIGURE 14-6

Probability distribution of ages of U.S. population, 1960 census

of teenagers, from 13.00 to 20.00, is an area under the curve of 0.11, or 11%. For a small enough range of ages about a certain value, we can find the area merely by multiplying the height of the curve by the range. The fraction of all people between 19.9 years and 20.1 years is (0.00113)(0.2) = 0.00022.

For large numbers of particles, as in a cubic meter of gas (2.7×10^{25} particles), the distribution graph becomes an almost perfectly smooth curve. A definite formula for the way in which speeds of the particles are distributed was developed by Maxwell in the latter part of the nineteenth century, using the expectation that there were equal numbers of particles moving in each of the three directions x, y, and z, and that speeds should be independent of direction in a uniform gas of large extent. A graph* of the distribution of molecules according to speed is shown in Figure 14-7. We note as expected that there are few particles that have very low or very high speeds, and that the curve has a peak, at what is called the *most probable speed* v_p. From the formula for the curve, this speed depends on

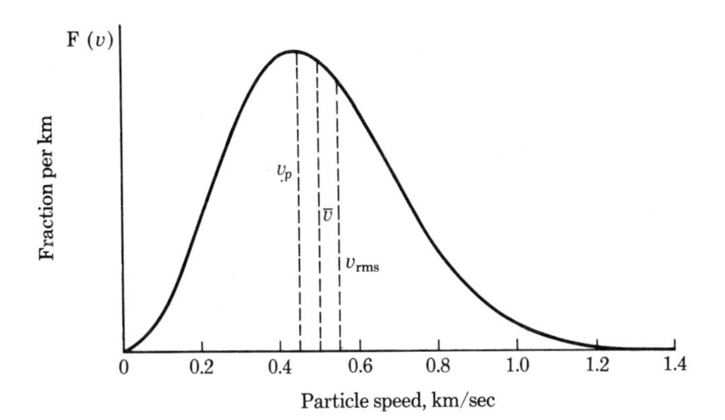

FIGURE 14-7
Maxwellian distribution of speeds in air

the temperature of the gas according to $v_p = \sqrt{2kT/m}$. We may also define an *average speed:* if in a collection of molecules there are n_1 particles of speed v_1, n_2 of speed v_2, etc., the average speed is simply

$$\bar{v} = \frac{n_1v_1 + n_2v_2 + \ldots}{n_1 + n_2 + \ldots}$$

From the formula for the curve, we find that $\bar{v} = (2/\sqrt{\pi})v_p$. The average kinetic energy is

$$\bar{E}_k = \frac{n_1(\frac{1}{2}m_1v_1^2) + n_2(\frac{1}{2}mv_2^2) + \ldots}{n_1 + n_2 + \ldots}$$

* The fractional number of particles per unit speed at v is given by the formula $F(v) = 4\pi Av^2e^{-mv^2/2kT}$, where T is the absolute temperature; k is Boltzmann's constant, 1.38×10^{-23} J/°K; and $A = \left(\frac{m}{2\pi kT}\right)^{3/2}$.

which is also $\frac{1}{2}m$ times the average *square* of the speed, $\overline{v^2}$, found from the curve to be $\frac{3}{2}v_p^2$. Combining,

$$\overline{E}_k = \tfrac{3}{2}kT$$

which we deduced earlier by elementary means, assuming all particles to have a common speed. The location of the various averages of speed is shown in Figure 14-7. The symbol v_{rms} stands for the *root-mean-square* speed, $\sqrt{\overline{v^2}}$.

It is interesting to investigate the effect on the speed distribution of raising the temperature of a gas. Since v_p depends on $T^{1/2}$, the larger T, the larger the value of v at which the peak of the curve $F(v)$ occurs. This implies that heating the gas shifts the curve generally toward higher speeds. We would expect this from our knowledge that the average particle energy is directly proportional to temperature.

In a molecular distribution, there are a few molecules with speeds above the escape speed from the Earth, $v_e = 11.2$ km/sec (recall Section 8-4). If one of these particles reached the top of the atmosphere and made no further collisions, it could leave the Earth completely. The fraction above v_e, as the area under the curve beyond that speed, is very small but finite. Had our planet a weaker force of gravity than it does, the air would have long ago been reduced below a density that could support life. In fact, our atmosphere has a lower concentration of helium (atomic weight 4) than would be expected from the known rate of generation of alpha particles by radioactive decay of heavy elements. The value of v_p for this very light gas is large and there are many high-speed particles; thus the rate of escape from the gravitational field is larger than for air molecules.

The temperature concept can be extended to an electrical discharge, in which there is a high degree of ionization. The positive ions, electrons, and neutral gas particles all have a temperature—that is, they have average energies given by $\overline{E} = \frac{3}{2}kT$. For illustration, the temperature of electrons of 1 eV average energy, 1.60×10^{-19} J is

$$T = \frac{2\overline{E}}{3k} = \frac{2}{3}\,\frac{1.60 \times 10^{-19}}{1.38 \times 10^{-23}} = 7730^\circ\text{K}$$

The atomic theory of gases provides us a description, in terms of molecular motion, of the relation of pressure, temperature, and volume of a substance. Such a description is analogous to information on a trip in which the cities we visit are known, but the means of travel from one to the other is not. For example, in pumping up a flat tire the gas law will tell us the relation between original and final values of P, V, and T, but will not reveal how much work was required, how much heat was developed, or the most efficient way to reach the objective. In order to complete the picture, we must bring in the law of conservation of energy, and seek other laws based on the detailed behavior of an assembly of particles that comprise a gas. In the next chapter, we shall examine the processes by which we go from one "state" of the gas to another, and develop the *second law of thermodynamics* from fundamental statistical concepts.

PROBLEMS

14-1 How many pairs of interactions are there in the nucleus of the oxygen atom, ${}_8O^{16}$?

14-2 Calculate the average translational energy, in joules and electron-volts, and the speed $v = \sqrt{3kT/m}$ for hydrogen molecules (H_2) at the surface of the Sun, $T = 6000°K$.

14-3 When a nucleus of ${}_{92}U^{238}$ absorbs a neutron, the resulting nucleus has an excess energy of about 6 MeV. What temperature would the nuclear "gas" seem to have?

14-4* Suppose that in a thermonuclear explosion involving 1 kg of deuterium (${}_1H^2$), 1% of the mass is converted into energy. What is the total internal energy released? What temperature and speed do the remaining gas particles have?

14-5 The tire of a jet plane is inflated at a temperature of 20°C to an absolute pressure of 3 atm. At the end of the runway the tire and its contained air become hot and the air temperature is measured as 140°C. What are the absolute pressure and the gauge pressure, assuming negligible expansion?

14-6 An automobile tire contains 0.005 kmole of air. How many grams of air and how many molecules does it contain? At a temperature of 37°C, what pressure in millimeters and in pounds per square inch does the air exert on the inner wall of the tire of volume 56 liters?

14-7 What is the number of collisions per square meter per second on the face of a piston containing He gas at pressure 1000 lb/in.^{2} and room temperature?

14-8 What mass of air strikes your face, area about 0.01 m^2, each hour?

14-9 How many molecules of air does 1 liter volume contain at standard conditions (1 atm, 0°C)? At a pressure of 10^{-3} lb/in.^{2} and -80°C?

14-10 Calculate the total prohibited space in a mole of gas molecules of radius 10^{-10} m. What fraction of the total space is this?

14-11 Inspection of Figure 14-4 shows that for a real gas at given pressure and temperature below T_c there are three possible volumes, corresponding to the fact that van der Waals formula is a cubic equation in V, with three roots. What do these roots become at the critical temperature?

14-12 Calculate the specific heat at constant pressure, in J/kg-°C, for the monatomic gas helium and for the diatomic gas oxygen.

14-13 Find the specific heat of the metal lead (Pb-206) in J/kg-°C, and the heat capacity of 1 kmole of Pb.

14-14 Develop a table of the probabilities of achieving various numbers for the sum of the spots on two dice, and thus verify Figure 14-5.

14-15 How many possible "hands" of two cards each are there in a 52-card deck? How many of these hands contain high cards (Ace, King, Queen, Jack)?

14-16 Calculate the probability in poker of dealing a flush (any five cards of the same suit), from a deck of 52 cards, none "wild."

14-17 Find the most probable speed of a "gas" of neutrons, $m = 1.67 \times 10^{-27}$ kg, in a nuclear reactor, at temperature 293°K.

14-18 Calculate the three characteristic speeds v_p, $\bar{v}$, and v_{rms} for air at the lowest recorded temperature in the United States, −76°F (Alaska, 1886).

14-19* Estimate the fraction f of all the molecules of air at 0°C, 760-mm pressure that have speeds faster than Mach-3 (3 times the speed of sound, 331 m/sec) if the area under the tail of the Maxwell curve is approximately $\frac{2x}{\sqrt{\pi}} e^{-x^2}$, where x is the ratio of speed to most probable speed, v/v_p.

14-20 What is the temperature corresponding to a plasma of heavy-hydrogen ions (as in an experiment to test the possibility of nuclear fusion) at average particle energy 0.04 MeV?

STATISTICS AND THERMODYNAMICS

15

The detailed description of all interactions in a substance with enormous numbers of particles is impossible, and statistical averages are required. This involves a serious loss of information content, but we still want to learn as much as possible about the behavior of the assembly. We thus seek general principles that will permit description of all systems, whether they be mechanical, electromagnetic, gravitational, or nuclear in character. The science of *thermodynamics*, based on statistics, will give us information on possible changes, directions of trends, and limitations on energy conversion for useful purposes. Its two laws are applicable whatever the details of the microscopic processes.

15-1 THE LAWS OF THERMODYNAMICS

The two basic principles that form the foundation of the science of heat energy in motion are simple:

FIRST LAW: ENERGY IS CONSERVED.

SECOND LAW: PROCESSES IN NATURE TEND TO GO FROM AN ORDERLY STATE TO ONE OF DISORDER.

No surprises or complications arise from the first law, which merely states that the total energy of an isolated system is constant, which is familiar to us by now. The second law has more subtle and fundamental meaning, and we must be careful in interpreting it. Our attention will initially be given to the first law. As usual, we use mathematical shorthand in particular applications of the laws. First, some notation is introduced.

Suppose that we add an amount of heat energy ΔQ to a substance, and an increase in internal energy ΔU takes place, while the system does work in amount ΔW. Conservation demands that

$$\Delta Q = \Delta U + \Delta W$$

which may be considered an algebraic statement of the first law of thermodynamics. We shall now proceed to apply this statement to a definite physical system such as the perfect monatomic gas, but keep in mind that the methods apply in general. Our attention will be focused on changes that are called *reversible*, those in which we can return to the original state by performing operations in the opposite order. For example, suppose we have the gas contained in a cylinder with a piston. An automobile tire pump will be adequate for illustration. Experience tells us that pushing the piston down compresses and heats the gas. On lifting the piston back, the gas expands and cools. If the previous conditions of pressure and temperature are restored, the initial (downward) movement was reversible. Normally, changes have to be slow and very small in order to meet this requirement. An explosion of a chemical mixture—say, hydrogen burning violently in oxygen—would not permit such reversal. The variables that give the "state" of the gas P, V, and T are seen to be important, along with the thermodynamic variables Q, U, and W. In principle, all six of these could change, but there are some interrelations that reduce the freedom of independent adjustment.

The first relation involves the definition of *work*. Again let us visualize the piston of area A being displaced by a force F through a distance Δx by expansion of the gas (Figure 15-1). The work done by the gas on the

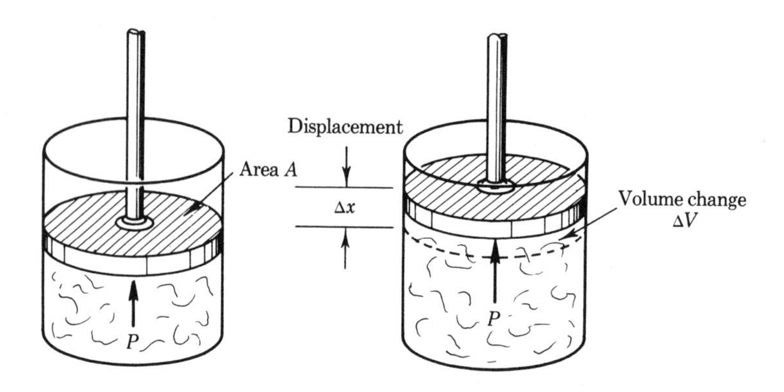

FIGURE 15-1

Expansion of a gas in a cylinder

piston is

$$\Delta W = F\Delta x$$

The pressure is force per unit area, $P = F/A$, and the change in volume of the gas is $\Delta V = A\Delta x$, and we find the connection

$$\Delta W = P\Delta V$$

Second is the meaning of *temperature* as a measure of internal energy. If a gas of mass M and specific heat c_v experiences a temperature change ΔT, the internal energy (translational, rotational, vibrational, and electronic) changes by

$$\Delta U = Mc_v\Delta T$$

In this formula the symbol c_v, *specific heat at constant volume*, is used instead of the symbol c, as in Section 14-4. The relation holds, however, for all processes, including those in which the volume is not constant.

15-2 APPLICATION OF THE FIRST LAW TO GAS PROCESSES

Out of the many possible processes one could imagine, with one or more variables being kept constant, we select five to illustrate the versatility of thermodynamics:

Quantity held constant	*Type of process*
V	Constant volume, isochoric
P	Constant pressure, isobaric
T	Constant temperature, isothermal
Q	No heat flow, adiabatic
Q, W	Constant internal energy, free expansion

In discussing each of these cases, we shall make reference to the trends of pressure and volume on the P–V diagram (Figure 15-2), which is one method of displaying a series of *states* characterized by P, V, and T. (For some of the processes the analysis is displayed, while in others in the interest of time only the results will be noted.)

1. Constant volume, $\Delta V = 0$**.** Suppose the position of the piston is fixed. From the definition of work, it is seen that ΔW is also zero. The first law reduces to $\Delta Q = \Delta U$, but from the definition of temperature, this is also $Mc_v\Delta T$. If c_v is not dependent on temperature, we may write the total change $Q = Mc_v(T_2 - T_1)$ and may compute the energy required to cause any rise in temperature—say, T_1 to T_2. For 1 kmole of a perfect gas, c_v is 12,500 J/°C, as shown in Section 14-4. To change it from 0°C to 100°C, the amount of heat to be added is

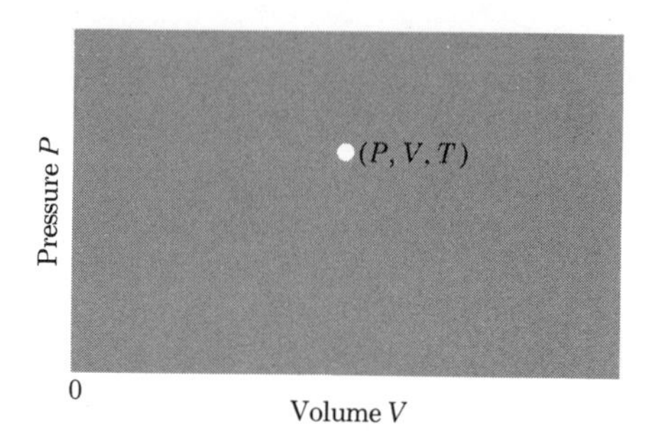

FIGURE 15-2
P–V diagram

$$Q = Mc_v(T_2 - T_1) = (1.25 \times 10^4 \text{ J/°C})(100\text{°C}) = 1.25 \times 10^6 \text{ J}$$

The absorbed heat of course goes to increase the pressure in the vessel. We are familiar with this effect in automobile tires, as they gain heat during a summer day.

***2. Constant pressure,* $\Delta P = 0$.** We heat the gas, allowing the piston to push against the constant atmospheric pressure plus that due to the weight of the piston and handle. The work done is $\Delta W = P\Delta V$, which may be extended to read

$$W = P(V_2 - V_1)$$

Recalling the perfect gas formula written as $V = NkT/P$, we find also that $W = Nk(T_2 - T_1)$. The first law then yields

$$Q = Nk(T_2 - T_1) + Mc_v(T_2 - T_1)$$

which can be forced into a form

$$Q = Mc_p(T_2 - T_1)$$

to make it resemble the temperature definition. The new *specific heat at constant pressure,* c_p, is introduced artificially. Comparing the two forms,

$$c_p = c_v + \frac{k}{m} \qquad \textit{(relation of specific heats at constant volume and at constant pressure)}$$

where $m = M/N$ is the mass of 1 particle.

***3. Constant temperature,* $\Delta T = 0$.** We apply heat to the gas, but insist that T be held constant, $\Delta U = 0$. The entering energy obviously must go into doing work against external pressure. Combining the first law, now $\Delta Q = P\Delta V$, where $P = NkT/V$ is variable, we obtain* the heat (and work done)

$$Q = NkT \ln \frac{V_2}{V_1}$$

* Since $\Delta Q = (NkT/V)\Delta V$, Q is the area beneath the curve NkT/V plotted against V.

for a definite change in volume. For example, suppose 1 kmole, $N = 6.02 \times 10^{26}/\text{m}^3$, is expanded by a factor of 10 ($V_2/V_1 = 10$), with T at 273°K. Then, in this *isothermal* process,

$$Q = (6.02 \times 10^{26})(1.38 \times 10^{-23})(273) \ln 10 = 5.22 \times 10^6 \text{ J}$$

4. *Zero heat flow,* $\Delta Q = 0$. To prevent any gain or loss of heat, the walls of the cylinder and the piston must be insulated. If then the gas is allowed to expand, pushing the piston up and doing work on it, we have, from the first law,

$$0 = \Delta U + \Delta W$$

or

$$\Delta U = -\Delta W$$

The negative sign merely means that the internal energy drops and the gas cools. Combining this with the gas law and the formula for c_p, and introducing $\gamma = c_p/c_v$, we find

$$P_1V_1^\gamma = P_2V_2^\gamma$$

This process is also labeled *adiabatic*. The ratio of specific heats γ can be found from previous information (Section 14-4). For a monatomic gas, $c_v = \frac{3}{2}(k/m)$, but $c_p = c_v + k/m = \frac{5}{2}(k/m)$. Then, $\gamma = \frac{5}{3} = 1.67$. It is easy to show that for a diatomic gas, $\gamma = \frac{7}{5} = 1.40$.

When we put these results on a kilomole basis (Section 14-4), we find the compact relation

$$C_p - C_v = R$$

and the specific heats for gases become as listed below.

	Monatomic	*Diatomic*
Constant Volume	$\frac{3}{2}R$	$\frac{5}{2}R$
Constant Pressure	$\frac{5}{2}R$	$\frac{7}{2}R$

5. *Free expansion* $\Delta Q = 0$, $\Delta W = 0$. We imagine our cylinder to be placed in a vacuum, and assume the piston to have negligible mass. When the piston that compresses the gas is released, expansion takes place freely. Looking at the system as a whole, we see that the gas does no work, $\Delta W = 0$, but also that no heat has been added, $\Delta Q = 0$, and from the first law $\Delta U = 0$. Thus no temperature change takes place.

In order to compare these various processes, we draw Figure 15-3, the P–V diagrams, showing initial and final states and the "paths" that the gas takes in going between them. Since the free expansion as a sudden transition is not a reversible process, we do not know the actual path. Thus we sketch only a dotted curve between the end points on the P–V plot.

All five of the above processes can also be illustrated by the behavior of air in a balloon with special properties ascribed to the containing wall

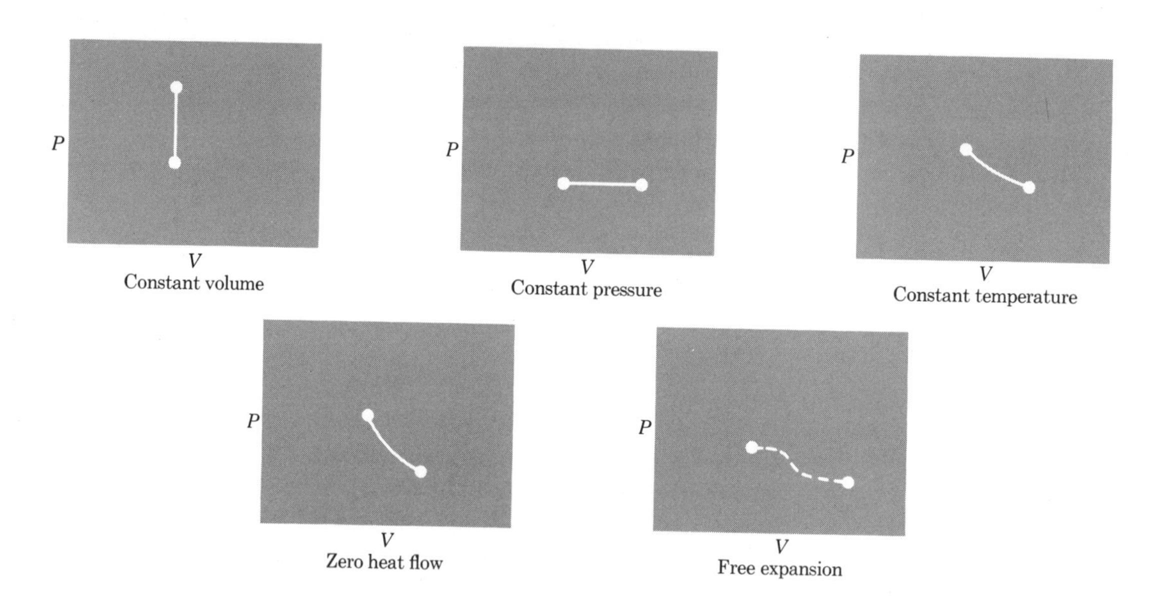

FIGURE 15-3

Several thermal processes in a gas

and under appropriate environments. If the walls are made of rigid plastic that conducts heat, exposure to the surrounding atmosphere on a hot day will result in an absorption of heat by gas at *constant volume*. If the walls are of a flexible rubber, the same exposure will allow the balloon to expand against the atmosphere at *constant pressure*. If the walls are highly conducting for heat and are flexible, absorption of heat from the surroundings will result in expansion *isothermally*. If the walls are composed of a very good insulator against heat, but flexible, expansion takes place *adiabatically*. Finally if the balloon bursts very high in the Earth's atmosphere, the air rushes out in a *free expansion*. We have by no means exhausted the possible means by which a gas can go from one state to another. Those cited are simplest and most often of interest.

In earlier chapters, when we studied the forces acting on a body to produce motion, it was necessary to define the *system* carefully. Since thermodynamics is based on mechanics, we would expect to take similar precautions. For example, let us heat a gas in a chamber, but allow some of the particles to escape into the surrounding atmosphere. The principle of energy conservation must be applied to the whole mass of gas, not merely to that in the container. Let us again examine the familiar process of putting air in a tire by means of a hand pump, to illustrate the need to account for the whole system. We shall assume the tire has a constant volume and any heat developed can readily escape, so the process is isothermal. Initially, all air is at atmospheric pressure, and a tire gauge would read zero. On the first stroke of the piston pump we do work on the *total* gas in the cylinder and the tire, $V_c + V_t$, compressing it to the tire volume V_t. On the next stroke, the system consists of the new air admitted to the cylinder plus that in the tire, with a larger number of particles than before.

15-3 ORDER AND DISORDER

Statistical mechanics, as its name suggests, is based on the theory of probability, which in turn had its origin in the study of games of chance. If the number of particles is enormous, detailed information on particle motion will not be obtainable; but we may hope to learn what changes are possible or likely. Our experience tells us that the natural course of events is toward a more disordered and chaotic situation. A few illustrations may be mentioned.

1. We place a drop of ink at a well-defined spot on the surface of water in a glass and wait a few minutes. Molecular motion causes the particles of ink to disperse throughout the whole volume.

2. We fire a rifle at a target. The bullet has *directed* energy while in flight. On impact, the energy is converted into the form of heat, which is *random* motion of the molecules of the target plus bullet.

3. We open a bottle of carbonated soft drink. Bubbles rise to the top of the liquid, and foam containing carbon-dioxide gas escapes into the air.

In each of these familiar examples, the trend is toward more random motion. A great deal of ingenuity and effort would be required to reverse these processes—to get the ink back into a drop, to send the bullet back to the muzzle of the gun, or to capture and concentrate the CO_2 again. These are clearly exaggerated examples of *irreversible* processes.

The science of statistical mechanics puts such events on a universal basis applicable to any system, through the concepts of *probability*, discussed here, and *entropy*, taken up in the next section. The second law of thermodynamics is evolved to supplement the first law, which is related to energy conservation. Imagine an evacuated chamber, isolated to prevent gain or loss of heat. A valve in the side is opened momentarily, letting a stream of air enter the container. Two important questions are: (1) How do the particles move within the space as time goes on? (2) How is the energy divided up among the particles at any instant?

It is conceivable but highly unlikely that every particle would retain its original speed and bounce back and forth indefinitely between the walls without collisions with other particles. We sense that it is even more improbable that by collisions all the energy would concentrate on one particle, with all others at rest. We seek to find just how implausible are these states of motion or, conversely, what the actual distribution is.

The appearance of laws of chance can be seen in a study of the *locations* of particles. Visualize an isolated container that has only 1 molecule of gas inside (Figure 15-4). It bounces around with a constant kinetic energy. The chance of it being in either half of the box at any instant is obviously $\frac{1}{2}$. Now, instead let us introduce 10 molecules of identical energy

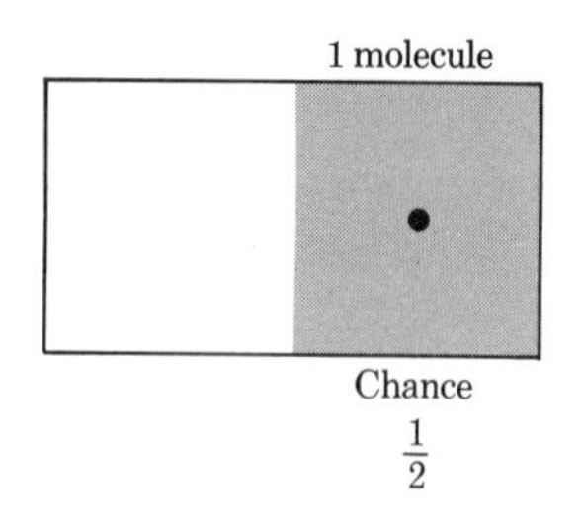

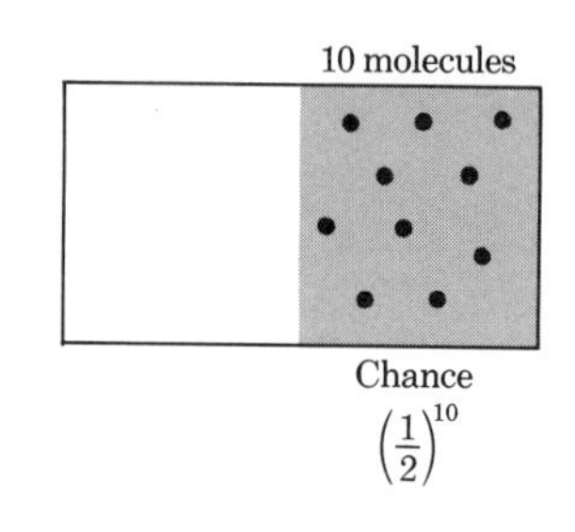

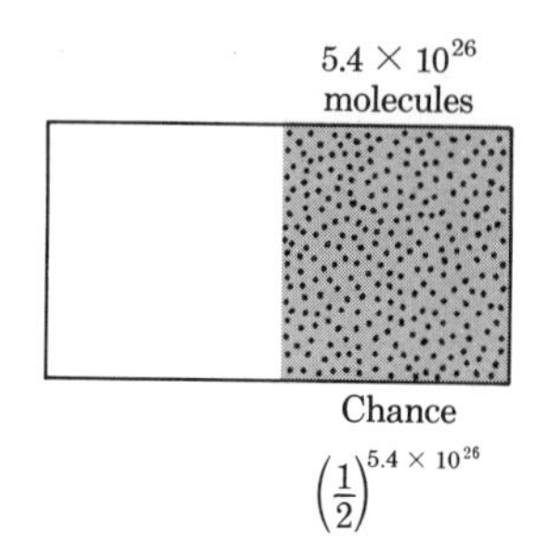

FIGURE 15-4

Probabilities of molecules being in one-half of a box

in the box. What is the probability that all of the particles will be in the half of the box on the right? The answer is $(\frac{1}{2})^{10} = \frac{1}{1024}$. Now, let us look at a room of volume, say, 20 m^3 and ask what is the chance that all molecules of air will be found in one half of the room, leaving a vacuum on the other side? The number of particles is $(20\ m^3)(2.7 \times 10^{25}/m^3) = 5.4 \times 10^{26}$. The probability is extremely low, $\frac{1}{2}$ raised to the *power* 5.4×10^{26}. We can see why an ordinary room has a very nearly uniform air density. The system tends to remain in a situation of high probability.

Now let us study the way in which *energy* is distributed among particles, seeking general statistical rules that will apply to gases in containers, the atmosphere surrounding the Earth, electrons found in atoms or free to move in metals, charges in vacuum tubes, or even to protons and neutrons in the nucleus. We can guess that the chances are very small that all the energy of such systems of many particles is concentrated on any one of the particles.

Substantial appreciation for the relation of probability and processes can be gained by examining simple systems. Whatever the particles are, we propose that they can occupy only a definite set of energy levels, say the

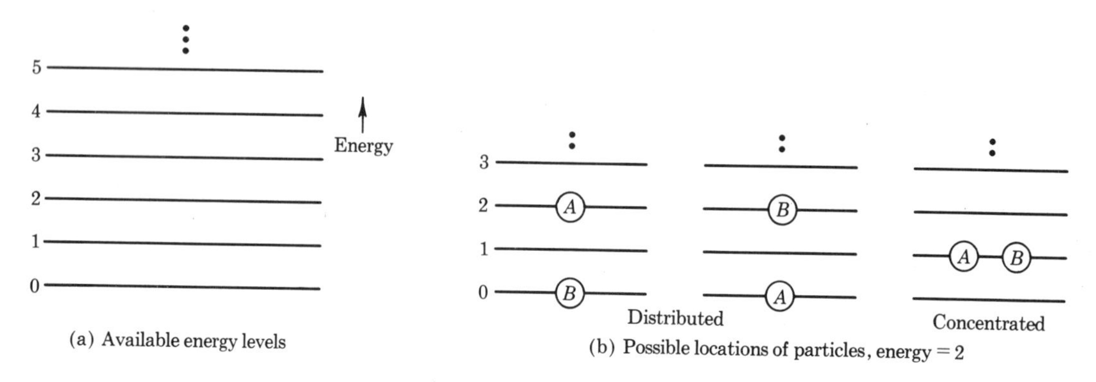

FIGURE 15-5

Energy distribution—two-particle system

lowest of zero energy, the next of 1 unit, 2 units, etc, as illustrated in Figure 15-5. Suppose, now, that the system contains only two particles, labeled A and B, and that the total energy in our arbitrary scale is 2. We ask the question, what are the possible "locations" of the particles? Figure 15-5 shows us there are only three. If these are equally likely, we see that there is a greater tendency by a factor of 2 for the energy to be distributed among particles at different levels than to have it concentrated on particles at the same level. To find out if this trend is continued by a larger number of particles, take a case with three particles, A, B, and C—i.e., $N = 3$—and let the total energy be $E = 3$. Ignoring labels on the particles, we see that the different types of arrangements or "states" are as shown in Figure 15-6.

Now let us look at state (a). To "fill" the lowest energy level with two particles, there are three possible combinations: AB, AC, and BC.

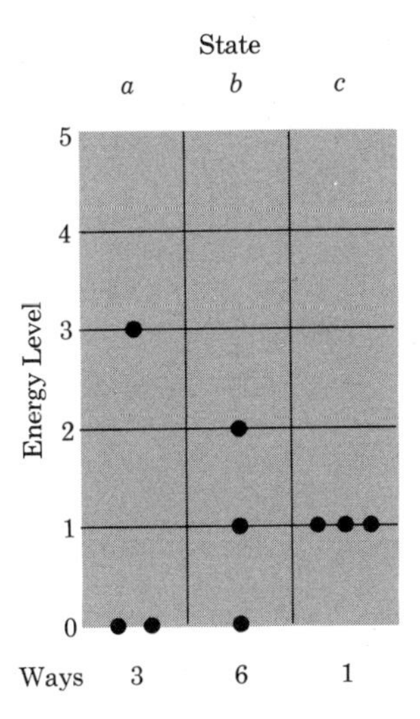

FIGURE 15-6

Energy distribution for three particles, total energy 3

Once one of these choices is made, there remains only one possibility to fill the remaining level. Thus the number of ways to achieve state (a) is 3. The theory of probability can be applied to such a problem. If there are n levels, N particles, and the numbers in level i is N_i, the number of ways w may be expressed using the notation of factorials*

$$w = \frac{N!}{N_0!N_1!N_2! \ldots N_n!}$$

For the case considered, $N = 3$, $N_0 = 2$, $N_3 = 1$, and all the other N_i are zero. Then, recalling that $0! = 1$, $1! = 1$, etc, we verify that $w_a = 3$. One can easily check that $w_b = 6$ and $w_c = 1$. Again we discover that the state that has particles more widely distributed in energy is more likely. The

* $N! = N(N-1)(N-2) \ldots (2)(1)$.

important fact that emerges from these exercises is that the *disordered* states, the ones with particles more widely dispersed over the levels, are more probable than the *ordered* states, with the particles more clustered about certain levels. Had we started with the system in state (a) or (c), it would have naturally tended to change to state (b). We generalize by saying that w seeks to increase, and when it reaches its maximum value, with no further change possible, the system is at equilibrium. The quantity w, which tells us how likely a state is, goes by the name *thermodynamic probability*. Thus our conclusion can be stated as follows: *equilibrium is the state of highest thermodynamic probability*. We shall pursue this powerful idea further as we proceed to discuss the idea of entropy. Our investigation here involved only two or three particles, whereas real systems involve very large numbers, requiring certain approximations to be made.

The continuation of the logical counting process we have started to the situation in which there are large numbers of particles leads to an answer to the next important question: What is the distribution of particles in the various energy levels at equilibrium? We can get a clue as to the dependence on energy by referring to our development of the exponential variation of pressure with height in the atmosphere (Section 13-4). Since the pressure is proportional to the number of gas particles per unit volume, the particle density at any height h, where the potential energy is mgh, is

$$n = n_0 e^{-mgh/kT}$$

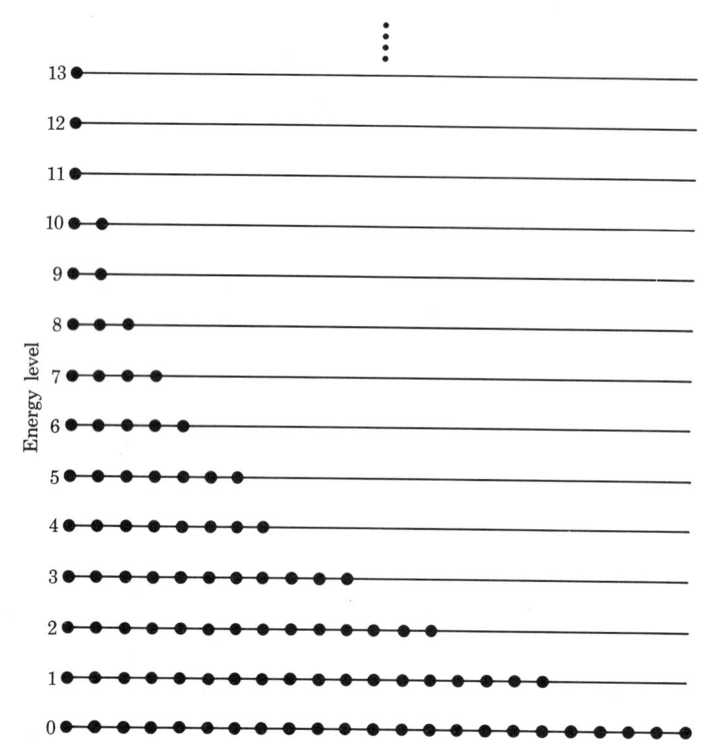

FIGURE 15-7
Boltzmann distribution of 100 particles

This expression is a special case of the *Boltzmann distribution*, a general law which states that in a system at equilibrium at temperature T, if there are N_0 particles in the lowest energy state, the number at the level l where the energy is E_l is

$$N = N_0 e^{-E_l/kT} \qquad \textit{(Boltzmann distribution)}$$

Figure 15-7 shows how a set of levels is occupied by particles using this relation. We have selected the total N to be 100, $kT \cong 5$ and $N_0 = 23$. The Boltzmann formula, derived from basic statistical principles, is also the origin of Maxwell's distribution of gas particles speeds (Section 14-5), and has many other important physical and biological consequences.

15-4 ENTROPY

We now introduce a new property of a system to supplement its state variables—volume, temperature, pressure, and internal energy. We have just seen that in natural processes the thermodynamic probability w tends to increase, corresponding to larger amount of disorder. Let us define the *entropy* S by

$$S = k \ln w$$

where k is Boltzmann's constant. As w increases, so does S, but more slowly, because of the logarithmic form. Thus we can state the second law of thermodynamics:

> PROCESSES IN NATURE TEND TO INCREASE THE ENTROPY, AS A MEASURE OF THE AMOUNT OF DISORDER OF THE PARTICLES; AT EQUILIBRIUM, AN ISOLATED SYSTEM HAS MAXIMUM ENTROPY.

Inspection of the Boltzmann formula and the diagram (Figure 15-7) shows that the higher the energy, the smaller is the occupation by particles. This relation, when used in conjunction with the definitions of w and S, leads to a simple but important thermodynamic expression, applicable to equilibrium states. When a small amount of heat energy ΔQ is added, with the system at temperature T, the entropy increases by an amount*

* To demonstrate this, we imagine that the added energy is given to one particle of the many in the system. It shifts from level m to a higher level n, so that $E_n = E_m + \Delta Q$. Thus there is a change in state from a to b, with probabilities w_a and w_b, whose ratio is easily seen to be

$$\frac{w_b}{w_a} = \frac{N_m!}{(N_m - 1)!} \cdot \frac{N_n!}{(N_n + 1)!} \cong \frac{N_m}{N_n}$$

The ratio of populations from the Boltzmann formula is

$$\frac{N_m}{N_n} = e^{(E_n - E_m)/kT}$$

Then the change in entropy between states is

$$\Delta S = S_b - S_a = k \ln w_b - k \ln w_a = k \ln \frac{w_b}{w_a} = k \ln (e^{\Delta Q/kT}) = k \frac{\Delta Q}{kT} = \frac{\Delta Q}{T}$$

$$\Delta S = \frac{\Delta Q}{T}$$

This can be considered a part of the second law, and provides fundamental supplementary information to the first law. It must be emphasized that this relation between changes in entropy and heat at a given temperature is valid only if the process is performed slowly and in small steps to assure that it is *reversible*. There is certainly a change in entropy in an irreversible process, but it is *not* governed by the expression $\Delta S = \Delta Q/T$. To verify this statement, consider the free expansion process (Section 15-2). There we found that ΔQ is zero, but it is clear that the released gas has become more disordered, and hence entropy has increased.

Properties of Entropy

We may investigate the behavior of physical systems by combining the two laws of thermodynamics. Several examples are given below to illustrate their power.

1. *Transfer of heat.* Two identical objects at different temperatures are brought into contact, so that heat energy flows from one to the other. Let us find the entropy change of this *system*, assuming it to be isolated from all sources of energy. Suppose that the temperatures are T_1 and T_2, with T_2 larger than T_1. Then the heat added to body 1 is ΔQ_1, which is equal to $-\Delta Q_2$ by the first law. The entropy change is

$$\Delta S = \frac{\Delta Q_1}{T_1} + \frac{\Delta Q_2}{T_2} = \Delta Q_1 \left(\frac{1}{T_1} - \frac{1}{T_2}\right)$$

The two factors in the last expression are both positive, hence ΔS is positive. We verify the statement that the entropy increases as the system tends to go from an ordered state (heat energy separated) to the disordered (heat energy mixed).

2. *Heating of a gas.* We take a perfect gas at some temperature and add heat in steps to bring the system up to a state characterized by P, V, and T. The sum of the entropy changes ΔS can be formed, and the result is a relatively simple expression

$$S = Nk(\tfrac{3}{2} \ln T + \ln V) = S_0$$

where S_0 is the original entropy. This verifies our statement that *entropy depends on the state of the gas.* It also shows that the *entropy will return to its original value in a cyclic process*—that is, if the original pressure, volume, and temperature are restored periodically.

3. *Mixing of gases.* By opening a valve between two identical containers of gas, we allow the contents to mix. We seek to calculate the change in entropy of the system. Initially, we have two sets of N particles in vol-

umes V, and finally we have one set of $2N$ particles in one volume $2V$. Calculating the entropy change (final minus initial) using the S formula for a gas, we find that

$$\begin{aligned}\Delta S &= S_f - S_i = 2Nk(\tfrac{3}{2}\ln T + \ln 2V) - 2Nk(\tfrac{3}{2}\ln T + \ln V) \\ &= 2Nk \ln 2\end{aligned}$$

This is a positive number, again illustrating that entropy increases as we go from the ordered situation (separate gases) to the disordered (mixed gases).

We have added entropy to our collection of variables that describe states and processes; the array now includes P, V, T, Q, U, W, and S. The science of thermodynamics in its fullest form consists of an elaborate set of equations employing the first and second laws. The first of these is a group called *Maxwell's relations*, which give the slopes of curves of P, V, T, and S plotted against each other. In each of the following the subscript notes which of the variables is held constant:

$$\left(\frac{\Delta V}{\Delta T}\right)_P = -\left(\frac{\Delta S}{\Delta P}\right)_T, \qquad \left(\frac{\Delta V}{\Delta S}\right)_P = \left(\frac{\Delta T}{\Delta P}\right)_S,$$

$$\left(\frac{\Delta P}{\Delta T}\right)_V = \left(\frac{\Delta S}{\Delta V}\right)_T, \qquad \left(\frac{\Delta P}{\Delta S}\right)_V = -\left(\frac{\Delta T}{\Delta V}\right)_S$$

These expressions are completely general; hence any conclusions deduced from them are rigorously applicable to any system. Let us consider a change of state from liquid to vapor. We recall that this involves an absorption of the heat of fusion L and a change in volume ΔV, both occurring at a constant temperature. The entropy law is $\Delta Q = T\Delta S$, where the absorbed heat is also L. Now $\Delta S = \left(\frac{\Delta S}{\Delta V}\right)_T \Delta V = \left(\frac{\Delta P}{\Delta T}\right)_V \Delta V$, by the third of Maxwell's relations. The change in volume ΔV is the difference between vapor and liquid volumes. Combining,

$$L = T\left(\frac{\Delta P}{\Delta T}\right)_V \Delta V$$

which is called *Clapeyron's equation*. Among its many uses is in deducing the latent heat from measurements on the change of vapor pressure with temperature.

15-5 IMPLICATIONS OF ENERGY AND ENTROPY

The fundamental statistical basis of the two laws of thermodynamics has now been established, and we are ready to look into the utility of the laws. We should note that the first law tells what process is *possible*, while the second law prescribes what is *probable*.

Our technological civilization has a tremendous need for energy in usable form, such as that of directed motion in automobiles, ships, aircraft, and rockets, or of industrial machinery, driven normally by electricity from rotating generators. Most of the usable energy (excluding water power) originates in the burning of some natural resource—oil, coal, or nuclear fuels. One may reasonably ask the question: Why can we not convert the boundless energy of molecular agitation in the Earth, the sea, or the air into energy of directed motion?

One possible answer might be that man has not yet proved ingenious enough to devise a converter that will tap this disordered motion, but this is not the correct answer. There is a fundamental limitation of nature, describable in terms of entropy. In order to appreciate this situation, let us consider the details of the operation of an engine that turns heat energy into work. The device is designed to develop the most work possible.

Let us imagine an absurdly simple experiment that will give us a start in the study of heat engines. A cooking pan with lid and containing a little water is placed on a hot stove burner. In a short time, the steam developed blows the lid off. Heat energy has been converted to work. Clearly this device is not very practical as an energy source since the motion is not repeated or maintained; but we may achieve continuous operation by utilizing the steam turbine. Water is vaporized by a source of heat, and the steam under pressure allowed to strike the vanes of a rotating turbine wheel. The alternative is to use the heat from a chemical reaction, as in the explosive burning of gasoline in an engine, where the *cycle* of operation appears.

We now study an idealized thermal cycle. Visualize a gas container equipped with a movable piston that may be set on a hot surface, or on a cold surface, or lifted far enough off either surface to effectively insulate the container. Four operations are performed in sequence, over and over, as follows (see Figure 15-8).

1. Place the container on the hot surface and allow the gas to absorb heat and, in expanding, to raise the piston. We receive useful work from

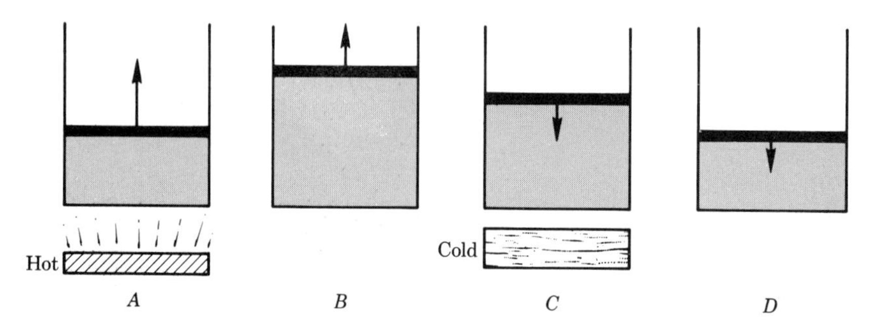

FIGURE 15-8

Steps in the Carnot cycle

the gas in this stage. We adjust the motion so that the heat absorbed, Q, matches the work done, W, so there is no change in internal energy or temperature.

2. Take the vessel off the hot surface, and allow the gas to expand further. This gives us additional useful work, at the expense of internal energy, since no heat can get in or out of the container walls.

3. Put the container on the cold surface, and push the piston down. Here we do work on the gas, and if the temperature is held constant, the energy we supply is delivered to the cold body.

4. Lift the vessel off and continue compressing until the original volume, pressure, and temperature are reached. Again we have supplied energy to the gas, and are back where we started.

What have we accomplished by these maneuvers? The net result is that more work was received from the gas than was delivered to it, and more heat taken from the hot surface than was sent into the cold one. Thus heat energy has been converted into available useful work. The processes described constitute a *Carnot cycle,* made up of alternating isothermal and adiabatic processes. We may now apply algebra to determine amounts of work available, using the first law of thermodynamics $\Delta Q = \Delta W + \Delta U$ and the perfect gas law $PV = NkT$ that relates the pressure, volume, and temperature.

Conservation of energy tells us that the *net* work W available is the difference the heats absorbed at the *high* temperature, Q_H, and delivered at the *low*, Q_L:

$$W = Q_H - Q_L$$

We make a distinction between these two quantities of heat. The quantity we should like to convert completely into work is Q_H, without the necessity of wasting the amount Q_L. If a perfect machine existed, Q_L would be zero, and W would be Q_H; but in a real machine we obtain only a fraction called the *efficiency*:

$$\epsilon = \frac{W}{Q_H}$$

The ideal efficiency value is 1 (100%). Alternately, this fraction can be written

$$\epsilon = 1 - \frac{Q_L}{Q_H}$$

which is less than 1 unless Q_L is zero.

All the ingredients for computing ϵ are available from our previous study of isothermal and adiabatic processes. Let us draw a P–V graph (Figure 15-9) of the steps in going from each state to the next and study it in connection with Table 15-1. By solving the adiabatic expressions for the ratios P_1/P_4 and P_2/P_3, and inserting them in the ratios of the isother-

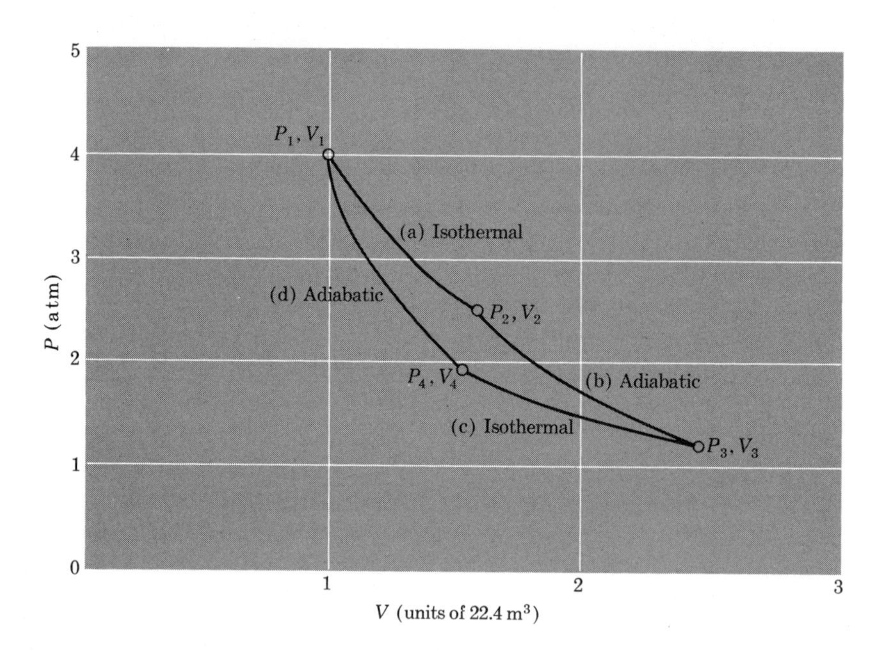

FIGURE 15-9

P–V diagram for Carnot cycle (monatomic gas, $\gamma = \frac{5}{3}$)

mal formulas, we obtain $(V_3/V_4)^{\gamma-1} = (V_2/V_1)^{\gamma-1}$. Hence $V_3/V_4 = V_2/V_1$. The expansion ratios are the same. Forming the ratio of the Q's,

$$\frac{Q_H}{Q_L} = \frac{W_H}{W_L} = \frac{T_H}{T_L}$$

and the efficiency becomes

$$\epsilon = 1 - \frac{T_L}{T_H}$$

TABLE 15-1

CARNOT CYCLE

Step		Between States		Formulas
Isothermal (T_H)	A	1	2	$Q_H = W_H = NkT_H \ln (V_2/V_1)$ $P_1V_1 = P_2V_2 = NkT_H$
Adiabatic	B	2	3	$P_2V_2^\gamma = P_3V_3^\gamma$
Isothermal (T_L)	C	3	4	$Q_L = W_L = NkT_L \ln (V_3/V_4)$ $P_3V_3 = P_4V_4 = NkT_L$
Adiabatic	D	4	1	$P_4V_4^\gamma = P_1V_1^\gamma$

We see a new fact here: the engine is perfectly efficient only if T_L is zero—that is, the region into which heat is exhausted is at absolute zero temperature.

The Carnot cycle is far removed from real engine cycles in that it involves ideal isothermal and adiabatic changes taking place under reversible conditions, which are by definition slow. The same mass of gas is used throughout. Consider how different is the case in a steam engine. Water in a boiler is heated and evaporated into steam, which enters an intake valve of a cylinder to fill and drive a piston. The valve is closed and further expansion causes cooling and condensation. Thus the *number of particles has changed during the cycle.* The rest of the steam is pushed out to a condenser, by the return stroke of the cylinder. In the engine of an automobile, the gasoline is mixed with air, compressed, and fired by a spark to drive the piston. Here the source of heat is *internal* to the system, being an explosive chemical, and the process is *nonreversible.* In each of these processes, definite amounts of friction are present, and there are heat-conduction losses to the surroundings. The Carnot cycle nonetheless tells us the optimum efficiency of a cycle.

The Carnot cycle has been discussed in terms of producing useful work from thermal energy. If engines are run *backward*—that is, if the P–V cycle is traversed in the opposite direction—work is done to provide desired heating or cooling. The "heat pump" provides heat to a room or building, while the refrigerator and air conditioner are designed to extract heat. Let us describe the cycle of operation of the heat pump, referring to the Carnot diagram of Figure 15-9. Starting at point 1, the working fluid is allowed to expand adiabatically to a temperature that is lower than the outdoors (air or ground). It passes through outside coils and receives heat, then is compressed to raise the temperature above room level. Heat is extracted to warm the room, and the cycle is repeated. For such systems, we are interested in a large value of ratio of the heat delivered at high temperature, Q_H, to the work required, W. From the Carnot relations we see that this coefficient of performance $C = Q_H/W$ is the reciprocal of efficiency, and thus

$$C = \frac{T_H}{T_H - T_L}$$

a number larger than 1, which means that the heat provided is larger than the work input.

The refrigerator can also be viewed as a heat engine, one that is run backward. It takes heat from a cold body (the air of the refrigerator cavity) at temperature T_L—say, 35°F—and delivers it to a hot body (the air in the room) at temperature T_H—say, 70°F. To do so requires work, provided by the electric motor. A liquid chemical such as Freon ($CClF_2$) is allowed to *evaporate* and thus absorb heat as it passes through the interior refrigerator coils. The pump driven by the motor compresses the vapor in the exterior coils where condensation releases the heat of vaporization.

The thermodynamic analysis of the familiar *internal-combustion* (Otto) *engine* is fairly simple. First we identify the steps in the cycle, as illustrated in Figure 15-10. The operations are as follows:

1. The intake valve I opens, the piston moves down, and a mixture of air and gasoline vapor enters the cylinder.

2. With both valves closed, the fuel mixture is compressed by the return stroke of the piston. A sparkplug fires, igniting the gasoline, and a large amount of heat Q_H is added.

3. The power stroke occurs, with the piston driven down by the expanding gases.

4. With the exhaust valve E open, the piston pushes out the residual gases, which still contains an amount of heat Q_L.

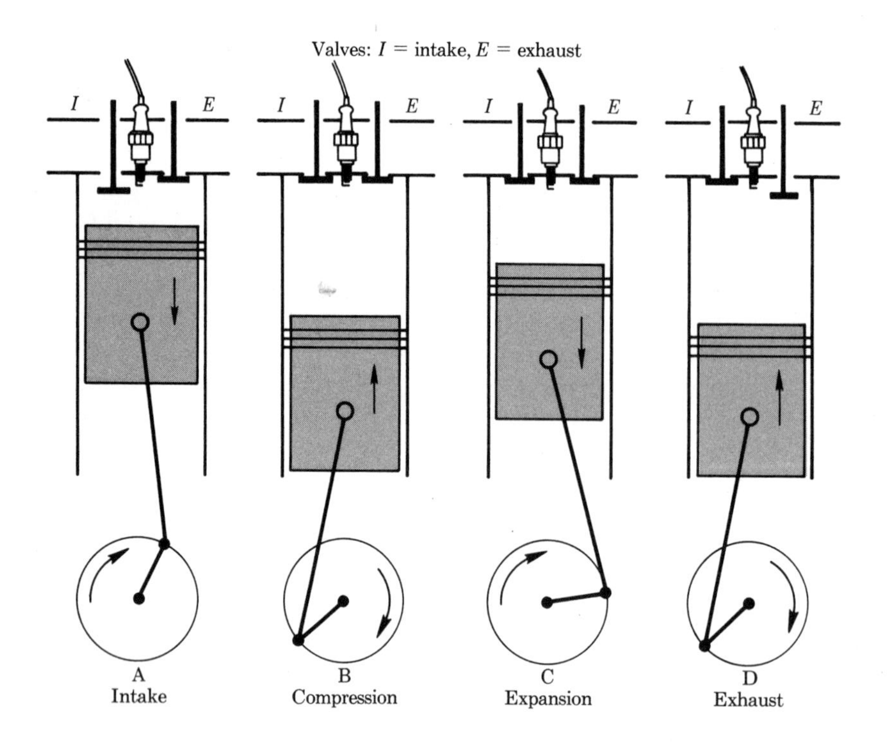

FIGURE 15-10

Steps of cycle in internal combustion engine

The cycle may be plotted on a P–V diagram (Figure 15-11). The intake step at constant pressure carries the system from point 0 to point 1. An adiabatic compression along 1 to 2 occurs, followed by a sudden rise in pressure from 2 to 3 as the gasoline burns. An adiabatic expansion carries the mixture from 3 to 4. The step from 4 to 1 corresponds to the loss of heat by discharge, which actually occurs at the same time as the step from 1 to 0 that sweeps out the gas.

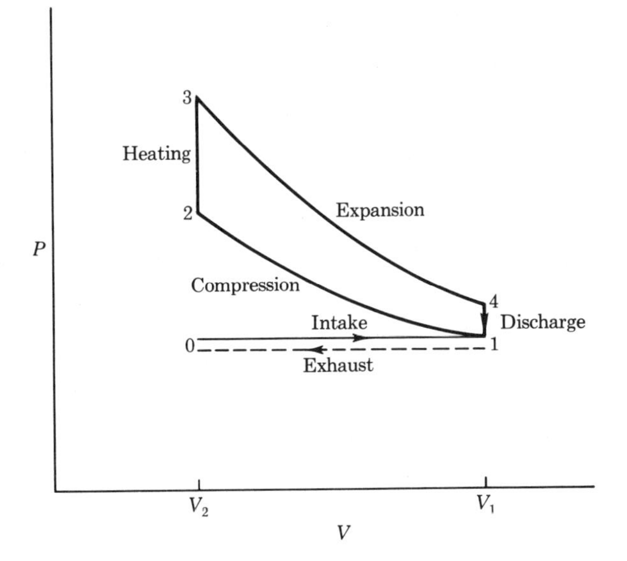

FIGURE 15-11

P–V diagram for cycle of internal combustion engine

The amounts of heat are

$$Q_H = cM(T_3 - T_2), \qquad Q_L = cM(T_4 - T_1)$$

from which the work done $W = Q_H - Q_L$ may be found.

The efficiency is

$$\epsilon = \frac{W}{Q_H} = 1 - \frac{Q_L}{Q_H} = 1 - \left(\frac{T_4 - T_1}{T_3 - T_2}\right)$$

The temperatures between the ends of the adiabatic steps are related by

$$T_4 = T_3 \left(\frac{V_3}{V_4}\right)^{\gamma-1}, \qquad T_1 = T_2 \left(\frac{V_2}{V_1}\right)^{\gamma-1}$$

Letting the compression ratio be $r = V_1/V_2$, a little algebra reveals that

$$\epsilon = 1 - \frac{1}{r^{\gamma-1}}$$

For example, using the value $\gamma = 1.4$ as for the diatomic gases O_2 and N_2 in air, and with a compression ratio of 5, the ideal efficiency would be $1 - 1/(5)^{0.4} = 0.475$, a little under 50%. The cycle as actually measured on an automobile engine is distorted somewhat from our theoretical diagram. The corners are rounded and the intake–exhaust part forms a loop rather than coincident parallel lines.

In the *diesel engine,* air only is admitted to the cylinder in step A (Figure 15-10); it is compressed to a very small volume in step B, at which point low-grade fuel is admitted. Ignition takes place because of the high air temperature. If gasoline were used in such an engine, premature burning would take place. Compression ratios are typically as large as 20:1 in the diesel engine.

The Meaning of Temperature

A survey of the trend toward improved precision of temperature measurements and scales over the course of history will show us how the present Kelvin temperature scale evolved. Until the 1700's, people presumably said that it was either "hot" or "cold," depending on how comfortable they were. Then *Fahrenheit* immersed a mercury bulb with glass tube in a selected mixture of ice and salt (which depresses the freezing point) and marked the tube at the level of mercury as "zero." The 100°F mark apparently was set by taking his body temperature (inaccurately, unless he had a fever). On the improved *centigrade* scale advocated by *Celsius*, the marks were set by immersion in ice water and steam, respectively labeled 0°C and 100°C. In each case, the size of the degree is merely the change in column divided by the number of degrees assigned to the range. The gas thermometer is preferable to one using mercury because of its accuracy. As the temperature rises, the pressure goes up linearly, according to the law $PV = NkT$. A barometer measurement for a fixed volume of gas gives the temperature. The curve plotted in Figure 15-12 is

$$P = CT$$

The *absolute* or *Kelvin* scale is based on the knowledge that real gases obey this law so long as the density is low—i.e., at high temperatures. The slope C of the curve of pressure vs. temperature is thus taken to be a known constant. If one pressure is measured with the gas thermometer in a selected standard environment, the curve can be drawn. Instead of fixing the zero, the temperature of a mixture of ice, water, and steam in equilibrium—the *triple point* of water—is arbitrarily said to be 273.16°K. The temperature where the pressure graph reaches zero (no molecular motion) is said to be *absolute zero*. This fixes the complete scale, including end points and size of degree. The triple point corresponds to 0.01°C, so that absolute

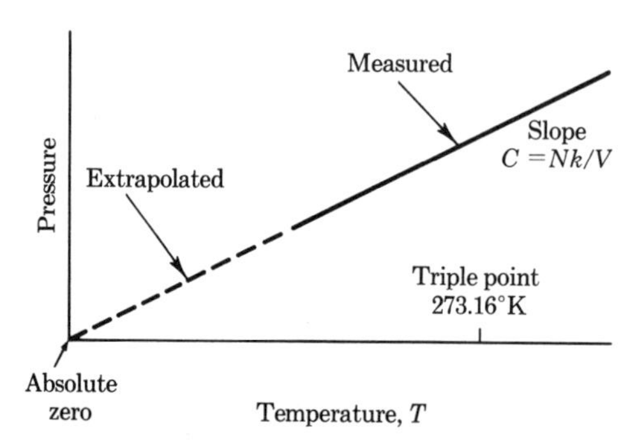

FIGURE 15-12

Determination of absolute zero

zero is $-273.15°C$. Real gases at temperatures near absolute zero deviate, of course, from the perfect gas law, but the linear part of the curve can be extended—i.e., extrapolated to zero.

Kelvin put this technique on a thermodynamic basis by expressing the relation in terms of the Carnot cycle:

$$\frac{Q_H}{Q_L} = \frac{T_H}{T_L}$$

This states simply that the heat transferred in an isothermal process is proportional to the absolute temperature:

$$Q \sim T$$

On cooling a substance, a point is reached when *no heat is transferred in an isothermal process*. Then, *the temperature is zero* (on the Kelvin scale). Experimental determinations of temperatures below 1°K, where all gases have become liquid or solid, make use of this result.

Perpetual-motion Machines

Many inventors and amateur scientists have conceived of "perpetual-motion" machines, which they expect to deliver work indefinitely with either no input (first kind) or by tapping the basic stored energy of the atmosphere, Earth, or sea (second kind). Such machines are based on hope rather than principles, and are often ingenious and complicated, which makes it difficult to demonstrate the fallacy in the idea, at least to the inventor's satisfaction.

Consider perpetual-motion machines of the first kind. If the machine delivers work, it must be at the expense of its initial kinetic or potential energy, which will soon be used up. If it does no work, the inevitable friction will eventually stop the device. Thus the law of conservation of energy (first law of thermodynamics) makes the machine impractical. The use of natural resources such as stored water from rainfall, or natural fuels such as coal, oil, or uranium is not prohibited by the first law, of course.

A perpetual-motion machine of the second kind is not as ambitious a concept. Basically, one would like only to take heat energy of molecular motion from a reservoir, such as the sea, at temperature T, and convert some of the energy into useful work in a cyclic fashion. The unused energy would be returned to the reservoir at the same temperature T. If it were possible, such conversion of existing energy into directed motion would be of inestimable value to mankind. The concept may be deflated by referring at the Carnot cycle. If the temperatures of the heat intake, T_H, and exhaust, T_L, are the same, then the efficiency ϵ is zero, and no work can be done. A heat engine demands *two* different temperatures, with the discharge of waste heat to the lower of the two. Since the Carnot cycle is based on gas theory, one might argue in favor of more complex cycles involving other phenomena, such as electromagnetism and mechanics, and fluids more exotic than a perfect gas. Unfortunately, there is no evidence that such systems would work. The second law of thermodynamics, as a generalization from a tremendous amount of experience, is considered to be applicable to *any* system, no matter how complicated.

A third type of machine is allowed by the second law of thermodynamics and thus does not fall in the category of "perpetual-motion" machines. It consists of a device that extracts heat from the substances of the Earth, such as equatorial air or water, or the hot interior of the planet; the waste heat must be discharged to a colder region, such as at the poles, and thus there would be a definite temperature difference. We may consider such systems feasible but probably impractical.

PROBLEMS

15-1 Compare the amounts of work done by a gas when heated (a) at constant pressure, and (b) at constant temperature, assuming initial conditions, $P_1 = 1.0 \times 10^5$ N/m², $T_1 = 300°$K, $V_1 = 0.6$ m³, and final volume 1.4 m³. (c) Discuss the difference in the results.

15-2 Show that for an adiabatic process,

$$P^{1-\gamma}T^{\gamma} = K \qquad (a\ constant)$$

15-3 Compute the pressure ratio P_2/P_1 for an adiabatic change in the diatomic gas Cl_2 if the volume increases by a factor of 10.

15-4 (a) If an isothermal process took place at constant volume, how much heat would be absorbed? (b) Two slowly combining chemicals in gaseous form are introduced into a closed, insulated vessel. As the atoms combine and release energy the pressure changes, but the volume occupied remains constant. Examine the role of the general gas law and the first law of thermodynamics in this process.

15-5 Compute the heat capacity at constant pressure per kilomole of He and O_2, taken to be perfect monatomic and diatomic gases, respectively, expressing results in J/°C and kcal/°C.

15-6 Explain the physical difference between a reversible and an irreversible process. Consider each of the following: (a) free expansion of a gas; (b) heating a closed gas–liquid system; (c) increasing the pressure on a gas with no heat flow; (d) condensation of a gas; (e) cooling by evaporation.

15-7 Find the chance that all the air molecules in a room of volume 20 m³ are found in one particular cubic meter.

15-8* Prepare a diagram showing all the possible energy distributions for 4 particles, with total energy 5 units, assuming equal spacing of the levels. Find the number of ways to achieve each state (the value of w) for each.

15-9 Calculate the thermodynamic probability w for a state in which $N_0 = 2$, $N_1 = 1$, $N_2 = 0$, $N_3 = 3$, $N_4 = 1$.

15-10 The thermodynamic probability of a certain state is $w = 10^{10}$. Find the entropy corresponding to the state.

15-11 Find the efficiency of a heat engine that operates on a Carnot cycle between the temperatures 100°C and 0°C. How much work is available per joule of heat?

15-12 A Carnot engine operates between temperatures 700°C and 100°C. The power absorbed is 5 kW. What is the efficiency of the engine? What is the amount of useful power?

15-13* Imagine a heat engine with "rectangular" power cycle on a P–V diagram—that is, (1) the volume is allowed to increase at constant pressure; (2) the pressure is reduced at constant volume; (3) the fluid is compressed at constant pressure; and (4) the pressure is raised at fixed volume (Figure P15-13). Find temperatures T_2, T_3, and T_4 in terms of T_1. Develop expressions for the heat absorbed or released in each step in the cycle and the amount of work done. Show that the latter is equal to the area enclosed on the P–V diagram by the cycle paths.

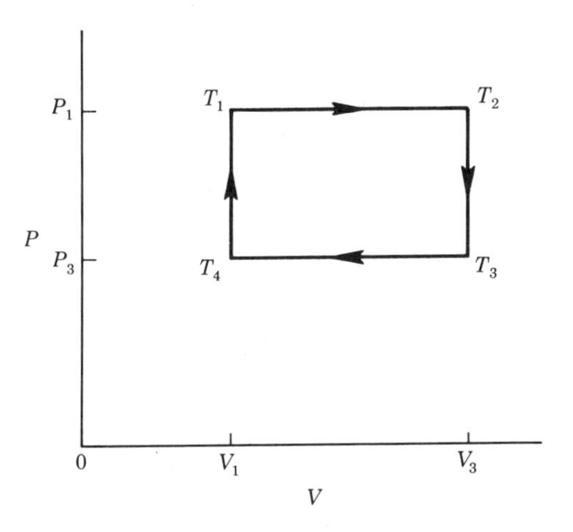

FIGURE P15-13

15-14 From the formula for an internal-combustion engine, determine if the efficiency is improved by increasing or decreasing the compression ratio. In the text example, what happens to ϵ if r is reduced to 2?

15-15* (a) We pump up a tire of volume 3000 cm³, initially at atmospheric pressure 14.7 lb/in.² and room temperature by the application of 25 strokes to an air pump with cylinder of diameter 2.5 cm, length 30 cm. What is the final pressure in lb/in.²? (b) Assuming the process to be isothermal, how much work have we done? (c) When the pressure is 20 lb/in.², how far down must the piston move before air is admitted through the valve on the tire? (See Figure P15-15.)

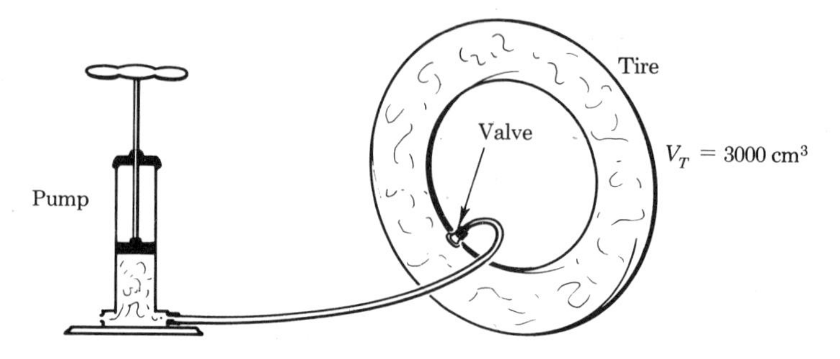

FIGURE P15-15

15-16 Diesel engines are more efficient than gasoline-operated internal-combustion types. Can you explain why?

15-17 What are the desirable properties of a heat engine? Which are the most difficult to obtain practically?

15-18 A helium-filled dirigible in outer space develops a leak and the gas escapes. Is there a change in entropy? If the dirigible were instead in the Earth's atmosphere, would the change in entropy be larger, smaller, or the same as before? Explain.

15-19 How much work by a heat engine acting as refrigerator would be required to reduce the temperature of an 0.8-kg mass of helium gas from room temperature (25°C) to one near absolute zero (−265°C)?

15-20 A heat pump operates between an outdoor temperature of 10°C and an indoor temperature of 25°C, and provides 2×10^4 J/sec of heat to a building. What is the smallest amount of power needed?

15-21 An inventor suggests a hydroelectric plant consisting of a long pipe thrust down in the ocean. He claims that water will continually flow in the top, drop down the pipe, and turn indefinitely an electric generating turbine at the bottom. Examine the practicality of this concept.

TRANSPORT PHENOMENA

16

The idea of the equilibrium state has played a dominant role in our study of thermal systems. The speed distribution in gases and the entropy–temperature relation required the assumption of such a situation.

True equilibrium implies complete uniformity on the macroscopic scale, and no changes with time. Only in a "homogenized universe" with the same temperature everywhere would the state be realized exactly. No matter how well we try to isolate smaller real systems, there will always be some transfer of heat energy through the insulation. Thus we are faced with a real world in which there are nonuniformities and flows, of both the steady and time-dependent type. The term *transport* refers to flow processes involving particles, momentum, energy, and electric charge in motion from one region to another. In this chapter* we shall develop descriptions of flows in terms of the variation in composition and properties of substances, thus accepting a certain amount of departure from equilibrium.

16-1 TRANSPORT PROCESSES

Let us consider several familiar processes or simple experiments to bring out the idea of the transport of various physical quantities.

* The chapter covers topics that are ordinarily beyond the scope of a basic text; therefore, the instructor may wish to omit certain sections.

The first example has to do with the process called *diffusion*. The tire on our car picks up a nail, is punctured, and air escapes. We know, of course, that the pressure and particle density of air inside the tire is larger than it is outside, and there tends to be a flow outward. A similar flow process would occur if a meteorite were to penetrate the wall of a spaceship moving in the vacuum of outer space. If we open a bottle containing a chemical with a strong odor, the foreign particles will soon make themselves known throughout the room. These are examples of particle diffusion.

The flow just described takes place in a gas at rest, but if the fluid is moving, new effects come into play. For example, when we pour cold motor oil from a can we notice that the flow is very sluggish. The oil has a large *viscosity*, meaning that each layer of fluid tends to hold back the other layers from sliding over it. As water flows along in the bed of a river, or from the sea onto a beach, or in the pipes of a building, viscous forces slow the motion. Similar effects are found in gas flow, because of molecular interactions.

Another "flow" process is that of *heat transfer* through materials—for instance, through the walls of buildings. Most houses are constructed with some kind of insulation to help keep the interior warm in winter and cool in summer. This may be a layer of rock wool, or a "dead-air" space between two walls. Heat energy must be carried by molecules across the poorly conducting gap. If we heat one end of a bar of metal, within a very short time the temperature at the other end will rise. Kinetic energy has been transferred by atoms or electrons, a process of heat transfer.

In addition to carrying heat energy, particles such as electrons and ions can transfer electricity, the process of *electrical conduction*. When we apply an electrical-potential difference between two surfaces with a gas between them, there is a flow of charge, primarily that of electrons. A few examples involving this process are the fluorescent lightbulb, the Geiger counter, the welding arc, the gas-filled "vacuum" tube, and the ion source for a particle accelerator.

The processes listed above have somewhat different physical mechanisms, but all are examples of *transport phenomena*, and can be studied by use of a common mathematical approach. These flows are different from those to which Bernoulli's principle (Section 12-6) applies. There, the principle of conservation of energy for bulk material was invoked. Now we must examine microscopic details to determine rates of transfer.

Molecules of a gas, liquid, or solid move from place to place in a random, chaotic manner. If the system is completely uniform, a particle that leaves a small region is soon replaced by another, and from a large-scale view no change has occurred. However, if there are more molecules at one spot than another, losses are not replenished, and there is a flow. The particles have various properties that they "carry" with them, such as mass, momentum, energy, and charge. Thus there can be a flow of any or all of these entities.

We should like to develop a unified description for such transfers.

The basic approach is to examine the nature of forces between particles and the mechanics of collisions, then to find out how far particles travel before colliding again. The gross effects are deduced by adding up the contributions of individual molecules, with their various speeds and directions of motion. The general mathematical analysis is beyond our present level or interest, but we can appreciate the processes and obtain some approximate relations using simplified derivations. First, let us look into the matter of flight and collision.

16-2 CROSS-SECTION AND MEAN FREE PATH

If we were able to view a gas with a powerful microscope, we would see a great deal of empty space, with a few particles moving freely about in straight lines until they came near one another. The chance of a collision depends on the sizes of the particles, or more precisely, on the target *area* presented to one particle as a projectile. The term *cross-section* is given to this area, symbolized by σ, in units of square meters. It is not easy to calculate cross-sections for molecules as complex structures of atoms, which are in turn composed of nuclei and electrons. Two colliding water molecules are shown in Figure 16-1. The effective area of interaction is sure to be different depending on the type of particles and on the way in which the particles happen to collide. The simplest model we can imagine involves two hard elastic spheres. Let the radius of the projectile be r_1 and that of the target r_2. They will just graze each other when the surfaces are in contact, as shown in Figure 16-2. If the separation of the lines between centers is less than $r_1 + r_2$, they collide; if it is greater, they miss. The "target" is thus a circle of radius $r_1 + r_2$, hence the area is $\sigma = \pi(r_1 + r_2)^2$. If the two spheres have the same radius r, the cross-section becomes $4\pi r^2$. (It is interesting but not significant that this is the total surface area of one sphere.) Realizing that molecules are not hard spheres, we seek ways to find the effective radii and cross-sections. One approach is to make detailed calculations of the forces between all the component particles of two

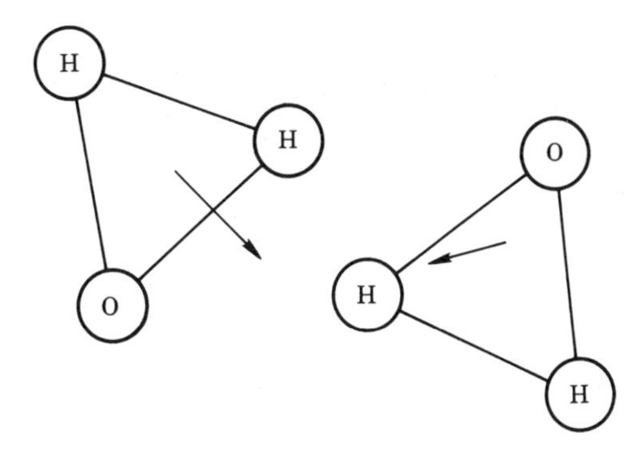

FIGURE 16-1
Collision of two water molecules

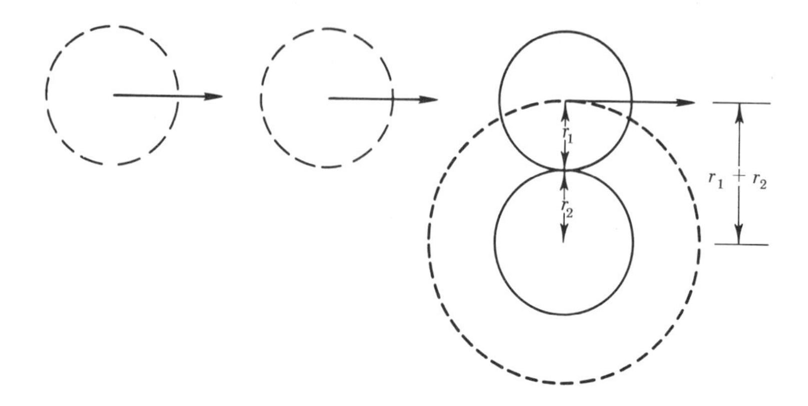

FIGURE 16-2

Grazing collision of hard spheres

molecules as they approach each other. Unfortunately, this is extremely complicated because of the number of component particles, the structure, and the need to apply quantum mechanics (discussed later in Chapter 22). Another indirect *empirical* method involves the study of much experimental data on molecular transport, equations of state, and phase changes. An

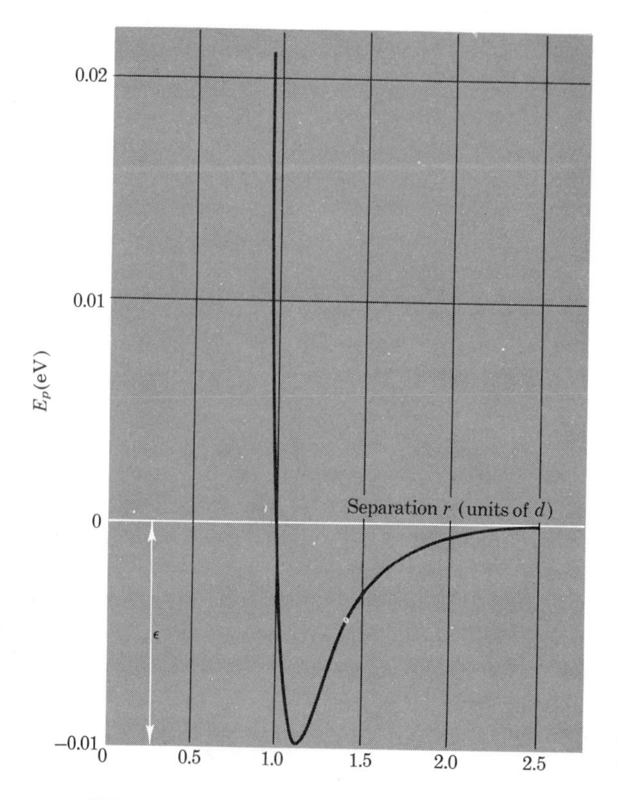

FIGURE 16-3

Potential energy diagram for colliding O_2 molecules

expression for the potential energy of a system of two molecules as it depends on their distance of separation is devised, to give the best fit to the observations when the theory of collisions is applied. Such relations display the lack of interaction at large distances, the attraction as the particles approach each other, and the very strong repulsion at small distances. We recall similar trends for *atoms* (Section 11-4). Figure 16-3 shows a graph of potential energy E_p against separation r for the interaction of two oxygen molecules O_2. It is clear that they do not interpenetrate far beyond a separation distance d, where E_p is rising rapidly. Thus $d/2$ can be regarded

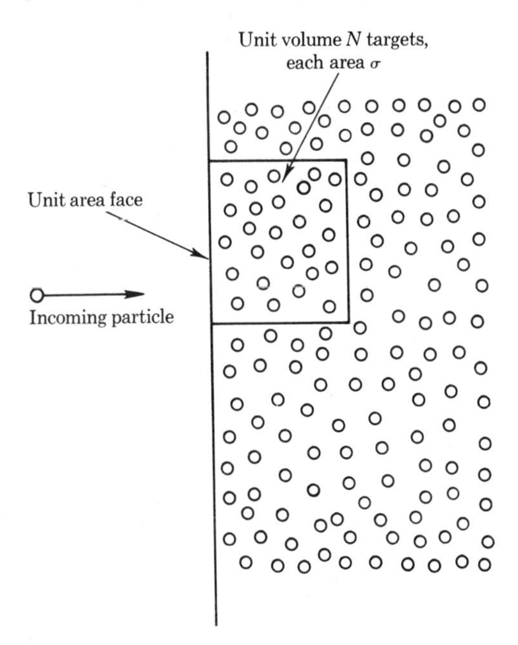

FIGURE 16-4

Penetration of a medium by atoms

as the effective radius of each molecule, $r_1 = r_2 = d/2$, and the cross-section become

$$\sigma = \pi d^2$$

Using the constants for oxygen, we would estimate

$$\sigma = \pi(3.6 \times 10^{-10}\ \text{m})^2 = 4.07 \times 10^{-19}\ \text{m}^2$$

With the knowledge of the cross-section, we are able to find out how far a particle will travel before it collides with another particle. We realize that chance determines the length of free path. Some particles will collide immediately, others will go great distances before striking another. We can deduce the average of the free paths, or *mean free path*, as follows. We allow a particle to cross a unit area of a boundary of a rarefied gas containing N targets per unit volume (Figure 16-4). The area seen in that volume is $N\sigma$, so the chance of colliding is the ratio of target area to total area, or $N\sigma/1$, assumed to be a small fraction. Now to be sure of colliding, probability 1, the projectile must go a distance of $1/N\sigma$, which is the mean free path λ. Thus

$$\lambda = \frac{1}{N\sigma} \qquad \textit{(mean free path)}$$

Our derivation, although it gives us the correct answer, is questionable if there are so many particles that they "shadow" each other.

Instead of using the unit volume, let us select a thin layer of targets of unit area, thickness Δx, at any location x in the medium. On the front face, the current is j, and the chance of collision is $N\sigma\Delta x/1$, so the number of collisions in the layer is $jN\sigma\Delta x$. The reduction in current is thus

$$\Delta j = -jN\sigma\Delta x$$

This equation, involving a change in a quantity that is proportional to the quantity itself, appeared before (Section 5-8). If j_0 is the current at the origin, $x = 0$, then

$$j = j_0 e^{-N\sigma x}$$

This attenuation formula, plotted in Figure 16-5, indicates that the stream drops off exponentially with distance of penetration. We can now find the average distance $\bar{x}$ that a particle goes by finding the total distance all particles travel and dividing by the original number of particles—i.e., j_0. As expected, this average distance turns out to be $1/N\sigma$. This is the mean free path found previously, being the average of all the free flights through the medium prior to collision. The result is more generally applicable than just to the specific experiment used to derive this formula. A particle moving an any direction through a substance will go a distance λ between successive collisions. Whether the particle started a free flight at $x = 0$ or at any position is immaterial. Notice that at depth $\bar{x} = 1/N\sigma$ the particle current has been reduced to $j(\bar{x}) = j_0 e^{-1}$ or $j_0/e = j_0/2.718$, as shown in Fig-

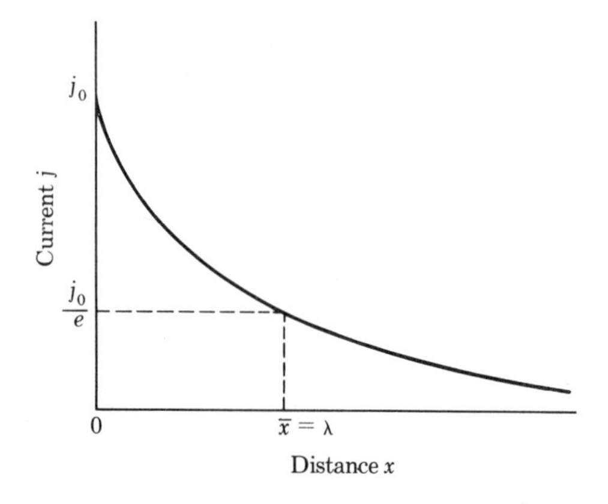

FIGURE 16-5

Exponential variation of current density

ure 16-5. In O_2 at normal temperature and pressure, $N = 2.69 \times 10^{25}/m^3$, $N\sigma = 1.10 \times 10^7/m$, and the mean free path is $\lambda = 1/N\sigma = 9.1 \times 10^{-8}$ m. This is seen to be very small—it would require 10 million such paths for a molecule to go a distance of 1 meter. From another viewpoint, λ is only about 250 times as large as a molecule itself.

We can now find how short a distance gas particles in a stream go without making collisions, using $j(x) = j_0 e^{-N\sigma x}$. For example, in a distance as small as 1 millimeter, $x = 0.001$ m, the exponent is $N\sigma x = (1.1 \times 10^7)(10^{-3}) = 1.1 \times 10^4$, and $j/j_0 \cong e^{-11,000}$ which is zero for all practical purposes.

16-3 PARTICLE DIFFUSION

We are now ready to look into the process of particle diffusion, by study of a simple experiment. If we blow through a drinking straw, there is obviously a flow of air through it because the pressure is higher at one end than the other. We have no control over the molecules within the tube, so it is their own motion that provides the flow. Picture a small area perpendicular to the tube, as shown in Figure 16-6. The air molecules in their chaotic motion cross the area in both directions, as indicated by the arrows. Let j_+ be the number of molecules each second crossing a unit area in the positive x-direction and j_- be the corresponding number in the negative direction. We can deduce what these currents depend on by examination of the details of the process. First, suppose that there was a completely *uniform* distribution of atoms in the space. With a particle-number density n and speed v, if all were moving freely in the positive direction, j_+ would be simply nv. However, some make collisions before they reach the plane, and also are traveling in various directions, so the current is smaller, actually* $nv/4$. The same number crosses from the opposite side, and so the net current $j = j_+ - j_-$ is zero. Now we return to the case in which there are more molecules on the left than on the right, as in Figure 16-6. Over a distance Δx, the particle-number density changes from n to $n + \Delta n$. It is the difference Δn that gives rise to a net flow. We can readily see that the larger the ratio of mean free path λ to the distance Δx the easier it is for particles to get to the area; and that the higher the speed v, the more will be crossing. When one takes account of the fact that particles are moving in many different directions (see Appendix E), the net current $j = j_+ - j_-$ is found to be

$$j = -\frac{\lambda v}{3} \frac{\Delta n}{\Delta x}$$

* To arrive at the factor $\frac{1}{4}$, a sum of all the particles that reach the plane without suffering a collision, and are aimed in the right direction to cross it, is formed by the process of integration.

The dependence on $\lambda/\Delta x$ and Δn agrees with our deduction. The negative sign merely guarantees that j is positive when $\Delta n/\Delta x$ is negative—that is, if the density drops as x increases. The name *diffusion coefficient,* labeled D, is applied to the factor depending on material properties:

$$D = \frac{\lambda v}{3} \qquad \textit{(diffusion coefficient)}$$

Let us estimate D for argon gas at temperature 273°K, using a molecular diameter $d = 3.4 \times 10^{-10}$ m, $\sigma = \pi d^2 = 3.6 \times 10^{-19}$ m^2. Then, with atom-number density $N = 2.69 \times 10^{25}/\text{m}^3$, the mean free path is $\lambda = 1/N\sigma = 1.02 \times 10^{-7}$ m. The speed is computed from $v = \sqrt{3kT/m}$ (Section 14-2), using $k = 1.38 \times 10^{-23}$ J/°K and $m = (39.4)(1.67 \times 10^{-27}) = 6.58 \times 10^{-26}$ kg. This gives $v = 415$ m/sec. Finally, $D = \lambda v/3 = 1.4 \times 10^{-5}$ m^2/sec.

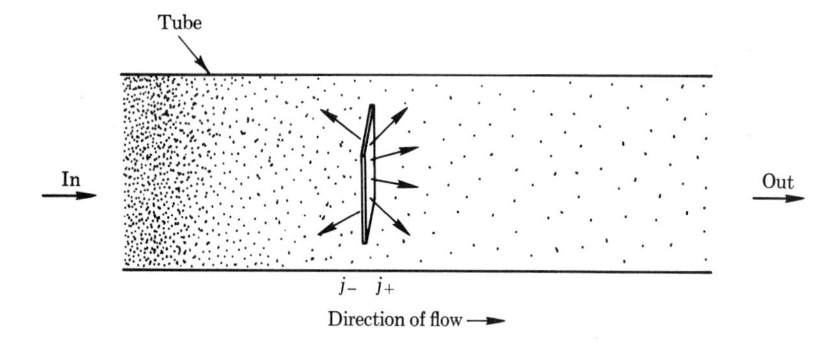

FIGURE 16-6

Density variation and diffusion of particles

For simplicity, we assumed in the above treatment that all particles had the same speed v. There is some logic in using this approximation, since the effects of particles of speed higher than the average of the distribution tend to compensate for the effects of those with lower speed.

16-4 TRANSFER OF ENERGY AND MOMENTUM

The diffusive flow of particles when the concentration of material varies with position is only one of several types of transport. The phenomena of heat conduction in gases, viscosity effects, and electric-charge conduction may be treated as extensions of the process of particle diffusion. Let us start by examining the situation in an insulating space between two walls of a house, as in Figure 16-7. Let the outside wall be at a low temperature, as in winter, and the inside wall be warm. Molecules of air tend to share the temperature of the nearest surface; hence there will be a temperature variation with position x. Recalling the connection between energy

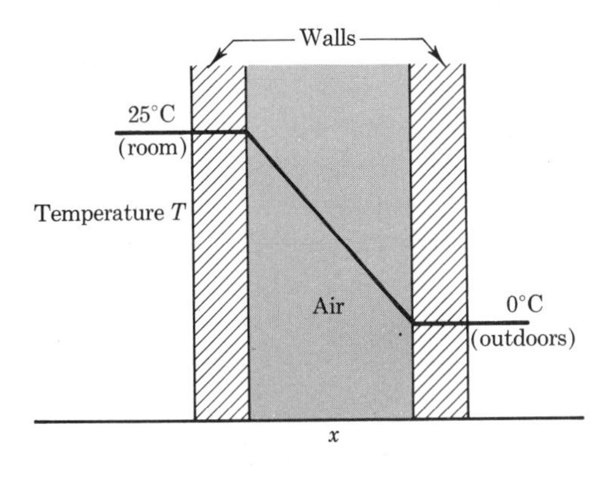

FIGURE 16-7

Temperature variation across insulating gas layer

and temperature (Section 14-2), we find a corresponding variation with position in the average energy of gas molecules.

Now let us examine flows due to random motion across a unit area parallel to the walls. Those crossing from left to right are of higher energy on the average than those crossing in the opposite direction, hence there is a net transfer, and heat is conducted across the gap.

As another example, imagine a flow of fluid such as air being forced at some speed v_f between two parallel walls, as in Figure 16-8(a). Such an arrangement might correspond to a method of cooling the two surfaces; also, it is similar to the flow in a pipe. Because of friction, we would expect the speed to be higher near the center of the space and smaller toward the walls. The vector arrows in the figure indicate this trend. In addition to the airflow speed v_f, there is the kinetic speed v of the particles, associated with the random thermal agitation of the molecules. The velocity of any particle is thus the vector sum of the two velocities $\mathbf{v}_f$ and $\mathbf{v}$. Consider a particle from the center of the stream that crosses the plane, as in Figure

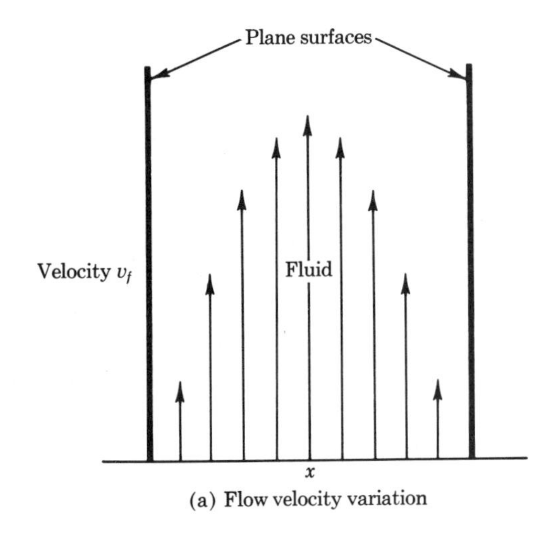

(a) Flow velocity variation

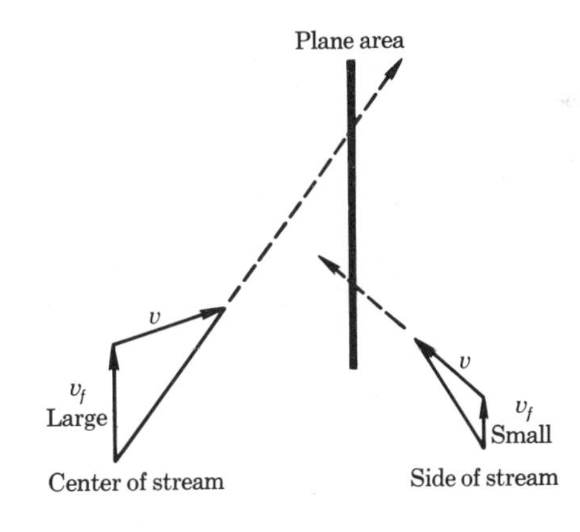

(b) Transfer of momentum

FIGURE 16-8

Flow of fluid between parallel plates

16-8(b), and compare it with another that crosses from the more slowly moving side of the stream. We see that more momentum is transferred to the right than to the left. Such a net transfer of momentum corresponds to a force, according to Newton's second law. This force is that of viscosity, where adjacent layers of a flowing fluid tend to exert a "drag" effect on each other.

The rates of heat conduction H and the viscous force F can be described by simple relations that are very similar in form to that for particle diffusion. There is a dependence on the rate of change of temperature T with position or the rate of change of flow speed v_f with position, each with a numerical coefficient. Extending the ideas in our previous derivation for diffusion, we visualize that each particle carries with it some quantity q as it crosses a selected plane area. For heat transfer, we let q be proportional to kT, which has the units of energy; for viscous effects, we let q be proportional to mv_f, the flow momentum. Instead of n as in the diffusion expression, we use nq, and propose that the net flow be written

$$Q = -\frac{\lambda v}{3}\frac{\Delta(nq)}{\Delta x}$$

The results are shown in Table 16-1, which is a summary of the several transport processes. The new symbols are: H, the rate of flow of energy per unit area; κ, the *thermal conductivity*; F, the force per unit area; and η, the *viscosity coefficient*.

TABLE 16-1

TRANSPORT OF PARTICLES, ENERGY, MOMENTUM, AND CHARGE

Process	Quantity Transferred	Variable	Formula	Units	Coefficient
Particle diffusion	Particles	Concentration	$j = -D\frac{\Delta n}{\Delta x}$	1/m²-sec	$D = \frac{\lambda v}{3}$
Heat conduction	Kinetic energy	Temperature	$H = -\kappa\frac{\Delta T}{\Delta x}$	J/m²-sec	$\kappa = \frac{\lambda vnk}{3}$
Viscous drag	Momentum	Flow speed	$F = -\eta\frac{\Delta v_f}{\Delta x}$	N/m²	$\eta = \frac{\lambda vnm}{3}$
Charge flow	Charge	Potential	$i = \frac{\sigma}{e}\frac{\Delta V}{\Delta x}$	C/m²-sec	$\frac{\sigma}{e} = \frac{\lambda nve}{3kT}$

For argon gas, using constants from a previous paragraph, we can readily compute

$$\kappa = \frac{\lambda vnk}{3} = \frac{(1.02 \times 10^{-7})(415)(2.69 \times 10^{25})(1.38 \times 10^{-23})}{3}$$
$$= 5.2 \times 10^{-3}\ \text{J/m-sec-°C}$$

and

$$\eta = \frac{\lambda vnm}{3} = \frac{(1.02 \times 10^{-7})(415)(2.69 \times 10^{25})(6.58 \times 10^{-26})}{3}$$

$$= 2.5 \times 10^{-5} \text{ kg/m-sec}$$

16-5 TRANSFER OF ELECTRIC CHARGE

Let us consider a space between parallel plates that contains a gas and some singly charged particles (Figure 16-9). When an electric field is produced in the region by a battery or other electrical supply, a flow of electrical current takes place that is the product of the particle current and the charge on each particle, $i = je$. Such an arrangement can serve as a detector for radiation or particles emitted in radioactivity.

Our transport viewpoint may be used to find the dependence of i on the strength of the electric field $\mathcal{E}$. Here we use an indirect approach, first finding a value of a hypothetical particle-concentration change that would yield the same flow as does the actual electrical-potential change between the plates. Suppose that the quantity q transported (see Section 16-4) is in general the combination

$$q = kT + Ve$$

where we note that both terms have the meaning of energy. Compare particle flows in two special cases:

1. Particle-density variation only; potential $V = 0$ and no temperature variation:

$$j = -\frac{\lambda vkT}{3}\frac{\Delta n}{\Delta x}$$

2. Potential variation only; no particle concentration or temperature variations:

$$j = -\frac{\lambda vne}{3}\frac{\Delta V}{\Delta x}$$

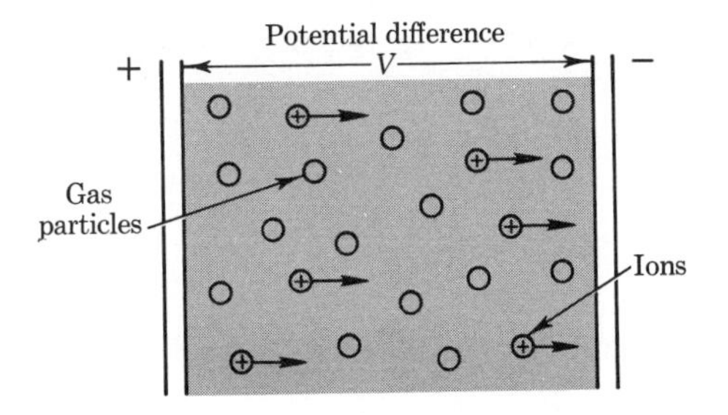

FIGURE 16-9
Charge flow in gas—particle detector

The particle-density variation $\Delta n/\Delta x$ that would give the same flow as provided by the potential variation is thus seen to be $(ne/kT)(\Delta V/\Delta x)$. Inserting that in the formula for diffusion (Section 16-3), and then letting $\mathcal{E} = -(\Delta V/\Delta x)$ and $i = je$, yields

$$i = \sigma\mathcal{E}$$

where the electrical conductivity* is

$$\sigma = \frac{\lambda v n}{3kT} e^2$$

(For completeness, we include these results in Table 16-1.) It should be noted that this expression holds for conduction in gases, but not for conduction in metals, where new interactions take place.

The current in the above formula is directly proportional to the electric field, as we might expect. The formula is one form of *Ohm's law*, which is very widely used in the study of electrical currents. For the description of electrical discharges in gases, a slightly different expression is convenient. The current density i can be written as

$$i = n_0 v_d e$$

where v_d is the "drift" speed of the n_0 particles per unit volume, each with charge e. Comparing formulas,

$$v_d = \frac{\sigma\mathcal{E}}{n_0 e} = \frac{De}{kT}\mathcal{E} \qquad \textit{(drift speed)}$$

Now, we define the ratio of speed to field as the *mobility* $\mu = De/kT$.

More than one force creates the motion of charges in gaseous electric discharges and in the semiconductors found in a transistor radio. Variations with position of both particle density and potential give rise to a combined force, and the resulting current is

$$i = -eD\frac{dn}{dx} + n\mu e\mathcal{E}$$

The flow of positive charges is implied here.

It is instructive to compare speeds in various situations, using the argon data of Section 14-3. First, in argon gas there is the kinetic speed, 415 m/sec. Next, let us find the final speed of argon ions accelerated in vacuum through a gap of 0.01 m with potential difference $V = 100$ volts. Then, the energy gain is $Ve = \frac{1}{2}mv^2$ and $v = \sqrt{2eV/m} = 2.2 \times 10^4$ m/sec.

* The symbol σ is used for surface-charge density (Section 6-2), for cross-section (Section 16-2), and for conductivity. This is another example of the inadequacy of notation.

Finally, the drift speed of ions through argon gas in a field $\mathcal{E} = 10^4$ V/m is

$$v_d = \frac{De}{kT}\mathcal{E} = \frac{(1.4 \times 10^{-5})(1.60 \times 10^{-19})(10^4)}{(1.38 \times 10^{-23})(273)} = 6.0 \text{ m/sec}$$

This is very small because the ion flights are continually interrupted by collisions, and little progress is made.

The similarities in mathematical forms for the transport of particles, thermal energy, momentum, and charge suggests that the constants D, κ, η, μ and σ for a substance may be related (see Problems 16-22 and 16-23).

16-6 HEAT TRANSFER IN SOLIDS, LIQUIDS, AND GASES

The energy of thermal motion is found to be transferred in solid and liquid bodies according to the same rule as for gases, $H = -\kappa(\Delta T/\Delta x)$, where κ is the thermal conductivity.

Consider this simple experiment: a thin metal rod of length L and cross-sectional area A is heated continuously at one end by the flame from a laboratory Bunsen burner, with the other end dipped in a large beaker of water (Figure 16-10). The heated end and the cooled end are at different temperatures T_1 and T_2, and there is a linear variation of temperature

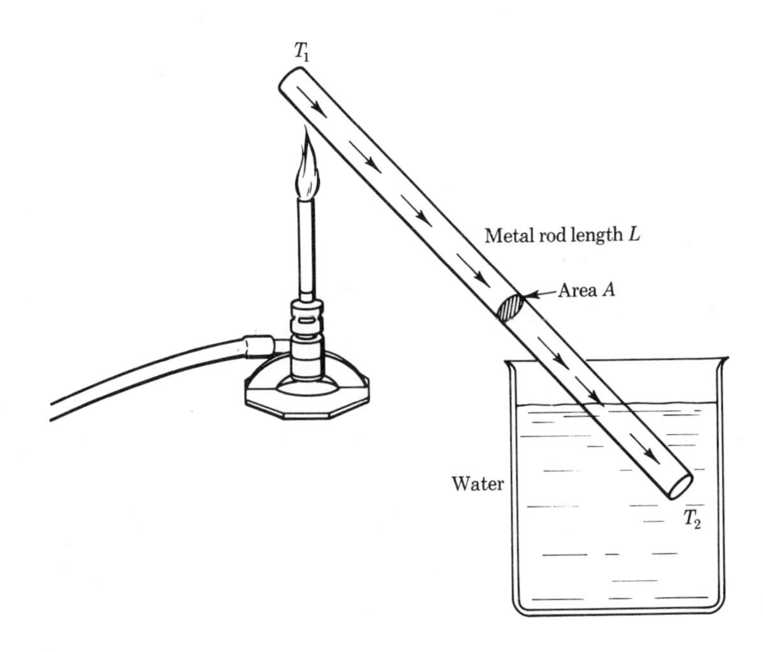

FIGURE 16-10

Conduction of heat in metal rod

along the rod. We assume that the rod is insulated by some nonconducting material or by a vacuum, so that heat only flows along the x-axis. The rate the temperature changes with position is $\Delta T/\Delta x = (T_1 - T_2)/L = \Delta T/L$, so that the amount of heat energy transferred by conduction each second through a unit area is $\kappa(\Delta T/L)$. In a time t, through the rod area A, the total amount of heat transferred is then

$$Q = \frac{\kappa A \Delta T t}{L}$$

The units of Q are

$$(\text{J/m-sec-°C})(\text{m}^2\text{-°C-sec/m}) = \text{J}$$

as expected. Table 16-2 gives values for κ for typical materials. Even though we used a thin rod in the above experiment, we would find the same dependence for a sheet of metal or a wall. The only requirement is that there be no flow perpendicular to the x-direction.

As a numerical example, find the heat transferred per minute through a silver rod of 0.5-cm diameter (area 2.0×10^{-5} m^2), 0.25 m long, with temperatures at the ends 100°C and 20°C:

$$Q = \frac{\kappa A \Delta T t}{L} = \frac{(420)(2 \times 10^{-5})(80)(60)}{0.25} = 160 \text{ J}$$

Let us compare the heat lost from a house per unit area through glass windowpanes ($\kappa = 0.8$; thickness, 0.003 m) and through concrete walls

TABLE 16-2

THERMAL CONDUCTIVITIES

SUBSTANCE	κ (J/m-sec-°C)
Air	0.023
Asbestos	0.17
Aluminum	200
Concrete	1.0
Copper	380
Glass	0.8
Ice	2
Iron	60
Rock wool	0.04
Silver	420
Water	0.60
Wood, pine	0.1

Conversions:
1 Btu/ft^2-hr-°F/in. = 0.144 J/m-sec-°C
1 cal/cm-sec-°C = 418.5 J/m-sec-°C

(κ = 1.0; thickness, 0.12 m). We assume that the temperature inside is a normal 20°C, and outside is freezing, 0°C. For glass, the loss per second is

$$Q = \frac{(0.8)(1)(20)(1)}{0.003} = 5300 \text{ J}$$

For concrete it is

$$Q = \frac{(1.0)(1)(20)(1)}{0.12} = 167 \text{ J}$$

From these results we can appreciate why the air is colder near a window than near a wall, and why two glass panes with an air space between them give good insulation.

16-7 CONVECTION OF HEAT

The conduction of heat is poor in liquids and gases, in which the mechanism of *convection* is responsible for most of the heat transfer. This process involves motion of large amounts of fluid due to temperature differences and buoyancy. *Natural* or *free convection* is a circulation that takes place automatically when a fluid is in contact with a hot surface; *forced convection* involves a flow of fluid over the surface that continually sweeps away heated particles.

We know that hot air rises and that cold air settles in a room. This is due primarily to the lower density of the heated gas, which causes it to be buoyed up. A circulation in a room takes place when air is heated on contact with a "radiator" and rises, comes in contact with walls and cools down, then becomes more dense and sinks toward the floor. We also know that there is a cooling of surfaces by flow of a fluid. A breeze on a hot day will cool our face, or a strong north wind in winter can remove body heat very rapidly. The passage of air over the radiator coils of our moving car takes away the heat developed in the engine.

From a molecular viewpoint, the process is easy to visualize but hard to describe precisely. Molecules of air strike the heated surface, gain energy (at the expense of the surface), and move on with the stream. We can suggest properties of the cooling agent and boundary that might be important in *increasing* the rate of heat transfer:

high velocity (heated fluid replaced quickly);
large specific heat (heat absorbed with small temperature rise);
large conductivity (ability to transfer heat well);
high degree of internal motion (much exchange of fluid);
large surface area (much contact with fluid);
large difference in temperature between surface and fluid (good local conduction).

The amount of heat transfer ΔQ between surfaces at temperatures T_1 and T_2 in a time interval Δt is written in the convenient empirical form

$$\Delta Q = hS(T_1 - T_2)\Delta t$$

where the direct dependence on surface area and temperature drop are displayed, and h accounts for all other factors in the above list. This proportionality constant, called the *heat-transfer coefficient,* is expressed in units of J/m^2-°C. It depends on the substance through its specific heat, conductivity, and density. It also varies with the fluid speed and turbulence, and the shape of the fluid passage. Experimental measurements lead to tables of data, graphs, or empirical formulas for h.

The proportionality to the temperature difference was observed by Newton, and the dependence is called *Newton's law of cooling:*

$$Q \sim \Delta T$$

Heat transfer may be steady or time-dependent. Imagine the following experiment. A solid of mass M is raised to a high temperature T_0, and allowed to cool in air, suspended in such a way as to prevent conduction loss (Figure 16-11). We know that its temperature T falls until it eventually reaches that of the air, T_a. Let us estimate T as it depends on time.

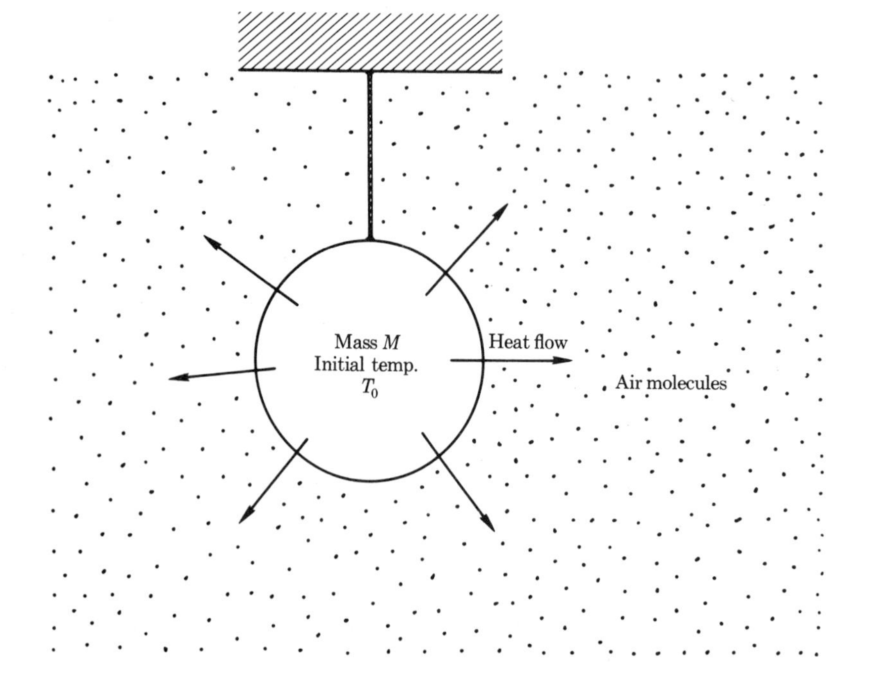

FIGURE 16-11

Heat transfer by convection

Conservation of energy requires that the flow of heat to the surrounding air be equal to the change in the internal energy of the solid. Then,

$$cM\Delta T = -hS(T - T_a)\Delta t$$

Suppose that we let the air temperature be zero, knowing that its actual temperature can be added to any value we obtain by solving the above equation. Letting the "time constant" τ be an abbreviation for cM/hS,

$$\frac{\Delta T}{\Delta t} = -\frac{T}{\tau}$$

We have solved similar equations several times before (e.g., in Section 5-8). If $T = T_0$ at $t = 0$, then

$$T = T_0 e^{-t/\tau}$$

The equations tell us that the rate of cooling is high initially when T is large, but that the rate gets quite small as T approaches zero, the temperature of the surrounding air. The quantity τ tells how long a time is required for T to fall by a factor of e. A plot of temperature is shown in Figure 16-12 (with the actual air temperature included).

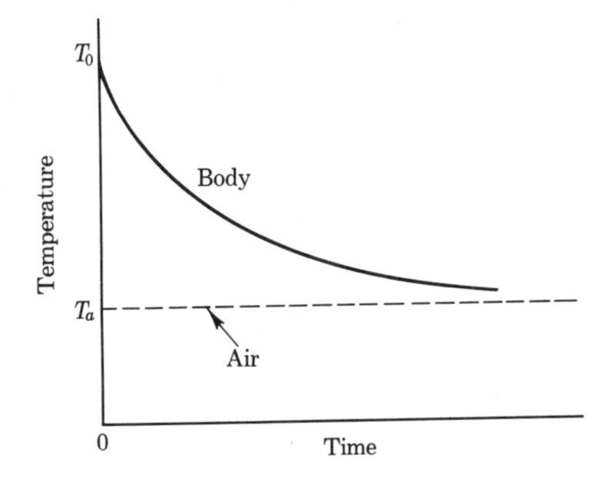

FIGURE 16-12

Cooling in air—Newton's law

If the amount of surface area per kilogram of material is large, as would be the case for a flat plate, the smaller is τ, which means that the exponential curve drops rapidly with time. The larger the heat transfer coefficient h, the smaller is τ, as expected.

Experimental comparisons between heat capacities $C = cM$ of two liquids (subscripts 1, 2) can be made by immersing identical hot objects (subscript 0) in them. The ratio of time t_1 to time t_2 for the liquids to reach a specified temperature is

$$\frac{t_1}{t_2} = \frac{C_1 + C_0}{C_2 + C_0}$$

Newton's law of cooling by convection is the dominant description so long as the temperature difference ΔT is relatively small. When the surface is raised to very high temperature compared to the surroundings, a new process of heat transfer by radiation becomes important. This mechanism involves electromagnetic waves, and we shall reserve discussion of it until Chapter 18.

16-8 THE SPEED OF VISCOUS LIQUIDS

The viscosity coefficient η determines how much pressure difference is required to create fluid flow through a pipe. The difficulty in pouring a thick liquid such as cold syrup from a bottle suggests that this pressure varies directly with η. Study of Table 16-3 reveals several points of interest.

TABLE 16-3
VISCOSITIES

MATERIAL	TEMPERATURE (°C)	η (kg/m-sec)
Hydrogen	0	0.84×10^{-5}
Carbon dioxide	0	1.40×10^{-5}
Air	−32	1.54×10^{-5}
	0	1.71×10^{-5}
	18	1.83×10^{-5}
Water	0	1.79×10^{-3}
	20	1.00×10^{-3}
Mercury	0	1.68
	20	1.55
Light machine oil	16	0.114
	100	0.0049
Glycerin	0	12.1
	20	1.49

Comparing η for H_2 and air, we see that the viscosity of gases increases with mass, as predicted by our formula (Section 16-4); water is much less viscous than liquids such as glycerin or oil; the viscosity decreases sharply with increasing temperature. The necessity in winter for the motor and the oil of a car to warm up before parts move freely is a good illustration of the latter point.

Picture the flow of a liquid in a horizontal circular tube of radius R and length ΔL, as in Figure 16-13. Familiar examples might be a section of garden hose, or household water piping, or oil pipeline. In order to maintain continuous flow, a pressure difference ΔP must be maintained between the ends. The fluid will move in streamlines, with the highest speed along the central axis of the tube and zero speed at the inner wall. The average speed of liquid is directly proportional to the pressure drop and to the pipe cross-sectional area, and inversely proportional to the length of pipe and

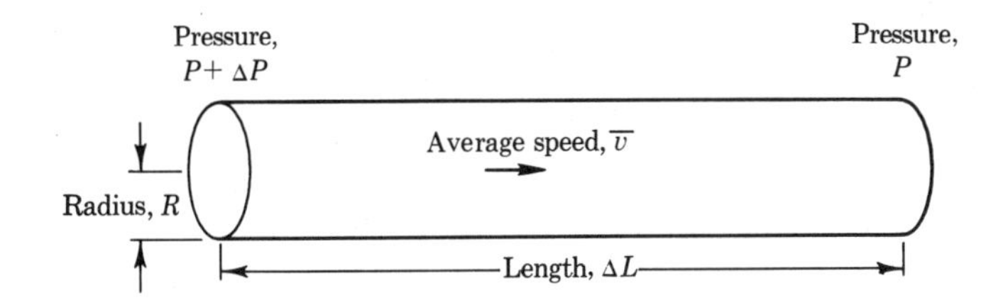

FIGURE 16-13

Viscous flow of liquid in a circular tube

to the viscosity. By the use of the calculus, as a slight extension of methods used in Section 2-3 to obtain distance as a function of time, it is found that the average speed of fluid in a circular pipe is given by the formula

$$\overline{v} = \frac{\Delta P}{\Delta L}\frac{R^2}{8\eta}$$

Similar relations can be derived for ducts of other shapes. Only the numerical factors are different. It must be noted, however, that if the speed of fluid is high, the phenomenon of turbulence sets in, and such streamline flow no longer occurs.

PROBLEMS

16-1 Compute the cross-section and mean free path of high-speed helium atoms that are injected as a stream into air at 100°C and atmospheric pressure. Assume that the diameters are $d_{\text{He}} = 2.6 \times 10^{-10}$ m and $d_{\text{air}} = 3.7 \times 10^{-10}$ m. How far, on the average, will the atoms go before colliding?

16-2 Calculate the potential energy of attraction between two oxygen molecules separated by a distance of 3.6×10^{-9} m, which is 10 times the diameter of one molecule. Compare with the value at $d = 3.6 \times 10^{-10}$ m.

16-3 To measure the air density in the rare upper atmosphere, a stream of charged particles is projected across a gap and the transmitted current is measured. Assuming a cross-section $\sigma = 4 \times 10^{-19}$ m² and a gap of 0.01 m, estimate the air-particle density if the current is reduced by a factor of 2, noting that $e^{-0.693} = 0.5$. How does this compare with the density at sea level?

16-4 Radioactive argon atoms are released from a plane surface at a definite rate and diffuse through argon gas to another parallel surface, where they are absorbed. If the measured drop in particle density is from $2.7 \times 10^{20}/\text{m}^3$ to zero in 0.015 cm, and the rate of flow is $6 \times 10^{12}/\text{m}^2\text{-sec}$, what is the value of the diffusion coefficient?

16-5 A container of volume V at initial pressure P_0 is immersed in a vacuum. If a hole of area A is opened, the inside pressure varies with time as $P = P_0 e^{-Avt/4V}$, where v is the average particle speed in the gas—e.g., 500 m/sec for air. If a meteorite punctures a 0.005-m²-area hole in a spacecraft of volume 100 m³, how long will it take the pressure to drop to $\frac{1}{2}$ atmosphere?

16-6 What would you expect the ratio of thermal conductivities of diatomic to monatomic gases to be?

16-7* Calculate the electrical conductivity for argon ions in argon gas at 2 atm pressure and 100°C, assuming particle diameters of 3.4×10^{-10} m.

16-8 The measured mobility of helium ions in helium at atmospheric pressure is 0.09 m²/V-sec. (a) Find its value at 1.5 atm. (b) Find the length of time it would take an alpha particle to cross a particle detector with potential difference 800 V across a gap of 3 cm, if the pressure were 1.5 atm.

16-9 Estimate the transport constants D, κ, and η for neon gas (atomic weight 20, $d = 2.95 \times 10^{-10}$ m) at 0°C and atmospheric pressure.

16-10 Using the diffusion coefficient for neon calculated in Problem 16-9, compute the mobility and then the drift speed of singly charged neon ions in a gap of width 2 cm across which a potential difference of 2500 V is maintained.

16-11* Compare numerically the conduction-heat loss through two types of wall—(a) solid concrete, and (b) equal layers of concrete, rock wool and concrete—assuming the total thickness and temperature differences to be the same. *Suggestion:* First show that the ratio of heat flows is $\dfrac{3}{2 + \kappa_c/\kappa_r}$.

16-12 Which is a better insulator for heat transfer—an air gap of 0.05-m width or a wooden panel 0.2-m thick?

16-13 The ends of a circular iron rod of radius 0.007 m and of length 0.5 m are maintained at 0°C and 100°C by contact with ice and steam, respectively. If the sides are insulated, how much heat flow per second takes place?

16-14 The interior of a spaceship chamber with aluminum wall 0.03-m thick is kept at 25°C, and the outside is at absolute zero. Estimate how many watts a heater would have to provide to balance the conduction loss from a total surface area of 24 m². Comment on the practicality of a space capsule of this construction.

16-15 The following temperature data were taken on a body in air at room temperature 25°C, of surface area 0.2 m² and heat capacity $cM =$ 6 kcal/°C:

t (sec)	T (°C)
0	250
60	200
120	165

Estimate the time constant for cooling τ and the heat-transfer coefficient for convection.

16-16 Hot water flows through thin metal tubing in a floor heating system. What temperature difference is required in order to deliver 15 watts per meter length of tube if the tube is 0.006 m in diameter and the heat transfer coefficient is 40 W/m²-°C?

16-17 When coffee is poured, you have a choice of adding cream and sugar at once or waiting until you are ready to drink it. Which will insure the higher final temperature? Why?

16-18* Investigate the following natural phenomena and relate them to principles discussed in this chapter and in Chapter 13: (a) jet stream, (b) gulf stream, (c) tornado, (d) hurricane, (e) flow from a nozzle.

16-19 A pump for oil is to be chosen for transferring oil in Antarctica, where temperatures go as low as −60°C and where the viscosity is very high, $\eta = 40$ kg/m-sec. Find the pressure drop in N/m² and lb/in.² required to obtain an average flow speed $\bar{v}$ of 0.5 m/sec in a pipe of 0.02-m radius and 3-m length, assuming that $\bar{v} = \dfrac{\Delta P}{\Delta L}\dfrac{R^2}{8\eta}$.

16-20* Stokes's law states that a small spherical object of radius r moving slowly with speed v through a viscous medium experiences a force of $6\pi\eta r v$. (a) Show that the limiting or terminal speed of fall of an object of density ρ (when the viscous force is equal to the weight of the sphere minus the buoyant force) in a medium of density ρ_0 is given by $v_{\text{max}} = \dfrac{2}{9\eta}(\rho - \rho_0)r^2 g$. (b) Calculate v_{max} in water at 0°C, for a steel ball of radius 0.006 m and density 7.8×10^3 kg/m³. (c) Find v_{max} in air at 0°C for sphere of density 1000 kg/m³ and volume 1.5 m³. How accurate would you expect this result to be?

16-21* Verify the following transport ratios: (a) diffusion coefficient to mobility, $D/\mu = kT/e$ (Einstein's relation); (b) electrical conductivity to mobility, $\sigma/\mu = ne$; (c) thermal conductivity to viscosity, $\kappa/\eta = k/m \cong c_v$; (d) thermal conductivity to electrical conductivity, $\kappa/\sigma = (k/e)^2 T$ (Wiedemann–Franz law).

16-22 Calculate the ratio of thermal to electrical conductivities at room temperature, using the Wiedemann–Franz law (Problem 16-21).

16-23 Transport concepts are often applicable to solids as well as to gases. If copper and porcelain are respectively good and poor conductors of electricity, discuss their predicted ability to conduct heat (note the results of Problem 16-21).

ELECTRICAL AND MAGNETIC PROPERTIES OF MATTER

17

The importance of electricity in our modern civilization is obvious—in the home it provides light and heat and operates many devices; in industry it drives machinery, controls automatic processes, and provides heat for chemical reactions; in communication it makes possible the operation of the telephone, telegraph, radio, and television; in scientific research, it permits the detection and measurement of signals of many kinds. During the many years before the atomic basis of electricity was discovered a variety of experimental rules and recipes was developed.

In previous chapters, we examined the concepts of electric and magnetic fields in a vacuum as produced by fixed or moving charges. When atoms of matter are present, with their charged electrons and nuclei, there will be an *interaction* between the charged particles and the fields due to atoms and the externally applied fields. The study of these interactions from the fundamental-particle viewpoint is the subject of this chapter.

17-1 MOTION OF ELECTRICITY IN MATERIALS—RESISTANCE

The main facts and ideas on which our study of electricity is based are the following:

1. The electron, which possesses 1 basic unit of electrical charge

(1.60×10^{-19} C) is the principal agent in electrical-charge flow. It is relatively free to move under the influence of electric and magnetic fields. Protons and other nuclear particles are much less mobile than electrons.

2. Electrons and other charges can have momentum and kinetic energy as individual particles. Work is done on and by charges as they interact with electric and magnetic fields.

3. Charges may be deposited electrostatically on surfaces of insulating and conducting materials, or they may stream through a vacuum, or migrate with many collisions through gaseous, liquid, or solid materials.

4. Magnetism is due to charged-particle currents, both on a large scale (as in the electromagnet) and on a submicroscopic scale (as in atoms and nuclei). Electrons are the main agents of magnetism, but positive and negative ions may participate also.

5. Changes in magnetic fields with time give rise to electric fields that can drive charges along certain paths.

As part of our study of transport of electrons in gases, we derived Ohm's law, which states that electric current is proportional to the electric field and depends on the material's conductivity. Early workers attempted to carry this analysis over to solid metallic materials, where a similar dependence had been noted. The similarity is only superficial, however, because conduction in metals is more correctly described by the use of wave mechanics.

The strength with which electrons are bound to atoms in a solid depends greatly on the type of material. Some materials, such as copper or silver, have many free electrons that can serve to carry current; while others, such as quartz or porcelain, have very few free electrons and do not conduct electricity appreciably. Materials at these extremes are called respectively *conductors* and *insulators*. Between them are the *semiconductors* that form the basis for the transistor, tunnel diode, and integrated circuits of modern electronics. The number of free electrons increases in all materials as the temperature of the substance is raised, since more energy is available for the electrons to escape from the bonds provided by the positive nuclei. We shall reserve further discussion of the distinctions between these classes of material until Section 17-6, where a more detailed atomic picture will be developed.

To reveal some practical aspects of electricity and matter, let us perform three basic experiments. In the first experiment, we connect a source of potential difference V—say, a flashlight battery—to a length of thin copper wire, as in Figure 17-1. By use of a meter we measure a steady current I. If we increase the voltage by adding more batteries, the current goes up proportionately. Ohm's law describes this effect:

$$I = \frac{V}{R} \quad \text{or} \quad V = IR \qquad (\textit{Ohm's law})$$

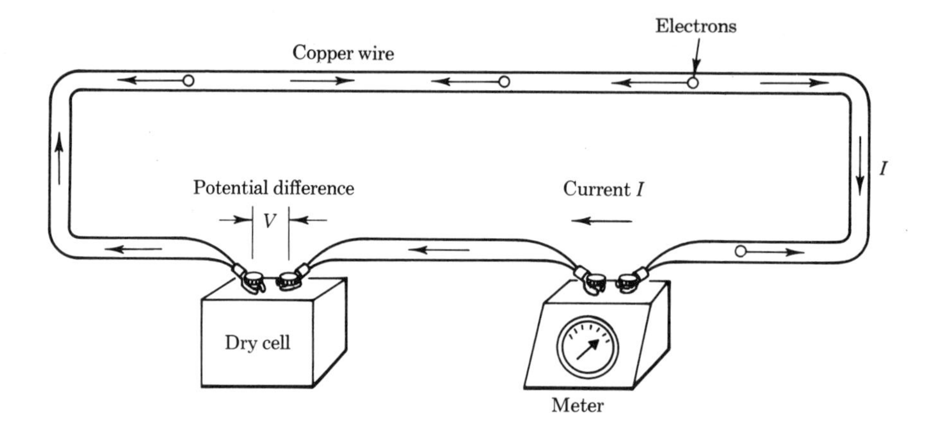

FIGURE 17-1

Application of Ohm's law, $V = IR$

where the *resistance* R is a measure of the difficulty in sending charges through the particular sample of wire. The unit of resistance is called the *ohm*, after G. S. Ohm, an electrical researcher of the early nineteenth century. For example, if $V = 1.5$ volts and $I = 0.03$ ampere, $R = V/I = (1.5\text{ V})/(0.03\text{ A}) = 50$ ohms. We find that if we double the length of the copper conductor, the current is halved; if we double the area through which charges flow, the current doubles; if we use another metal, the resistance is found to have a different value.

All these results can be explained by proposing that the ability of each substance to conduct electricity is measured by an *electrical conductivity* σ, and that the flow of electrons per unit of area is proportional to the electric field within the wire according to

$$i = \sigma \mathcal{E}$$

(This has the same form as the expression derived in Section 16-5.) Now, if the total current through the cross-sectional area A is $I = iA$, and if the field is uniform along the length of wire l, then $\mathcal{E} = V/l$. Combining,

$$I = \frac{\sigma A}{l} V$$

The direct and inverse dependence of current on A and l respectively is verified. Now, if we let $R = l/(\sigma A)$, we obtain Ohm's law, $I = V/R$. It is important to note the assumption that the field is uniform within the material, because in circuit elements such as the vacuum tube this is not true, and Ohm's law does not hold.

Using the formula $\sigma = i/\mathcal{E}$, the units of conductivity are found to be A/V-m = $(\text{ohm-m})^{-1}$.

The reciprocal of the conductivity, $1/\sigma$, is called the *resistivity* ρ, with units of ohm-meters. From measured data on ρ, such as in Table 17-1, the

TABLE 17-1
RESISTIVITY VALUES

	MATERIAL	ρ (ohm-m)
Insulators	Glass	9×10^{11}
	Mica	9×10^{13}
	Fused quartz	5×10^{16}
	Porcelain	3×10^{12}
	Hard rubber	1×10^{16}
	Water, distilled	5×10^{3}
Conductors	Aluminum	2.6×10^{-8}
	Copper	1.7×10^{-8}
	Iron	1.0×10^{-7}
	Platinum	1.1×10^{-7}
	Nichrome	1.0×10^{-6}

resistance of a wire of known length and diameter can be calculated. For example, No. 12 wire such as used in household circuits has a diameter d of 2.6×10^{-3} m, and thus an area $\pi d^2/4$ of 5.3×10^{-6} m^2. A 15-m length of copper wire, $\rho = 1.7 \times 10^{-8}$ ohm-m, thus has a resistance of

$$R = \frac{\rho L}{A} = \frac{(1.7 \times 10^{-8} \text{ ohm-m})(15 \text{ m})}{5.3 \times 10^{-6} \text{ m}^2} = 0.048 \text{ ohm}$$

Combinations of Resistances

An arrangement of resistance elements in a circuit can generally be viewed as a combination of groups in series or in parallel. The equivalent single resistance in each case is readily deduced. In Figure 17-2, the total voltage drop V in a *series* circuit is the sum of $V_1 = IR_1$ and $V_2 = IR_2$, the current being common. If we let $V = IR$ where R is the equivalent resistance, then $IR = IR_1 + IR_2$, or

$$R = R_1 + R_2$$

This is to be expected, since the longer the path, the greater the resistance and the greater the work to carry charge from one end to the other. For

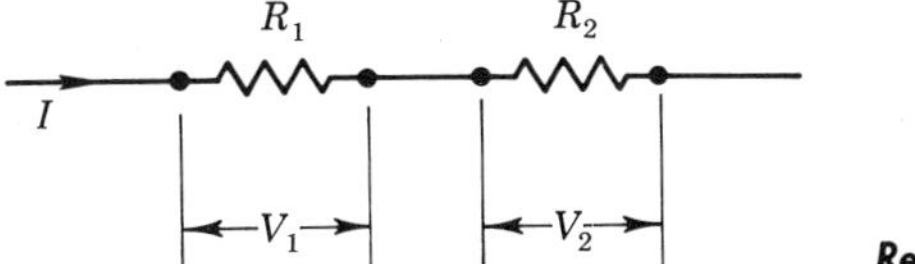

FIGURE 17-2
Resistors in series

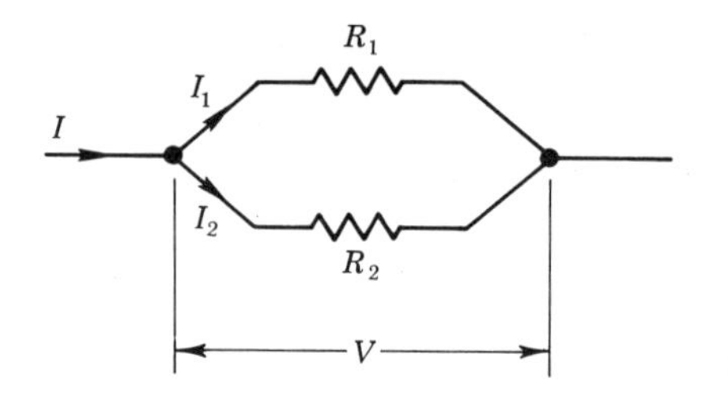

FIGURE 17-3

Resistors in parallel

any set, by extension,

$$R = \sum_i R_i \qquad (series)$$

For example, if a set of 6 lightbulbs, each 0.2 ohm, are attached in series for a Christmas tree, the total resistance is $6(0.2) = 1.2$ ohms. In Figure 17-3, the voltage drop V in a *parallel* circuit is common to both elements, hence is either $V = I_1R_1$ or $V = I_2R_2$. The total current is the sum $I = I_1 + I_2$ Again let $V = IR$, where R is the equivalent resistance; then $V/R = V/R_1 + V/R_2$, or

$$\frac{1}{R} = \frac{1}{R_1} + \frac{1}{R_2}$$

This is reasonable, since the more choices of path a charge has, the easier it can be transported between two points. For several resistors in parallel, by extension of this approach,

$$\frac{1}{R} = \sum_i \frac{1}{R_i} \qquad (parallel)$$

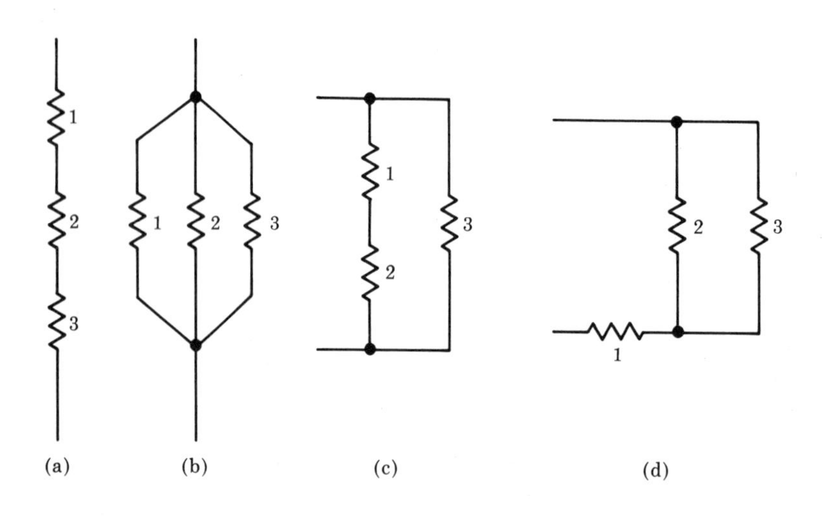

FIGURE 17-4

Combination of resistances

If the set of 6 bulbs in the example above were arranged in parallel, then $1/R = 6(1/0.2) = 30$, and $R = \frac{1}{30}$ ohm.

By the use of the series and parallel relations, we can calculate the total resistance of various circuits. For example, let us find R for several combinations of resistors of resistance 1, 2, and 3 ohms, as in Figure 17-4. For case (a), the total is $R = 1 + 2 + 3 = 6$ ohms. For case (b), $1/R = 1/1 + 1/2 + 1/3 = 1.83$ and $R = 0.55$ ohm. For case (c), the two resistances in series add to 3 ohms; then, $1/R = 1/3 + 1/3$ and $R = 1.5$ ohm. For case (d), the two resistances in parallel combine as $1/R_{23} = 1/2 + 1/3 = 0.833$, so that $R_{23} = 1.2$ ohm. Then, the total is $1 + 1.2 = 2.2$ ohms.

Experimentally, the resistivity of most conductors increases with temperature, expressed by

$$\rho = \rho_0(1 + \alpha\Delta T)$$

where ρ_0 is the resistivity at a specified temperature, and ΔT is the increase, and α is the *temperature coefficient of resistivity* in $(°\text{C})^{-1}$. (Note the analogy to expansion, Section 13-3.) Values are listed in Table 17-2. The very low

TABLE 17-2
TEMPERATURE COEFFICIENT OF RESISTIVITY
(*at* 20°C)

MATERIAL	α in $(°\text{C})^{-1}$
Aluminum	4.0×10^{-3}
Copper	4.0×10^{-3}
Iron	5.8×10^{-3}
Platinum	3.6×10^{-3}
Nichrome	1.6×10^{-4}
Constantan	2.0×10^{-6}

value of α for constantan makes this alloy of Ni and Cu ideal for precision resistors. The resistance of a given wire will increase by the same factor, $1 + \alpha\Delta T$, as does its resistivity. In our example of No. 12 copper wire, an increase from 20°C to 100°C yields $1 + \alpha\Delta T = 1 + (4 \times 10^{-3})(80) = 1.32$, and the new resistance is $R = (1.32)(0.048) = 0.063$ ohm.

Heat Developed in an Electrical Circuit

As current flows through a resistor, such as a length or coil of wire, electrical energy is turned into thermal energy. From the viewpoint of the circuit, this energy is lost or dissipated; from the outside it may appear as a source of useful energy—as, for example, in a room heater, an electric iron, or a kitchen range. The rate at which electrical energy is converted into thermal energy is easily found. The work done in carrying a charge Q

through a constant potential difference V is VQ. Thus the power, defined in Section 8-5 as the rate of doing work, is $P = V(Q/t) = VI$. By Ohm's law however, $V = IR$, or $R = V/I$; hence

$$P = VI = I^2R = \frac{V^2}{R} \qquad \textit{(heating by a resistor)}$$

We can view these relations as equivalent to and as basic as Ohm's law. When R is in ohms and I is in amperes, the power is in joules/second or *watts*. For example, suppose a flashlight bulb of resistance 1.2 ohms is powered by a 1.5-V battery. The current drawn is $I = V/R = 1.5/1.2 = 1.25$ A, and the power is $P = I^2R = (1.25)^2(1.2) = 1.88$ W.

The heating effect can be understood qualitatively on an atomic basis. Electrons accelerated by the electric field inside the conductor make frequent collisions with the atoms. The speed of drift of electrons in a given direction is very small in comparison with speeds achieved between collisions, which means that much of the electrical energy is converted into thermal motion of the atoms that constitute the conductor.

17-2 STORAGE OF ELECTRICAL CHARGE—CAPACITANCE

Let us now perform a second basic experiment. We connect a source of potential difference to a pair of parallel metal plates separated by a gap, as shown in Figure 17-5. We shall assume that there is a vacuum between the plates. This is the very best insulator, there being no material present to conduct charges. We close the switch for a moment, and then open it again. A potential difference V remains, as could be measured by an instru-

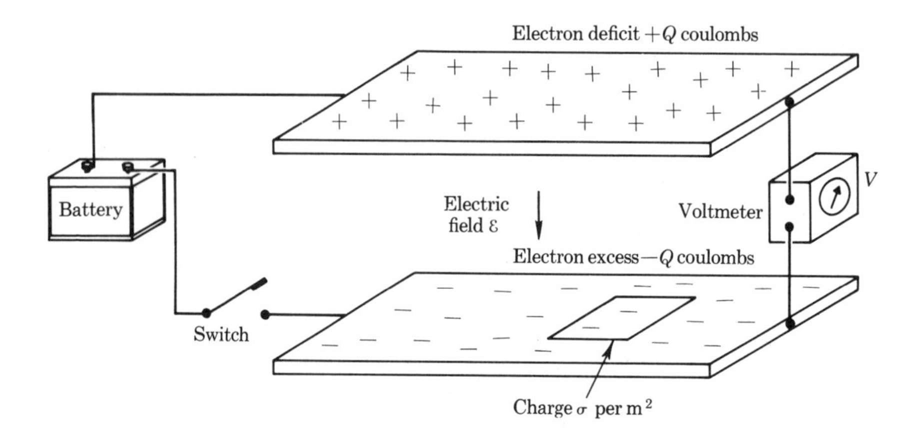

FIGURE 17-5
Parallel-plate capacitor

ment connected between the plates. We deduce that charges have been stored on the plates of this simple *capacitor* (or "condenser"). Our electron model will provide the explanation. Electrons attracted by the positive battery electrode have drained off the upper plate, leaving it positively charged by an amount Q coulombs. Electrons have deposited on the lower plate, to give it a negative charge of the same amount. We find experimentally that the larger V is, the larger is Q. This is reasonable, because the amount of force on the charges depends on V. Thus

$$Q = CV$$

where C is a constant, called the *capacitance*. It measures the "capacity" of the device for each unit of potential difference. The units of capacitance are coulombs per volt (C/V) defined as the *farad*, in honor of Michael Faraday. Since 1 farad is an extremely large capacitance, it is convenient to define the *microfarad* (μF) as 10^{-6} farads, and the *micromicrofarad* ($\mu\mu$F) or *picofarad* (pF) as 10^{-12} farads.

Experimentally, we also find that the capacitance is higher the larger the surface area. This is to be expected, since there is more area on which the charge can deposit. Also, if the gap between plates x is decreased, C and thus Q goes up. Let us try to understand this effect. In a space of width x an electric field is set up. Its magnitude, derived in Section 6-2 by use of Gauss's law, is

$$\mathcal{E} = 4\pi k_E \sigma$$

where the amount of charge per unit area is $\sigma = Q/A$ with units C/m². If we carry a *unit* positive test charge completely around this elementary circuit, we must supply work to lift it from the lower plate to the upper; but the energy will be recovered as the charge goes through the battery. The potential difference between the plates will be the same as that across the battery terminals. If the force on the unit charge is $\mathcal{E}$ and the distance moved is x, the work is

$$V = \mathcal{E}x$$

Thus we conclude that $V = 4\pi k_E \sigma x = 4\pi k_E (Q/A)x$. The ratio Q/V is the capacitance of this particular form of capacitor:

$$C = \frac{A}{4\pi k_E x} \qquad \textit{(parallel-plate capacitor)}$$

The above expression verifies the direct dependence on area and the inverse dependence on gap.

Let us compute C for a small capacitor of area 1 cm², separation 1 mm.

$$C = \frac{10^{-4}\ \text{m}^2}{4\pi(9 \times 10^9\ \text{N-m/C}^2)(10^{-3}\ \text{m})}$$
$$= 0.89 \times 10^{-12}\ \text{C}^2/\text{N}$$

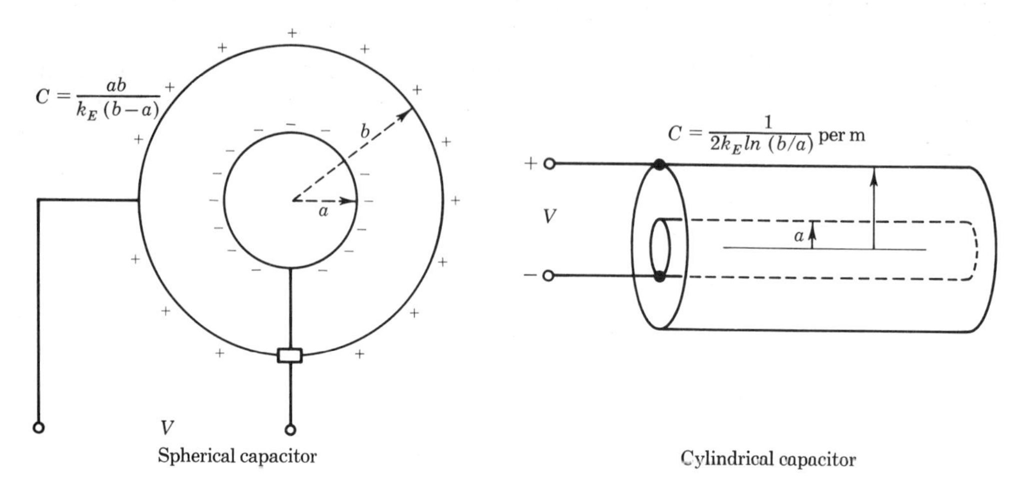

FIGURE 17-6

Capacitors

Since 1 volt = 1 newton/coulomb, the units of capacitance are again found to be coulombs per volt, which is the farad. For our example, C is also 0.89 $\mu\mu$F.

Two other types of capacitors are of interest—concentric spherical or cylindrical shells, as shown in Figure 17-6, with capacitance listed. Notice that if we let the radius of the outer spherical shell approach infinity, the capacitance of the isolated sphere is simply a/k_E, which is dependent only on the radius.

Combinations of Capacitances

The prescription for finding the equivalent capacitances of series and parallel arrangements of capacitors—Figure 17-7(a) and (b)—is *opposite* to that for resistances, as follows:

$$\frac{1}{C} = \sum_i \frac{1}{C_i} \qquad (series)$$

$$C = \sum_i C_i \qquad (parallel)$$

The two equations are easily derived. In Figure 17-7(a), the total voltage drop V is the sum of $V_1 = Q/C_1$ and $V_2 = Q/C_2$, where the amount of charge is common. If we let $V = Q/C$, where C is the equivalent capacitance, then $Q/C = Q/C_1 + Q/C_2$, which reduces to the series formula. In Figure 17-7(b), the total charge is the sum of $Q_1 = VC_1$ and $Q_2 = VC_2$, where the voltage drop is common. If we let $Q = VC$, where C is the equivalent capacitance, then $VC = VC_1 + VC_2$, which yields the parallel

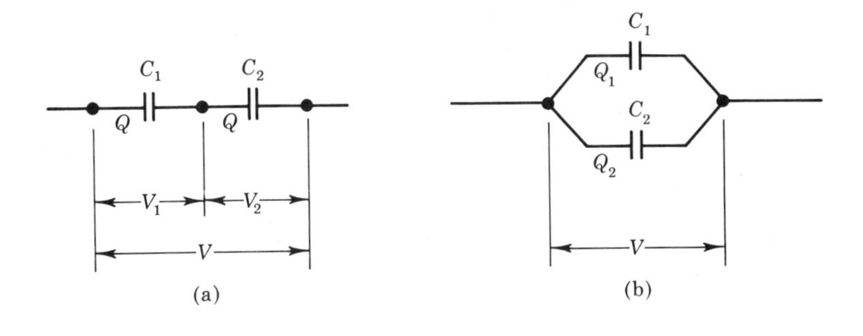

FIGURE 17-7

Capacitors in series and parallel

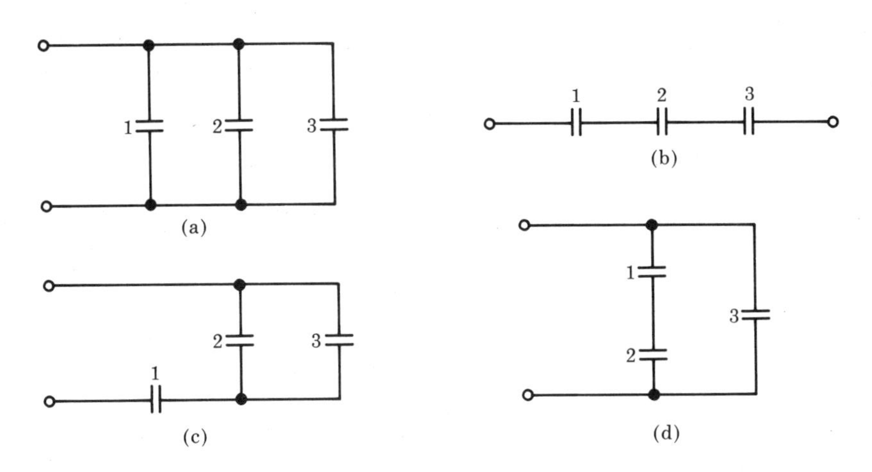

FIGURE 17-8

Addition of capacitances

formula. This result conforms to expectations: the capacitors in parallel provide a larger charge-storage area, hence addition is required.

Let us study a few numerical examples. First, we find the total capacitance for capacitors of 1 μF, 2 μF, and 3 μF arranged in several ways (Figure 17-8). For case (a), $C = 1 + 2 + 3 = 6\ \mu$F. For case (b), $1/C = 1/1 + 1/2 + 1/3 = 1.83$ or $C = 0.55\ \mu$F. For case (c), we first combine the parallel 2-μF and 3-μF capacitances to obtain 5 μF, then add again by the series formula $1/C = 1/1 + 1/5 = 1.2$ or $C = 0.83\ \mu$F. For case (d), we first combine the series 1-μF and 2-μF capacitances to obtain 0.67 μF; the total is then 3.67 μF.

17-3 EFFECT OF AN ELECTRIC FIELD ON INSULATORS—DIELECTRICS

Let us charge a capacitor of capacitance C_0 and then note the effect of introducing between the plates some nonconducting substance called a

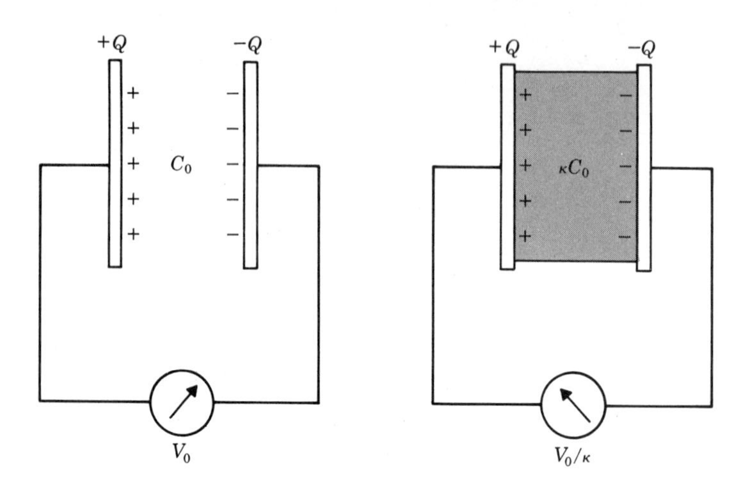

FIGURE 17-9

Effect of inserting a dielectric in a capacitor

dielectric (see Figure 17-9). This situation would be encountered if a vapor such as mercury were used in a vacuum tube, or if a gas such as argon were used in a detector for high-speed charged particles, or if air or a solid filled the gap of a capacitor.

Experimentally, we find that the potential difference between the plates drops from a value V_0 to a smaller value $V = V_0/\kappa$, where κ is called the *dielectric constant*. Since the charge Q is unchanged, we deduce that the capacitance has increased from C_0 to κC_0, or

$$\kappa = \frac{C}{C_0} \qquad \textit{(dielectric constant)}$$

The numerical value of κ depends on the material, as shown in Table 17-3.

Capacitors with large C for electric circuits are formed by rolling up sheets of thin metal foil with a layer of insulating material between them. The insulating material may be paper, mica, glass, or a metal oxide. Advantage is taken of the inverse dependence of C on gap width x.

TABLE 17-3

DIELECTRIC CONSTANTS

MATERIAL	κ
Vacuum	1.00000
Air	1.00054
Water	78.5
Oil	2.2
Mica	7.0
HCl gas	1.0046

A qualitative explanation of the effect of introducing dielectric material between the plates can be deduced. We shall let the inserted substance be atomic hydrogen gas in order better to visualize the processes. The positive charge on each of the atoms is attracted toward the negatively charged plate, the negative charge toward the positively charged plate. The bond between the charges is stretched, as shown for one atom in Figure 17-10(b). When there are many atoms present, as in Figure 17-10(c), some of the electrons near the edge of the dielectric are brought in close proximity to the positively charged plate, and effectively neutralize part of the charge. At the other plate a similar effect takes place. Thus the charge density is reduced from σ_0 to σ. Removal of the gas would restore initial conditions, of course. Recalling that electric field between parallel plates is proportional to the charge density (Section 17-2), and that for fixed separation of plates the electric field is proportional to the potential difference, we see that

$$\frac{\mathcal{E}_0}{\mathcal{E}} = \frac{V_0}{V} = \frac{\sigma_0}{\sigma} = \kappa$$

For the special case of a vacuum, the charge on the plates remains unchanged, and thus κ is exactly 1.

Now let us consider the effects on the atomic bonds. A separation of charge, s, is produced in each of the atoms, so the neutral atom becomes an electric dipole of moment $M = es$. We know, from Section 6-3, that a dipole produces an electric field, and thus the field between the plates is disturbed. If there are N atoms in each unit of volume, the total dipole moment per unit volume is NM, which we shall label P, called the *polarization.* Now, P depends mutually on the field strength and how susceptible the bonds are to stretching. This dual dependence can be expressed as

$$P = \frac{1}{4\pi k_E}\chi_E\mathcal{E}$$

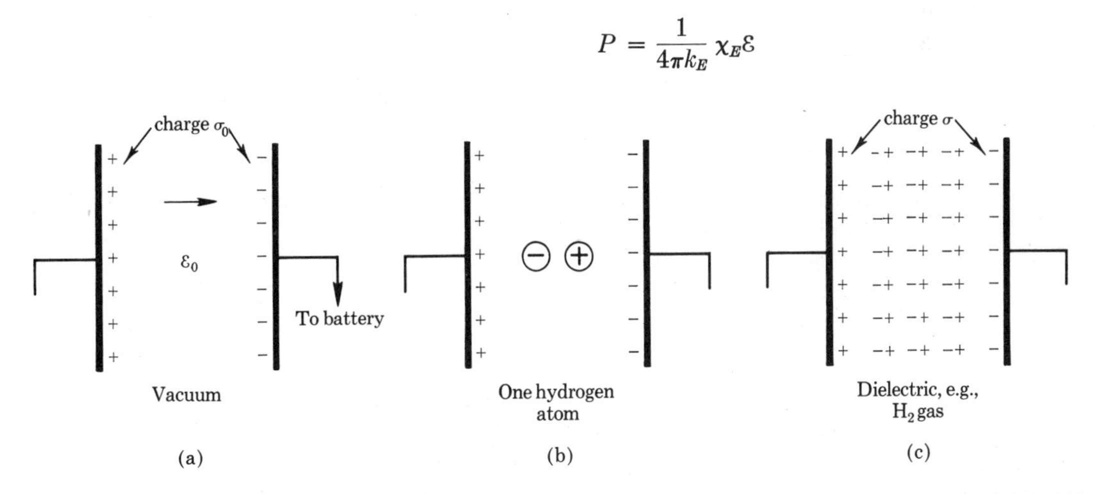

FIGURE 17-10

Displacement of charges in a dielectric

where χ_E is logically called the *electric susceptibility*. We would suspect some connection between it and the dielectric constant κ, since both are dependent on the substance. It is found* that $\kappa = 1 + \chi_E$.

Two types of substances must be distinguished: those with a permanent electric-dipole moment, called *polar* molecules—e.g., HCl, H_2O—and those without, called *nonpolar*—e.g., Ne, C_2H_2. The effect of an electric field is to *align* polar molecules parallel to the field so that there is a polarization. In competition with this effect is the thermal motion. The larger the field $\mathcal{E}$, the greater the alignment; the higher the temperature T, the more random are directions of electric moments; the larger the moments M or numbers per unit volume of molecules, N, the larger the polarization P.

17-4 MAGNETIC FIELDS IN MATERIALS

We now perform a third important experiment using the apparatus of Figure 17-11. We send a current I through the ring-shaped solenoid of n turns per unit length. As discussed in Section 7-3, the magnetic field inside is

$$\mathcal{B}_0 = k_M 4\pi n I$$

where the subscript zero implies that there is nothing inside the torus. We then turn the current off, fill the space inside with some substance, and turn the current on again. The magnetic field is now larger by a factor p, which we call the *relative permeability:*

$$\frac{\mathcal{B}}{\mathcal{B}_0} = p$$

An explanation of this magnetic effect in atomic terms involves the fact that a circulating current I in a loop of area A has a magnetic moment $M = IA$, as described in Section 9-7. Electrons in the atoms thus act as dipoles which produce a magnetic field that can modify that provided by the solenoid.

Let us suppose that each atom in the substance has an effective magnetic moment M, and that in each unit volume containing N atoms the total magnetic moment is NM, which we shall label $\mathfrak{M}$ and call the *magnetization*. Its value depends on the nature of the electron orbits in the substance and on the current in the coil. We introduce the symbol $\mathcal{H} = nI$ called the *magnetizing "force,"* and propose that the magnetization is given

* The number of charges that go from the dielectric to the plates per unit area is the number per unit volume times the separation produced, Ns, and the corresponding amount of charge $\sigma_0 - \sigma$ is Nse, which is also P. Thus

$$\kappa = \frac{\sigma_0}{\sigma} = 1 + \frac{\sigma_0 - \sigma}{\sigma} = 1 + \frac{4\pi k_E}{\mathcal{E}} P = 1 + \chi_E$$

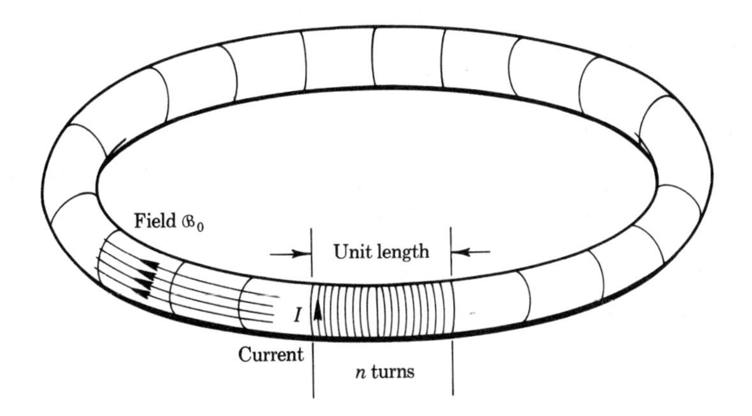

FIGURE 17-11

Solenoid in the form of a torus

by

$$\mathcal{M} = \chi_M \mathcal{H}$$

where χ_M is the *magnetic susceptibility*. (Note the analogy to electric polarization and field.)

We now wish to relate permeability to susceptibility. This will put our large-scale observations on a microscopic or atomic basis. To do so, we seek to achieve the same internal field $\mathcal{B}$ by a *different* experiment. If the current in the solenoid without any material inside is increased by an amount ΔI, the same field $\mathcal{B}$ can be reached as with material inside. Thus

$$\mathcal{B} = k_M 4\pi n(I + \Delta I)$$

and hence

$$p = \frac{\mathcal{B}}{\mathcal{B}_0} = 1 + \frac{\Delta I}{I}$$

This excess current causes a magnetic moment per turn of $(\Delta I)A$, and for n turns in a unit length it is $n(\Delta I)A$. The volume per unit length is A; hence the magnetic moment per unit volume (the magnetization) is $\mathcal{M} = n\Delta I$. We thus see that

$$p = 1 + \frac{\mathcal{M}}{\mathcal{H}} = 1 + \chi_M$$

Just as elements differ in electrical conductivity and dielectric character, they differ in magnetic properties. Some atoms (paramagnetic), such as oxygen and aluminum, have unfilled electron shells that give a permanent magnetic moment. Others (diamagnetic), such as helium and bismuth, respond to applied fields by setting up opposing moments. Finally, an exaggerated case of paramagnetism is found in elements such

TABLE 17-4
MAGNETIC SUSCEPTIBILITIES

Substance		χ_M
Diamagnetic	Helium	-2.59×10^6
	Copper	-0.11×10^6
	Bismuth	-1.7×10^6
Paramagnetic	Platinum	$+1.65 \times 10^6$
	Aluminum	$+0.82 \times 10^6$

as iron (ferromagnetism). The magnetic susceptibility χ_M is a measure of these properties (values are listed in Table 17-4).

Let us examine these types of magnetism in terms of electric orbits in the atoms of the substances. Recall from Section 9-7 that the magnetic moment of an electron moving with speed v in a circle of radius r is $M = evr/2$, and that the ratio of magnetic moment M to angular momentum $L = mvr$ is $M/L = 8.79 \times 10^{10}$ C/kg. According to quantum theory, L can take on only certain values that are integer multiples of $h/2\pi$, where $h = 6.63 \times 10^{-34}$ J-sec is Planck's constant. The smallest value of L is thus 1.05×10^{-34} J-sec, which corresponds to a magnetic moment of

$$(8.79 \times 10^{10})(1.05 \times 10^{-34}) = 9.27 \times 10^{-24} \text{ A-m}^2$$

This smallest amount of magnetic moment is called the *Bohr magneton,* and serves as a special unit.

Paramagnetic substances, such as oxygen, have a permanent magnetic-dipole moment due to the motion of certain electrons in the atoms. The directions of the moments are random in a gas or in a solid at high temperature, and there is no net magnetic effect. When a magnetic field is applied, however, there are torques on each dipole, as discussed in Section 9-7, and the dipoles tend to be aligned with the field. This is quite similar to the effect of an electric field on polar molecules. The larger the magnetic field $\mathcal{B}$, the greater the alignment; the higher the temperature, the more random the moments; the larger the atom-number density or the moment of each molecule, the greater the magnetization. Curie's law, which states that paramagnetism varies inversely with the absolute temperature, is an experimental verification of our reasoning.

Diamagnetic substances, such as helium, react in opposition to the application of a magnetic field. The classical explanation of this effect, given by Langevin, starts with the idea that an electron circulates at speed v in an orbit of radius r. When a magnetic field $\mathcal{B}$ is applied, directed normally to the plane of the orbit, a magnetic force on the charge of amount $qv\mathcal{B}$ is produced. To stay at the original radius, the charge must slow down, which reduces the magnetic moment $evr/2$. Thus the original magnetic field due to the circulating electrons is decreased, corresponding to a negative susceptibility.

Ferromagnetism, found in iron, cobalt, and nickel, is an extreme case of paramagnetism. The three elements mentioned have some incomplete electron shells that give very strong dipole moments. Average values per atom are, in Bohr magnetons, Fe 2.22, Co 1.70, and Ni 0.61. It is recognized that there are small regions or *domains* in these metals with all the moments aligned. In an unmagnetized state, the domains are randomly oriented; but application of a magnetic field tends to line them up. Above a certain temperature called the *Curie point*, this alignment is lost. Values of the Curie point for the three elements are: Fe 770°C, Co 1120°C, and Ni 358°C.

Let us find the increase in field due to an iron core in a solenoid if M, the magnetic moment per atom, is 2.22 Bohr magnetons, or

$$(2.22)(9.27 \times 10^{-24}) = 2.06 \times 10^{-23}\ \text{A-m}^2.$$

In iron with density 7.88×10^3 kg/m^3, N is $8.50 \times 10^{28}/\text{m}^3$. Hence the magnetization is

$$\mathfrak{M} = NM = (8.50 \times 10^{28}/\text{m}^3)(2.06 \times 10^{-23}\ \text{A-m}^2) = 1.75 \times 10^6\ \text{A/m}$$

The increase in field is thus

$$\Delta\mathcal{B} = 4\pi k_M \mathfrak{M} = (4\pi \times 10^{-7})(1.75 \times 10^6) = 2.20\ \text{Wb/m}^2$$

For ferromagnetic materials, the field $\mathcal{B}$ does not vary linearly with solenoid current as in the case of ordinary (paramagnetic and diamagnetic) substances. The domains become aligned in steps; once they have become oriented with the field, they tend to stay that way, and there is a *residual* magnetism even when the current is brought back to zero. This phenomenon, called *hysteresis*, is shown in Figure 17-12. The arrows indicate the direction in which the changes are made.

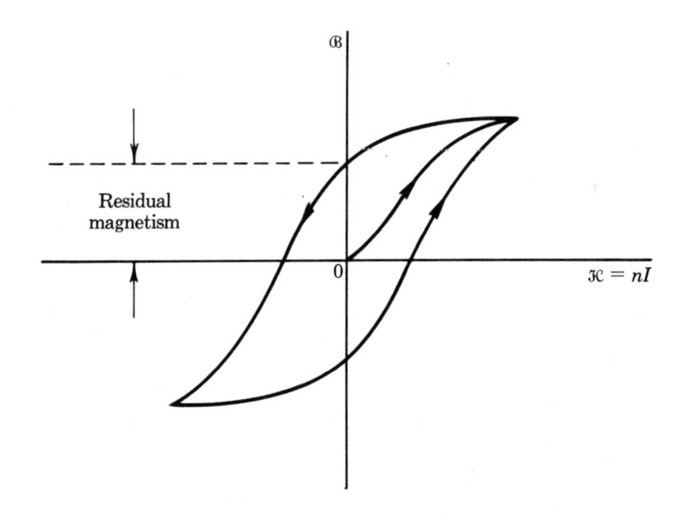

FIGURE 17-12

Hysteresis in magnetic materials

The magnetism of a bar magnet is also due to such alignment of domains. So long as no currents are present or the temperature is kept well below the Curie point, the magnetism can be considered *permanent*. Ferromagnetic materials are characterized by the relative permeability p, some values of which are listed in Table 17-5.

TABLE 17-5
RELATIVE PERMEABILITIES
(*maximum values*)

SUBSTANCE	p
Iron	5,000
78 Permalloy	100,000
Purified Iron	180,000
Supermalloy	800,000

In concluding our discussion of the atomic basis of electricity and magnetism, we note a great similarity between the roles played by the various fields and materials constants, $\mathcal{E}$ and $\mathcal{B}$, χ_E and χ_M, P and $\mathfrak{M}$, and so on.* This assembly of constants provides the basis for describing electric and magnetic effects from the bulk or molecular views, and with or without material present.

17-5 MAGNETICALLY INDUCED CURRENTS—INDUCTANCE

Faraday's discovery that changing magnetic fields create electric fields, as discussed in Section 7-4, forms the basis for the circuit element called the *inductor*. In simplest form, it consists of one or more loops of wire through which a changing current passes. A potential difference V between the ends of the wire is produced, proportional to the rate at which

* The complete description involves an additional electric vector called the *displacement*, defined by

$$D = \frac{\kappa\mathcal{E}}{4\pi k_E} = \frac{\mathcal{E}}{4\pi k_E} + P$$

The *permittivity* is then the ratio $\epsilon = D/\mathcal{E}$, which is also $\kappa/4\pi k_E$. For a vacuum, where $\kappa = 1$, ϵ becomes simply

$$\epsilon_0 = 1/4\pi k_E$$

called the *permittivity of free space*. Turning to magnetism, the *absolute permeability* is defined by $\mu = \mathcal{B}/\mathcal{H}$; but since $p = \mathcal{B}/\mathcal{B}_0$ and $\mathcal{B}_0 = 4\pi k_M\mathcal{H}$, we find that $\mu = 4\pi k_M p$. For a vacuum, $p = 1$, and μ becomes $\mu_0 = 4\pi k_M$, the absolute permeability of free space.

the current changes, $\Delta I/\Delta t$. The *self-inductance* L is defined as the constant of proportionality, or the ratio of magnitudes

$$L = \frac{V}{\Delta I/\Delta t}$$

It depends on the geometric features of the coil, not on current or voltage. For each shape, size, and arrangement of current loops there is a particular value of L, which can be measured experimentally or calculated. Let us find the self-inductance for a long straight solenoid (Figure 17-13). The magnetic field is $\mathcal{B} = 4\pi k_M nI$, where n is the number of turns per unit length. The flux through each turn is then $\Phi = \mathcal{B}A$, and a change in I gives a potential difference according to Faraday's law:

$$V = -\frac{\Delta\Phi}{\Delta t} = -4\pi k_M n \frac{\Delta I}{\Delta t}A$$

In the whole solenoid with length l there are nl turns, and since the loops are connected in series the potential differences add. We easily find the self-inductance to be $L = 4\pi k_M n^2 lA$ for this case. The units of L are volt-

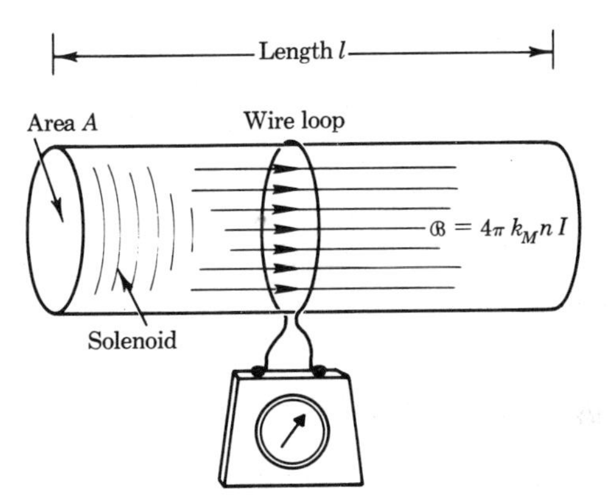

FIGURE 17-13
Self-inductance

seconds per ampere, which combination is called the *henry*, after Joseph Henry.

As an example, let us find L for a solenoid with $n = 5000$ turns per meter, length $l = 2.0$ m, diameter $= 0.03$ m, area $A = \pi d^2/4 = 7.1 \times 10^{-4}$ m^2. Then,

$$L = 4\pi k_M n^2 lA = 4\pi(10^{-7})(5000)^2(2.0)(7.1 \times 10^{-4}) = 0.045 \text{ henry}$$

If the current were changed at a rate of, say, $\Delta I/\Delta t = 60$ A/sec, the magnitude of the induced voltage would be

$$V = L\frac{\Delta I}{\Delta t} = (0.045)(60) = 2.7 \text{ V}$$

When several inductances are in series, but far enough apart so that their fields do not interact, their sum is given by $L = \sum_i L_i$; or, if they are in parallel, by $1/L = \sum_i 1/L_i$. This is the same rule as for resistances.

Mutual Inductance

When the current changes in one circuit whose flux is linked with another circuit, voltages are induced in the latter. Suppose we closely wrap two sets of wire about a single hollow cylinder (Figure 17-14), creating one solenoid within another. When a current I_1 passes through the inner one, it creates a flux through the loops of the outer circuit that can be written

$$\Phi_2 = MI_1$$

where M is called the *mutual inductance*, also measured in henrys. Its value is obtained by noting that $\mathcal{B}_2 = 4\pi k_M n I_1$; and if the outer coil has N_2 turns of area A, the total flux through it is $\Phi_2 = N_2 \mathcal{B}_2 A$ or $4\pi k_M n N_2 A I_1$. Hence, for this arrangement,

$$M = 4\pi k_M n N_2 A$$

Now, suppose the current I_1 changes at a rate $\Delta I_1/\Delta t$. This creates a potential difference in the outer coil according to Faraday's law:

$$V_2 = -\frac{\Delta\Phi_2}{\Delta t} = -M\frac{\Delta I_1}{\Delta t}$$

The role of M in these magnetically coupled circuits is the same as L in a

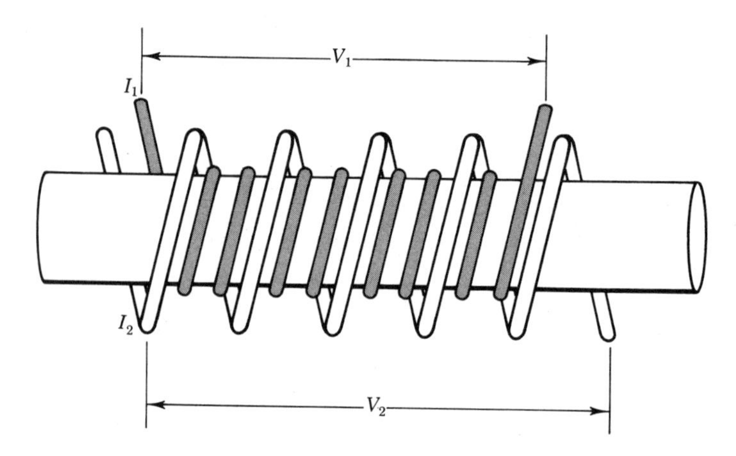

FIGURE 17-14
Mutual inductance

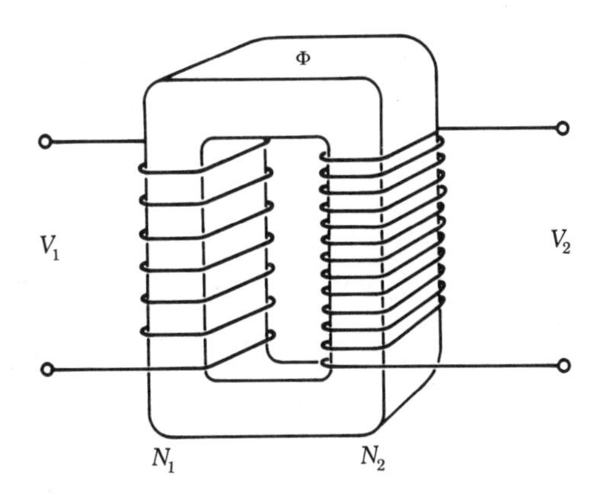

FIGURE 17-15

Simple transformer

single circuit. We can also reverse the point of view and set $\Phi_1 = MI_2$, so that $V_1 = -M(\Delta I_2/\Delta t)$ is the voltage induced in the inner solenoid by current changes in the outer.

Two circuits connected by a common flux serve as a voltage *transformer*. The flux is the same through an iron yoke with primary loops and secondary loops as shown in Figure 17-15. The voltage in the primary varies as $V_1 = -N_1(\Delta\Phi/\Delta t)$ according to Faraday's law, while the voltage induced across the secondary is $V_2 = -N_2(\Delta\Phi/\Delta t)$. Dividing equations, we obtain $V_1/V_2 = N_1/N_2$. For example, a transformer that "steps down" a voltage of 120 V to one of 6 V must have a ratio of turns $N_1/N_2 = 20$.

Energy in Electrical Components

The processes of charging a capacitor or changing the current in an inductor require that work be done by the electrical supply; at any instant, therefore, a certain amount of energy must be stored in these components. We may deduce the amounts by an examination of the processes.

An uncharged plane capacitor has zero potential difference between the plates, while a fully charged capacitor has a potential difference V. The work done to bring a unit charge onto the plates for these two cases is 0 and V, respectively. On the average, during accumulation, the work per charge is $V/2$; or, for a total charge Q it is $QV/2$. The stored internal or potential energy is thus

$$U = \frac{QV}{2} \qquad \textit{(energy in a capacitor)}$$

Alternate forms using $V = Q/C$ are $CV^2/2$ and $Q^2/2C$. We can find the energy density (energy per unit volume) by recalling that the capacitance of a parallel-plate capacitor with vacuum inside is

$$C = \frac{A}{4\pi k_E x}$$

If the volume is $v = Ax$ and the field is $\mathcal{E} = V/x$, a little algebra shows us that the energy density is

$$\frac{U}{v} = \frac{\mathcal{E}^2}{8\pi k_E} \qquad \text{(energy density in a capacitor)}$$

This relation holds for fields in general, even though we derived the relation for this particular type of capacitor. If there is a dielectric between the plates, the energy density can easily be shown to be κ times as large as given above.

A similar development yields the energy and its density for an inductor. Let us bring the current up from zero to a value I. The power, as the rate of doing work, is $\Delta W/\Delta t = P = VI$; but $V = L(\Delta I/\Delta t)$. Thus $\Delta W = LI\Delta I$. This expression is mathematically of the same form as those encountered earlier, such as in Section 8-4 (the work required to stretch a spring). By analogy, we write down the total work, which is the stored potential energy, as

$$U = \frac{LI^2}{2} \qquad \text{(energy in an inductor)}$$

Let us apply this relation to a particular inductor, the toroidal coil, for which the inductance is $L = 4\pi k_M n^2 lA$. The field is $\mathcal{B}_0 = 4\pi k_M nI$, and the volume $v = lA$. Combining, the energy density without material inside is

$$\frac{U}{v} = \frac{\mathcal{B}_0^2}{8\pi k_M} \qquad \text{(energy density in an inductor)}$$

If instead the coil is filled with material of relative permeability p, the energy density becomes p times as large as given above.

17-6 CONDUCTORS AND SEMICONDUCTORS—THERMOELECTRICITY

The conduction of electricity in crystals depends very much on the element and its degree of purity. Consider the fact that the metal copper is an excellent conductor, while diamond is a poor conductor or a good insulator. For a detailed understanding of the distinctions between conductors and insulators it is necessary to use the rigorous quantum-mechanical approach. However, we can combine our knowledge of potential wells, electron orbits, and energies to form a working picture. A crystalline solid consists of a regular array of atoms with the positive nuclei relatively fixed at certain locations. Some of the electrons are bound very strongly to these atoms and do not move about in the crystal. Figure 17-16 shows a few atoms in one plane of the three-dimensional structure. Some electrons are shown in orbits around individual nuclei, while others are shared by more than one atom.

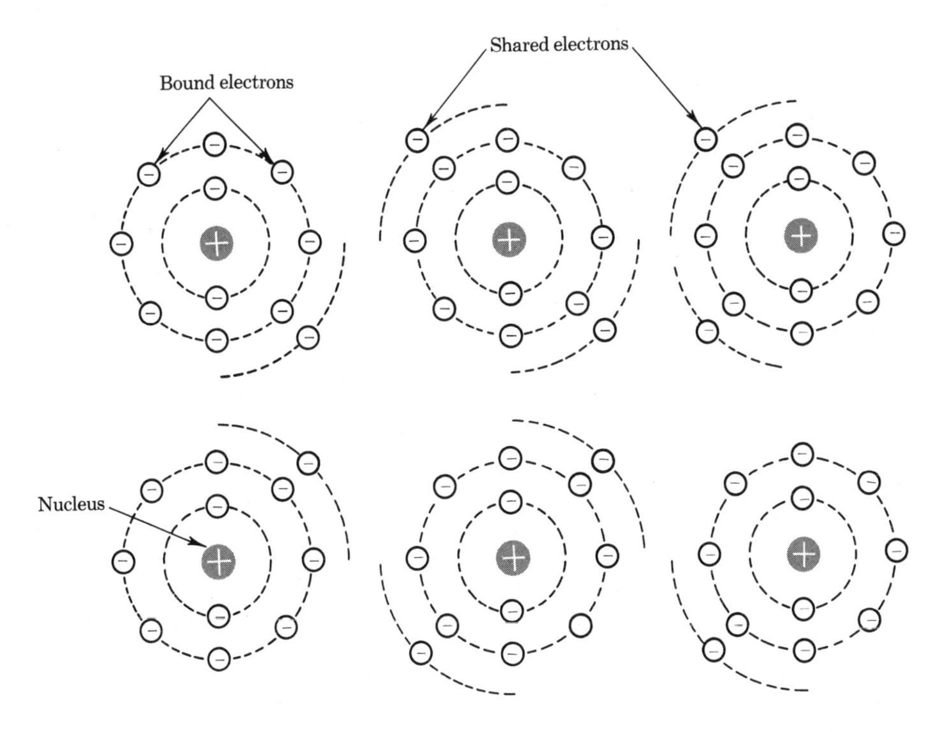

FIGURE 17-16

Atoms in a crystal lattice

Let us study the potentials in the material. First we draw, as we did in Section 8-3, the potential in the vicinity of a single isolated positive charge. It consists of the barrier, as shown in Figure 17-17(a). To an electron, with negative charge, this corresponds to a potential *well*, as in Figure 17-17(b). When we put many atoms together in a solid, the poten-

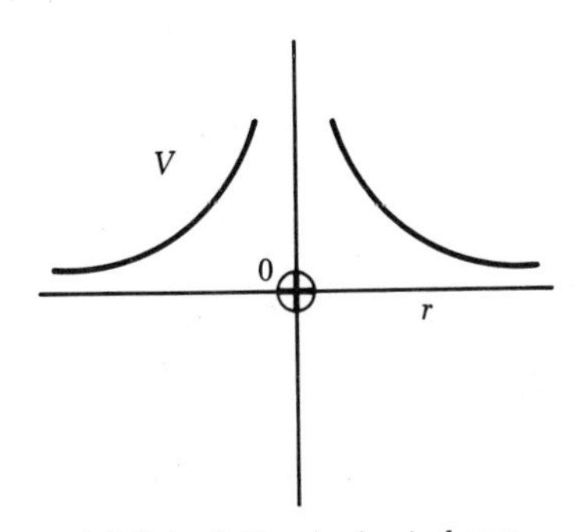

(a) Potential barrier for + charge

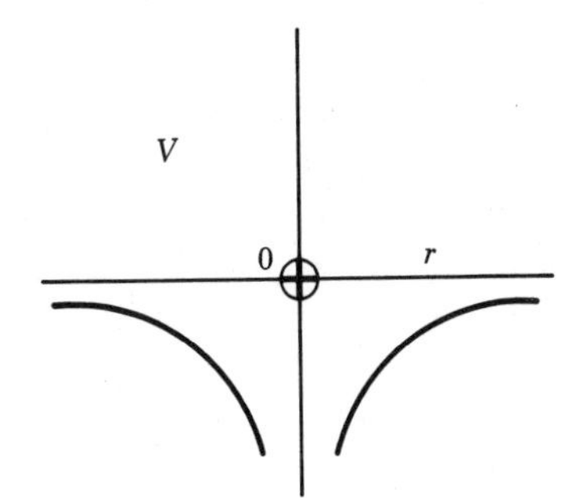

(b) Potential well for electron

FIGURE 17-17

Potential distributions near a positive charge

tial experienced by an electron is a combination of overlapping curves due to many nuclei, as shown in Figure 17-18. An electron of small kinetic energy remains bound to the nucleus, while one of high energy is relatively free. Recall our discussion in Section 12-5 of waves and energy levels of electrons in a potential well. In the system composed of many nuclei, there are so many discrete energy levels that there appears to be a band of energy in which electrons can reside. Figure 17-19 shows these imposed on the potential diagram of the lattice. In order to distinguish the types of material, we use the band concept. The upper band, labeled "conduction," can be used by electrons to travel from one atom to another in the crystal. The electrons normally occupy the lowest levels available, and there may not be any electrons in the conduction band unless the temperature is increased.* This type of material is a good electrical *insulator*.

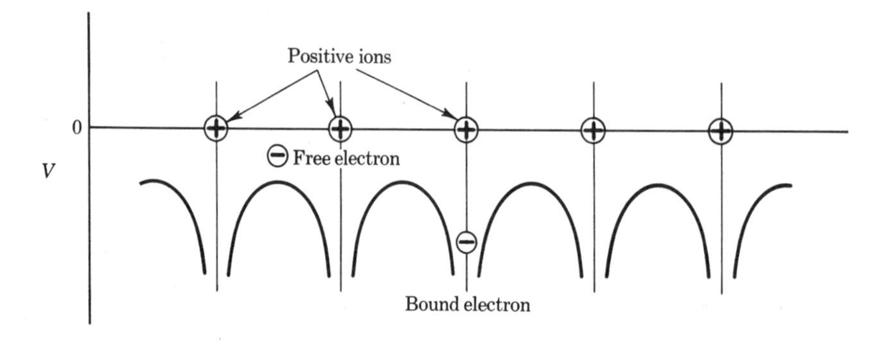

FIGURE 17-18

Atomic potential distributions in matter

If the conduction band normally has electrons in it, the transfer of charge can take place easily and we call the material a *conductor*. The distinction can be made by use of a simplified picture that includes only the energy bands, as shown in Figure 17-20. Here the filled valence bands, the conduction band, and the gap between them are shown for three characteristic materials. In a metal such as copper, there is no gap, and there are many conduction electrons; in an insulator, such as quartz, the gap is relatively wide, and practically no conduction can occur, except at high temperature. As expected, a *semiconductor*, such as germanium, has a gap of intermediate width—Figure 17-20(c).

The addition to pure germanium of a small amount of arsenic (As) as an impurity raises the conductivity because of an extra electron in the valence shell of arsenic. This extra electron enters the conduction band,

* Note that this is the *opposite* effect from that in conductors where the resistance increases with temperature. In insulators and semiconductors the resistance decreases with temperature increases.

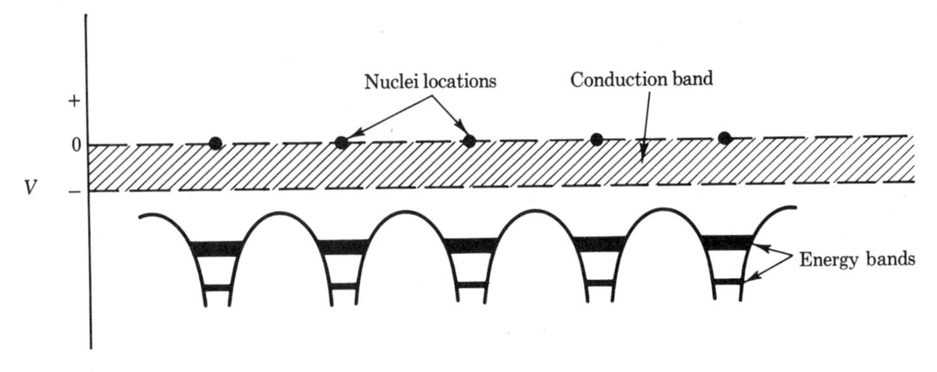

FIGURE 17-19

Potential for electrons in a lattice

adding to those already there, so that the arsenic atom serves as a "donor." On the other hand, the impurity gallium (Ga) extracts electrons from the conduction band, since the atom lacks one electron in the valence shell, and serves as an "acceptor." On taking an electron out of circulation, it creates a net positive charge that can be viewed as a positive "hole." These holes have the property of being able to be displaced, thus forming charge carriers.

Two types of special germanium semiconductors may be produced. The first is n-type, or "negative"—e.g., with As added. This type has more donors than acceptors and an excess of electrons. The second is p-type, or "positive"—e.g., with Ga added. This type has more acceptors than donors and an excess of holes. The construction of "solid-state devices" is accomplished by putting together p- and n-types in various combinations to make transistors, diodes, and other electronic components, as we shall discuss in Section 18-4.

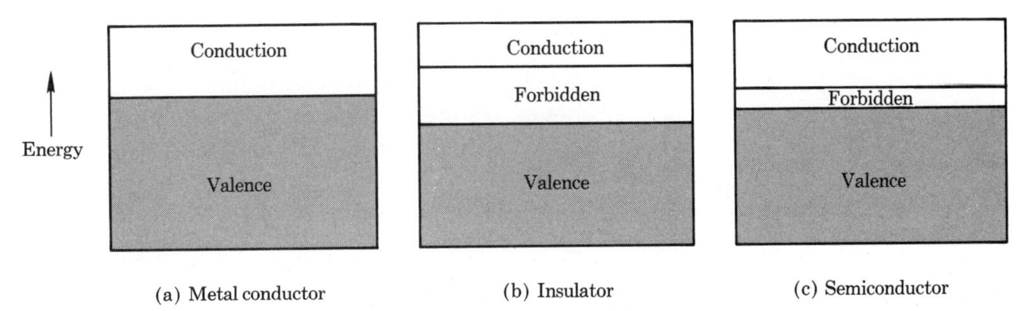

FIGURE 17-20

Energy band diagrams for various materials

Thermoelectric Effects

We can gain a better appreciation of the differences between n-type and p-type materials if we study temperature effects on them. Suppose we heat one end of a rod composed of n-type material and keep the other end cold. The addition of thermal energy excites electrons which then cross the forbidden gap and enter the conduction band. The particle density is high at the hot end, and by the transport concepts of Section 16-1 they tend to flow toward the cold end where the density is low. That end becomes negatively charged, while the hot end becomes positive. A potential difference and thus an electric field is produced, and a charge flow is set up. In the p-type material, there is a flow of "holes" from the hot end to the cold end; the latter charges positively, and an electric field is produced in the opposite direction to that of the n-type semiconductor. The use of these properties in a thermoelectric cell will be discussed in Section 18-1.

Electrical Circuits

The devices just described form the building blocks for a very large number of electrical circuits. By combining in different ways the elements that we have described—resistors, capacitors, inductors (with self- and mutual-inductance), and semiconductor elements—many functions are served. These include heating, control of equipment, measurements, power transmission and communication by telephone, telegraph, radio, and television. In the next chapter we shall analyze several such circuits of interest.

PROBLEMS

17-1 A choice of copper wire of diameter d is to be made in the electrical wiring of a house. The current in a 20-m length may be as high as 25 amperes. Calculate the appropriate value of wire diameter to be sure that the voltage drop in the wire is no more than 1.0 V.

17-2 What is the resistance of a cube of aluminum 5 mm on an edge?

17-3 Steel trolley-car rails that have a cross-sectional area of 7 in.^{2} and a resistivity of 6×10^{-7} ohm-m are used to carry the power used by the motors. What is the resistance of single track per mile of length?

17-4 A platinum wire of length 40 m, resistance 60 ohms/m at 20°C, is heated to 100°C. Compare the total resistances at the two temperatures.

17-5 At what temperature would the resistance of a piece of copper be the same as an identically constructed piece of aluminum at room temperature?

17-6* A heating coil *rated* at 400 watts is composed of 8 meters of steel wire of area 6×10^{-7} m^2. (a) At room temperature (20°C), what direct

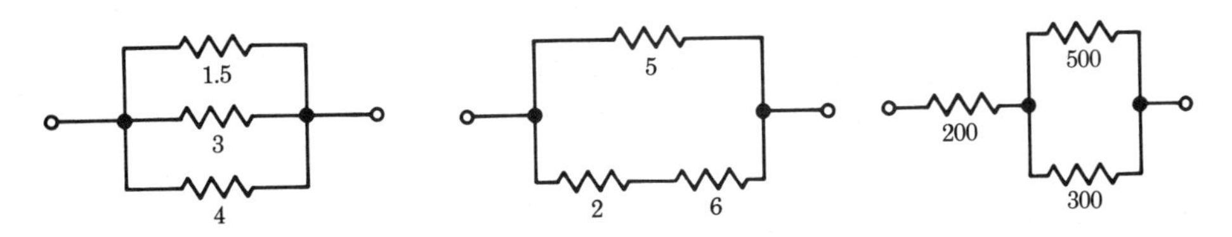

FIGURE P17-7

current would it draw at 125 volts potential difference, using the resistance of the coil? (b) What is the resistivity of the steel at its operating temperature of 1000°K? (c) How close to its rating is the power developed in the coil?

17-7 Calculate the total resistances of the combinations illustrated in Figure P17-7, where all values are in ohms.

17-8 A 12-volt dc car battery provides power for two 150-watt headlights in parallel. How much current is drawn by each bulb?

17-9* Twelve resistors, each with resistance R ohm, form the edges of a cube. Find the resistance between opposite corners.

17-10 By using only two resistors, either singly, in series, or in parallel, a student finds he can form a total resistance of 3, 4, 12, and 16 ohms. What are the separate values of the two resistors?

17-11 A set of eight Christmas-tree lightbulbs forms a parallel circuit (Figure P17-11). Each bulb has a power rating of 10 watts. When attached to a direct current supply of 120 volts, what value of current does each bulb and the whole circuit draw? If one bulb burns out, does the current through each remaining bulb increase, decrease, or remain the same?

17-12* A fuse in an electrical circuit at room temperature 25°C contains a link composed of tin of length 1.0 cm, width 0.25 cm, and thickness 0.5 mm. Using data listed below on tin, find (a) the resistance of the strip, (b) the energy required to melt it, (c) the time required for the fuse to blow at 50-ampere current.

density: 7.3×10^3 kg/m^3
resistivity: 2.0×10^{-7} ohm-m
specific heat: 0.06 kcal/kg-°C, 250 J/kg-°C
melting point: 232°C

17-13 The plates of a condenser for tuning a radio (see Figure P17-13 are separated by 0.8 mm. The area of one plate is 6 cm^2. (a) Find the capac-

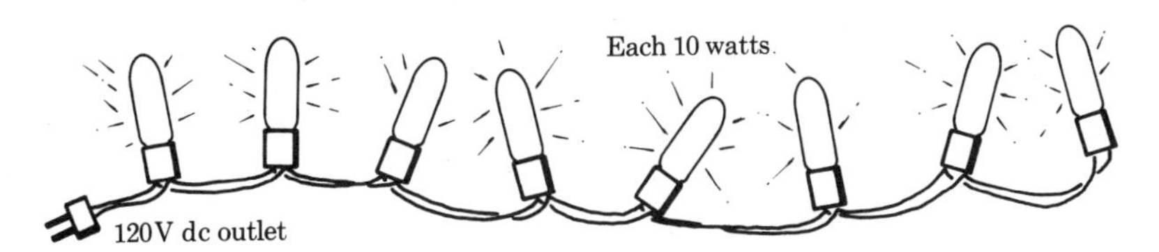

FIGURE P17-11

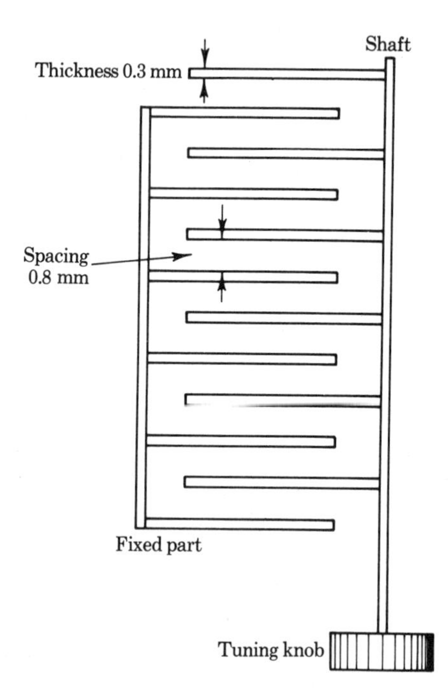

FIGURE P17-13

itance of two adjacent plates. (b) Find C for the whole parallel combination.

17-14 The capacitance of a cylindrical tube in a Geiger counter for the detection of gamma rays is to be computed (see Figure P17-14), using the relation for the value of C per meter length of $(2k_E \ln b/a)^{-1}$. The diameter of the tube is 0.025 m, the collector-wire diameter is 1.5×10^{-3} m, and the length is 0.35 m.

17-15 Find the total capacitance of the combinations shown in Figure P17-15, where all values are in microfarads.

17-16* A capacitance C_1 with charge Q_1 is suddenly connected with another uncharged capacitor C_2 by throwing the switch (see Figure P17-16). Find the final potentials and charges on both.

17-17 Sheets of mica and metal foil are wrapped up to form a compact

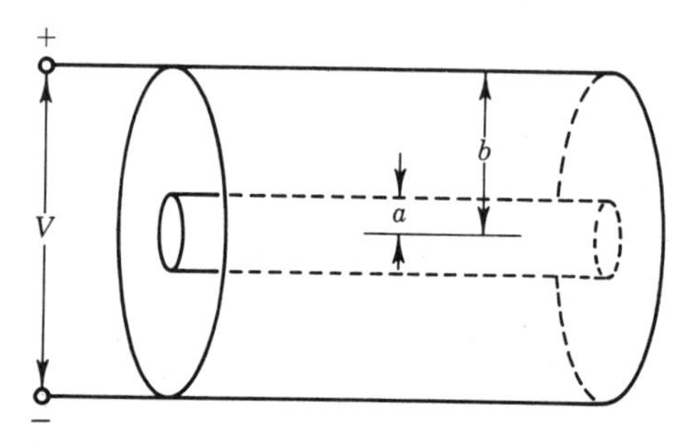

FIGURE P17-14

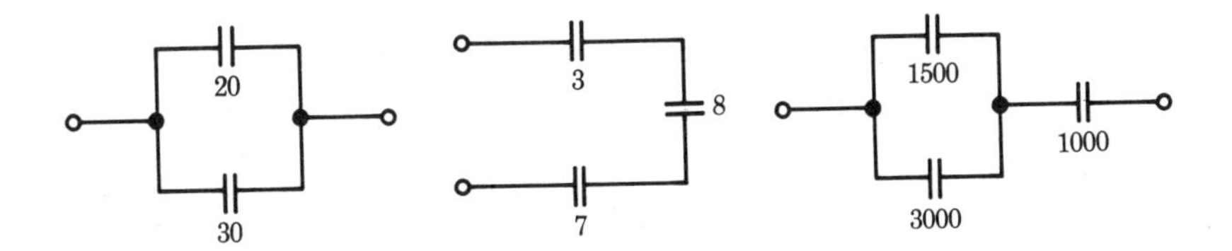

FIGURE P17-15

condenser of diameter 0.9 cm and length 1.4 cm. (a) Estimate the effective area of both the foil and mica back of which has thickness 5×10^{-5}-m. (b) What is the C-value of the capacitor?

17-18 A capacitor is connected to a battery with fixed potential difference V_0, and achieves a charge $Q_0 = C_0/V_0$. Then it is disconnected from the battery and immersed in oil of dielectric constant 2.2. What happens?

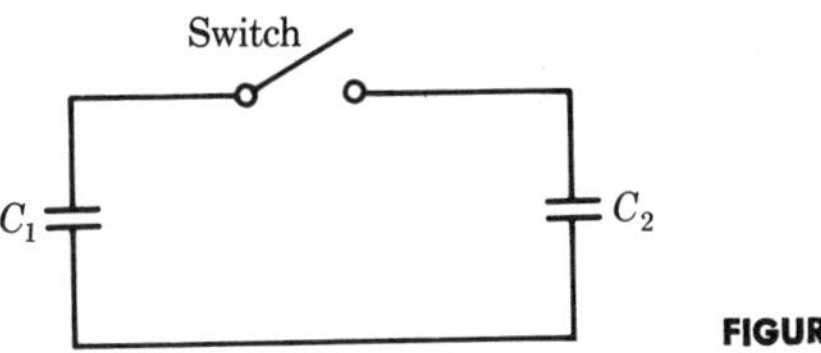

FIGURE P17-16

17-19 Derive a formula for the capacitance of a parallel-plate capacitor with two parallel layers of different dielectric materials and thicknesses.

17-20 Two plates are immersed in water with a slight amount of salt dissolved. The device acts as a combined capacitor with dielectric constant 78.5 and resistor with resistivity 4×10^3 ohm-m. If the area of each plate is 20 cm² and the separation is 0.5 cm, find C and R.

17-21 Assuming that the electric-dipole moment of H_2O is 6.2×10^{-30} C-m, calculate the susceptibility and dielectric constant of water vapor at 100°C temperature and atmospheric pressure, making use of the relation $\chi_E = \dfrac{4\pi k_E N M^2}{kT}$ applicable for polar molecules.

17-22 An infinitely long solenoid of diameter 0.02 m with 5000 turns per meter has a current of 0.7 amperes. (a) What is its magnetic field with and without iron of relative permeability $p = 400$? (b) What is its self-inductance per unit length with iron present?

17-23 What is the magnetic moment per unit volume of the iron core of the solenoid of Problem 17-22? What is the moment per iron atom if the atom-number density is $8.5 \times 10^{28}/\text{m}^3$? What fraction of the iron atoms are effectively participating in the magnetism of the core?

17-24 Calculate the self-inductance of a solenoid in the form of a toroid of diameters 0.41 m outside and 0.39 m inside, if there are 300 turns per meter.

17-25 A transformer has 400 turns on the primary and 2500 turns on the secondary. To what voltage can 110 volts be stepped up?

17-26 An electromagnet has a magnetic field of 4 Wb/m^2 permeating its coil which is composed of 300 turns 0.05 m^2 in cross-sectional area. If this field is reduced to zero in 2 sec, what potential difference is produced across the ends of the coil?

ELECTRICITY IN ACTION

18

The period from about 1850 to the present time could well be called the "electrical age." Almost all of the many inventions and applications involving electricity have appeared since then. The great strides of modern electronics in recent years have given us many practical devices as well as some of the most sophisticated tools available to the scientist. These systems range from diminutive radios to electronic data-processing apparatus to the accelerators of modern high-energy nuclear-physics research. In this chapter, we shall study the basic principles of electricity and magnetism by examining some of the more important of these devices.

First, we shall consider the primary sources of electrical potential and energy, such as chemical batteries, thermoelectric cells, and electromagnetic generators. These can be connected to electrical circuits which conduct and store charge and which transform electrical energy into other types, such as thermal or mechanical. Such circuits may contain resistors, capacitors, inductors, vacuum tubes, transistors, and motors. We shall also study the use of meters for measuring currents and voltages within circuits. The subject of alternating currents is then given special attention because of its role in the modern world. We then shall review the arrangement and operation of particle accelerators widely used in the exploration of the atomic nucleus.

18-1 SOURCES OF POTENTIAL

The electrical nature of matter makes it possible to use electrical energy to cause dissociation of a chemical compound, or to transform the energy of chemical binding to the electrical form. Similarly, mechanical or thermal energies can be converted into electrical energy. We shall first review the atomic mechanisms involved in the processes of electrolysis and electroplating, then analyze the battery, the electromagnetic generator, and the thermoelectric cell as sources of potential difference.

Electrolysis

When a potential difference is applied to two metal plates immersed in H_2O* (Figure 18-1), a measurable current is observed. There is an evolution of hydrogen gas at the negative electrode and of oxygen gas at the positive electrode. Such a process is an example of *electrolysis*, as we

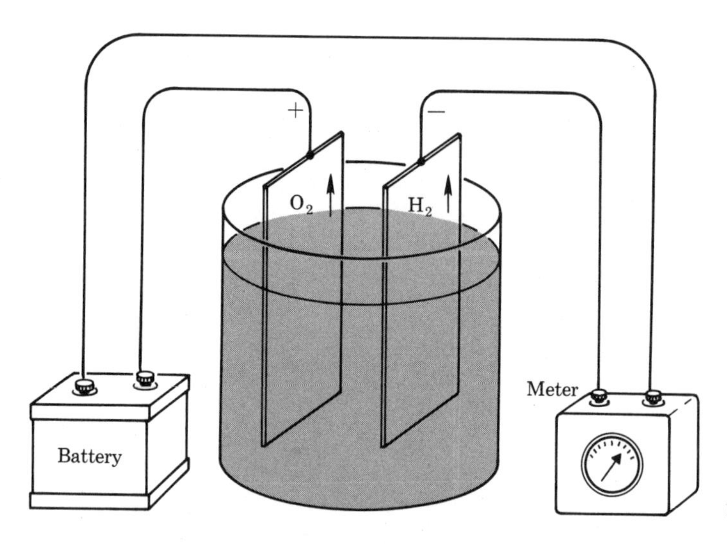

FIGURE 18-1

Electrolysis of water

have already studied in chemistry. There is a slight tendency even in pure water for the reaction

$$H_2O \rightarrow H^+ + OH^-$$

to take place. The application of a potential difference causes a field in the liquid between the plates; the positive hydrogen ion H^+ is attracted to the negative electrode (*cathode*) and the negative hydroxyl ion OH^- is attracted

* Usually the water is made more conductive by the addition of an *electrolyte*. The process occurs much more slowly without this addition.

to the positive electrode (*anode*). On arrival, H^+ picks up an electron to form neutral hydrogen gas according to

$$4H^+ + 4e^- \rightarrow 2H_2$$

while OH^- gives up an electron to yield oxygen gas by the reaction

$$4OH^- - 4e^- \rightarrow 2H_2O + O_2$$

In effect, two out of four water molecules have been *dissociated*. The process will continue so long as there is water to keep delivering the H^+ and OH^- ions.

The two quantities of interest in this process are the charge transferred and the energy required. Now, every singly charged ion has an excess or deficit of 1 electron, of charge $e = 1.60 \times 10^{-19}$ C, and there are 6.02×10^{26} atoms per kmole (Avogadro's number N_a). The total charge required to transfer 1 kmole is thus

$$N_a e = (6.02 \times 10^{26})(1.60 \times 10^{-19}\ \text{C}) = 9.65 \times 10^7\ \text{C}$$

This amount of electricity is called the *faraday*. When hydrogen and oxygen combine, as in a burning process, the reaction

$$H_2 + \tfrac{1}{2}O_2 \rightarrow H_2O$$

yields an energy per water molecule of around 3 electron-volts. The dissociation process, which is the reverse of combustion, requires the same energy.

Electroplating

The plating of a metal such as copper on another metal surface is a familiar chemical process, called *electroplating*. In a water solution of copper sulfate, the reaction that provides the ions is

$$CuSO_4 \rightarrow Cu^{++} + SO_4^{--}$$

The applied field between electrodes drives copper ions through the solution; the ions then collect electrons and deposit on the metal surface.

The Battery

In a cell or battery, a chemical reaction produces a potential difference. We are well aware of the main components of the lead–acid storage battery, as in an automobile. The two terminals or electrodes are connected to plates that are immersed in a water solution of about 30 per cent sulfuric acid, H_2SO_4 (Figure 18-2). The potential difference available, for example 12 volts, depends on the number of plates connected in series. The negative

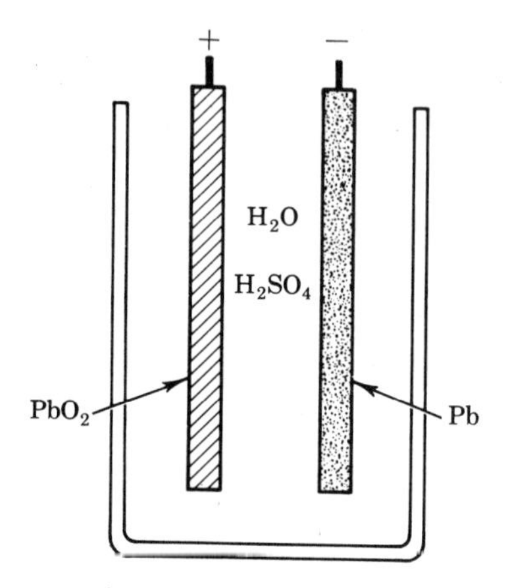

FIGURE 18-2

Pair of plates in a lead storage battery

plate is composed of the element lead (Pb) as a metal, while the positive one is lead peroxide, PbO_2. Within the solution some of the acid is dissociated according to

$$H_2SO_4 \rightarrow 2H^+ + SO_4^{--}$$

The reactions at the two electrodes which provide and collect electrons for an external circuit are

$$Pb + SO_4^{--} \rightarrow PbSO_4 + 2e^- \qquad (negative)$$
$$PbO_2 + 4H^+ + SO_4^{--} + 2e^- \rightarrow PbSO_4 + 2H_2O \qquad (positive)$$

As the battery operates, the concentration of sulfuric acid is gradually reduced. We test by measuring the solution's specific gravity, which should be about 1.3. The periodic addition of water merely makes up for evaporation, while the "charging" needed when the battery is "low" reverses the processes to bring the plates and solution back to their original conditions. In the nickel–cadmium cell used in space vehicles, voltages are provided by similar reactions at the electrodes with the net effect

$$Cd + Ni(OH)_4 \rightarrow CdO + Ni(OH)_2 + H_2O$$

The Generator as a Source of Electrical Energy

Most of the electricity that supplies our cities, homes, and manufacturing plants comes from electrical generators, which in turn derive their primary energy from falling water, or steam developed by burning some fuel. The physical principle involved is different from that of the battery, in that mechanical energy is converted into electrical energy. Most simply stated, forces applied to a conductor cause it to rotate through a magnetic field, causing a potential difference V to be produced between the ends of

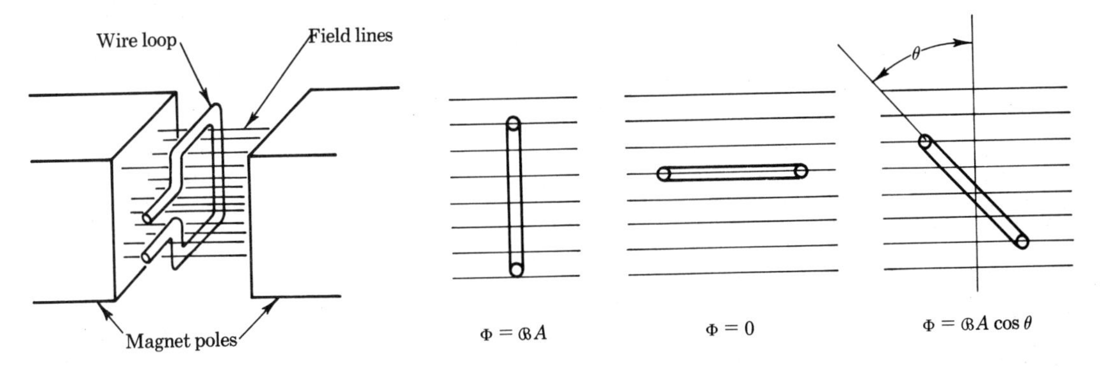

FIGURE 18-3

Elementary generator

the conductor. In accord with Faraday's law (Section 7-4),

$$V = -\frac{\Delta\Phi}{\Delta t}$$

where Φ is the product of field $\mathcal{B}$ and area A perpendicular to $\mathcal{B}$. As in Figure 18-3, let us move a rectangular loop of wire from the vertical to the horizontal. Initially there were $\mathcal{B}A$ lines through the loop; at the end of the motion there are none. Instead, let us rotate the loop with a constant angular speed ω. At any angle $\theta = \omega t$ from the vertical, the area seen by the field is $A \cos\theta$ and

$$\Phi = \mathcal{B}A \cos\omega t$$

Recalling the shape of the cosine curve, the rate of change of flux is zero at $t = 0$ and a maximum at $t = \pi/2\omega$. The potential difference produced is a sine curve (Figure 18-4),

$$V = V_0 \sin\omega t$$

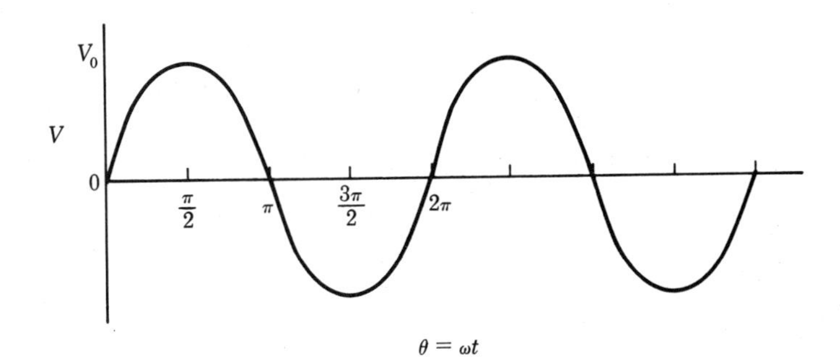

FIGURE 18-4

Potential output of generator

where $V_0 = \mathcal{B}A\omega$ is the peak potential difference. Note the analogy between this behavior and simple harmonic motion (Section 11-2).

In order to generate current of one polarity, it is merely necessary to change the contacts of the wire loops with the terminals each time V reaches 0—i.e., at $t = 0$, $t = \pi/\omega$, $t = 2\pi/\omega$, etc. If we put in more coils and connect them together, a nearly smooth potential may be generated.

The Thermoelectric Cell

The application of heat to a semiconductor was shown in Section 17-6 to yield a potential difference. The thermoelectric cell, as used for the conversion of combustion, solar, or nuclear heat into electrical power, exemplifies this effect. Suppose that an *n*-type and a *p*-type semiconductor are joined at one end, which is then heated, as in Figure 18-5. The cold end of the *n*-material becomes negatively charged by an accumulation of electrons; the cold end of the *p*-type is positive because of the flow of positive "holes." The two terminals thus have a potential difference of the order of a few volts. By connecting many such elements together, higher potentials can be generated to supply power to an external circuit. The virtue of such devices for electrical supplies can be seen by comparison with a conventional power plant where energy is transformed *twice*—from thermal (from a burning fuel) to mechanical (by a heat engine or turbine), and from mechanical to electrical (using a generator). In a thermoelectric cell, thermal energy is transformed into electrical energy in *one* step, called a *direct-conversion* process. One advantage is the elimination of moving parts. Unfortunately the process is not as efficient as one might hope because of thermal conduction that tends to counteract the thermoelectric effect.

We saw that the heat engine could be reversed to extract heat as a refrigerator (Section 15-5). The thermoelectric effect can be used in the same way in what is called the *Peltier effect*. By the passage of electricity, a temperature difference is created, and a flow of heat can be achieved.

A similar effect can be employed in the *thermocouple* to measure temperatures. Two dissimilar wires are connected at one end and the

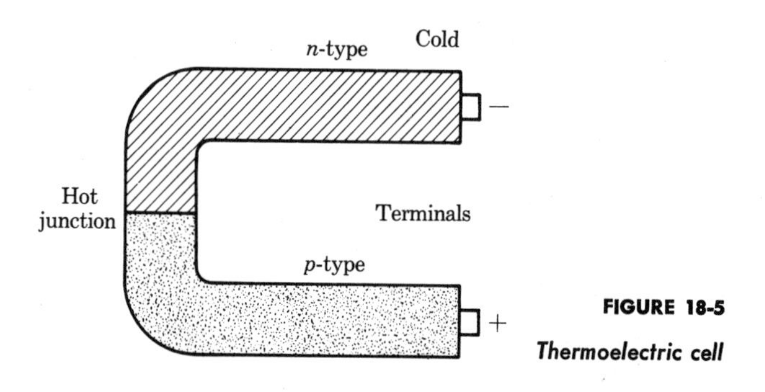

FIGURE 18-5
Thermoelectric cell

junction immersed in a hot liquid or attached to a hot metal structure. The potential difference between the free terminals, which are at lower temperature, can be used to measure the temperature difference.

18-2 ELECTRICAL CIRCUITS

The battery, generator, and thermoelectric cell just discussed are sources of what is called *electromotive force*. This quantity is not a force in the scientific sense, but a measure of the ability of the devices to transform energy. Electromotive force, abbreviated emf or V_E, is the amount of work the source can perform on a unit charge in bringing it from low potential to high. The unit of emf is the joule per coulomb, which is also the volt. If no current is delivered to an external circuit, the emf is the same as the potential difference between the terminals of the source, $V_T = V_E$. If there is a current I, however, work is done to carry charges through the internal resistance r of the battery or generator. This results in an internal potential drop of amount Ir, and the potential difference between the terminals is

$$V_T = V_E - Ir$$

(Note that when I is zero, the voltages become the same.) Thus it is useful to think of the emf as the *maximum terminal voltage*. Now, when the terminals are connected to an external resistance R, the current flow by Ohm's law is

$$V_T = IR$$

By combining equations, we find the current to be

$$I = \frac{V_E}{R + r}$$

Suppose that we turn on the headlights of a car equipped with a battery having 12-V maximum terminal voltage. Let the external resistance be 10 ohms, the internal resistance 0.5 ohm. The current is $I = V_E/(R + r) = 12/10.5 = 1.1$ A, and the terminal voltage is $V_T = V_E - Ir = 12 - (1.1)(0.5) = 11.4$ V.

Since a good source of emf has a low internal resistance, we often can neglect r in comparison with R in rough calculations, and make no distinction between emf and terminal voltage.

Direct Current Circuits

The simplest type of current is steady, constant, or *direct current* (dc). It involves a voltage source and various circuit elements that have been in operation for some time. Before reaching such steady conditions, the circuit may have carried *transient* currents—i.e., those varying with time. In a large number of devices however, the current is alternating (ac) with time according to the sine (or cosine) function. Currents and voltages that change with time in other ways are often used in research and control. We shall now concentrate on the direct currents, taking advantage of the fact that the large numbers of charged particles involved can be imagined to flow as a continuous fluid such as air or water. Several analogies between fluids and electrical currents are useful.

Two simple and reasonable principles can be employed to analyze any dc circuit. The first rule is that *the algebraic sum of currents to and from a point is zero.* This merely says that charge is not lost or gained. Formally, however,

$$\sum_n I_n = 0$$

where the sign of each current I_n depends on the direction of flow (i.e., into or out of the point). The second rule is that *the work done in carrying a charge around any closed loop or around the whole circuit is zero.* Physically this rule says that energy is conserved and mathematically it is written

$$\sum_n V_n = 0$$

Potential drops and rises must be treated as algebraically opposite, of course. These rules will have more meaning if we analyze the simplest possible circuit, a resistor connected to the terminals of a dry-cell battery. We let the longer of the two lines symbolizing a battery in Figure 18-6 be the + terminal (anode), taken at a higher potential than the − terminal (cathode). Electrons can now flow through the complete circuit; however, we usually visualize a flow of positive charges in the opposite direction, and call that the current. Thus we use this imaginary positive current for purposes of circuit analysis, keeping in mind that the true current is largely

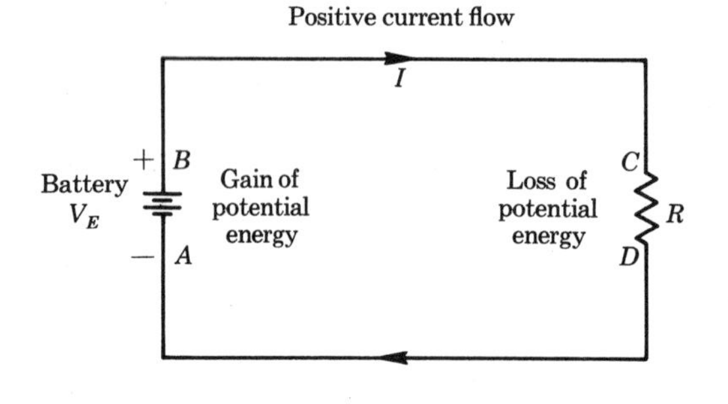

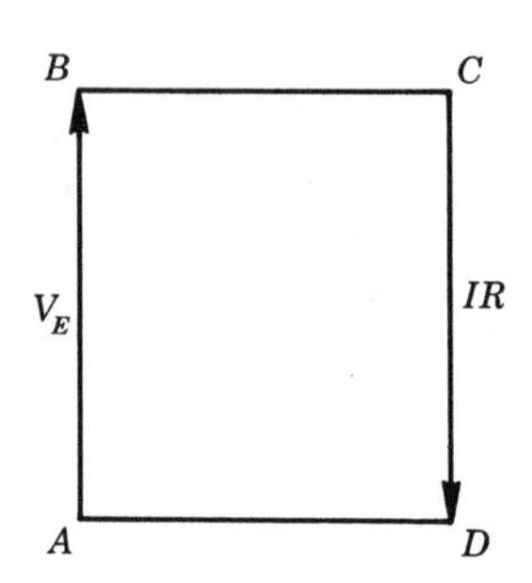

FIGURE 18-6

Potential rises and drops in simple circuit

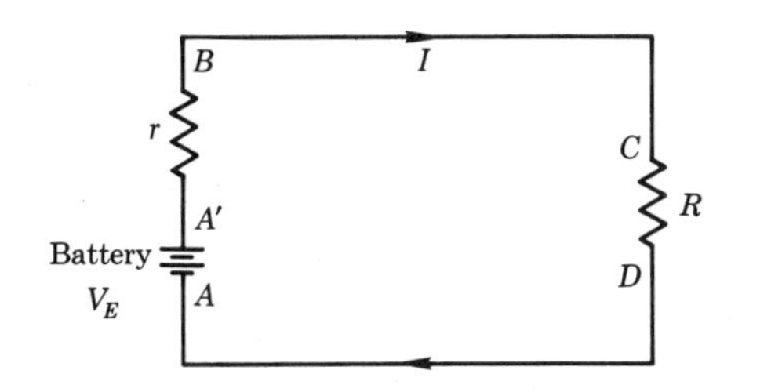

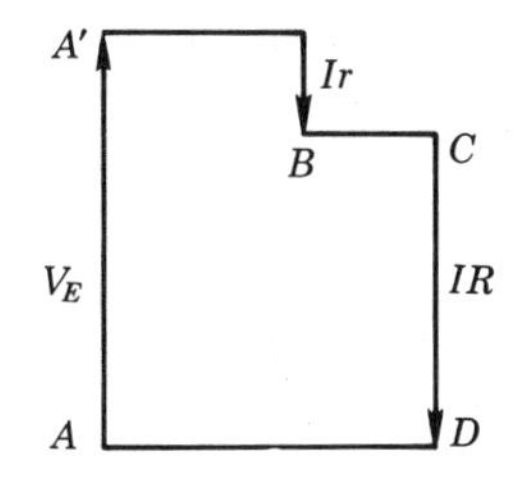

FIGURE 18-7

Effect of internal resistance

that of electrons, and in the opposite direction. The battery supplies the charges with potential energy, which is subsequently converted to heat in the resistor. We may follow a unit positive charge as it goes around the circuit. The potential increases in going from A to B by an amount V_E volts and drops by an equal amount IR volts in going from C to D, assuming that Ohm's law holds (Section 17-1). Thus, on going completely around the loop, the total potential change is zero, or

$$V_E - IR = 0$$

The current is then

$$I = \frac{V_E}{R}$$

It is most natural to trace the change in the direction of positive current flow because of the intimate relation of work and potential. Nothing prevents us, however, from going in the reverse order $DCBA$. To be consistent we should assign to the path from D to C a potential *rise* IR, followed by a potential *drop* $-V_E$ on going from B to A. Again we get

$$IR - V_E = 0$$

The process of adding potentials in a circuit containing a voltage supply with internal resistance is shown in Figure 18-7.

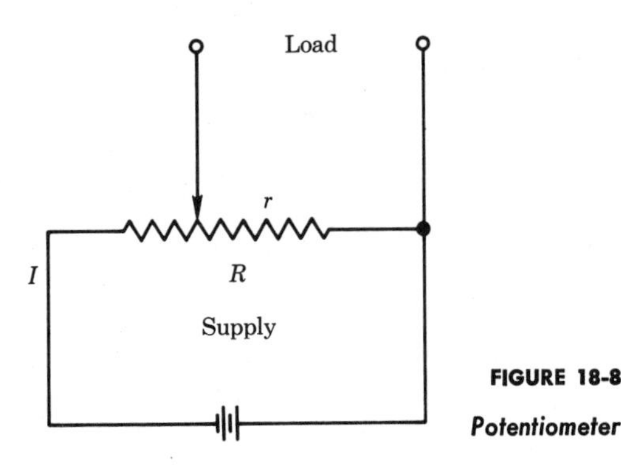

FIGURE 18-8

Potentiometer

As another illustration of the division of voltages, let us consider the *potentiometer*, a device for adjusting the voltage to an external circuit while employing a fixed supply. The circuit in Figure 18-8 is a simple version. The current I passes through the resistor R, which has a movable contact point. The potential drop through the portion r is $V_L = Ir$ compared with the total $V = IR$. By ratios, the voltage applied to the load is

$$V_L = \frac{r}{R}V$$

18-3 METERS AND MOTORS

Meters

Measurements in electrical circuits are achieved by proper insertion of galvanometers, voltmeters, and ammeters, each of which consists basically of a coil of wire in a magnetic field that is directed radially. Every conducting wire experiences a force, just as in the electric motor, and the whole coil experiences a torque. Because of an attached spring that provides a restoring torque, the coil does not spin, but will experience steady deflection at some angle. The amount of rotation as measured by a needle on a scale indicates the voltage or current. This basic meter is the *galvanometer*, shown in Figure 18-9. A very small current is sufficient to cause deflection. In order to convert a galvanometer into an *ammeter*, used to measure larger currents, it is necessary to add a parallel bypass or "shunt" for most of the current—Figure 18-10(a). Let us compute the fraction of the total current I that goes through the coil. First, $I_c + I_s = I$ and $I_cR_c = I_sR_s$, there being a common potential difference across the elements. Then, $\frac{I_c}{I} = \frac{R_s}{(R_c + R_s)}$, from which we can find the shunt resistance. For example, let us consider a sensitive galvanometer, with coil resistance 200 ohms and full-scale needle deflection at current 4×10^{-3} A. We wish to convert it to an ammeter that reads 1 ampere full-scale. To

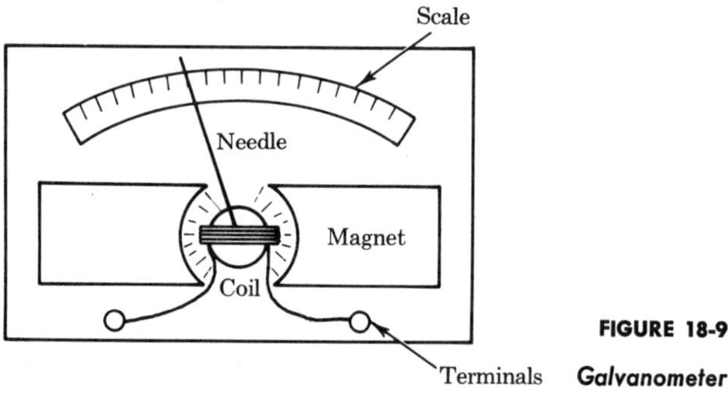

FIGURE 18-9
Galvanometer

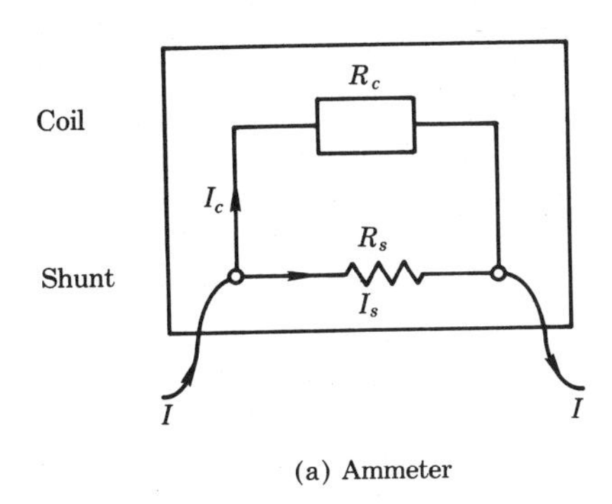

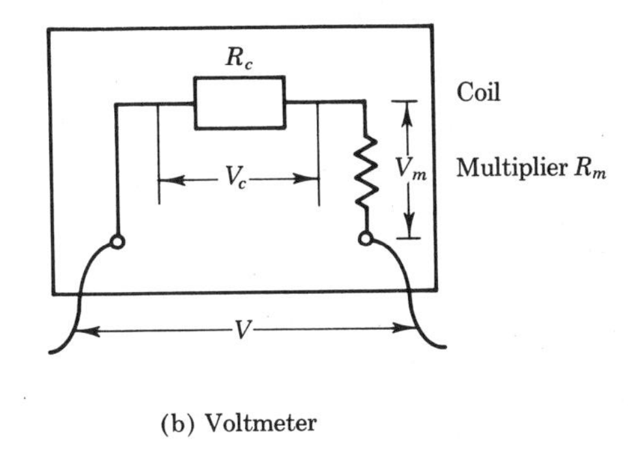

FIGURE 18-10

Meters for measuring current and potential

find the necessary value of the shunt, we solve for

$$R_s = \frac{R_c}{\frac{I}{I_c} - 1} = \frac{200}{\frac{1.0}{4 \times 10^{-3}} - 1} = \frac{200}{249} = 0.803 \text{ ohm}$$

In the *voltmeter*, a high resistance (*multiplier*) is placed in series with the galvanometer coil, as shown in Figure 18-10(b). The same current I_c flows through R_c and R_m. Its relation to the terminal voltage is found from Ohm's law

$$V = I_c(R_c + R_m),$$

from which we can find the multiplier resistance. We can convert the galvanometer into a voltmeter that reads 100-V full-scale, using R_m of value

$$R_m = \frac{V}{I_c} - R_c = \frac{100}{4 \times 10^{-3}} - 200 = 24{,}800 \text{ ohms}$$

Now let us consider a more complicated network, as in Figure 18-11, which contains several resistors and two sources of emf with internal resistance. The labeled quantities are either known or to be found, depending on the way the problem is stated. We apply the two basic rules of circuits (Section 18-2). At the junction at the top of the circuit,

$$I_1 - I_2 - I_3 = 0$$

where currents approaching are taken as positive, and those leaving are negative. (Note that if a current arrow happened to be attached incorrectly, the sign on that current would correctly come out negative.) At the junction on the bottom we find the same result. Going around the outside of the circuit, we find the voltage sum to be

$$V_T - I_1R_i - I_1R_1 - I_2R_2 - I_2R_m + V_B - I_1R_4 = 0$$

Going around the right loop,

$$-I_2R_2 - I_2R_m + V_B + I_3R_3 = 0$$

A positive sign was attached to I_3R_3 because the direction of path chosen was opposite to the anticipated current flow. The reader is left to verify that no new information is obtained by forming the potential equation for the left loop in Figure 18-11.

As the situation stands, we have *three* equations in *eleven* quantities. We see that at least *eight* of the latter must be given if we are to be able

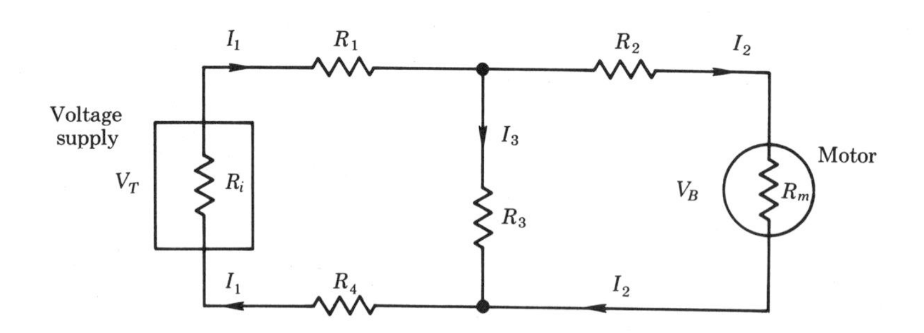

FIGURE 18-11

Electrical circuit with several elements

to solve the problem uniquely. As a numerical example, let us assume the following as given:

Voltages (volts)	*Resistances* (ohms)	
$V_T = 6$	$R_i = 0.1$	$R_2 = 1$
$V_B = 2$	$R_m = 0.5$	$R_3 = 2.5$
	$R_1 = 3$	$R_4 = 4$

Thus we need to solve for I_1, I_2, and I_3. Inserting numbers, we find that

$$I_1 - I_2 - I_3 = 0$$

$$6 - I_1(0.1) - I_1(3) - I_2(1) - I_2(0.5) + 2 - I_1(4) = 0$$

$$-I_2(1) - I_2(0.5) + 2 + I_3(2.5) = 0$$

We may carry out the simultaneous solution of the three equations in several ways, each yielding

$$I_1 = 0.90 \text{ A}, \qquad I_2 = 1.06 \text{ A}, \qquad I_3 = -0.16 \text{ A}$$

Note that I_3 is negative, indicating that it flows opposite our assumed direction. In problems of this type, care with signs and decimals is especially important, and the answers should be checked by substitution back into *all* of the original equations.

The Electric Motor

It is a well-understood fact that an ac supply can be used to provide power to drive motors in common appliances such as the refrigerator, vacuum cleaner, and washer. Also, machinery in most factories is powered by electric motors. Let us examine a simple motor operated on direct current, such as that used to start an automobile.

The basic principle of the direct current motor is that a current-carrying wire experiences a force when placed in a magnetic field. Recall from Section 7-1 that the force on a single moving charge is

$$\mathbf{F} = q\mathbf{v} \times \mathcal{B}$$

We also showed, in Section 9-7, that the torque on a rectangular loop of current-carrying wire in a vertical field is

$$N = IA\mathcal{B} \sin \theta$$

where θ was the angle between the plane of the loop and the horizontal. This torque tends to bring the plane of the loop perpendicular to the field. By use of a split ring (Figure 18-12), the direction of current can be reversed on each half-revolution, allowing the loop to rotate continuously as a rudimentary motor. The motor is thus similar to the electric generator,

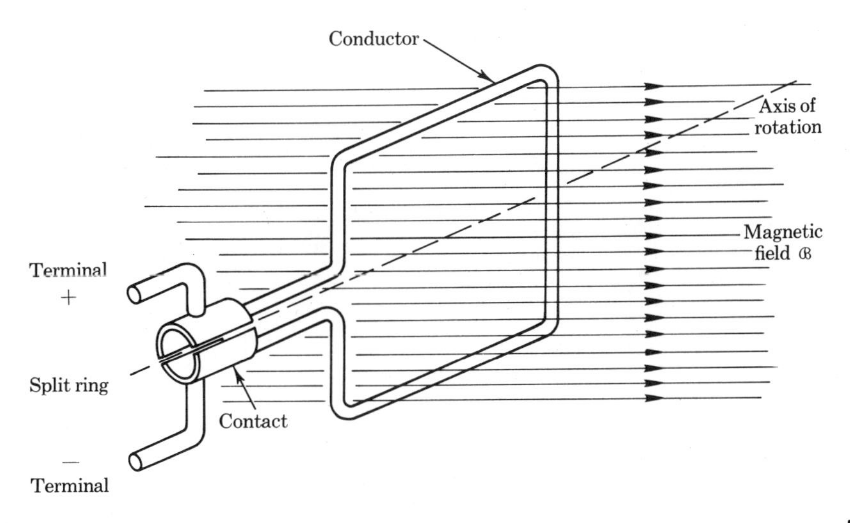

FIGURE 18-12
Elementary motor

except that motion in the former is produced by currents, instead of currents being produced by motion, as in the latter. In a real motor there are many such wires, mounted on a cylindrical iron *armature*. When the current is continuously switched by contacts of carbon *brushes* with a *commutator*, the sum of all the torques keeps the motor turning and doing work of rotation.

18-4 VACUUM TUBES AND TRANSISTORS

Many electrical operations not possible for resistors, capacitors, and inductors can be performed by a *vacuum tube*. Although the transistor has tended to take its place, the vacuum tube served to make radio and television possible, and still is used in many processes. Also, we find that part of the mathematical description of the tube carries over to that of the transistor.

Diodes

The simplest version of a vacuum tube is the *diode* (two electrodes), as in Figure 18-13. A pair of plates is sealed into an evacuated tube, with electrical leads coming out through the base. One plate, the cathode, is heated to incandescence by a fine wire "*heater*," which has a separate circuit. Electrons "boil out" of the metal if it is at high enough temperature, a process called *thermionic emission*. Thomas Edison was the first to observe this effect, during his study of the electric lamp. When the kinetic energy due to thermal agitation exceeds the work function of the substance (at that temperature), electrons can escape. There is a limit to the amount of current of electrons that can be drawn from a heated electrode at a given temperature. The higher the absolute temperature T, the greater the average kinetic energy of the electrons in the metal. How many of these can escape depends on the thermionic work function W (Section 8-8). The lower W is, the easier it is for electrons to escape from the surface.

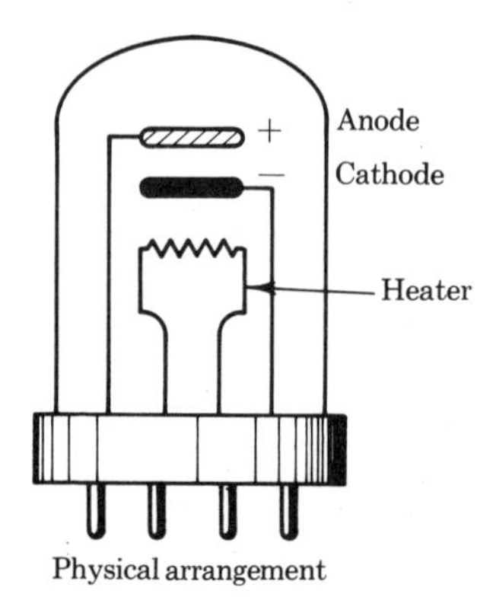

Physical arrangement

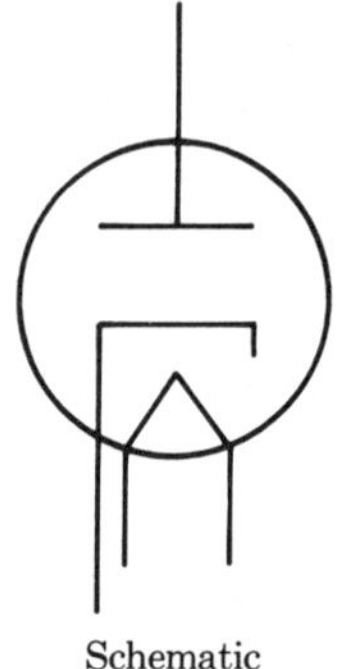

Schematic

FIGURE 18-13
Diode vacuum tube

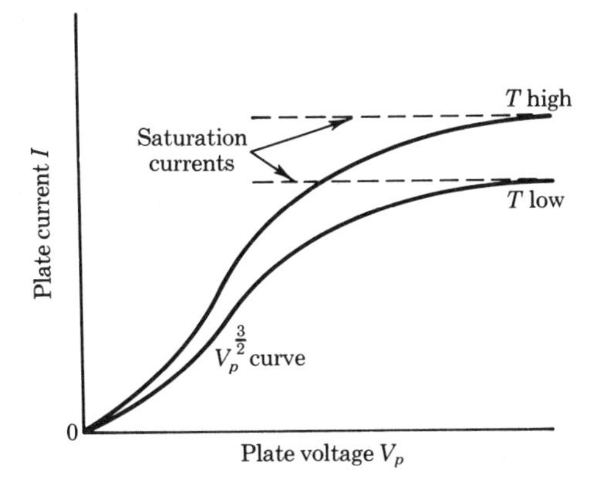

FIGURE 18-14

Current-voltage relation for diode vacuum tube (The current is proportional to $V_p^{3/2}$ and inversely proportional to d^2)

Between the plates of the diode, a potential difference V_p (plate voltage) is applied. The electrons emitted by the cathode are accelerated across the gap of width d, creating a plate current. Again a limitation on the amount of charge flow is experienced, since the charges in transit create a field of their own, and tend to retard the emission of electrons. The higher the plate voltage, or the narrower the gap, the greater the allowable flow, as we would expect. It is found that the current is proportional to $V_p^{3/2}$ and inversely proportional to d^2. If the potential difference V_p is continuously increased, the current reaches a "saturation" level, as shown in Figure 18-14. Every electron that is available is drawn into the gap and collected by the anode. Only by raising the temperature can current

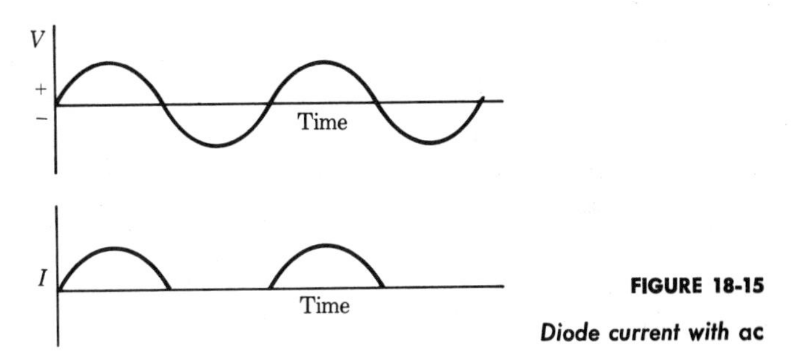

FIGURE 18-15

Diode current with ac

through the device be increased. The vacuum tube (or "valve," as it was formerly called) has the ability to convert alternating current (ac) into pulsations of only one sign. When V_p is positive, as is one half-cycle of an alternating supply, current is passed, but if V_p is negative, electrons are repelled by the plate, and no current flows. Figure 18-15 shows the supply voltage and the "*rectified*" current.

Triodes

Precise control of currents in the plate circuit is achieved by the addition of a third electrode, the *grid*, between the other two, as shown in

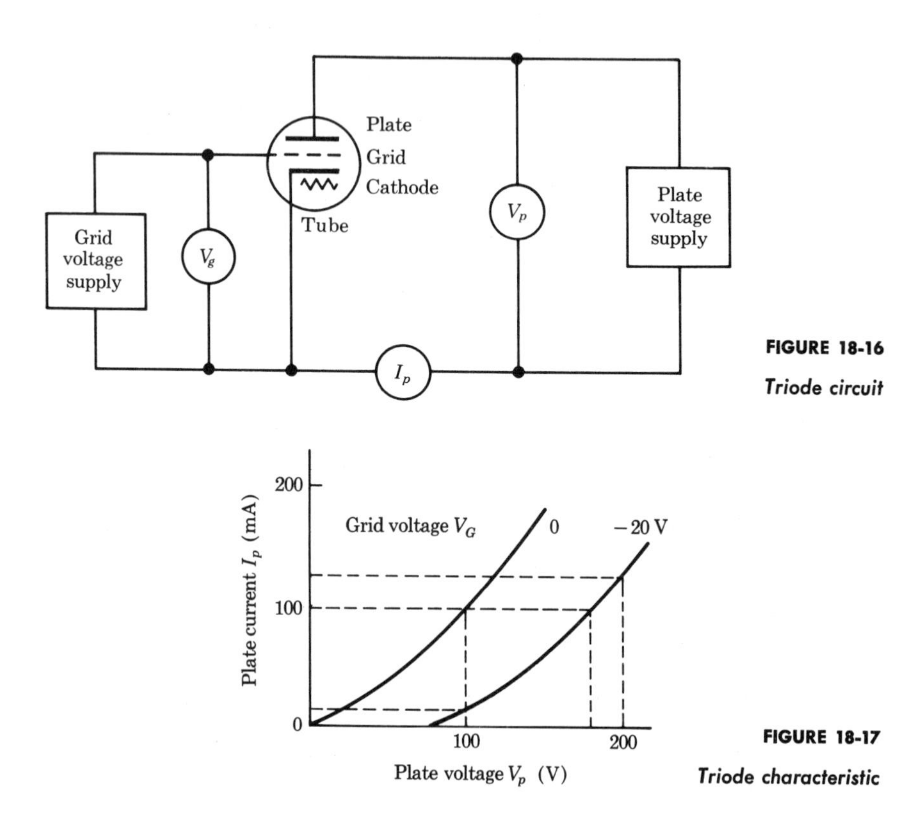

FIGURE 18-16
Triode circuit

FIGURE 18-17
Triode characteristic

Figure 18-16, with the plate-heater circuit omitted for simplicity. The grid is a screen that is much closer to the cathode than is the plate, and variations in its potential V_g affect the space charge and plate current more effectively than equivalent changes in the plate voltage. Figure 18-17 shows schematically the tube-current characteristics for two different grid voltages, zero and -20 V. The value of I_p can be dropped by a factor of 10 by changing V_g by only 20 V, while the same reduction requires 80 V of plate-voltage change. The triode thus serves as an *amplifier*. A small grid-voltage variation, as from a weak radio signal, causes a large change in plate current, and thus in voltage across a resistor in the plate circuit. Many stages can be placed in series, to multiply amplification factors.

Transistors

Important effects occur when n-type and p-type semiconductors (recall Section 17-6) are joined in the same crystal, as in Figure 18-18. The tendency toward an excess of one type of charge carrier over the other is indicated by the + and − symbols. Such a distribution favors the flow of electrons from right to left, and inhibits the flow of electrons from left to right. An n–p junction thus acts as a rectifier, and can replace a much more elaborate and expensive diode vacuum tube.

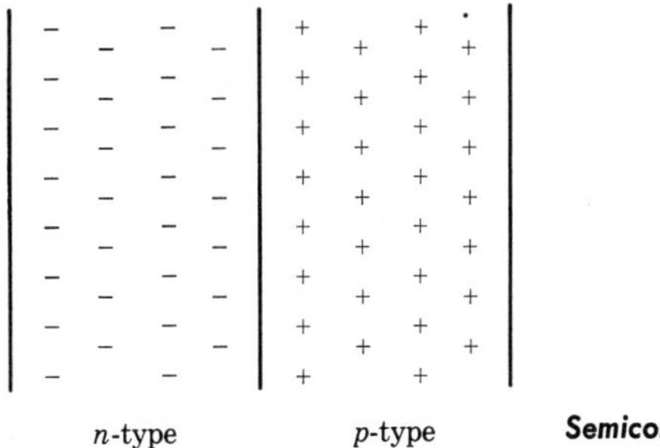

FIGURE 18-18

Semiconductor diode

When three layers, *n–p–n* or *p–n–p*, are constructed (Figure 18-19), the system is called a *transistor*, which behaves much like a triode vacuum tube, but is of very minute size. Among the unique features of transistors are their low power drain (since they do not have heaters), reliability, and long life. These small semiconductor devices have less natural capacitance

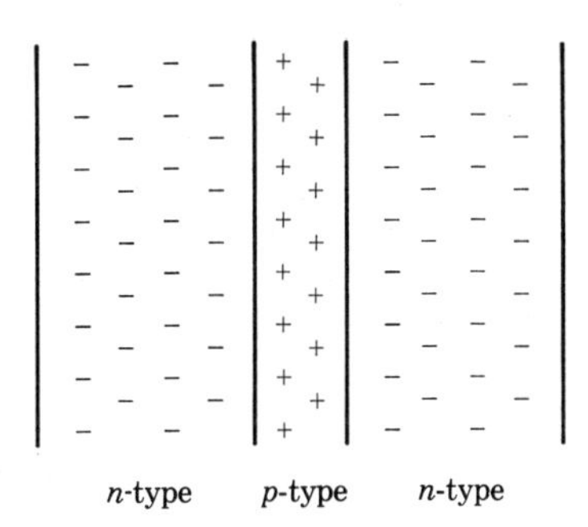

FIGURE 18-19

Transistor

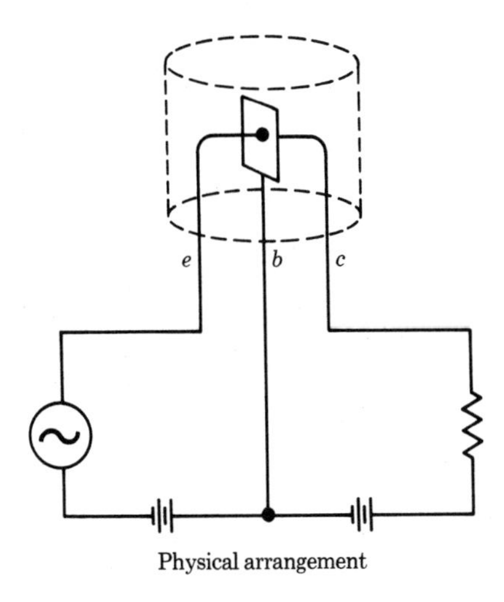

Physical arrangement

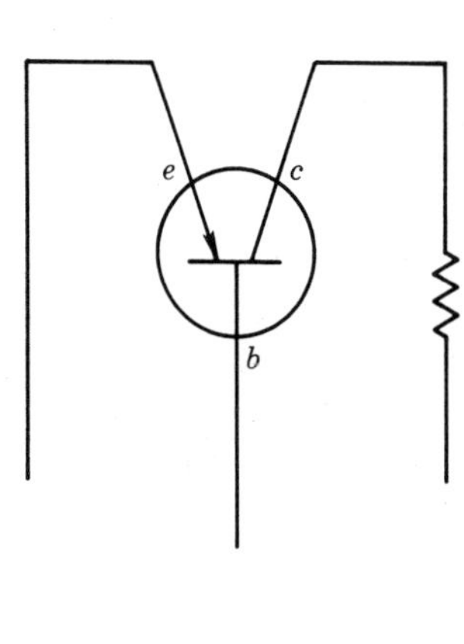

Schematic diagram

FIGURE 18-20

Transistor amplifier

and the shorter leads have less inductance, thus they can operate at higher frequencies than conventional tubes.

Since the transistor may be constructed either as an *n–p–n* or as a *p–n–p* device, it is able to use either electrons or "holes" as the charge carrier. The net effect is rather like having two types of tubes available, one with negative electrons and one with "positive" electrons. A typical transistor circuit is shown in Figure 18-20.

The three connections are labeled *emitter e*, *collector c*, and *base b*. These correspond roughly to the cathode, plate, and grid, respectively, of a vacuum triode. The two devices are not directly interchangeable, however, and the circuits must be designed specifically for one or the other.

18-5 ELECTRICAL OSCILLATIONS AND AC CIRCUITS

The most familiar example of electrical vibrations consists of those from the 60-cycle ac commercial supply. The original source is some remote generator, but for our purposes we may study the variation of voltage difference between the two terminals in a wall receptacle, as in the home.

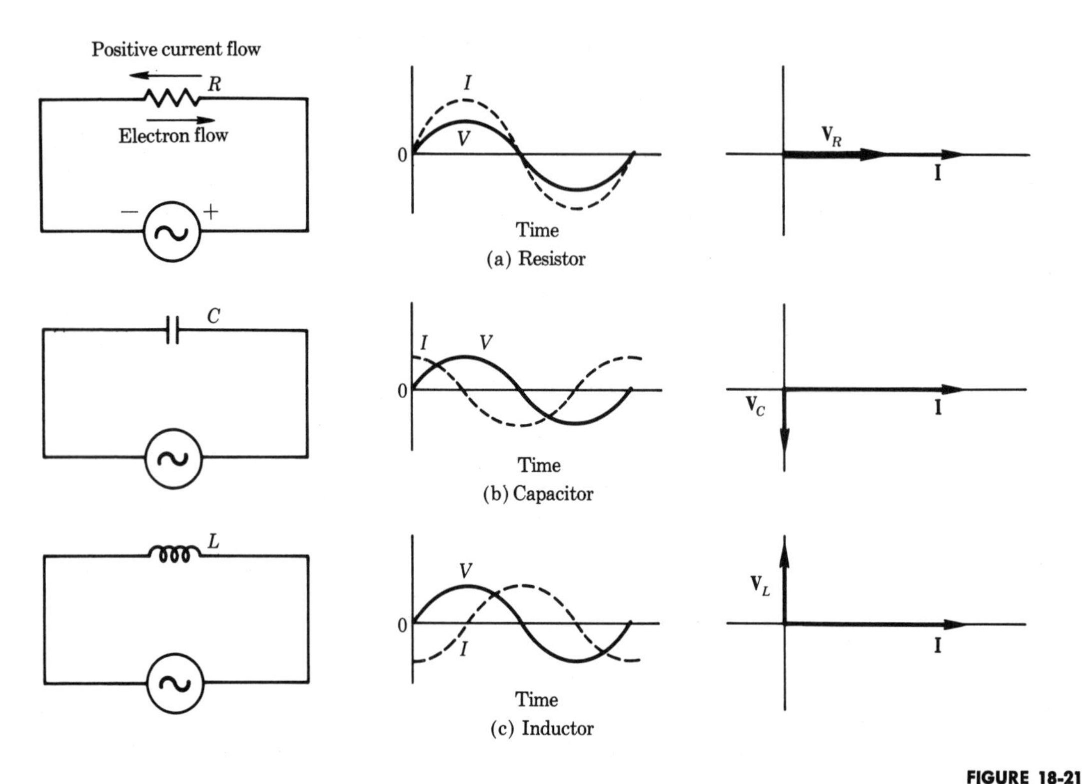

FIGURE 18-21

Elementary ac circuits

The supply voltage may be written

$$V_S = V_0 \sin \omega t$$

where V_0 is the maximum amplitude and $\omega = 2\pi\nu$ is the angular frequency, with ν the frequency, here 60 cps. At the terminals, the voltage alternates between + and −. Suppose that we connect a simple resistor to the outlet—Figure 18-21(a). A flow of electrons from left to right will occur in the conductor when the polarity is − +, and from right to left when the polarity is + −. By the convention of current, the positive current I flows oppositely. In order to find I, we start with the rule that the work done is zero, in carrying a positive charge about the closed circuit consisting of the resistor and the supply. Then at any instant

$$V_S - V_R = 0$$

where positive work must be done to bring the + charge from the negative terminal to the positive terminal in the supply, and the reverse in the wire. Now, by Ohm's law,

$$V_R = IR$$

Thus

$$IR = V_S$$

The current past any point in the circuit is, then,

$$I = \frac{V_0}{R} \sin \omega t$$

We note that I and V_R are always proportional—i.e., are *in phase*—Figure 18-21(a).

Suppose that instead of the resistor, we insert a capacitor as a charge storage element—Figure 18-21(b). We shall assume that it is capable of holding a total charge Q with potential difference V_C between its terminals:

$$Q = CV_C$$

where C is the capacitance. The equation of this new circuit is

$$V_S - V_C = 0$$

Thus

$$\frac{Q}{C} = V_S$$

or

$$Q = CV_0 \sin \omega t$$

As the potential difference rises from zero at $t = 0$ to a maximum of V_0 at $T/2$, the charge increases from zero to a maximum CV_0. Now, the current

I, as the rate of flow of charge $\Delta Q/\Delta t$, is greatest at the start of the cycle and least at $T/2$, as plotted in Figure 18-21 (b). By analogy with the relation of displacement y and speed $\Delta y/\Delta t$ in simple harmonic motion (Section 11-1), we find that the current is

$$I = \frac{\Delta Q}{\Delta t} = CV_0\,\omega \cos \omega t$$

This is easily understood on physical grounds. As the driving potential increases, the charge flow increases, and the capacitor begins to "fill up" with charges. As the maximum charge $Q_m = CV_0$ is approached, the potential developed tends to reduce the amount of flow to the capacitor. The current *leads* the voltage in phase by 90 deg, as we can verify by writing I in a different way*:

$$I = \frac{V_0}{X_c} \sin\left(\omega t + \frac{\pi}{2}\right)$$

where we have introduced $X_c = 1/\omega C$, called the *capacitative reactance*, with units ohms.

As a third possibility, we insert a coil as inductor—Figure 18-21(c)—which creates a magnetic field that interacts with the charge flow. The voltage V_L developed in such an inductance L is proportional to the rate of change of current:

$$V_L = L\frac{\Delta I}{\Delta t}$$

For this circuit,

$$V_S - V_L = 0$$

Thus

$$L\frac{\Delta I}{\Delta t} = V_S$$

or

$$\frac{\Delta I}{\Delta t} = \frac{V_0}{L} \sin \omega t$$

We may again use to advantage the analogy between the electric circuit and the harmonic oscillator. The current is of the form $I = -(V_0/L\omega) \cos \omega t$ or, in terms of a phase angle,

$$I = \frac{V_0}{X_L} \sin\left(\omega t - \frac{\pi}{2}\right)$$

where $X_L = \omega L$ is called the *inductive reactance*, also in ohms. This shows

* Note that $\sin\left(\omega t + \frac{\pi}{2}\right) = \cos \omega t$.

that the current *lags* the voltage—that is, reaches its maximum later, as in Figure 18-21(c).

The phase relations for the three circuit elements R, C, and L can be expressed simply by a vector scheme. Let the current be represented by a vector $\mathbf{I}$. Then, as also illustrated in Figure 18-21, the voltages across the elements R, C, and L are vectors respectively aligned with, 90 deg behind, and 90 deg ahead of the current vector.

Series RCL Circuits

The vector method is especially helpful in analyzing the behavior of RCL circuits, which contain all three elements in series, as shown in Figure 18-22. The alternating supply voltage is $V = V_0 \sin \omega t$. The currents that flow in this series circuit—through the resistor, through the inductor, and

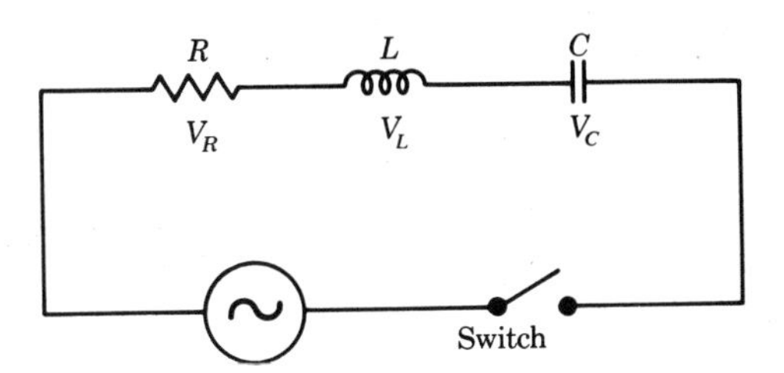

FIGURE 18-22
Series RCL circuit with ac supply

to the capacitor plates—are one and the same, and thus are in phase, but the voltages are out of phase with each other. We are especially interested in the current that will be drawn and the relation between the phase of the current and that of the supply voltage. At any instant, the supply voltage as a vector is the sum of voltages for the individual elements (in analogy with relations for dc circuits, Section 18-2):

$$\mathbf{V} = \mathbf{V}_R + \mathbf{V}_C + \mathbf{V}_L$$

We have already seen that the magnitudes of the voltages are $V_R = IR$, $V_C = IX_C$, and $V_L = IX_L$, but we must express the voltages as vectors. Let us introduce a vector $\mathbf{R}$, of magnitude equal to the resistance and directed along the current vector. Then, $\mathbf{V}_R = I\mathbf{R}$. Similarly, we let the reactances be vectors $\mathbf{X}_C$ and $\mathbf{X}_L$ at right angles to $\mathbf{R}$, as shown in Figure 18-23(a). The corresponding voltage vectors are thus $\mathbf{V}_C$ and $\mathbf{V}_L$, as shown in Figure 18-23(b). As an extension of Ohm's law, we then propose that

$$\mathbf{V} = I\mathbf{Z}$$

where $\mathbf{Z}$ is called the *impedance*, a generalization of resistance for ac circuits. We see at once that vectorially

$$\mathbf{Z} = \mathbf{R} + \mathbf{X}_C + \mathbf{X}_L$$

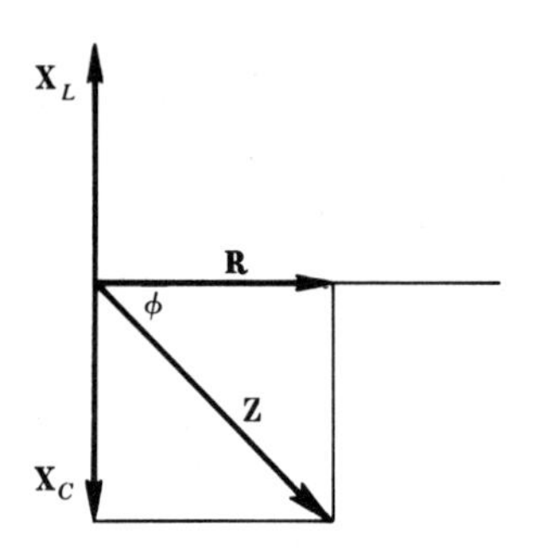

(a) Vector reactances

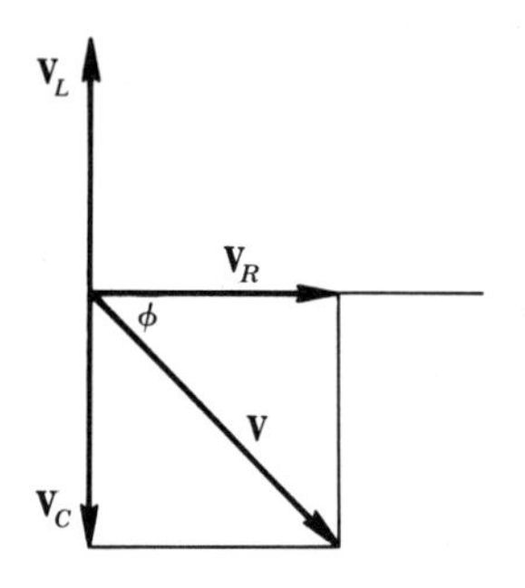

(b) Vector voltages

FIGURE 18-23
Vector relations in ac series RCL circuit

—also sketched in Figure 18-23(a)—and that its magnitude is

$$Z = \sqrt{R^2 + (X_L - X_C)^2}$$

The phase angle between **Z** and **R** is the same as between **V** and **I**, and is obtained from either

$$\cos \phi = \frac{R}{Z} \qquad \text{or} \qquad \tan \phi = \frac{X_L - X_C}{R}$$

The maximum value of the current, with peak applied voltage V_0, is easily found to be

$$I_0 = \frac{V_0}{Z}$$

We should note several features of the impedance. Recalling that $X_L = \omega L$ and that $X_C = 1/\omega C$, we see that Z depends on the frequency. The amount of current through the circuit will be low if ω is either very high or very low, and reaches a maximum when $X_L = X_C$—i.e., when $Z = R$, which says that the circuit acts as a pure resistor. The current and voltage are then exactly in phase, as seen from the expression for the phase angle ϕ. Such a condition is called *resonance* (also see Section 11-3). The particular frequency for resonance is found from

$$\omega L = \frac{1}{\omega C}$$

or

$$\omega = \frac{1}{\sqrt{LC}} \qquad \textit{(resonance condition)}$$

For example, if $L = 0.04$ henry and $C = 110\ \mu\text{F}$, then

$$\omega = \frac{1}{\sqrt{(4 \times 10^{-2})(1.1 \times 10^{-4})}} = 477/\text{sec},$$

or

$$\nu = \frac{\omega}{2\pi} = \frac{477}{6.28} = 76 \text{ cps}$$

The differences in phase play an important role in determing the way energy is shared among the elements at any instant. The ac supply transfers energy to the circuit at a certain rate, which is the input power. Since the capacitor stores charge half of the time and discharges it half of the time over a cycle, its power on the average is zero. Similarly, the magnetic field in the inductor builds up half of the time and drops off half of the time, with average power zero. We conclude that all of the power input goes into heating of the resistor. Thus, as in Section 17-1, the instantaneous power delivered to the circuit is $P = I^2R$. Since I varies periodically, its square varies as the square of the sine function. From the plots in Figure 18-24, we observe that the average value of the latter is $\frac{1}{2}$. The *average* power over a cycle is then

$$\overline{P} = \tfrac{1}{2}I_0^2R$$

Recalling, however, that $I_0 = V_0/Z$ and $\cos\phi = R/Z$, we find that

$$\overline{P} = \tfrac{1}{2}I_0V_0\cos\phi$$

The quantity $\cos\phi$ is called the *power factor*, determined by the relative sizes of the circuit elements and by the frequency.

We can invent an effective dc current I_e that would give the same amount of heating in the circuit as the actual ac current I. Let $\overline{P} = I_e^2R$ and equate it to $\overline{P} = \frac{1}{2}I_0^2R$, so that

$$I_e = \frac{I_0}{\sqrt{2}}$$

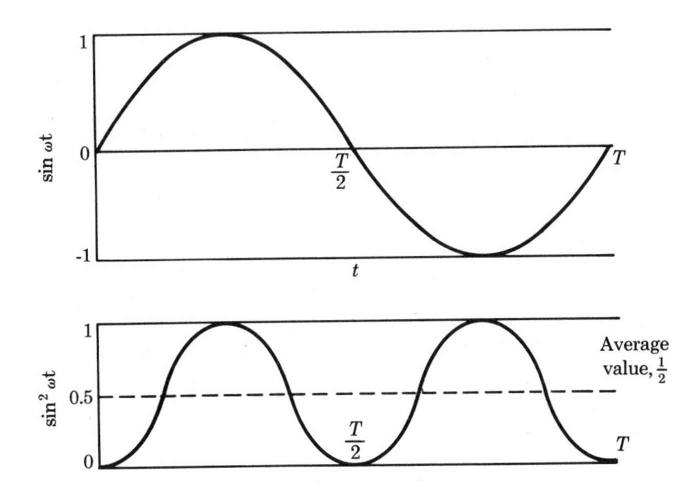

FIGURE 18-24
Average value of $\sin^2 \omega t$

Introducing also an effective voltage $V_e = V_0/\sqrt{2}$, we see that the average power is $\overline{P} = I_e V_e \cos\phi$. For maximum power delivery, ϕ should be zero, $\cos\phi = 1$, corresponding to the resonance condition.

Let us carry through a numerical illustration for an ac circuit. Suppose that $R = 20$ ohms and that $L = 0.04$ henry, $C = 110\ \mu$F. Let $V_0 = 165$ volts and $\nu = 60$ cps as in an ordinary commercial system. Now, $\omega = 2\pi\nu = 377$/sec and the reactance components are

$$X_L = \omega L = 377(4 \times 10^{-2}) = 15.1 \text{ ohms}$$

$$X_C = \frac{1}{\omega C} = \frac{1}{377(1.1 \times 10^{-4})} = 24.1 \text{ ohms}$$

Then, $X_L - X_C = -9.0$, and

$$Z = \sqrt{(20)^2 + (-9)^2} = 21.9 \text{ ohms}$$

The peak current is

$$I_0 = \frac{V_0}{Z} = \frac{165}{21.9} = 7.5 \text{ A}$$

and the effective current is

$$I_e = \frac{I_0}{\sqrt{2}} = 5.3 \text{ A}$$

To find the phase angle, we write

$$\tan\phi = \frac{X_L - X_C}{R} = -\frac{9.0}{20}$$

$$\phi = -0.42 \qquad \text{or} \qquad -24 \text{ deg}$$

The power factor is, then,

$$\cos\phi = 0.91$$

and the power delivered to the circuit on the average over the cycle is

$$\overline{P} = \tfrac{1}{2}V_0 I_0 \cos\phi = \tfrac{1}{2}(165)(7.5)(0.91) = 560 \text{ W}$$

18-6 PARTICLE ACCELERATORS

A charged-particle accelerator, with its many components and processes, is an excellent application of a wide variety of physics principles. We shall describe some simple versions of several accelerators. First, we must emphasize that any device that employs electric and magnetic fields in projecting ions or electrons can be considered an accelerator, and that the principles of operation of all have much in common.

The Linear Accelerator

The potential required to accelerate charges from zero energy directly into the million-electron-volt region would be excessive. To obtain 1-MeV protons, V must of course be 1 million volts. Wideröe in 1928 had the idea of the *linear accelerator* in which successive "kicks," each of low energy, can give a particle a high cumulative energy. Visualize a series of hollow tubes of varying length and spacing, to which an alternating radio frequency (rf) potential is applied as shown in Figure 18-25. Tube diameters are small compared to their lengths, giving a nearly uniform field across gaps and zero field inside the tube. By suitable choice of frequency, a particle such as a proton of charge q will cross a gap at a time when the potential difference is near its maximum, V_0, and will gain an energy approximately qV_0. It "drifts" down the field-free tube without change of speed, to reach the next gap when the potential difference is again right for acceleration. It follows that the tube lengths must be successively longer along the path.

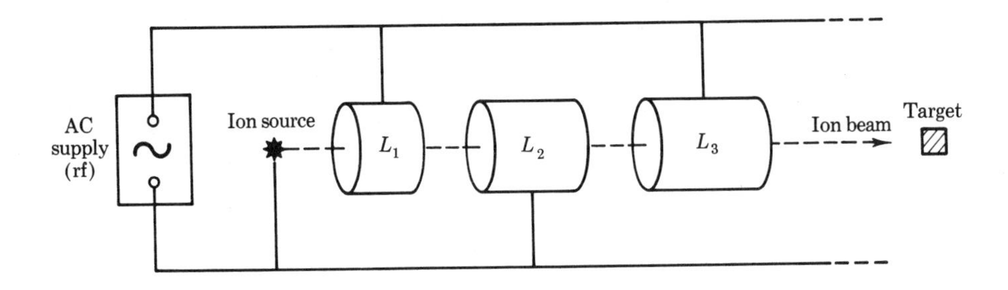

FIGURE 18-25

Simple linear accelerator

Let us examine the energy gains and the requirement on individual "drift-tube" lengths. Suppose the charge starts from rest. In the first gap, the gain in kinetic energy is qV_0 and its speed on entering the field-free tube is $v_1 = \sqrt{2qV_0/m}$. By the time it travels a distance L_1, the rf has gone through 1 half-period $T/2$; hence $L_1 = v_1T/2$. Similarly, the length of the nth tube is $L_n = v_nT/2$. However, the total energy gain by this time is $E_k = nqV_0$ and the speed is $v_n = \sqrt{n}v_1$; thus $L_n = L_1\sqrt{n}$. The total length of the accelerator is the sum of all these different lengths.

Let us calculate the length of the first tube in a linear accelerator with 10,000-volt peak and frequency $\nu = 10$ megacycles per second. The speed of a proton after crossing the first gap is

$$v_1 = \sqrt{\frac{2qV_0}{m}} = \sqrt{\frac{2(1.60 \times 10^{-19})(10{,}000)}{1.67 \times 10^{-27}}} = 1.39 \times 10^6 \text{ m/sec}$$

The length of the tube is

$$L_1 = \frac{v_1}{2\nu} = \frac{1.39 \times 10^6}{2(10^7)} = 0.07 \text{ m}$$

For the second tube, $L_2 = \sqrt{2}(0.07)$, for the third $L_3 = \sqrt{3}(0.07)$, and so on.

The Cyclotron

The particle accelerator called the *cyclotron*, invented in 1930 by E. O. Lawrence, makes ingenious use of electric and magnetic fields. In this device, charges are brought to high enough speeds to cause nuclear disintegration. We saw how alternating potential differences between gaps in the linear accelerator allow repeated additions of energy. The excessive length of the accelerator can be avoided if there is a magnetic field to keep the particles moving in circular paths while in the field-free region. Visualize a flat circular box that is cut in two D-shaped halves, as in Figure 18-26. To the gap between these "dees" is applied the alternating rf potential. An ion is accelerated across the gap, and enters the dee cavity, a space where the electric field is very weak. In the uniform magnetic field perpendicular to the dee faces, the ion now moves in a semicircle, coming out at a gap again. By this time, the potential has reversed polarity, and the charge is again given an additional impetus. We can show a variety of important features of the cyclotron by use of relations between potential difference V, speed v, magnetic field $\mathcal{B}$, and radius of motion, R. We shall assume the speed is low enough so that relativity corrections are not needed.

A particle of mass m (kg) having charge q (in coulombs) and speed v (in meters per second) will move on a circular path of radius r (in meters) in a field $\mathcal{B}$ (in webers per square meter) according to $v = \omega_c r$, where the cyclotron frequency (Section 7-1) is

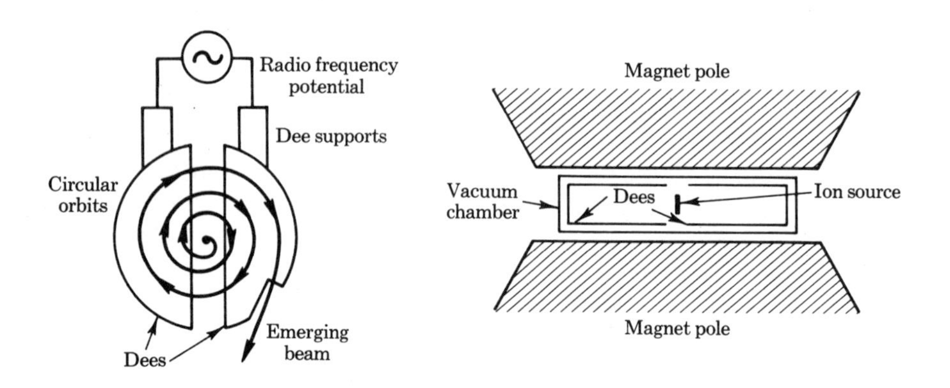

FIGURE 18-26
Cyclotron

$$\omega_c = \frac{q\mathcal{B}}{m}$$

In order for the particle to receive an increase in energy qV_0 each time it crosses the gap at peak potential difference V_0, the angular frequency of the electrical supply ω_e must be the same as the cyclotron frequency. Therefore, we shall label each ω. The total kinetic energy accumulated after N gap crossings will be $E_k = NqV_0 = \frac{1}{2}mv^2$. Thus

$$v = \sqrt{\frac{2E_k}{m}} = \sqrt{\frac{2NqV_0}{m}} = \omega r$$

This shows that r is proportional to $\sqrt{N}$, meaning that the radius of motion increases with the number of times the separate accelerations have been applied. The particle travels in a sort of *spiral* path from the center outwards. The total energy by the time the ion reaches some maximum radius R (determined by the size of the dee) is

$$E_k = \tfrac{1}{2}mv^2 = \tfrac{1}{2}m\omega^2R^2 = \frac{(q\mathcal{B}R)^2}{2m}$$

Usually, the design of a cyclotron is chosen to provide the largest possible final energy, achieved by choosing (1) a large machine radius, (2) a strong magnetic field, (3) a low mass particle, (4) a highly charged particle. There is no apparent dependence on the accelerating voltage. Its importance is tied in with the fact that the smaller total length of path, the larger is V_0, and hence the fewer chances there are for the ion to be deflected by collision with gas particles remaining in the near-vacuum.

Let us suppose that protons of energy 5 MeV are sought, with a 2000-volt electrical supply and a magnet that allows a maximum useful dee radius of 1.2 m (46 in.). The necessary radio frequency and magnetic-field intensity are to be found. The speed of the proton on completing its path is

$$v = \sqrt{\frac{2E_k}{m}} = \sqrt{\frac{2(1.60 \times 10^{-19})(5 \times 10^6)}{1.67 \times 10^{-27}}} = 3.1 \times 10^7 \text{ m/sec}$$

This is only $\frac{1}{10}$ the speed of light, which allows us to use the classical rather than relativistic relations. We then find the angular frequency of the electrical supply to be $\omega = \dfrac{v}{R} = \dfrac{3.1 \times 10^7}{1.2} = 2.6 \times 10^7$ rad/sec; the ordinary frequency is thus $\nu = \omega/2\pi = 4.1 \times 10^6$ cps. The magnetic field strength must be

$$\mathcal{B} = \frac{m\omega}{q} = \frac{(1.67 \times 10^{-27})(2.6 \times 10^7)}{1.60 \times 10^{-19}} = 0.27 \text{ Wb/m}^2$$

One can show (Problems 18-22 and 18-27) that the total path traveled by the proton in going from zero energy to the final energy is of the order of *miles*.

The Betatron

The linear accelerator and cyclotron use time-varying electrostatic fields to build up particle energies. In contrast, the *betatron*, invented by Kerst in 1940, employs electric fields induced by changes in the magnetic field with time (Figure 18-27).

We again start with the fact that a charge q in a uniform field $\mathcal{B}$ moves on a circle of radius R at angular speed $\omega = q\mathcal{B}/m$. Its momentum is $p = mv = m\omega R$ or

$$p = q\mathcal{B}R$$

The ions in a betatron are to be accelerated to high speed while remaining on a constant radius R. At any instant then, the force on the charges must be of magnitude $F = \frac{\Delta p}{\Delta t} = q\frac{\Delta\mathcal{B}}{\Delta t}R$. Faraday's law provides the basis for that force. Recall that a changing magnetic flux gives rise to a potential difference of amount $V = \frac{\Delta\Phi}{\Delta t}\cdot$ By suitable currents in the electromagnet, the magnetic field is caused to vary with position, with an average value $\overline{\mathcal{B}}$ over the region within the circle, corresponding to a flux $\Phi = \overline{\mathcal{B}}\pi R^2$. Now, the electric field experienced by the charges is $\mathcal{E} = V/2\pi R$, and the force is $F = \mathcal{E}q$. Combining, $F = \frac{q}{2}\frac{\Delta\overline{\mathcal{B}}}{\Delta t}R$. We see that the required force and that available by induction are the same only if $\overline{\mathcal{B}} = 2\mathcal{B}$—that is, if the average field is *twice* that at the orbit.

In a typical betatron, the acceleration from rest to full speed takes place in 1 quarter-cycle of a sine-wave variation of field with time. Suppose that we seek a final kinetic energy E_k of 0.49 MeV. At this energy, the electron speed is close to that of light, and it is necessary to use the rel-

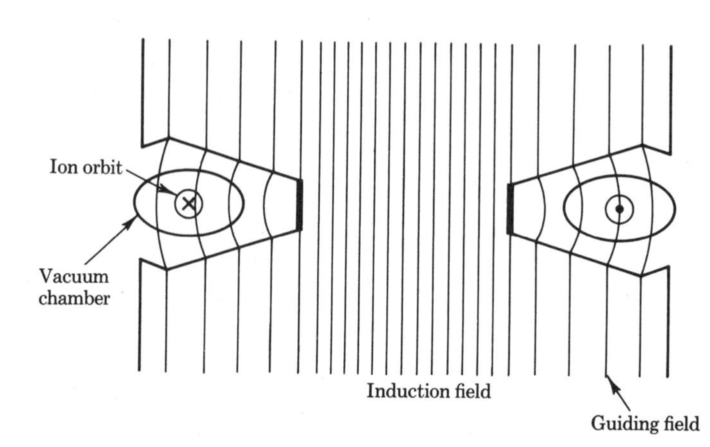

FIGURE 18-27
Betatron

ativistic formula for momentum:

$$p = \frac{\sqrt{E^2 - E_0^2}}{c}$$

where E_0 is the rest energy m_0c^2, which for an electron is $E_0 = 0.51$ MeV $= 0.82 \times 10^{-13}$ J. The total mass-energy is $E = E_k + E_0 = 1$ MeV $= 1.60 \times 10^{-13}$ J. Thus the momentum is

$$p = \frac{\sqrt{(1.60 \times 10^{-13})^2 - (0.82 \times 10^{-13})^2}}{3 \times 10^8} = 4.58 \times 10^{-22} \text{ kg-m/sec}$$

If the orbit radius is 0.5 m, the field at that radius for particles of final energy must be

$$\mathcal{B} = \frac{p}{qR} = \frac{4.58 \times 10^{-22}}{(1.60 \times 10^{-19})(0.5)} = 0.00572 \text{ Wb/m}^2$$

At that time, the value of $\bar{\mathcal{B}}$ must be twice this or 0.0114 Wb/m². Suppose that $\bar{\mathcal{B}}$ increases linearly from zero to its maximum in a time $\frac{1}{250}$ sec. Then, $\Delta\bar{\mathcal{B}}/\Delta t = (0.0114)(250) = 2.85$ Wb/m²-sec. The electric field is $\mathcal{B} = \frac{R\Delta\bar{\mathcal{B}}}{2\Delta t} = \frac{(0.5)(2.85)}{2} = 0.712$ V/m. The energy gain per revolution, of path length $2\pi R = 3.14$ m, is $(0.712)(3.14) = 2.24$ eV. The number of turns to reach 0.49 MeV kinetic energy is thus $\frac{0.49 \times 10^6}{2.24} = 220{,}000$, and total distance traveled is $(2.2 \times 10^5)(3.14) = 6.9 \times 10^5$ m.

The accelerators described in this section are popularly—and incorrectly—called "atom smashers"; in fact, the high-speed projectiles that they produce disrupt the internal structure of target *nuclei*. Details of such transmutation or disintegration processes will be discussed in Chapter 22.

18-7 TELECOMMUNICATIONS

A most rapid progress has been made in electronic-communication methods and techniques in the last quarter-century. The rapid growth of television, including the use of satellites for intercontinental "live" broadcasts, has been remarkable.

A good example of the complexity of present communications is found in the traffic-guidance system of a modern airport. The airplane naturally is in voice contact with the tower. The tower also has the incoming plane identified and located by ground-control equipment that gives the plane's location, speed, height, and angle of descent. This information is continuously available and is not under the pilot's control. For each flight, thousands of messages are needed each way to carry out the complicated flight plans and to negotiate the traffic patterns of an airport. Many simultaneous beams are employed to carry all this information back and forth.

This same situation exists in a telephone cable. With present techniques of time-sharing and scrambling, each pair of wires may carry four to six telephone conversations. This seems very wasteful, when we learn that modern electronics has the capability of admixing hundreds of thousands of telephone conversations on one microwave beam and then separating them again. To keep the beam localized and directed, we can enclose it in a small pipe called a waveguide.

Communications of the future are likely to be even more astounding to today's layman than our own would be to yesterday's untrained individual. We can anticipate the replacement of the present-day telephone, as a combination pleasure and business device. The possibility exists for a central communication console in each home or office, containing a single television screen replacing our present morning paper, telephone, radio, television, tape recorder and maybe even the neighborhood library and photocopy service. One could "call" a book out of a central storage library and have copies made for convenient reading or have tapes made for display on a screen.

Furthermore, the use of the laser in communications has only just begun and surely will one day be able to perform the tasks now assigned to microwave links.

PROBLEMS

18-1 Calculate the mass of lead (Pb^{206}) that is converted into lead sulfate in a storage battery if it draws 100 amperes for 5 minutes, as if one were trying to start a car that had run out of gasoline.

18-2 A single loop of wire of area 4 cm^2 is rotated at 60 rpm in a magnetic field of 3 Wb/m^2. What maximum potential difference is developed by this elementary generator?

18-3 Find the internal resistance of a battery rated at 6 volts if the terminal voltage at 2-amperes current is 5.7 volts.

18-4 Find the number of the setting of the contact in the potentiometer in Figure P18-4 to obtain an output potential difference of 4 volts, if each segment has a resistance of 5 ohms.

18-5 Find the unknown current and resistance in the circuit in Figure P18-5.

18-6 Show that if the current is zero in the line joining terminals A and B of the Wheatstone-bridge circuit in Figure P18-6, $R_1/R_2 = R_3/R_4$. Notice that if accurate values of three resistances are known, the fourth may be measured.

18-7 To measure the emf of auto batteries ($V = 12.6$ volts), a galvanometer of resistance 40 ohms and a sensitivity of 1 mA for full-scale deflection is to be used. If the meter is to read full-scale for 12.6 volts, what series resistance is needed?

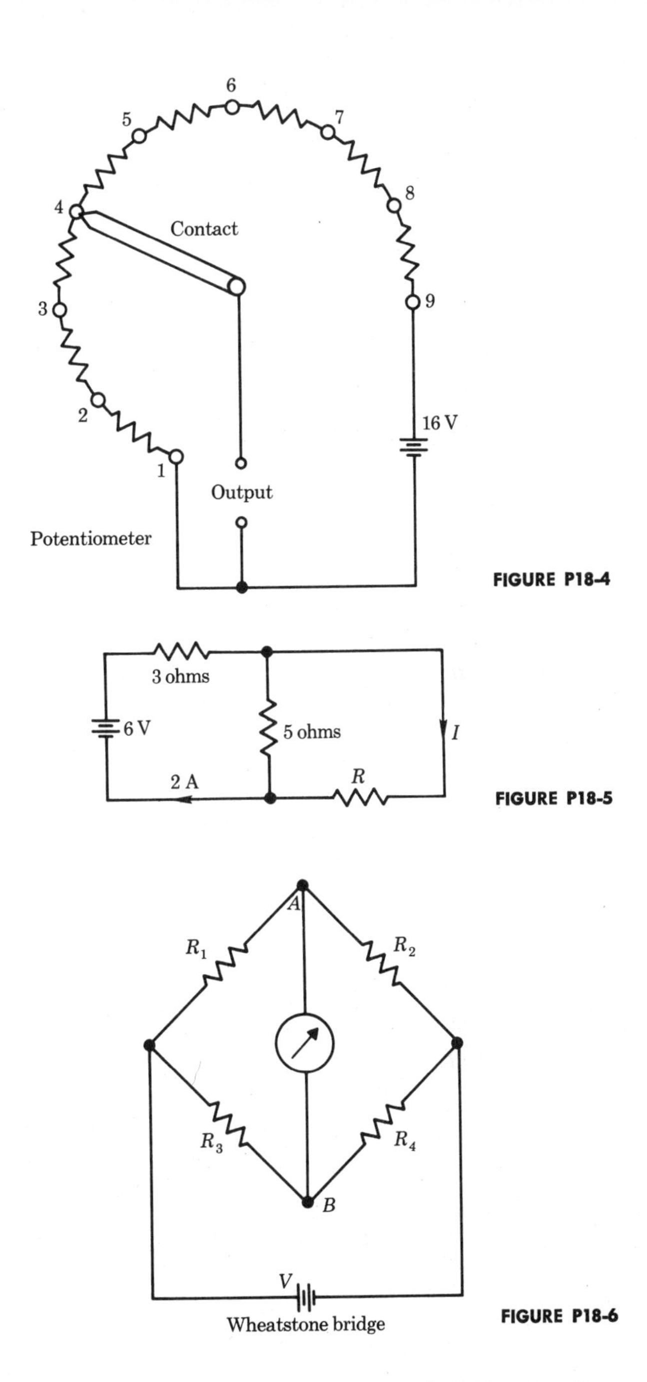

FIGURE P18-4

FIGURE P18-5

FIGURE P18-6

18-8 A galvanometer with coil resistance 10 ohms reads full-scale when the current through it is 0.02 amperes—i.e., with potential drop 0.2 volts. (a) What shunt resistance should be inserted to be able to use the device to read 5 amperes full-scale? (b) What multiplier resistance should be used instead to obtain 250 volts full-scale? Will the needle deflections be directly proportional to the current and voltage to which the meters are attached? Explain.

18-9 In the circuit shown in Figure P18-9, find the potentials at points labeled A, B, C, etc. *Suggestion*: Let $V_A = 0$ and proceed clockwise.

18-10* Develop, by use of the two circuit principles, a set of equations for the currents in the circuit in Figure P18-10 and find the values of the currents.

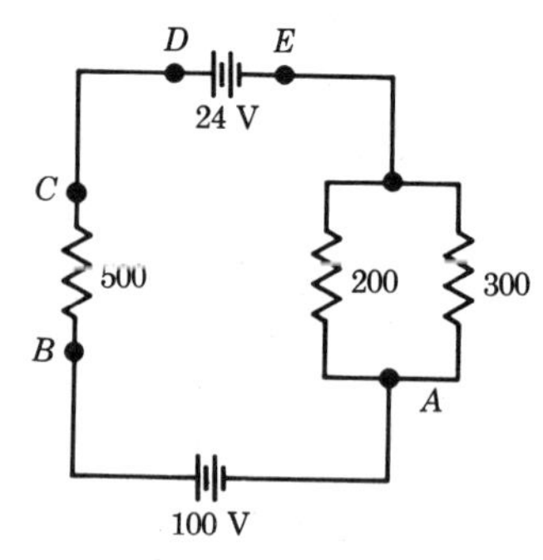

FIGURE P18-9

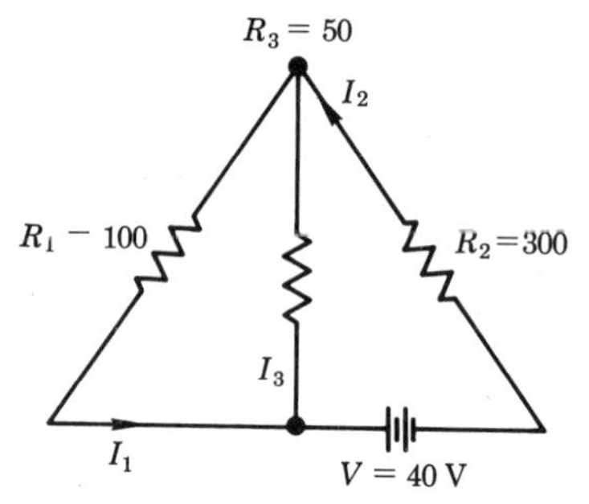

FIGURE P18-10

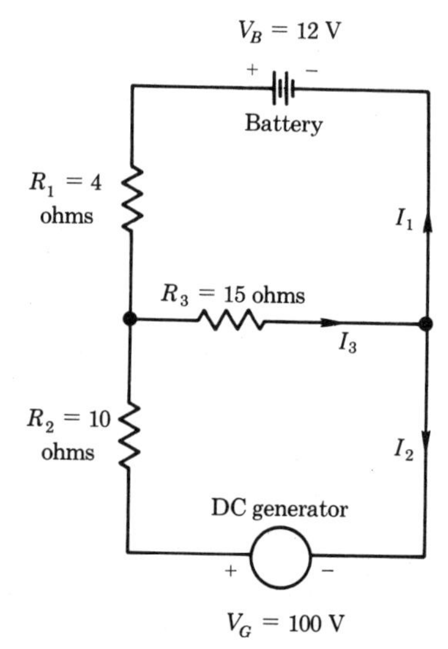

FIGURE P18-11

18-11 Solve for the currents in the dc circuit in Figure P18-11, which might be used to charge car batteries.

18-12* A motor consists of 1600 loops of wire of dimensions 15 cm by 20 cm, in a magnetic field of 30 Wb/m^2 that is always perpendicular to the longer dimension of the loops. (a) What total torque is developed in steady operation when the current in each loop is 1.2 A? (b) Suppose the starting current on a 150 V dc supply is 100 A and the motor is turning at 4 rpm, with 40% efficiency. How much torque is developed at this stage?

18-13 The switch is closed at time zero in the circuit containing a dc source of potential, a resistor, and a capacitor (Figure P18-13). The potential equation is

$$V_R + V_C - V_S = 0$$

where $V_R = IR$ and $V_C = Q/C$. (a) What is the current at time zero? (b) What is the charge after a long time has elapsed?

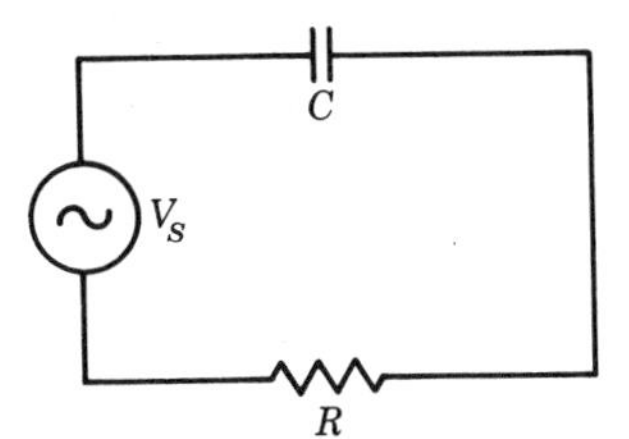

FIGURE P18-13

18-14 If the charge on the capacitor C drops off according to $Q = Q_0e^{-t/RC}$, when shorted through a resistor R, show that the length of time to fall from Q_0 to $\frac{1}{2}Q_0$ is $0.693RC$.

18-15 A 10-henry coil has a resistance of 180 ohms. What is its impedance to 60-cycle ac current? What value capacitor could be put in series with the coil to make the circuit have a minimum impedance?

18-16* We are given a 1.0-millihenry inductor and two capacitors of 5 and 10 microfarad. What resonant frequencies can be obtained by various combinations of these items in a circuit?

18-17 Spaceships of the future may be propelled by accelerated charged particles, for example cesium ions, Cs^+. By what factor would the electric current increase if the potential difference were increased by a factor of 4 and the gap between electrodes decreased by a factor of 2?

18-18 An electric blanket consists of many fine strands of insulated wire in parallel. What total resistance should a 100-watt blanket have with voltage $V_e = 120$ V? How much effective and peak current will it draw?

18-19 A simple series circuit consisting of a resistor of 500 ohms and a capacitor of 20 μF is connected to a 170-V, 60-cycle ac wall outlet. Compute the capacitative reactance, the impedance, the phase angle, and the peak current drawn.

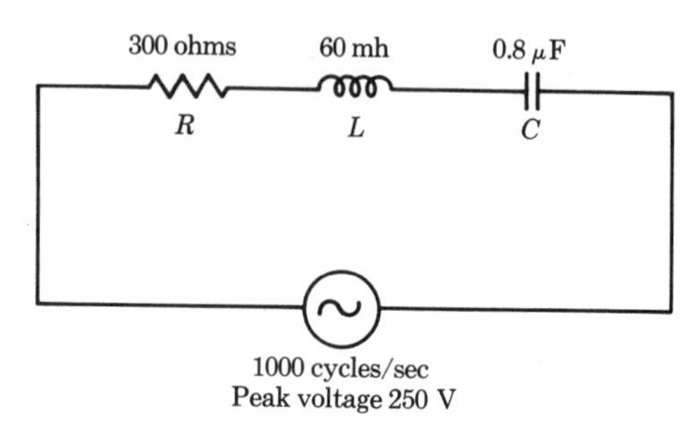

FIGURE P18-20

18-20* For the RCL circuit shown in Figure P18-20, calculate the following quantities: ω, X_L, X_C, Z, ϕ, I_0, V_e, I_e, and $\overline{P}$.

18-21 A linear accelerator for protons is designed with peak rf voltage 5000 V and radio frequency 200 Mc. (a) What is the speed at the end of the first acceleration and the length of the first tube? (b) How many gaps n does the proton have to cross to reach 5-MeV energy? (c) The total length of an accelerator with n gaps and tubes is

$$L = L_1 + L_2 + L_3 + \ldots = L_1(\sqrt{1} + \sqrt{2} + \sqrt{3} + \ldots) \cong \tfrac{2}{3}n^{3/2}L_1$$

Compute the length required for the proton machine.

18-22 Ions of charge q are accelerated through a potential difference V_0 each time they cross the gap between dees in a cyclotron. How many accelerations do they receive in arriving at final energy E_k? If the average path in the dees is $\pi R/2$, where R is the final radius, how far does the particle travel in getting up to full energy?

18-23 What magnetic field would be needed at the Earth's Equator so that a proton of speed 10^7 m/sec could just circulate around the Earth? Is this large or small compared with the Earth's field?

18-24 Show that the radius of the orbit of a charged particle moving in a magnetic field is proportional to its momentum. Does this relation hold for both classical and relativistic motion?

18-25 A proton, a deuteron, and an alpha particle are accelerated through the same potential difference and enter a region of uniform magnetic field perpendicular to their initial velocities. Compare the radii of the three orbits.

18-26 An electron travels in a circular orbit due to a magnetic field normal to the plane of motion. What happens to the orbit if the field is suddenly doubled? Halved? Turned off?

18-27 A cyclotron is designed to obtain 2-MeV protons with peak rf voltage 3000 V and maximum radius 0.75 m. Compute the rf frequency, the magnetic-field strength, and the total ion path (see answer to Problem 18-22, Appendix G).

18-28 A beam of 10-MeV alpha particles falls on a metal target. The beam represents a current of 20×10^{-6} A. How many deuterons strike the block each second? What is the heat power delivered to the target?

18-29 Protons are to be used to propel a spaceship by acceleration of the ions through 10,000-volts potential difference. (a) If the current is 250 amperes, what is the rate of discharge of particles per square meter? (b) What is the proton speed? (c) How much thrust per square meter is possible (see Section 10-3)? (d) What area of "nozzle" is needed to give one "g" acceleration?

ELECTROMAGNETIC RADIATION

19

In our daily lives we come across radiation in many forms. We experience light with its various colors by means of the sense of sight. We feel the heat from the Sun on our skins, and "sunburn" is an after-effect of too much exposure. We listen to radio, watch television, travel in airplanes guided by radar, have our teeth X-rayed, and worry about radiation from fallout. At first thought, we might tend to consider all these as unconnected phenomena. On a physical basis, however, they all are (or use) forms of *electromagnetic radiation,* differing only in their wavelength or frequency; and all these various forms have the same speed in a vacuum.

19-1 THE ELECTROMAGNETIC SPECTRUM

The description of the phenomena of electricity and magnetism, involving forces, fields, matter, and charges, was brought into a compact and elegant mathematical form by Maxwell in the latter part of the nineteenth century. He took into account all of the previous experimental and theoretical work of scientists such as Coulomb, Ampère, Ohm, Henry, and Faraday, and provided the missing part of the picture, the concept that changing electric fields induced magnetic fields. He translated the set of electrostatic and electromagnetic laws, which we presented in Chapters 6 and 7, into the simple but comprehensive mathematical statement that bears his name, Maxwell's equations. He then proceeded to predict the existence of radio waves long before they were discovered. As shown in Appendix F, electromagnetic waves in free space are proposed to consist of periodic variations of electric and magnetic fields that are directed perpendicularly to each other and progress through space with speed c. Their form is

$$\mathcal{E} = \mathcal{E}_0 \sin (kz - \omega t) \qquad (along\ x)$$

$$\mathcal{B} = \mathcal{B}_0 \sin (kz - \omega t) \qquad (along\ y)$$

The maximum amplitudes are $\mathcal{E}_0$ and $\mathcal{B}_0$, the angular frequency is $\omega = 2\pi\nu$, and the propagation constant is $k = 2\pi/\lambda$, such that $\omega/k = \lambda\nu = c$, the speed of the waves. By use of Faraday's law (Section 7-4), which states that the circulation of $\mathcal{B}$ is the negative of the time rate of change of magnetic flux, and by use of Ampère's law (Section 7-3), which states that the circulation of $\mathcal{E}$ is k_M/k_E multiplied by the rate of change of the electric flux, we derive (Appendix F) two important results:

$$\frac{\mathcal{E}_0}{\mathcal{B}_0} = c \quad \text{and} \quad c = \sqrt{\frac{k_E}{k_M}} = \sqrt{\frac{9 \times 10^9}{10^{-7}}} = 3 \times 10^8 \text{ m/sec}$$

The speed of visible light, radio waves, and all other forms of electromagnetic radiation is the same in a vacuum, $c = 3 \times 10^8$ m/sec; or, more exactly, 2.997925×10^8 m/sec.

The frequency and wavelength of electromagnetic waves depend on the type of source, which may be nuclei, atoms, molecules, heated solids, or accelerated charges in various electrical devices. Figure 19-1 shows the complete electromagnetic spectrum, with scales of wavelength λ and frequency ν given. The names used for the regions arise because of differences in the way the radiation is produced, detected, or employed. The distinctions are somewhat artificial when we consider that all are electromagnetic in nature. Let us review some of the features of the parts of the spectrum.

1. *Visible light.* The wavelengths that the eye can detect and identify as colors, ranging from violet to red, form a surprisingly small part of the spectrum. The eye is very sensitive, but unable to respond to most radiation.

2. *Ultraviolet light.* This radiation, just above the visible range in frequency, is abundant in sunlight. Small amounts of ultraviolet light are healthful through the vitamin D that it produces in the body, but large amounts are deadly. The Earth's atmosphere fortunately filters out the bulk of this radiation.

3. *Infrared light.* The heat from the Sun comes to us by these rays, of frequency lower than those in the visible range. The skin of the body will detect these rays through a warming effect, and by the use of sensitive heat detectors, infrared photographs can be made in complete darkness (at least in the visible sense).

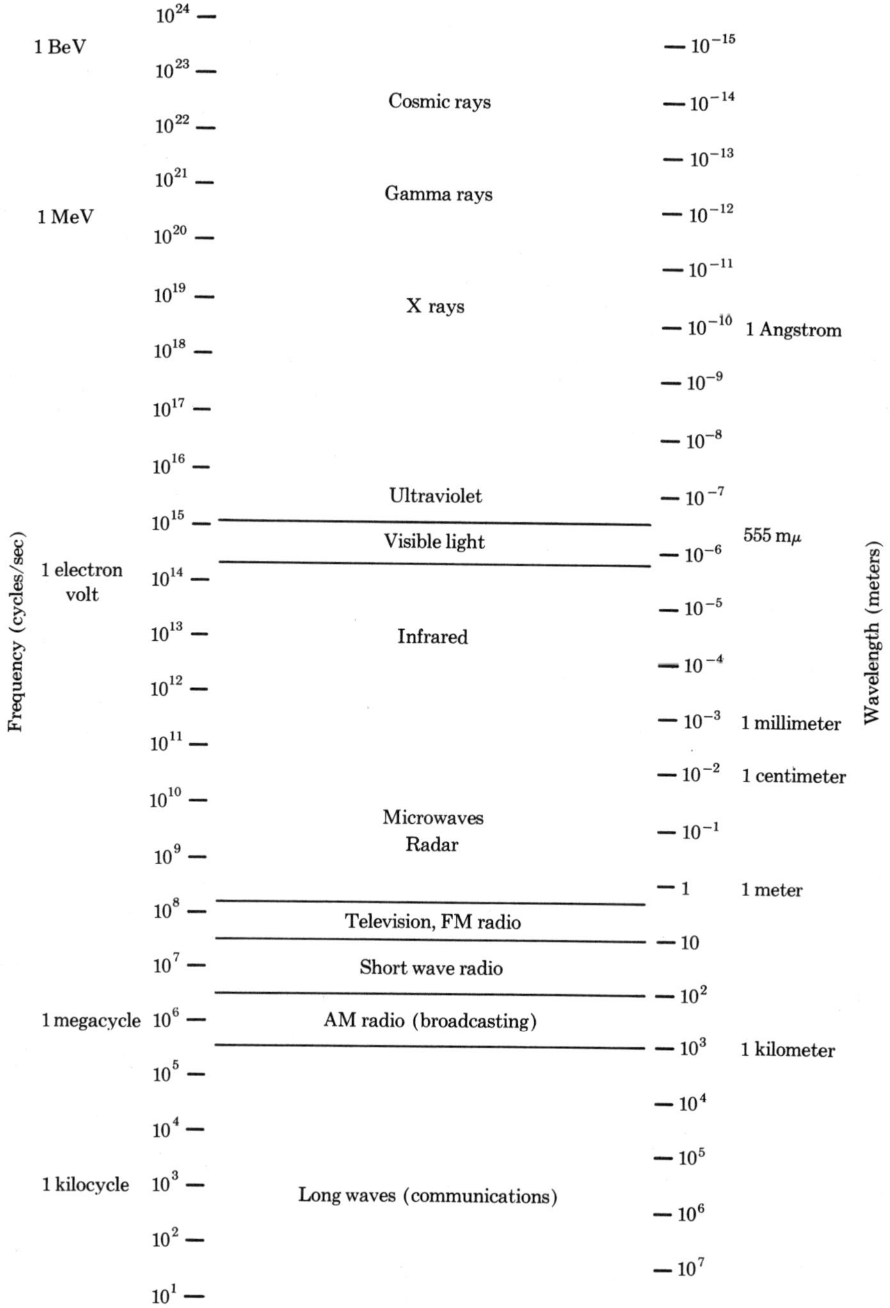

Frequency (cycles/sec)

10^{24} —
1 BeV
10^{23} —
10^{22} —
10^{21} —
1 MeV
10^{20} —
10^{19} —
10^{18} —
10^{17} —
10^{16} —
10^{15} —
1 electron volt
10^{14} —
10^{13} —
10^{12} —
10^{11} —
10^{10} —
10^{9} —
10^{8} —
10^{7} —
1 megacycle 10^{6} —
10^{5} —
10^{4} —
1 kilocycle 10^{3} —
10^{2} —
10^{1} —

Cosmic rays

Gamma rays

X rays

Ultraviolet

Visible light

Infrared

Microwaves
Radar

Television, FM radio

Short wave radio

AM radio (broadcasting)

Long waves (communications)

Wavelength (meters)

— 10^{-15}
— 10^{-14}
— 10^{-13}
— 10^{-12}
— 10^{-11}
— 10^{-10} 1 Angstrom
— 10^{-9}
— 10^{-8}
— 10^{-7}
555 mμ
— 10^{-6}
— 10^{-5}
— 10^{-4}
— 10^{-3} 1 millimeter
— 10^{-2} 1 centimeter
— 10^{-1}
— 1 1 meter
— 10
— 10^{2}
— 10^{3} 1 kilometer
— 10^{4}
— 10^{5}
— 10^{6}
— 10^{7}

FIGURE 19-1

Electromagnetic spectrum

4. *X rays, gamma rays, and cosmic rays.* These very-high-frequency radiations are of various origins. X rays are commonly produced by a machine involving the bombardment of atoms by electrons; gamma rays come from nuclear reactions; cosmic rays from outer space contain high-frequency electromagnetic components.

5. *Radio waves.* These low-frequency radiations, produced by electrical circuits, include the licensed bands used for ordinary radio, television, and "short-wave" communication—e.g., by police and amateurs. The radar commonly used for detecting distant objects is between the frequencies of infrared and television.

19-2 RADIATION FROM ELECTRICAL OSCILLATIONS

The sight of television antennas sprouting from rooftops is a common one. We are aware that such structures receive waves from a television station that has similar emitting devices, and that no wires are required to guide the signals from the transmitter to the receiver. We also have noted that the loudness and clarity of a radio with a built-in antenna, as in a transistor model, depend on the direction in which it is turned. The first improvement in our level of understanding is that the oscillation of charge and current produces an electromagnetic disturbance at a distance.

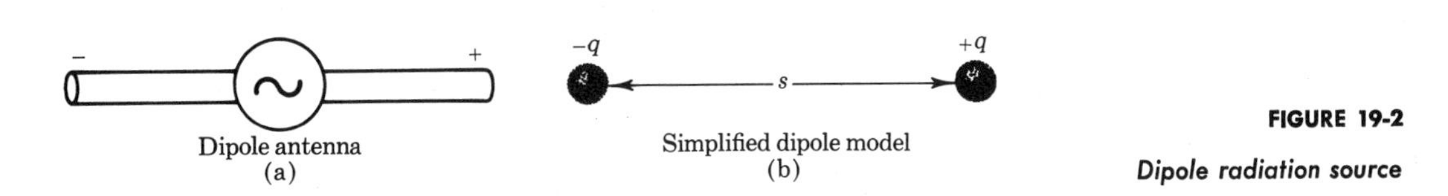

FIGURE 19-2
Dipole radiation source

The click in a radio when a nearby light switch is closed, or the crackling sound during an electrical storm give us evidence of this fact. Although the mathematics of sending and receiving these electromagnetic waves is complicated, we can appreciate the process by examining an elementary antenna. Picture a pair of straight wires attached to a source of alternating electric potential, as in Figure 19-2(a). At any instant, one arm is negative while the other is positive, but 1 half-cycle later, the signs are reversed. The electric field near the charged wire also reverses direction. The flow of current in the wire causes a magnetic field that also oscillates. We can simplify the situation by imagining that at any instant there are two unlike charges separated by distance s, charges $+q$ and $-q$, as in Figure 19-2(b).

As discussed in Section 6-3, this structure is an *electric dipole*, which produces an electric field in the vicinity. In our simplified model, the alternating voltage of frequency ν causes the charges to move back and forth with a sinusoidal motion. Let us study the variation with time of the electric field at various distances R from the oscillating dipole, referring to Figure 19-3. What we observe is found to depend greatly on the size of R.

First, let us suppose that R *is small* compared with the distance that light travels in 1 cycle of vibrations, and small compared with the wavelength of radiation λ, but still large compared with the separation of charges. The effect of a displacement of particles is detected almost instantly at the point P. The static electric field from a dipole (Section 6-3) is $k_E M/R^3$, where M is the dipole moment $M = qs$. The charge separation and thus the moment now vary periodically, $M = M_m \sin \omega t$, where M_m is its maximum value. Thus the field nearby is

$$\mathcal{E} = \frac{k_E M_m \sin \omega t}{R^3} \qquad \textit{(electric field near a dipole)}$$

which is rather strong at these distances because of the dependence on $1/R^3$.

We now turn to the case in which R *is large* compared with the distance light travels in 1 cycle. It takes a length of time approximately R/c for the effect of any disturbance of the charges to be experienced at P. Thus the electric field $\mathcal{E}$ there at time t is due to charge motions at an earlier time $t' = t - R/c$. A part of the total field that was unimportant at small distances now becomes dominant, and is given by $-k_M q a_y/R$, where a_y is the acceleration of the charges at time t'. Recalling that, in harmonic motion, acceleration is proportional to displacement, and that the latter is periodic, we obtain what is called the *radiation field:*

$$\mathcal{E} = \frac{A \sin \omega(t - R/c)}{R} \qquad \textit{(electric field far from a dipole)}$$

We see that the field is in the form of a *wave,* moving outward with speed c

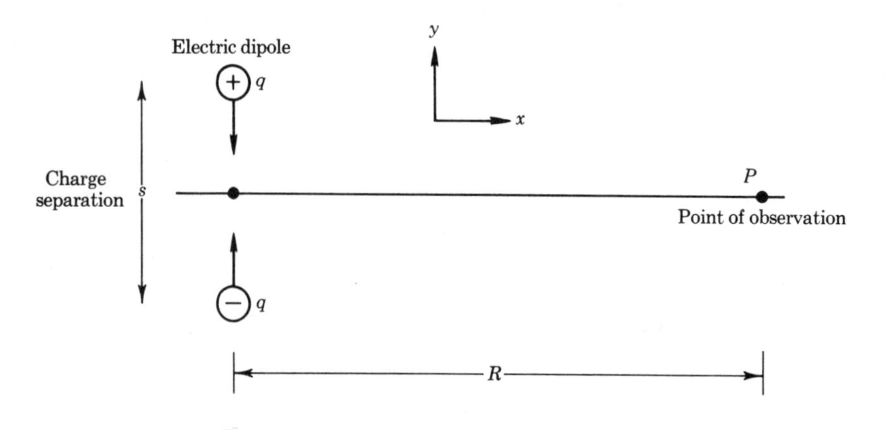

FIGURE 19-3

Production of electromagnetic waves by an oscillating dipole

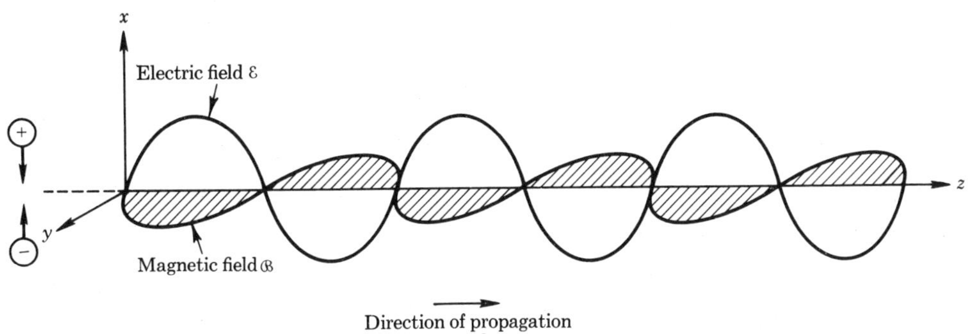

FIGURE 19-4

Electric and magnetic waves far from an oscillating dipole

across a curved surface of radius R. Since the radius is very large, the wave appears to be nearly of plane form at the point of observation. There is an accompanying magnetic field $\mathcal{B}$, perpendicular to $\mathcal{E}$, and the whole combination (Figure 19-4), is called the *electromagnetic wave*.

Electromagnetic waves are produced when charges are accelerated either in an electrical circuit or by encounter with other charges. The term *bremsstrahlung* ("braking rays") is applied to the latter, as when electrons interact with atomic particles in the atmosphere or in an X-ray generator.

Let us investigate the form of an electric field for electromagnetic signals in some particular situations. A typical radio transmitter operates at 1000 kc/sec. Thus the distance that signals travel in 1 cycle is $R = \frac{c}{\nu} = \frac{3 \times 10^8}{10^6} = 300$ m. Since most stations serve a region with radius much larger than 300 meters, the nearly plane radiation-field relation is dominant. On the other hand, visible light produced by atoms has a typical wavelength 5.55×10^{-7} m, in the yellow–green region to which the eye is most sensitive. For nearby atoms, of the order of 1 mean free path away, about 10^{-7} m, the distance R is of the order of λ, and the dipole field is important.

Summarizing, the properties of electromagnetic waves in a vacuum are the speed c, the frequency ν, the wavelength λ, the amplitude A, a phase angle, and a plane of vibration or polarization. This last property is usually referred to the electric-field direction. Radiation from a single dipole is said to be polarized, while that from many atoms with random vibration directions is unpolarized.

Communication by Radio Waves

Electromagnetic energy is transferred at a rate proportional to both the electric- and magnetic-field strengths, and to the speed of light. Since each field varies as $1/R$, the energy flow varies as $1/R^2$—that is, inversely

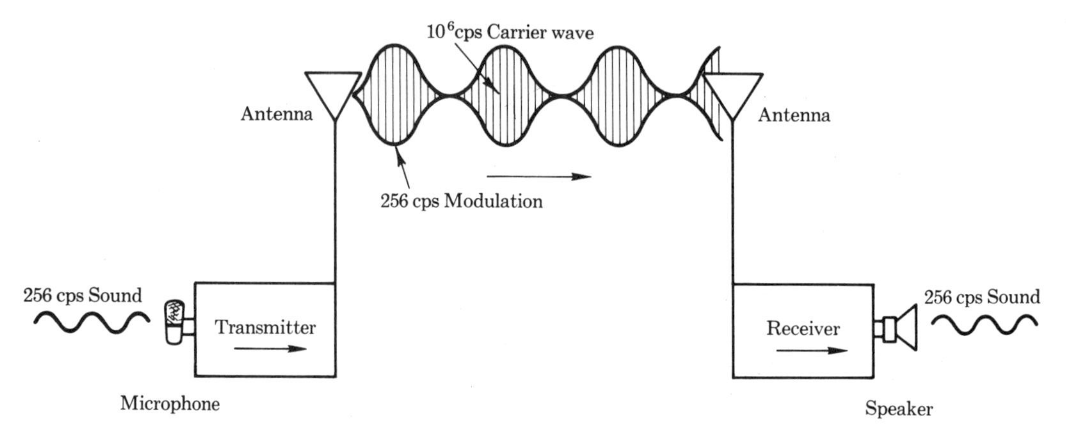

FIGURE 19-5

Amplitude-modulated radio wave transmission

as the square of the distance from the source, a fact that can be correlated with our knowledge that light from stars or radio signals from distant stations is weak.

The electromagnetic waves of radio are noted from Figure 19-1 to be in a frequency range near 1,000,000 cps. If these vibrations were converted into sound of the same frequency, they would be completely inaudible to the human ear. The radio waves serve, however, as a "carrier" of information such as speech, music, and other sounds. The signal that is transmitted and received consists of some change or *modulation* of the basic sine wave of radio. Suppose that a station wants to send out a note of 256 cps. The low-frequency variation is fed through a microphone to the transmitter, which, in the case of AM (*amplitude-modulated*) radio, varies the amplitude of the electromagnetic signal, as shown schematically in Figure 19-5. At the radio receiver, the variations are turned into electrical impulses that cause mechanical vibrations of the speaker cone. These agitate the air and the signal appears as audible sound again. More complex sounds are handled in the same way.

At the transmitter for FM (*frequency-modulated*) radio, the sound creates slight changes in the frequency of the radio waves. The changes are translated back into sound at the receiver.

The generation of electric and magnetic waves by oscillating charges was predicted on the basis of theory alone by Maxwell. This stands as one of the major scientific contributions of all time, comparable to Newton's discoveries in the field of mechanics.

19-3 SPECTRA FROM HEATED BODIES

A study of the laws of thermal radiation is important for two reasons. First is the interest in knowing how the amounts emitted, transmitted, and absorbed depend on temperature, materials, and the size and shape of the objects participating. The second is historical, since the quantum theory of radiation had its origin in the eventual correct interpretation of the above processes.

The variation of temperature and of the color of heated objects is familiar to us. The burner coil on an electric range or the filament of a lightbulb changes from black to red on heating up, and with high enough current becomes white.

We know from experience that there is a rapid change in the total amount of radiant-heat energy with temperature. Such energy is transmitted through a vacuum, in contrast to the mechanisms of conduction and convection (Sections 16-6 and 16-7). The radiation from the Sun that warms the Earth crosses empty space: the origin is in nuclear and atomic processes, the transmission is in the nature of electromagnetic radiation, and heating effects result from absorption by atoms of matter. Let us summarize the experimental facts. First, the amount of radiation from an ideal emitting surface in a unit range of wavelength at a certain λ is found to have a maximum value that depends on absolute temperature T, as shown in Figure 19-6. The peaks of these curves occur at wavelengths λ_m, where m refers to maximum (of the radiation rate) such that

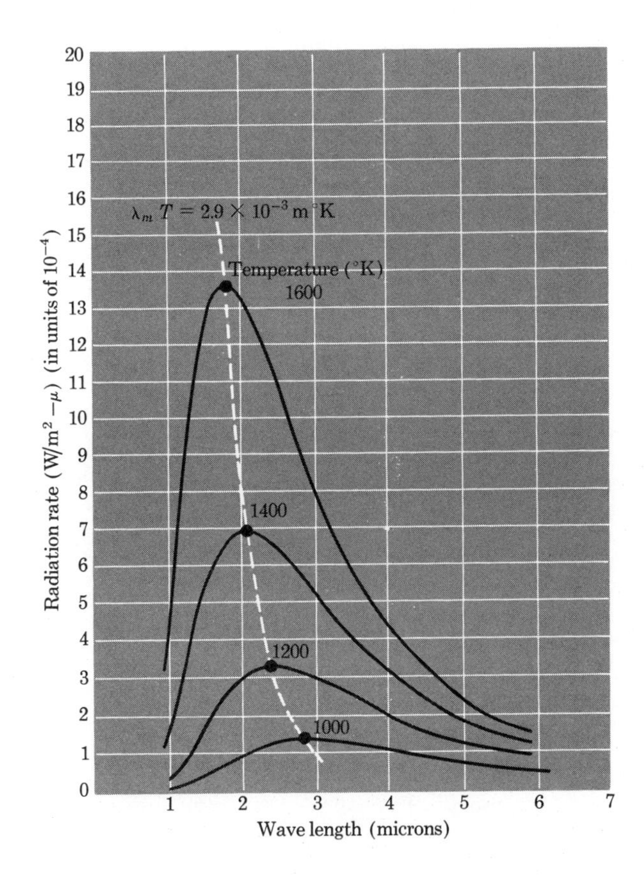

FIGURE 19-6
Radiation from an ideal heated object

$$\lambda_m T = \text{constant}$$

The dotted line on Figure 19-6 connects these points. Second, the total radiation energy at all wavelengths—the area under a curve—is proportional to the fourth power of the absolute temperature. It also depends on the nature of the surface: one that is rough and black is a much better radiator than one that is smooth and highly reflecting.*

The term *absorptivity* α is the fraction of the radiation that strikes an object that is absorbed by it. We know that objects such as dark clothing or black automobiles absorb more heat from the sun than do white ones. A perfect absorber, with $\alpha = 1$, is called a *"black body."* It has an ideal surface with which we can compare real surfaces.

Let us try to imagine the darkest possible object. There is very little reflection from dull-black paint. The keyhole of a closet is even better, because most of the light that we shine through it gets lost inside and little comes out again. The black body as an ideal absorber is just such a cavity with a very small aperture, Figure 19-7. We imagine that the interior can be heated to any temperature desired. Radiation is continually emitted and

* The word "radiator" as applied to heating equipment in buildings is a misnomer, since in these devices *convection* is the main mode by which heat is transferred.

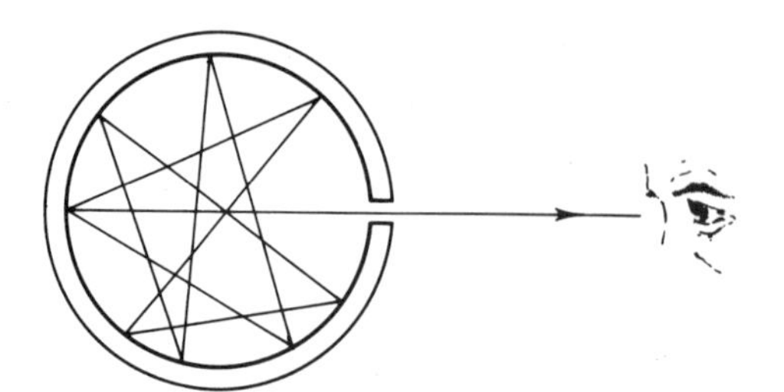

FIGURE 19-7

The black body—the perfect radiator and absorber

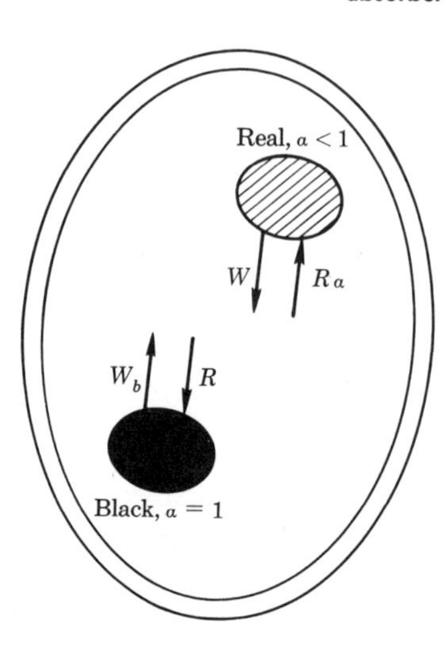

FIGURE 19-8

Geometry for Kirchhoff's law

absorbed by the walls of this furnace, and what comes out of the small hole in the side is a sample of the radiation inside.

We can derive a relation between the ability of objects to absorb and emit radiation by performing an imaginary experiment. We put two small objects in a large heated cavity (Figure 19-8) and allow the system to come to thermal equilibrium. One object is a real body with absorptivity α, the other is a black body with absorptivity 1. Examining unit areas of each, which receive radiation from the walls of the cavity of amount R, we find that absorption rates are $R\alpha$ for the real body and R for the black one, while we let the emission rates be W and W_b. Heat balances are written as follows:

$$R\alpha = W$$

$$R = W_b$$

Dividing equations yields

$$\frac{W}{W_b} = \alpha$$

The ratio W/W_b is called the *emissivity*, ϵ, which is the ratio of the radiation rate from a real body to that from a black body. We conclude that

$$\epsilon = \alpha$$

which is *Kirchhoff's law*, stating that *the emissivity is equal to the absorptivity for all surfaces.* Also, we see why a good absorber is a good emitter, and vice versa. A polished metal plate that reflects most of the incident radiation is a poor source of heat. If the radiation rate W_b from a black body at some temperature is known, we merely multiply by ϵ to find it for a real body. The value of ϵ depends on the surface of the material—the element, its chemical composition, and the degree of roughness. For instance, at 200°C, aluminum has a value of ϵ of about 0.1, while steel has 0.8. In general, ϵ varies also with wavelength. For example, as the radiation wavelength values proceed from 0.55×10^{-4} m to 0.65×10^{-4} m, ϵ for silver goes from 0.35 to 0.04.

The *Stefan–Boltzmann law* is the starting point for calculating rates of radiation. From a black body, the rate of radiation per unit area, including all frequencies, is

$$J = \sigma T^4$$

where $\sigma = 5.67 \times 10^{-8}$ W/m²-(°K)⁴, the *Stefan–Boltzmann constant.* We observe that the emission rate goes up very rapidly with temperature—as the fourth power. Doubling T causes J to increase by a factor of $(2)^4 = 16$. For example, let us find the total radiation rate per unit area from a black body at temperature 2000°K. In this case, $J = (5.67 \times 10^{-8})(2000)^4 = 9.1 \times 10^5$ W/m². If the emissivity were that of steel, $\epsilon = 0.8$, the rate of radiation would be smaller, 7.3×10^5 W/m².

Imagine a 100-watt lightbulb hanging in a room with nonreflecting walls at negligible temperature. To what temperature will the filament of area 25 (mm)^2 rise, if its emissivity is 0.9?

$$T^4 = \frac{J}{\sigma\epsilon} = \frac{(100\text{ W})/(25 \times 10^{-6}\text{ m}^2)}{[5.67 \times 10^{-8}\text{ W/m}^2\text{-}(°\text{K})^4](0.9)} = 78.4 \times 10^{12}\ (°\text{K})^4$$

or

$$T = 2970°\text{K}$$

The net amount of radiation that is transmitted from one body to another depends on the temperatures of both, since each is radiating to the other. Suppose that an object of area A_1 and emissivity ϵ_1 at absolute temperature T_1 is completely surrounded by a surface at lower temperature T_2. The net rate of radiation is

$$Q = JA = \epsilon_1 A_1 \sigma (T_1^4 - T_2^4)$$

Let us illustrate by considering the problem of heating communications satellites or spacecraft by solar radiation. We note that the Sun's temperature T_s is very high, about 6000°K. To determine its contribution of energy to a spaceship, let us visualize a sphere of radius equal to the distance to the Sun. The ratio of the Sun's area on that sphere, A_s, to the total sphere area A_t is very small, about 5×10^{-6}. By equating the actual radiation from the Sun to that received if the Sun were spread out over the whole sphere surface, we can find an effective temperature of space T_e. Neglecting constants of proportionality,

$$A_t T_e^4 = A_s T_s^4$$

or $T_e = (5 \times 10^{-6})^{1/4}\, 6000 = 284°\text{K}$, which is surprisingly low, only 11°C.

With the knowledge of the effective temperature of the surroundings, we can now find how the surface temperature of a space capsule depends on rate of radiation. Suppose that the equipment in the vehicle, area 100 m^2 and $\epsilon = 0.8$, produces 200 kW that must be continuously radiated to space. Then,

$$T_1^4 = T_e^4 + \frac{Q}{\epsilon_1 \sigma A_1}$$

$$= (284)^4 + \frac{2 \times 10^5}{(0.8)(5.67 \times 10^{-8})(100)}$$

from which we find $T_1 = 474°\text{K}$, or an uncomfortably warm 201°C.

The temperature of the surface of the spacecraft was assumed to be uniform in the above analysis. This would be true only if heat were conducted very readily in the hull or if the spacecraft were rotated, as was done in the Apollo missions to the Moon. Also, we ignored the radiation from all of the stars, which turn out to give an effective temperature of space of about 3°K.

19-4 THE ORIGIN OF QUANTUM THEORY

By the latter part of the nineteenth century, all of the discoveries and developments that we call classical physics had been made. Included were the sciences of mechanics, kinetic theory of gases, thermodynamics, electricity and magnetism, and electromagnetic-wave theory. Methods of making accurate measurement of physical quantities had been perfected as well. The agreement of theory and experiment was generally most satisfactory. When the subject of radiation from heated bodies was investigated, however, serious gaps in understanding remained. The spectrum of light from matter at various temperatures, such as we saw in Figure 19-6, was well known. But attempts on the basis of classical theory to explain the black-body radiation curves turned out to be highly unsuccessful, in spite of the efforts of many capable scientists. They assumed reasonably that the atoms in the solid heated material were responsible for the radiation, through vibration of charges in what were called atomic oscillators. Light was believed to be emitted and absorbed at the natural frequencies of the oscillators, each of which was assigned an average energy proportional to T

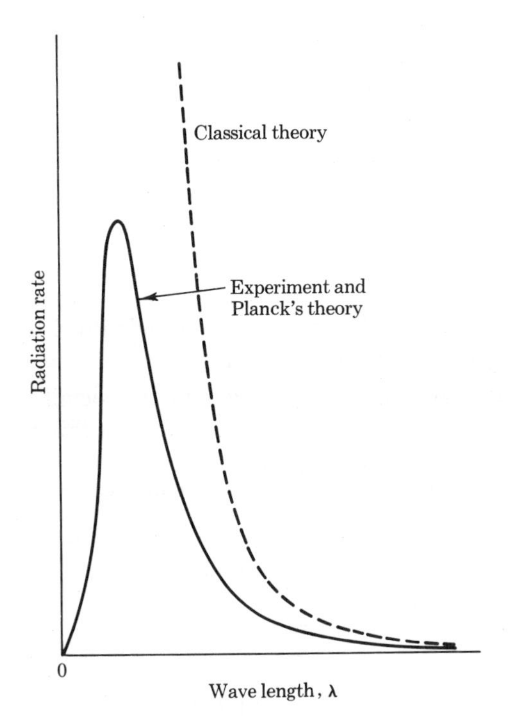

FIGURE 19-9

Radiation spectrum from heated body compared with classical prediction

on the basis of equipartition of energy. A typical *calculated* graph of the frequency distribution of intensity (Figure 19-9), we note, contrasts sharply with the *experimental* results. The theoretical curves fit very well at long wavelengths, but deviate drastically at short wavelengths.

This discrepancy became known as the "ultraviolet catastrophe." By calculating the area under a given curve, physicists could compare the theoretical total energy radiated at a given temperature with experimental findings. In this case, the theoretical curves were of no practical use at all, since the area under the curves was infinite for all temperatures.

In 1900, Max Planck, a German scientist, succeeded in deriving a formula* which fit the experimental data almost perfectly, and which also yielded the Stefan–Boltzmann fourth-power law. The assumption which he had to make was that the oscillating charges that absorb and emit the electromagnetic radiation can only do so in definite amounts and that these are in units of $h\nu$, where ν is the natural frequency of the oscillator and h, now known as *Planck's constant,* is 6.63×10^{-34} J-sec.

Planck actually regarded his *quantum hypothesis* as a calculational trick rather than a fundamental concept. However, the existence of discrete energy levels in atoms and the photon theory of light had their beginning with this assumption. Such ideas were revolutionary at the time.

The work of Planck is of historical importance in physics, in that it marked the end of the classical era and the beginning of modern physics, which is characterized by descriptions of all phenomena on an atomic and quantum-mechanical basis.

PROBLEMS

19-1 Find the wavelength λ, frequency ν, and propagation constant k for the following electromagnetic radiations: (a) yellow light from sodium, $\lambda = 5.89 \times 10^{-7}$ m; (b) an X-ray photon of energy 50,000 eV; (c) 850-kc radio waves.

19-2 Find the frequency and wavelength of a cosmic-ray photon of 1-Bev (10^9 eV) energy. If all the energy were converted into visible light at 555 mμ, how many photons would be produced?

19-3 A sodium-vapor lamp emits light in the vicinity of 590 mμ. Find how many photons are emitted per second from a 100-watt lamp.

* The radiant-energy flow per unit area per unit frequency at a given frequency ν, according to Planck, is

$$j = 2\pi \left(\frac{\nu}{c}\right)^2 \frac{h\nu}{e^{h\nu/kT} - 1}$$

and the Stefan–Boltzmann constant is

$$\sigma = \frac{2\pi k^4}{15h^3c^2}$$

19-4 The amplitude of an electromagnetic wave at great distances from a dipole can be shown to be $A = k_M 2qa$, with $a = \omega^2 r$. Find its value for an oscillator of atomic dimensions, comparable to the hydrogen atom, with $q = 1.6 \times 10^{-19}$ C, $r = 5.3 \times 10^{-11}$ m, and $v = 2.2 \times 10^6$ m/sec.

19-5* According to classical radiation theory, the centripetal acceleration of an electron results in radiation with a power $P = 2k_M(qa)^2/3c$. (a) How much power would be expected from the hydrogen atom (see Problem 19-4)? (b) How much power would come from an electron of speed 10^6 m/sec circulating in a magnetic field of 2 Wb/m^2?

19-6 Find the wavelength of light that is most prominent at a temperature of 1000°K. Is this in the visible range?

19-7 Calculate the rate of energy loss by radiation from a steel sphere of radius 0.05 m at temperature 1000°C if $\epsilon = 0.85$.

19-8 The net radiation rate between two parallel planes with different temperatures, areas, and emissivities is given by

$$Q = FA_1\sigma(T_1^4 - T_2^4)$$

where $F = (1/\epsilon_1 + 1/\epsilon_2 - 1)^{-1}$. Estimate the heat lost per second by the sides of a furnace of area 4 m^2, $\epsilon = 0.8$, $T = 600$°K, to an aluminum shield $\epsilon = 0.1$ at $T = 300$°K.

19-9 A Thermos bottle is constructed of two layers of silvered glass, with vacuum between them. Explain why it can maintain the temperature of either hot or cold liquids.

19-10 A steel pipe ($\epsilon = 0.8$) containing steam at temperature 120°C is suspended in a dark room at 40°C. The amount of heat lost per unit area by radiation plus convection is 5000 W/m^2. What fraction is radiation? Deduce the value of the overall heat transfer coefficient. (See Section 16-7.)

19-11 To what temperature would an unmanned satellite go if all power inside were shut off?

19-12* A nuclear-powered spaceship requires large radiating surfaces to eliminate the "waste heat" of the thermodynamic cycle. If the useful energy is 200 MW, the efficiency is 0.25, $\epsilon \cong 1$, and the radiator temperature is 1000°C, what area would be needed?

19-13* Show that Planck's formula (footnote, page 528) goes into the classical expression $j = 2\pi kT/\lambda^2$ for large wavelengths, small frequencies.

LIGHT WAVES AND RAYS

20

Regardless of the origin of electromagnetic radiation—from electrical devices, heated bodies, or atomic particles—the form of the disturbance is the same. It will thus be propagated through a vacuum and interact with materials in a manner that is determined by the geometry, the properties of the substances, and of course, the frequency and wavelength of the radiation. We shall use the terms light, radiation, waves, and electromagnetic radiation as essentially synonymous for the study of transmission through space and effects within material bodies.

20-1 DETERMINATION OF THE SPEED OF LIGHT

The speed of light or electromagnetic waves has been the subject of intensive experimental measurement over many years, because of the importance of that constant in physical theories. The ancient Greeks thought it to be infinite. Galileo attempted around 1638 to find the time required for light to travel a certain distance and back, using lanterns. Human reflexes being slow, the experiment failed and Galileo was able to conclude only that "it is extraordinarily rapid." The first indication that c was finite came from Roemer's astronomical observations in 1675. He found that times of eclipses of one of Jupiter's moons differed by about 22 minutes when seen at six-month intervals. Ascribing the discrepancy to the greater

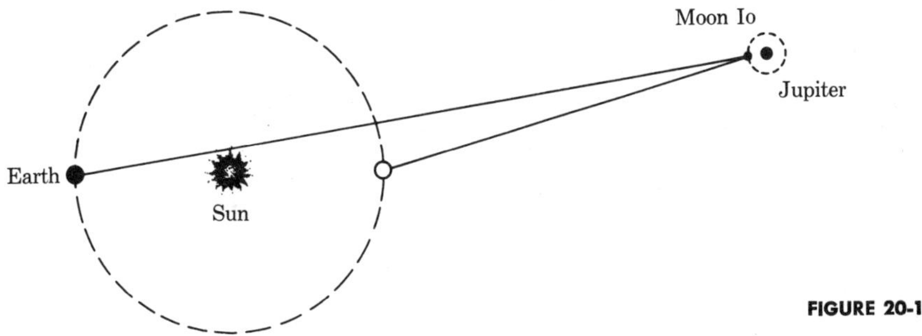

FIGURE 20-1

Roemer's estimate of the speed of light

distance light has to travel when the Earth and Jupiter are on opposite sides of the Sun (Figure 20-1)—twice 93 million miles—Roemer computed a value of c of a little over 2×10^8 m/sec. Measurements by others using a variety of techniques followed. Bradley found how much a telescope moving with the Earth had to be tilted forward for light to go down the telescope axis (somewhat similar to the tilt we must give an umbrella as we run in the rain). Fizeau set up a rotating cog wheel that allowed flashes of light to go to a distant mirror and back. Foucault substituted a rotating mirror for the wheel of Fizeau. The most precise measurements for many years were made by Michelson, who used a rotating mirror with as many as 32 faces. Figure 20-2 shows a simplified version of one of his arrangements. When the mirror gets to the right speed (around 528 revolutions per second), the detector receives full illumination from the source. In the time the mirror turns through $\frac{1}{8}$ revolution, $1/(8)(528) = 2.37 \times 10^{-4}$ sec, the light has traveled 44 mi, which gives a speed of 186,000 mi/sec. The value for the speed of light in vacuum as deduced by use of his later apparatus was 2.99774×10^8 m/sec. Since then, several improvements in the value of c have been obtained, using electromagnetic microwaves and

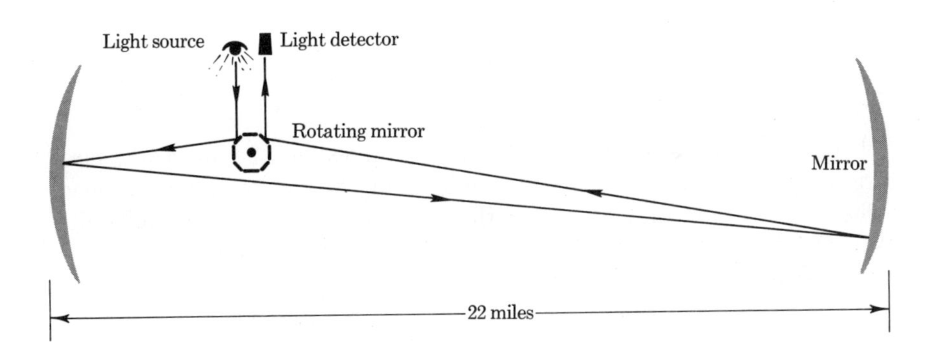

FIGURE 20-2

Michelson's rotating mirror method for measuring the speed of light

molecular spectra. The "best" value is currently $(2.997925 \pm 0.000002) \times 10^8$ m/sec. There has been some speculation that the speed of light changes slowly with time, but the trend is probably only due to coincidence as a result of the improved measurements over the years.

20-2 TRANSMISSION OF RADIATION IN SPACE

A radio announcer sometimes quotes the radiated power of his station—as, say, 50,000 watts. By this he means the total rate at which electromagnetic energy flows into the vicinity. We have previously encountered flows of particles, heat, momentum, and electric charge in Chapter 16. Electromagnetic energy also flows through space and materials—the rate of flow depends mutually on the electric and magnetic fields. The intensity of radiation I is defined as the energy flow per unit area and, as in the case of motion of particles or heat energy, is the product

$$\text{flow rate} = (\text{energy density})(\text{speed})$$

We may recall from Section 17-5 that the electrical-energy density within a capacitor is $\mathcal{E}^2/8\pi k_E$, while that within an inductor is $\mathcal{B}^2/8\pi k_M$. An electromagnetic wave involves both fields and, far from a dipole source in vacuum, they are related by $\mathcal{B} = \mathcal{E}/c$, as discussed in Section 19-1. Using $c = \sqrt{k_E/k_M}$, we find at once that the electric- and magnetic-energy densities are the same, and thus the total energy density is $\mathcal{E}^2/4\pi k_E$. Multiplying by the speed of light c, we find the intensity to be simply

$$I = \frac{\mathcal{E}^2 c}{4\pi k_E}$$

The units are J/m²-sec or W/m²—i.e., power per unit area. We can insert values of the constants to obtain

$$I = \frac{\mathcal{E}^2}{377}$$

where $\mathcal{E}$ is in volts per meter. To get an impression of how strong electric fields are, we estimate $\mathcal{E}$ for sunlight at the top of the atmosphere where $I = 2$ cal/cm²-min, called the *solar constant*. In our usual units, this is 1.4×10^3 J/m²-sec. Then, if the light were of one frequency, $\mathcal{E} = \sqrt{5.3 \times 10^5} \cong 730$ V/m.

Inverse-square Spreading of Radiation

When we are at distances from a source of light that are large compared with its size, it is convenient to think of the source as a *point*. Whether the radiation comes from an atom or from a radio antenna, the detailed

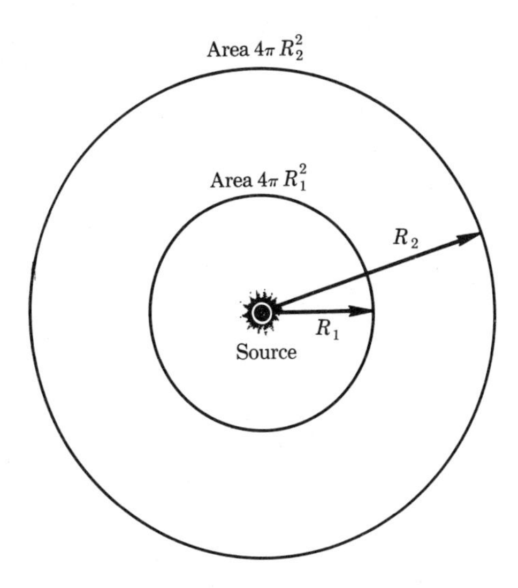

FIGURE 20-3

Flow of energy from a source of radiation

structure of the source is not important when the point of observation is far away. We saw in Section 19-2 that the electric field $\mathcal{E}$ from an oscillating electric dipole is inversely proportional* to R. Thus the intensity varies as $1/R^2$. This *inverse-square* dependence can be understood from the point of view of energy conservation. Suppose the energy is released by a point source in equal amounts in every direction. Let P be the power—that is, the energy emitted each second. In a vacuum, where there is no absorption, this same energy passes through any sphere of radius R and surface area $4\pi R^2$ (see Figure 20-3). There the intensity as the flow per unit area per second is clearly

$$I = \frac{P}{4\pi R^2}$$

Let us calculate the intensity I and the field $\mathcal{E}$ at 100 km from a 50-kW radio station.

$$I = \frac{P}{4\pi R^2} = \frac{5 \times 10^4 \text{ W}}{4\pi(10^5 \text{ m})^2} = 4.0 \times 10^{-7} \text{ W/m}^2$$

$$\mathcal{E} = \sqrt{377I} = \sqrt{(377)(4.0 \times 10^{-7})} = 0.012 \text{ V/m}$$

For any two radii R_1 and R_2, the ratio of intensities is

$$\frac{I_2}{I_1} = \frac{R_1^2}{R_2^2}$$

an inverse-square spreading of the radiation. Another way to view the spreading is shown in Figure 20-4. Imagine windows in two spherical shells about a point source of light. The area of window at radius R_2 is larger than

* Note that this field is quite different from the electrostatic field from a single charge, inversely proportional to R^2.

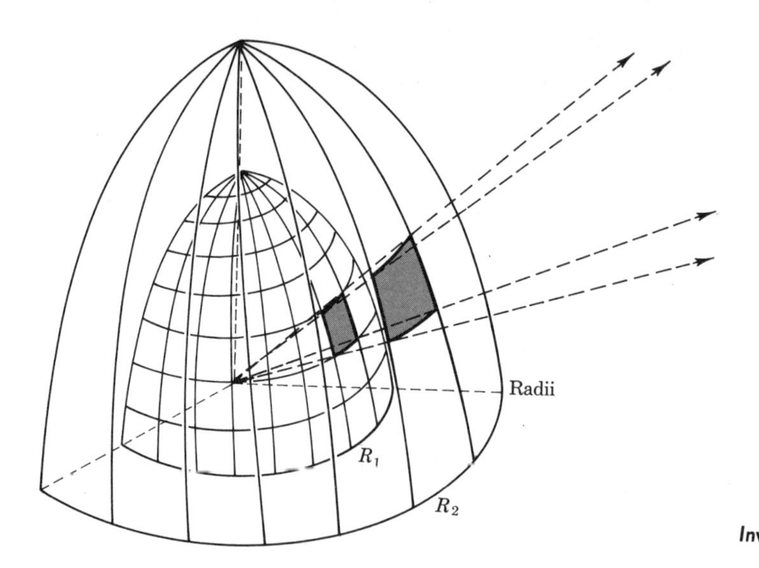

FIGURE 20-4

Inverse square spreading of radiation

that at radius R_1 in the ratio R_2^2/R_1^2, but since the same amount of light passes through each window, the intensities are in the ratio R_1^2/R_2^2.

Illumination

We are often concerned with the amount of light that is available to us for work or reading. For example, we may find that a 40-watt bulb is too weak for studying and must be replaced by a 100-watt bulb. When we take photographs, it is desirable to have a light meter in order to set our camera properly. The measurement of the amount of illumination of objects involves some specialized units. These were defined arbitrarily in the distant past, and now appear rather quaint. The *candle* was at one time a standard source of light (made of spermaceti, weight $\frac{1}{6}$ lb and burning at 120 grains per hour). The amount of light emission from other sources was then measured in *candlepower*. The *lumen* was then defined as the light falling on 1-square-foot area of a sphere of 1-foot radius with a one-candle source at the center. Although filament bulbs serve as standards nowadays, the units continue to be used. A typical 120-volt, 100-watt tungsten light bulb gives about 1600 lumens. Thus, at 1 foot, the intensity would be 1600 lumens/ft^2. If the bulb were 3 ft above our desk, the inverse-square spreading would reduce the intensity to $1600/(3)^2 = 178$ lumens/ft^2. Some photoflash bulbs give a peak radiation as high as 4 million lumens, but of course the flash lasts for a very short time, such as $\frac{1}{50}$ sec.

20-3 INTERACTION OF LIGHT WITH MATTER

We have seen that electromagnetic radiation can be produced by the controlled acceleration of electrons, as in a radio transmitter, or by heating of solid bodies to incandescence, or by the excitation of atoms in a gas. There is a great range of possible frequencies and wavelengths, but radiations have a common waveform and speed. Now we turn to the behavior of light as it passes through matter with its molecules, atoms, electrons, and nuclei. The nature of the interaction depends greatly on the structure and composition of the material, and thus our ability to generalize is somewhat reduced. Effects are highly dependent on the degree of freedom or binding of the charges in the substance. We can say, however, that the electric field of the radiation causes motion of electrons, which in turn become sources of supporting or competing radiation.

Conducting Material

First let us examine the interaction of waves in a metal, which was described in Section 17-6 as containing many free electrons. The electric field exerts forces on these charged particles, and if the material has a conductivity σ, as shown earlier, a conduction current density i will be produced by the electric field $\mathcal{E}$, of amount $i = \sigma\mathcal{E}$. The electric field is changing with time, hence there is a displacement current as well, in accord with Maxwell's discovery. These currents produce fields that combine with those inducing them. If the field outside is

$$\mathcal{E}_0 \sin \omega \left(t - \frac{z}{c}\right)$$

it is found that the field inside is of modified form

$$\mathcal{E}_0 \sin \omega \left(t - \frac{z}{v}\right) e^{-\gamma z}$$

Figure 20-5 shows graphs of these fields. Two features are to be noted—the speed v is different from $c = 3 \times 10^8$ m/sec, and there is an exponential reduction in field strength as the radiation penetrates the medium.

The quantity γ, called the *attenuation coefficient,* is dependent on the electrical conductivity of the material and the characteristics of the radiation. We can deduce the effects that take place in the medium: the electric

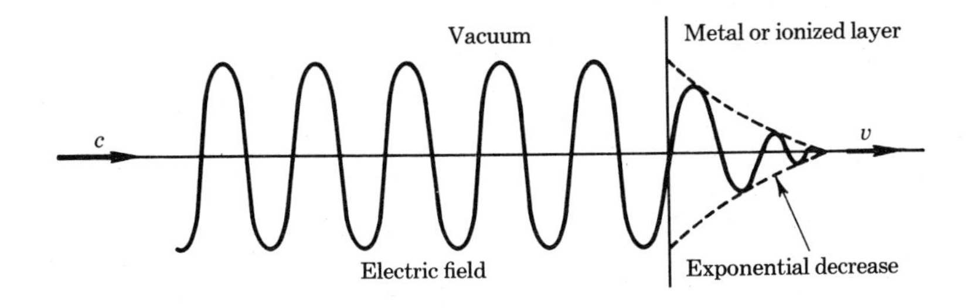

FIGURE 20-5

Attenuation of electromagnetic waves—skin effect

field of the incoming wave accelerates electrons; the energy of the wave is thus reduced and eventually approaches zero as the depth of penetration increases; the accelerated electrons produce electromagnetic radiation which is observed from outside the metal as a *reflected* wave. As expected, the higher the frequency of incident light and the larger the conductivity, the greater the energy loss by the wave and thus the shorter the distance of penetration. In a metal, some of the energy goes into heating a very thin layer of the surface, a process which goes by the name "skin effect." The thickness of the layer (measured by the distance for $\mathcal{E}$ to be reduced by a factor $1/e = 0.37$) is surprisingly small. Radiation of 1-cm wavelength in copper, for example, has a skin thickness of less than 0.001 mm.

The transmission of high-frequency currents in conductors is made difficult because of the skin effect. If the current is limited to a thin surface layer, the resistance and heating are excessive. A *coaxial cable* is one solution to this problem. It consists of a metal tube with a wire along its axis. Electromagnetic energy flows along in the space between the conducting surfaces. For very high frequencies (the *microwave* region), transmission is effected by a *wave guide*, which is a hollow rectangular box with no central wire. Fluctuating currents in the walls cause fields to combine with those in the cavity. The electric-field vector vibrates in the plane perpendicular to two faces and the magnetic field vibrates in the form of closed loops that are end to end along the length. These are special solutions of Maxwell's equations.

The reflection of radio waves from the ionized gases in the upper atmosphere (ionosphere) involves the same mechanisms of electron acceleration and radiation as take place in a metal, even though the electron density is much lower in this partly ionized atmosphere. Radio communication between widely distant points on the Earth makes use of this phenomenon. Another example of the effect is the temporary "blackout" of communication with a spacecraft that is in the stage of re-entry into the atmosphere. Frictional heating of the air by the high speed of the vehicle creates a plasma layer that reflects radio waves.

Insulating Material

Earlier (Section 17-6), we showed that the ability of a substance to conduct electricity depends on the degree of binding of electrons to the atoms. In the insulator, a material with tightly bound charges, the electric field of the electromagnetic wave can only set the electrons into vibration about their normal positions. We also saw (Section 17-3) that a steady electric field creates dipoles in a medium. The resulting polarization was described in terms of a dielectric constant κ. Similarly, an oscillating wave produces dipoles that have their own electric and magnetic fields. Imagine a large assembly of tiny radio transmitters induced to send out waves by the incoming light. Electric fields are superposed, the net effect of which is to change the speed of light in the medium from c, its value in a vacuum, to a value

$$v = \frac{c}{\sqrt{\kappa}}$$

This means that the higher the dielectric constant of the material, the lower the speed of light (corresponding to greater electrical polarization of molecules). The ratio of speed of light in a vacuum to that in the medium $c/v = \sqrt{\kappa}$ is designated as the *index of refraction,*

$$n = \frac{c}{v}$$

This index determines how much bending a ray of light experiences on entering the medium. For water, n is about $\frac{4}{3}$; in some glass about $\frac{3}{2}$; while for air, with its low particle content, n is very close to 1. There is some dependence of the speed of light—and hence the index of refraction—on the wavelength of light, as shown in Table 20-1. These differences permit

TABLE 20-1

INDEX OF REFRACTION OF QUARTZ

Wavelength λ (mμ)	Index of Refraction n
202.6	1.547
303.4	1.486
404.7	1.470
508.6	1.462
706.5	1.455

white light to be decomposed into its various colors by use of a prism (Section 21-2), a process called *dispersion.*

We are quite familiar with the fact that some nonconducting materials readily transmit visible light, while others are opaque. Such differences depend on the details of molecular structure and the frequency of the light. Some plastics pass no visible light, but are so transparent to infrared waves that they can be used as lenses in that range. In contrast, water is nearly transparent to visible light, but absorbs infrared rays strongly. High-frequency X rays tend to penetrate much more readily than visible light, which accounts for their use in medical diagnosis.

Let us consider a simple experiment. A plane wave of light originating in a region of vacuum is allowed to impinge on the flat surface of some transparent material. The interaction of the time-varying fields with the material of index of refraction n will create oscillations of the charges and a resulting reradiation of light. This secondary radiation supplements the original stimulating radiation. Such emissions can send waves backward that give the observer the impression that the original waves undergo *reflection*. The combined radiation will penetrate the surface and, depending on the angle of incidence, the radiation may experience a bending or *refraction*. The substance may transmit some of the wave vibrations and stop others, giving rise to the process of *polarization*. All of the above processes are shown to be closely related by electromagnetic theory.

20-4 REFLECTION AND REFRACTION

The action of light in many optical processes and devices is governed by certain experimentally deduced laws that are already familiar to us. For example, we know that light striking a plane mirror at an angle with the surface is reflected away at the same angle, and that light entering a substance such as water at an angle is bent into the medium. Electromagnetic-wave theory will now be brought to bear on these phenomena. First, let us suppose that a light wave moving in the z-direction in a vacuum is represented by its electric field

$$\mathcal{E} = \mathcal{E}_0 \sin (\omega t - kz)$$

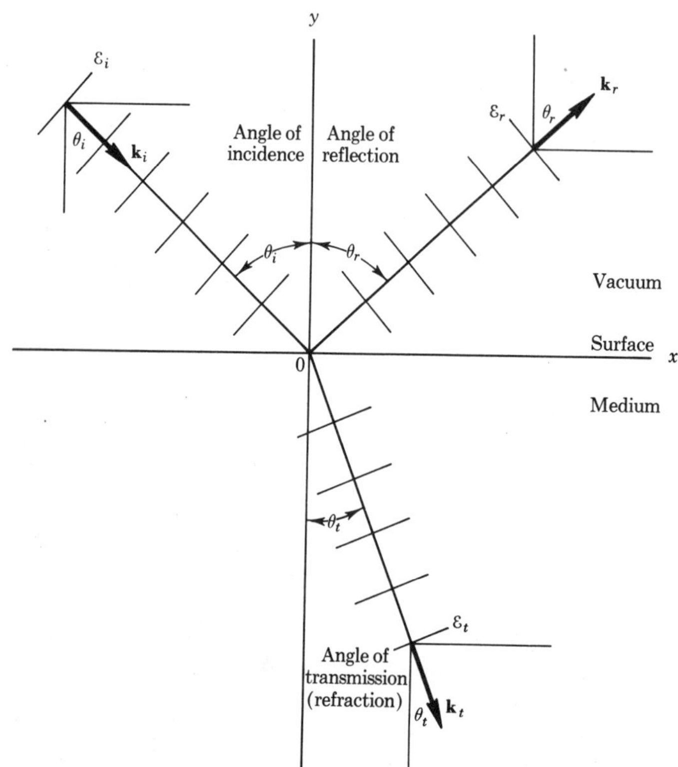

FIGURE 20-6
Incident, reflected, and transmitted waves

where k is ω/c. A wave that is moving in any direction in a medium in a three-dimensional coordinate system with axes x, y, and z can be characterized by a propagation vector **k** along the direction of motion and of magnitude ω/v. In place of kz in the wave expression above, the quantity $k_x x + k_y y + k_z z$ appears. Picture, as in Figure 20-6, the behavior of the electric fields when light from a vacuum strikes a plane surface of a substance at some angle. We can expect that the incident field $\mathcal{E}_i$ will give rise to a reflected field $\mathcal{E}_r$ and a transmitted field $\mathcal{E}_t$. The propagation vectors are seen to be all different. Electromagnetic theory tells us that the x-components of the vector **k**, those parallel to the surface, are all the same at the boundary between the two media. From the diagram, these components are found as

$$k_{ix} = \frac{\omega}{c} \sin \theta_i \qquad \textit{(incident)}$$

$$k_{rx} = \frac{\omega}{c} \sin \theta_r \qquad \textit{(reflected)}$$

$$k_{tx} = \frac{\omega}{v} \sin \theta_t \qquad \textit{(transmitted)}$$

Equating all three and canceling ω, we get a compact general expression relating speeds and angles:

$$\frac{\sin \theta_i}{c} = \frac{\sin \theta_r}{c} = \frac{\sin \theta_t}{v}$$

From the first two parts, we find $\sin \theta_i = \sin \theta_r$, or

$$\theta_i = \theta_r \qquad \textit{(law of reflection)}$$

which says that *the angle of incidence is equal to the angle of reflection.* This simple relation allows us to solve mirror problems. Then, using $c/v = n$, we see that

$$\frac{\sin \theta_i}{\sin \theta_t} = n = \frac{c}{v} \qquad \textit{(law of refraction)}$$

which states that *the ratio of the sines of the angles of the incident and transmitted (refracted) waves is the ratio of the speeds of light in the two media.* This second result, called *Snell's law*, forms the basis for solution of lens problems. The two basic formulas are depicted geometrically in Figure 20-7.

When light passes from one medium (say water), with index of refraction $n_1 = c/v_1$, at the speed of light v_1, into another (say glass) with index $n_2 = c/v_2$, speed v_2, the relation between angles of the rays to the normal is

$$\frac{\sin \theta_1}{\sin \theta_2} = \frac{n_2}{n_1}$$

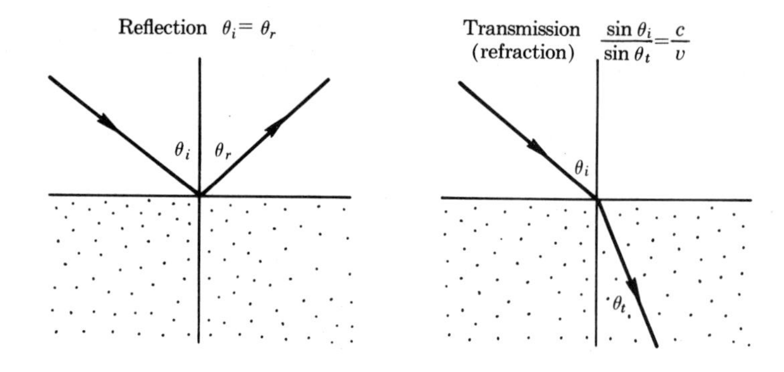

FIGURE 20-7

Laws of reflection and refraction

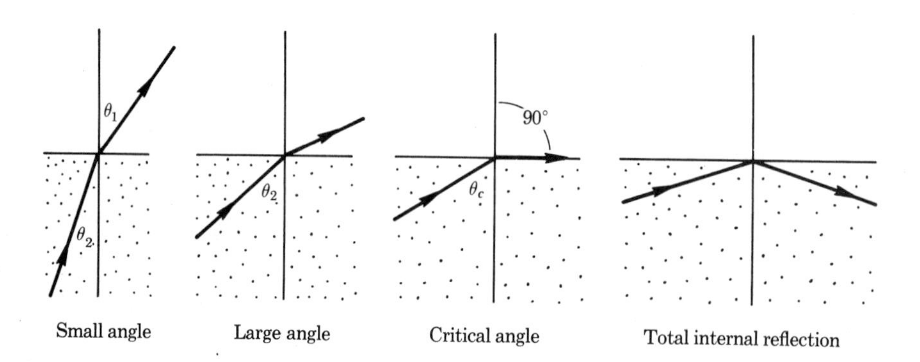

FIGURE 20-8

Approach to internal reflection

We can deduce an important rule from this general form of Snell's law. If light passes from a medium with smaller n to one of larger n it is bent, or refracted, *toward* the normal; if it goes in the reverse direction, it is bent *away* from the normal. To check, note that if $n_2 > n_1$, then $\sin\theta_1 > \sin\theta_2$ or $\theta_1 > \theta_2$. A peculiar effect occurs when one changes the angle of incidence as light passes from a dense to a rare medium. As seen in Figure 20-8, the angle of transmission becomes successively larger, until it reaches 90 deg. For larger angles of incidence, only the reflected wave remains, a condition called *total internal reflection.*

Let us compute the critical angle of incidence θ_c corresponding to refraction at 90 deg. Applying Snell's law,

$$n_1 \sin\theta_1 = n_2 \sin\theta_2$$

but since $\theta_2 = \pi/2$, $\sin\theta_2 = 1$. Thus

$$\sin\theta_c = \frac{n_2}{n_1}$$

When light goes from glass, $n_1 = \frac{3}{2}$, to water $n_2 = \frac{4}{3}$, then $\sin\theta_c = \frac{8}{9} = 0.889$, and $\theta_c = 62°44'$.

Reflected- and Transmitted-light Intensities

The special ability of glass to transmit most of the visible light incident on it is closely related to its dielectric properties. We can deduce the reflected and transmitted intensities from electromagnetic theory by application of the principle of conservation of energy. Suppose that light is incident *normally* on a plate of glass from air, such that the electric field $\mathcal{E}$ is parallel to the surface. Let the fields of the waves reflected back from the glass and transmitted through it be abbreviated for convenience as $\mathcal{R}$ and $\mathfrak{T}$. It is easy to show that the net electric fields on the two sides of the boundary are the same.* As in the case of transverse waves in a wire attached to a wall (Section 12-1), the reflected wave is 180 deg out of phase with the incident one. Thus we have the magnitudes of fields related by

$$\mathcal{E} - \mathcal{R} = \mathfrak{T}$$

Associated with each wave is a flow of energy, the *intensity*. Neglecting heating, the law of conservation of energy requires that the transmitted intensity equal the difference between those incident and reflected:

$$I_{\mathcal{E}} - I_{\mathcal{R}} = I_{\mathfrak{T}}$$

We know that the intensity of light in a vacuum (Section 20-2) is proportional to the square of the electric field $\mathcal{E}$ and to the speed of light c. We now inquire into its nature in a dielectric medium. Recalling that for a given electric field the energy density in a dielectric is $\kappa = n^2$ times as large as in a vacuum, but that the speed of radiation v is only $1/n$ times as large, we conclude that the intensity is n times as large. Our intensity balance can thus be written

$$\mathcal{E}^2 - \mathcal{R}^2 = n\mathfrak{T}^2$$

The pair of equations can easily be solved to find the ratios of the fields:

$$\frac{\mathcal{R}}{\mathcal{E}} = \frac{n-1}{n+1}, \qquad \frac{\mathfrak{T}}{\mathcal{E}} = \frac{2}{n+1}$$

For example, let us find the fraction of light that is reflected from the surface of glass of index of refraction $n = 1.5$. The ratio of intensities is

$$\frac{I_{\mathcal{R}}}{I_{\mathcal{E}}} = \frac{\mathcal{R}^2}{\mathcal{E}^2} = \left(\frac{n-1}{n+1}\right)^2 = \left(\frac{1.5-1}{1.5+1}\right)^2 = 0.04$$

* To show this on a static basis, we merely carry a charge along the outside a distance x, then go just inside the surface and come back. The net work is zero in the closed path, so the fields must be the same on both sides.

that is, only 4% of the incident energy is reflected, which tends to verify our experience. In the design of various optical instruments, the relative transmission and reflection are of considerable importance.

The Maxwell-Fresnel Equations

The amplitudes of the fields just found were especially simple in form for the special case of normal incidence. More general expressions, derived by Fresnel on the basis of Maxwell's theory, take account of light incident at an angle θ_i from the normal. The ratios are

$$\frac{\mathcal{R}}{\mathcal{E}} = \frac{\tan(\theta_i - \theta_t)}{\tan(\theta_i + \theta_t)}, \qquad \frac{\mathcal{T}}{\mathcal{E}} = \frac{2 \sin\theta_t \cos\theta_i}{\sin(\theta_i + \theta_t)}$$

Snells' law, $\sin\theta_i = n \sin\theta_t$, is required to find the ratios for a given substance.* As an interesting application of the equations, let us examine their

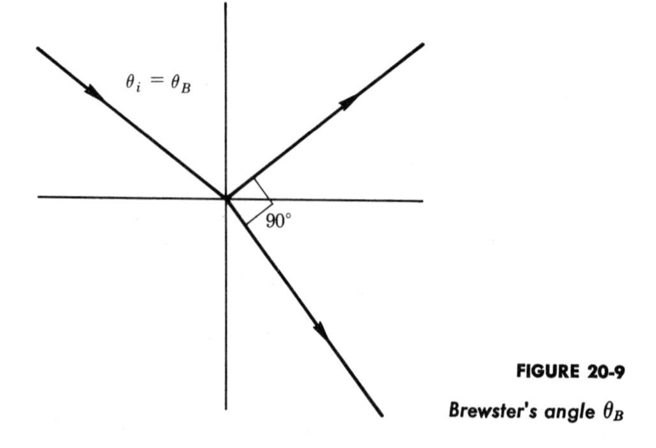

FIGURE 20-9
Brewster's angle θ_B

form as the angle of incidence θ_i is changed. There comes a point where the directions of reflection and transmission are 90 deg apart (see Figure 20-9). Since $\theta_i = \theta_r$, this may be written $\theta_i + \theta_t = \pi/2$. In our formula for $\mathcal{R}$, the tangent goes to infinity, and $\mathcal{R}$ must become *zero*. There is no reflection at all of light with its electric vibration in the plane of incidence. Light vibrating in other directions will still be reflected, however. This effect occurs at what is called *Brewster's angle* θ_B, an angle of incidence such that

$$\tan\theta_B = n$$

* The general relations can be shown to hold in the case of normal incidence. For small angles, the tangents and sines are both close to the angles. Thus $\frac{\mathcal{R}}{\mathcal{E}} \cong \frac{\theta_i - \theta_t}{\theta_i + \theta_t}$; and since $\theta_i \cong n\theta_t$, we find that $\frac{\mathcal{R}}{\mathcal{E}} = \frac{n-1}{n+1}$.

To prove this, we merely substitute $\sin\theta_t = \sin(\pi/2 - \theta_i) = \cos\theta_i$ into Snell's law $\sin\theta_i = n\sin\theta_t$. When light enters water, $n = \frac{4}{3}$, we compute $\theta_B = \tan^{-1}\frac{4}{3} = 53°8'$. We can guess the physical reason for the loss of reflected component when $\theta_r + \theta_t = \pi/2$. Dipoles near the surface are excited into oscillation by the refracted wave. They oscillate parallel to the direction where there should be a reflected wave. However, radiation from dipoles is zero in that direction and a maximum in the perpendicular direction—that is, along the refracted wave. The selection of waves according to their directions of vibration forms a part of the subject of *polarization* of light, to be discussed in more detail in Section 20-6.

Waveforms

The sound waves of music, speech, and noise in general were noted in Section 12-2 to be of complicated form, consisting of a mixture of waves of different frequencies and amplitudes. Similarly, many electromagnetic waves are also of a compound nature. It can be shown that pure monochromatic (single-color) light would exist only if the wave were infinitely long. Actual waves, being of finite length, must be composed of many frequencies. As a simple example, suppose that the aperture on our camera is opened and immediately closed. The light wave of one frequency from a distant object would be chopped at both ends, leaving the portion sketched in Figure 20-10(a). A form such as this can be decomposed mathematically (by *Fourier analysis*) into a sum of infinite wave trains with different pure frequencies. Our chopped light has a distribution in fre-

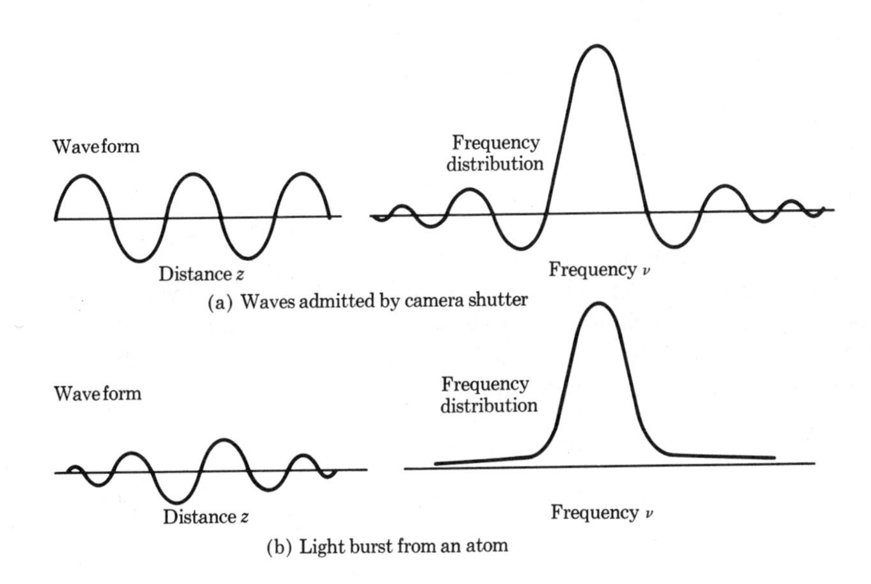

FIGURE 20-10

Waveforms and spectra

quency as illustrated in Figure 20-10(a). Such a plot is called a *spectrum* (plural, "spectra"). Another example of a group of waves is shown in Figure 20-10(b), corresponding to light from a collection of atoms at high temperature. Most of the spectrum is very near one frequency, as seen in a Figure 20-10(b), but again all possible frequencies appear in the distribution.

We have seen that the index of refraction and thus the speed of light in a medium often depend on the wavelength. Thus, when a beam of light containing many colors strikes a material substance, the various waves can move through it at different speeds. They add to form groups similar to those discussed above. In order to see this, let us add two waves only. They will be represented by $y_1 = \sin(\omega_1 t - k_1 z)$ and $y_2 = \sin(\omega_2 t - k_2 z)$, where it is assumed that the frequencies and speeds are quite close to each other.

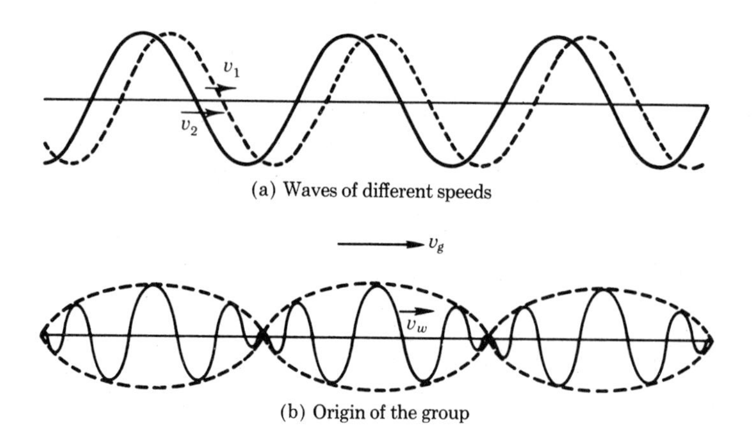

(a) Waves of different speeds

(b) Origin of the group

FIGURE 20-11

Wave and group speeds

The two waves and the sum $y = y_1 + y_2$ are shown in Figure 20-11. The envelope of the wave, shown as a dashed curve in Figure 20-11(b), represents the *group*, which moves along with a speed different from the speed of any of the *waves*.* Depending on how the index of refraction varies with wavelength in the medium, the group may move to the right, to the left, or stand still relative to the motion of the waves. It should be noted that energy is transmitted by the group, hence measurements yield its speed.

* Use of formulas of trigonometry yields the sum

$$y = 2 \sin(\bar{\omega}t - \bar{k}z) \cos\left(\frac{\Delta\omega}{2}t - \frac{\Delta k}{2}z\right)$$

where $\bar{\omega}$ and $\bar{k}$ are averages and $\Delta\omega$ and Δk are differences. The speed of the group is $v_g = \Delta\omega/\Delta k$.

20-5 THE PRINCIPLES OF FERMAT AND HUYGENS

To complete our review of the basic ideas of the interaction of radiation with matter, we shall state two generalizations that are consistent with the Maxwell–Fresnel theory.

Fermat's Principle

A theorem formulated by Fermat in 1650 is known as the *principle of least time.* It is useful for the study of geometrical optics, even though it is an approximation of Maxwell's rigorous theory. Fermat's statement is:

THE TIME REQUIRED FOR LIGHT TO FOLLOW AN ACTUAL PATH IS LESS THAN THE TIME FOR ANY OTHER NEARBY PATH.

The word "path" must be interpreted as being the *optical path,* defined as the product of distance and index of refraction. Three consequences are noted below:

1. Light travels between two points in vacuum along a straight line. The proof is simple: a straight line is the shortest distance between two points and since the speed of light c is constant, the time elapsed is shorter than for any other curve.

2. The angles of incidence and reflection from a surface are the same. To prove this, we draw rays AB and BC that give equal angles, as in Figure 20-12. Then we select some other paths AB' and $B'C$. Completing the geometric construction by dotted lines, with $BC = BD$ and $B'C = B'D$, we see that the path ABD (or ABC) is shorter than any other $AB'D$ (or $AB'C$) and thus is the actual path by Fermat's theorem.

3. Snell's law of refraction. For time to be smallest along the correct path, the change in time for slightly different paths should approach zero. We construct two sets of possible rays in Figure 20-13. Now, $\Delta t = \Delta l_1/v_1 -$

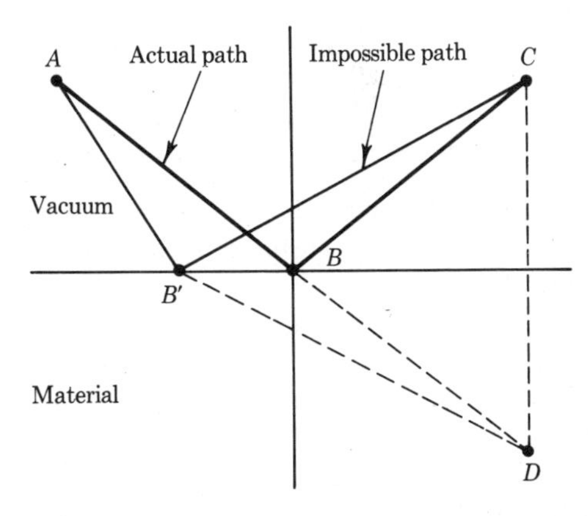

FIGURE 20-12

Applications of Fermat's principle

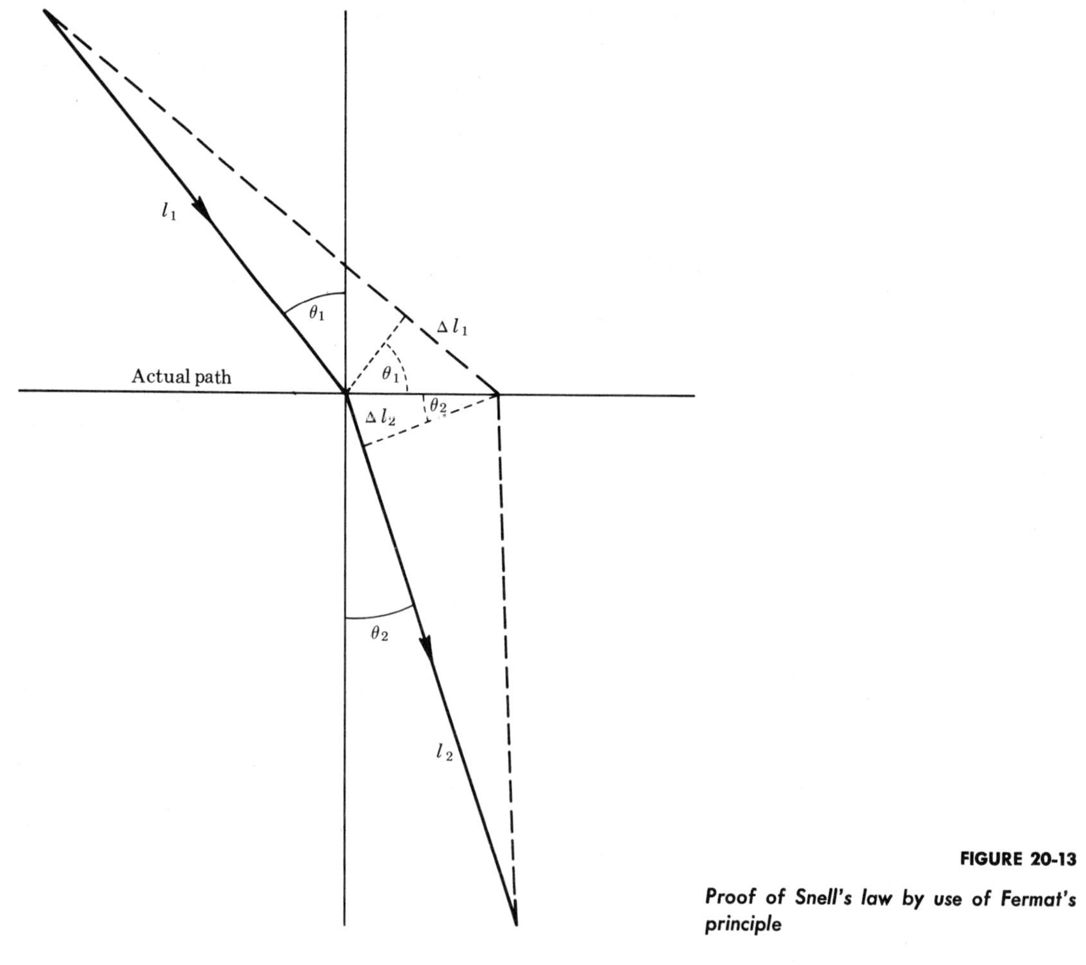

FIGURE 20-13

Proof of Snell's law by use of Fermat's principle

$\Delta l_2/v_2 = 0$, but $v_1 = c/n_1$ and $v_2 = c/n_2$ by definition, and from the small triangles, $\Delta l_1/\sin\theta_1 = \Delta l_2/\sin\theta_2$. Combining, we obtain

$$n_1 \sin\theta_1 = n_2 \sin\theta_2$$

Huygens's Principle

When light and other waves pass from one medium to another, or through an aperture in an absorbing plate, there are unusual effects on the wave. A powerful theorem formulated by Huygens, a Dutch scientist of the seventeenth century, can be used to help understand the process:

> ANY WAVEFRONT AT A CERTAIN TIME CAN BE CONSIDERED AS A SOURCE OF SPHERICAL "WAVELETS" THAT RADIATE FROM EVERY POINT. THE WAVEFRONT AT A LATER TIME IS THE ENVELOPE OF THE WAVELETS.

To visualize this, let us consider a plane wave. At time t_1, we replace the front by a set of points that give spherical waves. At a later time t_2, the

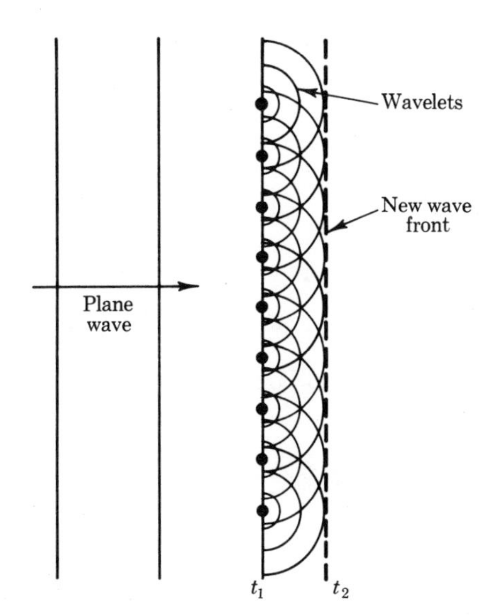

FIGURE 20-14

Illustration of Huygens's principle

new wave can be constructed by superposing the individual wavelets (Figure 20-14). Next, picture a plane wave passing from one medium into another in which the speed of light is lower, as in Figure 20-15. The same time interval is used in drawing the circles. We note the slower progress in the second medium of the plane wave. Snell's law of refraction can readily be derived by use of Huygens's principle. As in Figure 20-16, light is incident at an angle with the surface of a medium. We pick out two of the many wavelets that go to make up the plane wave. In a certain time inter-

FIGURE 20-15

Passage of light across a boundary between two materials

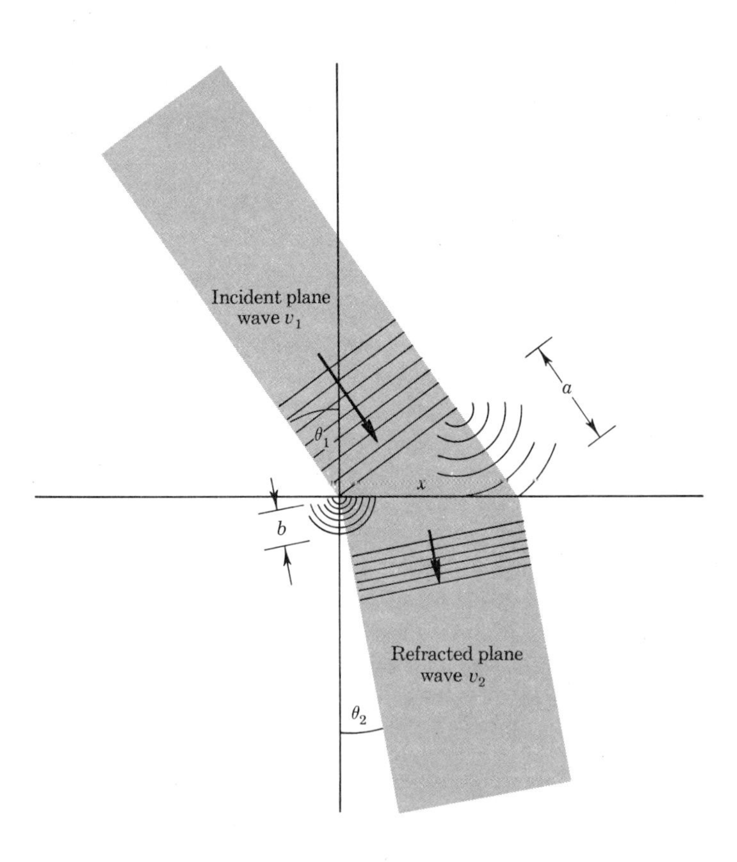

FIGURE 20-16

Application of Huygens's principle to refraction

val t, one wavelet has moved a distance $a = v_1 t$, while the other has moved a distance $b = v_2 t$. Thus

$$\frac{a}{v_1} = \frac{b}{v_2}$$

From the triangles, however, we see that

$$x = \frac{a}{\sin \theta_1} = \frac{b}{\sin \theta_2}$$

Dividing the two equations yields $v_1 \sin \theta_1 = v_2 \sin \theta_2$, as found previously. This example merely suggests the utility of Huygens's principle. We shall see its application later to the phenomena of interference and diffraction of light as it passes near boundaries and through openings.

With this collection of basic properties and principles of waves and rays, we are now able to investigate the detailed behavior of light as it is reflected and transmitted by different substances in various shapes, particularly those that assist the human eye in discerning the very small and the very far.

20-6 POLARIZATION OF LIGHT

Electromagnetic waves observed at great distances from dipole antennas were shown in Section 19-2 to consist of vibrations of the electric- and magnetic-field vectors in two perpendicular planes only. In contrast, the light from individual atoms that are oriented in random directions includes vibrations in all planes. Let us concentrate on the electric vector as the one that can create electron motion in materials and thus give the visible optical effects. If we examine an approaching beam of light, as in Figure 20-17(a), we see that it is composed of many electric vibrations. Such light is said to be *unpolarized*. Each of the $\mathcal{E}$ vectors can be resolved

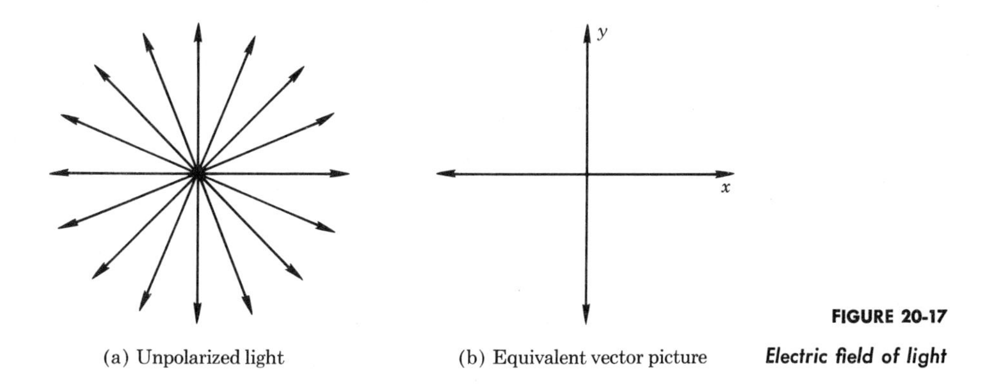

(a) Unpolarized light (b) Equivalent vector picture

FIGURE 20-17
Electric field of light

into x- and y-components. Combining the components of the vectors separately, we can draw an equivalent view of the disturbance, as in Figure 20-17(b). There are now electric vibrations in two perpendicular planes.

On striking a surface of a medium, the relative amount of reflection and transmission depends on the angle that the vibrations make with the plane of incidence. Thus, using appropriate reflection, it is possible to favor or eliminate some components and achieve *polarized* light, vibrating in one plane only. Light passing through certain natural crystals such as tourmaline also becomes polarized. The most commonly available material with this property is the artificial product Polaroid, invented by Land in 1935. It consists of transparent plastic sheets with embedded tiny crystals. In the manufacture, the sheets are stretched to line up the particles in one direction or axis. On passage through Polaroid, vibrations perpendicular to the axis are absorbed, while those parallel are transmitted.

An analogy may assist in visualizing the process. Suppose that we threaded a rope through a picket fence as in Figure 20-18(a), and oscillated the end vertically. Such waves would readily go through. If instead we vibrated it horizontally, the signals would stop at the pickets—Figure 20-18(b).

The effect on unpolarized light—of the form shown in Figure 20-17(b)—of a Polaroid plate with axis parallel to one of the electric directions is shown in Figure 20-19. If the plates were rotated through 90 deg, the

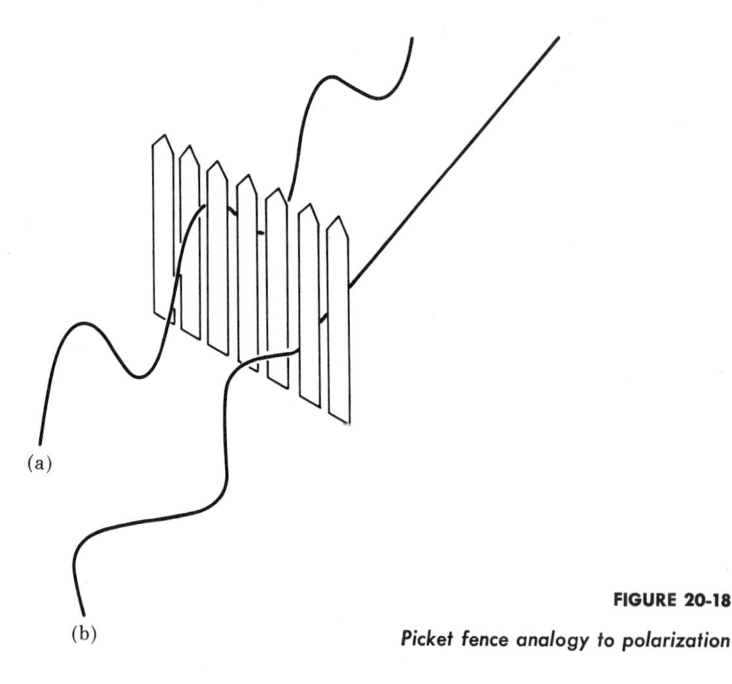

FIGURE 20-18

Picket fence analogy to polarization

horizontal component would be transmitted instead. Polaroid thus serves to "create" polarized light by this process of elimination.

Having achieved plane polarized light, we now ask what happens when it passes through a second Polaroid plate with its axis at any angle θ to the plane of vibration. The single incident electric-field vector $\boldsymbol{\mathcal{E}}$ may be resolved into components as in Figure 20-20. Only the part $\mathcal{E}_{\parallel} = \mathcal{E} \cos\theta$ is transmitted. The intensity (proportional to $\mathcal{E}_{\parallel}^2$) that gets through is $I_{\parallel} = I \cos^2\theta$.

Let us calculate the transmitted intensity for several angles:

$\theta = 0°$ (vibrations parallel to axis of Polaroid) $\cos\theta = 1$, $I_{\parallel} = I$. All of the light gets through.

$\theta = 45°$, $\cos\theta = \sqrt{2}/2$, $I_{\parallel} = I/2$. One-half of the light gets through.

$\theta = 90°$ (vibration perpendicular to axis) $\cos\theta = 0$, $I_{\parallel} = 0$. No light gets through.

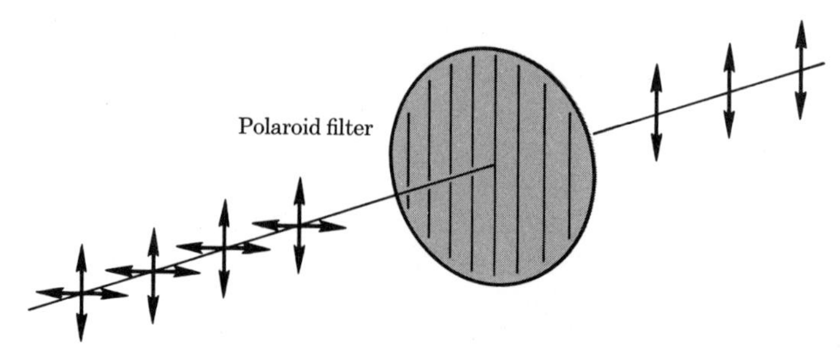

FIGURE 20-19

Polarization by use of Polaroid

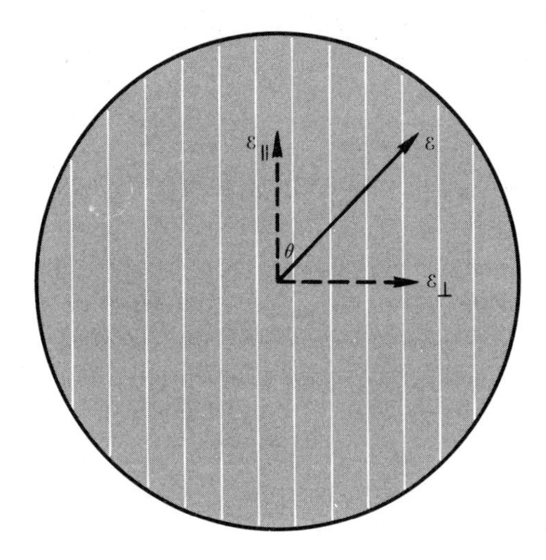

FIGURE 20-20

Resolution of polarized light

As we continue to still larger angles, the intensity builds back to a maximum, and repeats after $\theta = 180°$.

For a more complex case, we let unpolarized light fall on a stack of three plates with successive axes at 30 deg to each other. The incident light is viewed as equivalent to two perpendicular components, chosen so that one is parallel to the axis of the first plate. One-half the intensity is thus cut out immediately. Of the rest, the second plate transmits a fraction $(\cos 30°)^2 = (\sqrt{3}/2)^2 = 0.75$, and the third plate transmits a fraction 0.75 of that. Of the original light, the final fraction left is $(0.5)(0.75)(0.75) = 0.28$.

We can now understand the action of sunglasses with Polaroid lenses. Partly polarized light is reflected from a surface such as a highway, as discussed in Section 20-4. The axis of crystals in the glasses is arranged so as to screen out this undesired light. Another practical use of polarization effects is in the study of strains in materials. The extent to which light transmitted by a transparent solid or reflected from an opaque solid is polarized depends on the local distortions in the material. The pattern observed thus reveals points of structural weakness.

PROBLEMS

20-1 How long would it take light to travel between Galileo's points of measurement if they were 1 mile apart?

20-2 How long did it take TV signals to return from the Mariner spacecraft near Mars? (See Table 10-1 for data.)

20-3 How much radiant energy would strike a unit area if the electric field of the electromagnetic wave were 200 volts per meter?

20-4 What is the rate of radiation from the Sun at the planet Mercury (see Table 10-1), if the figure 1395 J/m^2-sec at the Earth is scaled up using the inverse-square law?

20-5 Compute the intensity of radiation at 2 meters from a 100-watt lightbulb. Also find the energy density and strength of the electric field there.

20-6 A lightbulb is lowered from height 1.5 m to 0.5 m above a desk surface. By how much is the illumination increased?

20-7 The attenuation coefficient γ for electromagnetic waves in metal is found to be $\gamma = \frac{2\pi}{c}\sqrt{k_E\sigma\nu}$. Compute the skin thickness $1/\gamma$ for radiation of frequency 3×10^4 Mc, incident on aluminum, resistivity 2.6×10^{-8} ohm-m.

20-8 Explain why one cannot be electrocuted on contact with a source of very high frequency.

20-9 The speed of light in water is measured to be 2.25×10^8 m/sec. Deduce its dielectric constant. Why might this differ from the value of 78.5 quoted in Table 17-3?

20-10 In Michelson's experiment, how much longer did light take to travel the 44 miles with air present than it would have in a vacuum? *Note:* for air, $n = 1.00002926$.

20-11 Calculate the angle of refraction of light that enters water, $n = \frac{4}{3}$, from air, $n = 1$, if the angle of incidence is 30 deg.

20-12 Find the critical angle of incidence for visible light that goes to air, $n_2 = 1$, from glass, $n_1 = 1.5$.

20-13 Find Brewster's angle for light entering glass from air.

20-14 What fraction of the *intensity* of light incident normally on a water surface is reflected?

20-15* Verify the formulas for reflected and transmitted intensity by applying the law of conservation of energy.

20-16 A parabola is defined as a curve with points equidistant from a line D and a point F, as shown in Figure P20-16. Show that Fermat's principle

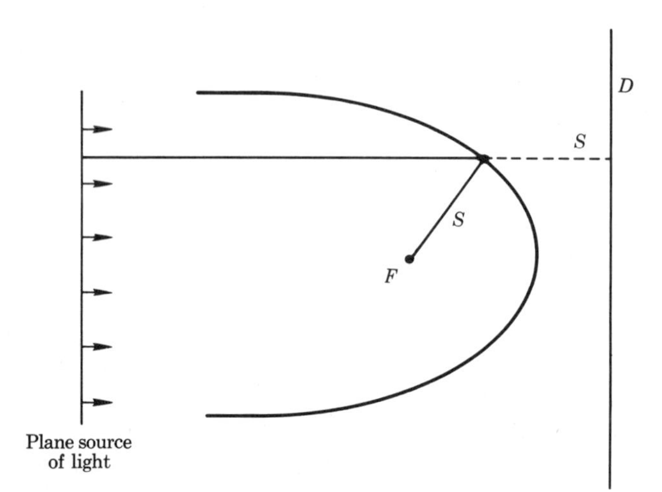

FIGURE P20-16

Application of Fermat's principle to focusing by parabolic mirror

is consistent with the fact that parallel rays are brought to a point in a parabolic mirror.

20-17 Polarized light is incident on Polaroid with the electric vibration at 30 deg to the axis of the sheet. What fraction of the original intensity is transmitted?

20-18* A stack of three Polaroid sheets with angles of axes at 45 deg to each other is illuminated with unpolarized light. How much of the original intensity gets through?

OPTICAL INSTRUMENTS

21

In man's search for full understanding of his surroundings, he finds that the unaided human eye is a rather poor instrument. To probe into the space outside the Earth, or into the details of material structure, devices such as the telescope and microscope are needed. Even to observe and record in his immediate vicinity, spectacles and the camera are valuable aids. We are now in a position to examine the function of such devices, by use of the principles and methods that come from the foregoing theory of radiation. Our new rules for reflection and refraction allow the interpretation of the action on light of plane and curved mirrors, lenses of various shapes, and combinations of these in optical instruments. The wave character of light can essentially be ignored in many instances, and light may be viewed as a set of rays that travel in straight-line segments. Optical studies involving reflection and refraction become exercises in geometry, hence the term *geometrical optics* is used. Our knowledge of the properties and principles of wave motion also allows us to examine optical processes in which the behavior is dominated by the wave aspect of light. This forms the subject called *physical optics*. Among these processes are the diffraction of light by edges, apertures, and lenses, and the polarization of light, as by Polaroid. For some optical systems, it is necessary to invoke the ray and wave views of light at the same time.

Let us first review how we see things in general. Light from a bright object such as a star or a lightbulb comes directly to us. Most objects are

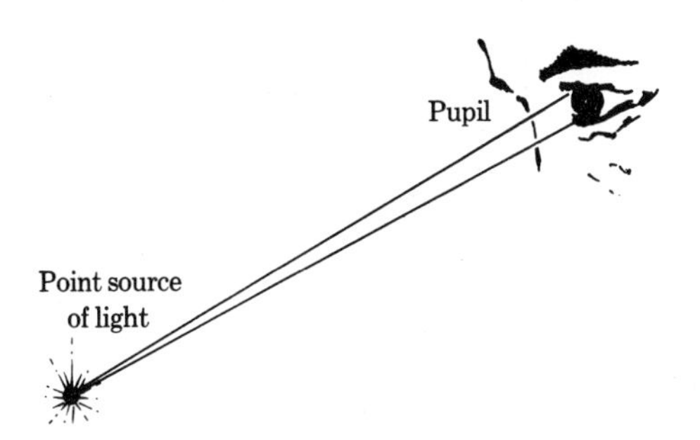

FIGURE 21-1

Rays from source of light to pupil of the eye

seen by reflected light, however. The pupil of the eye, though small—about $\frac{1}{8}$ in. in diameter—permits a small cone of rays from a point to enter. The light is focused on sensitive receivers connected to the optic nerve. Figure 21-1 shows two of the rays from a point that can be thus registered. We shall reserve further discussion of the eye until Section 24-3.

21-1 REFLECTION OF LIGHT

Plane Mirrors

The behavior of light used to view ourselves in a mirror is our first and simplest example. For instance, suppose one stands before a full-length mirror on the wall, as in Figure 21-2, and looks at the tip of a shoe, which reflects light from the room, and serves as a new source of light. The rays

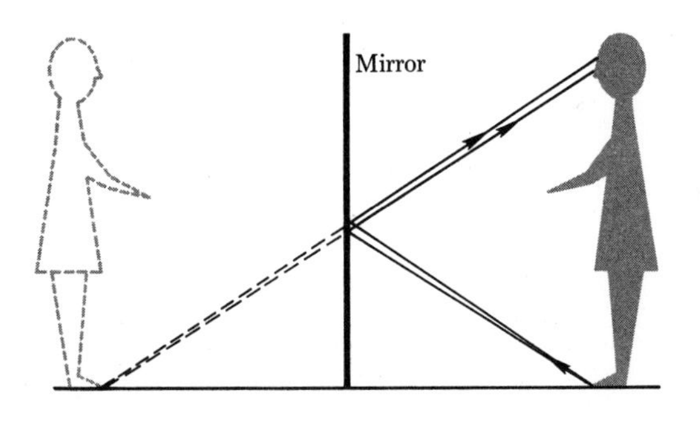

FIGURE 21-2

Wall mirror

are then reflected from the plane silvered surface, with the angle of incidence equal to the angle of reflection. As we well know, an object viewed by reflection appears to be behind the mirror. The distance of this image from the viewer is easily seen from Figure 21-3 to be twice his distance to the mirror. An extension of the rays to the eye backward will intercept the apparent or "virtual" image. We also note that a minimum of four rays are needed to obtain an indication of length.

An algebraic convention is used to define distance. Let p be the distance from the plane to the object being viewed and q the distance from

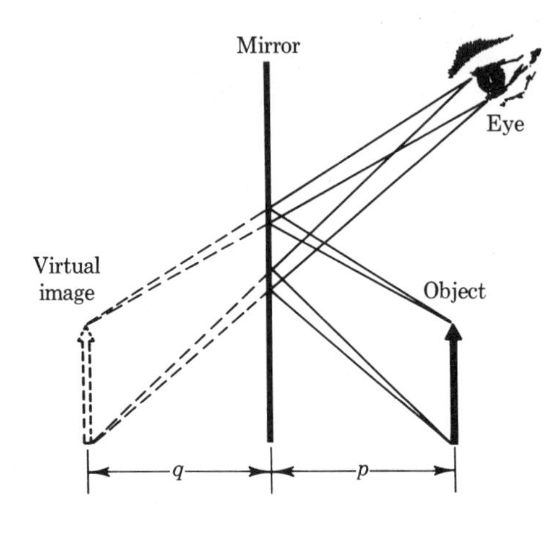

FIGURE 21-3

Image formation, plane mirror

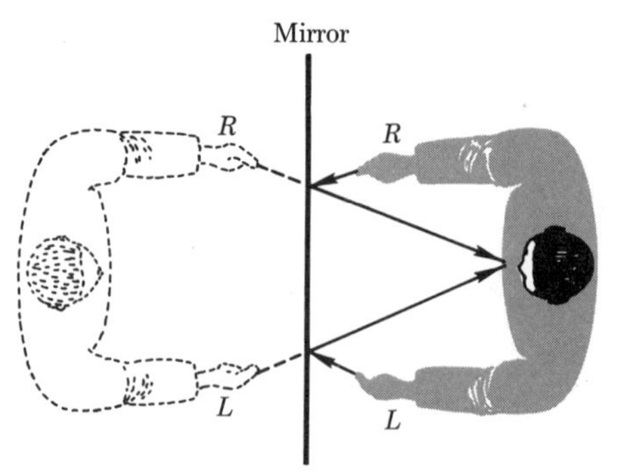

FIGURE 21-4

Reversal of image by plane mirror

the plane to the image. It is clear that from Figure 21-3 that

$$q = p$$

The size of the image and object are the same; there is no magnification in the case of a plane mirror. We are also aware that one's image in the mirror is reversed. The person that one sees in the mirror in Figure 21-4 has his right hand where his left ought to be, and vice versa. Since no one's face is quite symmetric, a mirror image looks different from a photograph. More complicated ray patterns are developed when one views himself in a system of two or more mirrors, as in a clothing store or in a barber shop.

Reflection at Curved Mirror Surfaces

Light rays from a distant source such as a star are parallel and are brought to an approximate *point* on reflection from a spherical surface. This makes possible the reflecting telescope used in astronomy. An image of the body is formed by the curved reflecting surface. Concave surfaces, as in automobile headlights, are also used to provide nearly parallel beams of light. These effects are easily demonstrated by the geometric approach. Let us construct, as in Figure 21-5, a ray diagram for a spherical mirror of

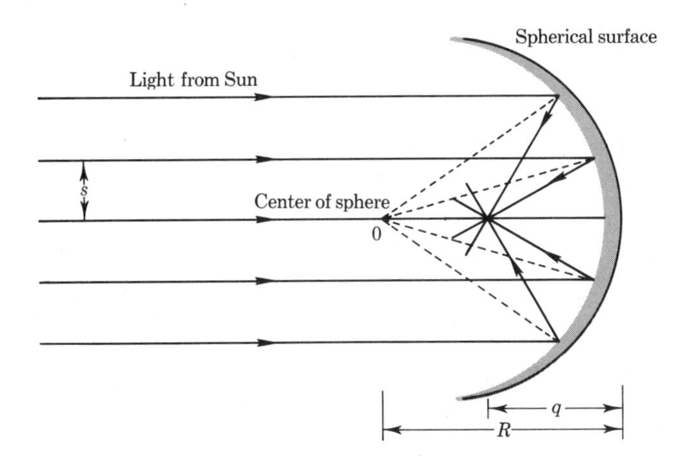

FIGURE 21-5

Focusing of light by spherical mirror

radius R, with light coming from a distant source, such as the Sun. The rays are taken to be parallel, and on reflection from the surface pass through about the same point on the axis. The image distance q is very close to $R/2$ if the incoming rays are near the axis. We can thus say that the light has been focused, and let $f = R/2$ be the *focal length* of this concave mirror.*

An orderly graphical procedure for finding images and deriving formulas is now presented. It is assumed that the object is small, so that s is negligible compared with R. First we draw a ray parallel to the axis, as in Figure 21-6(a). It is reflected through the point F or *focus*. A separate ray through the center of the sphere O passes through the same point on reflection along the radius—Figure 21-6(b). This pair of rays intersects to locate the image at a distance $q = QS$ from the mirror—see Figure 21-6(c). We attach distance labels to the diagram—Figure 21-6(d)—and find the sides of the similar triangles formed. The ratio of image height to object height can be written in two ways:

$$\frac{H}{h} = \frac{2f - q}{p - 2f} = \frac{q - f}{f}$$

Rearranging the second equation yields

$$\frac{1}{p} + \frac{1}{q} = \frac{1}{f} \qquad \textit{(concave-mirror formula)}$$

Substituting $1/f$ back in the original equation gives

$$\frac{H}{h} = \frac{q}{p} \qquad \textit{(magnification)}$$

* From the triangles in the figure, it is easy to show that q/R is more precisely $1 - \frac{1}{2}\left[1 - \left(\frac{s}{R}\right)^2\right]^{-1/2}$; thus, if $R = 0.2$ m and $s = 0.02$ m, then $q = 0.0995$, which is slightly different from $R/2$. A spherical mirror thus exhibits a small amount of *aberration*.

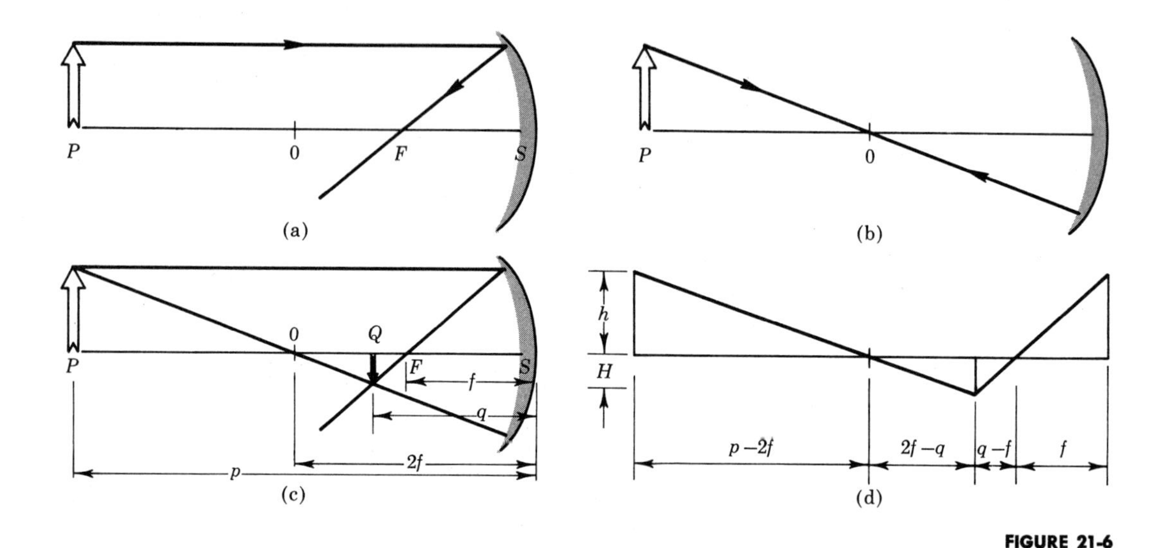

FIGURE 21-6

Graphical method for image location, mirror

which says that the magnification H/h is equal to the ratio of object distance to image distance.

Suppose the mirror radius is $R = 3$ m and $f = 1.5$ m. If we place an object at $p = 5$ m from the surface, then

$$\frac{1}{q} = \frac{1}{f} - \frac{1}{p} = \frac{1}{1.5} - \frac{1}{5} = \frac{7}{15}$$

or $q = 2.14$ m. The image height is $q/p = 2.14/5 = 0.428$ times the object height.

Hand mirrors, such as those used for shaving or applying makeup, have a concave surface that provides considerable enlargement. The object viewed must be nearer to the mirror than the focal point, which creates a large virtual image behind the surface, as shown in Figure 21-7. Suppose the radius of curvature is 0.8 m. Then, $f = 0.4$ m, and one's face at $p = 0.25$ m has an image found from $1/q = 1/0.4 - 1/0.25 = 2.5 - 4.0 = -1.5$. Then, $|q| = 0.67$ m, and the magnification is $|q|/p = 0.67/0.25 = 2.7$.

Parabolic Mirrors

A surface in the shape of a parabola has the special property of reflecting parallel light from distant objects to a single point, which is the focus of the parabola.* On the basis that light travels in either direction along the same path, this means that a lightbulb located at the focus will

* Strictly speaking, such a mirror should be called *paraboloidal.*

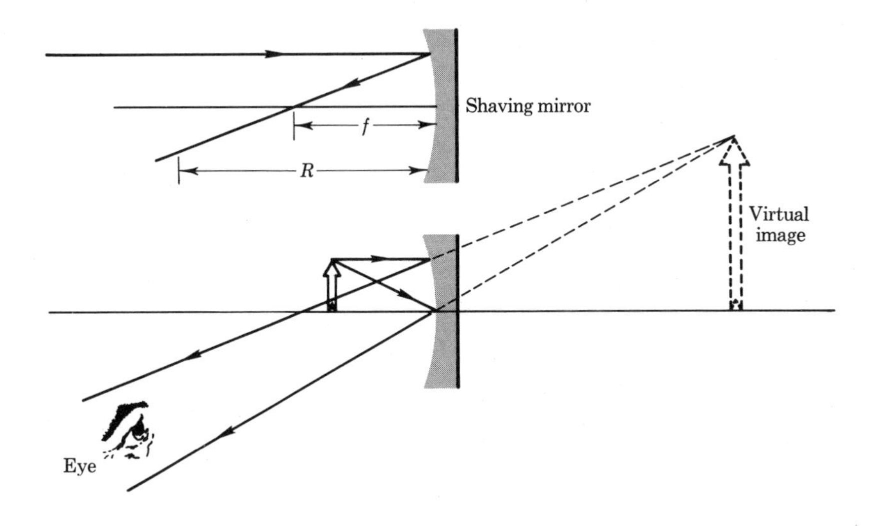

FIGURE 21-7

Virtual image formation by object inside focus of concave mirror

give rise to a parallel beam that will travel great distances without spreading out and thus losing intensity. Reflecting mirrors in searchlights are often constructed in this shape. A parabolic surface does not focus or project well, however, if the objects are off its main axis. The slight spreading that is desirable in automobile headlights is obtained by locating filaments off the focal point.

21-2 REFRACTION OF LIGHT

Refraction at Plane Boundary

Snell's law was derived (Section 20-4) on the assumption of plane waves, which by their very nature extend indefinitely far in two directions, while progressing in the third direction. We can also regard a beam or bundle of light rays passing through a small opening as having plane characteristics so long as the wavelength is much smaller than the aperture. It is convenient now to let the subscripts on indexes and angles refer to the *medium*. Thus, as a ray goes from water to glass or vice versa, we write

$$n_w \sin \theta_w = n_g \sin \theta_g$$

From these expressions, we see that it does not matter which direction the light is traveling; the ray lines are the same. In specific problems, we know part of the information and wish to find the rest. For example, let us measure the index of refraction of a certain transparent solid—say, a new type of glass—by light measurements near a water–glass boundary. Suppose (Figure 21-8) that if we set $\theta_w = 30$ deg, we find experimentally that $\theta_g =$

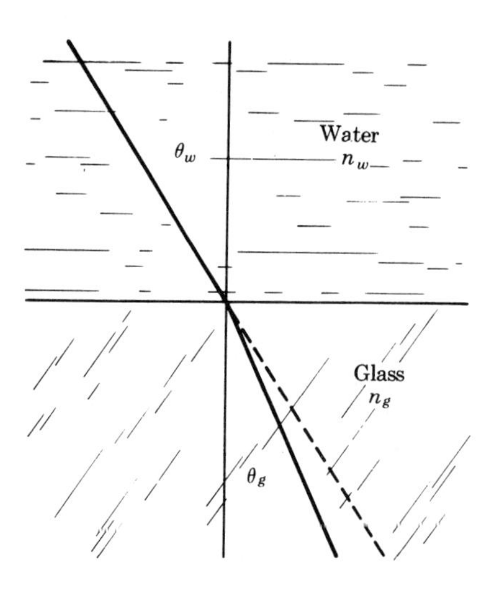

FIGURE 21-8

Refraction at boundary between transparent media

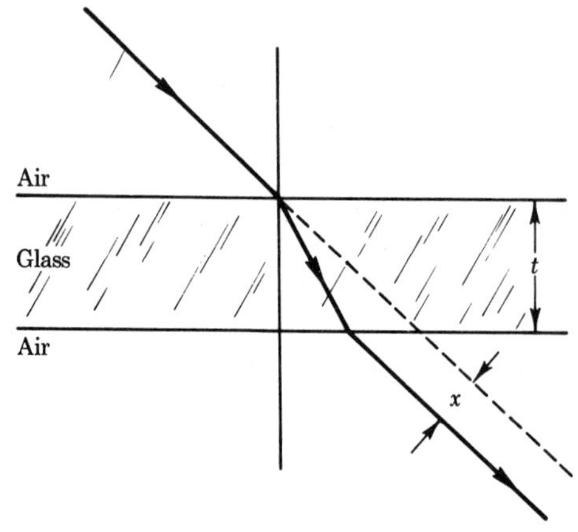

FIGURE 21-9

Displacement of ray by passage through a plate

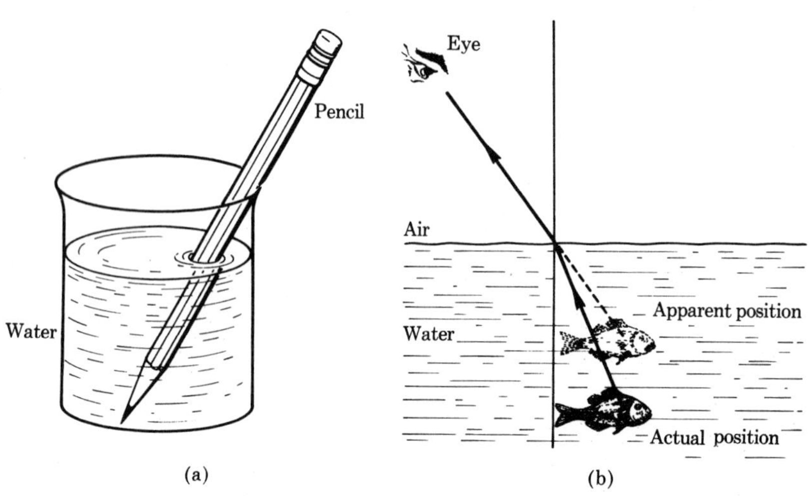

FIGURE 21-10

Optical illusions due to refraction

25 deg. If we assume that $n_w = 1.33$ is known previously, then

$$n_g = n_w \frac{\sin \theta_w}{\sin \theta_g} = \frac{1.33 \sin 30°}{\sin 25°} = \frac{(1.33)(0.5)}{0.423} = 1.57$$

A pencil of light will be shifted sideways on passage completely through a slab of material, as shown in Figure 21-9. The angles made with the normal of the original and final rays are the same. All objects seen through an ordinary windowpane in any direction that is not perpendicular to the surface are not quite where they seem. The displacement is small, however.

Some interesting optical illusions are caused by the refraction of light. A pencil partly immersed in a glass of water appears to have a sharp break at the surface—Figure 21-10(a). A fish in a pool seems to be at a higher level in the water than it really is—Figure 21-10(b).

Prisms

A triangular block of refracting material such as glass (Figure 21-11) has the property of separating white light into its various colors. This *prism* takes advantage of the variation of index of refraction with wavelength (see Table 20-1 and Problem 21-20). The longer wavelengths (toward the red end of the spectrum) with small n are bent less than the shorter wavelengths (toward the blue end). These differences, based on the interaction of light waves with atomic electrons (Section 20-3), give rise to the process of *dispersion*—i.e., a spreading of light according to color. By the

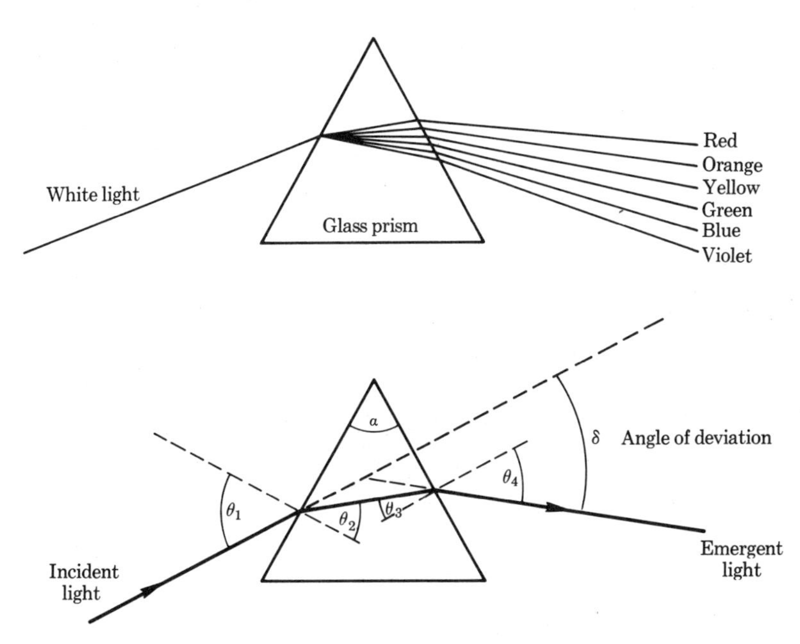

FIGURE 21-11

Refraction and dispersion of light by a prism, $\delta = \theta_1 - \theta_2 + \theta_4 - \theta_3$

application of Snell's law at each surface of the prism, we can obtain the *angle of deviation* δ using only simple geometry, trigonometry, and algebra. The angle δ depends on the incident angle θ_1 and the index of refraction. If the prism is rotated but the direction of incoming ray is fixed, a point is reached where the angle of deviation δ is a minimum.

Refraction by Curved Surfaces

In preparation for the study of optical instruments of various types, we now develop a set of relations for the refraction at a single curved surface and by a double surface or *lens*. We seek to trace the path of light in the medium and find the point where it crosses the axis. First, let a ray of parallel light strike a glass cylinder, as shown in Figure 21-12. The radial line OP is perpendicular to the surface, and serves as the normal. If we keep the distance s and the angles θ_i and θ_t small, we see that, approximately,

$$s = R\theta_i = q(\theta_i - \theta_t)$$

The sines appearing in Snell's law are practically the same as the angles, so the relative index of refraction is

$$n = \frac{n_t}{n_i} = \frac{\theta_t}{\theta_i}$$

Combining,

$$\frac{1}{q} = \frac{1}{R}\left(1 - \frac{1}{n}\right)$$

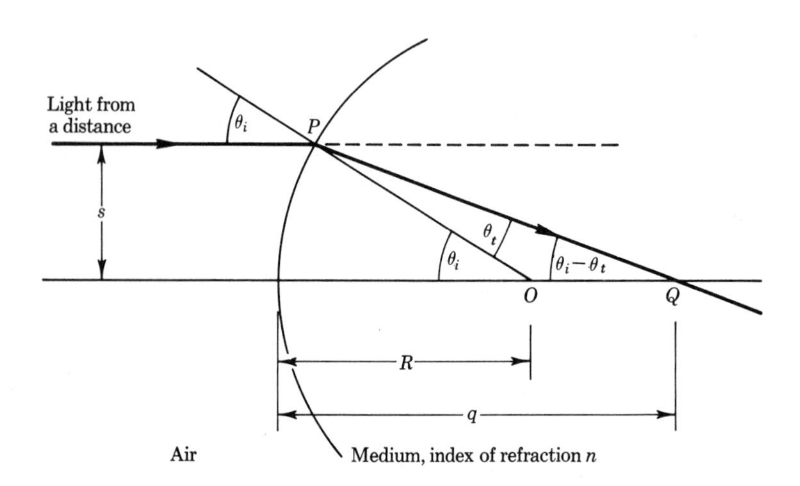

FIGURE 21-12

Refraction by a curved surface

For example, if $R = 0.2$ m and $n = 1.5$, as from air to glass,

$$\frac{1}{q} = \frac{1}{0.2}\left(1 - \frac{1}{1.5}\right) = \frac{1}{0.6}$$

or $q = 0.6$ m.

The refraction of light in a lens that has two surfaces can be analyzed by an extension of the above method. Two lens relations are sufficient for many applications.

The first is the focal length in terms of lens properties:

$$\frac{1}{f} = (n - 1)\left(\frac{1}{R_1} + \frac{1}{R_2}\right) \qquad \textit{(lensmaker's formula)}$$

The formula for f holds for a double-convex lens as written. If one of the surfaces is concave, a negative sign should be affixed to its radius. If one surface is plane, its R is infinity, with reciprocal zero, and the term is not present. For instance, let us find f for a convex–concave lens with $n = 1.5$, $R_1 = 0.2$, $R_2 = -0.25$:

$$\frac{1}{f} = (1.5 - 1)\left(\frac{1}{0.2} - \frac{1}{0.25}\right) = 0.5$$

or $f = 2$ m.

The second relation expresses the object and image distances q and p in terms of the focal length. A simple derivation of this expression can be performed that also yields the magnification, which is again the ratio of image height to object height. As in Figure 21-13, we draw a parallel ray (a) which goes through the focus. Then we draw (b) a ray that is undeflected through the center of the lens. Combining in (c), the image is beyond the focus and is inverted. From the geometry of (d), the magnification is

$$\frac{H}{h} = \frac{q - f}{f} = \frac{q}{p}$$

or

$$\frac{1}{p} + \frac{1}{q} = \frac{1}{f} \qquad \textit{(lens equation)}$$

We observe the fact that the lens equation is of the same form as that for a mirror. Let us apply it to a simple convex lens, such as the magnifying glass used for close reading, that has a focal length 2.5 cm, when the object is placed 3 cm from the lens. Then $1/q = 1/f - 1/p = 1/2.5 - 1/3 = \frac{1}{15}$ and $q = 15$ cm. The magnification is $q/p = \frac{15}{3} = 5$. If the object is placed between the focus and a convex lens, the image is virtual and enlarged, as shown in Figure 21-14. Formal solution of the lens equation for this case leads to a negative number for q. The *magnifying power* M is the ratio of

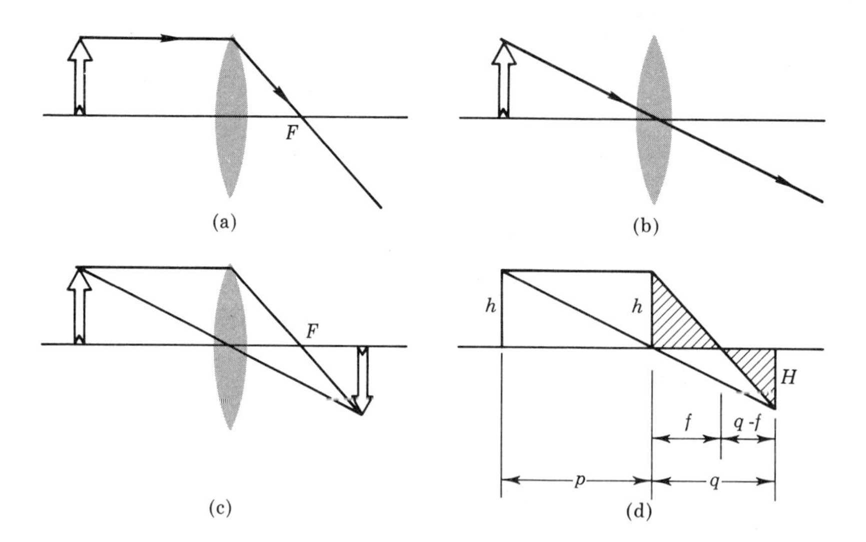

FIGURE 21-13

Ray tracing in lens

angles that the image and the object make with the eye. The image normally is adjusted to be at convenient distance from the eye, 25 cm, and M is about $25/f$.

We see from Figure 21-14 that the advantage of a lens for reading fine print or viewing small objects is thus confirmed. Before the microscope was fully developed, the Dutch merchant Van Leeuwenhoek was able to use an accurately ground single lens to observe sperm and blood cells, one-celled animals in stagnant water, and even some of the larger bacteria.

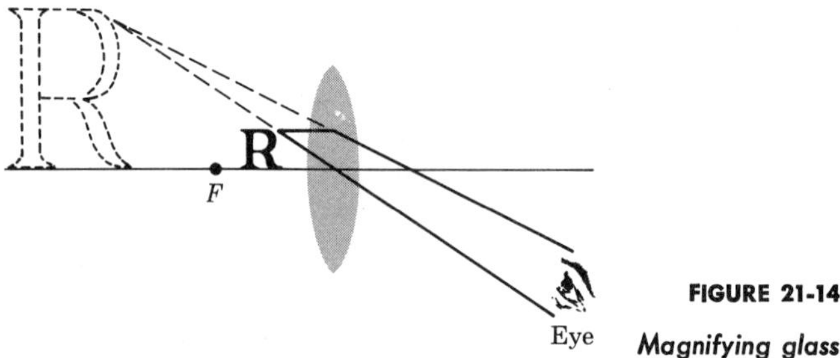

FIGURE 21-14

Magnifying glass

An object's image by a double concave lens is virtual—in the lens formula, the focal length is expressed as a negative number. For example, if the object is at $p = 5$ cm from a concave lens of focal length 4 cm, we write

$$\frac{1}{q} = \frac{1}{f} + \frac{1}{p} = \frac{1}{-4} + \frac{1}{5} = -\frac{1}{20}$$

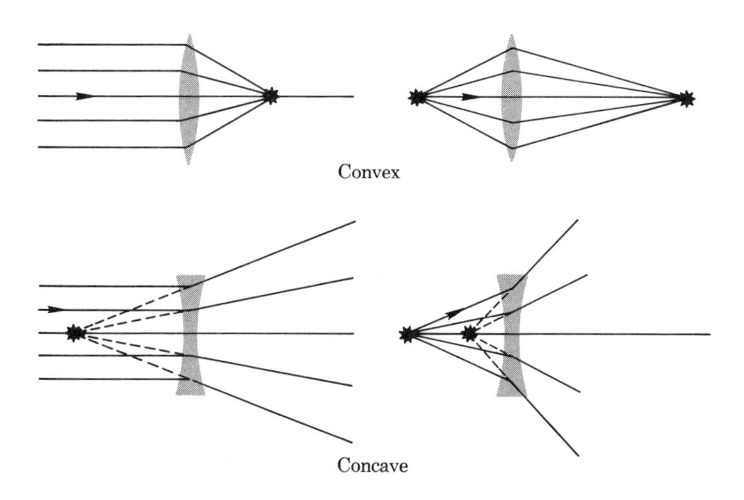

FIGURE 21-15

Lens images

and $q = -20$ cm, with the negative sign signifying the virtual nature of the image.

The behavior of light as it traverses lenses with two surfaces depends on the nature of the shape and the location of the object relative to the focus. Figure 21-15 shows the images of a bright spot or parallel light by various converging and diverging lenses.

21-3 OPTICAL INSTRUMENTS

The design of instruments to aid and extend the scope of the human eye is based on the geometrical ray treatment just discussed. The function of all devices such as spectacles, cameras, microscopes, and telescopes is to form appropriate images through refraction of rays from the objects.

The Camera

As we all know, this device permits the recording of a permanent image. Light from an object is focused by a convex lens or combination of lenses with focal length ordinarily about 5 cm (see Figure 21-16) on a film emulsion, in which chemical reactions take place in degree dependent on the intensity and color. We move the lens back and forth to obtain the clearest focus of the light on the photographic plate or film. The shutter admits light for a specified length of time and the size of the aperture is adjusted to limit the amount of light and prevent under- or overexposure. If the adjustable aperture diameter is D, and the focal length of the lens is f, the "F-number" is given by the ratio $F = f/D$. Now, the amount of light received by the film in a time t is clearly proportional to both the area of the opening and to the time—i.e., to $(\pi D^2/4)t$. The exposure thus varies as t/F^2. For constant exposure, if we increase F by a factor of 2, the time must be increased by a factor of 4. For example, a "shutter speed" of $\frac{1}{500}$ sec at

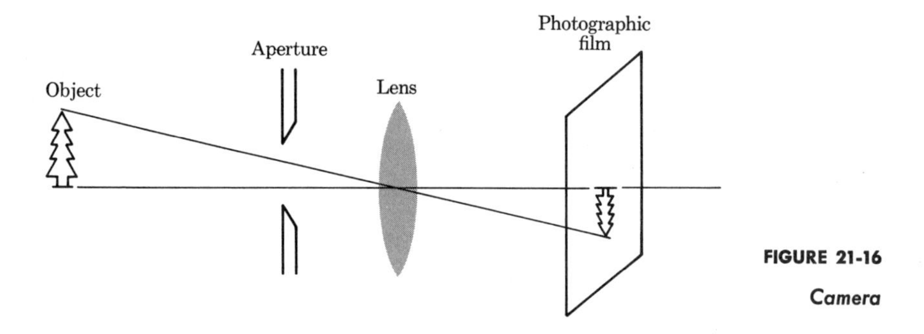

FIGURE 21-16

Camera

F-8 is comparable to $\frac{1}{125}$ sec at F-16. The role of the flashbulb is obvious: to assure adequate illumination of the object during the time the shutter is open. The "film speed" refers to the sensitivity of the film to light intensity, which in turn is related to the detail desired.

Astronomical Telescopes

The magnification of very distant objects, such as the Moon, planets, or stars, is achieved by two lenses. The first (*objective*) creates a real image I_1 in front of the second (*eyepiece*), as shown in Figure 21-17(a). The real image serves as an object for the eyepiece, which gives a large inverted virtual image I_2, as shown in Figure 21-17(b). The magnifying power M, as in the simple lens, is the ratio of the angles that the final image and original object make with the eye. To find these, let us consider (Figure 21-18) the focusing of light from the Moon. The first image of diameter d_1 is produced at a distance f_0 from the objective, which is at a distance approximately f_e from the eyepiece. The final image is of diameter d_2. From

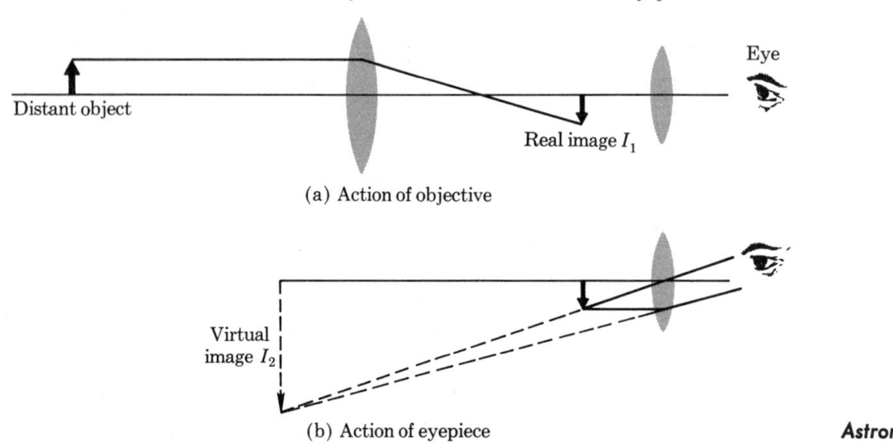

FIGURE 21-17

Astronomical telescope

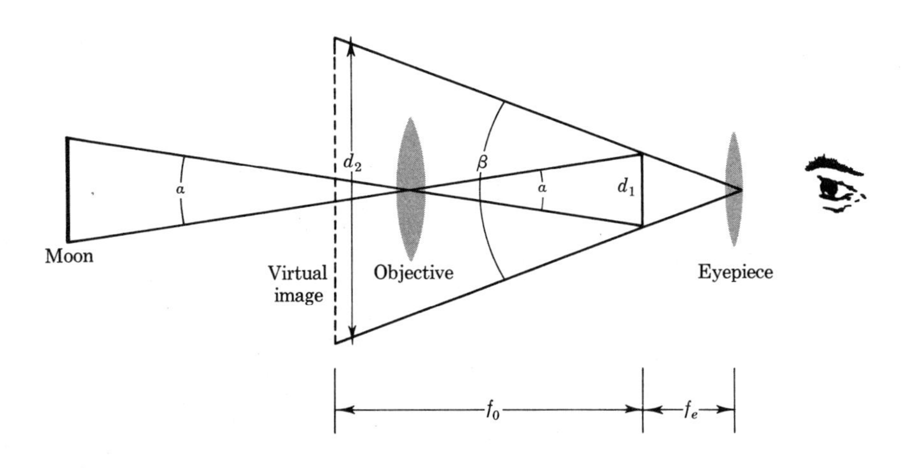

FIGURE 21-18

Angle relations and magnifying power of telescope

the triangles, $d_1 = \alpha f_0$ and also $d_1 = \beta f_e$, then $M = \beta/\alpha = f_0/f_e$. For example, if the focal lengths of the objective and eyepiece are respectively 1.00 m and 0.02 m, then the magnifying power is 50.

The Microscope

There is a great similarity in construction between the telescope and the microscope, as seen in Figure 21-19. If we place a very small object near the focal point of the objective, a real image I_1 is formed in front of the eyepiece. This in turn serves as object for the eyepiece, and a large virtual image is formed at I_2. The magnifying power M is the product of the magnifying powers of the two lenses M_0M_e. If the distances between lenses is x and the distance from eye to final image is X, then $M_0 = x/f_0$ and $M_e = X/f_e$. For example, if x is 20 cm (a typical distance) and X is a

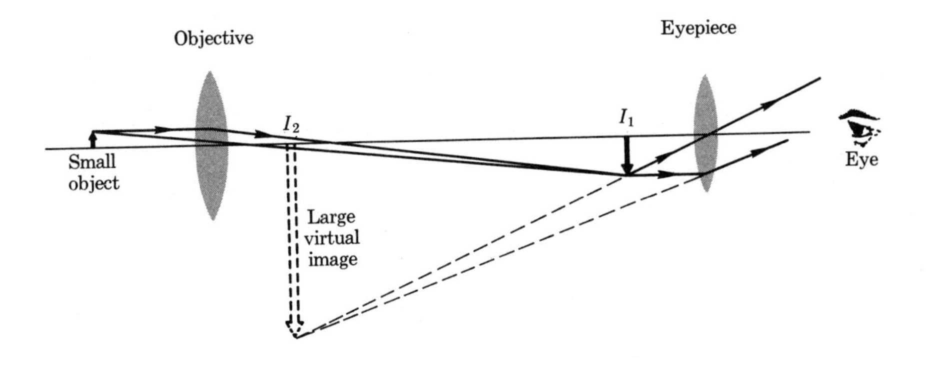

FIGURE 21-19

Microscope

normal 25 cm, and the focal lengths are $f_0 = 0.5$ cm and $f_e = 1.0$ cm, the magnifying power is $(20/0.5)(25/1.0) = 1000$.

Measurement of Astronomical Distances

The distance from the Earth to celestial bodies obviously cannot be measured by meter sticks. The only tools available are optical in nature. One is an extension of our visual method in which we compare the results of each eye's estimate. Suppose we hold a pencil about 1 ft from our face, and look at it with one eye closed and then with the other eye closed. The pencil seems to shift sideways an inch or so. If we move the pencil out farther from us, however, the shift is considerably smaller and at great distances imperceptible. This effect is given the name *parallax*, and the shift is a *parallactic displacement*. Astronomical distances, such as to stars, can be measured by comparing photographs of the star taken when the Earth is at each end of its orbit about the Sun (Figure 21-20). The angles from the vertical are measured at times six months apart to be α and α';

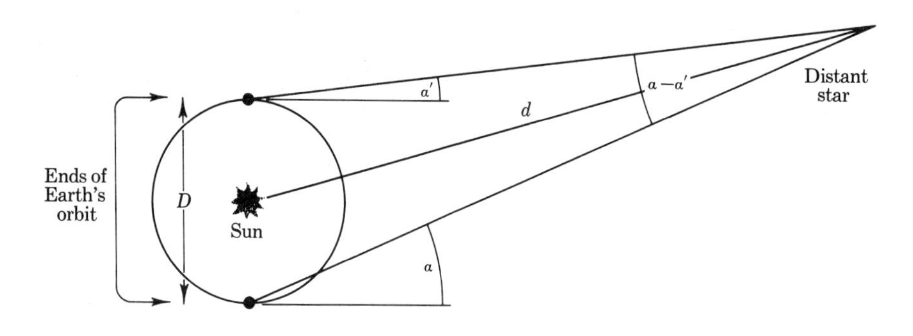

FIGURE 21-20

Use of parallax to measure astronomical distances

their difference is $\Delta\alpha = \alpha - \alpha'$. The desired distance d is then approximately

$$d = \frac{D}{\alpha - \alpha'}$$

The first such measurement was made by Bessel around 1840. Let us find how large is the angular difference $\Delta\alpha$ for the nearest star, Alpha Centauri, 4.3 light years away. The Earth's orbit has a diameter of about 16.6 light *minutes*. Since 1 year $= 5.26 \times 10^5$ min and 1 rad $= 57.3$ deg, the angle in degrees is

$$\Delta\alpha = \frac{(16.6)(57.3)}{(4.3)(5.26 \times 10^5)} = 4.2 \times 10^{-4}$$

which is less than 1.5 seconds of arc.

21-4 INTERFERENCE AND DIFFRACTION

The process called *interference* of light consists of the addition or cancellation of waves because of differences in phase. To study this phenomenon, we must shift over from a geometrical-ray viewpoint to the wave approach. In the following discussion, the term *diffraction* refers to an apparent bending process, while *interference* refers to the combination of light. There is really no fine distinction between the terms, since they are merely different aspects of wave behavior as described by the Huygens principle.

Young's Double-slit Experiment

Let us picture a simple experiment similar to that first performed by Young* in 1802. Three opaque screens labeled A, B, and C are lined up parallel to each other, as in Figure 21-21(a). A narrow slit is cut in A, two slits in B, and none in C. Sunlight is allowed to pass through the system. Let us *predict* the transmission of light by the geometrical optics we have learned. Since neither of the slits in B line up with the slit in A, we would expect no light to get to C. Were we to carry out the experiment, however, we would find C displaying an interesting pattern of dark and bright *lines*, as sketched in Figure 21-21(b). Being certain that screen B is really opaque, we conclude first that light has apparently bent around corners, and second,

* Thomas Young (1773–1829) was a medical doctor whose name is also attached to the modulus of elasticity (Section 13-2). Newton had been convinced that light consisted of particles ("corpuscles"). Young revived the wave theory by his experiment.

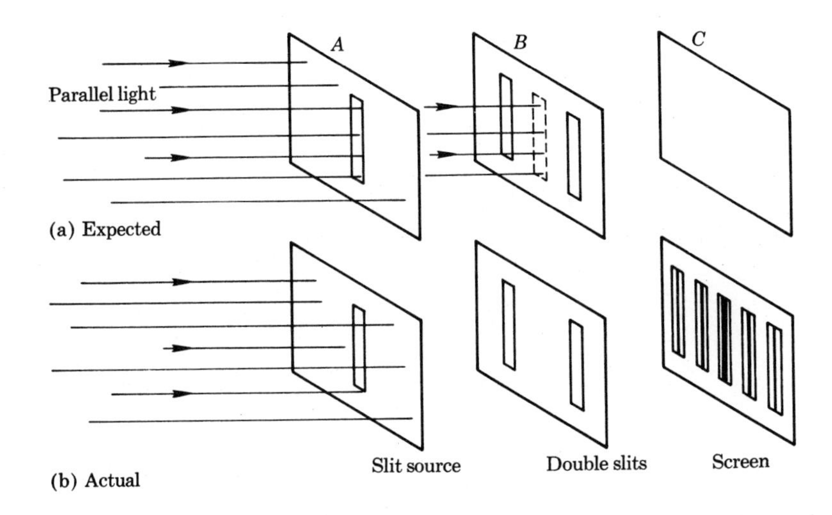

FIGURE 21-21

Double-slit interference experiment

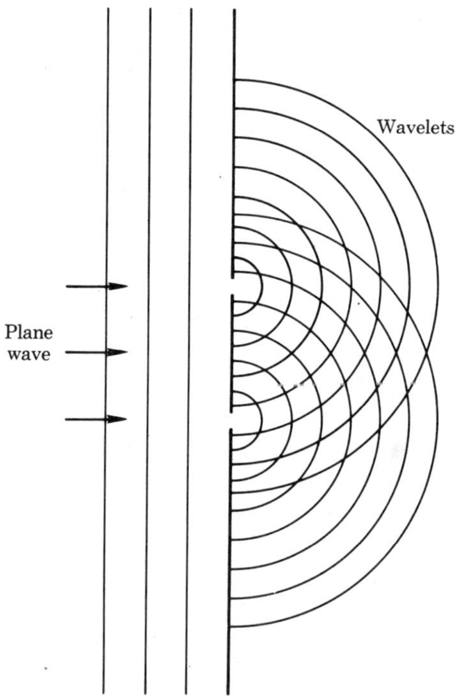

FIGURE 21-22

Application of Huygens's principle to Young's experiment

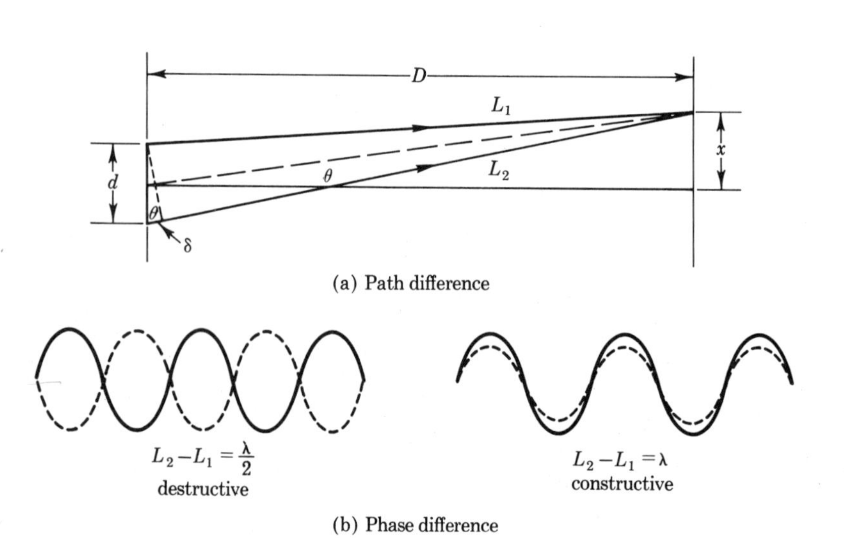

FIGURE 21-23

Interference in double-slit experiment

that it has combined in an unusual way. The explanation becomes simple if we apply Huygens's principle that each point on the plane wave sends out spherical wavelets. Figure 21-22 shows the pattern: the radii of circles represent the distances the light has traveled and thus the time since it started at the slit. Where circles of the same radius intersect, waves are in phase and add; elsewhere, phase differences cause waves to cancel partially or totally. We thus treat the two slits in B as new sources of essentially zero width. (The line of points that each slit forms gives by superposition a cylindrical wave.) Rays coming from the two slits to a point on screen C in general take different path lengths L_1 and L_2—Figure 21-23(a). The light may arrive there out of phase, in phase, or partly out of phase. As one moves along the screen C, there will be regions of reinforcement and cancellation of fields and thus of intensity. For example, suppose $L_2 - L_1$ is exactly *one-half* wavelength, $\frac{1}{2}\lambda$. On arrival, the two waves are as shown in Figure 21-23(b): the amplitudes cancel everywhere, giving zero light intensity. Note that the same result would occur if the path difference were $\frac{3}{2}\lambda$, $\frac{5}{2}\lambda$, etc. On the other hand, if $L_2 - L_1$ is exactly *one* wavelength, λ, they add together, giving twice the amplitude and 4 times the intensity of the original light. Addition also takes place for path differences 2λ, 3λ, etc. The interference for these two cases is called *destructive* and *constructive*, respectively. It is clear that for path differences that are some other fraction of λ there will be partial cancellation. The light pattern on screen C thus consists of the series of dark and bright lines. A little geometry tells us just where to find these maxima and minima of intensity. From Figure 21-23(a),

$$\frac{\delta}{d} \cong \sin\theta$$

where it is assumed that $\lambda \ll d$ and $d \ll D$. Let $\delta = n\lambda$, where n is any number; then,

$$n\lambda = d \sin\theta$$

For a maximum, n is an integer 0, 1, 2, etc.; while for a minimum, n is a half-integer $\frac{1}{2}$, $\frac{3}{2}$, $\frac{5}{2}$, etc. However, $x/D = \tan\theta$; and if θ is small, $\sin\theta \cong \tan\theta \cong \theta$; so

$$x \cong \frac{D\delta}{d} = \frac{n\lambda D}{d}$$

Let us compute the distance x at which the first minimum ($n = \frac{1}{2}$) occurs for a case in which $\lambda = 589\ \text{m}\mu$, one of the yellow lines from sodium, where d is 2 mm (0.002 meter) and D is 1.5 m; then,

$$x = \frac{n\lambda D}{d} = \frac{(\frac{1}{2})(589 \times 10^{-9})(1.5)}{0.002}$$

$$= 2.2 \times 10^{-4}\ \text{m} = 0.22\ \text{mm}$$

Quarter-wavelength Films

Advantage may be taken of interference phenomena to cut down on reflection from lenses. One coats a lens with a transparent material of index intermediate to that of air (1) and glass (1.5). The layer is made $\frac{1}{4}$-wavelength thick *optically*, $nd = \lambda/4$, as shown in Figure 21-24. Let us allow light of one frequency to fall normally on the film, but displace the rays so we can see the various reflections and refractions. Ray 2 has traveled exactly one-half wavelength farther, and is out of phase with Ray 1 and tends to cancel. Pairs 3 and 4 also are out of phase with each other, and

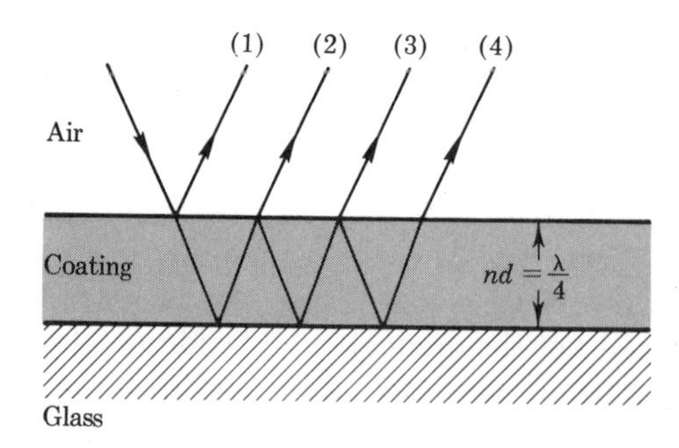

FIGURE 21-24

Rays in a nonreflecting layer on glass

so on. Some light is always reflected, since successive reflected rays are weaker and do not completely nullify their neighbors. The reflection is greatly reduced, however. Also, we note that for light of nearby frequencies the effect will not be quite so pronounced.

The Interferometer

Extremely accurate measurements of length are possible by use of the *interferometer*, one type of which (Fabry–Perot) is sketched in Figure 21-25.

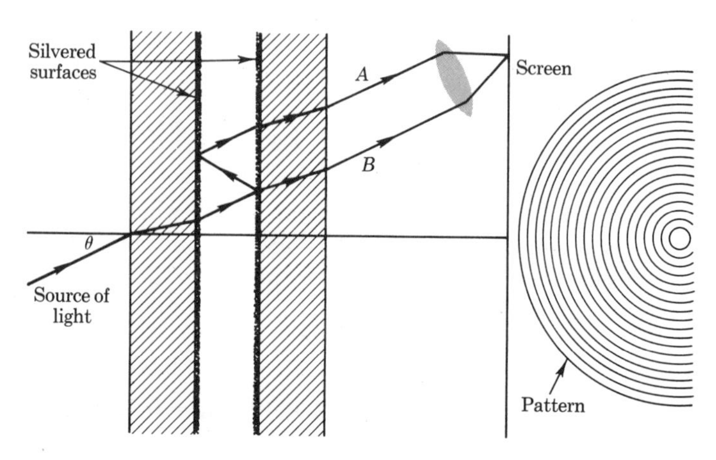

FIGURE 21-25

Interferometer

Parallel light is allowed to fall on a pair of partly silvered glass plates with an air gap between them. The transmitted light is the result of multiple reflections, only one of which is shown—for one incident ray. If the gap is width d, then ray B at angle θ has traveled a distance $2d \cos \theta$ farther than ray A. When this distance is equal to an even (or odd) number of half wavelengths, waves add to give a bright (or dark) spot. Thus a set of light and dark rings appears on the screen. The bright rings are quite sharp, permitting accurate comparisons of two wavelengths. By watching the locations of the rings shift as the gap is changed, very precise distance measurements are possible. This device served as the basis for the measurements of "ether drift" by Michelson and Morley (discussed in Section 2-5), and for comparative measurements of wavelengths of light. More recently, very accurate measurements of the distance to the Moon have been achieved using the laser beam reflector placed on the Moon by astronauts.

21-5 SINGLE-SLIT DIFFRACTION PATTERN—RESOLVING POWER

The two slits in Young's experiment were assumed to be extremely narrow compared with the wavelength of light, so that light emerged from points. We now ask: What happens if the slits are of some finite width? In this case, light *from different parts* of the slit can diffract and interfere. In fact, a single slit gives a diffraction pattern. We can verify this by the most elementary experiment. If one looks toward the light of a window through the narrow slit between two fingers held a few inches from his eye, he will see a series of dark and light lines. Let us analyze this simpler situation, and then see what effect the result has on the double-slit experiment.

Picture a slit of width a, with light from a distant source (or light from a point that is rendered parallel by a lens)—Figure 21-26. Consider light rays that come from points at the center and edge of the slit. The light

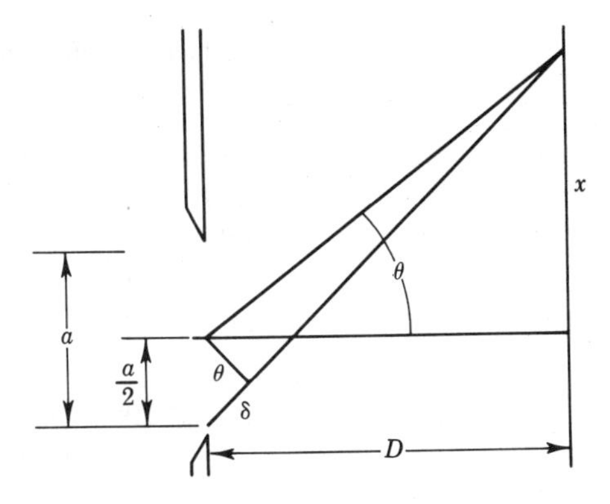

FIGURE 21-26

Interference between center and edge of slit

cancels if δ is $\lambda/2$. Light from points next to them will also cancel, and so on across the aperture. We can borrow the results from Young's experiment to find the minima, namely,

$$n\lambda = \frac{a}{2} \sin \theta$$

where n is a half-integer $\frac{1}{2}$, $\frac{3}{2}$, $\frac{5}{2}$, etc.

In order to find the exact pattern of light arriving at the screen, we must add up the field contributions from different parts of the aperture, taking proper account of the phase differences. A plot of the relative intensity is given in Figure 21-27, with $u = \pi a\theta/\lambda$. The angles shown on the graph refer to a slit whose width is 100 times the wavelength. Maxima of the light intensity occur near (but not at) values $u = 3\pi/2, 5\pi/2$, etc. Along the screen, at distance D from the slits, the positions of interest are

$$x = \frac{m\lambda D}{2a}$$

where $m \cong 0, 3, 5$, etc. for maxima, and 2, 4, 6, etc. for minima. (Note that $m = 1$ is missing.) The height of the first off-center maximum is only 4%

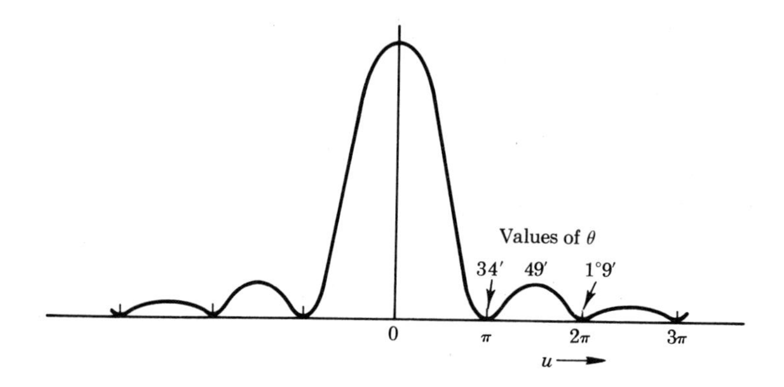

FIGURE 21-27
Single-slit diffraction pattern, $a = 100\lambda$

of the main peak, showing that the intensity falls off rapidly. To get an idea of the scale of this pattern compared to that of the double slit, we consider a light of 500 mμ passing through a pair of slits of width $a = 0.1$ mm, separated by $d = 1$ mm with the screen 1 m away. Young's interference pattern reaches its first minimum at

$$x = \frac{\lambda D}{d} = \frac{(500 \times 10^{-9})(1)}{0.001} = 0.5 \times 10^{-3} \quad \text{or} \quad 0.5 \text{ mm}$$

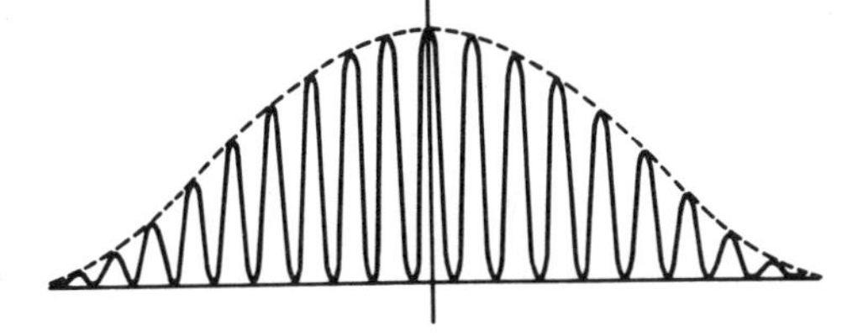

FIGURE 21-28

Details of double-slit interference pattern

The so-called single-slit diffraction pattern reaches its first minimum at

$$x = \frac{\lambda D}{a} = \frac{(500 \times 10^{-9})(1)}{0.0001} = 5 \times 10^{-3} \quad \text{or} \quad 5 \text{ mm}$$

This is 10 times as far out, which means that there are many interference peaks within a broad envelope of the diffraction pattern. Figure 21-28 shows this "modulation" effect.

Resolving Power of Lenses

The image of a point source of light, whether it be a distant star as viewed by a telescope or a bacterium as seen under a microscope, has a diffraction pattern. The ability to distinguish between two such objects by use of lenses depends on the distance between their diffraction peaks. The sum of intensities from two such sources is shown as dotted curves in Figure 21-29 for three separations—far, close, and very close. We see how the images become indistinguishable. *Rayleigh's criterion* states that two objects are just resolved if the maximum of one pattern falls onto the minimum of the other. The distance from the central peak of the single-slit

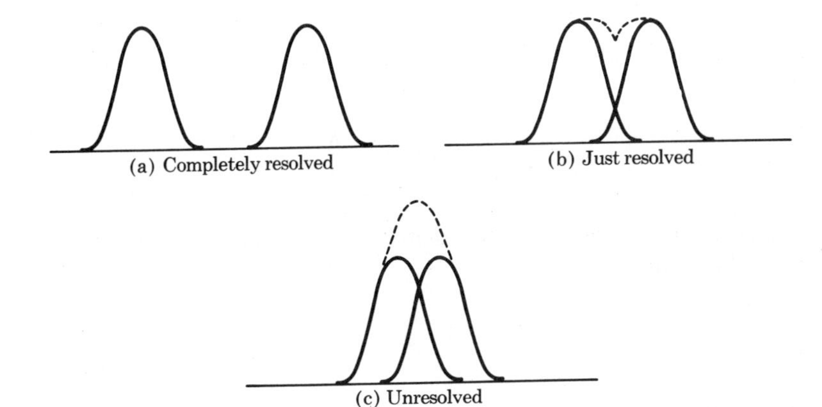

FIGURE 21-29

Rayleigh's criterion for resolution of images

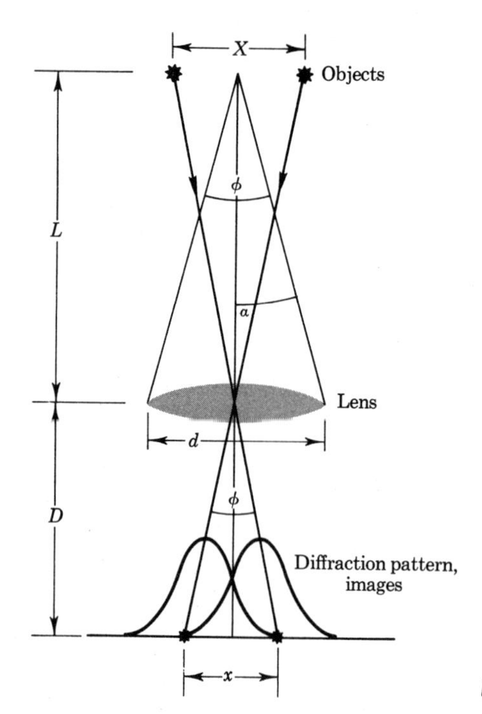

FIGURE 21-30
Resolution of distant objects

diffraction pattern out to its first minimum (where by Rayleigh's criterion the other peak falls) is $x = \lambda D/d$, as shown in a previous paragraph.

Let us look at a pair of distant objects, such as two stars, by means of a cylindrical lens, which acts like a slit of width d. From the similar triangles of Figure 21-30,

$$\phi = \frac{X}{L} = \frac{x}{D} \quad \text{and} \quad \sin\alpha = \frac{d}{2L}$$

Eliminating D and L, we obtain for the just-resolved linear separation of the stars

$$X = \frac{\lambda}{2\sin\alpha} \quad \text{or} \quad \phi = \frac{\lambda}{d}$$

For a circular aperture and lens of diameter d, the angular separation can be shown to be instead

$$\phi = 1.22\frac{\lambda}{d}$$

The factor 1.22 comes from the geometrical form of the opening.

For example, let us compute the separation of double stars 5000 light years away that can just be distinguished by a 200-in. telescope, assuming

$\lambda = 555\ m\mu$. Now,

$$d = (200\ \text{in.})(0.0254\ \text{m/in.}) = 5.08\ \text{m}$$

and

$$\frac{\lambda}{d} = \frac{5.55 \times 10^{-7}\ \text{m}}{5.08\ \text{m}} = 1.09 \times 10^{-7}$$

Then,

$$\phi = (1.22)(1.09 \times 10^{-7}) = 1.33 \times 10^{-7}\ \text{rad}$$

The linear separation of the resolved stars is then

$$X = (1.33 \times 10^{-7})(5000) \quad \text{or} \quad 6.65 \times 10^{-4}\ \text{light years}$$

which seems small but is actually four billion miles, i.e.,

$$(6.65 \times 10^{-4})(9.48 \times 10^{15}) = 6.3 \times 10^{12}\ \text{m}$$

21-6 OPTICAL SPECTROGRAPHS

Measurements of the frequency of visible light are made in the laboratory by devices known as *spectrographs* (or *spectroscopes*). A set of definite frequencies are observed, unique to each element in the periodic table. The total light, decomposed into a spectrum of colored bands or lines by the spectrograph, may be either viewed directly or photographed.

The basic component of the *prism spectrograph* is the triangular prism of quartz or glass. As shown in Figure 21-31, light from a distant object—such as the Sun or a star, or from an incandescent sample in an electric arc—experiences refraction and dispersion into colors by the prism. The refracted light appears on the screen or film as a background display of color, along with some intense lines, which differ in intensity according to how much light of each frequency is emitted.

The *diffraction-grating spectrograph* consists of sheets of glass or metal

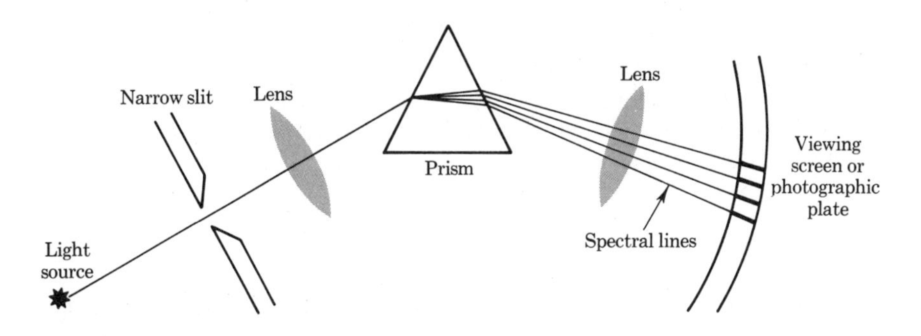

FIGURE 21-31

Prism spectrograph

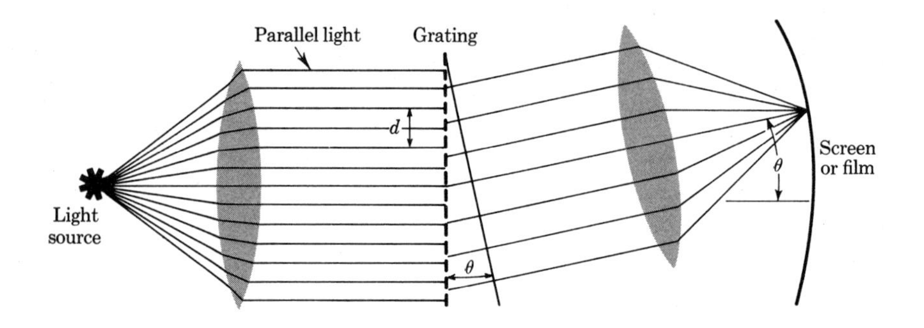

FIGURE 21-32

Diffraction grating

with thousands of closely spaced parallel lines cut in its surface. Light transmitted by the glass or reflected from the metal acting as a mirror exhibits diffraction patterns similar to those from a single slit. We may derive a relation for the location of maxima for the transmission type. Light of one wavelength is combined by means of a lens to a line on a screen (Figure 21-32). If the separation of slits is d, waves of light emerging at an angle θ from successive slits all differ in path by $d \sin \theta$, just as in Young's experiment. Constructive interference takes place when the path differences are an integer multiple of a wavelength:

$$d \sin \theta = n\lambda, \qquad n = 0, 1, 2, 3, \text{etc.}$$

This is the same interference condition as for a double slit. The integer n is called the *order* of the pattern.

The amplitude transmitted by a single-slit width a was found in Section 21-5. A grating consists of a large number N of such slits, and superposition of amplitudes is responsible for the overall effect. One can show that there is a very intense central maximum, so that images of light from the source having different wavelengths are distinctly separated on the viewing screen.

The *dispersion* of the grating, defined by

$$D = \frac{\Delta\theta}{\Delta\lambda}$$

is a measure of the ability to resolve wavelengths, being the change in angle of emergence per unit change in wavelength. In terms of the order of the pattern n, the line spacing d, and the angle θ,

$$D = \frac{n}{d \cos \theta}$$

To illustrate the effectiveness of a grating for spectral studies, let us calculate the location and separation of the two components of the sodium "doublet" at wavelengths 589.0 mμ and 589.6 mμ. We assume a grating with 500,000 lines per meter so that $d = 2 \times 10^{-6}$ m. Then, for first order on the shorter line,

$$\sin\theta = \frac{589 \times 10^{-9}}{2 \times 10^{-6}} = 0.294 \quad \text{and} \quad \theta \cong 17 \text{ deg}$$

The dispersion, with $\cos\theta = 0.956$, is

$$D = \frac{1}{(2 \times 10^{-6})(0.956)} = 5.23 \times 10^{5} \text{ rad/m}$$

For $\Delta\lambda = 589.6 - 589.0 = 0.6$ mμ $= 0.6 \times 10^{-9}$ m, the angle of separation is

$$\Delta\theta \cong D\Delta\lambda = (5.23 \times 10^{5})(0.6 \times 10^{-9}) = 3.1 \times 10^{-4}$$

On a screen 1 m away, the separation in space is $(1.0 \text{ m})(3.1 \times 10^{-4}) = 0.0003$ m or 0.3 mm.

21-7 X-RAY DIFFRACTION

A crystalline solid has a regularity of atom spacing that resembles a grating. X rays with wavelengths in the range of 10^{-10} m are scattered from the atomic centers, which have spacings also in that range. Experiments prompted by von Laue verified that such atoms acted as a grating. X rays were passed through a sample with a radiation-sensitive screen behind it, as in Figure 21-33, and a regular pattern of spots was observed. An alternate method developed by Sir William Bragg involves reflections from the crystal layers (Figure 21-34). Planes of atoms—say, Na and Cl, as in table salt—are separated by a distance d. Waves reflected from the second level of atoms interfere with those from the first level. When the total path difference $2d \sin\theta$ is equal to an integer number of wavelengths, a maximum intensity is found. This condition

$$2d \sin\theta = n\lambda$$

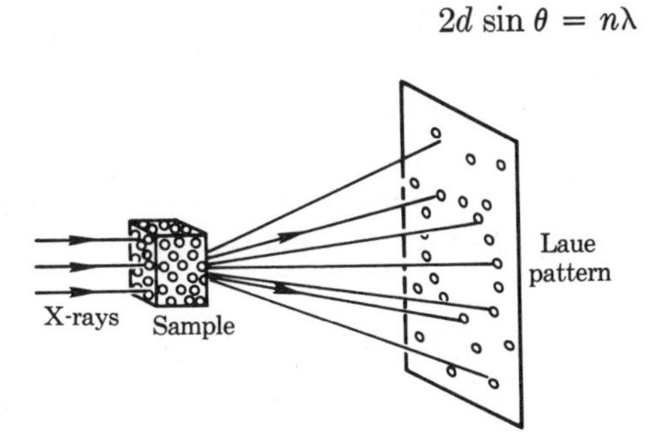

FIGURE 21-33

X-ray transmission through a crystal

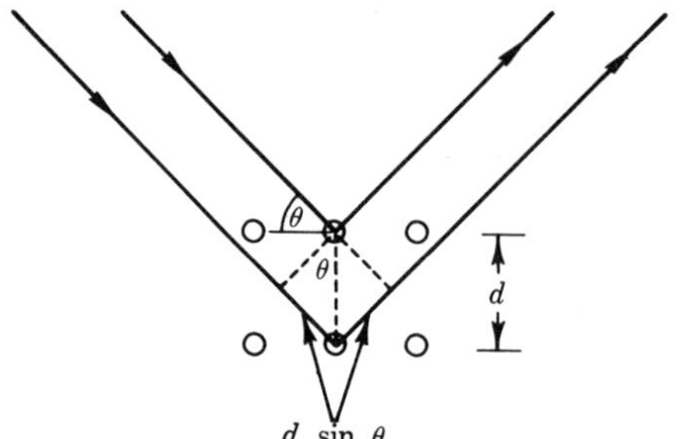

FIGURE 21-34

X-ray diffraction from a crystal plane

is called *Bragg's law*. Using X-ray spectrometers, with rays of one known wavelength, the spacing of atoms may be found, along with many other details of the structure of solids and location of particular atoms. From the knowledge of atom spacing d, we can compute Avogadro's number, as discussed in Section 4-5. It is not necessary to have a single smooth plane or a single crystal for the study of X-ray diffraction. A powder composed of randomly arranged crystals will suffice as a target.

X-ray diffraction serves as only one of the processes by which the detailed structure of solids and liquids is found. Electron-diffraction and neutron-diffraction devices for the study of submicroscopic phenomena are based on the discovery that particles such as electrons or neutrons in motion constitute matter waves, mentioned in Section 12-5. In the next chapter, we shall discuss the historical background and fundamentals of quantum mechanics, which encompasses the particle theory of light and the wave theory of matter.

PROBLEMS

21-1 Show that one needs a mirror that is as tall as one-half his height in order to view his whole body.

21-2 Mirrors on opposite walls, as in a barber shop, give multiple images that seem to recede into the distance. Investigate this effect by tracing rays from object to mirrors to the eye.

21-3 A light source is placed at 6 cm from the surface of a concave mirror of radius 4 cm. Where is the image located?

21-4 Find the nature, positions, and heights of images for a concave spherical mirror of radius 3 m when objects are placed at a distance from the surface of (a) 2 m, (b) 1 m.

21-5* Suppose that the headlight of a car is a spherical mirror of radius 0.12 m and diameter 0.10 m (see Figure P21-5). Where should a bulb be located in order to give a beam that is of width 20 m at a distance 50 m from the headlight?

21-6* Compute the horizontal displacement of a ray of light incident at 30 deg from the normal as it passes through a windowpane of thickness $\frac{1}{8}$ in., index of refraction 1.5.

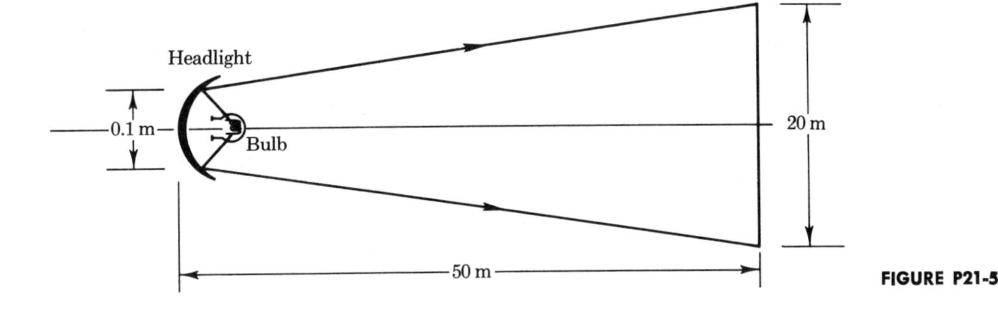

FIGURE P21-5

21-7 Suppose we are indulging in the easy sport of shooting fish in a barrel, as shown in Figure P21-7. The container of water ($n = \frac{4}{3}$) is a cube 4 ft on a side, half-full, with the fish actually 1 ft below the center of the tank. Find the apparent depth of the fish below the surface (a) using the approximation that the sine and tangent are equal, (b)* more accurately, using trial and error, graphical solution, or any other method.

21-8 The Moon appears to have risen above the horizon before it actually does. Explain, using diagrams.

21-9 (a) Draw a sketch of light rays for sodium light of frequency 589 mμ passing through a glass prism, $n = 1.52$, $\alpha = 30$ deg, when the angle of incidence is 45 deg. (b)* Show that the angle of deviations is $\delta = \theta_1 - \theta_2 + \theta_4 - \theta_3$ and calculate its value.

21-10 The minimum angle of deviation of light in a prism of angle α, index of refraction n, appears in the relation

$$n \sin \frac{\alpha}{2} = \sin \left(\frac{\alpha + \delta}{2} \right)$$

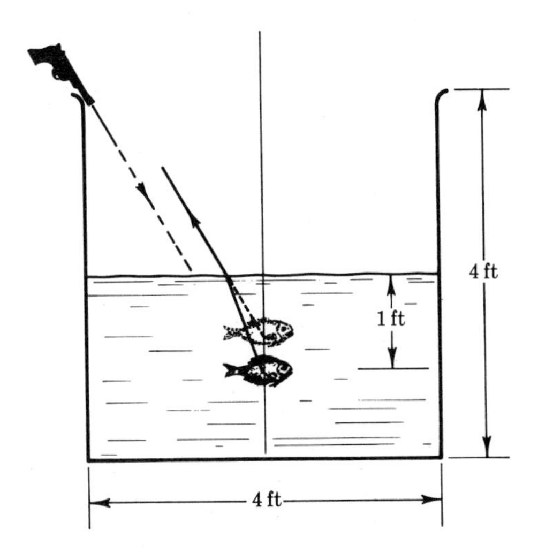

FIGURE P21-7

Suppose that δ is 30 deg for orange light of wavelength 606 $m\mu$ on passing through a prism of angle 60 deg. What is the index of refraction of the prism?

21-11 A "burning glass" is a double-convex lens with equal radii of curvature 10 cm, index of refraction 1.5. Find its focal length.

21-12 Sunlight is brought to a point 5 cm from a certain lens. Where would the image of a lightbulb 2 m away be located? How much magnification would be obtained?

21-13 A special lens is constructed with one surface cylindrical and the others as planes (see Figure P21-13). Continue the ray shown completely through the lens to the point where the ray crosses the axis, using geometry and Snell's law.

21-14* Show, by use of geometry only, that light reflected from the back surface of a spherical raindrop (with refraction on entering and leaving) experiences an apparent angle of reflection $\gamma = 2\theta_t - \theta_i$ (see Figure P21-14).

21-15 What radius R of a spherical surface of material of index $n = 1.4$ is needed to bring parallel light to a focus 15 cm from the surface?

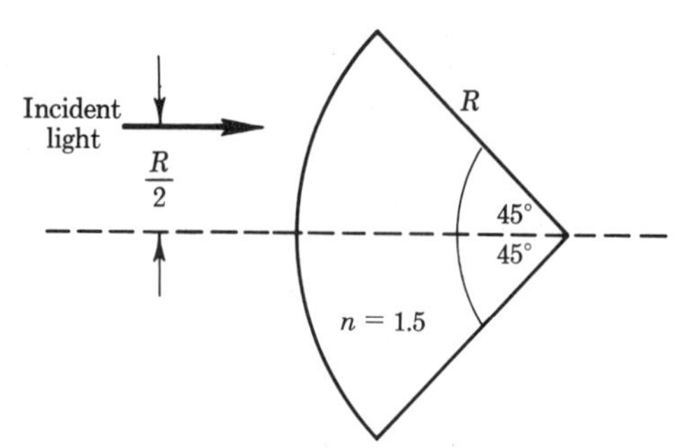

FIGURE P21-13

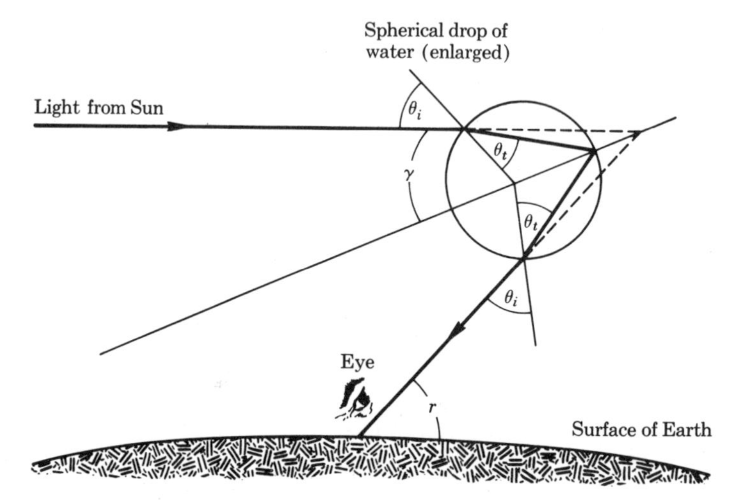

FIGURE P21-14

21-16 The glass of a thermometer tube acts as a cylindrical lens for the liquid inside, such that the magnification of the column of mercury is equal to the index of refraction of the glass. For a tube with $n = \frac{5}{3}$ and radius R, how large does the column of radius $\frac{2}{5}R$ appear?

21-17 A projector for "35mm" slides (2.3 × 3.4 cm) has a lens with focal length 5 cm. The distance between the slide and lens is 5.05 cm. Show why the slides must be put in upside down. How far away should the screen be placed? What is the smallest size of screen that would permit viewing of slides either vertically or horizontally?

21-18 Where in a spherical goldfish bowl full of water is a small tropical fish able to hide without being visible to an observer above the bowl?

21-19 Compute the focal length of a convex lens of flint glass, $n = 1.9$, with radii 0.05 m and 0.08 m.

21-20 A pair of lenses in contact are constructed of crown glass and flint glass respectively. The indexes of refraction of these materials vary with wavelength, as shown below:

λ	n	
(mμ)	*Crown*	*Flint*
434	1.528	1.675
486	1.523	1.664
589	1.517	1.650
656	1.514	1.644
768	1.511	1.638

Find the index of refraction of each substance for green light of 555 mμ.

21-21 A camera has a focal length of 2 cm. What is the diameter of the aperture at F-12? A picture is taken at $\frac{1}{125}$ sec and then the camera is reset to F-8. To get the same light exposure as before, what new shutter speed should be used?

21-22 Calculate the magnifying power of a telescope that has the following focal lengths: eyepiece 0.025 m, objective 1.25 m. When the telescope is trained on the Moon, which occupies a 0.52-deg arc, what arc does the Moon appear to occupy?

21-23 The star Sirius appears to shift by a maximum of 0.74 seconds of arc every six months. How far away, in light years, is the star?

21-24 Find the position on a screen of the second minimum of the double-slit diffraction pattern for two pinholes 4 mm apart in a piece of cardboard, for light of wavelength 600 mμ, with the screen 2.5 cm away.

21-25 Parallel light from a light source using the element mercury (λ = 254 mμ) passes through a slit formed by the edges of two razor blades separated by 0.15 mm. On a screen 0.2 m away, where would the first minimum of the interference pattern be located?

21-26 Suppose that astronauts set up on the Moon, 239,000 mi away, a pair of bright beacon lights of principal wavelength 555 mμ. How many feet apart must the lights be located in order to be resolved by a 200-in. telescope on Earth?

21-27 Calculate the angle for the first-order diffraction-grating pattern of green light $\lambda = 520\ m\mu$, when there are 8×10^5 lines per meter.

21-28 Find the dispersion of the diffraction grating in Problem 21-26.

21-29 An X-ray line is found at first order at an angle of 3 deg, when the target is a copper crystal with lattice spacing $d = 3.6 \times 10^{-10}$ m. What wavelength does the line have? By what factor is this larger or smaller than visible light, $\lambda = 555\ m\mu$?

QUANTA, WAVES, AND PARTICLES

22

A well-recognized goal of science is to represent natural processes in an orderly form. Some of the characteristics of *order* are completeness, simplicity, unity, and symmetry. Indeed, much of the progress in science has been based on hypotheses that were suggested by the belief or hope that the ultimate description of the universe had these properties. The wave theory of matter is an excellent example of such a far-reaching discovery. It originated in a hopeful guess in 1924 by de Broglie, a French scientist. The dual nature of light as waves and particles served as a background for his hypothesis.

In this chapter, we shall review some of the evidence used to arrive at this strange theory and then investigate the changes in fundamental concepts that are required by the new *quantum mechanics.*

22-1 LINE SPECTRA—BOHR'S MODEL OF THE HYDROGEN ATOM

For many years in the latter part of the nineteenth century, physicists studied the character of light from the Sun and from gases heated in electrical discharges, using the optical spectrographs based on the separation of light into colors by a glass prism. Well-defined bright lines of colors were observed within the continuous rainbow-like spectrum. Figure 22-1 shows one of the many series of lines from hydrogen gas. Various formulas

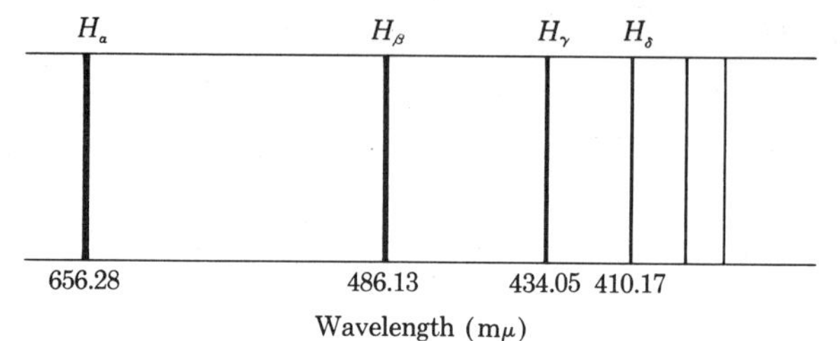

FIGURE 22-1

Spectral lines from hydrogen, Balmer series

relating the wavelengths of the different lines were devised by Rydberg, Balmer, and others. These spectroscopists observed that the wavelengths fit the formula

$$\frac{1}{\lambda} = \mathcal{R}\left(\frac{1}{n^2} - \frac{1}{m^2}\right)$$

where $\mathcal{R}$ is called the *Rydberg constant*, and m and n are integers. No physical explanation of the regularity was advanced, however, until 1913, when Niels Bohr of Denmark proposed his famous theory of the hydrogen atom. Bohr drew an analogy between the motion of a planet about the Sun and the motion of the electron about the proton. He assumed, on classical grounds, that the attractive electrostatic force provides the centripetal

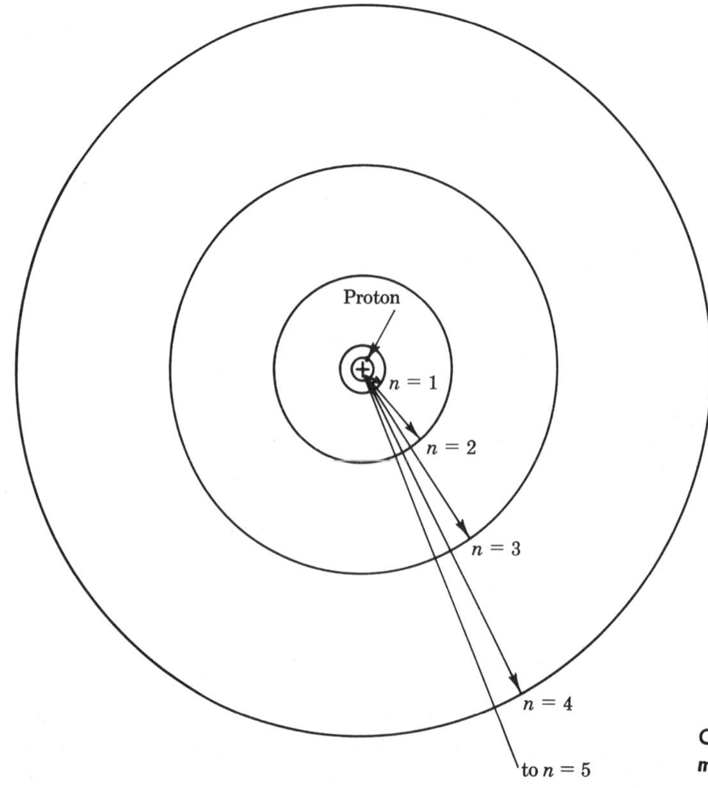

FIGURE 22-2

Orbits of electron in H atom by Bohr model

acceleration for circular motion,

$$\frac{k_E e^2}{R^2} = \frac{mv^2}{R}$$

His next contribution involved the radical idea that the angular momentum of the electron, $L = mvR$, can take on only definite or discrete numerical values. Specifically, these are integer multiples of the constant $h/2\pi$, where h is Planck's radiation constant 6.63×10^{-34} J-sec. Thus he proposed

$$L = n\left(\frac{h}{2\pi}\right) \qquad \textit{(Bohr's quantum condition)}$$

with $n = 1, 2, 3$, etc. The angular momentum was said to be "quantized," with n called the *quantum number.*

This restricts the electron motion to well-defined orbits, as illustrated in Figure 22-2. The increase in radius as the square of the number n can be demonstrated by eliminating speed between the preceding equations:

$$R_n = \frac{n^2(\hbar/e)^2}{mk_E}$$

where $\hbar = h/2\pi = 1.05 \times 10^{-34}$ J-sec. Let us calculate the radius of the smallest Bohr orbit, $n = 1$, from the basic constants:

$$R_1 = \frac{(\hbar/e)^2}{mk_E} = \frac{\left(\dfrac{1.05 \times 10^{-34}}{1.60 \times 10^{-19}}\right)^2}{(9.11 \times 10^{-31})(9 \times 10^9)}$$
$$= 5.3 \times 10^{-11}\text{ m}$$

The energy of the electron in *any* orbit can also be determined. From the force equation, the kinetic energy is

$$\tfrac{1}{2}mv^2 = \frac{k_E e^2}{2R}$$

The total energy of the system for given n is the negative of this result, borrowing from the analysis of planetary motion (Section 10-6),

$$E_n = -\frac{k_E e^2}{2R_n} = -\frac{1}{n^2}\frac{m}{2}\left(\frac{k_E e^2}{\hbar}\right)^2$$

These *energy levels* are plotted in Figure 22-3. The lowest energy level, corresponding to the smallest radius, $n = 1$, is

$$E_1 = -\frac{k_E e^2}{2R_1} = -\frac{(9 \times 10^9)(1.60 \times 10^{-19})^2}{(2)(5.3 \times 10^{-11})}$$
$$= -2.18 \times 10^{-18}\text{ J} \quad \text{or} \quad -13.6\text{ eV}$$

Then, for any orbit,

$$E = \frac{E_1}{n^2}$$

Bohr's third assumption was that light is absorbed or emitted in bursts of energy when the electrons "jump" from one orbit to another. This of course corresponds to Planck's idea. When an electron goes from a larger orbit labeled i (initial) to a smaller orbit labeled f (final), the energy lost, it was assumed, appears as light of frequency ν such that

$$E_i - E_f = h\nu \qquad \textit{(photon energy)}$$

Thus the energy released on transition between any two energy levels with quantum numbers n_i and n_f is

$$\Delta E = h\nu = E_1\left(\frac{1}{n_i^2} - \frac{1}{n_f^2}\right)$$

For example, the photon due to a jump between $n_i = 3$ and $n_f = 2$ is of energy

$$\begin{aligned}\Delta E &= -13.6(\tfrac{1}{9} - \tfrac{1}{4}) = 1.89 \text{ eV} \\ &= (1.89 \text{ eV})(1.60 \times 10^{-19} \text{ J/eV}) = 3.02 \times 10^{-19} \text{ J}\end{aligned}$$

The frequency is thus $\nu = \Delta E/h = 4.56 \times 10^{14}$ cps. Figure 22-3 also shows this radiation schematically. The wavelengths and frequencies predicted by Bohr agreed almost perfectly with the series of lines in hydrogen

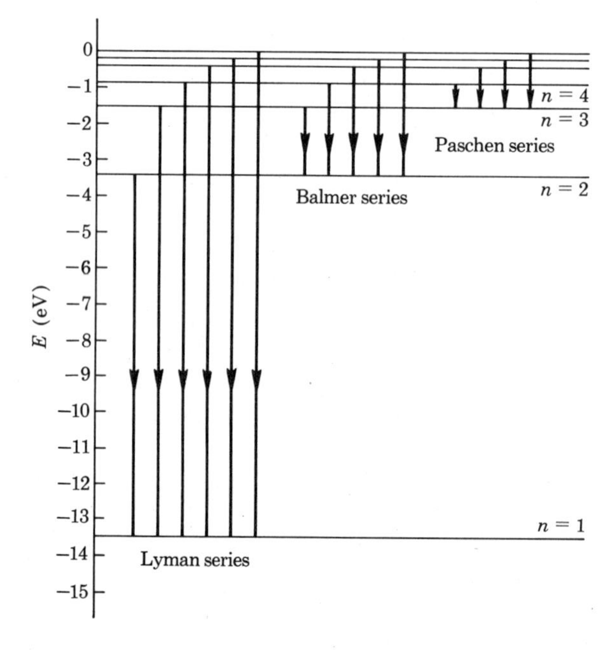

FIGURE 22-3

Energy levels and transitions in atomic hydrogen

identified by the spectroscopists. The presence of integers was accounted for, and calculations led to the correct value of the Rydberg constant:

$$\mathcal{R} = -\frac{E_1}{hc} = \frac{2.18 \times 10^{-18}}{(6.63 \times 10^{-34})(3.00 \times 10^{8})} = 1.10 \times 10^{7}\ \mathrm{m}^{-1}$$

In the theory there is an obvious defect, which was admitted by Bohr. Recall from Sections 19-2 and 20-2 that classical electromagnetic theory predicts a radiation of energy by a charge that experiences acceleration. Since the planetary model demands a continuous centripetal acceleration, there should be a radiation-energy loss that would cause orbits to collapse. This puzzling situation has been completely resolved by the newer wave mechanics, and Bohr's theory cannot be viewed as being correct mechanically and electrically. It has the virtue, however, of giving almost the right answer. Because of its pictorial simplicity, there is a tendency to take Bohr's theory literally and make the flat statement that electrons move in orbits about the nuclei. We must keep in mind that the *model is not correct*, even though it allows us to visualize atoms and thus be able to think about radiation processes more easily than is possible with the more abstract quantum-mechanical model. Bohr's theory had a profound effect on physics, for it stimulated others, such as Schroedinger and Heisenberg, to develop the theory of wave mechanics, to be discussed shortly.

22-2 MATTER WAVES AND QUANTUM NUMBERS

Let us begin our discussion of the newer quantum theory with a review of the experimental facts. The photoelectric effect in metals (Section 8-9) and the Compton scattering of photons on free electrons (Section 10-4) are excellent pieces of evidence for the particle model of light. On the other hand, the wave models of light refraction (Section 20-4) and diffraction (Section 21-4) are equally valid descriptions of these experimental observations. Although this duality had not yet been fully explained, de Broglie (see Section 12-5) conceived of the idea that if light waves acted like particles, then particles of matter might have wave character. He drew the following plausible analogy. The wave picture states that the wavelength of light is given by $\lambda = c/\nu$. According to Planck (Section 19-4) and Bohr (Section 22-1), the energy of a photon is given by $E = h\nu$, while Compton (Section 10-4) verified that its momentum is $p = E/c$. Combining these three relations, we find the wavelength of a *photon* to be $\lambda = h/p$. At this point, de Broglie made a simple but important conjecture, based on the concept of symmetry of nature. He proposed that a *particle of matter* of momentum p should have a wave associated with it of length h/p, so that

$$\lambda = \frac{h}{p} \qquad \text{(wavelength of a particle of matter—de Broglie)}$$

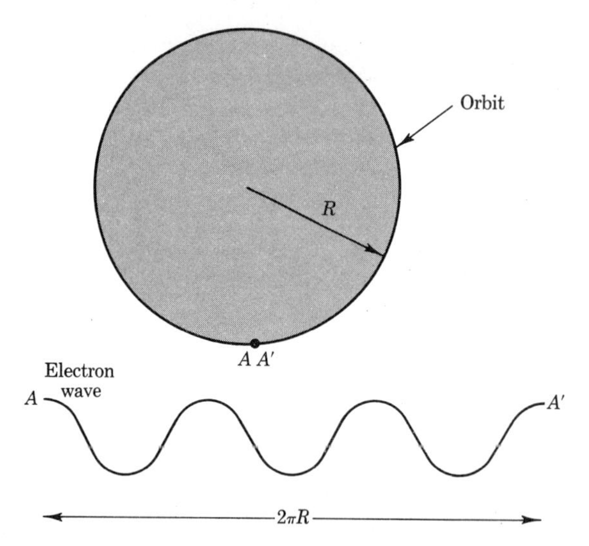

FIGURE 22-4

Relation of de Broglie waves to Bohr orbit, $n = 3$

When he looked at the then-known Bohr orbit of electron motion in hydrogen, he saw that waves of this type would fit perfectly into the circumference of the circle. As illustrated in Figure 22-4, if the circle is broken at a point and unwrapped, one could insert n full waves of length λ in the length $2\pi R$; hence

$$n\lambda = n\frac{h}{p} = 2\pi R$$

Since the angular momentum is $L = pR$,

$$L = n\frac{h}{2\pi}$$

agreeing with Bohr's quantum condition (Section 22-1).

Diffraction of Electrons and Neutrons

Great impetus to the acceptance of de Broglie's hypothesis that matter had wave properties was provided by the experiments of Davisson and Germer in 1927. They found that the reflection of beams of electrons from a metal crystal gave rise to *interference*, a wave effect.

Their experimental apparatus consisted of an electron "gun" with adjustable potential, a nickel metal crystal surface, a collector with narrow opening, and a current meter, as in Figure 22-5(a). As the angle θ was changed, the current went through a peak, as shown in Figure 22-5(b). With electrons of 54-eV energy, the angle was 50 deg. If de Broglie's theory were correct, the wavelength of an electron of kinetic energy E should be

$$\lambda = \frac{h}{p} = \frac{h}{\sqrt{2mE}}$$

or, if E is in electron-volts, the wavelength in meters is

$$\lambda = \frac{1.23 \times 10^{-9}}{\sqrt{E}}$$

In analogy with optical diffraction (Sections 21-6 and 21-7), the crystal should act as a ruled grating. For normal incidence $\lambda = d \sin \theta$ applies in first order, with d as the distance between crystal planes—2.15×10^{-10} m, in the case of nickel. We can check agreement in this early experiment by applying both formulas:

$$\lambda = \frac{1.23 \times 10^{-9}}{\sqrt{54}} = 1.67 \times 10^{-10}\ \text{m}$$

$$\lambda = (2.15 \times 10^{-10})(\sin 50°) = 1.65 \times 10^{-10}\ \text{m}$$

The agreement is too striking to be mere coincidence.

The *neutron* also exhibits wave properties, in accord with the expression

$$\lambda(\text{m}) = \frac{2.86 \times 10^{-11}}{\sqrt{E(\text{eV})}}$$

A beam of neutrons of low energy, around 0.025 eV, is allowed to leave a nuclear reactor and strike a calcite crystal at a glancing angle. Those reflected from the surface are of a single wavelength, which provides a means of selecting neutrons of one speed from a Maxwellian distribution. Since the cross-section for neutrons on hydrogen is large, the diffraction method may be applied to study crystal structure of compounds with

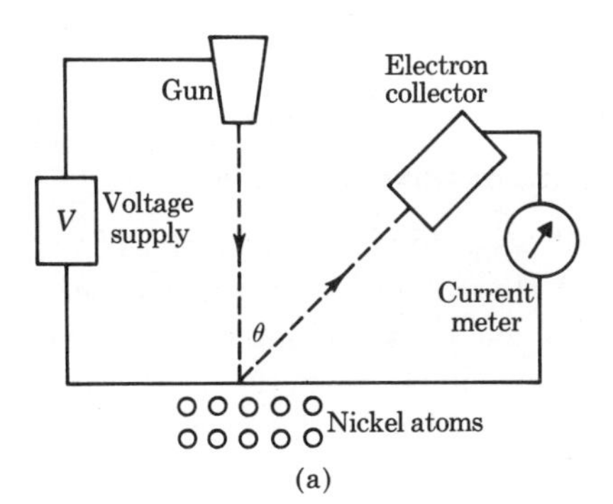

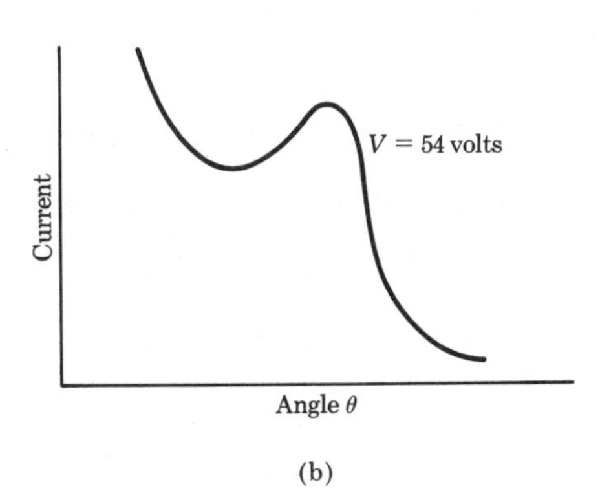

FIGURE 22-5

Electron diffraction apparatus and data (the current shown includes electrons reflected directly from the surface)

hydrogen atoms—e.g., H_2O and NaH. The method supplements X-ray diffraction which occurs best in heavy elements containing many electrons.

22-3 THE UNCERTAINTY PRINCIPLE

That chance is involved in the specification of location of a particle that has wave-like character was brought out by Heisenberg's *uncertainty principle,* which in words states:

> THE LOCATION x AND MOMENTUM p OF A PARTICLE CANNOT BE KNOWN SIMULTANEOUSLY WITH UNLIMITED ACCURACY.

We are of course familiar with the difficulty of making precise measurements. Use of a meter stick to find the distance from the origin to a point will give an answer $x \pm \Delta x$, where Δx measures the experimental error. Similarly, the speed of a particle will be found to be $v \pm \Delta v$ with the error Δv determined by the precision of the clock used to measure time. We would assume that by improving our apparatus, Δx and Δv (or $\Delta p = m\Delta v$) could be reduced to as small values as we wish. The inclusion in Heisenberg's principle of the word "cannot," however, means that no matter how elaborate or expensive or delicate our equipment, we can never achieve the goal of unlimited precision.

We naturally could counter with the remark that Δx and Δp can never be made *exactly* zero, but that we can certainly approach that limit. The uncertainty principle erects a definite barrier by stating, mathematically,

$$(\Delta x)(\Delta p) \geq \hbar \qquad \textit{(uncertainty principle)}$$

Although this product is small, 1.05×10^{-34} J-sec, it is still a long way from zero. We have three choices: (1) the two uncertainties can be made smaller together, but their product will never be less than $\hbar$; (2) the location can be found exactly, $\Delta x = 0$, but then Δp is infinite, meaning that we have no knowledge whatever of the momentum; (3) the converse, Δp can be zero, but we do not know where the particle is.

The uncertainty principle may be stated in even simpler terms:

> THE ACT OF MEASURING DISTURBS THE OBJECT BEING MEASURED.

A simple proof of this concept that also allows us to verify the degree of the limitation is now made. Imagine, as did Heisenberg, a "thought experiment" (*gedankenexperiment*) designed to locate a moving electron by reflecting light from it (Figure 22-6). After passing through the microscope, the photon of wavelength λ and frequency ν registers its position on a photographic film. The image, however, is really a diffraction pattern with a width of about λ/d, where d is the lens diameter. The ability to distinguish actual positions of the photon is then limited to

$$\Delta x = \frac{\lambda}{2 \sin \alpha}$$

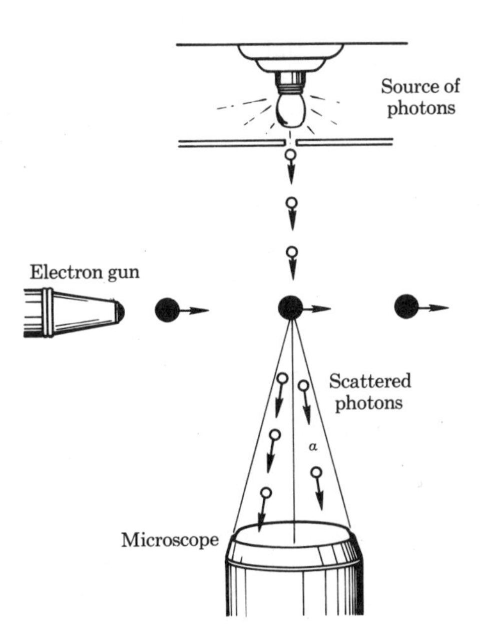

FIGURE 22-6

Apparatus for demonstrating uncertainty principle

as derived in Section 21-5. Now, in order to enter the lens, the photon of momentum $h\nu/c$ must have been deflected through angles in the range $\pm\alpha$. This means it was given a momentum component in the x-direction in the range $\pm(h\nu/c)\sin\alpha$. The electron being studied had the same amount of momentum imparted to it, but in the opposite direction. The total uncertainty in electron momentum is thus

$$\Delta p = \frac{2h\nu \sin\alpha}{c}$$

Recalling $c = \nu\lambda$, we find the product of the two uncertainties to be $\Delta x \Delta p = h$. The more precise expression differs by only the factor 2π, being $\Delta x \Delta p \geq h/2\pi$. By reducing the wavelength λ, we gain better position resolution, but then the frequency ν increases, which disturbs the electron more and we lose definition of momentum.

For large-scale objects, the effect is negligible. Let us suppose that we require the fractional errors $\Delta p/p$ and $\Delta x/x$ to be the same, labeled f. Then $f = \sqrt{\hbar mvx}$ for a particle of mass m and speed v. For example, a football weighing 0.5 kg is passed to a receiver 10 m away with a speed 20 m/sec. How large are the fractional errors in location or speed? Inserting numbers,

$$f = \sqrt{(1.05 \times 10^{-34})(0.5)(20)(10)} \cong 10^{-16}$$

In this example, the uncertainty principle has nothing to do with ability to catch the football. On the level of atomic-particle motions however, it becomes all-important.

Other physical quantities exhibit mutual uncertainty—for instance, the energy E of a system and the time t at which it is measured. Here we find

$$\Delta E \Delta t \geq \hbar$$

To show this quickly, suppose we try to measure the frequency of a photon, by counting the number of peaks of waves arriving in a time Δt. Our measurement will have an uncertainty of 1 cycle, the gap that lies between two maxima. Thus $\Delta \nu = 1/\Delta t$. However, $E = h\nu$ and the error in E is $\Delta E = h\Delta \nu$. We form the product $\Delta E \Delta t = h$, again within a factor 2π of the correct result. In the study of energies in atoms and nuclei, this form of the principle is significant.

The discovery of the uncertainty principle had a great impact on the science of physics, since it upset the idea that all motions were determined by precise laws of cause and effect. There is little doubt that the principle is valid, but some scientists such as Einstein were unwilling to accept it on philosophical grounds. Efforts have been made at the other extreme to elevate the uncertainty principle to a basis for moral and theological thought. Most scientists would not subscribe to this extension.

The Interpretation of Matter Waves

The explanation of the phenomenon of particle diffraction (Section 22-2) required the use of waves. The experiments of Davisson and Germer, however, did not reveal any information as to the character of the waves that either represented or accompanied the electrons. Interpretations by Schroedinger and Born led to our present theoretical basis of wave mechanics, involving the concept of *wave functions*. In particular, they gave some physical meaning to what is labeled the ψ*-function*. To help understand these waves, let us perform a thought experiment with a stream of electrons, a pair of very narrow parallel slits, and a detecting screen, as shown in Figure 22-7.

We suppose that we can release various numbers of electrons and that the detector is appropriate for counting individual particles. We start the experiment by accelerating and noting the arrival of single electrons, finding that they seem to strike the screen in a rather random fashion. Some are found in locations they could not reach according to classical physics. Then, if we either increase the flow or allow the process to go on for a long time, a pattern of detected charges begins to emerge, one which is very reminiscent of diffraction fringes of light in Young's experiment (Section 21-4). Electrons travel from the source to some points on the screen and not to others, giving a distribution that suggests that waves from the two slits interfered destructively at certain places. In the case of light, we could understand such effects in terms of Huygens's principle, but it is more difficult to see how one electron could be going through *both* slits. We are led to the alternative that the wave is somehow indicative of the *probability* of a given electron being found at a point in space. Following Schroedinger, let us propose that $\psi(r, t)$ represents the wave associated with the electron. Since a wave of the form $y = A \sin (kx - \omega t)$ goes positive and negative, while probability must always be positive, we are led to the idea that ψ^2 measures the probability of finding the electron at a point. Thus the square* of the wave function is in the nature of a distribution function discussed in Section 14-5.

The diffraction pattern on the screen is interpreted by saying that the wave there due to slits 1 and 2 consists of two parts, ψ_1 and ψ_2, which add to form ψ. A completely different pattern is found on the screen if we alternately cover over one slit and then the other, a fact explained by the contrast between $(\psi_1 + \psi_2)^2$ and $|\psi_1|^2 + |\psi_2|^2$. The wave-function concept is compatible with Heisenberg's uncertainty principle. Whereas classical mechanics predicts *exact* positions, speeds, and energies of particles under

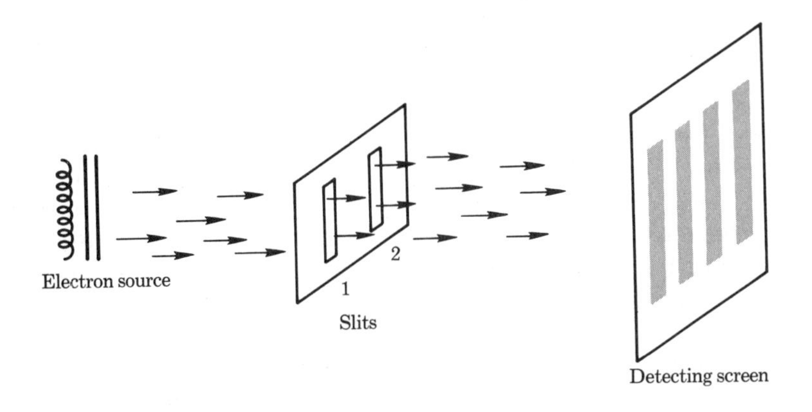

FIGURE 22-7

Electron diffraction and probability

the influence of forces, wave mechanics yields only the *probable* values. An electron in the hydrogen atom is not *at* the radius 5.3×10^{-11} m, as the Bohr model would indicate, but if many atoms could be observed, the electron distances would on the average be that value. In the determination of the detailed mathematical form of the ψ-function for a given physical situation, use is made of what is designated as *Schroedinger's equation.* We shall not undertake discussion of its form, but merely note that it includes the nature of the forces acting, and may roughly be considered as a generalization of the principle of conservation of energy, which was derived from Newton's second law.

* It turns out that the ψ-function is complex, of the form $\psi(x, t) = \psi_0 e^{i(kx - \omega t)}$ for a simple plane wave; hence we must use the square of the absolute value $|\psi|^2$ as the measure of probability.

22-4 THE RELATIONSHIP BETWEEN CLASSICAL AND QUANTUM PHYSICS

Regardless of how willing we are to accept the correctness of the wave-mechanical view of nature, it tends to remain vague and remote from reality. Psychologically, we think in terms of our large-scale classical world where objects, positions, and speeds are rather well-defined quantities. The concept of smallness is easy to grasp, and there is sufficient proof that atoms exist. However, since we cannot see or feel particles of atomic size, we tend to use our classical ideas in that domain. Were our form of life of the dimensions of viruses, with capability of thought, we might get accustomed to thinking wave-mechanically.

The Wave Packet and Newton's Laws

Although the full appreciation of quantum-mechanical phenomena depends greatly on mathematics, we can help bridge the gap between the two views of nature by use of the concept of the *wave packet*. Recall from Section 20-4 that complicated waveforms are composites of pure frequencies with various amplitudes. The group was shown to have a speed v_g that is different from the speed v of the individual waves of which it was formed. This idea is now extended to the matter waves for a particle in quantum physics. We combine such waves in a range $\Delta\nu$, to form a single wave packet of length $\Delta x = 2\pi/\Delta k$, with no disturbance outside that range of position, as shown in Figure 22-8. The assembly moves along with its average coordinate $\bar{x}$ having the group speed, here labeled $\bar{v}$, so that $\bar{x} = \bar{v}t$. The amplitude of the wave at a point in space is related to the probability of finding the object there. Thus there is a close analog between the average

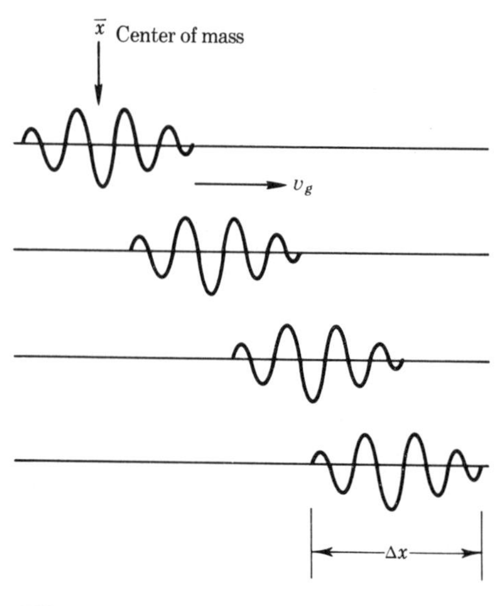

FIGURE 22-8

Motion of a wave packet representing a particle

coordinate of a material object (its center of mass) and the average position of the group (its "center of probability"). By advanced methods, the wave group is found to obey the equation

$$F = m\frac{\Delta\bar{v}}{\Delta t}$$

which we see to be of the form of Newton's second law. In words, then, the "center of mass" of the wave packet representing the particle moves in accordance with the laws of mechanics. The precise location of the particle is ill-defined within the length of the packet, Δx. Recall de Broglie's expression $p = h/\lambda$. However, $k = 2\pi/\lambda$; hence $p = hk/2\pi$. The error in definition of momentum is thus $\Delta p = h\Delta k/2$. Then, since $\Delta x = 2\pi/\Delta k$, we again obtain $\Delta x\Delta p = h$, which agrees with the uncertainty principle except for a factor of 2π.

The connection with classical mechanics is now more firmly established. Recall that if the mass of an object is large, then the fractional error in position and momentum is extremely small. For practical purposes, this means that we can set h equal to zero in the uncertainty principle:

$$\Delta p\Delta x = 0 \qquad \textit{(classical)}$$

Now, if h is set equal to zero in de Broglie's relation

$$\lambda = \frac{h}{p}$$

then $\lambda = 0$, the packet is a point, $\bar{x}$ becomes x, and Newton's law of classical mechanics holds exactly. This connection is one facet of Bohr's *correspondence principle*, which states that the principles of quantum mechanics must agree with the laws of classical mechanics when applied to macroscopic bodies. In other words, quantum theory is the more general, applicable to both large and small bodies, and thus embraces classical theory as a special case. On a practical basis, however, Newton's laws are sufficiently accurate for many purposes. This situation is quite analogous to the relation of relativistic physics and classical physics. Whereas Einstein's special theory of relativity is required to describe rigorously the motion of bodies of all speeds, large or small, it is approximated by the simpler useful classical form if the speed is low. In passing, we note that relativistic quantum mechanics is an even broader generalization of classical mechanics.

Wave and Particle Concepts of Light

We observed the duality of light as particles and waves (Section 22-2) to be the starting point of the new quantum mechanics. Not until after de Broglie advanced his hypothesis was the relation of quanta of light and waves of light fully understood. In fact, it was the newly derived wave

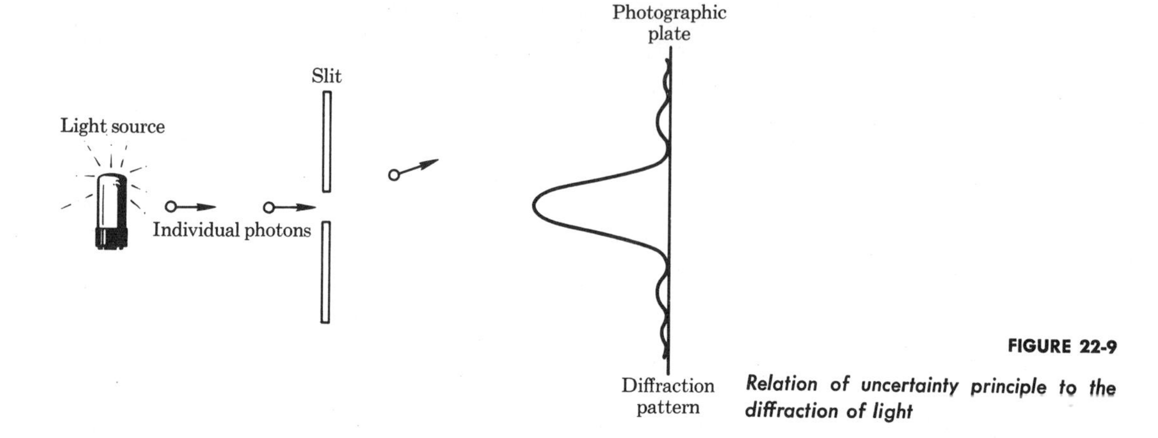

FIGURE 22-9
Relation of uncertainty principle to the diffraction of light

mechanics with its uncertainty principle and probability view that provided the explanation. Let us consider an experimental arrangement as in Figure 22-9. We allow light to enter a slit and be registered on a photographic film. The beam is adjusted to be so weak that we are certain that no more than one photon goes through at a time, which eliminates the possibility that particle interaction can occur. We would expect from geometric optics that after a long time a line will form on the film exactly opposite the slit. Instead, we find a diffraction pattern with alternating bands that extend far beyond the direct line of sight. The results of this experiment must be interpreted by saying that light is composed of photons of very small but definite energy $h\nu$. The wave is a representation of the probability of finding a photon at a point. The very presence of a screen with a slit cut in it, as an obstruction to the wave, disturbs the probability distribution of photons. As each photon enters the aperture, it has a certain chance of going anywhere. However, over a long time, as many particles come through, they distribute themselves over the photographic film in accordance with the new probability, and the pattern coincides with the interference pattern of the wave model. The uncertainty principle serves as a guide to the process, in that it tells us that the interposition of the slit is an act of measurement that disturbs the situation. It also gives the approximate size of the diffraction pattern, but it does not have the power to specify the details of the photon distribution.

22-5 VECTOR MODEL OF ATOMS AND SPECTRA

The full description of a hydrogen atom that has an energy larger than the lowest level (ground state) requires additional quantum numbers besides n. Their origin is mathematical, being the result of solutions of Schroedinger's equation. There are three degrees of freedom corresponding to the coordinates r, θ, and ϕ; hence three quantum numbers are required. As in the Bohr theory, the value of n, called the "total" quantum number, indicates the radius R and the total energy E. Corresponding to θ-motion is the "orbital" quantum number l; to ϕ-motion, the "magnetic" quantum number m. It can be proved that l can have several values 0, 1, 2, and so on, up to largest value $n - 1$. For a given l-value, m can be 0, ± 1, ± 2, and so on, up to a maximum of $\pm l$.

We can construct a partial table (up through $n = 3$) of possible combinations of these three quantum numbers of hydrogen. The classification of these different *states* wherein the electron can exist consists of writing first the n-value 1 or 2 or 3, etc., then the l-value according to the

TABLE 22-1

SOME STATES OF THE HYDROGEN ATOM

n	l	m	CLASSIFICATION
1	0	0	1s
2	0	0	2s
	1	+1, 0, −1	2p
3	0	0	3s
	1	+1, 0, −1	3p
	2	+2, +1, 0, −1, −2	3d

"code" $0 = s$, $1 = p$, $2 = d$, $3 = f$, etc., and finally the possible m-values.

Before the wave theory was developed, graphical or *vector models* were devised, with angular momenta drawn as vectors. These are still employed for ease in visualization. In our model of the hydrogen atom, we let the origin of coordinates be at the proton. The use of r to indicate radial position of an electron is straightforward. However, the angles θ and ϕ have meaning only if there are some axes from which they are to be measured. To a free hydrogen atom there is no up, down, right, or left that suggests any reference frame. A magnetic field can supply the needed frame. If we place the atom in a magnetic field that is along the z-axis, there is now a definite direction even if the field is so weak that it has no appreciable effect on the motion. The angular momentum of the electron in orbit about the proton is taken to be a vector $\mathbf{L}$, with magnitude $l\hbar$. Since l is an integer, L has "quantized" lengths. Its *direction* is also restricted to some definite values relative to the field lines—specifically, such that its *component* in the z-direction is $m\hbar$, with m the magnetic

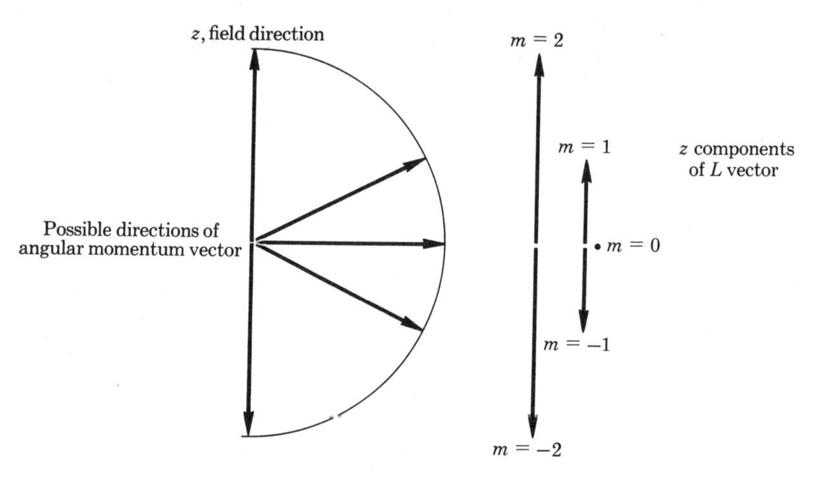

FIGURE 22-10

Quantized directions of **L**

quantum number. Figure 22-10 shows the possible direction of L-vectors for the case $l = 2$—for example, in the $3d$ electron state. Experimental measurements of light from hydrogen gas at high temperature were used to check this picture. Continual transitions of the electrons provide spectral lines from which states are deduced.

Zeeman Effect

The frequencies of light from an incandescent gas such as hydrogen are changed slightly on application of a magnetic field. The shifting of energy levels and resulting changes in the energy transitions that give rise to light goes by the name *Zeeman effect,* after its discoverer. We recall from Section 9-7 that a current loop (e.g., an electron in orbit) forms a magnetic dipole, whose moment M interacts with a magnetic field. If we turn a dipole from a direction perpendicular to the field to a direction making an angle θ with the field, the potential energy is found to change by an amount

$$\Delta E = M\mathcal{B} \cos \theta$$

We also saw in Section 9-7 that the ratio of magnetic moment to angular momentum for a circulating electron is $M/L = q/2m$. If we insert this in the relation for ΔE and note that the grouping of constants $q\mathcal{B}/2m$ has the form of an angular frequency (recall Sections 7-1 and 18-6), and can be labeled ω_L, then

$$\Delta E = \omega_L L \cos \theta$$

The important feature to note is that $L \cos \theta$ is the component of angular momentum parallel to the field, as in Figure 22-10. Its value according to quantum mechanics is also $m\hbar$, where m is here the magnetic quantum number (not the mass). Thus the energy change on application of a magnetic field to an atom is

$$\Delta E = m\hbar\omega_L$$

Figure 22-11 shows how the energy levels for two adjacent levels $3p$ and $3d$ are "split," and how transitions of the electron between states with different n give three spectral lines. Since ΔE is directly proportional to m, and m is an integer, groups of three transitions all yield the same frequency, and only three distinct lines are seen. Experimentally, such lines are observed in the light from many atoms, if the magnetic field is *very strong*. Discrepancies still exist for some atoms, however, if the field is weak. Only by proposing still another degree of freedom called *electron spin* can the problem be resolved. This innovation by Uhlenbeck and Goudsmit was based on simple classical reasoning. They extended the analogy between planetary and electron motion, arguing that if there is an angular momentum due to the circulation of an electron in an orbit about the nucleus, there should also be an angular momentum due to the rotation or spin of the electron on its own axis.

The model worked, but we do not take it literally, any more than we do the Bohr planetary orbits. Instead, we accept the fact that there is a *basic property of particles that we call spin*. An electron has an extra angular momentum of magnitude $\frac{1}{2}\hbar$, as if there were a special quantum number s of value always *one-half*. This new spin momentum adds vectorially with the orbital momentum to give the total for the atom. Letting the vector angular momenta be **L** (orbital) and **S** (spin), the total is

$$\mathbf{J} = \mathbf{L} + \mathbf{S}$$

As we are prepared to expect, the vector **J** can have only discrete values $j\hbar$

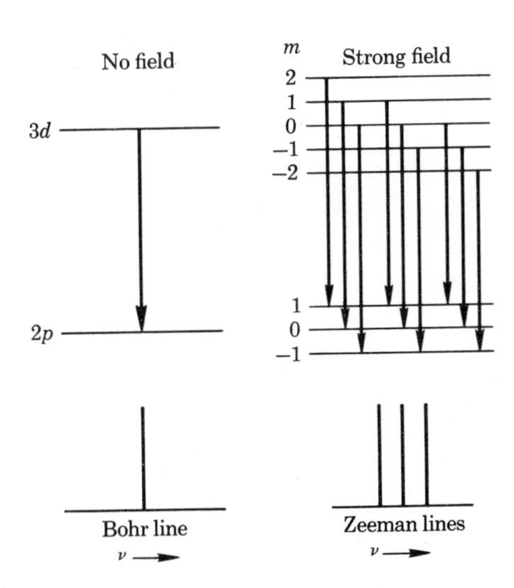

FIGURE 22-11

Effect of a magnetic field on the spectrum of hydrogen

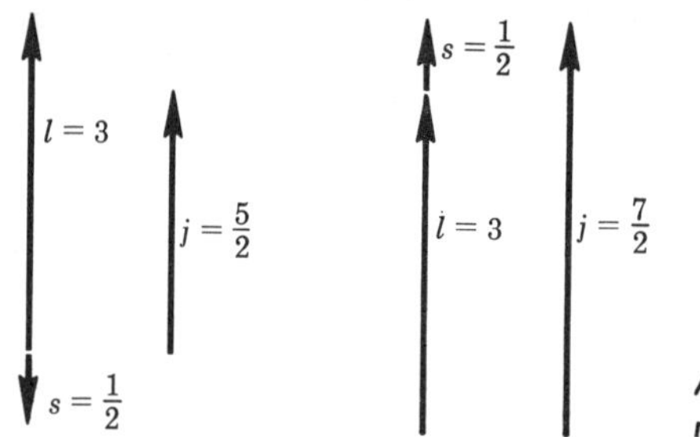

FIGURE 22-12

Addition of spin and orbital angular momentum vectors

(where j is a total-angular-momentum quantum number) and components parallel to the field of value $m_j\hbar$. For one electron, **L** and **S** can be either parallel or opposite each other, as shown in Figure 22-12 for a case in which $l = 3$. The quantum numbers j are either $3 + \frac{1}{2} = \frac{7}{2}$ or $3 - \frac{1}{2} = \frac{5}{2}$. If instead $l = 2$, we find $j = \frac{5}{2}$ or $\frac{3}{2}$, and so on.

At this point, complications arise. The electron in orbital motion gives rise to a magnetic field that is *internal* to the atom. The magnetic moment that is related to spin experiences this field and a sort of Zeeman effect inside the atom takes place, giving a further split of energy levels.

When we come to atoms with two or more electrons—for instance, helium—there are many new interactions and ways in which spin and orbital momenta add. In general, one must account for nuclear effects and relativity. Although the theory is available, the actual labor involved is large in the interpretation of spectra of complex atoms such as uranium $_{92}U^{238}$, with 92 electrons.

We have been able to provide only a brief indication of the power of the methods of quantum mechanics to explain the dynamics of atomic and molecular processes. Among the phenomena to which quantum mechanics has been successfully applied are the nature of the chemical bond in molecules, the variation of specific heat with temperature, the energies involved in chemical reactions, and the emission and absorption of light in complex gases. As we shall see in the next chapter, the principles of quantum mechanics hold as we probe still deeper into the submicroscopic realm, that of the nucleus.

PROBLEMS

22-1 The wavelength of H_γ line of hydrogen is 434 mμ. Noting that the Rydberg constant is 1.10×10^7 m^{-1}, determine the initial and final electron orbits. Are there any others that would give the same wavelength?

22-2 In what way is the singly ionized helium atom He^+ similar to and different from the hydrogen atom? Compute the electron energy in the smallest orbit of He^+ by use of ratios. How do the electron-orbit radii in the two "atoms" compare?

22-3 Find the frequency of the light emitted in the hydrogen-atom transition $n_i = 4$, $n_f = 2$.

22-4 Make a diagram, to scale, of the first six stable states in the hydrogen atom, using radius, energy, and angular momentum as variables.

22-5 Show that the linear speed of an electron in the nth orbit of hydrogen is given by

$$v = \frac{k_E e^2}{n\hbar}$$

22-6 The frequency of the light from hydrogen corresponding to a transition between Bohr orbits $n_i = 3$ and $n_f = 2$ was found to be 4.56×10^{14} cps (see Section 22-1). Find the wavelength and determine which line of the Balmer series it represents.

22-7 Verify that the wavelength in meters of an electron of energy E, in electron-volts, is $\lambda = \dfrac{1.23 \times 10^{-9}}{\sqrt{E}}$.

22-8 Find the energies, in electron-volts, for electrons, X rays, and neutrons to be used in diffraction studies of atomic structure, assuming that distance between atoms are 3×10^{-10} m and $\theta = 3$ deg, a *glancing* angle. (We should note that the condition for constructive interference, $\lambda = d \sin \theta$, refers to normal incidence, where θ is the angle from the normal; the relation $\lambda = 2d \sin \theta$ involves glancing incidence, at angle θ with the surface.)

22-9 Calculate the de Broglie wavelengths of the following objects: (a) the Earth in its orbit about the Sun (see Table 10-1); (b) a racing car of mass 1000 kg moving at speed 200 km/hr; (c) a 2-gram bullet of speed 1000 m/sec; (d) an air molecule, mass 4.8×10^{-26} kg, speed 400 m/sec; (e) a neutron in a nuclear reactor, mass 1.67×10^{-27} kg, speed 2200 m/sec. Discuss the significance of these wavelengths.

22-10 If the position of a proton is located within a range of 1 millimicron, how much uncertainty is there in its speed?

22-11 The position of a particle confined to a region is uncertain by the width of the latter. Place an electron in a "box" of side 10^{-10} m, comparable to the size of a hydrogen atom, and find the uncertainty in its momentum. Assuming the actual momentum to be at least this large, estimate the lowest energy in eV the electron could have in such a box.

22-12 A typical time required for an electron that has been excited to higher level to return to the ground state is 10^{-8} sec. What amount of uncertainty in frequency would be expected? If the frequency is 2×10^{15} cps, what fractional error is implied?

22-13 (a) Calculate the characteristic angular frequency $\omega_L = q\mathcal{B}/2m$ for an electron in a magnetic field of 4 Wb/m². (b) Find the *maximum* energy change due to the field in the hydrogen atom when it is in the $3p$ state (see Table 22-1), using $\Delta E = m\hbar\omega_L$.

NUCLEAR PHYSICS

23

We began our account of the science of physics with a factual description of molecules, atoms, and nuclei, without presenting experimental proof or theoretical evidence. Having developed an understanding of forces and energy, we are ready to examine the important details of nuclear processes. All of us are familiar with the impact on our technical and political world of the discovery of nuclear energy.

A great deal of understanding of nuclear events can be gained simply by considering three key ideas:

1. The nuclei of atoms are composed of very small clusters of positively charged protons and uncharged neutrons, together called nucleons.

2. Mass and energy may be intertransformed in accordance with Einstein's relation $E = mc^2$. Conservation of mass-energy is applicable to all processes.

3. Reactions involving nuclei have many analogies to the more familiar chemical reactions—processes such as dissociation, combination, and displacement.

23-1 THE STRUCTURE OF NUCLEI

The apparent dimensions of nuclei depend on the process that is used to measure; hence the radius R is not a well-defined quantity. One model of the nucleus is a compact spherical array of protons and neutrons, where the nuclear matter is not compressible. The volume $\frac{4}{3}\pi R^3$ is roughly proportional to the number of nucleons A, which means that the radius R is proportional to $A^{1/3}$. A useful approximate relation based on scattering data is

$$R \cong 1.4 \times 10^{-15}\, A^{1/3}\ \text{m}$$

Since nuclei contain charged particles, we would expect the nucleus to be highly unstable because of electrostatic repulsion. Since it is not, we conclude that other strong forces exist that attract nuclear particles to each other. We recall that the coulomb (electrostatic) force is proportional to $1/R^2$, which is very large at distances of the order of 10^{-15} m, the approximate size of a nucleus. There is good evidence of a force of attraction between pairs of nucleons that has the necessary strength at very short range, but is negligible at large separations. It can be considered as essentially independent of charge—that is, the force between nucleons is about the same, whether they are neutrons or protons. The fact that nuclear constitutents are tightly bound together means that energy must be expended

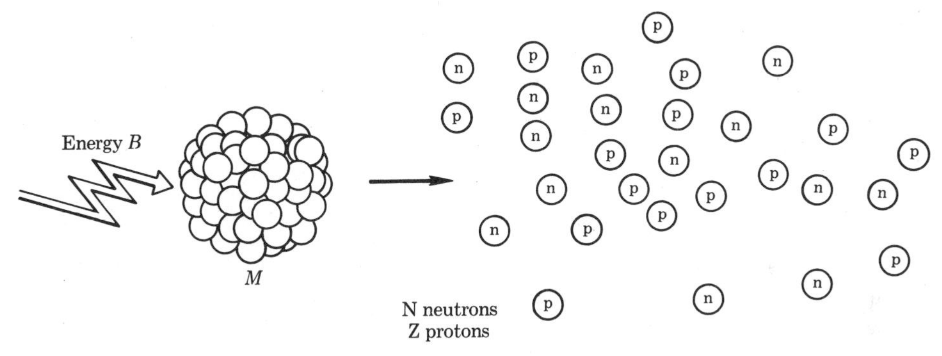

FIGURE 23-1

Dissociation of a nucleus

to disrupt a nucleus. In other words, to cause a nucleus to dissociate, we must inject an amount of energy equal to the *binding energy* of that nucleus.

There is an intimate relation between nuclear mass M and binding energy B. Suppose, as in Figure 23-1, that we had a means for supplying an amount of energy equal to B to the nucleus, causing it to completely dissociate into Z protons, each of mass M_p and N neutrons of mass M_n. Let us apply the law of conservation of mass-energy to this spectacular event. Before the reaction, the total is the sum of Mc^2 for the nucleus and B for the binding energy. After the reaction, the total is $(\text{Z}\, M_p + \text{N}\, M_n)c^2$. Equating and solving for B/c^2,

$$\frac{B}{c^2} = (\text{Z}\, M_p + \text{N}\, M_n) - M$$

Conversely,

$$M = (Z\,M_p + N\,M_n) - \frac{B}{c^2}$$

The binding-energy equation expresses the fact that the mass M of a nucleus is smaller than the total mass of its nucleons, $Z\,M_p + N\,M_n$. It also shows that the amount of binding energy can be calculated if the masses of the nucleus and all the nucleons are known.

Atomic Masses

For the study of nuclear structure, we must take very precise account of differences between masses. To the accuracy needed for calculating chemical reactions, atomic weights to three or four significant figures are adequate, for example 1.008 for hydrogen, 12.01 for carbon, 55.85 for iron, etc. Several additional significant figures are required for calculating nuclear reactions. As noted in Section 1-2, on the *physical scale* of atomic masses the isotope carbon-12 is taken as having exactly mass 12. A few values for other particles are then

M_p	*proton*	1.007277
M_e	*electron*	0.000549
M_H	*hydrogen atom*	1.007825
M_n	*neutron*	1.008665
M_d	*deuterium atom*	2.014102

These are expressed in terms of the *atomic-mass unit* (amu), which has a numerical value of 1.660420×10^{-27} kg.

To check, we note that Avogadro's number of particles $N_a = 6.02257 \times 10^{26}$, each of mass $m = 1.660420 \times 10^{-27}$ kg, has a total mass of 1 kg, $N_a m = 1$, exactly.

The energy corresponding to 1 amu may be computed from Einstein's formula:

$$E = mc^2 = (1.660420 \times 10^{-27}\ \text{kg/amu})(2.997925 \times 10^8\ \text{m/sec})^2$$
$$= 1.492312 \times 10^{-10}\ \text{joule/amu}$$

However, since 1 MeV $= 1.602095 \times 10^{-13}$ joule,

$$E = \frac{1.492312 \times 10^{-10}}{1.602095 \times 10^{-13}} = 931.476\ \text{MeV/amu}$$

The energy equivalent of the proton's mass is thus

$$(1.007277)(931.476) = 938.254\ \text{MeV}$$

The *mass spectrograph* is a device with which masses can be measured relatively accurately. One type, invented by Bainbridge, is shown in Figure 23-2. Its action is similar to that of a cyclotron. An element is

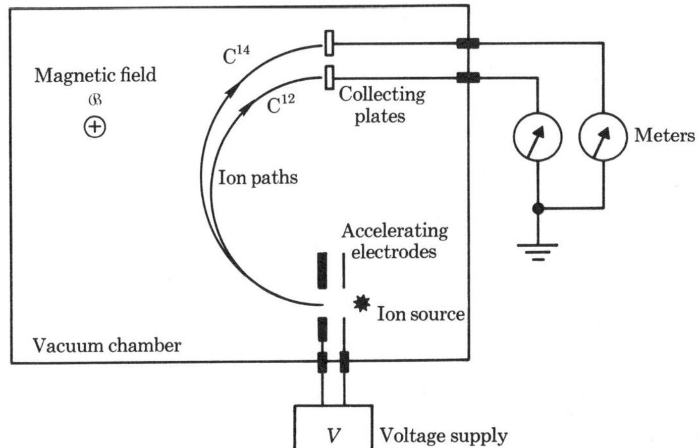

FIGURE 23-2

Spectrograph for measuring masses of isotopes

vaporized and its atoms are ionized by bombardment with electrons. The resulting ions of charge q are then accelerated through a potential difference V into a uniform magnetic field $\mathcal{B}$, in a vacuum. They move in a circular path of radius R and strike a plate connected to a current meter. The speed is $v = \sqrt{2qV/m}$, while the angular speed is v/R and also $\omega = q\mathcal{B}/m$; hence

$$R = \frac{\sqrt{2Vq/m}}{\mathcal{B}}$$

Now, if we use an element such as carbon, ions of the two isotopes C^{12} and C^{14} will move on different radii, since their masses are different, and we can measure separate currents corresponding to their ions. The ratio of masses for two ions is equal to the ratio of squares of the radii:

$$\frac{m_2}{m_1} = \frac{R_2^2}{R_1^2}$$

Comparisons of masses are thus made possible by accurate knowledge of ion paths. For example, if m_1 is exactly 12, the standard, then we may compute m_2, the mass of C^{14}.

Let us calculate the binding energy for the nucleus of helium-4, containing Z = 2 protons, N = 2 neutrons, with A = Z + N = 4 nucleons. The measured mass of the *atom* ${}_2He^4$ is $M = 4.002604$ amu, and the total mass of components, including the electron in hydrogen, is

$$\mathrm{Z}\, M_{\mathrm{H}} + \mathrm{N}\, M_n = 2(1.007825) + 2(1.008665) = 4.032980$$

The difference is 0.030376 amu, or

$$\begin{aligned} B &= (0.030376 \text{ amu})(931.476 \text{ MeV/amu}) \\ &= 28.29 \text{ MeV} \end{aligned}$$

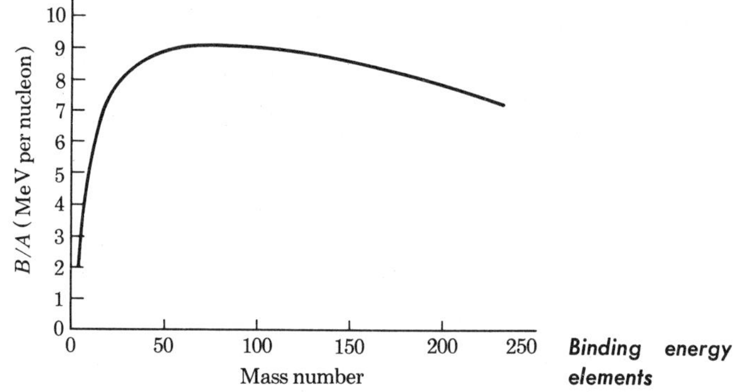

FIGURE 23-3

Binding energy per nucleon in the elements

Next, let us consider the heavy element uranium-238, with Z = 92, N = 146, A = 238. Its binding energy is $B = 92(1.007825) + 146(1.008665) - 238.0508 = 1.934$ amu or $B = 1801$ MeV.

Comparison of the binding energies *per nucleon* gives us valuable information about the nature of the forces between nucleons. For He^4, this is roughly $B/A = 28.29$ MeV$/4 = 7.1$ MeV. For U^{238} it is $B/A = 1801/238 = 7.6$ MeV. Figure 23-3 shows B/A for many nuclei. We see that over a wide range the binding energy is almost proportional to the number of nucleons:

$$B \sim A$$

Is this to be expected? If each particle interacted with every other particle, the number of bonds would be $\frac{A(A-1)}{2}$, as shown in Section 14-1. For large A, this would yield a value of B that is proportional to A^2, not to A. If instead, each nucleon only interacted with one other particle, the number of bonds would be A/2, and B would be proportional to A. We must conclude that the latter type of cohesive force is more nearly correct. Our *first approximation* to the dependence of binding energy on nuclear composition is

$$B \cong \alpha A$$

where α is some constant.

Opposing the cohesive effect is the electrostatic repulsion. Visualize the process of forming a collection of protons Z, each of charge e, packed together in a sphere of radius R. We can make a rough estimate of the total work required to do this. Suppose that we brought one-half the particles, A/2 in number, up from infinity against the repulsive force of the other half, and located them at a separation R. Then the work would be

$$W = \frac{k_E\left(\frac{Ze}{2}\right)^2}{R}$$

using the results of Section 8-4. It turns out that if, instead, the charges are collected one by one, the work is correctly evaluated as

$$W = \frac{3k_E(Z\,e)^2}{5R}$$

differing from our estimate only in the numerical factor.

The binding-energy formula is now improved to the form

$$B \cong \alpha\,A - \frac{3k_E(Z\,e)^2}{5R}$$

where the negative sign means that electrostatic repulsion tends to separate the nucleons. These two terms dominate the binding energy, but there are several other effects that give additional terms to the general formula.

The first is a surface correction. Nucleons at the surface do not have their full share of binding since there are no particles outside to attract them. This suggests a negative term proportional to the area $4\pi R^2$. We saw earlier that the radius was proportional to $A^{1/3}$, hence the surface term is proportional to $A^{2/3}$. Second, there is an observed tendency for the neutron–neutron or proton–proton force to exceed slightly the neutron–proton force. Thus any excess of neutrons beyond the value $A/2$ (equal number of protons and neutrons) weakens the average binding. This excess is $N - A/2 = (A - Z) - (A/2) = (A/2) - Z$. To account for this effect, we introduce a negative term proportional to $\dfrac{(A/2 - Z)^2}{A}$. Finally, any unpaired particles give a weakening of the total binding. In combining these effects into a single formula for mass, we find it convenient to work with the mass of the *atom*, including all electrons, instead of just the nuclear mass. In the expression for M, this requires that we replace the proton mass M_p with the hydrogen-atom mass M_H. The mass of an atom according to a semiempirical formula of Fermi and Weiszacker is

$$M = Z\,M_H + N\,M_n - \alpha\,A + \beta\,A^{2/3} + \epsilon\,Z^2A^{-1/3} + \frac{\gamma(A/2 - Z)^2}{A} + \delta$$

This is seen to contain all the terms mentioned above. Let M be in millimass units for convenience. The constants in the formula that fit expermental data are $M_H = 1007.825$, $M_n = 1008.665$, $\alpha = 15.04$, $\beta = 13.35$, $\epsilon = 0.627$, and $\gamma = 83$. The semiempirical quantity δ is $36a\ A^{-3/4}$, where a is 1 if Z and N are both odd, zero if Z or A are odd, and -1 if Z and N are both even.

23-2 RADIOACTIVE DECAY

As we discussed in Section 1-3, radioactivity is a process by which a nucleus is spontaneously transformed into another species, with the emission of a particle or ray. We now present the two basic ideas that underlie the radioactive process.

1. *A nuclear (or atomic) system will tend to go to its lowest possible energy state.* This is similar to the tendency of an electron in an outer orbit of the Bohr atom to jump to an unfilled smaller orbit. If there is some arrangement of nucleons that has lower energy, a nucleus will seek that state, even if it involves dissociation. Let us consider the simplest radioactive disintegration, that of a neutron into a proton plus an electron (we reserve discussion until Section 25-4 of the third particle produced—the neutrino):

$$_0n^1 \rightarrow {}_1H^1 + {}_{-1}e^0$$

The mass of the neutron in amu is 1.008665, while the sum for the proton and electron is smaller: 1.007277 + 0.000549 = 1.007826. By the general rule, the decay can occur, with the excess mass 0.000839 appearing as kinetic energy of the products, in amount 0.78 MeV. The most common particles released in radioactive decay are electrons (*beta* particles), photons (*gamma* rays), and helium nuclei (*alpha* particles).

2. One might expect that unstable isotopes would never be found in nature if decay were possible. *An individual nucleus can disintegrate at any time, but it is a matter of probability as to* when *it will.* Thus a natural radioactive nucleus or one produced by some reaction may decay in the first microsecond or may wait for many years before disintegrating. One cannot tell when the event will occur for an individual nucleus, but by observation of very large numbers of particles we can make a statistical prediction.

Suppose that we have a sample of a given radioactive substance in the laboratory. Measurements reveal that no matter when we start observing the process, one-half of the particles will have disintegrated and one-half will remain after a time lapse t_H. This time for decay to half-value is called the *half-life*, a quantity that is a property of the nuclear type. For example, let us take 1000 atoms of an isotope which has a half-life of 5 sec. At the end of 5 sec there will be 500 left, at 10 sec 250, at 15 sec 125, and so on. Such a behavior can be described by the relation between the number of atoms N at any time t and the number at whatever time we choose as zero, N_0:

$$\frac{N}{N_0} = \left(\frac{1}{2}\right)^{t/t_H}$$

Figure 23-4 shows this curve.

We can demonstrate that such a trend is consistent with the idea that radioactive decay is a statistical process. Suppose that there are N particles present at time t. After a short time Δt has elapsed, the number will be changed to $N + \Delta N$. If we assign the symbol λ to the chance per second that a given particle will decay, then the number on the average that will actually decay in time Δt is $N\lambda\Delta t$, or

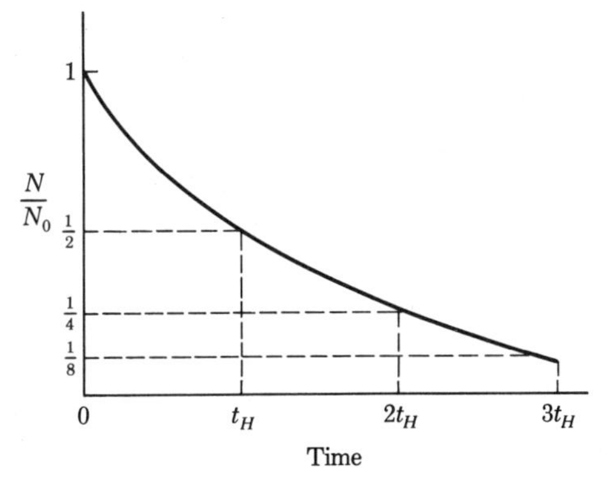

FIGURE 23-4
Radioactive decay

$$\Delta N = -N\lambda\Delta t$$

where the negative sign is attached to denote the reduction. Thus recalling similar equations (e.g., in Section 5-8), the number of atoms at any time t in terms of N_0, the number at time zero, is the exponential

$$N(t) = N_0 e^{-\lambda t}$$

This is identical to the formula involving half-life* if we let the *decay constant* λ be

$$\lambda = \frac{\ln 2}{t_H}$$

where $\ln 2 = 0.693$.

We can deduce the half-life of an isotope from experimental measurements of the rate of decay or *activity* A, defined as the magnitude of $\Delta N/\Delta t$. From our derivation, we see that

$$A = N\lambda = \lambda N_0 e^{-\lambda t}$$

At time zero, the activity is $A_0 = \lambda N_0$. Taking the logarithm of both sides of the equation, we get

$$\ln A = \ln A_0 - \lambda t$$

which shows that a plot of activity on semilog paper is a straight line, as seen in Figure 23-5. The slope of the best straight line through the observed points gives λ directly; i.e.,

$$\lambda = \frac{\ln A_0 - \ln A}{t}$$

The half-life is then calculated from $t_H = 0.693/\lambda$.

* Note that $N/N_0 = e^{-\lambda t} = e^{-(\ln 2)t/t_H} = (\frac{1}{2})^{t/t_H}$.

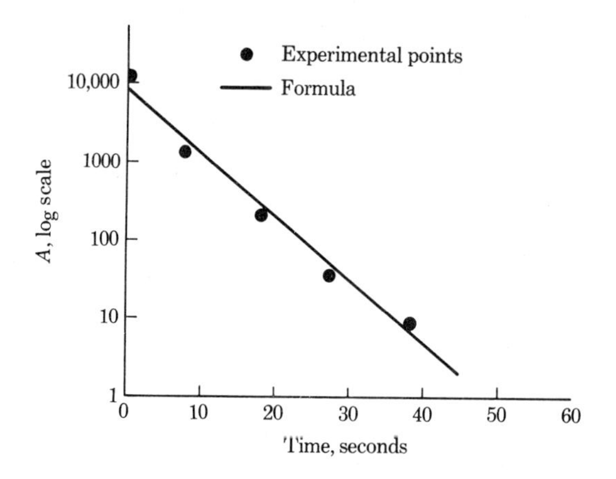

FIGURE 23-5

Activity of a radioactive substance (semi-log scale)

The radioactive isotope cobalt-60 has many applications in medicine and industry. Its half-life is 5.27 years, and the decay reaction is

$$_{27}\text{Co}^{60} \rightarrow {}_{28}\text{Ni}^{60} + {}_{-1}\text{e}^{0} + 2\gamma$$

where the energies of the products are: electron 0.3 MeV, gamma rays 1.17 and 1.33 MeV. Let us start with $N_0 = 10^{16}$ atoms (about 1 μg) at time zero and find how many are left at $t = 1.5$ years, and what the activity is then. (It will be convenient to leave the time units in years.) The decay constant is

$$\lambda = \frac{0.693}{t_H} = \frac{0.693}{5.27} = 0.131 \text{ yr}^{-1}$$

At $t = 1.5$ yr,

$$N = N_0 e^{-\lambda t} = (10^{16})e^{-(0.131)(1.5)} = 0.822 \times 10^{16} \text{ atoms}$$

The activity is

$$A = N\lambda = (0.822 \times 10^{16})(0.131 \text{ yr}^{-1}) = 1.08 \times 10^{15}/\text{yr}$$

or, since 1 yr = 3.16×10^7 sec, $A = 3.41 \times 10^7$/sec. It is rather surprising to find so many disintegrations, considering the long time of more than 5 years for half of the nuclei to decay. We must keep in mind, however, that the number of atoms, even in a microgram, is enormous.

The isotope phosphorus-32 is useful for the study of the effects of fertilizer on plant growth. Suppose that the activity of a sample is measured at time zero to be 2.00×10^6 disintegrations per second (d/sec) and 48 hours later to be 1.82×10^6. Let us deduce its half-life and the amount initially present:

$$\lambda = \frac{\ln (A_0/A)}{t} = \frac{\ln (2/1.82)}{48} = \frac{0.0943}{48} = 0.00196 \text{ hr}^{-1}$$

and

$$t_H = \frac{0.693}{0.00196} = 354 \text{ hr} = 14.7 \text{ days}$$

The actual half-life is 14.2 days.

Let us return to the study of the energy requirements for decay. Imagine a variety of nuclear species of the same atomic mass number A, but differing in the relative number of neutrons and protons. These nuclei are *isobars* (equal mass number) for all of which A = Z + N is the same. Their *masses M* are not identical, however, since the binding energy varies with Z and N. A typical graph of *M* as it depends on Z for A = 15 is shown in Figure 23-6. It is of the shape of a parabola with vertex at element Z = 7—i.e., nitrogen. Following the rule that nuclei seek the lowest mass, we would expect radioactive disintegrations along the directions of the arrows.

An increase in Z is achieved by an electron emission, while a decrease in Z is due to a positron emission. The isotopes to the right of the minimum at N^{15} have too much positive charge for stability, while those to the left have too little charge. We would predict the following reactions, among others:

$${}_6C^{15} \rightarrow {}_7N^{15} + {}_{-1}e^0 \qquad \textit{(electron emission)}$$

$${}_8O^{15} \rightarrow {}_7N^{15} + {}_{+1}e^0 \qquad \textit{(positron emission)}$$

Such disintegrations do indeed occur.

Several heavy elements in the crust of the Earth are naturally radioactive. Starting with ${}_{92}U^{238}$, ${}_{92}U^{235}$, and ${}_{90}Th^{232}$, three different chains of decay proceed, called the uranium, actinium, and thorium series, respectively. Eventually, after many alpha- and beta-decays, stable isotopes of lead ${}_{82}Pb^{206}$, ${}_{82}Pb^{207}$, and ${}_{82}Pb^{208}$ are reached. Mass numbers of the isotopes

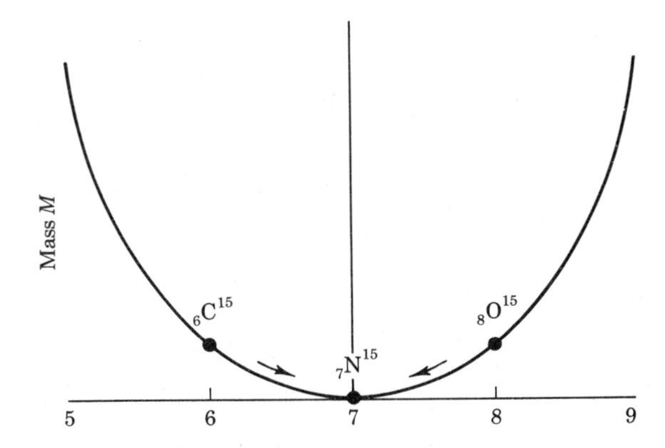

FIGURE 23-6

Decay of radioactive isotopes. Masses are ${}_6C^{15}$, 15.010600; ${}_7N^{15}$, 15.000108; ${}_8O^{15}$, 15.003072

in the series are respectively $4n + 2$, $4n + 1$, and $4n$, where n is an integer. The half-lives in the chains vary from billions of years to fractions of a second. For example, in the uranium series, t_H for $_{92}U^{238}$ is 4.5×10^9 yr, with an alpha-particle emission to become $_{90}Th^{234}$, which disintegrates by beta emission with a half-life of only 24 days. Down the chain we find radium $_{88}Ra^{226}$, half-life 1620 yr. This element is well known because of its discovery by Madame Curie, and for its long use in the treatment of disease and as the luminous coating on the dials of watches. After the last emission of an alpha particle by 138-day $_{84}Po^{210}$, the stable $_{82}Pb^{206}$ is formed.

Estimates of the age of rocks in the Earth or Moon are made from present abundances of various isotopes. The relative amount of lead and uranium in a sample of rock depends on how long ago the sample was formed—say, by volcanic eruption. The longest half-life in the chain definitely being U^{238}, we can write approximately

$$N_{\mathrm{U}} = N_{\mathrm{U}0}e^{-\lambda t}$$

and

$$N_{\mathrm{Pb}} = N_{\mathrm{U}0} - N_{\mathrm{U}}$$

Solving these two equations, and introducing the half-life, we find

$$t = \frac{t_H}{0.693} \ln \left(1 + \frac{N_{\mathrm{Pb}}}{N_{\mathrm{U}}}\right)$$

Suppose we find 10^{-4} atoms of lead for each atom of uranium in a certain ore. Then the age of the sample is estimated as

$$t = \frac{4.5 \times 10^9}{0.693} \ln (1 + 10^{-4}) = 6.5 \times 10^5 \text{ yr}$$

An interpretation of the decay of elements such as radium will give much insight into the phenomena of radioactivity, and show how quantum mechanics is used. If an alpha particle ($_2He^4$ ion) bombards a nucleus, it is deflected by the coulomb potential, as discussed in Section 10-6. If, instead, it is inside the nucleus already, it is strongly bound to other particles. We thus construct, as did Gamow, the potential-energy diagram, as in Figure 23-7. The nuclear radius is taken to be $R = 1.4 \times 10^{-15}A^{1/3}$, and since A is 226 for radium, $A^{1/3} = 6.09$ and $R = 8.5 \times 10^{-15}$ m. Now, the electrostatic part of the potential energy curve is

$$E_p = \frac{k_E\, Z_1\, Z_2\, e^2}{R}$$

where for the alpha particle $Z_1 = 2$, and the radium nucleus (minus the alpha particle) is $Z_2 = 86$. At the nuclear radius, then,

$$E_p = \frac{(9 \times 10^9)(2)(86)(1.6 \times 10^{-19})^2}{8.5 \times 10^{-15}} = 4.7 \times 10^{-12} \text{ J} = 29 \text{ MeV}$$

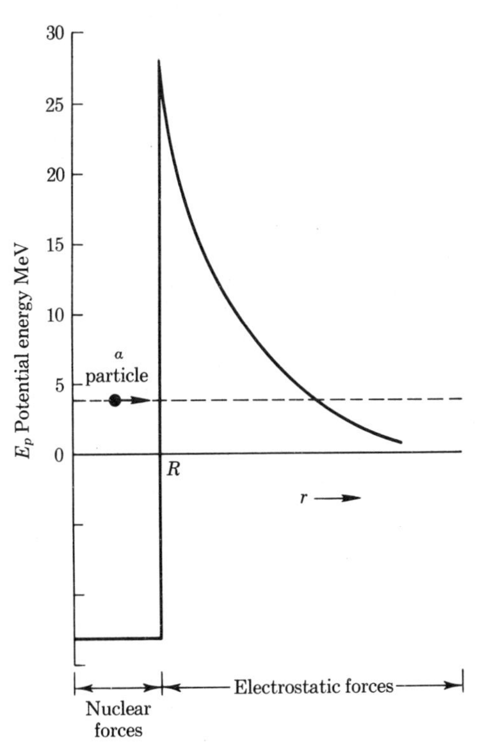

FIGURE 23-7

The potential barrier experienced by alpha particles in a nucleus

Experiments show that alpha particles from radium have an energy of about 5 MeV. Classically, an alpha particle of that energy in the nucleus could never climb the 29-MeV barrier and there would be no such thing as radioactive radium.

Quantum-wave mechanics (Section 22-2) tells us that a particle has a probability ψ^2 of being at a point. The wave function for the alpha particle is found to be large inside the nucleus; but it extends outside as well. Thus there is a small but finite probability of the particle's escape, by a process called "tunneling" through the barrier. There is a rough analogy between the reflection and transmission of light waves on a surface (Section 20-4), and the behavior of an alpha particle as it bombards the barrier.

TABLE 23-1

RELATION OF HALF-LIFE AND ENERGY FOR POLONIUM ISOTOPES, Z = 84

Nucleus	Half-Life (seconds)	Alpha Energy (MeV)
Po^{212}	3.0×10^{-7}	8.8
Po^{214}	1.6×10^{-4}	7.7
Po^{216}	1.6×10^{-1}	6.8
Po^{218}	1.8×10^{2}	6.0
Po^{210}	1.2×10^{7}	5.3

From the alternate particle viewpoint, the higher the energy of the alpha particle in the nucleus, the more frequently it encounters the barrier and thus the greater the chance of penetration. Therefore, we would expect that the higher the alpha-particle energy, the shorter the half-life of the isotope would be. This trend is verified experimentally, as can be seen in Table 23-1, which shows data for some polonium isotopes that appear in the heavy-element radioactive series.

23-3 NUCLEAR REACTIONS

A chemical reaction equation is balanced if two requirements are met—that the number of atoms of each species on both sides are equal, and that the valences of the reacting atoms add up correctly. Also, we recall that some reactions give up energy (*exothermic*) while others require energy to make them go (*endothermic*). For example, we write

$$(Na^{+}O^{--}H^{+}) + (H^{+}Cl^{-}) \rightarrow (Na^{+}Cl^{-}) + H_2^{++}O^{--} + Q$$

where Q represents an energy release.

An analogous nuclear reaction represents the result of bombarding tritium with deuterium:

$${}_1H^2 + {}_1H^3 \rightarrow {}_2He^4 + {}_0n^1 + Q$$

We see that the total nuclear charge is 2 on each side, and that the mass numbers add to 5. The energy release Q, here a positive number, is associated with the excess of mass contained in the reactants over that in the products.* We may abbreviate this reaction by the expression

$${}_1H^3\ (d, n)\ {}_2He^4$$

where (d, n) means in words, "deuteron in, neutron out." Note that the light particles ${}_1H^2$ and ${}_1He^3$ have been "fused" into a heavier particle ${}_2He^4$. The term *fusion* applies to such processes occuring at the low-mass end of the periodic table. We may compute Q by a mass balance

$$M_d + M_t = M_{He} + M_n + Q$$

where the subscripts are obvious. Transposing,

$$Q = (M_d + M_t) - (M_{He} + M_n)$$

$$Q = (2.014102 + 3.016049) - (4.002604 + 1.008665) = 0.018882 \text{ amu}$$

In terms of energy, using 1 amu = 931.5 MeV, we find the value of Q to be

* If there is an excess of mass in the products over that in the reactants, Q is negative. In such event, energy must be provided to the reactants (as kinetic energy) in order for the reaction to take place.

17.6×10^6 eV. We see that the available energy for such a nuclear reaction is *millions* of times greater than for a chemical reaction. We supposed that we could actually bring the deuterium and tritium nuclei (deuteron and triton) together for the fusion event. Recall however, that the particles are both charged and thus repel each other. It will be necessary to provide a high speed of approach to overcome this coulomb barrier. A rough estimate of the necessary deuteron energy, assuming the triton at rest, is obtained by the use of the potential energy, $E_p = k_E e^2/R$. In radioactive decay, the alpha particle is inside the nucleus looking out. Here, the deuteron is outside looking in. For the deuteron of charge $e =$

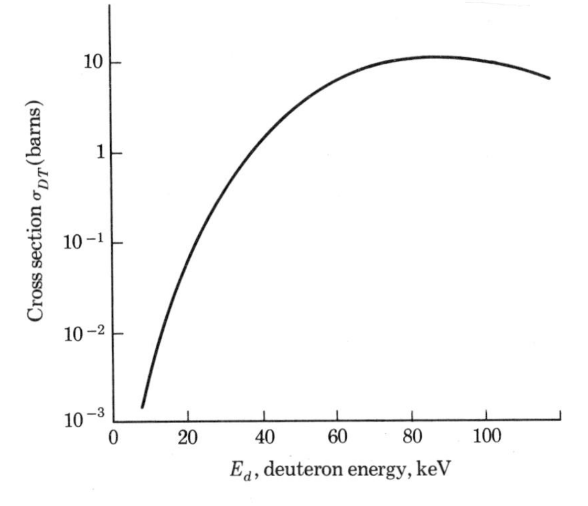

FIGURE 23-8

Cross-section for fusion of deuterium and tritium as it depends on energy, 1 barn = 10^{-24} cm^2

1.60×10^{-19} C to get as close as the nuclear radius of the triton, $R \cong 2 \times 10^{-15}$ m, the energy must be of the order of

$$E_p = \frac{k_E e^2}{R} = \frac{(9 \times 10^9)(1.60 \times 10^{-19})^2}{2 \times 10^{-15}} = 1.15 \times 10^{-13} \text{ J} = 0.72 \text{ MeV}$$

Again, according to quantum mechanics, the incident particle has a probability for penetrating the barrier. The reaction starts at considerably lower projectile energy than 0.72 MeV, with the probability of fusion measured by σ—the cross-section—increasing as E_d increases. Figure 23-8 shows this trend. The hydrogen bomb is based on the fusion principle. Light elements are brought together under conditions of high temperature so that a violent reaction takes place, with release of kinetic energy and radiation.

Particle energies sufficient to cause fusion events involving individual particles are readily attainable by many types of accelerators. However, in order for fusion to be harnessed on a large scale for practical use, it will be necessary to obtain more energy from the reaction than is supplied by the accelerating process.

The Carbon Cycle in Stars

The most important charged-particle reactions so far as life on Earth is concerned are those occurring in the Sun. On the basis of knowledge of stellar composition and energy release, it appears very likely that the processes involve conversion of hydrogen into helium. The elements nitrogen and carbon also enter into the chain of events known as the *carbon cycle*. Figure 23-9 shows the steps compactly. Ordinary carbon ${}_6C^{12}$ is bombarded by high-speed protons ${}_1H^1$, creating ${}_7N^{13}$ and a gamma ray. The ${}_7N^{13}$ decays radioactively giving a positron and ${}_6C^{13}$. The latter reacts with another proton to form ${}_7N^{14}$. The eventual products of the further reaction shown are ${}_2He^4$ and ${}_6C^{12}$, the original material. We can view the whole process as the reaction

$$4H^1 \rightarrow He^4 + \text{energy}$$

where carbon serves as a "nuclear catalyst." Energy in the form of particle motion and light causes the reaction to continue (at temperatures in the

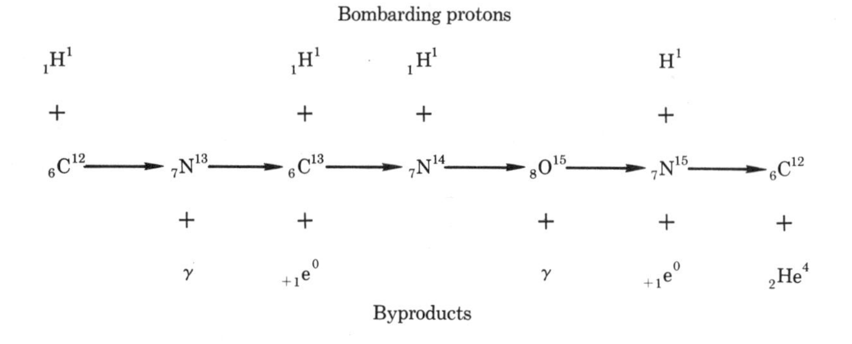

FIGURE 23-9

Diagram of the carbon cycle in the Sun

vicinity of 20,000,000°K), and radiation escapes into space. We should also note that the time for the cycle is about 5 million years because of the low probabilities of reaction, and that it takes a typical photon of light thousands of years to migrate by collisions out from the center of the Sun to the surface.

Carbon-14 Dating

The field of nuclear physics would seem to be little related to archeology, anthropology, history, and religion. Nevertheless, the method of determination of dates using the radioactivity of carbon-14, developed by Libby, serves as a powerful tool in correlating information on ancient documents and artifacts. First, we note that the formation of the radioactive isotope is by cosmic rays in the atmosphere, with reactions such as

$$_7N^{14} + _0n^1 \rightarrow _6C^{14} + _1H^1$$

The amount of C^{14} in the air stays constant, since its rate of formation and loss by decay is in equilibrium. A plant takes in carbon as CO_2 during its lifetime, and a certain equilibrium amount is maintained in the tissues. Animals partake of the isotope indirectly. Following the death of the organism, the C^{14} decays, with a half-life of around 5600 years, emitting a low-energy electron. At any later time, the C^{14} fraction (as determined by detecting the counting rate) is less than normal for carbon, allowing the date when the being was living to be computed from the ratio $e^{-\lambda t}$. Among the accomplishments using this method have been the dating of the Dead Sea Scrolls and of prehistoric animals and plants.

23-4 FISSION AND NUCLEAR CHAIN REACTIONS

The absorption of neutrons by certain isotopes—notably U^{235} and Pu^{239}—gives rise to a splitting of the resultant nucleus into two heavy fragments. This process, called *fission* (Figure 23-10) is accompanied by two important byproducts. The first is kinetic energy in tremendous quantities as judged from an atomic scale. The two charged fission fragments separate at high speed, bearing a total of about 160 MeV of energy.

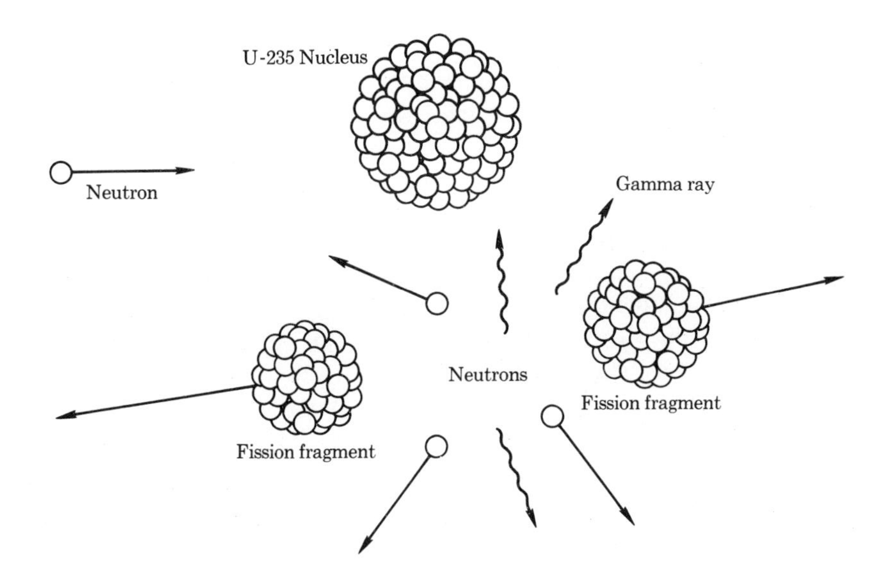

FIGURE 23-10

The fission process

This in turn is transmitted by collisions to particles of material in which the uranium is imbedded. Eventually, most of the kinetic energy is converted to energy of thermal agitation, providing a source of heat superior in many ways to oil or coal. The major difference between heat from nuclear processes and that from atomic processes is in the amount per atom involved in the reactions. The most energetic chemical reaction—neutralization of ionic hydrogen—yields 13.6 eV per amu. Fission, including all products, gives $\dfrac{2 \times 10^8 \text{ eV}}{236} = 8.5 \times 10^5$ eV per amu. Thus U^{235} gives 6.3×10^4 times as much energy per kg of material used. The second product of a fission is a burst of high-speed neutrons, ranging in number from 1 to 7, but averaging around 2.5. These are available for two purposes: (1) to bombard other uranium nuclei and create additional fissions, allowing a self-sustaining chain reaction to proceed; and (2) to be absorbed in normally nonfissionable nuclei, such as U^{238} or Th^{232}, converting these respectively into the fissionable variety, such as Pu^{239} or U^{233}. This "breeding" process extends the ultimate resources of fuels considerably, since the natural-abundance ratio of U^{235} to U^{238} is only 1 to 138. The two most important sets of reaction equations are

$$
\begin{aligned}
&{}_{92}U^{235} + {}_0n^1 \rightarrow {}_{92}U^{236} \\
&{}_{92}U^{236} \rightarrow F_1 + F_2 + 2.5\,{}_0n^1 + Q \qquad (\textit{fission}) \\
&{}_{92}U^{238} + {}_0n^1 \rightarrow {}_{92}U^{239} \\
&{}_{92}U^{239} \rightarrow {}_{94}Pu^{239} + 2\,{}_{-1}e^0 \qquad (\textit{plutonium production})
\end{aligned}
$$

In fission, F_1 and F_2 are two nuclear fragments. The utilization of neutrons in these cases is illustrated by some simple numerical examples. Suppose that one average fission event in uranium gives 2.5 neutrons. One neutron is needed to continue the chain reaction by splitting another uranium atom. This requirement is met if the shape, size, and materials in the system permit no more than 1.5 neutrons to be lost by capture or escape. The reaction will go on as long as uranium is supplied as a fuel. If an improvement in the arrangement is made so that only 0.5 neutron is lost, 1 neutron will be left over to be absorbed in an otherwise inert but "fertile" material to create a new fissionable atom. This replaces the one that was used up, and breeding will have been achieved.

The device in which the controlled chain reaction is made to proceed is the *nuclear reactor* (Figure 23-11). It consists typically of a mixture of water, U^{235}, and a metal such as aluminum or zirconium. The water is called the *moderator*, being composed of light elements H and O that are weak absorbers for neutrons and are good targets for fast neutrons from fission. After many collisions, the neutrons arrive at the energy of thermal agitation, where they are very effective in causing further fission. The metal serves as a *structure* to hold the U^{235} as *fuel*. An alloy of U^{235} and Al or Zr is coated with thin layers of the same metal to form fuel plates in the form of "sandwiches." The coating prevents radioactive-fission products

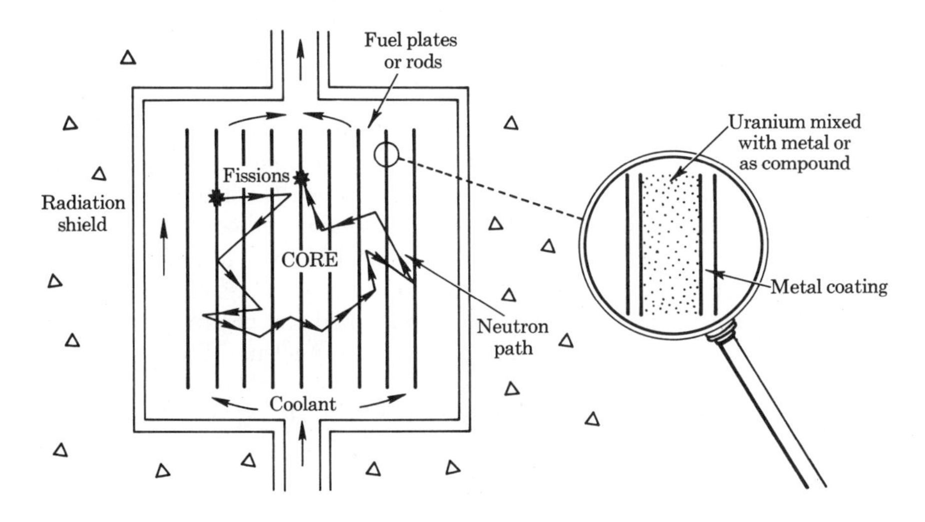

FIGURE 23-11

The nuclear reactor

from escaping into the water, which also serves as *coolant* to remove fission heat. The central part of the reactor is called the *core*; surrounding layers of materials are the *reflector* for conserving neutrons and a *shield* for radiation protection. The size of the system must be large enough to prevent excessive neutron escape through the boundary. The reactor is designed to minimize extraneous neutron capture in nonfissionable materials. Simple arithmetic governs the behavior of a reactor. One may define a *multiplication factor* k as the following product:

$$k = \mathcal{L} f \eta$$

For each fast neutron from fission, the fraction that remains in the fuel-bearing core without escaping ("leaking") from it is $\mathcal{L}$. Of each $\mathcal{L}$ neutrons left, a fraction f is absorbed in U, in competition with water and metal. For each $\mathcal{L}f$ neutrons absorbed in U, the number of new fission neutrons is η. The product $\mathcal{L}f\eta = k$ represents the number produced at the end of a cycle. When k is greater than 1, neutrons accumulate, and the system is said to be *supercritical*; when k is less than 1, the population declines to the *subcritical* state. For practical operation, k should be exactly 1, the *critical* state.

Typical values of these constants for a reactor such as the one that operated the first nuclear submarine "Nautilus" are $\eta = 2.1$, $f = 0.8$, and $\mathcal{L} = 0.6$, whose product is close to 1.

The so-called atomic bomb is an assembly of fissionable U^{235} or Pu^{239} that has a large value of multiplication factor, so that the chain reaction takes place in a few microseconds. The weapon is kept together long enough for much of the material to be converted into energy of explosion. Although the bomb and a reactor may use the same fuels, one is violent and uncontrolled, while the other is peaceful and controlled. In a typical fission reaction

$$_{92}U^{235} + {_0n^1} \rightarrow F_1 + F_2 + 2{_0n^1} + Q$$

the F_1 and F_2 are two atoms whose mass adds up to 234, as seen from the balance of A-values. They are called *fission fragments* when viewed as particles coming out of the reaction, or *fission products* when considered in the context of the problem of their use or disposal as radioactive materials. Since there are many combinations of A that add to 234, many different fission products appear in the "burning" of a certain mass of uranium, all highly radioactive because their nuclear mass and charge are far from the values that yield a stable isotope. Suppose the nucleus $_{92}U^{236}$ formed by capture of a neutron by $_{92}U^{235}$ releases two neutrons and splits in the likely ratio of about 3:2. Possible fragments have $Z_1 = 56$, $A_1 = 144$, $Z_2 = 36$, $A_2 = 90$, which correspond to the isotopes $_{56}Ba^{144}$ and $_{36}Kr^{90}$. By means of electron emission (beta decay) the krypton isotope is converted to zirconium. This takes place in a series or chain of reactions. The one that produces the dangerous isotope strontium-90 is

$$_{36}Kr^{90}(33\text{ sec}) \rightarrow {_{37}Rb^{90}}(2.7\text{ min}) \rightarrow {_{38}Sr^{90}}(28\text{ yr}) \rightarrow {_{39}Y^{90}}(64\text{ hr}) \rightarrow {_{40}Zr^{90}}$$

We see that the half-lives (in parentheses) get longer at the start. Each reaction involves the emission of a beta particle and one or more gamma rays.

23-5 PENETRATION OF MATTER BY HIGH-SPEED PARTICLES

We are interested in the interaction with matter of the various high-speed particles, since they must be detected, identified, and their numbers measured for scientific purposes. They must be stopped to protect those who work with radioactive materials, X-ray machines, charge accelerators, and nuclear reactors. They may heat and cause damage to the substance through which they pass. If they come into contact with the human body, there is a hazard to tissue and cells. Three classes of radiation

TABLE 23-2
RADIATIONS

Charged particles
electrons (β), protons, helium nuclei (α)
Neutral particles
neutrons, neutrinos
Electromagnetic radiation (photons)
X rays, gamma rays (γ)

can be formed according to their nature and mechanism of penetration or stopping, as shown in Table 23-2. Charged particles generally lose energy by the process of ionization on passage through matter. The electric field of the moving charge exerts a strong force on electrons and either excites them to a higher energy level or removes them from their atoms.

Let us examine this effect by finding the kinetic energy developed in an electron by a passing positive charge at closest distance b, as shown in Figure 23-12(a). The ion, of charge Ze, moves in the x-direction with a speed v, and exerts an electrostatic force on the electron of charge e in amount

$$F = \frac{k_E e(Ze)}{R^2}$$

We suppose that the transit is rapid enough so that the x-components of force give no net effect. However, $F_y = F \cos\theta$, and from Figure 23-12, $\cos\theta = b/R$, $R = \sqrt{x^2 + b^2}$, and $x = vt$. A plot of F_y as it depends on time is shown in Figure 24-12b. Recalling from Section 10-2 that the momentum

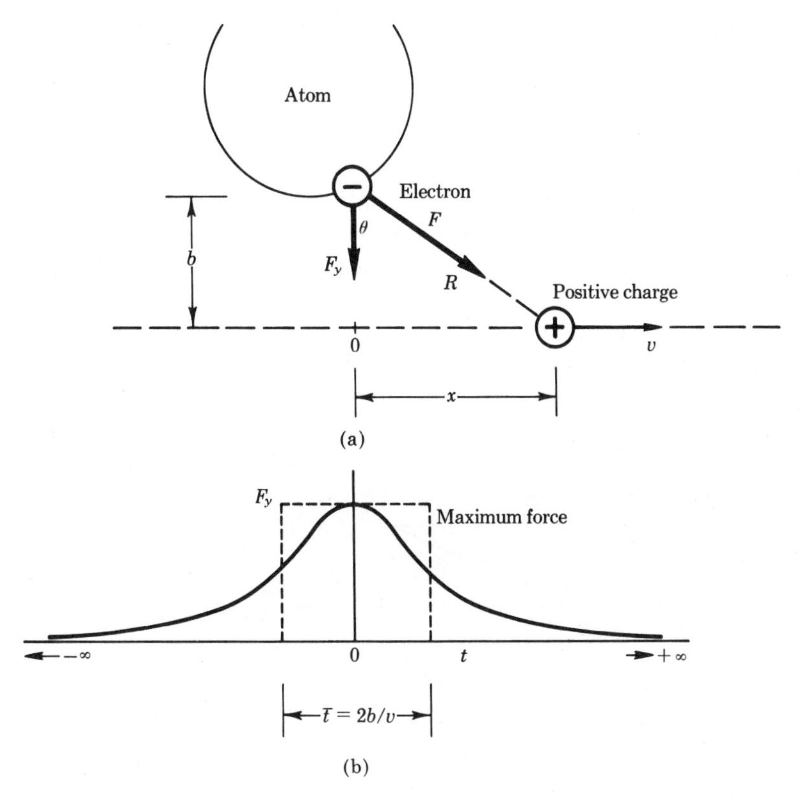

FIGURE 23-12

Interaction of a moving positive charge and an atomic electron

gain with an impulse force is $p = \bar{F}t$, we can replace the curve by the dotted one with constant maximum force acting over a shorter time $\bar{t}$, which turns out to be just $2b/v$; hence the y-component of momentum is

$$p_y = 2F\frac{v}{b} = \frac{2k_E Z e^2}{vb}$$

and the corresponding kinetic energy is

$$E_k = \frac{p_y^2}{2m} = \frac{2}{m}\left(\frac{k_E Z\, e^2}{vb}\right)^2$$

The energy supplied to the electron must of course come from the passing particle, which itself slows down. We can proceed to find out its energy loss per unit length of path through the medium, in terms of the number of atoms per unit volume n and the average energy I required to excite or ionize each atom. The *stopping power* is defined as this ratio $\Delta E/\Delta x$, and the *range* $\mathcal{R}$ is the distance it goes before losing all of its energy. We find, as expected, that the higher the atom number density, the higher the Z of the projectile, and the lower its speed, the shorter is the range. Roughly, $\mathcal{R}$ for an alpha particle in air at ordinary density is 0.5 cm per MeV of initial energy. We see immediately that the range in a liquid or solid is extremely small, merely because of the large number of electrons. A sheet of writing paper will stop most alpha particles. We can observe the paths of charges in the *cloud chamber*, where a small excess of vapor such as water or alcohol is present in air. The electrons that are torn from their atoms provide centers of condensation for drops, which grow in size, giving visible "tracks" of the projectile.

We can measure the particles electrically as well in the *ionization chamber*, which is most simply a capacitor with applied potential and with gas between the plates (Figure 23-13). A single alpha particle entering the space will cause ionization. The electric field sweeps out electrons and positive ions which are registered as a pulse of current. If there is a steady flow of charges into the space, a continuous current will be measured.

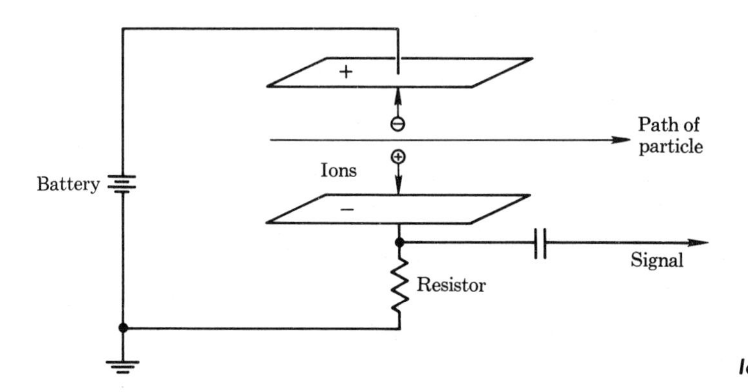

FIGURE 23-13

Ionization chamber

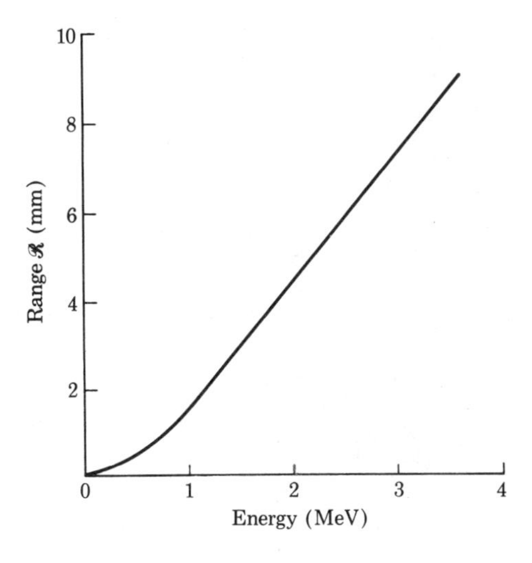

FIGURE 23-14

Range of electrons in metal as a function of energy

The mechanism of the stopping of beta particles in matter is similar in some ways to that of alpha particles. The electrons cause ionization by what amounts to collision, but account must be taken of the effects of relativity. There is also a loss of energy in the form of radiation due to acceleration of electrons. These *bremsstrahlung,* mentioned previously in Section 19-2, play an important role when the electron speed is very high. By interposing more and more layers of material in a stream of beta particles and observing the number transmitted, the range can be found experimentally. Figure 23-14 shows the range in millimeters for beta particles of different energies in aluminum. For small energy, $\mathcal{R}$ varies as E^2, for large energies as E. A straight-line fit of the curve for E greater than about 0.8 MeV leads to the formula

$$\mathcal{R}\left(\frac{\text{g}}{\text{cm}^2}\right) = 0.55E - 0.16$$

where E is in MeV. To find the range in cm, we divide this by the density in g/cm³. For example, the range in Al of 1.5 MeV beta particles is $\mathcal{R} = (0.55)(1.5) - 0.16 = 0.66 \text{ g/cm}^2$, or $\mathcal{R} = (0.66 \text{ g/cm}^2)/(2.7 \text{ g/cm}^3) = 0.24$ cm.

A convenient rule of thumb is that the energy loss per pair of ions formed in a gas is 32 electron volts. This is a good estimate for various particles, energies, and gases.

Three processes cause the attenuation of gamma rays in matter. The first is the *Compton effect,* where photons are scattered from electrons, as discussed in Section 10-4. This takes place even if the gamma energy is not high enough to cause the second process, the *photoelectric effect* (see Section 8-8), which is actually ionization. Gamma rays of higher than 1-MeV energy can be converted into electrons and positrons in the *pair-production process.*

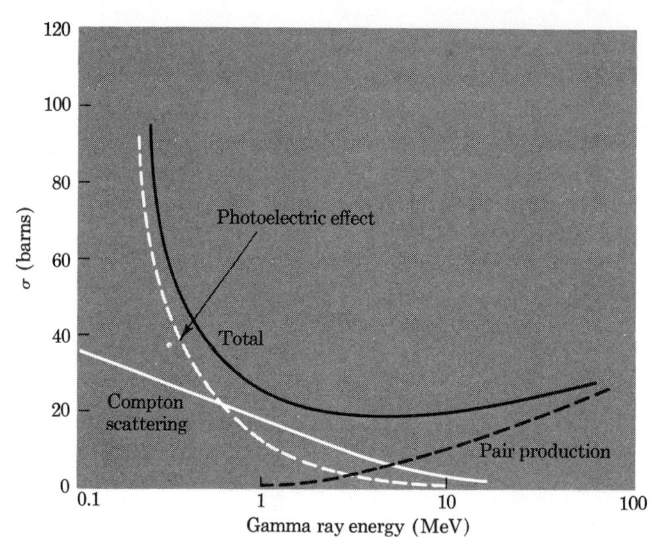

FIGURE 23-15

Cross-section for gamma-ray interactions in lead, 1 barn = 10^{-24} cm^2

We can also define a cross-section for collision of a gamma ray with a target, in close analogy with that for atomic-particle collisions (Section 16-2). First, we deduce the value of total cross-section σ from experimental measurements. Then, since there is a certain chance of each of the three types of interaction, we may assign separate cross-sections σ_c, σ_i, and σ_p. These are related to the total by

$$\sigma = \sigma_c + \sigma_i + \sigma_p$$

Figure 23-15 shows these component cross-sections and the total. The attenuation of a stream of such radiation follows the exponential law

$$j = j_0 e^{-N\sigma x}$$

where N is the number of atoms per unit volume. We saw earlier that alpha and beta particles have rather definite ranges, depending on their energy and the medium. The exponential nature of attenuation of gamma rays implies that there is some chance that radiation will penetrate to any depth, and the idea of range is thus not applicable. Instead, the thickness t_H of a layer that reduces the intensity to one-half its incident value is used:

$$t_H = \frac{0.693}{N\sigma}$$

The analogy between radioactive decay and gamma-ray attenuation comes from the common exponential behavior. We can compute t_H for lead, with density $\rho = 11.3 \times 10^3$ kg/m^3. In an atomic mass 207 kg, there are 6.02×10^{26} atoms; hence in 1 m^3 there are $N = \frac{11.3 \times 10^3}{207}(6.02 \times 10^{26}) = 3.28 \times$

10^{28} atoms. At 1-MeV energy, the cross-section from Figure 23-15 is 24 barns, and since 1 barn = 10^{-24} cm^2 = 10^{-28} m^2, $\sigma = 2.4 \times 10^{-27}$ m^2. The product $N\sigma$, labeled μ and called the *attenuation coefficient,* is

$$\mu = N\sigma = (3.28 \times 10^{28})(2.4 \times 10^{-27}) = 79 \text{ m}^{-1}$$

The half-thickness is thus

$$t_H = \frac{0.693}{79} = 0.0087 \text{ m}$$

or a little less than 1 cm.

Since the neutron is electrically neutral, no electrostatic force is exerted on it by the charge of atomic electrons or nuclei. Only when the neutron comes within the range of nuclear forces, which are of the order of 10^{-15} m, will there be a reaction. For fast neutrons, a hard-sphere model gives fair results. The cross-section for scattering is of order $\sigma_s = \pi R^2$, where the nuclear radius is about $R = 1.4 \times 10^{-15} \text{ A}^{1/3}$. Low-speed neutrons are captured by the nucleus, with a cross-section for absorption that varies inversely with speed:

$$\sigma_a = \frac{\sigma_0 v_0}{v}$$

where σ_0 is a value at a reference speed v_0 = 2200 m/sec. The values of σ_0 vary greatly from one element to another, as seen in Table 23-3. The

TABLE 23-3

CROSS-SECTIONS FOR NEUTRON CAPTURE, 2200 m/sec

Isotope	σ_0 (barns*)
$_2He^4$	$\cong 0$
$_1H^2$	5.7×10^{-4}
$_1H^1$	3.3×10^{-1}
$_{92}U^{235}$	6.8×10^{2}
$_5B^{10}$	4.0×10^{3}
$_{54}Xe^{135}$	2.7×10^{6}

* One barn = 10^{-28} m^2.

attenuation of a stream of neutrons obeys the exponential rule as discussed in Section 16-2. Let us compute the mean free path λ for neutrons of speed 2200 m/sec, $\sigma_0 = 4 \times 10^3$ barns $= 4 \times 10^{-25}$ m^2, with $N_B = 6 \times 10^{26}$/m^3, in a mixture of boron and weakly absorbing aluminum. Then, $N\sigma = (6 \times 10^{26})(4 \times 10^{-25}) = 240$ m^{-1} and the mean free path is

$$\lambda = \frac{1}{N\sigma} = \frac{1}{240} = 0.0042 \text{ m}$$

Radiation Hazards

When charged particles, gamma rays, and neutrons bombard the human body, damage to tissue occurs. Such radiation may come from an improperly shielded machine or an exploding atomic weapon, or from disintegration of radioactive materials that have entered the body. Energy is released in the cells with resultant harm to them. The term *dose* refers to the amount of radiation absorbed. The *roentgen* is a measure of dose for electromagnetic radiation. It gives an energy release of 9×10^{-6} J in a gram of air at NTP. One roentgen per hour would result from exposure to a stream of about 5.8×10^9 gamma rays of 1 MeV energy striking an area 1 square meter per second. The maximum permissible dose for workers in atomic plants is 0.1 roentgen per week, and the general population should be limited to less than 1 per cent of that value.

The *curie* is a measure of the activity or rate of decay of an isotope, being defined as 3.7×10^{10} disintegrations per second (d/sec). The maximum amount of radioactive material in the air that may be permitted is quoted in μcuries/cm^3. For example, 3×10^{-10} μcurie/cm^3 of Sr^{90} corresponds to the "tolerance" dose. We can show that this corresponds to a very small mass of strontium. The half-life is 28 years, hence the decay constant is

$$\lambda = \frac{0.693}{t_H} = \frac{0.693}{(28 \text{ yr})(3.16 \times 10^7 \text{ sec/yr})} = 7.8 \times 10^{-10} \text{ sec}^{-1}$$

The number of disintegrations per second, the activity, is

$$A = (3 \times 10^{-10}\,\mu\text{curie})(3.7 \times 10^4 \text{ d/sec-}\mu\text{curie}) = 1.1 \times 10^{-5} \text{ d/sec}$$

Since $A = N\lambda$, the number of atoms of strontium is

$$N = \frac{1.1 \times 10^{-5}}{7.8 \times 10^{-10}} = 1.4 \times 10^4$$

Since 90 grams contains 6×10^{23} atoms, this number corresponds to only 2×10^{-18} grams—i.e., a billionth of a billionth of a gram.

PROBLEMS

23-1 Estimate the radius of the nucleus of $_{92}U^{238}$.

23-2 Calculate the mass difference in atomic-mass units corresponding to the atomic-binding energy in hydrogen, 13.6 eV. Discuss the distinction between "atomic energy" and "nuclear energy."

23-3 If the radius of the circular path of the C^{12} ion in a mass spectrograph is 0.15 m, how far apart at the collecting plates are the ions of C^{12} and C^{14} (assumed to be of mass 14.00)?

23-4 Calculate the binding energy for the nucleus of tritium $_1H^3$, M_t = 3.016049.

23-5 Find the total electrostatic energy in MeV of the nucleus of iron $_{26}Fe^{56}$.

23-6 *Isobars* are particles that have the same mass number but different atomic number—e.g., $_7N^{15}$, $_8O^{15}$—while *isotones* have the same number of neutrons—e.g., $_7N^{14}$, $_8O^{15}$. By a study of the semiempirical mass formula, deduce which is the more stable, N^{15} or O^{15}.

23-7 A 1-microgram sample of radioactive iodine I^{131}, half-life 8.1 days, is deposited in a patient undergoing treatment. What fraction is left after 14 days? What activity does it have at that time?

23-8 A nuclear explosion produces 250 grams of a fission product with a half-life of 4 months. How much will be left after 1 year? 2 years? 10 years?

23-9 The abundance ratio of the two main isotopes of uranium is $N_{238}/N_{235} = 138$. The half-lives are: U^{238}, 4.5×10^9 yr; and U^{235}, 7.1×10^8 yr. What was the ratio 1 billion years ago?

23-10 In addition to a neutron, what particles are produced by bombarding (a) N^{14} with alpha particles? (b) Li^7 with protons? (c) O^{17} with protons? (d) C^{13} with alpha particles?

23-11 Complete the following nuclear equations:

$$_2He^4 + {}_4Be^9 \rightarrow {}_6C^{12} + (\quad)$$

$$_{88}Ra^{226} \rightarrow (\quad) + {}_2He^4$$

$$_{92}U^{238} + {}_0n^1 \rightarrow {}_{56}Ba^{144} + 2{}_0n^1 + (\quad)$$

23-12 Rutherford discovered the possibility of nuclear transmutation in 1919 by bombarding nitrogen with alpha particles. Find the energy Q in his reaction

$$_2He^4 + {}_7N^{14} \rightarrow {}_8O^{17} + {}_1H^1 + Q$$

(The mass of $_7N^{14}$ is 14.00307; that of $_8O^{17}$, 16.99914.) What minimum amount of energy must the alpha particles have to cause reaction?

23-13 The alchemist's dream of producing gold is possible by the (n, p) reaction

$$_{80}Hg^{198} + {}_0n^1 \rightarrow {}_{79}Au^{198} + {}_1H^1$$

Calculate the energy released or absorbed in the process, assuming that the mass of Hg is 197.9668, and that the mass of Au is 1.37 MeV larger.

23-14 The activity of C^{14} in a sample of wood from an Egyptian mummy case is only $\frac{2}{3}$ of its modern value. How old is the wood?

23-15 Plutonium undergoes fission with neutron bombardment according to the reaction

$$_{94}Pu^{239} + {}_0n^1 \rightarrow F_1 + F_2 + 3{}_0n^1$$

Assuming the fission fragments have the ratio 3:2 both in nuclear charge and mass, what isotopes are they?

23-16* Neutrons from the fission process have a high energy, 2 MeV, while those that produce fission are of low energy 0.02 eV. In a reactor that contains water as a moderator the neutrons on the average lose one-half their energy by each collision with hydrogen atoms. How many collisions are required to slow neutrons from 2 MeV to 0.02 eV? Estimate how far they travel if the average distance of flight between impacts is about 1 cm. How would you go about computing how long in time that the slowing process takes?

23-17 By interposing successive layers of aluminum foil in a beam of electrons, the maximum range is found to be 0.9 cm. What is the particle energy?

23-18 How thick should a lead shield be for a 75 keV X-ray machine, attenuation coefficient 3850 m^{-1}, in order to reduce the radiation to 0.1% of its incident value?

23-19* Air containing radioactive strontium is breathed continuously for a day. If the average beta-energy release per disintegration is 0.18 MeV, estimate the concentration of Sr-90 in air that would give a hazardous dose of 500 roentgens.

23-20* A metal sphere in a vacuum is painted with a radioactive compound that emits beta particles. Discuss the behavior of particles and potential, if the half-life of the element is t_H and the particle energy is E.

23-21 Alpha particles of 4-MeV energy arise from a nuclear reaction occurring in a cloud chamber. If the air is at 1.5 atm at temperature 20°C, what is the approximate range of the charges?

23-22* A spherical lead shield is erected to protect workers from the 1.2-MeV gamma rays of a 70 curie cobalt-60 source. (a) If the cross-section for Pb^{206}, specific gravity 11.3, at that energy is 25 barns, what is the attenuation coefficient? (b) Estimate, taking account of both inverse-square spreading and exponential attenuation of radiation, how thick the shield should be to limit the dose during an 8 hr work day to less than 0.1 roentgen.
Suggestion: Start with a one-foot thickness as trial value.

PHYSICS AND BIOLOGY

24

The question "What is life?" has preoccupied man since his beginning. Through the concepts and methods of modern chemistry, physics, and biology, some progress is being made in the understanding of the distinction between living organisms and nonliving substances. The subject of life has profound implications in terms of philosophy, sociology, and medicine. Most scientists believe that the processes that go on in living beings—metabolism, growth, reproduction, and adaptation—are governed by physical laws. On this premise, they seek to find logical and consistent descriptions of biological processes.

The subject of biology is far too vast for us to attempt any comprehensive discussion of its physical basis. Instead, in this chapter we shall merely select a few topics in which application of the facts and principles of physics provides some insight. In particular, we shall discuss the relation of molecular structure and the statistical character of living beings, describe the function of various physiological systems in physical terms, and draw analogies between thought and the operation of an electronic computer.

24-1 MOLECULES AND STATISTICS

Man is considered to be the most complex and highly developed of all living things. It is interesting to examine the relation between man's size and structure with those of the atom and molecule. His body, of the order of a meter in height and 75 kilograms in mass, contains some 2.5×10^{27} molecules of size 10^{-10} meters and mass around 10^{-26} kg. It is reasonable to assume that many types of molecules are required to permit the variety of specialized function, but it is not evident that the number of molecules need be so enormous. The concepts of statistics help us understand the situation. An organism composed of a very few molecules would be subject to a very chaotic environment, being bombarded by other particles in thermal agitation. It would be difficult for such a being to function in an orderly manner. Indeed, man would be incapable of concentrating to the extent needed to even develop the concept of an atom! We have seen, in our previous study of statistics, that the distributions of particles in space or in speed were uniform and smooth if the number of molecules was very large, and that steady flows of fluid take place by diffusion processes, even though the individual particles are in erratic motion. When large numbers of molecules operate in concert, the effects of thermal agitation are smoothed out. The physical boundary of any substance is fuzzy and fluctuating when considered on the submicroscopic scale of the atoms, but quite sharp and well-defined from the macroscopic viewpoint of man. Forms of life of near-molecular size such as viruses and bacteria have life processes that are simple and crude compared to that of higher animals. There is not enough complexity to sustain thought, memory, and control. Such simple organisms however do have functions which require that we identify them as living beings.

Much conjecture centers about the environment that could have favored the original formation of molecules of the type that sustain life. The atmosphere following the primeval cooling of the Earth may have included significant amounts of water (H_2O), ammonia (NH_3), and methane (CH_4). The necessary energy to convert these simple molecules into more complex forms may have been provided by ultraviolet light from the Sun, or by radiation from radioactivity in the Earth's minerals, or by electrical currents in the form of lightning. It has been demonstrated in the laboratory that electrical discharges, on passing through such elementary gases, yield compounds like formaldehyde (CH_2O), acetic acid (CH_3COOH), and simple amino acids. Modern biology tells us that the latter are the building blocks of the molecule DNA (deoxyribonucleic acid) that is the key to reproduction, in that it carries the instructions for the formation of new cells and tissues. The DNA structure is viewed as two long, linked chains of molecules, in the form of a spiral, the basic components of which are the smaller molecular units arranged as stair steps, A (adenine), G (guanine), C (cytosine), and T (thymine). For example, the chemical formula for unit G is $C_5H_5ON_5$. The DNA molecule is thus only a relatively small step away from the familiar hydrocarbons that are produced at will in the chemical laboratory. If such "chemical evolution" is the method by which life developed on Earth, it is easy to imagine that there is life on other planets elsewhere in the universe. Other life cycles, involving elements such as silicon in place of carbon, can be visualized. We can then surely speculate that other intelligent forms exist on some of the billions of planets of our galaxy. To determine this, the first step will be to ascertain whether Mars or Venus have simple plants or animals (in this respect, it is important that the first space vehicles to land there not carry Earth life as a contamination).

Evolution

The theory of *biological evolution,* usually ascribed to Darwin, is recalled as involving the gradual transition, over millions of years, of elementary one-celled biological entities into the higher forms of life. The method by which evolution is assumed to proceed is *natural selection,* in which individuals that are better suited to a particular environment tend to survive more often and reproduce their kind more successfully. The origin of differences in the character of a species is the process of *mutation.* A distinction is made between small variations in features that are not inherited and the significant "jump" in features involving mutated genes—the similarity to the transitions between energy levels as described by wave mechanics is worth noting. Once a mutation has occurred, the new property is inherited in accordance with the rigorous laws of genetics. The large-change events are, of necessity, rare; otherwise, those leading to disadvantageous characteristics might quickly wipe out a species. It might at first appear that the race could not change appreciably if the advantageous mutations were rare. However, if the improvement in survival were only 1 per cent, then 101 mutants would live for every 100 nonmutants, and over many generations the mutant would almost completely dominate the species.

The work of Muller revealed new information about the genes as carriers of heredity. He bombarded fruit flies with X rays and produced a host of mutations, demonstrating that high-energy photons can create local ionization that affects the genes greatly. The increase in the rate of mutation was found to increase with the radiation dosage—i.e., the amount of energy absorbed by the flies; the rate of mutation is not significantly dependent on the particular energy (or wavelength) of the radiation. The types of mutation were found to be similar to those occurring in nature.

One can compare mutation in genes produced by light with the excitation of electrons in the atom by photons. If sufficient energy is injected into either structure, a discrete transition can take place. It would seem that a change in a single molecule out of the myriad present in an organism could not have the profound effect that mutations involve. Strong evidence indicates, however, that genetic changes are very highly localized molecular displacements. To illustrate, it has been found that the mere substitution of one amino acid for another in an ordered array of nine such compounds is sufficient to change the gene related to hemoglobin in blood cells and to induce inheritable anemia. Thus the changing of the genes in a reproductive cell can be transmitted readily to the second generation.

24-2 THE DYNAMICS OF ORGANISMS

The physical principles of energy conservation and entropy production, and the mechanisms of transfer of electricity and heat, are applicable to descriptive and quantitative models of organs and systems. We shall briefly review here some of these applications of physics concepts.

Energy Conservation in Photosynthesis

The mechanisms by which a plant grows, using food, water, and sunlight, have long been recognized as involving energy transfer and conservation, but the details of the processes were not fully understood until isotopic tracers were used. As discussed in Section 23-3, isotopes may be stable or radioactive, and their presence may be detected and followed through a system. For example the stable isotope O^{18}, measured by electromagnetic forces on ions, can help trace oxygen through an organism; and the radioactive isotope C^{14}, detected by its emitted beta particles, is especially valuable in finding where carbon goes. As now viewed, *photosynthesis* consists of the following sequence: a plant takes in air, using the contained carbon dioxide CO_2, along with water H_2O and nitrogen from the soil, and produces tissue containing the elements C, H, O, and N, releasing oxygen as O_2 back to the atmosphere. The energy for the cycle comes from sunlight; the energy carried by photons is absorbed by the molecules of H_2O, causing them to split into hydrogen and oxygen atoms, which then carry the energy in chemical form; the green compound *chlorophyll* serves as a catalyst for the reactions that follow. Measurements using O^{18} show that the oxygen given off by plants comes from the absorbed water and not from the carbon dioxide. The nature of the process involving carbon was revealed by the tracer C^{14}. In a cyclic fashion, C is taken up by ribulose phosphate, a 5-carbon molecule; the product splits into two 3-carbon molecules, which eventually combine to form the single sugar glucose phosphate, and provide more molecules of ribulose phosphate to repeat the cycle. A start is thus made toward the formation of more complex and specialized molecules and structures in the plant.

The Relation of Metabolism and Heat Loss

The concepts of heat transfer in air, as discussed in Section 16-6, provide some ideas about the living habits and processes of various organisms. For example, we know that small animals with high metabolic rates, such as songbirds, may eat more than their weight each day, while man exists on a few pounds per day. A simple geometric analysis will give some insight to this distinction.

Let r be the rate of metabolism or heat generation per unit volume and V the volume of the body, while S is the skin surface area through which the heat is lost. If the temperature is to remain constant, then $rV \sim S$, which says that metabolism must be proportional to the ratio of surface to volume. Since S/V is inversely proportional to size*, we deduce that the

* For a sphere, $V = \frac{4}{3}\pi R^3$, $S = 4\pi R^2$, $S/V = 3/R$; for a cube, $V = L^3$, $S = 6L^2$, $S/V = 6/L$.

smaller the animal, the larger the amount of food it must consume in comparison with its weight.

The Thermodynamics of Organisms

The first law of thermodynamics—conservation of energy—is easily recognized as applicable to life processes, but the role of the second law of thermodynamics—the trend in entropy—is somewhat more subtle.

According to the second law, the natural trend that all systems experience is from a state of order to one of disorder. A nonliving isolated system will come to rest because of internal friction. Differences in potential are eliminated, and the temperature becomes uniform. The entropy continuously increases if the system is left alone. On the other hand, living organisms maintain a high degree of order over a very long period. Without life, an organism would decay in a very short time, just as corrosion, or rust, or the dispersal of inanimate substances takes place when exposed to the environment.

Whereas a nonliving entity gains entropy, an active living being loses entropy, or what is the same thing, draws negative entropy from the surroundings, thus causing an increase in the entropy of that environment. It achieves order at the expense of the world in which it lives. Eating, drinking, and breathing are the means by which animals and plants "extract" order from the surroundings to compensate for the tendency for decay and disorder that would normally result without the life processes. The complex compounds that are assimilated bring with them a well-ordered state of matter, then are returned in a degraded form. Plants of course take negative entropy from the basic compounds CO_2 and H_2O, as well as from the "well-ordered" sunlight.

In a broader sense, man as a species creates order by converting raw materials into objects for his use. Even population growth can be viewed as an ordering process. We conclude that man's actions cause a more rapid increase in the entropy of his surroundings than if he did not exist. However, he has ample opportunity to preserve and improve his environment for his own benefit and that of other living beings, keeping in mind that he lives in an essentially isolated system.

Physical Size and Structural Strength

Our description of elastic forces (Section 13-2) provides us with some indication of the appropriateness of man's size, and in fact of the range of sizes of animals in general. The smallest mammal is the shrew, weighing a few ounces, and the largest is the whale, weighing many tons, but this range is still only a few orders of magnitude.

Let us start with the fact that bone structure has a certain molecular arrangement and thus a fairly well-defined elastic constant. Bone is subject

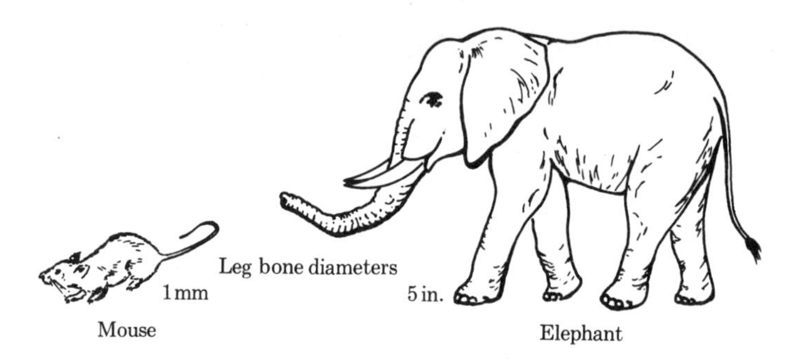

FIGURE 24-1

Comparison of structures of small and large animals

to Hooke's law—stress is proportional to strain, or

$$\frac{l}{L} = E_y \frac{F}{A}$$

We assume that the fractional change in height L of the organism l/L is limited to some small value—otherwise, the bones of the body would be crushed under the gravitational force. This in turn requires that the stress F/A be about the same for all animals. Now, since the force is roughly equal to the weight, proportional to the cube of a linear dimension, and A is the cross-sectional area of supporting bones, proportional to the square of their diameter, we find that

$$\frac{F}{A} \sim \frac{L^3}{d^2} \cong \text{a constant}$$

We conclude from this argument that bone thickness increases rapidly with size: $d \sim L^{3/2}$. This is in qualitative but not exact agreement with the fact that, relative to length, the mouse has very delicate thin bones while the elephant has very thick bones (see Figure 24-1). A land animal very much larger than an elephant would have to be composed of solid bone.

The Physics of the Heart and Circulatory System

Our previous study of fluid flow (Section 13-6) permits us to appreciate much of the action of the heart and the flow of blood. Roughly speaking, the system is an essentially closed circuit of constant volume containing an incompressible fluid—blood. Bernoulli's principle is aptly applied to account for variation in the pressure and speed of the liquid. (The circulatory system is illustrated in Figure 24-2.) The heart is basically a pump, transferring blood to and from the lungs, where carbon dioxide waste is discharged and fresh oxygen is taken in. It transfers the purified blood to the body by means of *arteries*, which in turn divide into fine *capillaries* that are in intimate contact with organs and tissues. The capillaries combine into *veins* that carry the blood back to the heart. Since the total area for flow in the fine blood vessels is very large, the blood speed is low there compared with that in the larger tubes. The pumping action is stimulated by electrical impulses, of frequency about 78/min and of the order of 30 millivolts, that spread over the surface of the heart, causing muscle fibers to contract. The average pressure in the circulatory system is around 100 mm of Hg, and the flow rate of blood ranges from 3.5 to 35 liters/min, corresponding to rest and activity.

Let us use fluid-flow principles (Section 13-6) to calculate the heart power at rest. Now, the power is $P = Fv = (F/A)(vA) = pJ$, where the pressure is

$$P = \left(\frac{100 \text{ mm}}{760 \text{ mm/atm}}\right)(1.0 \times 10^5 \text{ N/m}^2\text{-atm}) = 1.3 \times 10^4 \text{ N/m}^2$$

and the volume flow rate is

$$J = \left(\frac{3.5 \; l/\text{min}}{60 \text{ sec/min}}\right)(10^{-3} \; l/\text{m}^3) = 5.8 \times 10^{-5} \text{ m}^3/\text{sec}$$

Thus the heart power is estimated to be

$$P = (1.3 \times 10^4 \text{ N/m}^2)(5.8 \times 10^{-5} \text{ m}^3/\text{sec}) = 0.75 \text{ W}$$

Neglected in this calculation is the power associated with kinetic energy of

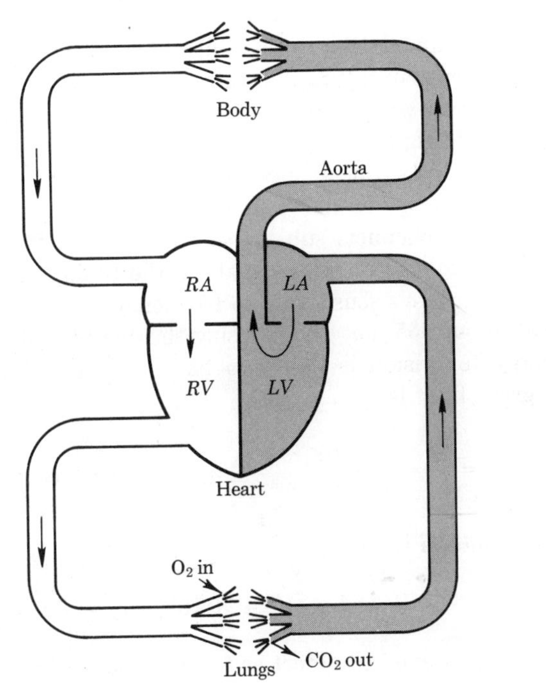

FIGURE 24-2

The circulatory system

the blood. Under active conditions this becomes the dominant component and the heart power must increase to the order of 100 watts.

24-3 THE PHYSICS OF THE SENSES

The Process of Vision

The human eye, illustrated in Figure 24-3, is a remarkable optical system, capable of perceiving and distinguishing near and distant objects, responding to a variety of colors, and being sensitive over a wide range of light intensities. The *cornea* is about 12 mm in diameter and has a radius of curvature of about 8 mm. It provides much of the refractive effect, supplemented by the convex *lens*. Light entering the *pupil* is focused on

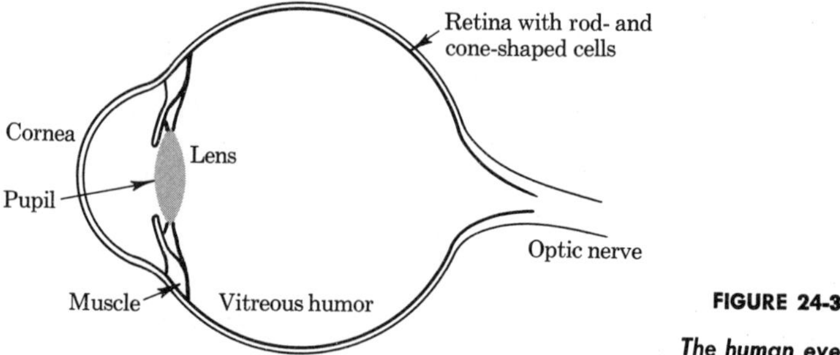

FIGURE 24-3
The human eye

the *retina*, with accommodation to the object distance provided by muscle adjustment of the lens curvature. Objects ranging from infinity to within about 0.25 m (the *near point*) can be clearly seen by the normal eye, with focusing as shown in Figure 24-4.

The retina, with its cone-shaped and rod-shaped cells, converts light energy to electrical energy on nerve fibers. It is believed that the various types of cells contain chemical substances that are selective in their light absorption. The *optic nerve* carries signals to the brain, where the images are interpreted. The eye's sensitivity to frequency has a maximum in the green region at about 555 mμ. By exposing the eye to flashes of light, the threshold energy for vision is found to be about 10^{-18} joule. Now, the frequency of green light is

$$\nu = \frac{c}{\lambda} = \frac{3 \times 10^8 \text{ m/sec}}{555 \times 10^{-9} \text{ m}} = 5.4 \times 10^{14}/\text{sec}$$

The energy of 1 photon is, then,

$$E = h\nu = (6.6 \times 10^{-34} \text{ J-sec})(5.4 \times 10^{14}/\text{sec}) = 3.6 \times 10^{-19} \text{ J}$$

Thus we find that the eye can detect as few as (about) *three* photons. The

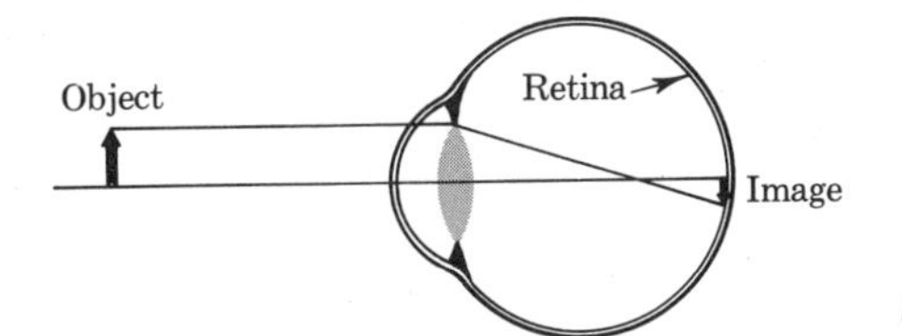

FIGURE 24-4

Image formation in the eye

optical–muscular–nervous system has many special features to enhance vision: depth perception is achieved by the superposition of the slightly different images from the two eyes; the iris of the eye opens and closes to admit more or less light as the occasion demands; certain cells that are more sensitive to blue color become dominant as night approaches.

In spite of the excellent features of the eye, defects are common, and there is a widespread need for spectacles to aid vision. Loss of muscle flexibility in controlling the lens occurs as a person ages, and the near point tends to shift outwards. The eye defects of *nearsightedness* (*myopia*, focusing in front of the retina), and *farsightedness* (*hypermetropia*, focusing behind the retina) may be corrected by spectacles. Inspection of Figure 24-5, along with study of the lens formula (Section 21-2),

$$\frac{1}{p} + \frac{1}{q} = \frac{1}{f}$$

reveals some of the effects. From our knowledge of the fixed size of the eyeball, we know that the image distance q is a definite number. Then, for large p (distant objects), f must be made large—that is, the radii of curvature are large and the lens is relatively flat. For small p, the lens must be made more nearly spherical. Whenever there is physiological limit on the value of f, there will be a focusing as shown. When an auxiliary spectacle lens of focal length f_s is used, the composite f is, in accord with optical

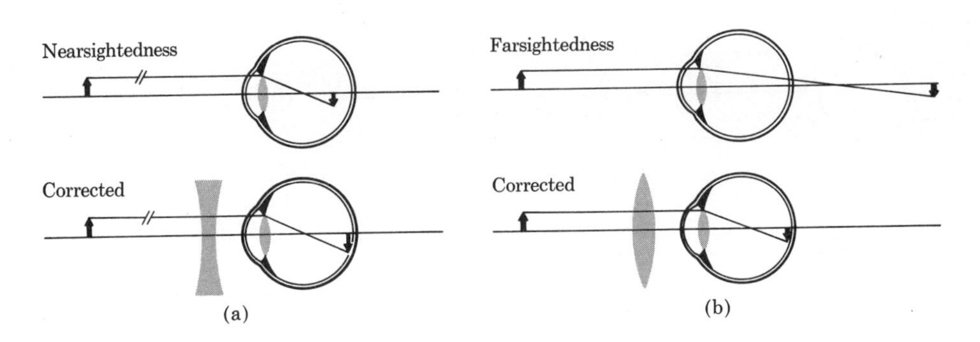

FIGURE 24-5

Defects in vision and corrections by spectacles

theory,

$$\frac{1}{f} = \frac{1}{f_s} + \frac{1}{f_e}$$

The case of nearsightedness—Figure 24-5(a)—requires a concave or divergent lens, for which f_s is negative, and f is larger than f_e. For farsightedness—Figure 24-5(b)—the glass must be a convex or convergent lens such that f is shorter than f_e.

Suppose that a farsighted person finds that the print on a page goes out of focus at 75 cm or nearer. Let us find what focal length his glasses should be, so that he can read well at a convenient 25 cm. Without glasses, $1/p + 1/q = 1/f_e$, where $p = 75$; with glasses, $1/p' + 1/q = 1/f_e + 1/f_s$ where $p' = 25$. Subtracting equations gives $1/f_s = 1/p' - 1/p = 1/25 - 1/75 = 2/75$, and $f_s = 37.5$ cm, a converging lens.

The Physics of Hearing

We are so familiar with hearing that it is hard to realize how complex are the structures and processes that permit us to hear. The following rough description—in conjunction with the schematic diagram of Figure 24-6, which omits a number of anatomical details—will serve to illustrate the physics of the ear. The *outer ear* is basically a funnel filled with air, ending in the membrane that we call the *eardrum*. The ear canal is analogous to a closed-end organ pipe, as discussed in Section 12-2. The eardrum responds to the molecular vibrations of the air that we call sound, and transmits them mechanically to the *middle ear*. The latter consists of a set of three connected bones that act as levers to amplify the very small displacements of the eardrum, to distinguish the signals from vibrations coming through the bone tissue of the head, and to protect the mechanism from

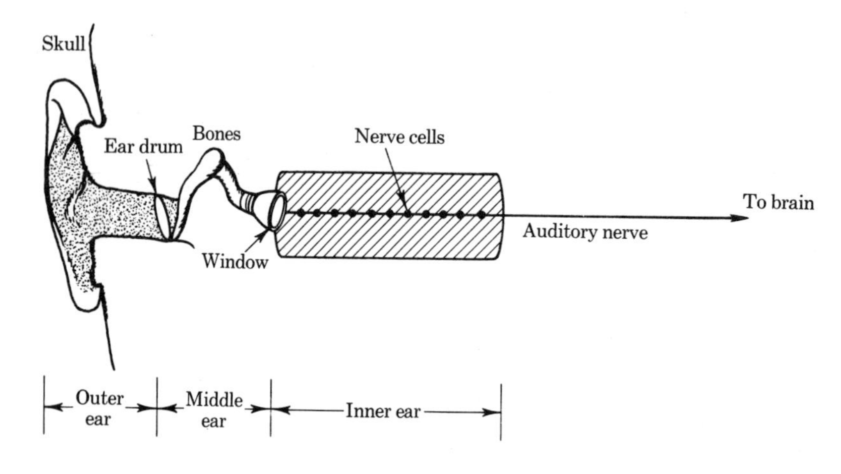

FIGURE 24-6

Functional diagram of the human ear

extremely loud sounds. A second membrane called a *window* separates the air-filled middle ear from the liquid-filled *inner ear* (*cochlea*), and transmits vibrations from one medium to the other. The cochlea consists of a spiral arrangement of parallel tubes, through which liquid waves travel. A (basilar) membrane separating two of these tubes contains a complex organ with various nerve cells. As the membrane is distorted by these hydrodynamic waves, certain hair-like cells are bent and trigger the nerve cells by complex mechanical, electrical, and chemical reactions. The signals then go to the brain via the *auditory nerve*.

The system has the surprisingly excellent ability to respond to a wide range of sound pressures, corresponding to $P_0 = 2 \times 10^{-5}$ N/m²,* which is the threshold of hearing, to around 100 N/m² at the threshold of pain. It is therefore convenient to use a logarithmic scale in expressing pressure amplitudes. The pressure level in decibels (db) is defined as $L = 10 \log (P^2/P_0^2)$. On this scale, L is about 60 db for ordinary speech, while the noise of a jet engine is around 130 db. Sensitivity to frequency varies greatly from one individual to another, but it is found that generally the ear can identify frequencies from a few cycles per second to some 10,000 cps, and is most sensitive at a few thousand cycles per second.

Senses in Animals

It is well known that various senses are particularly acute in some animals—for example, the sharp sight of birds of prey and the ability of dogs to detect scents and to hear very-high-frequency sounds. As an interesting illustration of the relation of physics to animal life, let us consider the behavior of the bat. In flight, it makes an unusual use of sound echoes. It emits a sequence of short complex pulses from its mouth or nose that are reflected from objects and return to the bat's sensitive ears. The process is similar to acoustic sonar or electromagnetic radar. The location, distances,

* The pressure variation from the average in a sound wave is related to both the wave speed v and the particle speed u according to

$$\Delta P = \rho u v$$

Consider air with normal density 1.22 kg/m³ and sound speed 343 m/sec. The particle speed corresponding to the sound pressure threshold for hearing of $\Delta P = 2 \times 10^{-5}$ N/m² is thus

$$u = \frac{\Delta p}{\rho v} = \frac{2 \times 10^{-5}}{(1.22)(343)} = 5 \times 10^{-8} \text{ m/sec}$$

Now, the speed and displacement y of a particle appear in $u = \omega y$, where $\omega = 2\pi\nu$ and $y = u/2\pi\nu$. At a frequency of 1000 cps, the distance a molecule moves is

$$y = \frac{5 \times 10^{-8}}{(2\pi)(1000)} = 0.8 \times 10^{-11} \text{ m}$$

This is of the order of magnitude of the radius of an atom, which shows that the motion of the eardrum is exceedingly small.

and sizes of obstacles or insects are computed in the bat's brain, apparently using information from both ears. Pulse lengths range from 1 to 100 milliseconds, with frequencies within the pulse going as high as 100,000 cps. The bat increases the repetition rate of the signals as it approaches its prey to take account of the shorter distance the sound has to travel. The time between generation and detection of a pulse of an object that is a distance d away is, of course, $t = 2d/v$, where v is the speed of sound, 343 m/sec. For example, if a moth were 1 meter away, this time would be about 6 milliseconds. Apparently, each bat uses its own frequency pattern that permits it to distinguish its echoes from those of many other bats nearby.

Insects such as ants and bees demonstrate an especially good sense of direction. Experiments indicate that bees make use of the location of the Sun in traveling from a feeding place to the hive, rather than relying on their "memory" of the surroundings. Bees communicate the location of new food sources to others in the hive by a complicated dance pattern, in which the amplitude of their vibration tells the time of flight, and the direction reveals the angle between the Sun's rays and the flight path. If a bee is caught and confined for a few hours, it starts off (erroneously) along a path that makes the same angle with the Sun's rays as when it was caught.

Much is yet to be learned about the homing or migratory abilities of birds. Various theories involve such factors as familiarity with the terrain; or use of the Sun or stars, combined with an excellent time sense; or use of the vertical component of the Earth's magnetic field; or the Coriolis force.

Even the force of gravity can be sensitively detected by certain organs in fish. These consist of discs of tissue supported by a large number of fine hair cells. If the animal tips from the vertical, the hairs are bent and nerve cells fire electrically.

24-4 THOUGHT AND THE COMPUTER

The principles of physics help us to appreciate the function of the nervous system. Here we shall find it instructive to draw analogies with electrical circuits, and to compare or contrast thought processes with actions in electronic computers.

Let us briefly review the generally accepted model of the nervous system. The sensory nerve endings or *receptors* receive stimuli of various types —pain, pressure, temperature, odor and flavor, and light. Impulses of electrochemical nature are produced and transmitted along the *nerve fibers* to the *spinal cord*, and finally to the *brain* as supervisory center. The signals are interpreted either automatically or consciously, and impulses are sent back along other nerves to the muscle fibers or chemical-producing devices in the body. The speed of these impulses ranges from 1 to 350 ft/sec. Nerve cells (*neurons*) are segmented, with "nodes" separating their sections. Inside the cell there is a slight excess of potassium ions and outside an excess of sodium ions, with a potential difference of about 0.1 volt across the cell membrane, the interior being negative. A stimulus causes the membrane

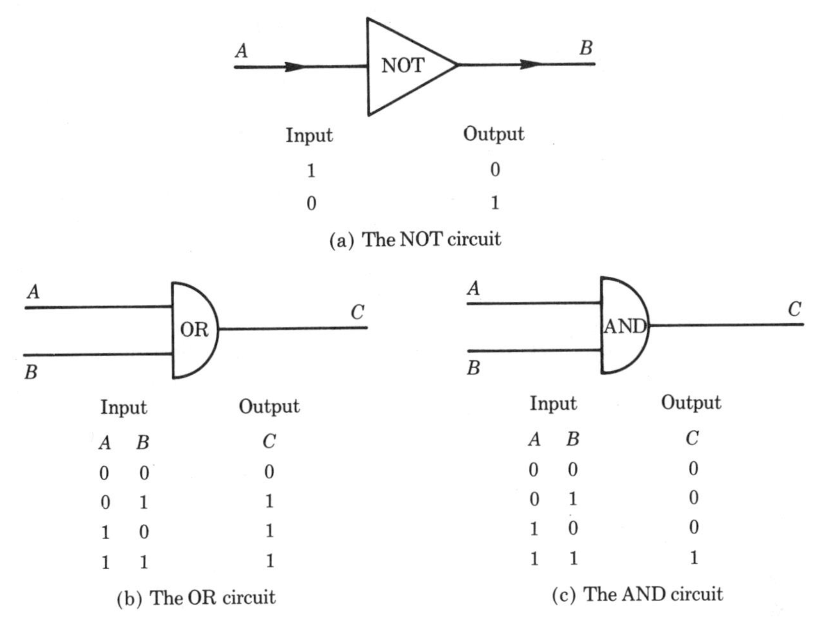

Input	Output
1	0
0	1

(a) The NOT circuit

Input A	Input B	Output C
0	0	0
0	1	1
1	0	1
1	1	1

(b) The OR circuit

Input A	Input B	Output C
0	0	0
0	1	0
1	0	0
1	1	1

(c) The AND circuit

FIGURE 24-7

Computer circuits

to admit extra Na ions, the potential drops to zero, and the cell is said to have "fired." The signal is transmitted from one segment to another along the length of the fiber, with each section recovering in a time of around 1 millisecond.

Most of the functions of the body, such as blood circulation, digestion, and breathing, are automatic. Within the body are many self-regulating devices which display the process of *feedback*. For example, if the temperature of the air goes up, a signal from the skin is sent to the hypothalamus in the brain. This organ serves as a thermostat to command more blood circulation to the skin to remove the heat, and to stimulate the release of perspiration for cooling. Other actions are conditioned responses, learned by experience, such as reading or typing. At the highest level are thought processes involving the coordination of stored information and creative activity.

The electronic computer imitates many of the functions of the nervous system. In addition to the wires that carry electrical impulses, it has circuit elements that "fire" in analogy to nerve cells. Its electronic circuits are capable of performing elementary logical processes such as arithmetic. For example, the NOT circuit shown in Figure 24-7(a) is designed with an output that is opposite to the input. Let us think in terms of volts of applied potential. If the input is 1, the output is 0; if the input is 0, the output is 1. In the OR circuit—Figure 24-7(b)—the output C is arranged to be 1 if either or both of the inputs are 1. In the AND circuit—Figure 24-7(c)—the output C is the result of both the inputs, and is only activated if both are sensed. Arithmetic operations can thus be performed by such simple circuits containing resistors, diodes, and transistors, in conjunction with memory elements. The basic unit of information is the "bit." A device such as a transistor or a ferrite core that can be in either one of two states is able to store a signal of 0 or 1, depending on polarity or direction of magnetization. Arithmetic operations are thus carried out in *binary* mode rather than the decimal mode with which we are more familiar.

Binary Arithmetic

An example of binary counting is as follows: each "place" in the number can only be 0 or 1 and thus counting 0, 1, 2, 3, 4, . . . becomes, 0, 1, 10, 11, 100, ... Table 24-1 lists some numbers in both decimal and binary coding.

TABLE 24-1
COMPARISON OF DECIMAL AND BINARY NUMBERS

Decimal	Binary	Using Powers of 2
1	1	2^0
2	10	2^1
4	100	2^2
8	1000	2^3
16	10000	2^4
32	100000	2^5

To express any decimal-system number in the binary mode, it must be broken down into powers of 2 (as is done with powers of 10 in the decimal mode). For example, 2746 may be expressed in powers of 10 as $(2 \times 10^3) + (7 \times 10^2) + (4 \times 10^1) + (6 \times 10^0)$. Similarly, the number 676 may be written as $(1 \times 2^9) + (0 \times 2^8) + (1 \times 2^7) + (0 \times 2^6) + (1 \times 2^5) + (0 \times 2^4) + (0 \times 2^3) + (1 \times 2^2) + (0 \times 2^1) + (0 \times 2^0)$, or more conveniently, 676 is 1 0 1 0 1 0 0 1 0 0 in binary. It is seen that many more places are required for this method, but it only requires circuit elements that are either "on" or "off"—such as a switch—for storage and use of numbers.

We now add two numbers in binary arithmetic, noting that when two 1's appear they are equal to the binary number 10. Thus

$$\begin{array}{r} 10101 \\ +10010 \\ \hline 100111 \end{array}$$

which is $32 + 4 + 2 + 1 = 39$. As a check, we write

$$\begin{array}{r} 21 \\ +18 \\ \hline 39 \end{array}$$

Carrying in binary is exactly the same as in decimal; for example,

$$\begin{array}{r} 111 \\ +101 \\ \hline 1100 \end{array} \quad \text{or} \quad \begin{array}{r} 7 \\ +5 \\ \hline 12 \end{array}$$

This is accomplished in a circuit by a *feedover* of signal. Whenever a 1 is changed back to a 0, a signal is sent to the next higher register to change it to a 1. Once basic addition can be accomplished, all the other arithmetic operations become available.

Implications of the Computer

The computer is regarded by many thoughtful persons as constituting a major development that will revolutionize man's existence. Let us examine what such devices can and cannot do, and what implications they present for the future.

1. *The expanded capability for memory.* An individual develops in the course of his lifetime an enormous resource of experience—that is, he programs his supply of electrical circuits. The problem, however, is that he cannot always call up information from his mind reliably and completely. By use of the paraphernalia of the computer system—cards, tapes, etc.—modern machines can store and release almost instantly billions of figures, facts, procedures, and information in general.

2. *The increased speed of accurate manipulation of information.* As a simple example, a person can multiply two 10-digit numbers together in a period of some five minutes using pencil and paper, but a computer performs the task in an interval of a few microseconds. There are several ways to view this improvement in terms of the complex mathematical tasks required in modern society. The computer can accomplish work that would require a lifetime of effort by one person, or would demand the attention of thousands of people, or would be humanly impossible in terms of the chance of one error that would nullify the results.

3. *The ability of the computer to direct the action of mechanical, chemical, and electrical devices that eliminate work.* The machines of the Industrial Revolution, although they increased man's productivity and strength, did little to eliminate the drudgery of daily occupation. The automated equipment that computers control, however, releases man from repetitive and monotonous work, and permits greatly increased free time for more satisfying activities.

Accompanying these benefits are a few technical problems. Computers can certainly be programed to perform many of the processes that can be classified as "reasoning." They undoubtedly can be designed to provide instructions on how to build new and better computers. However, they are incapable of the unique type of intelligence exhibited by man, who can accommodate his thinking to entirely unforeseen situations; this point is exemplified in the wise decision to give astronauts the option to control their spacecraft manually. Being inanimate, the computer can do no more than it is told to do. This fact uncovers a new problem. The speed and capacity of computers far outruns man's ability to come up with proper instructions. A disproportionate amount of time and effort is required to translate man's intention into terms with which the device can operate. For instance, computers can be programed to translate, after a fashion, from one language to another. Unfortunately, they perform an "automatic" job without reason. For example, "The spirit is willing, but the flesh is weak," once came out, in Russian, "The vodka is strong, but the meat is rotten"; and "out of sight, out of mind" was rendered as "blind idiot"!

In spite of these defects, it is generally believed that the coupling of electronic computers and labor-saving machinery will render much work obsolete and open up opportunity for man to devote his attention fully to recreational, creative, and cultural accomplishments.

PROBLEMS

24-1 Compute the total number of *atoms* in animals of mass 1 μg and of mass 10 kg if the density of each is approximately 1.0 g/cm^3 and water is the main compound present.

24-2 How many ions would be created by a 1.6-MeV gamma ray, if the amount of energy needed to create one pair is 32 eV? If each is singly charged, how many coulombs of each sign of charge are produced?

24-3 Estimate the amount of 1-million-volt lightning charge in coulombs that would have to be passed through a gas composed of equal parts of CO_2 and H_2O in order to produce a protein molecule of molecular weight 30,000, assuming it to be mainly carbon, with a binding energy of 1.5 eV per atom.

24-4* Suppose that a mutant strain that has a 10% higher survival rate than homo sapiens appears today, and that the remaining world population of 3 billion remains constant. How many years would it take for the total population (the sum of both species) to increase by 1%, assuming that one generation is 25 years, and that the birth rate in mutant families is the same as in the rest of the population? List any other assumptions that must be made.

24-5 In a study of photosynthesis, leaves of plants that have absorbed radioactive carbon (half-life 5600 years) in the form of CO_2 are burned and the gases introduced into a radiation detector. If the smallest observable counting rate is 2 per second, what minimum mass of C^{14} must be present in the leaves?

24-6 Estimate the smallest size animal capable of warm-bloodedness if the daily limit on food intake is 4 times its body weight, and if an animal of 1-m^3 volume consumes a weight equal to 2% of its body weight daily. Assume specific gravity 1.0 as needed. Is this a realistic result? What other factors must be considered?

24-7 The *horse* can be regarded as a heat engine. Discuss this idea and prepare an analysis that would relate horsepower developed to other thermal and thermodynamic features of the system.

24-8 Suppose that the bone diameter of an animal of height 5 cm is 1 mm. On the basis of elastic properties what would the diameter be for one of height 2 m? Is this a realistic result? What other factors must be considered?

24-9 If the average heart power of an individual is 5 watts, how many joules of work does it do in each of 80 beats per minute? How many Calories (kilocalories) of food energy per day is required for that organ, assuming 100% efficiency?

24-10 A certain farsighted person cannot see well at distances greater than 5 m. What type and focal length lens should be prescribed?

24-11 A certain person has difficulty seeing at distances greater than 1 m or less than 0.20 m. What is the focal length of a single lens that will enable him to read at 0.25 meter and also to watch a football game?

24-12 (a) A bat approaches its prey at rest and the emitted sound signals are reflected. What information does the bat obtain from use of the Doppler effect (Section 12-4)? If the sound emitted, of speed 343 m/sec, is of frequency 5000 cps, what frequency does a bat of speed 2 m/sec hear? (b) Airplanes make use of a "Doppler ground-speed indicator" by sending out and receiving reflected signals from the surface of the Earth. Investigate the operation of such a device, assuming that an electromagnetic beam is sent down at a 45-deg angle. If the plane speed were 424 m/sec, what fractional frequency change would be observed?

24-13 Write the following numbers in binary form:

8; 33; 127; 313; 682; 1000

24-14 Write the following binary numbers in decimal form:

100101; 1011011001; 100010001; 1010101; 1111111111

24-15 Convert the following simple addition problems to binary and solve them; then check your results:

$$\begin{array}{r}32\\+\ 8\\\hline\end{array}\qquad\begin{array}{r}20\\+\ 4\\\hline\end{array}\qquad\begin{array}{r}122\\+\ 32\\\hline\end{array}\qquad\begin{array}{r}101\\-\ 16\\\hline\end{array}$$

MAN AND THE UNIVERSE

25

The foregoing chapters of this book have presented only an introduction to the foundations of physics. We have omitted many interesting and useful topics, have discussed others in oversimplified terms, and have stopped short of the frontier of investigation in still others. On the basis of the concepts covered, however, the student will be in a good position to appreciate the role of physics as an area of scientific study, and its importance as a basis for new technical developments.

Research in physics and other sciences is progressing on a broad front at a rapid pace. Extensions of the principles we have studied are to be expected, but radically new viewpoints may have to be developed as scientists probe deeper into the microscopic and macroscopic realms.

A few examples of areas in which there are important problems yet to be solved are reviewed in this concluding chapter. The selection is intentionally diverse, ranging from the atomic to the cosmic and from the exotic to the practical, and is intended to convey to the reader some idea of the challenge and excitement that physics holds for the present and future generations.

25-1 THE UNIFICATION OF KNOWLEDGE

It has often been said that the objective of science is to find the smallest number of simple principles to explain the greatest number of phenomena. Much progress has obviously been made in this direction. Concepts of conservation permit the understanding of all types of diverse processes; the electromagnetic equations of Maxwell cover many of the phenomena of waves, fields, and charges; and the quantum–atomic view of the structure and behavior of materials is comprehensive in nature. There still remain, however, great gaps in our ability to correlate such entities as mass, gravity, charge, and light. In this section we shall elaborate on the efforts of Einstein and others to generalize and unify the relation between mass, gravity, and space, to illustrate the type of unusual thinking that is required.

Einstein's General Theory of Relativity

We found in Section 3-4 that application of Einstein's theory of special relativity was required if the two frames of reference were moving at high uniform speed relative to each other, and that the Lorentz transformation must be substituted for the Galilean transformation.

Einstein, after having completed his theory of special (or restricted) relativity, turned to the question "How are inertia and gravity related?" Throughout our studies to this point, we have used the concept of mass as a measure of inertia and also as the quantity that determines the amount of gravitational attraction. Since these phenomena appear quite different in character, there is no apparent reason that the two masses should be the same. Extremely accurate measurements show them to be identical, however. In simplest terms, the question reduces to asking if Newton's $F = ma$ and $F = k_G mM/R^2$ are merely two aspects of a more general statement. Einstein's *theory of general relativity* was the outcome.

One of the key features of this latter theory is the *principle of equivalence* of acceleration and gravitation. Let us picture a simple experiment

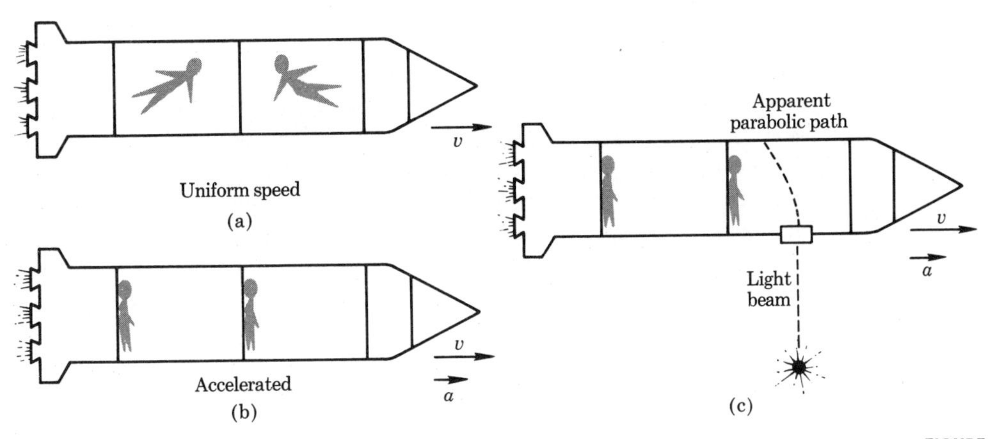

FIGURE 25-1

Equivalence of acceleration and gravitation

that will help reveal its meaning. A spaceship is moving uniformly in free space, far from any planets or stars—Figure 25-1(a). All objects within the ship float freely. Now, if the rocket engines are started up and the ship accelerates forward—Figure 25-1(b)—objects within the ship rush to the rear of the vessel, where they are pressed against the wall with a force that is indistinguishable from gravity. How can the crew members tell whether the ship accelerated or whether a new attracting mass has suddenly appeared? A second experiment brings out some additional ideas. A beam of light from a distant sun streams through a porthole in the side of the spaceship. Assuming that precise instruments are available, the path of the light across the interior of the accelerating ship would be found to be parabolic, as illustrated in Figure 25-1(c). The observers inside again cannot tell whether or not a gravitational force is acting on the light or whether the ship is accelerating. The new question, however, is: How can gravity act on photons which have no mass, as ordinarily defined?

Einstein's theory provides an answer to these questions, being a successful effort to put the principle of equivalence on a firm mathematical basis. The complications in the theory are such that for years after it was presented, the story went that only a half-dozen people understood it. Because of the unusual character of general relativity, there has been much allusion to it in popular literature, especially in science fiction, where the terms "fourth dimension" and "curved space" are freely used. We can give these phrases some meaning by simple illustration and by analogy with our own familiar space of three dimensions. Let us consider first an automobile moving along a straight highway. An event takes place—for example, a tire blows out and the car is wrecked. The description of the event in the newspaper would include both where and when it happened—i.e., the distance along the highway (x-coordinate) and the time of day (t-coordinate). Now, let us look at any particle in the whole world. Its position must be described by four coordinates: x, y, and z for spatial location, and t for time. The distance between any two such "world points" can be expressed by an extension of the Pythagorean theorem of ordinary geometry. Let us start by finding the distance, on a plane surface, from the origin (0, 0) to a point (x, y). This diagonal is obviously $s = \sqrt{x^2 + y^2}$ or its square is $s^2 = x^2 + y^2$. Next, we consider making the measurement of distance along the x-axis from either a vehicle moving with speed V, or from the ground. Using Newton's relativity, which relates the measurements by $X = x - Vt$, $x = X + Vt$ (as in Section 3-3), the displacement or distances are the same $\Delta X = \Delta x$. In three dimensions, this can be written $(\Delta S)^2 = (\Delta s)^2$, which states that displacements are the same or *invariant*, regardless of coordinate system used. It is very useful to introduce a new time variable $w = ict$, where c is the speed of light and i is the imaginary number $\sqrt{-1}$. A point in "space-time" is specified by all four coordinates (x, y, z, w) and (X, Y, Z, W) in the two frames moving with respect to each other. Displacements or distances between points turn out to be the same according to observers in either frame, using the Lorentz formulas. Another way of stating this result is that the laws of physics are the same on all frames moving uniformly with respect to each other. We have now a little appreciation for the view that nature is describable in *geometric* terms.

The Curvature of Space

Einstein went on to propose that gravitation is a geometric effect, suggesting that the presence of a planet or star causes a distortion of the space in the vicinity. To understand his theory in detail requires very complex mathematics, but we can use analogies to establish its qualitative meaning. In ancient times, it was generally believed that the Earth was a flat disc, with the sky forming a dome. The idea that the Earth was round was surely greeted with disbelief. The contrast between flat and curved space, and between Euclidean and non-Euclidean geometry, is readily understood by consideration of the Earth's surface. According to the plane geometry of Euclid, the sum of the angles in a triangle is 180 deg, the hypotenuse of a triangle is the square root of the sum of the squares of the sides, and the shortest distance between two points is a straight line. Although such theorems hold in a very small region on the surface of the globe, they break down when applied to extensive regions. Natural coordinates for spherical geometry are r, θ, and ϕ as shown in Figure 25-2(a). To illustrate, let us consider the locations of two large cities, as in Figure 25-2(b). New York City (NY) is at longitude $\phi_N = 73°59'39''$ (measured from the Greenwich Meridian) and latitude $\theta_N = 40°45'06''$ (measured

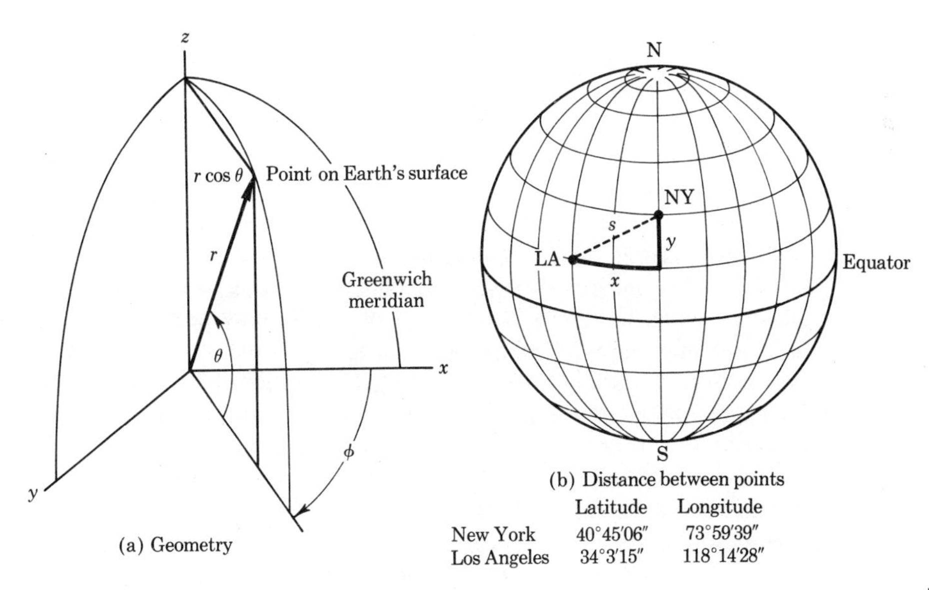

	Latitude	Longitude
New York	40°45′06″	73°59′39″
Los Angeles	34°3′15″	118°14′28″

FIGURE 25-2

The Earth's surface

from the Equator). Los Angeles (LA) is at $\phi_L = 118°14'28''$ and $\theta_L = 34°3'15''$. Thus NY is 6.70° north and 44.25° east of LA. Now, 1 deg of latitude is 69.1 mi and 1 deg of longitude at the location of LA is 69.1 $\cos \theta_L = 57.25$ mi. Thus the height and base of the spherical triangle formed are respectively 462.8 mi and 2533 mi. If the Pythagorean theorem were applicable, the distance between cities would be 2575 mi. The actual distance as part of a great circle, calculated by use of spherical trigonometry is instead 2447 mi. The sum of the three angles in the triangle is 207 deg, not 180 deg.

Let us now ask, What is it that characterizes a particular type of space? A generalized Pythagorean theorem can be developed relating changes in the coordinates. For example, if we start at a city located on the globe at coordinates θ and ϕ and travel to one nearby, making changes $\Delta\theta$ and $\Delta\phi$, it is easily shown that the distance between cities is

$$(\Delta s)^2 = (r \cos \theta)^2(\Delta\phi)^2 + r^2(\Delta\theta)^2$$

where r is the Earth's radius. Coefficients $(r \cos \theta)^2$ and r^2 multiply the squares of displacements in this coordinate system. These are called *metric coefficients*, labeled g, and depend on the nature of the surface.

In the non-Euclidean geometry employed by Einstein to represent displacements in four-dimensional world space, the generalized Pythagorean theorem is still more complicated. Let x, y, z, w represent small displacements, and s be the resultant. Then,

$$s^2 = g_{11}x^2 + g_{22}y^2 + g_{33}z^2 + g_{44}w^2 + 2g_{12}xy + 2g_{13}xz + 2g_{14}xw + 2g_{23}yz + 2g_{24}yw + 2g_{34}zw$$

where there are 10 of the g's. Einstein proposed that the presence of a chunk of matter causes local variations in the values of these coefficients, thus disturbing the space from flatness for which $g_{11} = g_{22} = g_{33} = g_{44} = 1$, and the rest are zero. Particles move on paths in warped space along *geodesics*, which are the shortest distances between pairs of points (the *great circle* on a spherical surface). We interpret this motion as that of a particle in a gravitational field created by the matter.

To get some further feeling for this effect, let us consider an analogy. A thin rubber sheet is stretched out flat, supported on all edges. We drop an object, such as a golfball, on the sheet, causing a depression. If the sheet had grid lines marked on it, the stretching would cause the coordinate system in the region of the depression to be distorted in such a way that the Pythagorean theorem and other aspects of Euclidean geometry would no longer hold in that vicinity: the space will have become curved because of the presence of the object. Carrying the analogy further, let us compare the motion of a marble rolling on the sheet when it is flat and then on the curved sheet. In the latter case, the marble rolls into the depression. By the same token, free space is considered to be "flat," while the space near a gravitational mass is "curved," and the acceleration of bodies in gravitational fields is due to that curvature.

Tests of the Theory

Experimental verifications of the theory of general relativity are few in number, because the effects are extremely small. One of Einstein's predictions was that light from a distant star would be deflected on passing a large gravitational body. The light should follow the straightest path, a *curve* in the distorted space. During a solar eclipse by the Moon, distant stars are visible, and a comparison can be made of the apparent star position and the location if light is undeflected. Displacements in the range 1.7–2.0 seconds of arc are observed, comparing very favorably with predicted 1.75 seconds. Another proposed experiment involves bouncing high-frequency radar off the planet Mercury as it passes behind the Sun and observing the shift in time for the waves to return to the Earth.

A second prediction is related to the behavior of the planet Mercury itself as it moves about the Sun. According to the theory of general relativity, the major axis of its elliptical orbit changes direction in time. Measurements show that the change is 43.6 seconds of arc each century, compared with a predicted 43.0 seconds. Consideration is being given to sending a gyroscope up in an orbit around the Earth in order to detect shifts in the axis of rotation of the gyroscope.

The third prediction has been only partially verified. Einstein stated that light coming to us from massive stars should experience a decrease in frequency, shifting the color toward the red end of the spectrum. There is evidence of such a gravitational "red shift," but the observed amount does not agree with theory in all cases.

A terrestrial check of the principle of equivalence has been made possible by the discovery of new means of producing and detecting light. We recall from Section 22-1 that the light from atoms such as hydrogen is emitted at unique frequencies related to internal energy changes. In an incandescent gas with the atoms moving at high speeds in all directions and serving as moving sources, the observed light has a significant spread of frequency due to the Doppler effect. A similar situation exists for *gamma* rays emitted by moving nuclei. Mössbauer found that extremely sharp spectral lines would be produced by radioactive elements embedded in a solid crystal, where the emitter could be considered as "frozen." Relative motions of emitter and detector composed of materials that exhibit this *Mössbauer effect* are readily distinguished by observing the shift in frequency. In an experiment conducted by Pound and Rebka in 1960, radiation was sent downward from a height of 22 meters above the detector at ground level. The difference in gravitational field between the levels produced an observable change in frequency of the light. The excellent agreement between theory and experiment provided a direct verification of Einstein's principle of equivalence.

The success of Einstein's efforts to unify mechanics and gravity encouraged him to seek still further unification, that of gravitation and electromagnetism, with the hope of including an explanation of elementary particles and the results described by quantum mechanics. Although he did not realize this ambition, his vision has continued to inspire his successors to continue the work.

Recently, a rival theory has been advanced by Dicke and Brans. It conceives of gravity having two components—a scalar field with no directional aspects, and a tensor field that is directional in a generalized vector sense. A combination of 5 per cent and 95 per cent of the two types is estimated to be correct. New sensitive measurements are needed on the deceleration of light as it passes by massive objects in order to test the difference between the Einstein and Dicke views.

Ultimately, a truly general field theory may be developed, as a mathematical concept and description that encompasses mechanics, electromagnetism, gravity, nuclear forces, relativity, particles, fields, and quantum mechanics. Although we are presently far from this goal, it must be remembered that science is still only about a thousand years old and that most of its progress has taken place in the last century.

25-2 ASTRONOMY AND COSMOLOGY

The character of processes within distant stars and galaxies must be inferred from limited information, gathered at terrestrial or satellite observatories. We shall first review some of the methods of observation and deduction. Naturally, the status of astrophysics and cosmology changes as new information is gained by means of ever-larger optical and radio telescopes through refinements in detection methods, and the development of space observatories. We shall discuss qualitatively the various theories that have been developed to render consistent the measurements of electromagnetic waves and cosmic particles that come to the Earth. Of special interest is the mechanism by which the universe was formed, if indeed the concept of origin is meaningful.

Star Characteristics

Let us first note the properties that can be observed or deduced. The distance from the Earth can be obtained for a few hundred stars in our vicinity by the method of parallax (Section 21-3). Generally, however, distances are found by indirect means. Certain near stars, the Cepheids, of which the pole star is an example, were observed to vary in brightness in a regular cycle with the period directly related to the distance away. This information greatly extends the range of measurements of distances to stars in other galaxies besides ours. The light from a given star as a radiator is distributed in wavelength as discussed in Section 19-3. The higher the temperature, the shorter the characteristic wavelength—i.e., the bluer its color. Because of the elemental composition of the star, it will exhibit discrete spectral lines as well. The fourth-power law of black-body radiation provides a direct connection between the temperature and the rate of radiation of energy per unit area. The apparent brightness or *luminosity*—that observed by detectors on Earth—can be translated into absolute brightness at the star surface by taking account of the inverse-square reduction with distance. Then, the knowledge of the rate of radiation per unit area and the total rate of radiation permit the calculation of the star surface area, and hence its dimensions. Masses can be computed for a number of double stars that revolve around each other, by applying the principles of mechanics and gravitation to the period. Through this sequence of measurements and deductions the properties of distance, temperature, luminosity, size, and mass are established. Generally, stars are divided into the "normal" and "abnormal" classes. In the former group—the main sequence—the true brightness decreases as one proceeds from large hot stars down to small cool stars. In the latter category are *white dwarfs, red giants,* and *supergiants.* The interpretation of the traits of stars involves a very involved combination of the theories of gravitational contraction, nuclear fusion, thermal radiation, and general relativity, and remains far from being a closed subject. One fact with which any cosmological explanation must deal is the existence of *galaxies* or *star clusters,* in contrast with a uniform distribution of stars throughout space. Our particular galaxy has a diameter of about 100,000 light years and contains about 200 billion stars.

The most unusual phenomenon to be explained is the apparent *expansion of the universe.* The wavelengths of spectral lines from stars are found in general to be longer—i.e., toward the red end of the spectrum—than those observed on Earth. On the basis of the Doppler effect (Section 12-4), this implies that the stars are receding from the observer. It is found that the lower the intensity of light from a star, the greater the shift. If we presume that the lower intensity is associated with greater distance of the stars away from us, it must be concluded that the universe is expanding. If d is the estimated distance away and v is the speed inferred from the Doppler shift, then d/v is a time t, which turns out to be about the same for all galaxies, and is about 10 billion years. It is tempting to conclude that the universe was formed 10 billion years ago by an explosion in a small region of space, setting all of the cosmic bodies into motion.

Another important observed feature is that the night sky is relatively dark—or, conversely, that there is remarkably little light from a universe known to be of such tremendous size. If one assumes that space on the average of large regions is equally populated by stars, then the total light we receive from every spherical layer of space should be the same. The inverse-square reduction in intensity as we go outward is just balanced by the direct-square area increase with distance. The total light received from the universe, we might then reason, should be nearly infinite. The flaw in this deduction is that it neglects the possibility of expansion, which would reduce the density of stellar material and thus the intensity from the more distant regions of space.

Theories of the Universe

Of the many theories of cosmology, there are two that deserve special discussion. One model is called an *evolving universe,* the other a *steady-state universe.*

The evolving model, conceived by Lemaître, a Belgian, is based on general relativity. It involves the concept that the universe is unbounded but finite, meaning that if one goes outward continuously from the Earth he eventually returns to it, in analogy to circumnavigation of the globe. This strange situation has been described by Sir James Jeans as analogous in a four-dimensional sense to a soap bubble, the surface of which is the universe. The theory also supposes that a primeval nuclear explosion of a super-body took place many billions of years ago. The popular term *"big-bang theory"* is applied to this view. The initial expansion was presumably slowed by gravitational attraction, but a second effect peculiar to general relativity may have taken over, that of repulsion increasing with distance. The formation of galaxies is attributed to the cooling and condensation under local gravitational forces.

The steady-state model states that if the universe is viewed on a large enough scale, it is seen as uniform in space and unchanging in time. However, if the universe is expanding, the average density of matter is decreasing indefinitely, which would be counter to the requirement of steady conditions. Thus the idea is introduced that there is a continual creation of matter out of nothing. This would seem to be an apparent gross violation of our conservation laws, but it has been proposed that the energy of expansion plays a role in the formation of matter. Within the cosmic steady state, it is presumed, galaxies are created by condensation and move away from each other as they age. At any time, however, the universe looks the same. A variant on these theories is the model of a pulsating universe that cyclically collapses and expands with a period of billions of years.

The determination of the correctness of any theory of cosmology obviously rests on observations rather than controlled experiments. Strictly speaking, it is the incorrectness that should be investigated, since the failure to explain some well-established fact can eliminate a given theory. Let us briefly cite some of the tests to be applied in such evaluations. We note first that what we see of distant parts of the universe, millions to billions of light years away, are conditions as they existed when the light was emitted. According to the evolutionary theory, there should thus be marked differences in the appearance of galaxies at various distances, but with the steady-state theory little difference would be expected, since continous creation would favor uniformity. Also, if the steady-state concept is to be considered reasonable, some consistent theoretical mechanism must be found to explain the formation of new galaxies in the presence of old ones.

The recent discovery of *quasars* ("quasi-stellar" sources) has opened a new chapter in the subjects of astronomy and cosmology. These are dim stars that are very powerful sources of radio waves, well localized in space at distances that are billions of light years away, as determined by their high speeds. Quasars number about a hundred, randomly scattered about the universe. To be visible at such distances from the Earth, the objects would have to give as much light as 50 galaxies each containing some 100 billion stars. They have an unexplained variation in intensity of as much as one per cent per month, which implies that they are small in comparison with the distance light travels in a month—i.e., far less than a light year. Various conjectures have been advanced, such as the idea that quasars are superstars that collapse with the speed of light, or involve negative mass, or are supernovae in a state of explosion. There is some possibility that the tremendous release of energy is related to the annihilation of matter and antimatter.

25-3 THE INVESTIGATION OF SPACE

A great new interest in the space outside the Earth and its atmosphere has arisen in the scientific world in recent years. There had been for centuries a continuing study of the motion of stars and planets—the subject of astronomy. The composition and radiations from stars, including the Sun, formed the more modern field of astrophysics. Interest developed also in cosmic rays—the primary particles from space, and the secondary radiation due to processes in the atmosphere. The achievement in the 1950's of long-range military missiles, using rockets for propulsion, led naturally to consideration of travel into the space outside the Earth. Extensive research programs were established by several countries, especially the United States and the U.S.S.R. The goal of such studies is to learn more of the mysterious regions of space and planets than can be observed from the Earth, using vehicles with instruments and carrying men as needed.

The Radiation Belts

The existence of enormous zones of high-energy radiation near the Earth was one of the new discoveries of recent years. By the use of balloons, rockets, and artificial satellites, it was found that there were two "belts" of radiation encircling the planet at about 1 and 3 Earth radii from the surface. Figure 25-3 shows a map of these regions, where the radiation, composed of protons and electrons, is so high that space travelers going through them would receive significant dosages of radiation. The shape of these rings suggests that they are caused by the interaction of the magnetic field of the Earth (Section 6-4) and moving, charged cosmic particles, as from the Sun. Charged-particle motion in corkscrew paths take place along field lines of the Earth. The loops are large at the level of the Equator, where the field is weak, and tight near the poles, where they

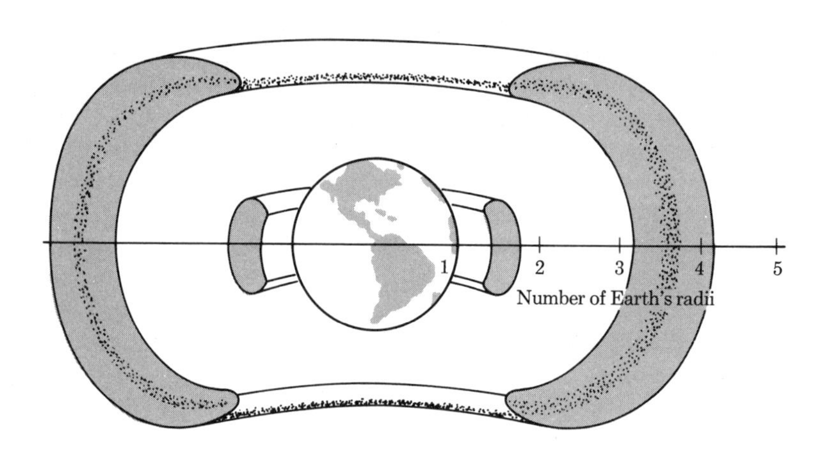

FIGURE 25-3

Radiation belts around the Earth

collide with the high-density atmosphere. There ionization and radiation is produced, explaining the aurora borealis or "northern lights." The reason for two belts instead of one is not well understood.

Satellite Orbits

The first artificial satellite of the Earth was Sputnik I, put into orbit by the Soviet Union on October 4, 1957. Soon after, in 1958, Explorer I was launched by the United States. The latter was a small sphere of mass 30.8 lb, which achieved a 106-min orbit at altitude ranging from 217 to 1100 mi. Since that date, several scores of instrumented satellites have been successfully placed in orbits around the Earth and sent to the Moon, to the nearby planets, and out farther into space. Physics has played an important role in these projects in several ways: (1) in the determination of the mechanics of motion under the influence of rocket thrust and the gravitational forces of celestial bodies; (2) in the invention of instruments and detection devices to control the vehicles, protect them from damage, and to provide technical information to the Earth-based tracking stations; and (3) in the observation and interpretation of space measurements of temperatures, pressure, charge density, energy and composition, solar radiation, meteorite frequency, and magnetic fields. Three practical applications of spacecraft are discussed below.

1. *Meteorology*. The science of weather prediction has long remained inexact because of insufficient information. Views of the Earth from out in space as provided by weather satellites yield new sources of data on cloud formation, beginning hurricanes, and radiation supply and loss. Long-range weather forecasts may become possible, using the satellite Aeros with 24-hour period such that it remains fixed in space over a point on the

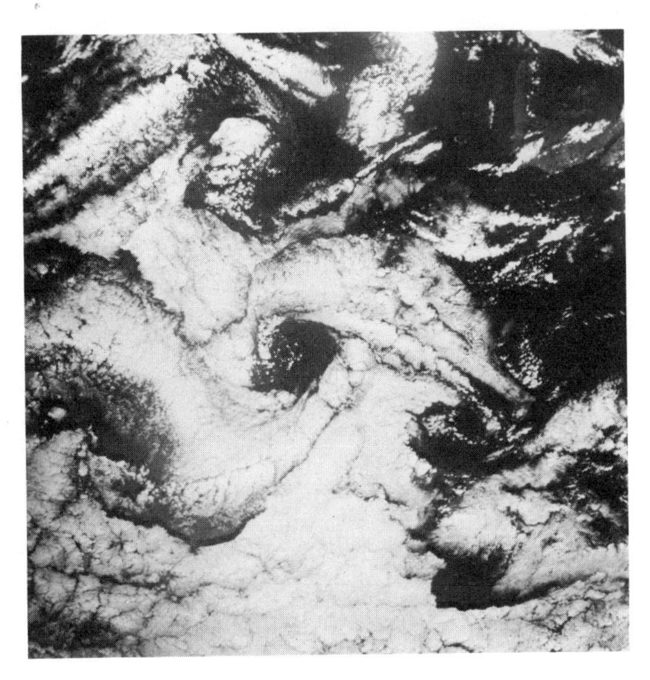

FIGURE 25-4

Weather patterns viewed from Earth satellite

Equator. Figure 25-4 shows photographs of cloud formations taken from the satellite Gemini VI (1965) at a height of 177 miles.

2. *Communications.* The direct transmission of electromagnetic waves between widely separated points on the Earth's surface is not possible because of the intervening Earth mass. In the case of radio, reflection from the charged upper atmosphere (*ionosphere*) is possible. Television, however, with its shorter wavelengths is not so reflected. A satellite in the form of a thin metal balloon 135 feet in diameter (Echo) has been found capable of providing a suitable surface for reflection. The relaying satellite Telstar, solar-powered, is far more effective in that signals to receiver are transmitted with greatly-amplified strength to distant points. Figure 25-5 shows the schematic operation of this communications satellite. Radio control of the electronic circuits is achieved from the Earth.

3. *Planetary investigations.* Conditions on nearby planets can be explored remotely by the use of *space probes.* Guided through space by radio waves and electronics, the probe Mariner sent back information on the planet Venus from a distance of a few thousand miles. Among the early findings were that the surface temperature is extremely high (600°F), probably because of a dense cloud cover; that there is a weak magnetic field; and that there is little daily rotation of the planet. Photographs of Mars were taken by television cameras on Mariner spacecraft and

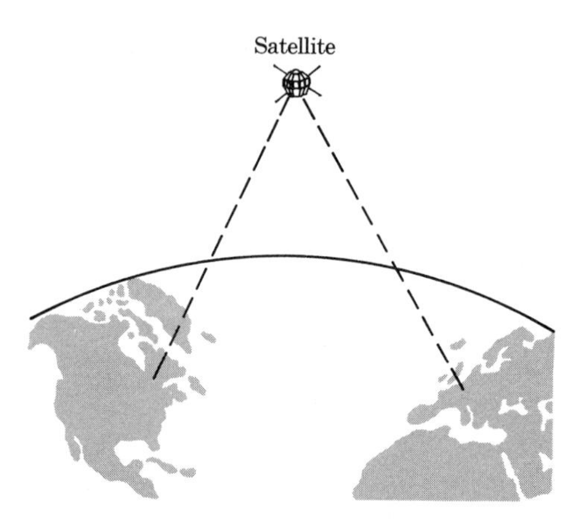

FIGURE 25-5

Television relaying satellite

transmitted by radio back to Earth by coded signals of intensity. The pictures showed that the surface was covered with craters, and resembled that of our Moon.

Space Travel

Success in achieving manned flight from the surface of the Earth to the surface of the Moon, as in the Apollo-11 voyage, can be regarded as a first step toward the exploration of the solar system and the space beyond. Using the rocket principle, particles having mass are given energy by some fuel to provide the needed thrust. The requirements depend on the "mission" selected—orbit about the Earth, escape from Earth to the Moon and return, travel to the nearby planets Mars and Venus, trips to the limits of our planetary system, and finally to planets of the stars.

Propulsion based on chemical reactions is adequate for missions of relatively short distance and small "payload." For manned exploration of Mars, however, a nuclear rocket is regarded as necessary. The engine of such a spacecraft would consist of a nuclear reactor (Section 23-4) in which *fission* serves to heat hydrogen gas, which is then expelled to provide thrust.

It is interesting to imagine a rocket for very long voyages using the best *fusion* reaction, in which protons combine:

$$4\,{}_1\mathrm{H}^1 \rightarrow {}_2\mathrm{He}^4 + 2\,{}_{-1}\mathrm{e}^0 + 26.7\ \mathrm{MeV}$$

If all the energy goes into the resulting alpha particles, each of mass 4.002 amu = 3728 MeV, the propellant speed v_p found from

$$\tfrac{1}{2}mv_p^2 = (\Delta m)c^2$$

is close to $\frac{1}{8}$ the speed of light. Relativity tells us that the ratio of initial to final rocket mass for a flight that involves launching, coasting at speed V, and landing is

$$\frac{m_0}{m} = \left(\frac{1 + V/c}{1 - V/c}\right)^{c/v_p}$$

For a trip with $V = 0.9c$, using the above fusion reaction, the ratio is $(19)^8 = 1.7 \times 10^{10}$. Even if we neglect the weight of the rocket completely, to send a small man of mass 50 kg would require a fantastic mass of 8.5×10^{11} kg or about 1 billion tons. However, at lower speeds, say $V = 0.2c$, the mass ratio is only $(1.5)^8 = 25.6$, which is much more easily obtained.

The most serious limitation to distant exploration is *time*. The nearest star, Alpha Centauri, is 4.3 light years away, and since speeds of material objects are less than the speed of light, any voyage there will require a time of $4.3(c/V)$ years. The fact that clocks and all biological processes including aging would be slower on a moving ship is helpful to the occupants only at high speed. According to the Lorentz transformation (Section 3-4), the elapsed ship time is

$$T = k\left(t - \frac{VX}{c^2}\right)$$

but since the distance traveled in Earth time t is $X = Vt$, we find that

$$\frac{T}{t} = \sqrt{1 - \left(\frac{V}{c}\right)^2}$$

At $V/c = 0.9$, the Earth time is $t = 4.3/0.9 = 4.78$ years, and since $T/t = \sqrt{1 - (0.9)^2} = 0.436$, we find the astronauts will have aged only 2.08 years. However at $V/c = 0.2$, $t = 4.3/0.2 = 21.5$ years, $T/t = \sqrt{1 - (0.2)^2} = 0.98$, and $T = 21.1$ years. The conclusion is that we save time by using mass or save mass by using time. The nearest planet whose climatic conditions can support life might accompany the nearest star. On the other hand, it may be necessary to travel for centuries to find such a planet. Only by the use of hibernation or colonization methods could such voyages be possible.

25-4 ELEMENTARY NUCLEAR PARTICLES

Four elementary particles—the electron, proton, neutron, and photon—are sufficient to describe a very large part of the phenomena of physics: the structure of nuclei, atoms, molecules, and bulk matter; the flow of electricity; the emission and absorption of light; and many nuclear reactions.

A host of new particles have been found in the last half-century. Their discovery was made possible for various reasons: (1) methods of measurement became more precise; (2) higher-energy nuclear projectiles were available from accelerators; (3) advanced theories of matter and fields predicted their existence. Below, we mention and point out the role of only a few of these.

Positron. The *positron* has the same mass as the electron, but is positively charged. Predicted by Dirac in his quantum theory that took account of relativity, it was the first example of "antimatter," a rare and unusual substance that has properties opposite to ordinary matter. Positrons were first observed by Anderson in 1932 in cosmic-ray studies, and since then many examples of their production in radioactive decay have been found. When a positron collides with an electron, both particles are annihilated, with two photons of light appearing in their place. In reverse, the energy of a gamma photon can be converted into mass of an electron–positron pair in a process referred to as *pair production.*

Antiproton. Existence of a negatively charged particle like the proton had been expected by analogy to the electron–positron relation. Only until accelerators of high enough energy were built could a successful observation be made by Segré and Chamberlain in 1955. When the *antiproton* collides with the proton, annihilation occurs, giving radiation in the form of gamma rays. The *antineutron* is the opposite of the neutron, presumably having the opposite internal-charge distribution. It has been conjectured that some stars and planets may be constructed of antimatter, which thus might be called "other-world" substance. One can imagine the catastrophic result of landing a spaceship from the Earth on such planets.

Neutrino. In the radioactive decay of nuclei that give rise to beta particles (electrons), a peculiar discrepancy of energy was noted. The simplest illustration is observed in the reaction for neutron disintegration:

$$ {}_0\mathrm{n}^1 \rightarrow {}_1\mathrm{H}^1 + {}_{-1}\mathrm{e}^0 + \nu^0 $$

The same reaction takes place in β-decay in many isotopes, where the measured kinetic energy of the products is not as large as the mass difference between original and final particles. Rather than conclude that the law of energy conservation was wrong, Pauli and Fermi proposed a neutral particle—the *neutrino* ("little neutral one")—that has zero rest energy, but at its speed c has the requisite energy. The neutrino is extremely difficult to detect, since it penetrates miles of solid matter without a collision. Reactions involving β-decay are classified as "weak" since the forces that come into play are far smaller than the binding forces between nucleons.

Pion. The *pion,* of rest mass 273 times that of an electron (therefore lighter than a proton), was predicted before it was observed. An analogy was used. Fermi's theory of the electrostatic field stated that photons were continually emitted and reabsorbed by the electron. Yukawa reasoned

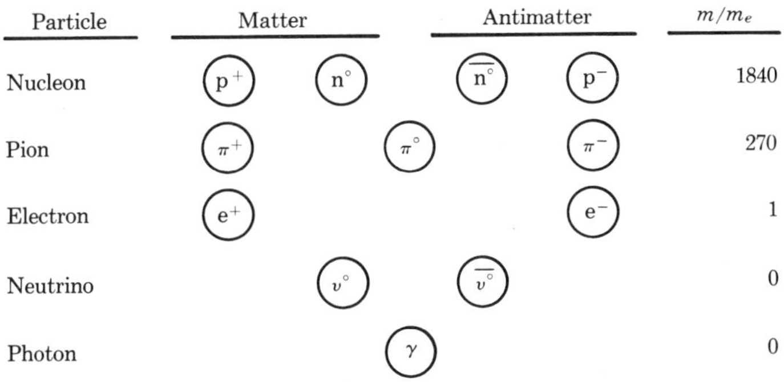

Particle	Matter			Antimatter		m/m_e
Nucleon	p^+	n°		$\overline{n}^\circ$	p^-	1840
Pion	π^+		π°		π^-	270
Electron	e^+				e^-	1
Neutrino		ν°		$\overline{\nu}^\circ$		0
Photon			γ			0

FIGURE 25-6

Twelve elementary particles

that the nuclear force was due to similar processes employing a *meson*, which since has been given the name *pi meson* or *pion*. Its escape from and capture by a nucleon occurs so rapidly that no energy change is detectable. If sufficient energy (135 MeV) is supplied, the pion can be observed. Three types are known, positive π^+, negative π^-, and neutral π^0, with the latter being its own antiparticle. Figure 25-6 shows the particles just discussed and the photon. The ratios of their masses to that of the electron are listed.

Other particles. Among other particles that have been identified in high-energy cosmic-ray and accelerator-produced reactions are the *mu meson* or *muon*, of mass 207 times that of the electron. The *negative muon* μ^- is a decay product of the pion, according to

$$\pi^- \longrightarrow \mu^- + \overline{\nu}^0$$

where $\overline{\nu}^0$ stands for the antineutrino. The negative muon also disintegrates to yield an electron and two neutrinos:

$$\mu^- \longrightarrow {}_{-1}e^0 + \nu^0 + \overline{\nu}^0$$

A *positive muon* is also known. Heavier than protons (2100–2600 electron masses) are the *xi*, *sigma*, and *lambda particles;* lighter ones (about 970 electron masses) are the *K-mesons*. All properties of these new particles (shown in Figure 25-7) are not yet fully known, nor is their exact role in the structure or processes of nature.

In a qualitative sense, the process of "atom smashing" by the injection of particles of ever-increasing energy has brought us to our present state of knowledge of nuclear structure. The early cathode-ray tube of Thomson showed the existence of the electron and its role in ionization and the production of light from atoms. Rutherford's bombardment of matter with energetic alpha particles first revealed the existence of a concentrated nucleus, and showed that transmutation of elements was possible. The

Particle	Matter		Antimatter		m/m_e
Muon	μ^-			μ^+	207
K meson	K^+	K°	$\overline{K^\circ}$	K^-	970
Lambda		Λ°	$\overline{\Lambda^\circ}$		2200
Sigma	Σ^+	Σ°	$\overline{\Sigma^\circ}$	$\overline{\Sigma^-}$	
	Σ^-			$\overline{\Sigma^+}$	2300
Xi	Ξ^-	Ξ°	$\overline{\Xi^\circ}$	Ξ^+	2600

FIGURE 25-7

Eighteen other elementary particles

electron-scattering experiments of Compton and of Davisson and Germer led to new concepts of particles and waves. The neutron was discovered by Chadwick as a byproduct of the interaction of alpha particles with beryllium, and later the neutron was found to be the projectile that initiated the fisson process. The pi meson (pion) was discovered by Powell as a rare component of cosmic rays, and modern high-energy particle accelerators now produce pions in large quantities. Antimatter was found to exist by the use of a *bevatron,* an accelerator giving particles energy in the vicinity of 6 BeV. The particles of mass greater than that of the proton (Figure 25-7) were also identified as products of high-energy reactions.

Some of the questions yet to be resolved are: What is the relation of mass and charge? Are heavy particles merely excited states of more basic particles? What are the truly basic forces and what is the smallest number of conservation laws required to describe all nuclear processes? Exploration to give answers continues along two lines. A consistent theoretical explanation is sought for the roles of the many particles and various types of force. According to Gell-Mann, there are only three basic building blocks, named "quarks".* At the same time, new accelerators are being planned or developed that provide higher energy and more copious numbers of product particles. The continued study of the properties of the new rare particles and their interactions with matter may well give new insight into the structure of the universe.

25-5 APPLICATIONS OF THE DISCOVERIES OF PHYSICS

The growth of science may turn out to be the most important historical characteristic of the twentieth century. The concept has evolved that a

* From a word in James Joyce's *Finnegan's Wake.*

nation's progress is intimately related to the proportion of its resources devoted to scientific investigations. It is noted that many technological innovations are the outgrowth or "spinoff" of basic research, and that the application bears no predictable relation to the subject of study. For example, the modern computer is partly the unexpected consequence of research on semiconductors. Similarly, the nuclear reactor is partly the result of studies on the interaction of neutrons with matter.

The process of translation of discovery into practicality is still underway; we shall illustrate this process by reviewing three interesting subjects: *cryogenics,** the *laser*, and *nuclear fusion.*

Low-temperature Phenomena

As discussed in Section 17-1, the resistivity of most metals tends to decrease gradually as the temperature is lowered. Qualitatively, this effect can be correlated with the reduction in freedom of motion of the electrons and atoms of which the substance is composed. In 1911, Onnes discovered that cooling of a mercury wire to the very low temperature of 4.2°K, the boiling point of helium, resulted in a sudden remarkable reduction of resistivity. It has since been found that below some transition temperature T_c some alloys have a resistance that is at least 10^{-17} times that at room temperature. For all practical purposes, this is zero, which means that a current once started in a wire will flow practically forever. A *superconductor* tends to deflect lines of magnetic field. Also, *superconductivity* can be destroyed by the application of a magnetic field that is as large as some critical value $\mathcal{B}_c$. Such a field can be produced by the current in the wire itself. Adapting relations discussed in Section 7-2, we find that the current in a wire of radius r required to achieve a field $\mathcal{B}_c$ at its surface will be $I_c = r\mathcal{B}_c/2k_M$. The explanation of superconductivity requires the application of quantum mechanics to the motion of electrons and atomic-lattice vibrations and will not be attempted here.

The liquefaction of gases in general is done by an adaption of the refrigerator. The gas to be cooled is first compressed; the heat developed is removed by a colder liquid surrounding the pipes containing the gas; the gas expands which cools it still more, and eventually it liquefies. The modern helium *cryostat* invented by Kapitza produces liquid He by allowing the gas to drive an engine, with the heat going to provide mechanical motion. Just as steam condenses to water in the steam engine, helium gas condenses to liquid helium.

Very unusual effects appear as liquid He is cooled from its boiling point 4.2°K to its λ-*point*, 2.19°K. There, the liquid flows without friction through the smallest openings, a phenomena known as *superfluidity*. The substance has gone into a new state, called Helium II. Its ability to flow is remarkable. Picture a test tube dipped in He II as in Figure 25-8. A very rapid flow along the surface causes the tube to fill until the levels are equal.

* Greek *kryos*, "icy cold."

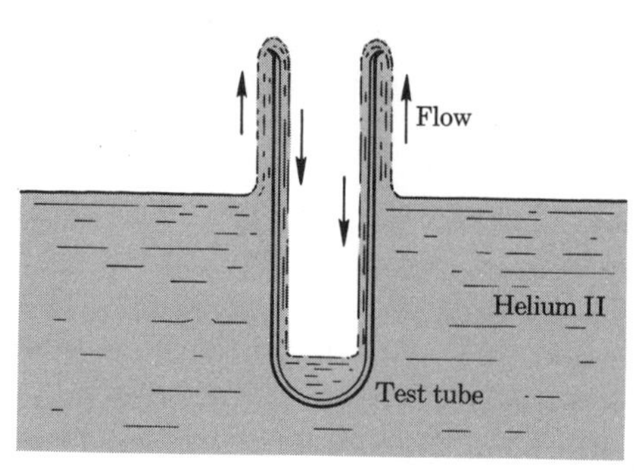

FIGURE 25-8

Flow of Helium-II at the λ-point

The substance is visualized as actually a mixture of two forms of helium. One is the superfluid, composed of He atoms in their lowest energy or ground state, the other a normal liquid, consisting of atoms with higher-energy levels. The two fluids tend to act independently except that energy exchange takes place. The superfluid conducts no heat, since it is already at its lowest energy. The two fluids can flow in opposite directions to each other.

There are many present and future uses of superconductivity. Magnets containing alloys such as niobium–titanium exhibit no resistance heating. Studies are underway on the possibility of using persistent currents for memory cores in computers. Other possible applications are (1) the creation of very high magnetic fields; (2) motors without friction and heating effects due to resistance, thus reaching 100% efficiency, (3) the transmission of electricity over great distances without loss. Unfortunately, it is difficult to reach the low temperatures needed for these purposes.

The Laser

We have seen in Section 19-2 that the forced oscillation of electrons gives rise to *electromagnetic waves* in the radio, television, and microwave regions. These are rather pure sine waves that can be emitted in phase with each other and travel in a single direction. The wavelengths are limited, however, to the order of the physical dimensions of the source, which for practical reasons can be no smaller than about 1 mm. On the other hand, the shorter wavelength of *visible light* has a great mixture of frequencies, the waves are out of phase with each other (incoherent), and the direction of radiation is random, because the light is due to chaotic atomic processes. By discovery of the *laser* (*l*ight *a*mplification by *s*timulated *e*mission of *r*adiation), the excellent properties of radio waves have been obtained for visible light, opening up a great new field of communication, electronics, and research.

To understand its operation, only one basic quantum idea is needed. Visualizing the Bohr atom model, suppose that energy is supplied to an atom to raise its electron from ground to a higher state, as in Figure 25-9(a). We now *send in* light of a frequency ν corresponding to the energy differ-

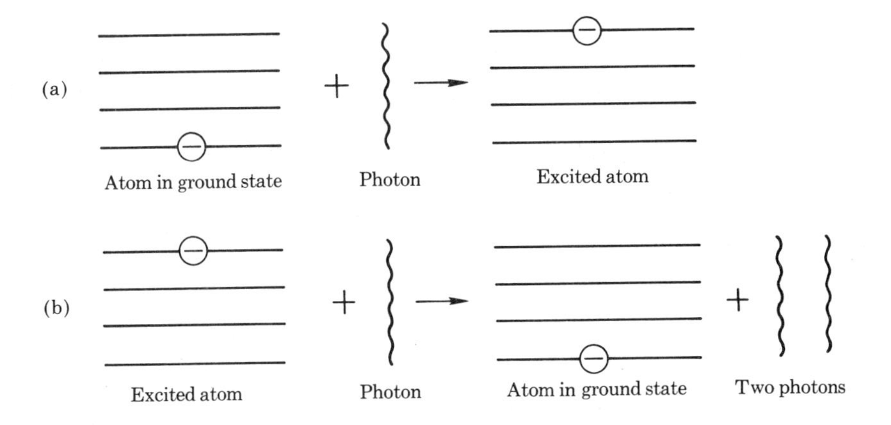

FIGURE 25-9

Atomic reactions in the laser

ence, as in Figure 25-9(b). The jump back to ground is stimulated to take place prematurely. *Two* photons of frequency ν are now available. In the laser, a great deal of energy is "pumped" into a medium containing many atoms. At a certain stage, a photon produced spontaneously can trigger a chain reaction of light emissions. It strikes an atom with stored energy, giving rise to 2 photons; these strike 2 other atoms and 4 photons result, etc. The burst of light is of single frequency. By an ingenious arrangement, the other desired features are obtained. A typical laser (Figure 25-10) is composed of the gem known as ruby, which is basically aluminum oxide containing a few chromium atoms (which give it the red color). The ends of a ruby rod, typically $\frac{1}{4}$ in. thick and 2 in. long, are polished smooth, flat, and as nearly parallel to each other as possible. Coatings of silver permit light to reflect back and forth inside many times before escaping through one partially silvered end. If the chain reaction starts at one end, a buildup in number of photons takes place along the length of the rod (Figure 25-11). They are reflected at the end and sweep back, stimulating

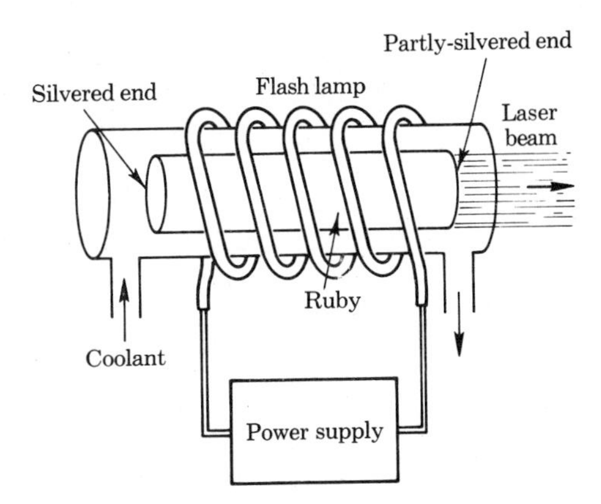

FIGURE 25-10

Ruby laser

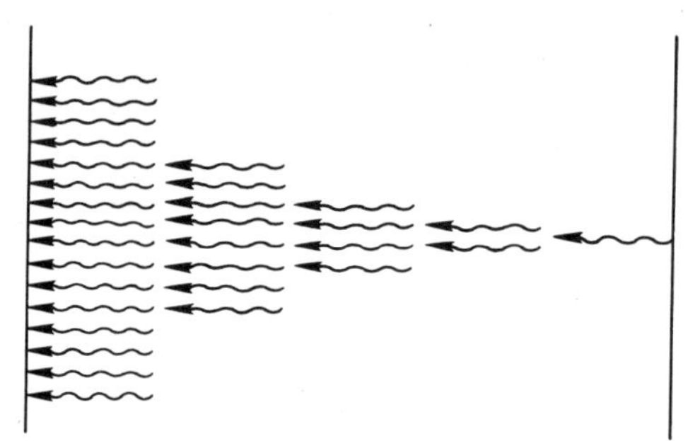

FIGURE 25-11

Cascade of photons in the laser

other atoms to emit. Out of the partly transparent end comes a wave of light that is in phase and is extremely close to being parallel. It spreads only about 5 feet per mile—only $\frac{1}{20}$ as much as a searchlight beam. By the use of lenses it can be focused to give a spot on the Moon only 2-mi across. A laser can act as a super "burning-glass." The large power can be focused to a spot 0.1-mm wide, giving a density of 100 billion watts per m^2, which will vaporize any known material. The agreement in phase of all parts of the wave is readily demonstrated by putting the two slits of Young's experiment (Section 21-4) almost on the end of the laser. The interference pattern is very clear, showing the beam to be highly coherent and parallel. There are many possible applications to astronomy, signaling, amplification of weak light, induced chemical reactions, and new frequency generators.

Control of Thermonuclear Energy

The process of *fusion* of light nuclei, if it could be made practical, could provide the world with power for ages to come. Deuterium from the water of the vast oceans could serve as the fuel for the process. The fraction of deuterium in the hydrogen of water is small, 0.00015, but on an energy basis, 1 gallon of water is equivalent to about 250 gallons of gasoline. In terms of the Q-unit of energy, which is 10^{18} Btu (or 1.05×10^{21} joules), the heat consumed in the world is about $0.1Q$ per year. The mass of sea water is so great that there is enough D_2 to last for a billion years even at an energy consumption of $15Q$ per year.

Several nuclear reactions provide a large energy release, as discussed in Section 23-3. Those most favorable involve deuterium H^2, tritium H^3, and a helium isotope ${}_2He^3$:

$$ {}_1H^2 + {}_1H^2 \rightarrow {}_1H^3 + {}_1H^1 + 4.03 \text{ MeV} $$
$$ {}_1H^2 + {}_1H^2 \rightarrow {}_2He^3 + {}_0n^1 + 3.27 \text{ MeV} $$
$$ {}_1H^2 + {}_1H^3 \rightarrow {}_2He^4 + {}_0n^1 + 17.6 \text{ MeV} $$
$$ {}_1H^2 + {}_2He^3 \rightarrow {}_2He^4 + {}_1H^1 + 18.3 \text{ MeV} $$

All of these reactions can be accomplished on a laboratory scale. Because of the electrostatic barrier of a nuclear target, the ions that act as projectiles must have a large energy for there to be a good chance of reaction, as mentioned in Section 23-3. One approach is to contain the particles in an electrical discharge, where part of the energy of nuclear reaction is used to accelerate the particles. The ultimate goal is a continuously operating machine that has enough energy left over for useful purposes. There are serious losses, however, due to *bremsstrahlung* (Section 19-2). As electrons come within the field of an ion, radiation proportional to the square of the acceleration takes place. Only if the temperature is very high, above 400 *million* degrees Kelvin, will the energy provided by fusion outweigh the energy loss by radiation.

Realizing that even the most heat-resistant materials disintegrate at temperatures of a few *thousand* degrees, we see the need for nonmaterial "containers." The only possible way to confine ions is by *electromagnetic forces*—i.e., by specially arranged electric and magnetic fields. The latter might be provided by superconducting magnets. Recall that charges can only move in spiral paths along magnetic field lines, with radii inversely proportional to the magnitude of $\mathcal{B}$. This suggests that by increasing $\mathcal{B}$, the volume of reacting ions can be made as small as desired.

Imagine a deuterium ionic "gas" or plasma, produced perhaps by a strong electrical discharge in which all the neutral atoms have been ionized. Collisions of ${}_1H^2$ particles having energies corresponding to the temperature ($\overline{E} = \frac{3}{2}kT$) result in nuclear reactions that release energy in the form of kinetic motion of the products such as ${}_1H^3$. The latter then react further with deuterium, or by collisions share the new energy with ${}_1H^2$. There is a tendency for particles to diffuse away from the center of the discharge and be lost from the plasma. However, streams of like charges moving in the same direction tend to be forced toward each other (see Section 5-4), a phenomenon labeled the "pinch effect." This magnetic constriction acts to counteract the electrostatic repulsion of charged particles. The resulting dense column of ions favors the needed higher temperature.

Ions having any motion other than perfectly straight paths can generate opposing and competing electromagnetic fields. Thus an electric discharge tends to be very unstable and can "blow up," which stops the reaction just as a wind can extinguish a match flame. The solution of the problem of releasing practical thermonuclear energy demands new and ingenious concepts of the interaction of particles and a combination of external and self-induced electromagnetic fields, in the presence of electromagnetic radiation. If a self-sustaining reaction is achieved, there still remains the practical matter of extracting energy in a useful form.

25-6 THE RELATION OF PHYSICS AND TECHNOLOGY

Recognizing that the science of physics is basic to the understanding of natural phenomena and manmade devices, we now wish to describe its professional aspects. The processes of discovery, invention, and application are considered, along with the views held by the people about their profession. The urge to analyze and to classify has resulted in distinctions both between kinds of activity and between types of motivation. Below, we shall note the conventional description of these various classes, keeping in mind that they are artificial, since there is a continuous variation of tasks to be done and of personal attitudes. First, we can compare "pure" physics and "applied" physics.

The objective of *pure physics* is to extend the boundaries of basic knowledge by the use of theory and experiment. An example is the measurement of the properties of some new particle and the interpretation of its role in the structure of matter. Whether the particle is of any practical use or not is immaterial to the pure physicist—he regards the advancement of knowledge as adequate motivation. Naturally, he would be pleased if some material benefit to man resulted from his studies, but he is not searching for applications.

The purpose of *applied physics* is to develop better understanding of mechanisms related to useful processes and devices. Examples are the measurements of the characteristics of solids that may lead to improved transistors or nuclear reactors or aircraft. Of course, the applied physicist would be pleased to make a discovery of fundamental import, but is not aiming toward that goal.

There is an overlap between the two types of scientist in that both are motivated by the satisfactions coming from creative effort. The investigations of pure science occasionally result in major new applications. A classic example is the relativity theory of Einstein that was the forerunner of the discovery that mass can be converted into nuclear energy. Conversely, work directed toward utility occasionally leads to major improvements in knowledge—the microwaves that were the outgrowth of wartime development of radar are widely used for the study of the structure of molecules.

The function of the *engineer* is closely allied to that of the scientist, and in some ways is indistinguishable. The engineer has the goal of designing and developing practical, reliable, and economically useful devices and processes. The technological needs of the consumer and, more generally, those of society tend to be uppermost in his mind. In general, the engineer is said to use the results of existing scientific information. This stereotyped view does not take into proper account the engineer's use of highly scientific techniques, or the abstract studies that often must precede design and development.

Serendipity

The role of *accident* further complicates the relation between the various technical investigators. Many discoveries have been unexpected, and in fact unrelated to the objective of the studies underway.* For example, Count Rumford was led to the concept that work and heat energy were related while watching the process of boring of cannon. While casually looking at light reflected from windows, Malus in 1808 discovered that electromagnetic waves were polarized by reflection (a process closely related to the action of Polaroid sunglasses). Galvani in 1792 discovered current electricity by noting the contraction of frog's legs placed in contact with dissimilar metals. Roentgen in 1895 discovered X rays by observing that light came from the surface of certain chemicals at considerable distance from an electron-accelerating tube. Becquerel in 1896 discovered radioactivity while attempting to study X rays, noting that uranium itself gave off radiation that affected photographic plates. Although purely accidental discoveries are important, it is generally conceded that usually it is the combination of a good background and hard work that puts a scientist in a position where he can properly interpret unusual observations. Such accidents happen only to those who deserve them, as Lagrange said of Newton.

Discoveries may not be recognized as such by the worker. While investigating filaments for electric light bulbs, Edison found that the introduction of another electrode would permit conduction of current in one direction only. He had discovered the principle of the vacuum-tube diode, but did not examine the phenomenon further. The credit for the invention is given to deForest, who found the effect again, but years later.

The time lapse between a discovery and its application is closely related to the availability of resources and the degree of need for the product. For instance, the rocket was invented many centuries ago by the Chinese, but it was not until recent times that it was used as a weapon and as a vehicle for space flight. Television was technically possible as early as the 1920's, but a waiting period of some twenty-five years elapsed before there was a demand for large-scale commercial production. It is a sign of our accelerated age that the time between the discovery of a scientific relationship and its incorporation into a useful device has grown progressively shorter. It required only about five years for the laser to be transformed from a laboratory curiosity into a powerful new tool of research and industry.

Finally, it must be noted that the ever-widening scope of man's activities requires the cooperative effort of scientists from all disciplines and engineers of every branch, making use of industrial capability and economic resources. Major national projects—such as the development of useful nuclear energy and the investigation of space—demand every type of professional talent available. In the nuclear energy program, there is a need for engineers of many types—chemical, electrical, mechanical, metallurgical, civil, and biological—as well as for scientists who specialize in physics, chemistry, biology, and medicine. In the space program, physicists study

* The gift of discovery by accident is often labeled "serendipity" after the heroes of *The Three Princes of Serendip,* who often found things they did not seek.

the character of radiation in space, doctors test the reactions of the pilots of spacecraft, biologists are concerned with possible types of life on other planets, and psychologists investigate reactions to weightlessness. Specialists do not yet exist in some subjects—for example, interplanetary law and lunar geology.

25-7 SCIENCE AND SOCIETY

The tremendous growth of science and technology in the twentieth century has brought us a host of devices that reduce the need for manual effort, that increase our mobility, and that increase the speed of communication. It has provided means for maintaining health and prolonging life. Accompanying these benefits to our standards of living are many new problems and challenges to society and its technical manpower. Even in the United States, which is generally regarded as the wealthiest and most highly developed nation in the world, there are many improvements to be made, by a combination of scientific study, engineering development, and commercial and governmental enterprise, led and supported by people who seek a better society. A few of the challenges to our ingenuity and humanity are now examined in light of our technical capabilities.

The decay and congestion of the cities. The transition from an economy based on farming to one based on industry has followed the mechanization of agricultural processes. The consequence has been mass migration to cities, resulting in overcrowding, inadequate housing, substandard employment conditions, overloaded transportation systems, and problems of air and stream pollution by wastes. The renovation of the cities is possible with the help of ingenious manufacturing and construction techniques, careful community planning, and adequate economic support.

Unemployment accompanying automation. Ever since the advent of the Industrial Revolution, there has been a temporary displacement of workers accompanying the installation of new labor-saving machinery. It has been repeatedly demonstrated that more jobs eventually became available, because of economic expansion and the need for increased service functions. No one knows whether or not this process of accommodation to change will persist indefinitely, especially with the rapid adaption of computer techniques. The matter is of much concern to organized labor, to industrial management, and to government. Solutions appear to include the introduction of new industry in depressed areas, massive organized retraining programs, the acceptance of shorter working days or weeks, or earlier retirement, with the expansion of commercial activities geared to increased leisure time.

The depletion of natural resources. The history of man's careless exploitation of his environment is familiar to everyone. The indiscriminate destruction of the forests, the plowing of the prairies, and the exhaustive cultivation of the soil resulted in the drought and duststorms of the Midwest in the early part of this century. The complete extinction of the passenger pigeon and near-disappearance of the American bison are well-known facts. The profligate use of irreplaceable coal, oil, and natural gas is believed by many to foretell serious future shortages of natural materials. The substitution of nuclear fuels such as uranium and thorium for fossil fuels has the effect of greatly stretching out the time span before energy resources are exhausted, and permits the conversion of fossil materials into vitally needed chemicals. Even the use of nuclear reserves in the crust of the Earth requires control, since the only natural fissionable isotope U^{235} constitutes less than 1 per cent of all uranium, and since the cost of fuel increases as the readily-available deposits are depleted. It is necessary to pursue actively the development of practical breeder reactors for conversion to useful form of the inert isotopes U^{238} and thorium yet available. The practicality of widespread use of the energy of sunlight has yet to be demonstrated, but the potential yet exists for harnessing the same nuclear fusion processes in terrestrial machines. The waters of the Earth would contain an almost inexhaustible source of deuterium as fuel for thermonuclear devices, if and when they are developed. A nominal effort is now being devoted to research toward such devices, on the grounds that much yet remains to be learned about the basic processes before development is attempted.

The needs for food and water. The absolute minimum requirement for human subsistence is an adequacy of food and water. These are related, of course, in that the growth of plants or animals for food depends on water supply. Only in the last century has agricultural and commercial fertilizer machinery been used on a large scale to provide an abundance of food in America, and many areas of the world still employ primitive farming methods. Inadequate distribution systems for water and the growing needs for agricultural, human, and industrial consumption conspire to create a growing problem, compounded by inadequate control of purity in streams and lakes. Water is abundant in many areas of the world, and the sea is an inexhaustible source, if only salt could be removed. Solutions of the problem include the development of mammoth diversion and distribution systems; economical methods for desalinization of sea water, including the likely use of nuclear energy; and careful control of wastes.

The information explosion. The increased emphasis on scientific investigation in the twentieth century has resulted in the phenomenon called the "information explosion." New discoveries lead to more powerful research tools, which collect data of greater scope and extent. These in turn lead to a deeper knowledge of life processes, our environment, and of manmade products and devices. Our standard of living is greatly enhanced by the commercial application of such information. The finiteness of man's available time and capacity to absorb and retain knowledge has caused scientists to become more highly specialized in their work, to the point that communication is limited to a single group of experts in each field. Also, the number of years devoted to preparation for a professional career tends to increase, and the amount of information to be gained requires a more intensive effort. The rapidity with which the technological world changes requires that education of all workers be continuous rather than terminating with the formal study period. A solution of these problems may well lie in the greater use of the technical developments themselves, such as a combination of telecommunications—telephone, radio, and television—with the power of the high-speed electronic computer to store, manipulate, and analyze information.

The inequitable distribution of progress. During the period of rapid technological development, there has been an uneven distribution of the fruits of scientific discovery and application throughout the world, such that only part of the nations have benefited significantly. In most of South America, Asia, and Africa, the standard of living of the people has been little improved, and the gap between conditions in advanced and developing countries tends to widen. Factors that have hindered progress are the lack of knowledge, of financial resources, and of the will to change. A solution appears to be possible only through a deliberate effort on the part of "have" nations to assist "have-not" nations to break the cycle of defeat.

The population explosion. A problem that is believed by many to overshadow all others is the phenomenal rise in world population in recent years. Throughout history, disease and famine have kept down the rate of population growth, but the advent of modern drugs and sanitation practices have greatly reduced infant mortality and increased life expectancy. As a result, the world population is increasing at a rate faster than technical advances can be assimilated to provide adequate food and standards of living. It is possible that new discoveries may change this relation, but many people are convinced that the only answer is planned parenthood based on birth control. This solution is very complicated on an international basis because of the many factors that must be considered—technical, educational, economic, political, religious, and cultural.

The threat of nuclear warfare. The discovery of nuclear weapons that have the potential to annihilate all of mankind has led to a pervading sense of fear, and to many a sense of futility. One approach to the situation is to provide full physical protection against radiation and blast accompanying a nuclear attack. Although this is feasible if a sufficient fraction of national wealth is allocated to such construction, the strategic and psychological consequences are unknown. It is debatable whether defense or retaliatory systems serve as a deterrent or a stimulus for another nation to act, and whether or not their existence is beneficial to the spirit of the people. International agreements to avoid the use of nuclear weapons can be sought, to provide some assurance of safety. Ultimately and ideally, it is through the development of fullest understanding and good will among nations that the hazard can be eliminated.

Technology in the future of society. Every new discovery, invention, and technical development opens up the possibility of further improvements in man's condition in two ways. First, the new knowledge and equipment is available for application to the accumulation of additional information on control of the environment. Second, opportunities are provided for increased economic strength to exploit material resources. We now have the power to accomplish almost any technical feat if we choose to devote sufficient manpower, time, and money to its achievement. Evidence for the truth of this concept has come from the rapid application of new discoveries. Examples are the dominance of automobile and air transportation, the universality of television in the home, the advent of nuclear-fueled electrical plants, the essential elimination of polio as a disease, and the amazing accomplishments in space exploration. It is believed that the question is not one of feasibility, but one of choice as to which are the preferable objectives.

All problems, challenges, and opportunities that face the world today have technical aspects of great complexity, but more important are the underlying human factors of physical and spiritual need and aspiration. The necessity for communication, understanding, and cooperation is evident at every level of our society—between persons, communities, and nations. To accomplish the task of adjustment to new realities, followed by progress toward the goals within our reach, the best efforts are needed from all segments of our society, including scientists, engineers, industrialists, teachers, ministers, and political leaders, as well as the general public. Such responsibilities can be met through understanding, capability and willingness to make decisions, and creative action, by citizens and leaders who have appreciation of both the inherent technological implications and the fundamental cultural and human values. In this era of increased specialization, the demands to broaden our horizons of thought are ever-increasing. The adjustment to the changing world—indeed, survival itself—depends on our success. Most of us believe, however, that man can do more than exist, and has the potential of developing toward a state of perfection. We can look hopefully toward a near-utopian civilization in which man is no longer subject to the uncertainties of his environment and has become master both of his world and himself.

PROBLEMS

25-1 Estimate the deflection of a 1-MeV photon as it traverses a spacecraft of diameter 2 m, under 3-g acceleration.

25-2 Show that the fractional changes in the frequency of light $\Delta\nu/\nu$ due to the gravitational red shift are (a) gz/c^2 for small displacements z near the Earth's surface; (b) $k_G M_S/R_S c^2$ for photons coming to us from the Sun's surface.

25-3 Using the masses (in atomic-mass units; 1 amu = 931 MeV) of the H atom, the neutron, and the deuterium atom—respectively 1.007825, 1.008665 and 2.014102—find the energy release when a neutron is captured.

25-4 How many tons of hydrogen fuel per year would supply the world's heat needs of $0.1Q$ per year if the reaction $4\,_1\mathrm{H}^1 \rightarrow {}_2\mathrm{He}^4 + 2\,_{-1}\mathrm{e}^0 + 26.7$ MeV could be used?

25-5 Find the ratio of initial mass to final mass for a relativistic rocket with speed $V = 0.3c$ and propellant speed $v_p = \frac{1}{8}c$.

25-6 Compare the Earth time and plane time in a flight of 3000 mi across the U.S. at 600 mph.

25-7 Compute the energies, in MeV, of the neutrinos emitted on decay of the pion π^- and the subsequent decay of the muon μ^-, assuming that the two from the latter are directed oppositely.

APPENDICES

A PHYSICAL CONSTANTS *

(For many calculations three significant figures are sufficient.)

Speed of light, c	2.997925×10^{8}	m/sec
Avogadro's number, N_a	6.02252×10^{26}	/kmole
Electronic charge, e	1.60210×10^{-19}	C
Electron rest mass, m_e	9.10908×10^{-31}	kg
Proton rest mass, m_p	1.67252×10^{-27}	kg
Neutron rest mass, m_n	1.67482×10^{-27}	kg
Planck's constant, h	6.62559×10^{-34}	J-sec
$\hbar = \dfrac{h}{2\pi}$	1.054494×10^{-34}	J-sec
Energy of 1 atomic-mass unit (amu)	931.478	MeV
Energy of electron mass	0.511006	MeV
Bohr orbit, $R_1 = \dfrac{\hbar^2}{me^2}$	5.29167×10^{-11}	m
Stefan–Boltzmann constant, σ	5.6697×10^{-8}	W/m²-(°K)⁴
Boltzmann's constant, k	1.38054×10^{-23}	J/°K
Gas constant, R	8.31434×10^{3}	J/°K-kmole
Bohr magneton, $\mu_0 = \dfrac{e\hbar}{2m}$	9.2732×10^{-24}	A-m²
Volume of kilomole of gas, V_0	22.4136	m³/kmole
Faraday constant, $F = N_a e$	9.64868×10^{7}	C/kmole
Rydberg constant, $\mathcal{R}$ (hydrogen)	1.0967759×10^{7}	m^{-1}

* E. R. Cohen and J. W. M. Du Mond, *Reviews of Modern Physics*, October, 1965, p. 590.

B SYMBOLS

A
activity, d/sec
a constant
area, m^2
amplitude, m
a
acceleration, m/sec^2
semimajor axis of ellipse, m
a distance, m
B
binding energy, J or MeV
$\mathcal{B}$
magnetic induction field, N-sec/C-m or Wb/m^2
b
semiminor axis of ellipse, m
C
circumference, m
heat capacity, kcal/°C
electrical capacitance, C/V or F
a constant
c
specific heat, kcal/kg-°C
speed of light, m/sec
torsion constant, m-N/rad
a constant
c
"centi-" (prefix)
D
diameter, m
a distance, m
diffusion coefficient, m^2/sec
electric displacement, V/m
d
distance, m
E
a modulus of elasticity
energy, J
E_p potential energy, J
E_k kinetic energy, J
$\mathcal{E}$
electric field, V/m or N/C
e
electronic charge, C
base of natural logarithms, 2.718 . . .
"electron" (subscript)

F
force, N
F-number
f
a function
a fraction
focal length, m
G
"gravitational" (subscript)
$\mathcal{G}$
gravitational field, N/kg
g
acceleration of gravity, m/sec^2
H
a height, m
"half" (subscript)
$\mathcal{H}$
magnetizing force, A
h
height, m
Planck's constant, J-sec
$\hbar = \frac{h}{2\pi}$
I
intensity of radiation, W/m^2-sec
moment of inertia, kg-m^2
electrical current, A
i
current density, A/m^2
$\sqrt{-1}$
"index" (of one of several items—subscript)
J
particle current, sec^{-1}
rate of radiation per unit area, J/m^2-sec
j
a particle-current density (m^2-sec)$^{-1}$
K
force constant, N/m
k
"kilo-" (prefix)
propagation constant, m^{-1}
Boltzmann's constant, J/°K
Einstein's $[1 - (V/c)^2]^{-1/2}$
a constant, such as k_G, k_E, and k_M
"kinetic" (subscript)
L
angular momentum, J-sec

self-inductance, henrys
a length, m
heat of change of state, kcal/kg
l
a length, m
M
mutual inductance, ohms
mass of an object, kg
dipole moment—electric, N-m^2/V; magnetic, N-m^3/Wb
m
mass of an object, kg
m
"milli-" (prefix)
N
torque, m-N
number of atoms per unit volume, m^{-3}
number of turns of wire
number of particles in a sample
N_a
Avogadro's number, (kmole)$^{-1}$
n
number of particles per unit volume, m^{-3}
number of turns of wire per unit length, m^{-1}
a quantum number
"neutron" (subscript)
index of refraction
n
"nano-" (prefix)
P
power, W
pressure, N/m^2
polarization
p
"potential" (subscript)
"proton" (subscript)
momentum, kg-m/sec
object distance, m
"photon" (subscript)
pole strength
relative permeability
p
"pico-" (prefix)
Q
amount of charge, C
amount of thermal or nuclear energy, J
q
amount of charge, C

image distance, m

R

a radial distance, m

resistance, ohms

gas constant per mole, kcal/°K-kmole or J/°K-kmole

$\mathcal{R}$

Rydberg constant, m^{-1}

range of particles, m

central-force constant, $\frac{L^2}{mC}$

r

radial coordinate, m

rate of discharge, kg/sec

resistance, ohms

S

entropy, kcal/°K or J/°K

surface area, m^2

s

arc or path length, m

T

temperature, °F, °C, or °K

period, sec

t

time, sec

thickness, m

U

internal energy, kcal or J

u

speed or velocity, m/sec

V

potential or potential difference, V

speed or velocity, m/sec

v

volume, m^3

speed or velocity, m/sec

W

work, N-m

a weight, N

work function, eV

w

a weight, N

thermodynamic probability

X

a distance, m

x

coordinate, m

Y
a distance, m
y
coordinate, m
Z
impedance, ohms
z
coordinate, m

GREEK SYMBOLS

α (*alpha*)
coefficient of linear thermal expansion, $(°C)^{-1}$
an angle, rad or deg
angular acceleration, rad/sec^2
β (*beta*)
coefficient of volume thermal expansion, $(°C)^{-1}$
an angle
γ (*gamma*)
ratio of specific heats, $\frac{c_P}{c_V}$
a constant
an angle
δ, Δ (*delta*)
a difference or change
an angle
ϵ (*epsilon*)
emissivity
eccentricity
efficiency
permittivity
η (*eta*)
viscosity coefficient, kg/m-sec
number of neutrons
θ (*theta*)
angular coordinate
κ (*kappa*)
thermal conductivity, J/m-sec-°C
dielectric constant
λ (*lambda*)
wavelength, m
mean free path, m
μ (*mu*)
"micro" (prefix)
attenuation coefficient
micron
coefficient of friction
mobility, m^2/V-sec
permeability
ν (*nu*)
frequency, sec^{-1}
Poisson's ratio
π (*pi*) = 3.14159 . . .
ρ (*rho*)
mass density, kg/m^3

charge density, C/m^3
resistivity, ohm-m
σ (*sigma*)
electrical conductivity, A/V-m or $(\text{ohm-m})^{-1}$
cross-section, m^2
surface-charge density, C/m^2
Σ (*sigma*)
summation symbol
τ (*tau*)
a time constant, sec
ϕ (*phi*)
an angle
Φ (*phi*)
flux
χ (*chi*)
compressibility, m^2/N
susceptibility—electric or magnetic
Ψ (*psi*)
quantum-mechanics wave function
ω (*omega*)
angular speed, rad/sec

C ATOMIC WEIGHTS

(Based on the atomic mass of $_6C^{12}$ as 12)

Atomic Number	Name	Symbol	Atomic Weight	Atomic Number	Name	Symbol	Atomic Weight
1	Hydrogen	H	1.00797	53	Iodine	I	126.9044
2	Helium	He	4.0026	54	Xenon	Xe	131.30
3	Lithium	Li	6.939	55	Cesium	Cs	132.905
4	Beryllium	Be	9.0122	56	Barium	Ba	137.34
5	Boron	B	10.811	57	Lanthanum	La	138.91
6	Carbon	C	12.01115	58	Cerium	Ce	140.12
7	Nitrogen	N	14.0067	59	Praseodymium	Pr	140.907
8	Oxygen	O	15.9994	60	Neodymium	Nd	144.24
9	Fluorine	F	18.9984	61	Promethium	Pm	[145]
10	Neon	Ne	20.183	62	Samarium	Sm	150.35
11	Sodium	Na	22.9898	63	Europium	Eu	151.96
12	Magnesium	Mg	24.312	64	Gadolinium	Gd	157.25
13	Aluminum	Al	26.9815	65	Terbium	Tb	158.924
14	Silicon	Si	28.086	66	Dysprosium	Dy	162.50
15	Phosphorus	P	30.9738	67	Holmium	Ho	164.930
16	Sulfur	S	32.064	68	Erbium	Er	167.26
17	Chlorine	Cl	35.453	69	Thulium	Tm	168.934
18	Argon	A	39.948	70	Ytterbium	Yb	173.04
19	Potassium	K	39.102	71	Lutetium	Lu	174.97
20	Calcium	Ca	40.08	72	Hafnium	Hf	178.49
21	Scandium	Sc	44.956	73	Tantalum	Ta	180.948
22	Titanium	Ti	47.90	74	Tungsten	W	183.85
23	Vanadium	V	50.942	75	Rhenium	Re	186.2
24	Chromium	Cr	51.996	76	Osmium	Os	190.2
25	Manganese	Mn	54.9380	77	Iridium	Ir	192.2
26	Iron	Fe	55.847	78	Platinum	Pt	195.09
27	Cobalt	Co	58.9332	79	Gold	Au	196.967
28	Nickel	Ni	58.71	80	Mercury	Hg	200.59
29	Copper	Cu	63.54	81	Thallium	Tl	204.37
30	Zinc	Zn	65.37	82	Lead	Pb	207.19
31	Gallium	Ga	69.72	83	Bismuth	Bi	208.980
32	Germanium	Ge	72.59	84	Polonium	Po	210
33	Arsenic	As	74.9216	85	Astatine	At	[210]
34	Selenium	Se	78.96	86	Radon	Rn	222
35	Bromine	Br	79.909	87	Francium	Fr	[223]
36	Krypton	Kr	83.80	88	Radium	Ra	226
37	Rubidium	Rb	85.47	89	Actinium	Ac	227
38	Strontium	Sr	87.62	90	Thorium	Th	232.038
39	Yttrium	Y	88.905	91	Protactinium	Pa	231
40	Zirconium	Zr	91.22	92	Uranium	U	238.03
41	Niobium	Nb	92.906	93	Neptunium	Np	(237)†
42	Molybdenum	Mo	95.94	94	Plutonium	Pu	(239)
43	Technetium	Tc	[99]*	95	Americium	Am	(243)
44	Ruthenium	Ru	101.07	96	Curium	Cm	(244)
45	Rhodium	Rh	102.905	97	Berkelium	Bk	(249)
46	Palladium	Pd	106.4	98	Californium	Cf	(252)
47	Silver	Ag	107.870	99	Einsteinium	Es	(253)
48	Cadmium	Cd	112.40	100	Fermium	Fm	(257)
49	Indium	In	114.82	101	Mendelevium	Md	[258]
50	Tin	Sn	118.69	102	Nobelium	No	[255]
51	Antimony	Sb	121.75	103	Lawrencium	Lw	[263]
52	Tellurium	Te	127.60	104			[261]

* [ ] = atomic weight of most stable artificial isotope.
† ( ) = atomic weight of most abundant artificial isotope.

D DERIVATIONS IN THE SPECIAL THEORY OF RELATIVITY

Some of the mathematical manipulations involved in Einstein's special theory, Sections 3-4, 4-7, 8-7, and 12-4, are collected here for reference.

The Coordinate (Lorentz) Transformation

The linear relations between coordinates proposed by Einstein are

$$x = k(X + VT)$$
$$X = k(x - Vt)$$

Divide the first equation by T and the second by t to obtain

$$\frac{x}{T} = k\left(\frac{X}{T} + V\right)$$
$$\frac{X}{t} = k\left(\frac{x}{t} - V\right)$$

Measurements of the speed of light in each of the two frames of reference give the same value, $C = c$. Then introduce $x = ct$ and $X = cT$ and multiply the equations together:

$$\frac{ct}{T} \cdot \frac{cT}{t} = k^2(c + V)(c - V)$$

Simplifying,

$$k^2 = \frac{c^2}{c^2 - V^2}$$

or

$$k = \frac{1}{\sqrt{1 - (V/c)^2}}$$

The Velocity Transformation

Let the speeds of an object as observed from the plane and ground be $U = X/T$ and $u = x/t$, respectively. In the first of these, insert the Lorentz transformation equations

$$X = k(x - Vt)$$
$$T = k\left(t - \frac{Vx}{c^2}\right)$$

Thus

$$U = \frac{x - Vt}{t - \frac{Vx}{c^2}}$$

Solve for

$$\frac{x}{t} = u = \frac{U + V}{1 + \dfrac{UV}{c^2}}$$

Doppler Effect for Light

The observed frequency of light emitted by a moving source is derived using relativistic ideas. The light reaching an observer from an approaching object such as a jet plane (used as an example in Sections 3-4 and 4-7) appears to be a wave of the form

$$\sin\left[2\pi \frac{\nu_o}{c}(x + ct)\right]$$

that is, moving in the negative x-direction, while the plane as a source describes it as

$$\sin\left[2\pi \frac{\nu_s}{c}(X + cT)\right]$$

It is the same light in reality, and the wave expressions must be the same. Thus

$$\nu_o(x + ct) = \nu_s(X + cT)$$

Insert the Lorentz transformation

$$X = k(x + Vt), \qquad T = k\left(t + \frac{Vx}{c^2}\right)$$

where $k = (1 - V^2/c^2)^{-1/2}$ and the sign before V is positive since the plane is moving in the negative x-direction. For a perfect match, coefficients of x and t must be equal on both sides of the equation. A little algebra shows that the ratio of observed and source frequencies is

$$\frac{\nu_o}{\nu_s} = \frac{\sqrt{1 + V/c}}{\sqrt{1 - V/c}}$$

Relativistic Formulas

A set of formulas that allow one to work back and forth between physical quantities is listed for reference. For example, E is given in terms of E_k by $E_k + m_0c^2$.

	v	p	E_k	E	m
v	v	$\dfrac{pc}{\sqrt{p^2 + (m_0c)^2}}$	$\dfrac{c\sqrt{\dfrac{2E_k}{m_0c^2}\left(1 + \dfrac{E_k}{2m_0c^2}\right)}}{1 + \dfrac{E_k}{m_0c^2}}$	$c\sqrt{1 - \left(\dfrac{m_0c^2}{E}\right)^2}$	$c\sqrt{1 - \left(\dfrac{m_0}{m}\right)^2}$
p	$\dfrac{m_0v}{\sqrt{1 - \left(\dfrac{v}{c}\right)^2}}$	p	$\sqrt{2m_0E_k\left(1 + \dfrac{E_k}{2m_0c^2}\right)}$	$\sqrt{\left(\dfrac{E}{c}\right)^2 - (m_0c)^2}$	$c\sqrt{m^2 - m_0^2}$
E_k	$m_0c^2\left[\dfrac{1}{\sqrt{1 - \left(\dfrac{v}{c}\right)^2}} - 1\right]$	$c\sqrt{p^2 + (m_0c)^2} - m_0c^2$	E_k	$E - m_0c^2$	$(m - m_0)c^2$
E	$\dfrac{m_0c^2}{\sqrt{1 - \left(\dfrac{v}{c}\right)^2}}$	$c\sqrt{p^2 + (m_0c)^2}$	$E_k + m_0c^2$	E	mc^2
m	$\dfrac{m_0}{\sqrt{1 - \left(\dfrac{v}{c}\right)^2}}$	$\sqrt{\left(\dfrac{p}{c}\right)^2 + m_0^2}$	$\dfrac{E_k}{c^2} + m_0$	$\dfrac{E}{c^2}$	m

E PARTICLE TRANSPORT

A formula for the rate of flow of particles in a gas is developed. Visualize, as in Figure A-1, a plane unit area across which flow is to be found. A particle of speed v that collides at P has a chance $e^{-N\sigma r}$ of reaching the plane without making another collision, and a chance $\dfrac{\cos\theta}{4\pi r^2}$ of being aimed correctly. Now, in a volume ΔV, if the particle and target-number densities are respectively n and N, and the cross-section for their interaction is σ, then the rate of collision in that volume is $nvN\sigma\Delta V$. Each of the

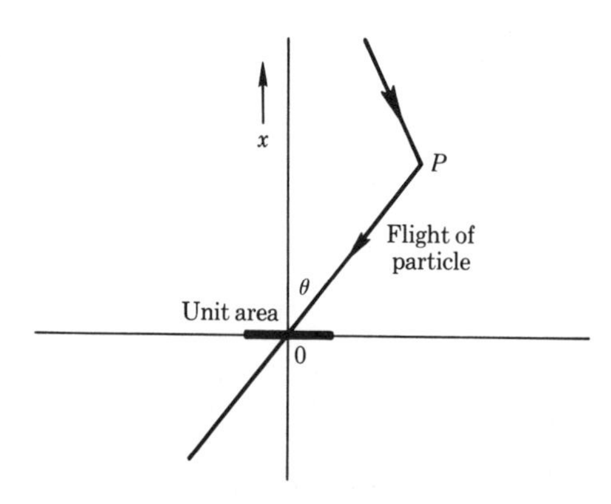

FIGURE A-1

scattered particles experiences the chances listed above. Assume that n varies only in the x-direction, so that all particles originating in a small ring-shaped volume (Figure A-2)

$$\Delta V = (r\Delta\theta)(\Delta r)(2\pi r \sin\theta)$$

behave similarly. Finally, if the mean free path is short, most of the particles crossing the area come from near the plane, and linear variation of n can be assumed, $n(x) = n_0 + n_0'x$, where n_0 and n_0' are respectively

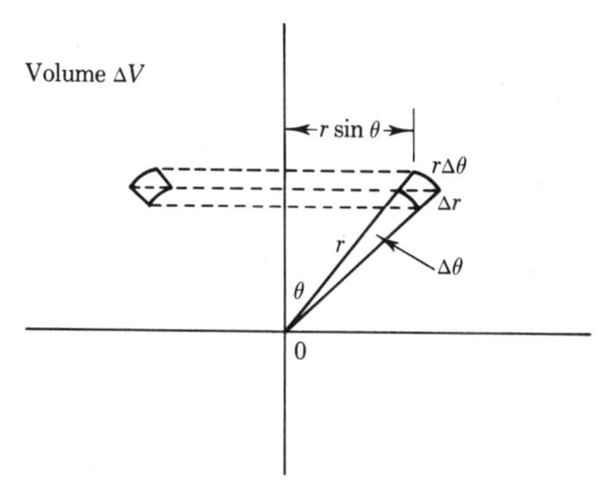

FIGURE A-2

the number density and its slope at the origin. Combining all these relations and adding up contributions from all of the space above the plane, the number of particles per second per unit area that flow in the negative x-direction is obtained. Using integral calculus

$$j_- = \int_{r=0}^{\infty} \int_{\theta=0}^{\pi/2} (n_0 + n_0' r \cos\theta)(vN\sigma)(r\,d\theta)(dr)(2\pi r \sin\theta)\left(\frac{\cos\theta}{4\pi r^2}\right)e^{-N\sigma r}$$

which reduces to

$$j_- = \frac{n_0 v}{4} + \frac{n_0' v}{6N\sigma}$$

A similar derivation shows that for the flow in the positive x-direction the sign between terms is negative. Thus the net current j is

$$j_+ - j_- = -\frac{n_0' v}{3N\sigma}$$

Since $\lambda = 1/N\sigma$ and $n_0' = dn/dx$, we obtain

$$j = -\frac{\lambda v}{3}\frac{dn}{dx}$$

which is the derivative form of the expression in Section 16-3.

F ELECTROMAGNETIC WAVES

A demonstration that waves are consistent with the principles of electromagnetism is provided. Assume that the electric and magnetic fields point in the x- and y-directions, respectively, and are governed by the equations of a wave propagated in the z-direction:

$$\mathcal{E} = \mathcal{E}_0 \sin (kz - \omega t)$$

$$\mathcal{B} = \mathcal{B}_0 \sin (kz - \omega t)$$

Faraday's law, discussed in Section 7-4, could be written in terms of the circulation of $\mathcal{E}$:

$$\mathcal{E}C = -\frac{\Delta\Phi_M}{\Delta t}$$

For the loop around which $\mathcal{E}$ is summed, select a rectangle in the x–z plane of any height h and base Δz, lying along the z-axis, as shown in Figure A-3(a). Also shown is the form of the electric wave at an instant. As one proceeds around the loop, the fields cancel except for the small difference $\Delta\mathcal{E}$, and thus the circulation reduces to $h\Delta\mathcal{E}$. Now, the magnetic field $\mathcal{B}$ is perpendicular to the loop, with area $h\Delta z$, and thus the flux is $\mathcal{B}h\Delta z$. Applying Faraday's law yields

$$\frac{\Delta\mathcal{E}}{\Delta z} = -\frac{\Delta\mathcal{B}}{\Delta t}$$

By analogy with the relation of position x and speed $\Delta x/\Delta t$, form

$$\frac{\Delta\mathcal{B}}{\Delta t} = -\mathcal{B}_0\omega \cos (kz - \omega t) \qquad \text{and} \qquad \frac{\Delta\mathcal{E}}{\Delta z} = \mathcal{E}_0 k \cos (kz - \omega t)$$

Thus

$$\mathcal{E}_0 k = \mathcal{B}_0 \omega$$

However, since $k = 2\pi/\lambda$ and $\omega = 2\pi\nu$, where the speed of the electromagnetic wave is $c = \lambda\nu$, the ratio of amplitudes of the electric fields is

$$\frac{\mathcal{E}_0}{\mathcal{B}_0} = c$$

Next, turn to *Ampère's law as generalized by Maxwell*, involving the circulation of $\mathcal{B}$ (Section 7-3). When there are no true currents, as in free space,

$$\mathcal{B}C = \frac{k_M}{k_E}\frac{\Delta\Phi_E}{\Delta t}$$

The loop around which $\mathcal{B}$ is summed is taken to be a rectangle in the y–z plane, height h, base Δz along the z-axis, as shown in Figure A-3(b). The

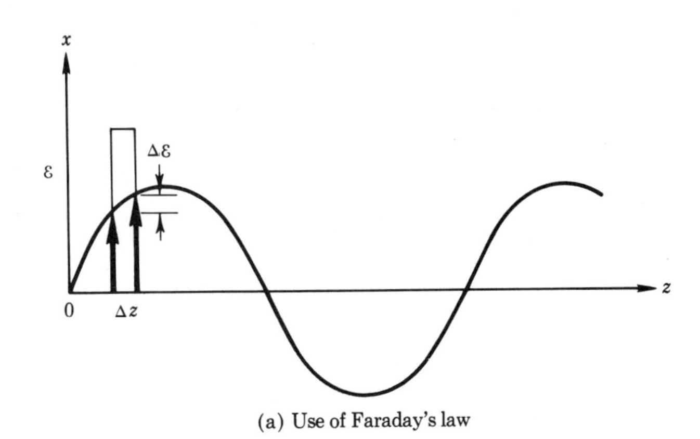

(a) Use of Faraday's law

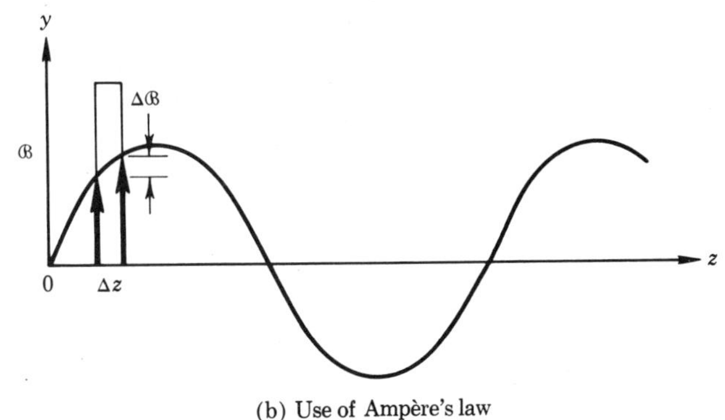

(b) Use of Ampère's law

FIGURE A-3

Derivation of electromagnetic relations

circulation of the magnetic field is $-h\Delta\mathcal{B}$, while the electric flux is $\mathcal{E}h\Delta z$. Applying Ampère's law,

$$\frac{\Delta\mathcal{B}}{\Delta z} = -\frac{k_M}{k_E}\frac{\Delta\mathcal{E}}{\Delta t}$$

Now, $\Delta\mathcal{B}/\Delta z = \mathcal{B}_0 k \cos(kz - \omega t)$ and $\Delta\mathcal{E}/\Delta t = -\mathcal{E}_0\omega \cos(kz - \omega t)$, so that

$$\mathcal{B}_0 k = \frac{k_M}{k_E}\mathcal{E}_0\omega$$

However, $\mathcal{E}_0/\mathcal{B}_0 = c$ and $\omega/k = c$, and thus

$$c = \sqrt{\frac{k_M}{k_E}}$$

that is, *the speed of light is the square root of the ratio of the electric and magnetic constants of proportionality*. Inserting numbers yields

$$c = \sqrt{\frac{9 \times 10^9}{10^{-7}}} = 3 \times 10^8 \text{ m/sec}$$

in agreement with the measured speed of light in free space.

G ANSWERS TO SELECTED PROBLEMS

1-2	8.64×10^{12}; 3.16×10^{15}
1-4	1.1×10^{30}
1-5	1.45×10^{17} kg/m^3; 5.23×10^{12} lb/in.^{3}
1-8	16 lb
1-10	4.55×10^{22}
1-14	7.1×10^{49}
1-17	VE 25.8×10^6 mi; ME 48.8×10^6 mi
1-18	10^{-5}
2-1	3:35 min
2-3	2.57 sec
2-6	200, 3.46×10^7 m
2-9	-31.2 ft/sec^2; 4.82 sec
2-11	159 ft, 3.12 sec; 101 ft/sec
2-12	1.75×10^{-3} rad/sec, 1.46×10^{-4} rad/sec; 2.8×10^4 m/sec, 1.75×10^{-5} m/sec
2-15	800 mi/hr, 200 mi/hr; 400 mi/hr
2-17	195 mi; 31.5 deg
2-20	7.34×10^4 mi/hr; 1.88×10^{-2} ft/sec^2
2-21	24.2 ft/sec; 0.76
2-23	788 m/sec^2
3-1	2.99793×10^8 m/sec; 0.000002×10^8 m/sec; 0.00002×10^8 m/sec
3-3	-0.0001, -0.0001, -0.01; -0.01, -0.009, -0.9; -8.1, -0.81, -81
3-6	1.55×10^{-7}; nonrelativistic
3-7	11 ft
3-9	4.32 sec
3-12	3 yr
4-2	2.4×10^4 kg-m/sec, 4.2×10^4 kg-m/sec; 1.8×10^4 kg-m/sec; 4.5×10^3 N
4-5	5.39×10^7 m/sec^2
4-7	1000 m/sec; 2.60×10^{-3} m/sec^2; 1.92×10^{20} N
4-10	5900 N; 1.80 m/sec^2
4-13	$w\sqrt{h(2r - h)}/(r - h)$
4-15	2.06×10^4 N
4-17	216 lb
4-18	They accelerate upward the same; 827 N
4-21	71.4 ft/sec^2
4-23	4.73×10^{28}; 2.76×10^{-10} m
4-24	1.93 lb/ft^3; 0.031
4-27	1.08 tons
4-29	1660 lb
4-31	1.22 m

4-33	2.04×10^{-29} kg; 6.11×10^{-21} kg-m/sec
4-35	1.45%
5-2	0.90; 22,800 mi
5-4	1/3%
5-6	6.82×10^{-7} N
5-8	600 ft
5-11	5.74×10^{13} C; 6×10^{5} kg
5-13	0.118 m
5-15	3×10^{8} m/sec
5-18	6; 2×10^{-8} C
5-19	44,700 ft
5-22	1.4 hr
5-23	5.1×10^{4} m; 204 sec; 3.5×10^{5} m
5-26	$x = R_0 \Big/ \left(\frac{K}{m\omega^2} - 1\right)$; 0.152 m
5-28	24.9 m/sec
5-30	0.07
5-31	0.27
6-1	6.1 N/kg
6-3	9.56 N/kg
6-5	830 N/C
6-7	5.4×10^{-15} N
6-8	2.16×10^{7} N/C
6-11	46 deg from x-axis; 6.08×10^{11} N/C
6-13	$8.64 \times 10^{-9}/R^2$; 5.4×10^{10} N/C
6-15	1.58×10^{7} N/C
6-16	1.88×10^{10} N/C
6-18	0.686; 120 deg from x-axis
6-23	6.8×10^{-6}
7-2	proton 2.09×10^{3} m; alpha 4.18×10^{3} m
7-3	1.41×10^{-6} N; ratio 1.8×10^{-8}
7-6	5.21 cm
7-7	0.10 cm
7-10	8.0×10^{6} A/m^2; 500 C, 3.1×10^{21}
7-11	3.6×10^{-4} N
7-15	1.25×10^{-6} C; 7.8×10^{12}
7-16	1.91 A
7-18	Toward the wire
7-20	4.1 cm
7-22	0.02×10^{11} C/kg (high)
8-1	2130 ft-lb
8-3	5.88×10^{6} J
8-5	1.88×10^{-4} J; 532

8-8	7.52×10^{-19} J
8-10	-4.35×10^{-18} J; -27.2 eV
8-12	420 m
8-14	980 J; 2
8-18	5.9×10^{5} m/sec; 0.2%
8-19	9000 V
8-21	-52.7 V
8-23	1.30×10^{9} J, 361 kWh; 1.23×10^{6} Btu
8-25	2.55×10^{7} m
8-26	24.9 m/sec^2; 59.6 km/sec
8-28	1.05×10^{4} lb
8-30	0.558 hp; 896,000 cal; 45%
8-32	1.76×10^{9} ft-lb/sec; 3.20×10^{6} hp; 2.39×10^{6} kW
8-34	-40°F, -40°C; 11.4°F, -11.4°C
8-36	0.47°C
8-40	5.63×10^{-21} J; 0.0352 eV
8-42	3.0 eV
8-43	1.57×10^{11} J
8-44	400 kg
8-47	5.16 yr
8-48	1.5×10^{-10} J, 9.4×10^{8} eV; 3.4×10^{-9} J, 2.1×10^{10} eV
8-51	7.5×10^{19}
9-2	5.11×10^{4} mi
9-5	0.0520 nm from H-H band
9-7	1.67 grams high
9-10	2.5 m; no
9-12	$\theta = \tan^{-1}\left(\frac{v^2}{gR}\right)$; $\theta = 70$ deg
9-14	2.96×10^{-47} kg-m^2
9-16	$\frac{1}{2}MR^2$
9-18	450 N-m; 3750 N
9-20	21.1 ft-lb
9-21	7.86×10^{3} lb-ft
9-26	51.4 ft-lb
9-27	105 rpm
9-28	1.0 rad/sec^2 counterclockwise
9-29	2.0 rad/sec
9-31	85 N-m; 1700 N; 382 lb
9-33	4.08×10^{-5} Wb/m^2
9-36	7.18×10^{33} kg-m^2/sec; 4.32×10^{-12} C/kg; 1.1×10^{17}
10-1	2.92 m/sec
10-4	1640
10-6	0.005 m
10-8	1.8 J
10-11	1110 kg/sec

10-12	0.632
10-13	0.252; 0.48 m_0
10-17	102 mi/hr, 8 mi/hr; 50.9 mi/hr, 4.1 mi/hr
10-18	0.0022
10-20	1.52×10^7 m/sec; 2.75×10^5 m/sec
10-23	Ratios of final to initial neutron energy 0.64 for Be^9, 0.11 for H^2
10-25	2.28×10^{-22} kg-m/sec; 9.7×10^3 m/sec
10-27	$\dfrac{6\,mv}{Ml}$
10-32	4.75 hr
10-33	14.5 deg; 3870 mi/hr
10-36	2.52×10^{12} kg-m^2/sec
10-37	Hyperbolic; 3.6×10^{12} kg-m^2/sec
10-38	1.53×10^{-16} sec
10-39	8.56×10^8 mi^2/sec
11-2	2.83
11-5	-6.99 m/sec; 2.1 sec later
11-7	7.96 cps
11-10	$\dfrac{T}{8}$
11-13	4.94×10^{-6} N-m/rad
11-14	0.83 sec
11-17	0.065 Wb/m^2
11-20	2.96×10^4 N
11-21	4.26 sec
11-23	9.91 m/sec^2; 0.13 m/sec^2
11-25	0.0622 m; 0.26 mm
11-28	9.78 m/sec^2
11-31	5.54 sec
11-32	3.23 cps
11-34	536 N/m; 1.28×10^{14} cps
11-36	83.3 m; 377 rad/sec; 0.0754 /m
11-37	1.43×10^9 cps
11-40	39.6 m/sec
11-42	17.9 m/sec; 19.9 cps
12-1	11.4 m, 0.0214 m
12-3	$2A \cos \dfrac{\phi}{2} \sin \left(\omega t - kz - \dfrac{\phi}{2}\right)$
12-5	3750 cps
12-6	20; 540, 560
12-9	0.00054 sec; 465 sec; 29.3 sec
12-11	3.7
12-13	Error 0.3%
12-14	3.10 kg
12-16	140 cps

12-20	20; 49, 9, 31; 38, 2, 42; etc.
12-21	882 cps
12-23	780 cps, 1300 cps, 1560 cps, etc.
12-25	1330 cps, 385 cps
12-27	6.0 min
12-29	6.36×10^6 m/sec
12-31	37.8 eV, 151 eV
13-1	5.52×10^3 N
13-5	3.63×10^{-4} cm^3
13-7	0.36 in.
13-9	278°C
13-11	1.4 deg
13-13	1.44×10^{-7} m^3; 1.44×10^{-7} m^3; 1.19×10^6 N
13-15	1.37 atm; 2.69×10^{25} /m^3
13-17	0.765 m; 0.28%
13-18	20.6 km
13-22	0.57 cm
13-23	0.008%; 2.9¢
13-27	25.5 N; 0.51 m/sec^2; 3.44 sec
13-28	2.77 ft/sec; 34.6 ft/sec; 42 lb/in.^{2}
13-31	1.08 kg/m^3; 6.12 km/sec; 1.62×10^5 N
13-34	0.1 cal/g-°C; iron
13-36	1.29×10^4 kg
14-2	1.24×10^{-19} J; 0.78 eV; 8.62×10^3 m/sec
14-4	9×10^{14} J; 1.5×10^{11} °K; 4.2×10^7 m/sec
14-5	4.23 atm; 3.23 atm
14-7	3.83×10^{29} /sec
14-9	7.8×10^{23}; 7.5×10^{19}
14-12	3100, 650 J/kg-°C
14-15	2652; 1392
14-17	2200 m/sec
14-19	0.00523
15-1	8.0×10^4 J; 5.08×10^4 J
15-3	0.040
15-5	2.08×10^4 J/°K, 4.96 kcal/°K; 2.94×10^4 J/°K, 6.94 kcal/°K
15-7	10 to the power (-7×10^{26})
15-9	420
15-11	0.268; 0.268 J
15-14	Increasing; drops to 0.242
15-15	32.7 lb/in.^{2}; 538 J; 7.95 cm
15-19	7.19×10^5 J
15-20	1 kW
16-1	3.11×10^{-19} m^2, 1.63×10^{-7} m; 1.63×10^{-7} m

16-3	$1.73 \times 10^{20}/m^3$; ratio 6.4×10^{-6}
16-5	111 sec
16-7	2.25×10^3/V-m-sec
16-8	0.06 m^2/V-sec; 1.87×10^{-5} sec
16-10	1.12×10^{-3} m^2/V-sec; 125 m/sec
16-11	Ratio 1:9
16-13	1.85 J/sec
16-16	20°C
16-19	175 lb/in.^{2}
16-20	298 m/sec; 6.42×10^7 m/sec
17-1	3.3 mm
17-4	2400 ohms, 3090 ohms
17-6	15.6 A; 3.06×10^{-6} ohm-m; within 17 W
17-7	0.8 ohm; 3.08 ohms; 388 ohms
17-10	4 ohms, 12 ohms
17-12	1.6×10^{-3} ohm; 4.73 J; 1.18 sec
17-13	6.62 $\mu\mu$F; 72.8 $\mu\mu$F
17-15	50 μF; 1.66 μF; 818 μF
17-17	4.48×10^{-3} m^2; 5540 $\mu\mu$F
17-20	277 $\mu\mu$F; 10^4 ohms
17-23	1.40×10^6 A/m; 1.64×10^{-23} A-m^2; 79%
17-24	1.12×10^{-3} henry
17-26	30 V
18-1	31.8 grams
18-3	0.15 ohm
18-5	0.4 A; 7.5 ohms
18-7	12,560 ohms
18-8	0.0402 ohm ; 12,490 ohms
18-11	$I_1 = -4.8$ A, $I_2 = 6.88$ A, $I_3 = 2.08$ A
18-12	17.3 N-m; 1500 N-m
18-15	3775 ohm; 7040 $\mu\mu$F
18-16	710, 505, 411, 874 cps
18-19	$X_c = 133$ ohms; $Z = 518$ ohms; $\phi = -14.4$ deg; $I_0 = 0.328$ A
18-21	9.8×10^5 m/sec, 2.45 cm; 1000; 774 m
18-22	$\dfrac{\pi R E_k}{2qV_0}$
18-23	1.63×10^{-11} Wb/m^2; much smaller
18-27	4.16×10^6 cps, 0.272 Wb/m^2, 785 m
18-28	6.25×10^{13}; 100 W
19-2	2.4×10^{23}sec; 1.2×10^{-15} m; 4.45×10^8
19-4	2.92×10^{-3} V/m
19-5	4.7×10^{-8} W; 7.0×10^{-19} W
19-7	3960 W
19-10	0.129; 62.5 W/m^2-°C

19-12	4050 m^2
20-2	760 sec
20-3	106 W/m^2
20-6	By a factor of 9
20-7	4.69×10^{-7} m
20-10	6.94×10^{-9} sec
20-11	22 deg
20-14	0.0204
20-17	$\frac{3}{4}$
21-3	3 cm in front
21-4	6 m behind object, real, inverted; 3 m behind mirror, virtual, erect
21-6	0.024 in.
21-10	1.41
21-11	10 cm
21-12	0.0513 m from lens; 0.026
21-15	4.3 cm
21-17	5.05 m; 3.4 m square
21-19	3.4 cm
21-21	0.22 cm; $\frac{1}{281}$ sec
21-23	8.8 light years
21-24	5.6×10^{-6} m
21-26	168 ft
21-28	8.8×10^5/m
22-1	2, 5; no
22-3	6.2×10^{14} cps
22-6	658 mμ; H_α
22-8	1.53×10^3 eV; 0.83 eV; 3.96×10^4 eV
22-10	62.9 m/sec
22-12	1.59×10^7 cps; 8×10^{-9}
23-3	0.024 m
23-4	7.97 MeV
23-7	0.70; 1.38×10^9 d/sec
23-9	61
23-11	${}_0n^1$; ${}_{86}Rn^{222}$; ${}_{36}Kr^{92}$
23-12	-2.14 MeV; 2.14 MeV
23-14	3300 yr
23-15	${}_{38}Sr^{95}$, ${}_{56}Ba^{142}$
23-17	4.7 MeV
23-19	0.45 mg/m^3
23-22	82/m; 24 cm
24-3	3.75×10^{-3} C
24-5	1.2×10^{-5} μg
24-6	0.167 m

24-10	0.264 m; convergent
24-12	5500 cps; 10^{-6}
24-13	1000; 100001; 1111111; 100111001; 1010101010; 1111101000
24-14	53; 729; 37; 137; 85; 1023
25-1	6.5×10^{-16} m
25-5	144
25-7	32 MeV; each 52.5 MeV

INDEX

A

B

C

F

N

O

P

Q

R

W

X

Y

Z